Biomedical Micro-Nanotechnologies Toward Translation

Alessandro Grattoni • Tejal Desai
Editors

Biomedical Micro-Nanotechnologies Toward Translation

A Tribute to Dr. Mauro Ferrari

Chapter 2 and Chapters 9 through 27 were originally published in *Biomedical Microdevices* in the Topical Collection "Biomedical Micro-Nanotechnologies Toward Translation"

Springer

Editors
Alessandro Grattoni
Houston, TX, USA

Tejal Desai
San Francisco, CA, USA

ISBN 978-1-0716-0186-0

This Springer imprint is published by the registered company Springer Science+Business Media, LLC part of Springer Nature.
The registered company address is: 233 Spring Street, New York, NY 10013, U.S.A.

Preface

Fig. 1 Dr. Mauro Ferrari. *Image courtesy* Houston Methodist Hospital

Advancements in micro- and nanotechnologies have continued to enrich human life. Beyond their prominence in various industries such as automotive, renewable energy, water filtration, and agriculture, these technologies have progressively transformed the biomedical field. This holds true across a broad spectrum of medical applications including diagnostic, therapeutic, and theranostics technologies. While "nanotechnology and nanomedicine" are common terminology today, 25 years ago these terms were infrequently used. The tremendous developments and technological advances we experience today were prompted by a handful of individuals who pioneered the field and showed its promise to the scientific and industrial community. Engineering molecules and materials at the nanoscale has proven to be of great importance for leveraging unique physicochemical properties and transport phenomena not accessible at the macroscopic scale.

Among these visionary pioneers is Dr. Mauro Ferrari, who in the early 90s at the University of California at Berkeley leveraged his background in mathematics, mechanical engineering, and composite materials to explore the world of nanoscale channels and its biomedical applications. That marked the beginning of "nanofluidics" era. Dr. Ferrari had the intuition that the molecular transport of confined fluids in channels of the same order of magnitude as the molecules themselves is dominated by their interactions within the channel walls. He also realized that precise nanochannels could be created with silicon by adopting advanced technologies used in the semiconductor industry. This led to the development of nanochannel membranes for molecular sieving, cell immunoisolation, and sustained drug delivery among other nanotechnologies for diagnostics and injectable therapeutics.

On the occasion of his 60th birthday, this book *Biomedical Micro-Nanotechnologies Toward Translation* is a tribute to Dr. Mauro Ferrari's vast contribution to the field of nanotechnology and nanomedicine. Covering the multifaceted areas of biomedical research, this book presents aspects of technological innovation, ethics, patient advocacy, and clinical translation, all of which are fundamental components to the development of biomedical technologies. The book includes the contributions of numerous scientists, bioethicists, patient advocates, and renowned experts who have worked closely and collaborated with Dr. Ferrari, leaving an even greater footprint in the field. Particular emphasis is given to recent advanced technologies on their path to clinical translation. Since the inception of his career, the development of nanotechnologies for the benefit of patients has remained a clear focus of Dr. Ferrari's efforts. He has pursued this goal as the principal investigator of a laboratory at UC Berkeley and Ohio State University, during his service at the National Cancer Institute, as the president of the Texas Alliance for NanoHealth, as a Department Chair at the University of Texas Health Science Center at Houston, and as The President and CEO of the Houston Methodist Research Institute. At all of these institutions, not only has he made pioneering contributions to science and engineering but has also demonstrated a true commitment to mentoring the next generation of scientists. Now, in a giant leap forward, Dr. Ferrari will be pursuing this goal from higher ground, serving as the President of the European Research Council. With his new role, we eagerly await the movement toward science for global justice on par with his influence and impact on nanomedicine and nanotechnology.

We would like to thank all authors who have contributed to this work and made this book possible Fig. 1.

Houston, USA Alessandro Grattoni
San Francisco, USA Tejal Desai

Contents

Part I
From Ethics to Innovation: The Multiple Aspects in Nanomedicine Development and Translation

Bridging the Death Valleys of Medical Innovation: One Person's Herculean Efforts to Develop and Drive Innovative Translational Research and Educational Programs

Hanh H. Hoang[1] • Tong Sun[1]

Author's Personal Note In the past 13 years, I have had the fortune to work closely with Mauro. From that first day in the summer 2006 until now, I have always been and continue to be inspired by his passion toward cancer research, in particular, for metastatic breast cancer and translational research, with the singular aim to reduce human suffering and eliminate incurable diseases. High profile publications and large grants were never his goals, and he would never claim success until the research discoveries become medical products that truly help patients. Recently, I heard a seminar by Mauro entitled "My 26 Years of Failures". It has been a great honor for me to be by his side and share the second half of those 26 years of "failures". I sincerely wish Mauro true success in the pursuit for an effective treatment for metastatic breast cancer, in the next 13 years!
Tong Sun

The definition of translation research is generally accepted as the process to bring basic scientific discoveries toward a human endpoint, or clinical application. This process, on average, requires vast investments of time and money—almost two decades and well over $2 billion, respectively—to bring a new drug from idea to clinical use. In addition, translational research is impeded by multiple potential "valleys of death", pitfalls where a medical innovation (i.e., new drug, device, or clinical process) remains imperiled by lack of or insufficient financial investments, regulatory hurdles, manufacturing capacity, or market complexities, among others (Fig. 1). In recognition of these immense challenges, Dr. Mauro Ferrari not just envisioned, but also implemented, a coordinated and multipronged approach to address each of these potential roadblocks along the path of translational research.

1 Funding Support from Conventional and New Sources

In relatively more recent years, the National Institutes of Health, the US Department of Defense, and others have begun to offer funding opportunities for translational research, including the Clinical and Translational Science Awards/National Center for Advancing Translational Sciences and the Breakthrough Awards for product development, respectively (https://ncats.nih.gov/funding; https://cdmrp.army.mil/funding/bcrp). However, the Federal and State funding agencies still largely support the basic sciences or clinical trials studies, but little for the intervening stages. The for-profit private sector, including pharmaceutical and medical device companies and venture-funded biotechnology startups, have traditionally accepted the role of advancing scientific discoveries through the development pipelines into medical products. However, the failure rate of developing and commercializing a medical product is abysmally high. For example, approximately 250 candidate drugs out of 5000–10,000 lead compounds will progress to preclinical development, and only five of these 250 will reach Phase I clinical studies. Only one will ultimately gain regulatory approval (http://www.fastercures.org/assets/Uploads/VOD-TranslationalResearch2.pdf). Given such extraordinary odds for failure and low return on investment, for-profit companies have become more risk-averse and reluctant to invest in the translation of medical products in the early phases of development. Rather, these companies have set incredibly high bars for investment and have become more apt to enter licensing agreements or acquiring somewhat de-risked medical products that have already surpassed at least Phase I clinical studies.

Dr. Ferrari has recognized the need and urgency for additional, paradigm-shifting avenues of research funding support, particularly for the potential "death valleys" where translational research often resides and funding from traditional sponsors remain insufficient. He conceptualized and catalyzed the Translational Research Initiative (TRI), currently being implemented at Houston Methodist Research Institute, to advance medical breakthroughs toward clinical utilities—much more quickly in a few years rather than decades. The TRI is sourced by philanthropy support and is intended for research projects that have moved beyond the laboratory research stage and are poised

H. H. Hoang · T. Sun (✉)
Houston Methodist Research Institute, Houston, TX 77030, USA
e-mail: tongsun2007@gmail.com

A. Grattoni and T. Desai (eds.), *Biomedical Micro-Nanotechnologies Toward Translation*,
https://doi.org/10.1007/978-1-0716-0187-7_1

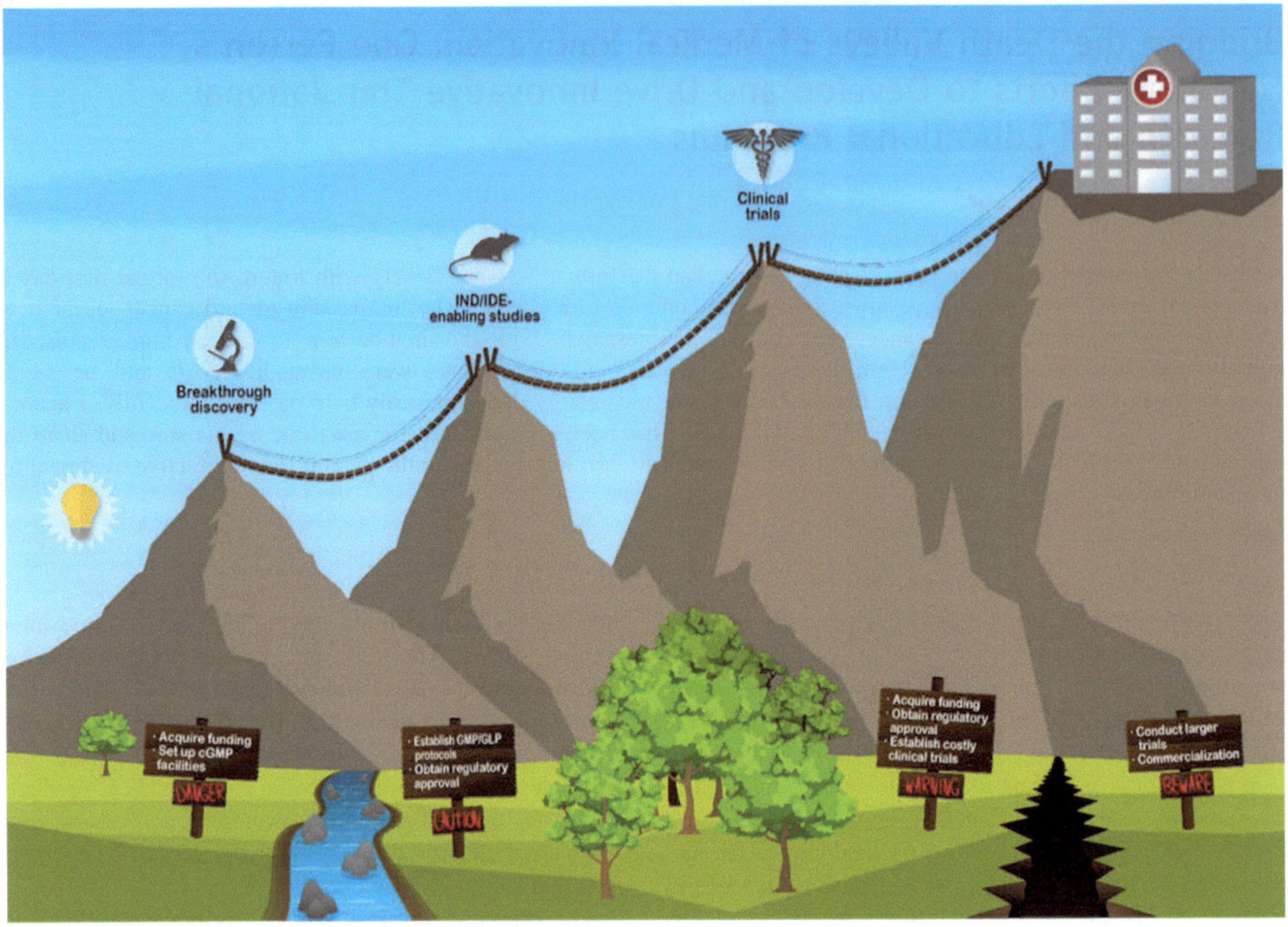

Fig. 1 The pathway from idea to medical solutions for patients is fraught with challenges, many often too monumental to overcome. Thus, many innovations remain in the so-called death valleys of medical research. Those challenges, including insufficient funding, complex infrastructure, regulatory approval, clinical trials, and commercialization, may be de-risked with investments in regulatory and manufacturing resources and expertise, development of unique funding streams, and positioning the innovations for eventual hand-off to potential industry partners

for product development or early-phase clinical trials. Highly promising projects that meet the criteria for review are vetted by a panel of experts in entrepreneurship, regulatory affairs, product development, manufacturing, marketing, clinical trials, and commercialization. Successful researchers are mainly those who have articulated and demonstrated true feasibility and potential commercial value of the medical inventions. Projects supported through the TRI are milestones-driven and closely managed throughout its TRI cycle. With ultimate success of the inventions or technologies (i.e., licensing to industry partners), the returns on investment from the TRI contribute back to the TRI in an "evergreen" fund that will help to support additional future projects in the TRI pipeline.

To successfully establish and implement such a vision requires the buy-in and dedication of a galvanized community of business leaders, patients, teachers, and other community advocates and leaders, who all not only believe in research and big ideas, but also in the power of rapid transformation of those ideas into medicine. To our knowledge, this type of research funding support has no replica in the world of academic medicine. With appropriate community and institutional support, and a visionary leader such as Dr. Ferrari, the TRI can serve as a model elsewhere for rapid advancement of life-saving and life quality-enabling medical solutions to patients.

2 Regulatory and Manufacturing Infrastructure

In order to bring new drugs, devices, or clinical practices to patients, needed are the stringent review by and eventual regulatory approval from the federal governing bodies (i.e., the United States Food and Drug Administration, the European Medicines Agency, others) (https://www.fda.gov/Drugs/DevelopmentApprovalProcess/default.htm; https://www.ema.europa.eu/). However, before regulatory approval can even be considered, development of the new medical technologies (i.e., drugs, devices, clinical processes or information databases) must progress through controlled and regulation-compliant processes including current Good Manufacturing Practice (cGMP) protocols and Good Laboratory Practice (GLP) preclinical studies for safety and toxicology testing of the medical product. The GLP preclinical experiments in relevant animal models (e.g., rodents, non-human primates, others) serve as Investigational New Drug (IND)- or Investigational Device Exemption (IDE)-enabling studies that prepare the medical innovation for examination by the regulatory agencies for approval of in-human applications. Once this approval has been obtained, the candidate drug, device, or clinical practice may proceed through clinical trials and eventual progression through a New Drug Application or Biologics License Application, or Premarket Approval Application or Premarket Notification [or 510(k)] for medical devices. The infrastructure, including human resources and expertise, needed for these vast undertakings is often prohibitive for many academic organizations and investigative teams and is precisely one of the potential death valleys.

Undeterred by the challenges because he recognizes the urgency of success and ramifications of failure, Dr. Ferrari passionately advocated for additional institutional investments in the GMP manufacturing capabilities (including for first-in-class drug candidates) at Houston Methodist, capabilities that also significantly raise the competitive profile of the institution in the eyes of funding agencies and other sponsors. Along with the physical infrastructure, he also bolstered the quality assurance, regulatory affairs, product development, and commercialization teams to support the simultaneous implementation and advancement of translational projects at Houston Methodist, including those that are funded through the TRI mechanism. At present, these investments position the medical innovations to potentially surpass the hurdles toward at least early-phase clinical trials (i.e., Phases I and II), to the point at which hand-off to industry partners are more likely to be unencumbered. Such a composite of infrastructure and capacity in one institution is very rare, if not unprecedented, in the domain of academic medicine, and it is a true testament to the passion and commitment to making a difference in patients' lives.

3 Educating Next-Generation Leaders in Business Development for Translational Research

Understanding and navigating the translational research landscape requires knowledge and expertise about many domains of medical research, product development, clinical trials, and commercialization. Education and training in conventional academic settings have typically partitioned to either the basic sciences or clinical realm, but little on the transitional areas or the business development en route to commercialization and market distribution. The new faces of next-generation learners and eventual experts will need these additional, critical knowledge bases and skill sets. Dr. Ferrari's identification of this fundamental gap spurred the vision and creation of the first-in-kind degree, the Master in Clinical Translation Management (MCTM), in collaboration with colleagues at the University of St. Thomas in Houston, Texas (https://www.stthom.edu/Academics/School-of-Business-Cameron-School-of-Business/Business-Programs/Business-Graduate-School/Master-in-Clinical-Translation-Management/Index.aqf). The MCTM program offers students convenient didactic learning, opportunities for international travel, access to some of the most innovative research for the formulation of capstone projects, and broadened potential to establish careers with research, healthcare, pharmaceutical, or technology-based industries. In just two to three short years since its inception, the MCTM program is already among the top 10 programs in the nation for Best Master's in Clinical Research (https://www.collegechoice.net/rankings/best-masters-in-clinical-research-degrees/), stacking up against powerhouses such as Boston University, University of Virginia, University of California San Diego, University of North Carolina at Chapel Hill, to name just a few. With this success, there are opportunities for the creation of additional degree-granting and training programs that aim to better prepare our students and trainees in scaling the complexities of translation research and management.

4 Concluding Remarks

With too many novel medical discoveries languishing in the valleys of death and far too many patients in need of medical breakthroughs, it is our moral duty to invest in the intellectual, human, and capital resources to pave the translational research pathway with feasible and effective strategies. Not only does Dr. Ferrari envision those strategies, he has created and cultivated the ecosystems and mechanisms to bring them to fruition, for the eventual benefit to patients and their families. As a servant leader, he certainly feels that it is his moral duty to devote and to effect change. Indeed, these are some of the legacies he will endow to current and future custodians of academic medicine.

References

https://ncats.nih.gov/funding. Accessed 26 Feb 2019
https://cdmrp.army.mil/funding/bcrp. Accessed 26 Feb 2019
http://www.fastercures.org/assets/Uploads/VOD-TranslationalResearch2.pdf. Accessed 21 Feb 2019
https://www.fda.gov/Drugs/DevelopmentApprovalProcess/default.htm. Accessed 26 Feb 2019
https://www.ema.europa.eu/. Accessed 26 Feb 2019
https://www.stthom.edu/Academics/School-of-Business-Cameron-School-of-Business/Business-Programs/Business-Graduate-School/Master-in-Clinical-Translation-Management/Index.aqf. Accessed 26 Feb 2019
https://www.collegechoice.net/rankings/best-masters-in-clinical-research-degrees/. Accessed 26 Feb 2019

Biomedical Microdevices (2019) 21: 32
https://doi.org/10.1007/s10544-019-0360-6

NCI Alliance for Nanotechnology in Cancer – from academic research to clinical interventions

Piotr Grodzinski[1] · Christina H. Liu[1] · Christopher M. Hartshorn[1] · Stephanie A. Morris[2] · Luisa M. Russell[1]

Published online: 23 March 2019

Abstract
The National Cancer Institute (NCI) of National Institutes of Health has funded and operated the NCI Alliance for Nanotechnology in Cancer - a large multi-disciplinary program which leverages research at the intersection of molecular biology, oncology, physics, chemistry, and engineering to develop innovative cancer interventions. The program has demonstrated that convergence of several scientific disciplines catalyzes innovation and progress in cancer nanotechnology and advances its clinical translation. This paper takes a look at last thirteen years of the Alliance program operations and delineates its outcomes, successes, and outlook for the future.

Keywords Cancer nanotechnology · Nanomedicine, multi-disciplinary research · Federal funding

1 Introduction

Nanotechnology has been increasingly finding new and successful applications in the medical arena. Moreover, it has been contributing to novel approaches which are significantly improving the way we diagnose and treat cancer. Nanoscale materials inhabit the same size scale as biological materials, enabling unique interactions with cells and proteins that can be harnessed for efficient delivery of drugs and imaging agents to target sites in the body. Nanotechnologies, which are capable of highly sensitive, specific, and versatile recognition of biological materials, can also be integrated into devices for use in disease detection and characterization applications.

The National Cancer Institute (NCI) recognized these unique research and clinical opportunities at the crossroads of biology, oncology, and technology early and established the Alliance for Nanotechnology in Cancer (ANC) program in 2005. The purpose of the program was to exploit nanotechnology's potential within cancer research and to the improvement of cancer care. ANC aimed to establish projects ranging from discovery-based research to translation, with a focus on innovative, clinically relevant technologies and it was the first federally-funded program to fund large scale cooperative research in cancer nanotechnology. In its original incarnation, the program focused on the development of technology platforms which were seeking appropriate cancer applications. Since the initial years, the program has matured and evolved into defining relevant biological and clinical problems, which serve as a driver for the implementation of suitable nanotechnologies. Subsequently, several technologies developed under the Nanotechnology Alliance funding have reached a level warranting the initiation of clinical trials.

✉ Piotr Grodzinski
grodzinp@mail.nih.gov

[1] Nanodelivery Systems and Devices Branch, Cancer Imaging Program, Division of Cancer Treatment and Diagnosis, National Cancer Institute, 9609 Medical Center Drive, Rockville, MD 20850, USA

[2] NIH Office of Strategic Coordination, National Institutes of Health, Bethesda, MD 20893, USA

2 ANC program background and history

Recognizing the potential of nanotechnology in cancer, NCI assembled a working group and held a series of workshops to structure programmatic initiatives that could capitalize on nanotechnology innovations in 2003–2004. Dr. Mauro Ferrari, spent two years at NCI at that time (away from his 'real' job as a faculty of the Ohio State University), and was instrumental to coordinating and shaping the strategies which were behind the launch of the Alliance for Nanotechnology in Cancer (ANC) in 2005 (Ferrari 2005). The goal of the program was to establish a network of interdisciplinary research teams that had the requisite collective expertise to develop and validate nanotechnologies applicable to cancer. ANC investigators were tasked to work in three broad research areas: i) early diagnosis using *in vitro* assays and devices or in vivo imaging techniques; ii) multifunctional nanotherapeutics, including nanoparticle-driven immunotherapies (introduced in Phase III); and iii) devices and techniques for cancer prevention and control. The ANC's development model called for the most promising ANC strategies developed in academia, to be handed off to for-profit partners for effective clinical translation and commercialization.

2.1 Program structure and network participants

The ANC Network was initially funded in 2005 through a set of Funding Opportunity Announcements (FOAs) released by the NCI. Internal and external evaluations of the program, as well as input from the extramural community (Zamboni et al. 2012), guide development of the ANC program through consecutive phases. NCI has renewed the program twice since its beginning in 2005, with Phase II and Phase III launched in 2010 and 2015, respectively. Each phase lasts five years and involves multiple synergistic funding initiatives for large research centers, smaller research projects, multidisciplinary training awards, and support of the Nanotechnology Characterization Laboratory (NCL) (Fig. 1). Figure 2 shows geographical distribution of different awards in all 3 phases of the program.

The Centers of Cancer Nanotechnology Excellence (CCNEs) are focused on integrating nanotechnology and cancer research to develop solutions that are clinically relevant. They provide infrastructure and translational support to the ANC network. The ANC program also includes smaller multidisciplinary research projects - Innovative Research in Cancer Nanotechnology (IRCN) awards, which are focused on building fundamental knowledge on interactions of nanomaterials with biological systems and mechanisms of nanotechnology-based in vivo delivery with the ultimate goal of contributing to efficient and well-informed nanotechnology translation. Multidisciplinary research training and team development initiatives were also funded to foster cross-disciplinary training of graduate students and postdoctoral fellows in nanotechnology and cancer biology. These efforts were a mix of individual post-doctoral awards; Pathway to Independence awards, allowing promising graduate students to transition to faculty positions; and Cancer Nanotechnology Research Training Centers, which conduct multi-mentor training at the cross-roads of several disciplines. The Nanotechnology Characterization Laboratory is also part of the ANC Network. NCL is an intramural laboratory serving as a centralized resource to characterize nanomaterials developed by ANC researchers and researchers from the broader academic, government, and industry research communities.

ANC researchers leverage program network resources through trans-network collaborations. These collaborative efforts are encouraged via trans-Alliance challenge projects which usually last 12–18 months, with 2–3 rounds of them being funded over the period of one phase. Many of these projects led to active collaborations and peer-reviewed publications. ANC also launched Grand Challenge competitions within the program network to fund projects in areas collectively identified by the investigators as those being most critical, yet requiring several research groups to work together to provide an adequate solution.

To chart forward strategies for the fast evolving field of cancer nanotechnology, NCI holds strategic workshops every five years (Nagahara et al. 2010; Grodzinski and Farrell 2014) and publish their outcomes in the form of a Cancer Nanotechnology Plan (https://www.cancer.gov/sites/nano/research/plan, Hartshorn et al. 2018).

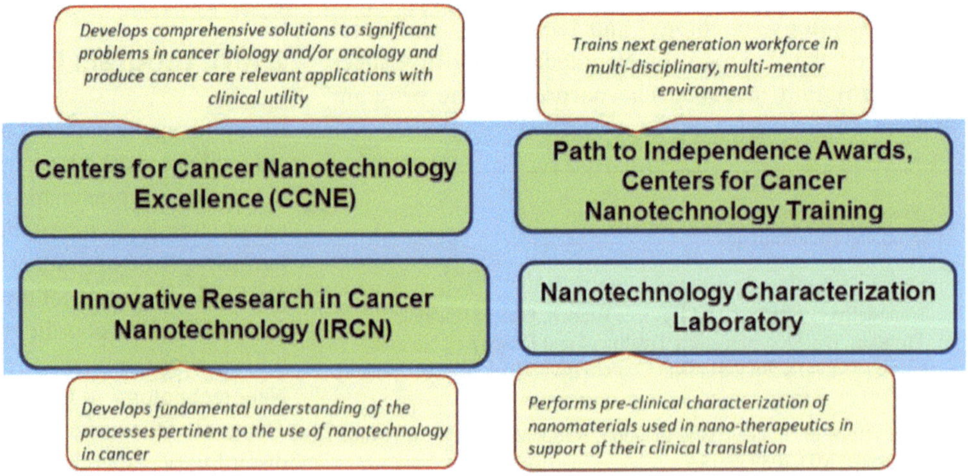

Fig. 1 ANC Program organization

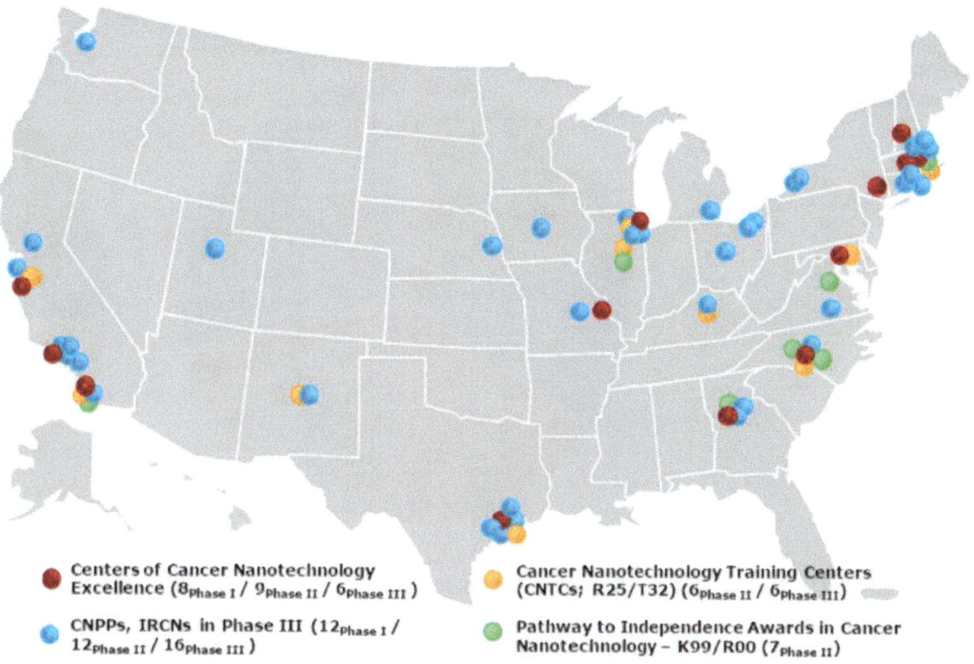

Fig. 2 Distribution of ANC awards among different academic institutions in Phases I, II, and III of the ANC program

2.2 Evolution of ANC objectives: Program Phases I to III

The ANC began with the goal of exploiting promising new technologies to improve cancer research and care. The program has since distinguished itself by pursuing additional goals of developing a strong interdisciplinary network of researchers as well as recruiting and training both physical scientists and engineers to focus their efforts on problems in cancer. The ANC was an early adopter of a multiple principle investigator (PI) model for NIH awards, and applicants have always been advised to recruit one PI from the physical sciences or engineering and another PI from the biological or clinical sciences. This has resulted in the deep integration of cancer biology and oncology practitioners into the technology design process within ANC projects.

The initial ANC awards supported development and early testing of cutting edge technology platforms in cancer settings. In some cases, these platforms adapted processes and materials from other industries, such as micro/nanofabrication techniques from the semiconductor industry (Farrell et al. 2010; Farrell et al. 2011). Many CCNE PIs were already accomplished physical scientists, and the ANC program provided support for dedicated inquiry into the most biologically-relevant applications for their work and promoted their collaborations with clinicians and cancer biologists. These Phase I projects were largely focused on technology development and the potential cancer applications were mostly used as prototypical testing platforms. Thus, most of the investigators operated in a "technology push" environment, where technology was defined first and applications for it were identified later. Phase I projects considered a wide variety of tumor types as application test beds, with many projects focused on "solid tumors," in general, rather than a specific disease or genetic sub-type. Increasing association with clinicians in Phases II and now III has focused efforts into specific interventions aimed at defined tumor types, molecular pathways, and disease markers that could benefit from a nanotechnology-based approach. Funded treatment and diagnostic modalities have evolved with time (Phase II vs III, Fig. 3), showing increasing prominence of immunotherapy applications, for example, with interest in gene therapy decreasing, while the number of projects on diagnostic tools has remained relatively constant.

3 Scientific productivity and output

Overall Alliance investigators have been highly productive and produced close to 3500 publications over the period of Phases I and II (2005–2014). To assess the impact of these publications, rather than counting individual citations, we used the metric Relative Citation Ratio (RCR) (http://www.metrics-toolkit.org/relative-citation-ratio/, Hutchins et al. 2016). Recently developed at NIH, RCR is a field-normalized citation indicator calculated based on co-citation networks which represent an article's field, as the reference set. A paper with RCR equal to 1 has an RCR higher than 50% of NIH-funded papers. The Alliance awards (both CCNEs and IRCNs) hold higher median RCRs than NCI NANO R01 awards (nanotechnology-specific projects submitted to NCI via 'parent' R01 solicitation), which already had median RCR values greather than 1 (Fig. 4).

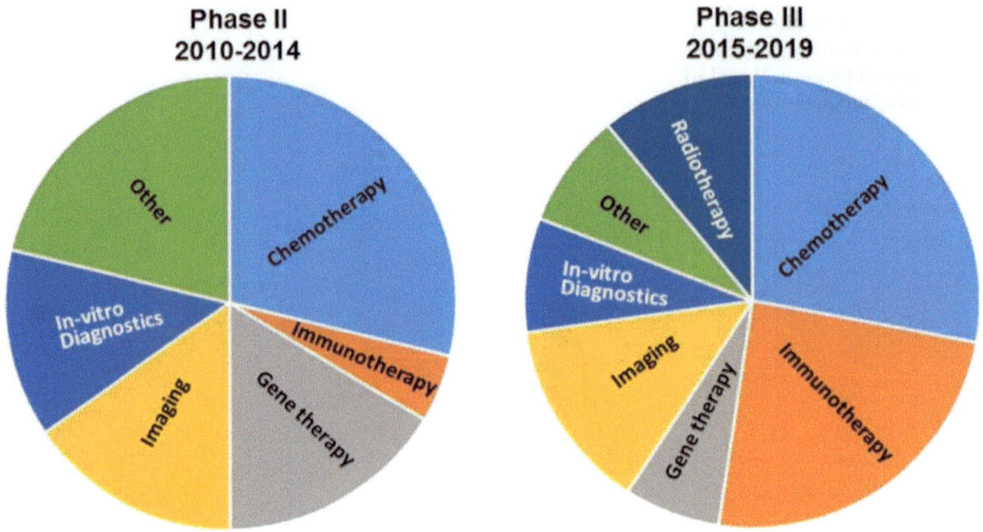

Fig. 3 Evolving focus of ANC program – change in distribution of different research topics between Phase II and Phase III

A publication-based analysis was also conducted to estimate the translational readiness of Alliance projects. Translational readiness measures were extracted using a machine learning tool (iTrans, https://itrans-beta.od.nih.gov/) and used to generate 'Triangle-of-Medicine' plots (Fig. 5, red color indicates a higher frequency of cell-line, animal, or human studies, respectively). The translation-readiness analysis showed that publications within Alliance (Phase II) resulted in significantly more animal studies and also progressed further towards studies with human samples as compared to the non-Alliance NANO R01 projects. This difference (as compared to NANO R01s) was more pronounced for CCNEs than for IRCN awards.

4 Model for translation

The translation of research discoveries to clinical application is an important goal for the ANC program. Alliance funding supports pre-clinical studies only, but members of the CCNEs and broader Alliance are expected to aggressively pursue advancements of their technologies to the clinic by leveraging ANC funding and resources. They are encouraged to seek other sources of funding for clinical trials and to partner with industry and/or establish start-up companies. Through the course of three program phases, the ANC awardees have established a path allowing for the development of innovative technologies in academic centers, allowing their maturation there, and then transitioning them to the translational stage in the commercial sector. This became a unique approach that allows for continuous input of new creative technologies, selecting those that are clinically viable, and moving them forward to the clinical trial environment. Often students and post-docs who were involved in the initial technology development within an academic group, move with the technology to start-up companies that are initiated by themselves or by their PI. Over 130 companies (Chapman et al. 2012; Lenoir and Herron 2015) were formed or associated with the ANC program since 2005. Most often these companies are funded using venture capital funding, but

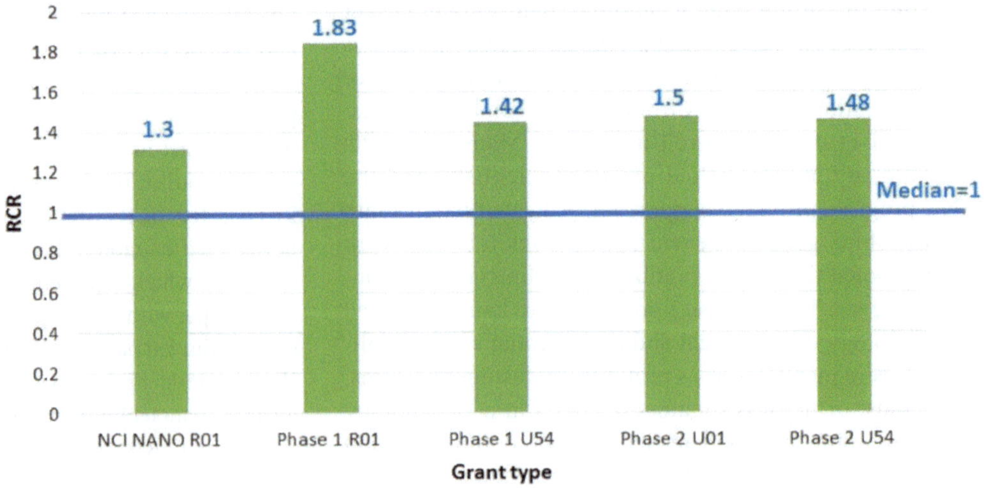

Fig. 4 RCR indices of publications produced by different grant mechanisms supporting nanotechnology at NCI

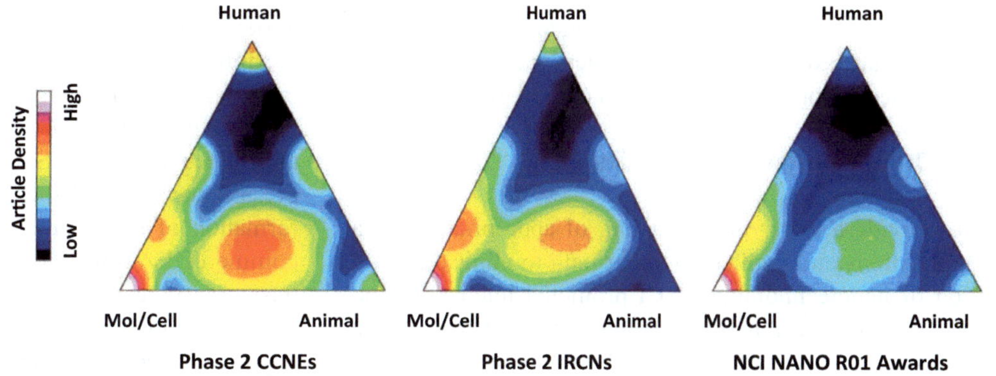

Fig. 5 Triangle-of-medicine plots for Phase II Alliance grants in comparison with NANO R01s. Heat maps of the individual cohorts are displayed with the apex of the triangle as human studies, the lower left corner as molecular/cellular studies, and the lower right corner as animal studies

many of them have also benefited from federal Small Business Innovative Research (SBIR) grants and contracts from NCI. Furthermore, several technologies that emerged from the ANC program have entered clinical trials (over 20) via these start-up companies. Majority of these trials are not funded by NCI; investigators or companies associated with them secured funds to support the trials and further technology development from other sources.

5 NCI nanotechnology translational resources

5.1 Nanotechnology Characterization Laboratory

The Nanotechnology Characterization Laboratory (NCL, https://ncl.cancer.gov/) was founded in 2005 as part of the ANC program to provide "pharmaceutical mentorship" to investigators working in cancer nanomedicine. NCL's mission was to develop an 'Assay Cascade' of scientific tests that would help determine the reproducibility, safety, and efficacy of nanotechnology-based cancer drugs and diagnostics and to provide investigators with additional tools and information required to meet regulatory requirements to move their technology towards clinic.

NCL's Assay Cascade assays had been previously tested on a variety of nanomaterial types and were developed with concurrent input from the FDA. This battery of tests and protocols has been used ever since to evaluate nanomaterials submitted to NCL by extramural investigators with the aim of generating data to support future Investigational New Drug (IND) or Investigational Device Exemption (IDE) filings to FDA. NCL has assisted more than 100 extramural investigators, tested over 300 nano-formulations in its Assay Cascade, and now has 14 collaborators with nanomedicine products in clinical trials. NCL also published over 150 scientific papers describing important trends in nanomedicine development.

Nanomedicine has been maturing, with several formulations in clinical trials, and growing interest from the pharmaceutical and biotech industry. Nanotechnology formulations are employing not only small molecule drugs and are also being considered for the delivery of peptides, proteins, siRNAs, mRNAs, plasmids, and CRISPR/Cas9-based approaches. To stay current with these new developments, NCL has been developing new capabilities in the area of drug re-formulation to reduce drug toxicities and widen their therapeutic windows through the use of nanoparticles. NCL has also participated and remains actively engaged in coordinating the formation of a sister laboratory in Europe, which has a similar mission and is funded under a European Union grant.

5.2 Data sharing

As part of the first phase of the ANC program, the Cancer Nanotechnology Laboratory data portal (caNanoLab; https://cananolab.nci.nih.gov/) was initiated as a collaborative effort with the NCI Center for Biomedical Informatics and Information Technology (CBIIT) to capture nanomaterials data generated by the research community. caNanoLab serves as a data repository that allows researchers to submit and retrieve information on well-characterized nanomaterials including compositions; physico-chemical, *in vitro*, and in vivo characteristics; associated publications; and assay protocols. caNanoLab has become an established resource designed to address the needs of the broader cancer, biomedical, and nanotechnology communities. Currently, this database holds 70 assay protocols, data on over 1200 curated nanomaterial samples, and almost 2000 publications that are available for public use. The majority of these data were curated from ANC and NCL publications by an in-house curator; however, individual users are strongly encouraged to directly submit their data to caNanoLab. Members of the NCI caNanoLab team actively work with other nanotechnology database teams, community-based programs, and federal initiatives, such as the National Nanotechnology Initiative's Signature Initiative on Nanotechnology Knowledge Infrastructure and the National Cancer Informatics Program Nanotechnology Working Group, to develop exchange standards and deposition guidelines for data submission and sharing. Through these interactions, NCI is in the process of building the framework for an organized nanoinformatics system that may lower barriers to technology development and clinical translation.

5.3 Other translational activities

To provide a better understanding of appropriate clinical applications and facilitate a better path to clinical translation of nanotechnologies, NCI promoted interactions between ANC researchers and representatives from pharma/biotech industry and established two opportunities: the Translation of Nanotechnology in Cancer Consortium (TONIC) and the Nanotechnology Startup Challenge in Cancer (NSC^2).

Translation Of Nanotechnology In Cancer consortium Established in 2011 (https://consortiapedia.fastercures.org/consortia/tonic/), with the purpose to facilitate communication and collaboration between early stage technology developers in academia and industry experienced in late pre-clinical and clinical development of cancer nanotechnology products. The main mission of TONIC was to create a consortium of the public, private, and academic sectors to accelerate the translation and development of nanotechnology-based solutions to the early detection, diagnosis, and treatment of cancer. TONIC membership includes companies currently pursuing or considering future nanomedicine applications (mostly nanomedicine startups) and several large pharmaceutical and biotechnology companies considering expansion of their pipeline to nanomedicines. The goals of this partnership were to provide ANC researchers with insight into industry needs in technology platforms and drug targets; promote collaborations between ANC investigators and industry partners on pre-competitive research programs; provide TONIC members with the opportunity to interact with regulatory authorities; and serve as a sustained forum for nanotechnology idea exchange. Since its inception, several round table discussions and a workshop (Prabhakar et al. 2013) have been organized to share opinions among different stakeholders of TONIC.

Nanotechnology Startup Challenge in Cancer In October 2015, NCI Nanotechnology Program Office and the NCI Technology Transfer Office partnered with the Center for Advancing Innovation (CAI) to launch the Nanotechnology Startup Challenge (https://www.nscsquared.org/). The goal of this Challenge (Currell and Bellringer 2016, Truman and Locke 2016) was to advance clinical translation of biomedical nanotechnology by developing business plans around selected portfolio of attractive nanomedicine intellectual property portfolio. Nanomedicine inventions from intramural NIH investigators were chosen to seed the Challenge using a model developed by CAI and NCI for previously operated Breast Cancer Startup Challenge (https://www.breastcancerstartupchallenge.com/). CAI then invited the formation of teams around these technologies to pursue commercialization. The teams mostly recruited from business students, currently enrolled in MBA program, who were interested in pursuing entrepreneurship and starting technology companies. Teams accepted into the Challenge received coaching from CAI and highly regarded entrepreneur mentors on best practices in business planning, formation, and development as they progressed through the stages of the Challenge – elevator pitch, business plan, and finally formation of a startup. At the end of the first two stages, judges drawn from the business and biotechnology sectors selected winning teams to advance to the next stage. In July 2016, ten teams were chosen from the initial 28 to enter the final stage—Startup—in which they received additional advisory support from CAI on how to formally start companies, seek licensing rights for their chosen inventions, and raise funds. The Challenge was highlighted in the National Nanotechnology Initiative Strategic Plan 2016, as 'a model for engaging industry in effective technology transfer'.

6 Conclusion

ANC has become a transformative program which has delivered a strong level of innovation to medicine (Dickherber et al. 2015). This was accomplished by combining several research communities representing nanotechnology innovation, biology, and clinical practice. The program continues to foster an environment that supports the development of innovative technologies in multi-disciplinary academic centers; allows for maturation of these technologies; and their subsequent transition to the clinical stage.

Although, many metrics of success can be cited relative to this program, one of particular interest to its core mission of translation has been in the ability of ANC's investigators to attract additional funds to supplement NCI funding thus expanding upon the opportunities which the NCI seed funds provided. This is true for both additional funds raised by ANC investigators to support their academic research, as well as translation efforts pursued by companies commercializing technologies initially developed by ANC-funded academic groups. This

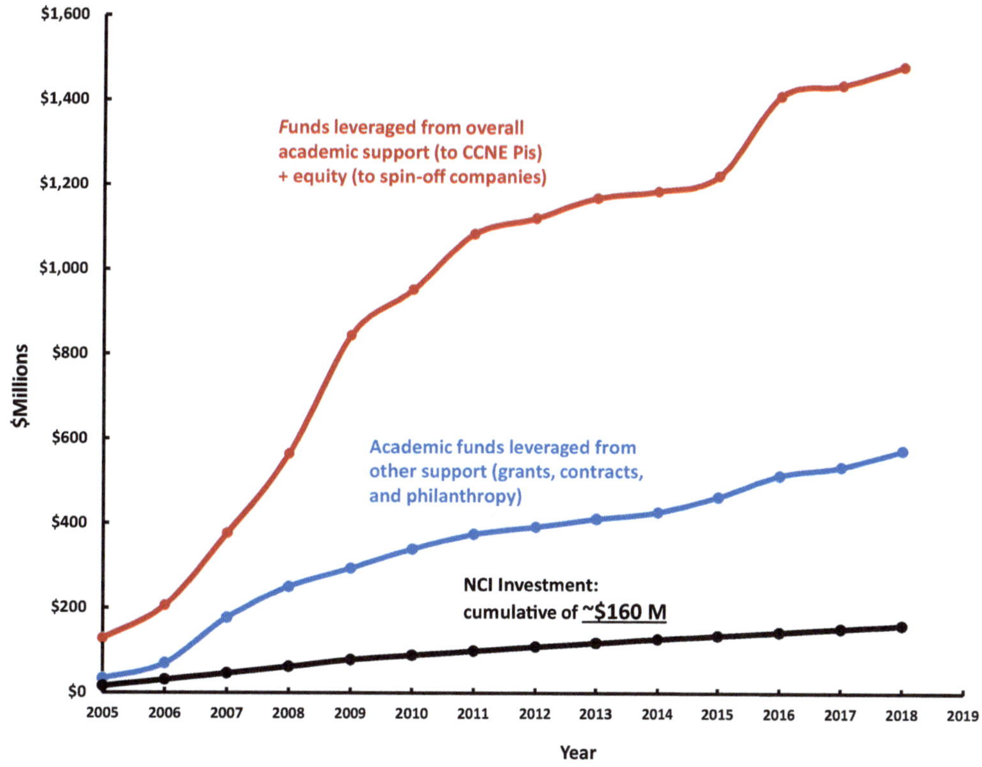

Fig. 6 Leveraging of initial NCI investment with additional academic grants and support for spin-off companies for four CCNEs funded in all 3 Phases: Stanford, Caltech/UCLA/ISB, Northwestern, UNC

'leveraging effect' of ANC awards can be seen in Fig. 6, where cumulative amounts of additional funding raised, in the form of academic grants or equity funding, are depicted for four CCNEs. Each funded through all three phases of the program (13 years), they include Stanford, Caltech-UCLA-Institute of Systems Biology, Northwestern, and the University of North Carolina CCNEs. We estimate that for each $1 invested by NCI into these four ANC awards, they generated additional ~$3 via academic grants and additional ~$5 via equity funding to companies originating from them. This very strong 8-fold leveraging factor, displays the amplification to the value of federal funding invested as well as the expansion to the scope of research and clinical translation resulting from this investment. This leveraging is particularly important in the era of tight federal funding and demonstrates a valuable model to support academic research and subsequent transition to the clinical stage. Moreover, the seed NCI award is a key enabling factor to raising of follow-up funding. The ANC investigators continue to stress the point, that having a peer-reviewed large NCI center grant award has been a critically important prerequisite to obtaining additional funding.

ANC can be also credited with increasing an overall interest of research community in using nanotechnology to address cancer. During the course of the program, we observed continuous increase of NCI NANO R01 applications (Fig. 7) reaching over 700 applications in fiscal year 2018.

Overall, the first thirteen years of the ANC make a powerful case for the influence of early investment of federal government into new research fields and technologies. The program brought several talented technologists to cancer research; they may have never used their talent and innovation for cancer applications without becoming part of CCNE teams. The initial technology-centric focus evolved and ANC-funded groups have become well integrated into biology and oncology and are capable of identifying compelling problems in these areas, while providing effective nanotechnology solutions to them. Furthermore, these teams are highly entrepreneurial and effective in raising additional funds to setup start-up companies, thus enabling technology translation and commercialization. In addition, the program became also a powerful training ground for graduate and post-doctoral students who complete their initial work in one of the ANC institutions and then move-on to obtain a faculty position in another or to join one of the start-up companies originating from ANC.

Overall, the ANC program is viewed as an example of a highly successful, long-lasting venture, which started as a science endeavor, but evolved into a complex organism involving academic investigators representing several synergistic fields working collaboratively, while being closely connected to industry.

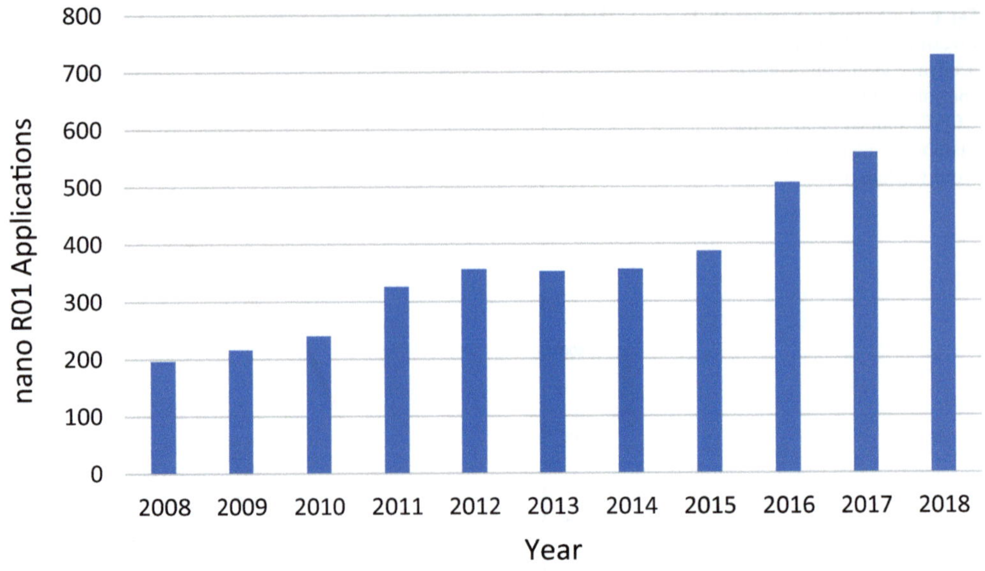

Fig. 7 NCI nano R01 applications. The number of nano R01 applications submitted to NCI increased from 200 in fiscal year 2008 to over 700

Publisher's note Springer Nature remains neutral with regard to jurisdictional claims in published maps and institutional affiliations.

References

S. Chapman, N.J. Panaro, G.W. Hinkal, S.S. Hook, U. Prabhakar, K. Ptak, D. Farrell, P. Grodzinski, Kindling translational cancer nanotechnology research. Nanomedicine (London) **7**, 321–325 (2012)

F. Currell, M. Bellringer, Cancer nanotechnology startup challenge: A new way to realize the fruits of innovation. Cancer Nanotechnology **7**(2) (2016)

A. Dickherber, S.A. Morris, P. Grodzinski, NCI investment in nanotechnology: Achievements and challenges for the future. Wiley Interdiscip Rev Nanomed Nanobiotechnol **7**, 251 (2015)

D. Farrell, J. Alper, K. Ptak, N.J. Panaro, P. Grodzinski, A.D. Barker, Recent advances from the National Cancer Institute Alliance for nanotechnology in Cancer. ACS Nano **4**, 589 (2010)

D. Farrell, K. Ptak, N.J. Panaro, P. Grodzinski, Nanotechnology-based cancer therapeutics-promise and challenge-lessons learned through the NCI Alliance for nanotechnology in Cancer. Pharm. Res. **28**, 273–278 (2011)

M. Ferrari, Cancer nanotechnology: Opportunities and challenges. Nat. Rev. Cancer **5**, 161 (2005)

P. Grodzinski, D. Farrell, Future opportunities in cancer nanotechnology–NCI strategic workshop report. Cancer Res. **74**, 1307 (2014)

C.M. Hartshorn, M.S. Bradbury, G.M. Lanza, A.E. Nel, J. Rao, A.Z. Wang, U.B. Wiesner, L. Yang, P. Grodzinski, Nanotechnology strategies to advance outcomes in clinical Cancer care. ACS Nano **12**, 24 (2018)

B.I. Hutchins, X. Yuan, J.M. Anderson, G.M. Santangelo, Relative citation ratio (RCR): A new metric that uses citation rates to measure influence at the article level. PLoS Biol. **14**, e1002541 (2016)

T. Lenoir, P. Herron, The NCI and the takeoff of nanomedicine. J Nanomedine Biotherapeutic Discovery **5**, 2 (2015)

L.A. Nagahara, J.S. Lee, L.K. Molnar, N.J. Panaro, D. Farrell, K. Ptak, J. Alper, P. Grodzinski, Strategic workshops on cancer nanotechnology. Cancer Res. **70**, 4265 (2010)

U. Prabhakar, D.C. Blakey, H. Maeda, R.K. Jain, E.M. Sevick-Muraca, W. Zamboni, O.C. Farokhzad, S.T. Barry, A. Gabizon, P. Grodzinski, Challenges and key considerations of the enhanced permeability and retention effect (EPR) for nanomedicine drug delivery in oncology. Cancer Res. **73**, 2412 (2013)

R. Truman, C.J. Locke, Gazelles, unicorns, and dragons battle cancer through the nanotechnology startup challenge. Cancer Nanotechnol **7**, 4 (2016)

W.C. Zamboni, V. Torchilin, A.K. Patri, J. Hrkach, S. Stern, R. Lee, A. Nel, N.J. Panaro, P. Grodzinski, Best practices in cancer nanotechnology: Perspective from NCI nanotechnology Alliance. Clin. Cancer Res. **18**, 3229 (2012)

The Impact and Relevance of Patient Advocacy

Anne Meyn[1]

Author's Personal Note Mauro was applying for the Department of Defense Innovator Award. Dr. John Mendelsohn had written a letter of recommendation. At the time, I was one of the patient advocates on the M.D. Anderson Cancer Center Breast Cancer SPORE and was their patient advocate on the Translational Breast Cancer Research Consortium which I assume is the way he got my name. I was asked to meet with Mauro regarding his application and to possibly serve as his patient advocate. I assumed it would be a few minutes, but time flew by. He was electric—so full of ideas. There was a whiteboard on the back of his office door. Before I knew it, we had covered the door exchanging ideas and mapping our plan. As I was soon to learn, Mauro didn't just want a name on a grant application, he named me executive director of the DoD Innovator Award. He has a true passion for the power and importance of patient advocates. I have always felt appreciated and that my ideas are honored. We have mutual respect. Mauro and Paola have been extremely supportive through Ray's surgeries, treatment, and eventual death. I will never forget their kindness.
Anne Meyn

The National Cancer Institute defines a research advocate as: *A person who serves as a link between patients and scientific researchers. Research advocates help patients understand scientific information and research findings that may help them. They may also bring a patient perspective on research activities to scientific advisory boards and committees. For example, research advocates use their own experiences and the experiences of patients to work with researchers to help develop clinical trials that are safe and meet the needs of patients. They also help teach patients and their families about clinical trials and recruit patients to clinical trials.*[1]

Cancer survivors as well as their friends, family members, and caregivers who have become breast cancer advocates add the patient perspective to all types of research from basic science to clinical research. It is important to emphasize that these research advocates both as individuals and groups of highly motivated volunteers are actively engaged in grassroots efforts to seek support for breast cancer research. As such, advocates eagerly seek partnerships with researchers in pursuing answers to the causes and prevention of breast cancer, and better targeted treatments to combat all types of breast cancer and thus to save lives. To serve as an effective advocate one must have good communication skills, be able to work collaboratively and possess the ability to separate their personal "story" from a representation of all patients.[1]

Initially, a novice advocate can feel overwhelmed by the science, but thankfully there are excellent resources. The National Breast Cancer Coalition conducts a weeklong intensive training called Project LEAD. The Research Advocacy Network provides many print and online courses and materials. Research Advocacy Network's Advocate Institute: http://researchadvocacy.org/advocate-institute. The American Association for Cancer Research, AACR, offers a Scientist↔Survivor Program during their annual conference. The AACR Scientist↔Survivor Program is designed to build enduring partnerships among the leaders of the scientific, cancer survivor, and patient advocacy communities worldwide. The program provides advocates with special lectures using lay language, small group discussions, and other opportunities for the exchange of information on key aspects of cancer research, survivorship, advocacy, and public policy. The American Society of Clinical Oncologists, ASCO, has programs and opportunities for patient advocates who attend the ASCO Annual Meeting. Information may be found here: https://www.cancer.net/research-and-advocacy/patient-advocates/patient-advocate-programs-asco-annual-meeting. Although ASCO does not have a formal research advocacy training program in place, they are working with the Research Advocacy Network (Elda Railey and Mary Lou Smith) to provide research advocacy education and training.

Other trainings include: Force (Facing Our Risk of Cancer Empowered) provides research advocacy training: http://www.facingourrisk.org/get-involved/how-to-help/volunteer-fundraise/volunteer.php#FRAT; YSC (Young Survival Coalition) RISE: Respected Influencers Through Science and Education: https://www.youngsurvival.org/get-involved/be-an-advocate/be-a-ysc-advocate#become-a-rise-advocate;

Susan G. Komen's Advocates in Science: https://ww5.komen.org/GetInvolved/Participate/BecomeanAdvocate/

A. Meyn (✉)
Houston Methodist Research Institute, MD Anderson Cancer Center, Houston, TX, USA
e-mail: ahmeyn@comcast.net

A. Grattoni and T. Desai (eds.), *Biomedical Micro-Nanotechnologies Toward Translation*,
https://doi.org/10.1007/978-1-0716-0187-7_3

BecomeanAdvocateinScience.html. In addition, Komen created a specific program for working with researchers called the ARMs Initiative…How do we build & cultivate meaningful, long-term advocate<=>researcher relationships. The goal is not about finding an "advocate" just as a name to submit in a grant proposal. The goal is to establish an ongoing relationship. The Call to ARMs webinar link is https://www.youtube.com/watch?v=wUEi-RN2kwc&feature=youtube.

During the San Antonio Breast Cancer Symposium, The Alamo Breast Cancer Foundation created Hot Topics that are nightly panels of renowned clinicians and researchers who identify what they feel are the most critical presentations of the day in a manner that the advocates can comprehend. It is extremely beneficial and well-received. The question and answer period can go almost as long as the presentations. The panelists are completely giving of their time and expertise. Not only are these sessions live-streamed, they are available for viewing at a later date. Kudos to the Alamo Breast Cancer Foundation members for their valuable contribution and hard work.

Advocates can help researchers understand that early-stage and advanced-stage breast cancer patients have different research needs. To have a metastatic breast cancer advocate involved in research is a constant reminder to the scientist of the urgency of positive results. Clinicians and researchers should be aware that research advocates can help reduce the barriers patients face in gaining access to clinical trials; provide resources to patients and their families; provide a link to the community; increase awareness about clinical trials; develop/review patient education materials; and match prospective clinical trial participants with a survivor who was a participant. The idea is to establish a rapport between the scientist/clinician and the advocate. For the young investigators one might have them give an "elevator" speech that is an explanation of his/her research in three minutes or less in terms that a lay audience could understand. It can frequently be very difficult for the graduate student, postdoctoral fellow or even a more veteran scientist. It is helpful for the advocate to visit the lab and attend an occasional lab meeting. Usually an institution will have departmental retreats. It is informative for advocates to attend.

Patient advocates who become research advocates enhance clinical trials by contributing to the study design, reviewing protocols, serving on study sections, participating in formal programs, and conducting as well as participating in focus groups. Many advocates serve on Institutional Review Boards and Data Safety Monitoring Boards as a nonmedical community representative. Research advocates have developed surveys to understand better the patients' needs, and their perceptions of research and clinical trials. The results have been used for protocol design. When an advocate has worked with a scientist on a successful research program, or with a clinician at the conclusion of a trial, the advocate can help to disseminate correct information regarding the conclusions. This might take the form of a seminar, or educational materials. Most advocates are involved with both online and face-to-face support groups and organizations. There are several that have been formed, for instance, for the advanced breast cancer community: The Metastatic Breast Cancer Network; The Metastatic Breast Cancer Alliance; The Lobular Breast Cancer Alliance (LBCA); Metavivor; and Metup.

A Patient Advocate's Perspective

When one survives breast cancer for 30 years, one thankfully sees much progress toward curing the diseases. As learned over time, and with much research, the cancer of any organ site is actually a compellation of numerous subtypes, breast cancer being no exception. Trite as the phrase is, it does seem that the more we learn, the more there is to learn. Although a drug regiment may be successful for one subtype, unfortunately it may very well not translate into shared success with other subtypes. There will not be one answer to all, which means bucking the current drug development trend and most notably addressing the novel emerging technologies, in respect to drug delivery systems, drug treatments, drug development to target cell-surface receptors, to identify potential for metastasis and resistance to therapeutics.

The surgery, chemotherapy, clinical trial of two more drugs and 6 weeks of radiation may have been a success story for this writer, but it cannot be said for thousands of others who were diagnosed and treated during the same period or in the subsequent years. Many were diagnosed with later-stage disease, or developed metastasis, and some patients did not respond to any treatments. Certainly, some could have been saved today with the discovery of HER2 and Herceptin (trastuzumab), but even today over 40,000 lives are still lost to metastatic breast cancer alone annually. Therefore, delivery of new and current drugs needs to reach the target whether it is a primary or metastatic site, in order destroy cancer cells. Deaths from breast cancer are not caused by the primary site, but in sites where the cancer spreads.

If the identification of the cancer and its particular cell types and pathways could be determined using small amounts of blood to test, this would certainly be a cheaper, more efficient method to establish the target pathways. With the development of molecular-specific contrast agents, it is now possible to facilitate the noninvasive detection and visualization of morphological and biochemical changes that influence disease and predict response to therapy. Physicians are better able to understand the molecular signatures of cancer cells enabling them to target abnormally activated pathways. When combined with more conventional diagnostic imaging, one could expect to have a major impact on the detection, diagnosis, and decision-making for personalized treatment. One attainable goal is to discover new biomarkers in order to verify cell types, and specific pathways to tailor treatments to the individual and to prevent unwarranted treatment. The times of one shot for all, big box, are over.

Hopefully, immunotherapy will have positive results in breast cancer, as it has in other cancers, but to date breast cancer patients have not numbered among them. Several new trials, however, suggest that the outlook for immunotherapy in breast cancer treatment may brighten soon. Recent trial results may very well prompt the first FDA approval of an immunotherapy treatment for breast cancer. Research advocates are most interested in this approach as well as many other strategies to boost the immune system to fight the cancer. Some trials are looking at radiation to stimulate the immune system prior to surgery. There are also trials where vaccines have been developed.

Using nanotechnology, nanoporous silicon particles, to deliver the drugs in a multistage approach allows these MSVs to carry their specific payload to the target increasing their therapeutic efficacy. This multistage approach along with implantable miniature devices to release the drugs further provides a controlled drug delivery of the predetermined, effective drug over a prescribed period of time. Rather than using a cumbersome pump, this miniature device could be implanted, relieving the patient of the burden of returning to a clinic for multiple infusions. The device might handle the drug delivery for months. This would save time and money for the patient and caregivers. The benefit to rural patients would be incredible. The side effects of the drugs should be significantly reduced, and therefore, this system should give the patient the ability to lead a more normal life during treatment. Hopefully, the drug will reach the target with limited invasiveness and without effecting surrounding organs and tissue.

Funding is always an issue for research. The competition is fierce for the limited money available. Nearly all the grants require a patient advocate to serve on the grant and some require this advocate to be part of a mentor team. Furthermore, advocates serve on the review panels and have equal voting rights as the scientific reviewers to fund new research. Dr. Mauro Ferrari completely believes in the patient advocate as a powerful voice in all endeavors within the realm of research. So much so when I originally worked with him in 2008 on his application for the Department of Defense Innovator award, he named me as Executive Director of the grant. This was not in name only; I attended weekly research update meetings. In addition, all breast cancer grant applications, publications, and new projects required my approval. Dr. Ferrari has backed this belief in patient advocates by budgeting money to pay his advocates for time spent on tasks for the department and the grants. Pursuit of knowledge has been high on his priorities. He has reimbursed advocates for attending symposiums, conferences, and classes. He has asked that I present during conferences he has hosted for the Physical Science Oncology Center. During brainstorming sessions for new projects, he has requested my impressions and ideas. For subsequent grant proposals, he has included additional advocates and myself. He has named me on the executive committee for the new NIH PSON Center grant. At one point, I had the opportunity to chair the members of CanCare for a day of talks and tours. Our team recently lost a patient advocate member of our Education Committee for PSON to metastatic lobular cancer. At our next meeting, she was honored, and her husband was sent an award in remembrance of her.

Research advocates desire more information on any method to successfully treat all types of breast cancer. Many new technologies provide much hope for patients. Through research and clinical trials, many lives will be saved, and the quality of life will be improved during treatment. The research advocates will distribute the successful results to the larger breast cancer community.

References

https://www.cancer.gov/publications/dictionaries/cancer-terms/def/research-advocate
Research Advocacy Network, www. http://www.researchadvocacy.org/
Breast Diseases Quarterly

Ethics Big and Small: From Communitarianism to Nanotechnology

Joseph J. Fins[1] • Susan Miller[1]

The paradox couldn't be greater: the pursuit of ethics on both a nano- and macroscale. When this sort of contrast is presented in discussions of ethics, the parallel is often metaphorical: scholars whose studies range from the microenvironment of the doctor–patient relationship all the way up to societal questions of distributive justice. But in the case of the life and work of Mauro Ferrari, the nano versus macro comparison is not metaphorical. It is quite literal. His work brilliantly examines the biochemistry, physics, and translational biomedical applications of the tiniest particles which are measured at 10^{-9} m while also seeking to transform STEM education and bring economic development to an impoverished town in southern Italy through *NanoGagliato*.

Co-founded by Mauro and his wife Paola Ferrari in 2008, *NanoGagliato* is an ambitious effort to bring first-rate science education to a small town in Calabria (2018). Once the center of the world, located on ancient shipping lanes on the Ionian Sea between Greece and Italy, Gagliato is now a decaying town. You feel steeped in the past and in the Western tradition which took root on this soil. But it is bittersweet because it seems that the town's better days are long past. Old elegant homes are in varying states of disrepair, and it is not uncommon to hear of a renovation that has to contend with an errant tree that has grown in what had once been a living room. But it is an enchanting setting, a sun-drenched place. At midday, the shutters on your window get so hot you could cook a frittata.

But the real warmth comes from the town folks, a welcoming community who embraces a band of scientists who come each summer to deliberate about nanoscience. When you arrive in town you are greeted by a banner across the main street. And all the people you meet know why you are here and make you feel part of the community.

We have attended as medical ethicists, scientists, physicians, and students, invited to address the normative implications of the science that is presented (Fins 2015). It is truly a magical voyage. We meet in the town's lone restaurant perched above the Mediterranean, cooled by a sea breeze that comes in off the balcony. Amidst pasta and pizza, some of the world's leading scientists give the sort of talks that could only be given in an intimate space. Work in progress, hypotheses, and speculations round out the usual data presentations one usually hears. It is all a very fertile mix and by the end of the day, your head is spinning full of ideas and hopes for interdisciplinary collaboration based on Mauro's academic style of generosity, passion, multidisciplinary cross-pollination, creative inquiry, and most importantly, skepticism. This crucible connects disparate subjects and expands the horizons of scientific knowledge by actualizing a mechanism to explore the ambiguities of the successes and the *failures* of the most recent scientific research. Self-correction occurs within the context of collegiality.

There are many highlights of the week but two stand out. The first is the night time festival when we present our conclusions in the town piazza. It's rather amazing as the entire community turns out for the show. For academics who usually operate in obscurity, the buzz is electric. You feel like a rock star when you talk about that little corner of academe that you inhabit. And the folks love it and applaud as if you were Dylan on tour, even if many don't understand what has been said. Afterward, there is a celebration. It's an Italian version of a massive block party, and it speaks to the power of community.

J. J. Fins (✉) · S. Miller
Division of Medical Ethics, Weill Cornell Medical College, 435 East 70th Street, Suite 4-J, New York, NY 10021, USA
e-mail: jjfins@med.cornell.edu

S. Miller
Department of Family Medicine, Houston Methodist Hospital, 6550 Fannin, Suite 1001, Houston, TX, 77030, USA
email: smmiller@houstonmethodist.org

A. Grattoni and T. Desai (eds.), *Biomedical Micro-Nanotechnologies Toward Translation*,
https://doi.org/10.1007/978-1-0716-0187-7_4

As good as that is, the true high point of the week is *Academia di Gagliato* which brings science education to children in Calabria who otherwise would not be exposed to first-rate STEM education. It is an innovative program that makes science fun, through a mix of summer camp and college that creatively makes learning accessible and enduring. And beyond the high schoolers who participate in the Academy, there is the *Piccola Academia* for the littler ones, children aged 5–14 where they can learn about nanoscience and its applications. Dreams begin there each summer. One can only imagine who these children will become because of this formative exposure to science and inquiry.

Paola and Mauro have been incredibly creative in their pedagogical approach, overcoming the one-room schoolhouse challenge. For example, before the visiting scientists and kids are brought together, the visitors answer a number of questions about our work and lives and we are asked to send a photo in advance. When we arrive, we discover that they have made personalized baseball cards. These are distributed to the kids and when we meet them they search us out and ask for our autographs and more importantly about our work. It is very affirming for the scientists and empowering for the kids who are told that we will be disappointed if no one asks for our autograph. So everyone feels welcomed and the kids have a pathway to engagement.

So what does this have to do with ethics? In their regard for Gagliato and the future of the children of Calabria, Mauro and Paola have given practical expression to communitarianism. Communitarianism is a philosophical school which stresses the needs of the community and social solidarity over individualism. It has roots in both the social justice mission of Catholicism and the European moderate left which has sought to minimize dramatic social inequalities and to improve quality of life standards enjoyed by all citizens. Fundamentally, communitarianism asserts that the relationship between the individual and his or her community is of central importance (Kuczewski 2001).

The work of *NanoGagliato* can be best understood as operating within this ethical and philosophical tradition. Here the collective well-being of the community, and that of each child, is best addressed by promoting economic development of the town *writ large* and ensuring that the children of Calabria have personalized access to improved educational opportunities. The approach is one that promotes solidarity and fosters community engagement all directed to a shared notion of the common good, economic development, and a better future for the children of the region.

In bringing *NanoGagliato* to Calabria at the southern tip of Italy, the Ferraris have planted seeds of hope in receptive soil. Diego Gracia Guillén, Spain's leading bioethicist and a distinguished historian of medicine, has observed that the geography of modern bioethics in Europe has ancient roots dating to the extent of the Holy Roman Empire (Gracia 1993). In areas of Southern Europe controlled by the Empire, Catholicism and a communitarian ethic has predominated. In Northern Europe, where Protestantism took root, rights language has predominated over social solidarity. And so it continues in the Mediterranean bioethics epitomized by the community building of *NanoGagliato.*

Closer to home, Daniel Callahan, a philosopher and co-founder of The Hastings Center, has counterposed communitarianism against the traditional liberalism of American bioethics which shares Anglo-Saxon roots with the bioethics of Northern Europe (Kuczewski 2001). In contrast to communitarianism which places the individual into the context of families and communities, American bioethics has favored notions of self-determination and the independence of the individual. Callahan has observed that while autonomy is important, viewing it as the primary ethical good can lead to distortions about the common good (Callahan 1984). For example, in the context of health care, it can lead prizing the negative right to refuse care over the positive entitlement to receive it.

Advancing communitarianism as central to his bioethics, Callahan has critiqued how American bioethics understands the relationship of the individual to community. Using a metaphor that will certainly appeal to Ferrari, the nanoengineer, Callahan has argued that American bioethicists have had an atomistic view of the individual (Callahan 2003). By that he meant one in which the individual is unconnected to others and operating independently from the forces of family and society. In Callahan's view, this is a moral fiction as we are socially constructed and that our individual choices and actions have consequences for others.

No one would imagine that Ferrari had an atomistic approach. Indeed, his work in nanotechnology is not at the atomic level but rather about atoms working together. Although neither of us are chemists, it is our understanding that nanoparticles are a *collection or community* of atoms. Indeed, the number of atoms in a nanoparticle is governed by the atom as well as the *shape* of the particle (Mori and Hegmann 2016), making the calculation of the number of atoms actually one that depends on how these atoms *relate to each other.* Without overstating the metaphor, we would assert that this is akin to a *communitarian nanochemistry.*

Ever attentive to the ethical implications of his work, Ferrari and colleagues have addressed ethical issues in nanotechnology. In one of the first articles on the subject (Hall et al. 2012), they summarized the historical precedents of nanotechnology and its transition from science fiction to Nobel awards in Chemistry and Physics. These authors linked the nascent field to the technology prediction summarized as "Moore's Law" and provided insights on the computational uses of nano-electron messaging. They reviewed the earliest clinical application of these initial nanomedicines which received FDA approval based on an understanding of the *mode of transport* rather than a molecular mode of action. This leads to subsequent approval of a "drug" that works through a thermal ablative effect rather than a more traditional pharmacological mechanism of action, a truly revolutionary development.

With this background, the essay considers the bioethical concerns associated with these innovations and the, as yet, unknown impact of these technologies on medicine and society. They note that while the controverted issues of risk/benefit, biodistribution, environmental contamination, weaponization, and economic effects of nanotechnology are not unique to this nascent field, the skillful application of oversight and regulations surrounding this area of scientific exploration remain transitory. However, based on the potential novel clinical applications of this technology, Dr. Ferrari and colleagues argue for a proactive approach to minimize regulatory delays while concurrently protecting public health. To make these arguments the authors contextualize the future of nanotechnology within the principlism articulated by Beauchamp and Childress (Hall et al. 2012). Dr. Ferrari argues that utilizing these precepts as a foundational approach to nanotechnology ethics will serve as an essential protective buttress while ensuring future societal benefits.

Even more fundamentally, Ferrari brings a prudential ethic to medical advancement. Although his work is at the forefront of an important and radically disruptive technology, he urges caution and deliberation, raising questions about the implications of these innovations. Indeed, he has called for "… careful scrutiny of the potential impact of nanomedicine on society, so that the community can guide its deployment in keeping with the fundamental tenets of medical ethics" (Ferrari et al. 2009). The roots of this appeal reside in his implicit communitarianism but also something deeper that he has explicitly acknowledged. And that is the legacy of medicine and the Holocaust.

In January 2012, the Center for Medicine After the Holocaust, led by Sheldon Rubenfeld MD, held an international conference on human subjects' research after the Holocaust at the Houston Methodist Research Institute (HMRI), (Rubenfeld and Benedict 2014). As president of the HMRI, Mauro hosted the gathering and gave his support. His hospitality as the leader of this Institute would have been testimony enough to declare his historical memory and ethical sensibility about mid-century horrors we should never forget, but he did more. He made his views explicit.

In an elegant and highly personal essay that is part confessional and part personal philosophy of doing good against the prospect of evil (Ferrari 2014), he begins, in a typically Mauro manner, with modesty and humility noting that:

> I am neither a medical ethicist nor a historian. I have no direct personal or Family relationship to the Shoah. I am not Jewish. If anything, because of my Italian origins, I am, in a broad sense, on the wrong side of the conversation… (Rubenfeld and Benedict 2014)

He explicitly argued that the lessons of the Holocaust must not be ignored in any consideration of medical ethics that has "any real depth or societal significance" (Rubenfeld and Benedict 2014). And more to the point, he warned of potential harm emerging from the dual use of therapeutic nanoparticles as weapons of mass destruction. Because of these concerns he made a decision to abandon "… nanomedical research for the oral delivery of biological pharmaceuticals and, at the same time, to publish my concerns on this type of research" (Rubenfeld and Benedict 2014).

Ferrari's decision to discontinue potentially dangerous research speaks to the view that science must promote human flourishing, a stance taken by the Presidential Commission for the Study of Bioethical Issues in (2011). His commitment to transparency points to need for science to be governed and guided by democratic deliberation. The promotion of human flourishing and social engagement are both are communitarian values. Each places science into a larger social context.

This is precisely what happens each summer at *NanoGagliato*, a name which itself embodies the work of Mauro Ferrari: regard for ethics big and small and attention to the interplay of science and society. Simply put, it is as the *NanoGagliato* mission statement asserts a place *where science and people meet* (2018). A very special place indeed.

References

D. Callahan, Autonomy: a moral good, not a moral obsession. Hast. Cent. Rep. **14**(5), 40–42 (1984)

D. Callahan, Individual good and common good: a communitarian approach to bioethics. Perspect. Biol. Med. **46**(4), 496–507 (2003)

M. Ferrari, M.A. Philibert, W.R. Sanhai, Nanomedicine and society. Clin. Pharmacol. Ther. **85**(5), 466–467 (2009)

M. Ferrari, No exceptions, no excuses: a testimonal, in *Human Subjects Research after the Holocaust*, ed. by S. Rubenfeld, S. Benedict (Springer International Publishing, Switzerland, 2014), pp. 283–297

J.J. Fins, Nanotechnology, neuromodulation & the immune response: discourse, materiality & ethics. Biomed. Microdevices: BioMEMS Biomed. Technol. **17**(2), 9934 (2015). https://doi.org/10.1007/s10544-015-9934-0

D. Gracia, The intellectual basis of bioethics in Southern European countries. Bioethics **7**(2/3), 97–107 (1993)

R.M. Hall, T. Sun, M. Ferrari, A portrait of nanomedicine and its bioethical implications. J. Law Med. Ethics **40**(4), 763–779 (2012)

M.G. Kuczewski, The epistemology of communitarian bioethics: tradition in the public debates. Theor. Med Bioeth. **22**(2), 135–150 (2001)

T. Mori, T. Hegmann, Determining the composition of gold nanoparticles: a compilation of shapes, sizes, and calculations using geometric considerations. J. Nanopart. Res. **18**(10), 295 (2016)

NanoGagliato. Accessed on 22 October 2018 at: http://www.nanogagliato.com/index.php?lang=en

Presidential Commission for the Study of Bioethical Issues, *"Ethically Impossible" STS Research in Guatemala from 1946 to 1948* (U.S. Government Printing Office, Washington, D.C., 2011)

S. Rubenfeld, S. Benedict (ed.), *Human Subjects Research After the Holocaust* (Springer International Publishing, Switzerland, 2014)

Medicine and Bioethics After the Holocaust

Sheldon Rubenfeld[1]

Author's Personal Note Just before he stepped down as president and chief executive officer of the Methodist Hospital system in 2011, Ron Girotto introduced me to Mauro Ferrari. Ron thought Mauro and I, as founding Chairman of the Center for Medicine after the Holocaust (CMATH), would work well together to advance the study, teaching, and researching of medicine after the Holocaust.

"As he was stepping down from his leadership role, Ron expressed his enthusiastic appreciation for CMATH's mission and recommended with great emotion that I continue supporting it. It has been my great privilege to do so ever since, and a true gift for my soul…" is how Mauro reflects upon this event in his typically gracious manner in the 2014 book that resulted from our first conference, *Human Subjects Research after the Holocaust.*

Drawing on his personal and family history, Mauro derived five insightful lessons about medical ethics that he recorded in his chapter in that book, "No Exceptions, No Excuses: A Testimonial." They are:

1. Any discussion of medical ethics that ignores the lessons from the Holocaust cannot have any real depth or societal significance.
2. We cannot make a real, positive difference, unless we realize that we are all part-bad guys—it is not "them evil vs. us good."
3. There are major elements of the path to the Holocaust in today's medicine and culture.
4. The future may be scarier than the past.
5. Forgiveness is the sole victor.

Mauro learned this last lesson from the searing and profound words of Eva Kor, a survivor of Josef Mengele's experiments on twins at Auschwitz. Mauro wrote, "As a father of two sets of twin girls and five children in total, it was impossible for me to sit through her presentation at the CMATH conference and not be moved to tears, many times."

He concluded that medical ethics pertains to everyone involved in healthcare, including medical researchers like himself, and ended his chapter with: "We are all deeply involved in medicine; we all have the power to bring relief from suffering and the power to cure. We are all healers. The tenets discussed here pertain to all of us. No exception. No excuses."

True to his word, Mauro has continued to support and advance the study, teaching, and researching of medicine after the Holocaust. In 2015 Mauro and HMRI co-hosted with CMATH the First International Scholars Workshop on Medicine after the Holocaust, which set the stage for a larger Second International Scholars Workshop on Medicine after the Holocaust in Israel in 2017.

Again in 2017 Mauro and HMRI co-hosted with CMATH a conference on Bioethics after the Holocaust. Mauro and I then chose to focus our efforts on one particularly thorny bioethical issue about which he is ardently concerned, physician-assisted suicide (PAS) and euthanasia after the Holocaust.

The Nazi euthanasia programs were the fulcrum of the Holocaust and the medical profession played a central and indispensable role in the design and implementation of those programs as well as the Holocaust. More than 70 years after the Holocaust, PAS and euthanasia are permitted in several European countries for infants, children, terminally ill adults, patients with psychiatric illnesses, and for those "ready to give up on life." While euthanasia is not yet legal in the United States, PAS is legal in nine states and Washington, D.C. and many other states are considering similar legislation.

To stimulate a debate about PAS and euthanasia after the Holocaust in the United States, Mauro and I invited a dozen American and Canadian academic physicians with experience in palliative care and bioethics to Europe in May 2018 for visits to medical sites relevant to the Holocaust and a conference presented by European scholars on PAS and euthanasia before, during, and after the Holocaust. The scholars processed this transformative experience and presented their manuscripts at another conference co-hosted by HMRI and CMATH in January 2019 at HMRI in anticipation of publication of a book entitled *Physician-Assisted Suicide and Euthanasia after the Holocaust.*

Mauro Ferrari is a man with broad experience, passionate interests, consummate talents, and an animated personality. It has been a pleasure and an honor to know him, to work with him, and to call him a colleague and a friend.

Sheldon Rubenfeld

S. Rubenfeld (✉)
Clinical Professor of Medicine, Baylor College of Medicine, Houston, TX, USA
Executive Director, Center for Medicine after the Holocaust, Houston, TX, USA
e-mail: sheldonrubenfeld@mac.com

A. Grattoni and T. Desai (eds.), *Biomedical Micro-Nanotechnologies Toward Translation*,
https://doi.org/10.1007/978-1-0716-0187-7_5

1 Introduction

The Nuremberg Medical Trial (NMT) was a watershed event and a missed opportunity for Western medicine. In 1947 an American three-judge panel declared 16 of 23 Nazi defendants guilty of medical war crimes, wrote "Permissible Medical Experiments" better known as the Nuremberg Code, and found one of Western civilization's most celebrated professions wanting. The opening statement for the prosecution states:

> The thanatological knowledge, derived in part from these experiments, supplied the techniques for genocide, a policy of the Third Reich, exemplified by the 'euthanasia' program and in the widespread slaughter of Jews, Gypsies, Poles, and Russians. This policy of mass extermination could not have been so effectively carried out without the active participation of German medical scientists (Schmidt 2004).

The NMT's judgment cries out for an in-depth examination of the causes, rationales, methods, policies, and underlying moral principles that led German physicians to play an indispensable role in the design and implementation of a uniquely medical genocide, the Holocaust. And yet, as we approach the 75th anniversary of the NMT in 2021, Western medicine has disregarded and dismissed the behavior of Nazi physicians, and of American eugenicists who provided moral, legal, and philanthropic support to them.

Self-reflection about these events is required because contemporary healthcare and medical education in Western democracies were built upon German models that made their medicine the most advanced in the world when Hitler became chancellor in 1933. Shortly after Bismarck unified Germany in 1871, he proposed centralized social welfare programs including universal medical insurance that were emulated by many European countries and Japan. When former Republican President Theodore ran for president as the nominee of the Progressive Party in 1912, he included the Bismarck model of health insurance in his political platform. Although his presidential bid was unsuccessful, Roosevelt's progressive, centralized healthcare proposals ultimately prevailed. As President Obama's said in his address to Congress in 2009, "I am not the first President to take up this cause, but I am determined to be the last. It has now been nearly a century since Theodore Roosevelt first called for health care reform. And ever since, nearly every President and Congress, whether Democrat or Republican, has attempted to meet this challenge in some way" (Nelson 1787).

The US, encouraged by philanthropists and the 1910 Flexner report, emulated the German pedagogic model of scientific medical education. Full-time medical faculty dedicated to scientific research and teaching with minimal clinical responsibilities became and remains the dominant American academic model. Clinicians such as Osler, Cushing, and Peabody voiced their fears that patients could be harmed in a medical model that favors science over caring for patients and the sanctity of life itself (Duffy 2011). Their fears became reality with the Third Reich's euthanasia programs.

Sanctity of life was given by Moses on Mt. Sinai, moored in medicine by Hippocrates on the Isle of Kos, and conveyed by Christianity to twentieth-century Western physicians. The NMT documented what happens when medicine rejects "Judeo-Christian Hippocratism" and chooses science over sanctity of life, relief of suffering over healing, and the state over the individual. This history challenges physicians to personally confront the medical ethics of the Holocaust and to critically examine the moral values and the mission of modern medicine (Cameron 2001).

2 German Eugenics

Hoping to improve the genetic stock of the *volk* (people), the German medical profession enthusiastically promoted eugenic policies long before Hitler became chancellor. In 1920 psychiatrist Alfred Hoche and jurist Karl Binding stated in their influential eugenic treatise that "The right to life must be earned and justified, not assumed" and that "Those incapable of human feeling—'empty human husks' in the mental hospitals—should be destroyed as a humane act, even involuntarily, by the state." They realized that their proposals were contrary to religious medical ethics and the Hippocratic Oath:

> There was a time, now considered barbaric, in which eliminating those who were born unfit for life, or who later became so, was taken for granted. Then came the phase, continuing into the present, in which, finally, preserving every existence, no matter how worthless, stood as the highest moral value. A new age will arrive – operating with a higher morality… (Binding and Hoche 1975)

and "The young physician enters practice without any legal delineation of his rights and duties—especially regarding the most important points. Not even the Hippocratic Oath, with its generalities, is operative today" (Binding and Hoche 1975).

After he read *Human Heredity and Racial Hygiene* in Landsberg prison, Hitler incorporated eugenics into his worldview (Proctor 1988). He wrote in *Mein Kampf*, "Whoever is not bodily and spiritually healthy and worthy, shall not have the right to pass on his sufferings in the body of his children" (Proctor 1988). Figure 1 from a high school biology textbook conveys the scientific and economic rationale for the eugenic Nazi healthcare policy of "Applied Biology" (Proctor 1988).

Physicians eagerly supported Hitler's healthcare policy because they themselves had pioneered it. In 1929, four years before Hitler became chancellor, they formed the

Fig. 1 The caption reads: "You Are Sharing the Load! A Hereditarily Ill Person Costs 50,000 Reichsmarks on Average up to the Age of Sixty."

Nazi Physician League to "coordinate Nazi medical policy and to purify the German medical community of the influence of Jewish Bolshevism" (Proctor 1988). At the end of WW II nearly 50 percent of Germany's physicians were members of the Nazi party and seven percent were members of the SS, both percentages much higher than the general population or other professions (Proctor 1988).

German physicians voluntarily and enthusiastically joined Hitler on the slide down this slippery slope

Applied Biology
    Sterilization Law
        Nuremberg Laws
            Child Euthanasia
                Adult T4 Euthanasia
                    Medical Expts
                        Final Solution

that began with the involuntary sterilization of 400,000 German men and women between 1934 and 1939. In accordance with the 1935 Nuremberg Laws, physicians examined all betrothed couples to ascertain that they were "fit to marry" (Proctor 1988).

The lynchpin of the Nazi healthcare policy was the child euthanasia program that began with an autonomous, unsolicited request from the family of infant Knauer that Hitler bring about its death in 1938, which was granted. In the subsequent nationwide program, physicians and midwives submitted a registration form for children with disabilities that was reviewed by three expert referee physicians without seeing or examining the children. Physicians at 28 institutions slowly poisoned, starved, and exposed to the elements 5,000 German children. Jews were initially excluded because they were not considered worthy of "mercy killing" (Proctor 1988).

It is apparent that if German physicians were willing to kill 5,000 of their countrymen's disabled children, they would have no qualms about experimenting upon or killing others in the rapidly enlarging sphere of the genetically inferior such as Jews, blacks, homosexuals, Sinti and Roma, and Jehovah's Witnesses.

The adult involuntary euthanasia or T4 program, named after its headquarters at Tiergartenstrasse 4, was modeled after the children's euthanasia program and began with this authorization signed by Hitler in October 1939:

> Reichleiter Bouhler and Dr med Brandt are charged with responsibility to extend the powers of specific doctors in such a way that, after the most careful assessment of their condition, those suffering from illnesses deemed to be incurable may be granted a mercy death. (Lifton 1986)

The "most careful assessment" was review of a registration form without seeing the patients and then, as in the children's euthanasia program, marking them with a red "+," a blue "–," or an occasional "?" (Lifton 1986) (See Fig. 2). Simultaneously physicians worked with chemists and engineers to design and install gas chambers and crematoria at six euthanasia centers to eliminate most of Germany's mental patients. The mass gassings ended after 70,000 deaths but physicians then murdered another 130,000 victims by poison, starvation, and exposure to the elements. It is important to note that physicians were empowered, not ordered, to commit these merciful murders and did so voluntarily and often enthusiastically (Proctor 1988).

On January 20, 1942, Reinhard Heydrich and Adolf Eichmann and 13 other Nazi Party and German government officials met in Wannsee to implement the "Final Solution of the Jewish Question." They planned to medicalize mass murder by employing T4 physicians to train more of their colleagues to kill Europe's 11 million Jews in bigger and better gas chambers and crematoria. While the Nazis did not accomplish their goal, they did gas approximately 4,500,000 Jews (Lifton 1986).

German physicians believed that relieving the suffering of lives "not worth living" had greater moral value than healing patients. When Nazi physicians took the SS oath, they transformed the Hippocratic Oath's doctor–patient relationship to a state-*volkskörper* (the people's body) relationship, thereby transferring their loyalty from their patients to the state as represented by Adolf Hitler:

> I swear to you, Adolf Hitler—as the Führer and Chancellor of the Reich—loyalty and bravery. I pledge to you and to my superiors, appointed by you, obedience unto death, so help me God. (Lifton 1986)

Karl Brandt, head of the T4 program and lead defendant at the NMT, justified his behavior:

> I am deeply conscious that when I said 'Yes' to euthanasia I did so with the deepest conviction, just as it is my conviction today, that it was right. Death can mean relief. Death is life-just as much as birth. It was never meant to be murder. (Schmidt 2004)

German physicians believed that healing by killing was ethical and noble, and that eugenic benefits to the *volk* and economic benefits to the state justified their medical genocide or "eugenicide."

3 American Eugenics

Many prominent Americans enthusiastically supported eugenics including Presidents Theodore Roosevelt and Woodrow Wilson; David Starr Jordan, MD, president of Stanford University; Nobel laureate Dr. Alexis Carrel; Alexander Graham Bell, inventor of the telephone; Margaret Sanger, founder of Planned Parenthood; the Carnegie Institution, the Rockefeller Foundation, and philanthropists Mary Harriman and John Harvey Kellogg, inventor of corn flakes; and United States Supreme Court Justice Oliver Wendell Holmes Jr. (Black 2003).

Indiana passed the world's first involuntary eugenic sterilization law in 1907. In 1927 the Supreme Court of the United States declared in an 8-1 decision in *Buck v. Bell* that the Virginia involuntary sterilization law, the model for the Nazis 1933 sterilization law, was constitutional. The infamous majority opinion ended with, "Three generations of imbeciles are enough." Thirty states sterilized approximately 60,000 Americans between 1907 and 1981 when Oregon performed the last legal involuntary sterilization. Most of the sterilization laws were never repealed (Black 2003).

Madison Grant, conservationist, lawyer, and friend to American presidents, extolled the virtues of the Nordic race in his 1916 *The Passing of the Great Race*. Hitler sent him a fan letter stating, "The book is my Bible" and then distributed the German translation of the book to his staff and

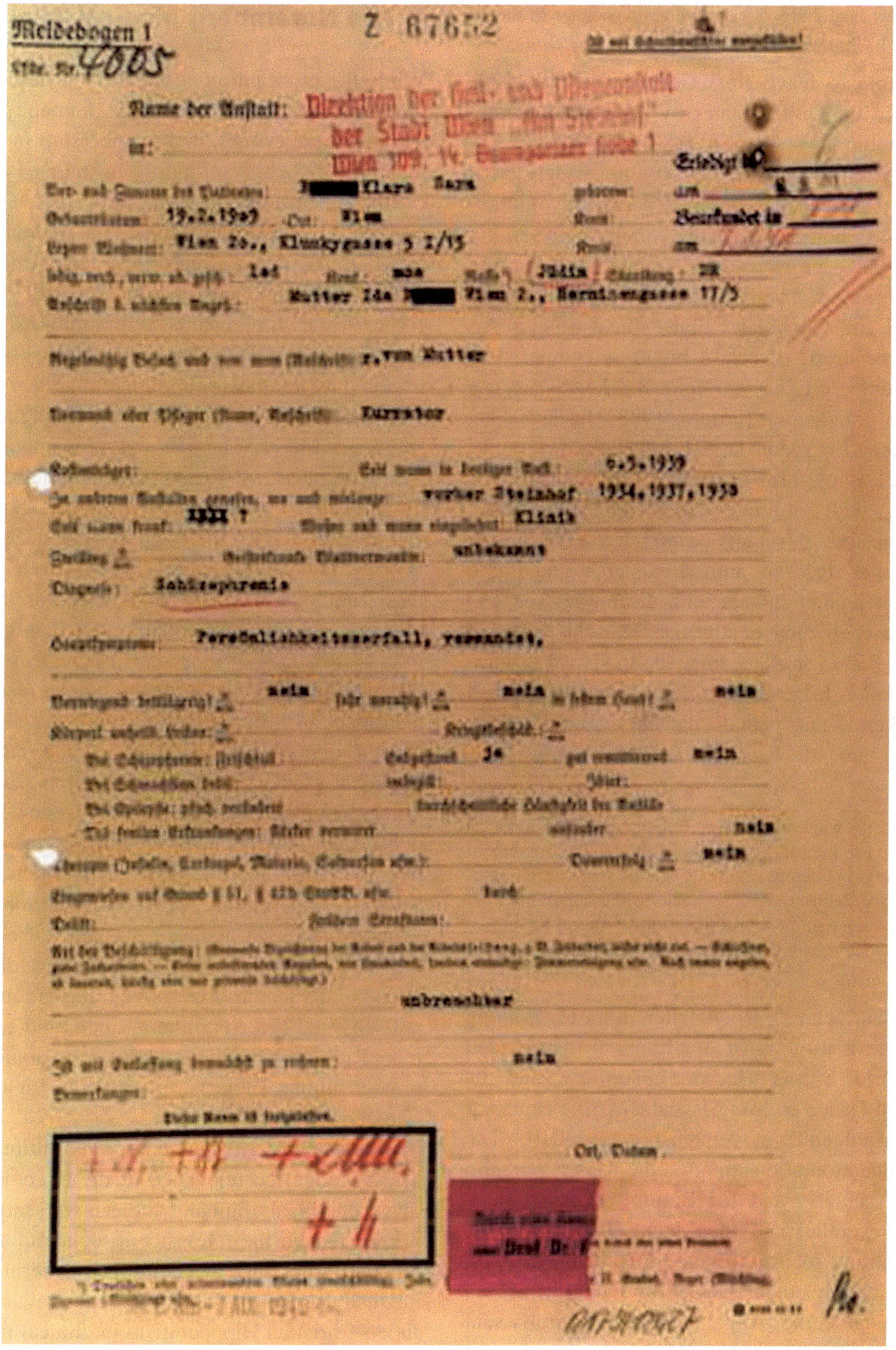
Meldebogen 1

Lfde. Nr. 4005

Name der Anstalt:

[redacted] Klara Sara

19.2.1909 Wien

Mutter Ida [redacted]

Kurator

6.5.1939

Klinik

unbekannt

Schizophrenie

nein

unbrauchbar

nein

Fig. 2 Registration form for patient Klara B (1909–1940) marked for death with a red "+" in the T4 euthanasia program

generals (Black 2003). In *Mein Kampf* Hitler praised American eugenic legislation passed in the early 1920s to prevent all Japanese and Chinese and most Jewish and Italian immigration:

> There is today one state in which at least weak beginnings toward a better conception are noticeable. Of course, it is not our model German Republic, but the American Union, in which an effort is made to consult reason at least partially. By refusing immigration on principle to elements in poor health, by simply excluding certain races from naturalization, it

> professes in slow beginnings a view which is peculiar to the folkish state concept. (Black 2003)

Jews attempting to leave Nazi Germany in the 1930s were barred from the United States by harsh enforcement of these xenophobic laws (Proctor 1988).

While the Nazi's sterilization law was not widely criticized—at least a dozen countries had passed involuntary sterilization laws before Germany—their anti-Semitic policies were. Nazi propagandists countered by citing America's centuries of slavery, racism in medical school admissions, and Jim Crow laws that defined a Negro as anyone with "one drop" of Negro blood as compared to Germany's Nuremberg Laws that defined a Jew as anyone with one Jewish grandparent (Proctor 1988; Black 2003).

In 1915 Chicago surgeon Dr. Harry Haiselden publicized his refusal to perform life-saving surgery on Baby Bollinger born with multiple congenital deformities, claiming that successful surgery would leave the infant with nothing better than "an animal existence and imbecility." He had let other infants with disabilities die and occasionally hastened their deaths. Dr. Haiselden played himself in *The Black Stork*, a film about his experience with Baby Bollinger. Eugenicists in Britain and the US had previously spoken about "lethal chambers" and Madison Grant expressed their ideology well in *The Passing of the Great Race*:

> Mistaken regard for what are believed to be divine laws and a sentimental belief in the sanctity of human life tend to prevent both the elimination of defective infants and the sterilization of such adults as are themselves of no value to the community. The laws of nature require the obliteration of the unfit and human life is valuable only when it is of use to the community or race. (Black 2003)

Mary Harriman, John D. Rockefeller, John H. Kellogg and the Carnegie Institution of Washington all provided funding for Cold Spring Harbor Laboratory and its Eugenics Record Office on Long Island, NY, the center of the international eugenics movement (Greely 2010). The Rockefeller Foundation also supported the Kaiser Wilhelm Institutes for Psychiatry and for Brain Research (Black 2003). First Rockefeller and then James Loeb provided funds to the Institute for Anthropology, Human Heredity, and Eugenics for twin studies headed by Otmar von Verschuer. Josef Mengele, the "Angel of Death," dutifully sent eyes, body parts, sera and paperwork from his experiments on twins at Auschwitz to his mentor Verschuer in Berlin (Lifton 1986; Black 2003).

4 The Nuremberg Medical Trial

While the prosecution at the NMT accused the defendants of war crimes, crimes against humanity, conspiracy to commit these crimes, and membership in a criminal organization, it focused primarily on human subjects research conducted on concentration camp prisoners without their consent. The defense argued that there was no binding international law distinguishing legal from illegal human subjects research and that the medical experiments performed by the defendants were similar to those of American scientists. The three-judge panel was sufficiently troubled by this line of argument that they included in their 1947 verdict ten basic principles defining legitimate human subjects research. The first and most important principle of the Nuremberg Code is: "The voluntary consent of the human subject is absolutely essential" (Schmidt 2004).

These principles were ignored for decades. In 1964 the World Medical Association (WMA) released its Helsinki Declaration, a less restrictive code for the protection of human research participants that appeared to put the interest of society in medical progress over the interest of the individual. Concern for the human research participant resurfaced in 1966 following Henry Beecher's article in the *New England Journal of Medicine* cataloguing many unethical American human subjects experiments. The article had little impact until 1972 when newspaper articles publicized the US Public Health Service's denial of curative penicillin to 400 poor black men in its Tuskegee Syphilis Experiment, which caused more than 100 preventable deaths. Caught in the glare of adverse publicity, the US government ended the experiment, passed the National Research Act, issued the Belmont Report that rediscovered the Nuremberg Code, and created the Common Rule and Institutional Review Boards or "IRBs" to protect human subjects volunteering for medical research (Greely 2010).

While Western medicine was ultimately forced to acknowledge the relevance of the Nuremberg Code to its own unethical human subjects research, it has not acknowledged the relevance to contemporary medicine of broader issues such as centralized governmental control of medical economic and political power, disregard for Judeo-Christian Hippocratism in clinical medicine, and the empowerment of physicians to kill their patients. Physicians in Western democracies dismiss these issues, in part, because they believe they are incapable of committing eugenic crimes similar to those perpetrated by their Nazi colleagues. The history of the American eugenics movement demonstrates that this comforting assumption is inaccurate.

When it comes to euthanasia, the lynchpin of the Holocaust, modern medicine notes that Nazi programs were involuntary while contemporary PAS and euthanasia are voluntary and originate in autonomous requests from patients or their surrogates. While this distinction is true it is a false moral premise—a medical profession committed to healing rather than relieving suffering should reject both involuntary and voluntary requests for PAS and euthanasia.

5 Medicine After the Holocaust

In 1948 the newly formed World Medical Association issued its Declaration of Geneva, a reworking of the Hippocratic Oath that opened the door for many other revisions that dramatically altered the intent of the original (Cameron 2001). For example, the Oath that Bears the Name of Hippocrates, which is recited by the graduates of Geisel School of Medicine at Dartmouth (Duncan 2017), changes two key paragraphs of the original from:

> I swear by Apollo the Physician, by Asclepius and by Hygieia, by Panacea, and by all the gods and goddesses, making them witnesses, that I will carry out, according to my ability and judgment, this oath and this indenture. (Cameron 2001)

to:

> I do solemnly swear, by whatever each of us holds most sacred. (Duncan 2017)

and:

> I will use treatment to help the sick according to my ability and judgment, but I will never use it to injure or wrong them. I will not give poison to anyone though asked to do so, nor will I suggest such a plan. Similarly I will not give a pessary to a woman to cause abortion. But in purity and in holiness I will guard my life and my art. (Cameron 2001)

to:

> That I will exercise my art solely for the cure of my patients, and will give no drug, perform no operation for a criminal purpose, even if solicited; far less suggest it. (Duncan 2017)

These two revisions alone change medicine from a supernaturalistic covenantal oath with specific prohibitions into a secular contractual vow with statements that leave patients wondering what kind of ethics their doctors profess. By revising a few universal principles easily understood by patients, medicine has abdicated its professional responsibilities and created a moral vacuum that a new cadre of experts—primarily lawyers, judges, biomedical ethicists, and politicians—has occupied.

The authors of *Principles of Biomedical Ethics*, a standard bioethics textbook, dismiss religious and Hippocratic medical ethics, stating in the book's opening pages:

> Although major writings in ancient, medieval, and modern health care contain a rich storehouse of reflection on the relationship between the professional and the patient, these writing are inadequate for contemporary biomedical ethics...To avoid a similar narrowness, we begin with philosophical reflection on morality and ethics that is removed from the history of professional ethics. (Beauchamp and Childress 1994)

They then restate the four principles of contemporary secular biomedical ethics articulated in the Belmont Report about human subjects research—autonomy, nonmaleficence, beneficence, and justice—and extend, without adequate justification in my view, these principles from the primarily contractual doctor–patient relationship in human medical experimentation to the primarily covenantal doctor–patient relationship in clinical medicine (Beauchamp and Childress 1994).

The Supreme Court of the United States in its 1973 Roe v. Wade decision similarly dismissed both religious and Hippocratic objections in its finding of "the right to privacy, including a woman's qualified right to terminate her pregnancy." Section VI.1 states, "Ancient religion did not bar abortion." While this may be technically true, it does not accurately capture the spirit of religious prohibitions against abortion. Traditional Judaism, for example, obligates a physician to perform an abortion when a pregnancy threatens the mother's life, but it bars abortion in most other circumstances. In section VI.2, the justices acknowledge the enduring influence of the Hippocratic prohibition of abortion but then dismiss it based upon Dr. Edelstein's suggestion that it is "a Pythagorean manifesto and not the expression of an absolute standard of medical conduct" (Wade 1973).

Roe v. Wade also contributed to a dangerous concentration of medical decision-making in the federal government by overruling state laws on abortion. Holocaust historian Henry Friedlander notes, "Until World War I, German eugenics paralleled the eugenics movement in the United States..." Why then didn't America, the world's leader in eugenics, go down the same path as German medicine? Friedlander points out that German eugenics was "more centralized" while in the United States "federalism and political heterogeneity encouraged diversity even within a single movement" (Friedlander 1995). While the concentration of political and economic medical power in the federal government might improve American healthcare, the results of well-intentioned medicine during the Third Reich highlights the potential for a centralized government's bad healthcare policy prescriptions to harm large numbers of patients. Prosecution witness Dr. Leo Alexander testified at the NMT about the harmful collaboration between the German government and willing members of his own profession:

> All of them have in common a callous lack of consideration of, and human regard for, and an unprincipled abuse of power over the poor, unfortunate defenceless creatures who had been deprived of their human rights by a ruthless and criminal government. All of them violated the Hippocratic commands which they had solemnly sworn to uphold and abide by—such as the fundamental principle never to do harm. (Schmidt 2004)

6 Healing by Killing

Reflection upon contemporary end-of-life practices highlights some disconcerting trends in Western medicine after the Holocaust. Just as scientists wishing to experiment on humans felt constrained by the Nuremberg Code's explicit language to protect human research participants from harm, decision-makers in clinical medicine and healthcare policy may feel constrained by the Hippocratic Oath and traditional religion's explicit language to protect the lives of patients. The danger is that experts may once again promote relief of suffering and put the interests of society and cost saving over the interests of patients and physicians to justify healing by killing.

Permission for PAS and euthanasia rests upon the perceived right of an autonomous patient or a surrogate to request medical assistance with dying to relieve suffering from a poor quality of life, and upon the physician's subsequent obligation to either assist with the patient's suicide or actively end the patient's life. Before the relatively recent introduction into clinical medicine of a patient's right to die, patients and physicians shared a duty to heal the patient, a duty that explicitly proscribed suicide by the patient and murder by the physician.

Two years after the end of the NMT Dr. Alexander reflected upon the potential danger of accepting quality of life as more important than life itself:

> The beginnings at first were merely a subtle shift in emphasis in the basic attitudes of physicians…It started with the acceptance…that there is such a thing as a life not worthy to be lived…Gradually the sphere of those to be included in this category was enlarged… (Alexander 1949)

Contemporary bioethics has slowly but surely shifted its emphases from life to quality of life, from healing to relief of suffering, and from Judeo-Christian Hippocratic covenantal duties to secular philosophical contractual rights with a worrisome legislative enlargement of the "sphere of those included in this category."

These changes are particularly problematic for the US because it was founded on Hebrew scripture as well as enlightenment principles, making it the only Judeo-Christian and arguably the most religious of any Western democracy (Shalev 2013). Figure 3 portrays a common view of America at her founding as the "Second Israel" in Benjamin Franklin's initial design for the first Great Seal of the United States.

Fig. 3 Benjamin Franklin described his first design for the Great Seal of the United States as: "Moses standing on the shore, and extending his hand over the sea, thereby causing the same to overwhelm Pharaoh who is sitting in an open chariot, a crown upon his head and a sword in his hand. Rays from a pillar of fire in the clouds reaching to Moses, to express that he acts by Command of the Deity." The motto read: "Rebellion to Tyrants is Obedience to God."

In 1989 Oregon developed a prioritized list of health services ranked in order of "clinical effectiveness and cost-effectiveness" for its citizens covered by Medicaid and the State Children's Health Insurance Program. In 1997 the Oregon Death with Dignity Act allowed terminally ill Oregonians to end their lives with voluntary PAS. In 2008 Barbara Wagner and Randy Stroup each received letters denying expensive treatment for their cancers but offering to pay for their inexpensive "comfort care" including PAS (James 2017).

In 1991, some time after Holland legalized PAS and euthanasia, the Remmelink Report documented that physicians *involuntarily* euthanized 1,000 patients (van Delden et al. 1993). Belgium legalized euthanasia in 2002 for adult patients and emancipated minors with full mental capacity and constant, unbearable physical or mental suffering, even if they are not terminally ill. In 2014 Belgium extended the law to terminally ill children with consent of their parents (Siegel et al. 2014). Legally enshrining relief of suffering as important as, if not more important than, healing opens the door for the identification of suffering everywhere and in everyone.

Given that autonomy is the dominant principle of contemporary secular biomedical ethics and that rights granted by the state are the method for guaranteeing that autonomy, we must consider the arguments in favor of infanticide by Peter Singer, Ira W. DeCamp Professor of Bioethics in the University Center for Human Values at Princeton University. He argues that quality of life is a sound moral basis for making life and death decisions in medicine. If spina bifida, for example, will give a child a "miserable" life, then that life is "not worth living;" if the death of a defective infant would lead to the birth of an infant with better prospects for happiness, the total amount of happiness will be greater if the defective infant is killed. The loss of happy life for the first infant is outweighed by the gain of a happier life for the second (Weikert 2016).

If society and politicians follow Singer's prescriptions to their logical conclusions, we risk transforming one of America's foundational principles enshrined in the Declaration of Independence:

> We hold these truths to be self-evident that all men are created equal, that they are endowed by their Creator with certain unalienable Rights, that among them are life, liberty and the pursuit of happiness.

into a life-threatening prescription:

> We hold these truths to be self-evident that all men are created equal, that they are endowed by their State with certain unalienable rights, that among them are quality of life, autonomy, and happiness.

Physicians must resist these changes if they are to protect the lives of their patients and preserve the moral integrity of their profession.

7 Conclusion

The Nuremberg Code could have alerted Western physicians to the potential for corruption of medical morality when healing is subjugated to relief of suffering for political, cultural, or philosophical reasons, but the behavior of Nazi physicians was either ignored or dismissed as an aberration not worthy of serious deliberation. By promoting the dominance of quality of life over life, autonomy over all other principles, and rights over duties, philosophers, lawyers, bioethicists and politicians have created a new medical morality that includes an unacceptable acceptance of physician involvement in killing patients. Physicians must study their history if they wish to renew their commitment to first do no harm and restore moral integrity to medicine after the Holocaust.

I conclude with a tale told by Dr. William Seidelman about

> a timeless symbol of medicine and the ethical spirit of the profession: the Greek Island of Kos. On Kos, there exists the remains of a major temple to the Greek god of medicine, Aescelapius. Kos is also the mythical birthplace of Hippocrates, who created the paradigm, perhaps mythical as well, of the ethical physician. An oft-visited site in the town of Kos is an ancient plane tree where, legend has it, Hippocrates taught under its branches. Seeds from the plane tree of Hippocrates have been distributed around the world as part of an effort to disseminate the Hippocratic spirit.
>
> In the summer of 1944, Kos was occupied by the German military. On July 23, the 120 Jews of Kos were assembled at the harbor, near the plane tree of Hippocrates. A small vessel arrived to pick them up, afterward joining other vessels containing the last Jews of Rhodes. From there, they were transported to the Greek mainland and from there they were conveyed by train to Auschwitz.
>
> Upon arrival at the rail siding in the Birkenau complex, the Jews of Hippocrates' birthplace were met on the ramp by the professional descendants of Hippocrates. Those licensed SS physicians who had been selected to select made a diagnosis on each of the Jews of Kos that he or she was a "useless life" and should receive the 'Great Therapy of Auschwitz,' which was death in 'The Central Hospital.' There, they all perished.
>
> Today the island of Kos and the empty synagogue of Kos, which adjoins the plane tree of Hippocrates, symbolize the spiritual crisis of medicine arising from the Holocaust; a crisis that medicine has failed to recognize, let alone resolve (Seidelman 1992).

References

L. Alexander, Medical Science under Dictatorship. N. Engl. J. Med. **241**, 39–47 (1949)

T.L. Beauchamp, J.F. Childress, *Principles of Biomedical Ethics*, 4th edn. (Oxford University Press, New York, 1994)

K. Binding, A. Hoche, *The Release of The Destruction of Life Devoid of Value* (Robert L. Sassone, Santa Ana, 1975)

E. Black, *War Against the Weak: Eugenics and America's Campaign to Create a Master Race* (Dialog Press, Washington, DC, 2003)

Cameron NM de S, *The New Medicine: Life and Death after Hippocrates* (Bioethics Press, Chicago, 2001)

T.P. Duffy, The Flexner report—100 years later. Yale J. Biol. Med. **84**, 269–276 (2011)

M. Duncan, *Pinto-Powell: Geisel School of Medicine at Dartmouth Student Policy Handbook 2016–2017*, (2017). https://geiselmed.dartmouth.edu/faculty/pdf/geisel_student_policy_handbook_public.pdf

H. Friedlander, *The Origins of Nazi Genocide: From Euthanasia to the Final Solution* (The University of North Carolina Press, Chapel Hill, 1995)

H.T. Greely, From nuremberg to the human genome: the rights of human research participants, in *Medicine After the Holocaust: From the Master Race to the Human Genome Project and Beyond*, ed. by S. Rubenfeld (Palgrave Macmillan, New York, 2010), pp. 185–200

SD. James, *Death Drugs Cause Uproar in Oregon* (2017). http://abcnews.go.com/Health/story?id=5517492&page=1

R.J. Lifton, *The Nazi Doctors: Medical Killing and the Psychology of Genocide* (Basic Books, Inc., New York, 1986)

M. Nelson, *The Evolving Presidency. Landmark Documents, 1787–2015*, Fifth edn. (CQ Press, Thousand Oaks, 2016)

R.N. Proctor, *Racial Hygiene: Medicine Under the Nazis* (Harvard University Press, Cambridge, 1988)

U. Schmidt, *Justice at Nuremberg: Leo Alexander and the Nazi Doctors' Trial* (Palgrave Macmillan, New York, 2004)

W.E. Seidelman, Medspeak for murder: the nazi experience and the culture of medicine, in *When Medicine Went Mad: Bioethics and the Holocaust*, ed. by A.L. Caplan (Humana Press, New York, 1992) pp. 271–280

E. Shalev, *American Zion: The Old Testament as a Political Text from the Revolution to the Civil War* (Yale University Press, New Haven, 2013)

A.M. Siegel, D.A. Sisti, A.L. Caplan, Pediatric Euthanasia in Belgium: disturbing developments. JAMA **311**, 1963–1964 (2014)

J.J.M. van Delden, L. Pijnenborg, P.J. vander Maas, The Remmelink study two years later. Hast. Cent. Rep. **23**, 24–27 (1993)

R. v. Wade, 410 U.S. 113 (1973)

R. Weikert, *The Death of Humanity and the Case for Life* (Regnery Faith, Washington, 2016)

Lessons Learned and Forecasts in Nanomedicine Startups and Commercialization

Amir Nashat[1]

Author's Personal Note Sometime around 1996, just before I left Mauro's lab at UC Berkeley to move to Boston, Mauro and I were walking around Hearst Mining Circle on a Saturday (discussing some results for our my thesis, I suspect). Mauro's eldest, Giacomo, who was probably 6 at the time, was with us also and was playing around the grass and the pond in Hearst Circle. At one point, Giacomo made an observation while staring at the pond, and relayed a big burst of words in Italian to Mauro, in a pretty excited way. Mauro suddenly took on this look of dazzled pride, and had his characteristic smile, and replied "good, good, very good" in Italian. I asked Mauro what the commotion was all about, and he replied "Well, Giacomo basically just explained to me, in his own words, that flies walk on water because of surface tension." Mauro had this glowing, childlike happiness on his face, and that cemented for me an impression I have always had about Mauro. As smart and creative as he is himself, Mauro's true joy and excitement comes from watching talented and smart people do truly remarkable things. He seems to live in this constant state of optimism and anticipation for the next big revelation from the people he surrounds himself with, and these moments are the happiest I have ever seen him.
Amir Nashat

Nanotechnology is aiming to bring a new dimension to medicine. About 14 years in, the field of nanomedicine is still young in drug development terms, but the past two years have proved exciting with a number of FDA approvals and positive data reported for nano-enabled treatments.

While discoveries that were foundational to the field of nanotechnology date back to the 1950s, commercialization of nano-driven products boomed in the 1990s, thanks to Sumio Iijima's discovery of carbon nanotubes and the subsequent backing of numerous research and industry groups. Many of the products we all encounter in our daily lives have been made possible by nanotechnology, from the zinc and titanium dioxide nanoparticles used in sunscreens to the engineered quantum dot technology that has enabled smartphones displays. Nanotechnology is employed in clothing to make garments more durable and water- and stain-resistant, in agriculture and food, where nanocapsules aid in pesticide delivery or in food packaging. It increases efficiency in solar panels and makes for lighter, stronger automotive materials.

In many ways, the application of nanotechnology in medicine and drug development is a natural one. The power of nanotechnology lies in its ability to hardness physical phenomena happening at the nanometer scale, and much of biology occurs in this size range. The commercial trajectory for medicine is quite different from the other industries, though, as the required FDA approval process means it often takes 10 to 15 years to bring emergent technologies to market. Other therapeutic modalities, from antibodies to antisense to gene therapy to gene silencing, all took two decades to get on the market and become important parts of the therapeutic landscape. So, against that backdrop, nanomedicine is following that classical timeline in novel drugs.

A major period of enthusiasm for the incorporation of nanotechnology into medicine began around 2005, stimulated by academic work from the labs of Mauro Ferrari, Omid Farokhzad of Harvard, Samir Mitragotri (then at UC Santa Barbara), Bob Langer of MIT, Joe DeSimone of the University of North Carolina, and many others. The work of these early pioneers led to the first wave of nanomedicine-focused startups. My firm Polaris Partners backed a number of these companies, including Selecta, BIND Therapeutics, Cerulean Pharmaceuticals, T2 Biosystems, and Kala Pharmaceuticals.

Most of the initial nanomedicine companies focused on improving drug delivery to the site of disease or relevant biology. NanoMedical Systems, co-founded by Dr. Ferrari, focused on developing nano-enabled subcutaneous implants for long-term, constant release of a variety of drugs and biologics. BIND and Cerulean sought to improve the effectiveness and lower the toxicity of various chemotherapies with the help of nanotechnology. Selecta harnessed nanoparticles for the delivery of antigen and adjuvant to dendritic cells. Kala employed nanoparticles to deliver steroids to the eye to treat dryness and pain. Another Polaris-backed company, Alnylam Pharmaceuticals, used lipid nanoparticles to deliver a new therapeutic modality, namely, RNAi, to cells and tissues.

A. Nashat (✉)
Polaris Partners, Boston, USA
e-mail: anashat@polarispartners.com

A. Grattoni and T. Desai (eds.), *Biomedical Micro-Nanotechnologies Toward Translation*,
https://doi.org/10.1007/978-1-0716-0187-7_6

A few companies from this era tapped into the sensitivity of nanoparticles to changes in the environment for use in diagnostic applications. T2 Biosystems is focusing on rapid diagnosis of pathogens causing sepsis, where speed has a dramatic impact on patient survival. This trend continues, with a new company called Glympse Bio just recently emerging out of the lab of Sangeeta Bhatia at MIT to use nanoparticles for the detection of diseases such as NASH.

Across the board, this first wave of nanomedicine startups has been able to raise healthy rounds of private capital, and several have gone public. Their success in bringing their respective products to markets has been mixed, and a few key lessons emerge when following their trajectories.

First, the good. Alnylam made history in 2018 in securing FDA clearance for Onpattro, its aforementioned drug for the rare disease hereditary transthyretin amyloidosis (hATTR). The approval, won 16 years after Alnylam started, marked the first FDA approval for an RNAi drug, but also bodes well for the advancement of nano-enabled medicines, as Onpattro employs lipid nanoparticle delivery. Also in 2018, T2 secured FDA clearance for its T2Bacteria panel for sepsis, building upon its previous clearance for its fungal diagnostic product. Kala's nanoparticle corticosteroid Inveltys received FDA in August 2018 and commercially launched in January 2019. Selecta recently presented Phase 2 results for its lead candidate SEL-212, showing the drug was effective in reducing incidence of gout flares.

In contrast, a number of companies hit hurdles in bringing their products to market and had to abandon development of their nanotechnology, however. The key lesson here is perhaps a bit counterintuitive. While nanotechnology improved existing products across a number of other industries, nanotechnology in medicine has seemed to stand the best chance by focusing on more revolutionary use cases and the hardest problems, with wide-open markets. Companies seeking to use nanotechnology to improve upon existing products, typically in crowded fields, have faced much rockier journeys. Tackling diseases that have high unmet medical need, with little or no alternative treatments on market, has proven to be a smoother path.

In the case of both Cerulean and BIND, they struggled in trials to prove their nanotechnology-driven approach was better than existing chemotherapies on the market, in the highly competitive field of cancer. They had the high burden of being compared to alternative treatments and proving they could lengthen survival rate substantially. In contrast, T2's target of early sepsis detection promises a huge potential impact; sepsis is third leading cause of death in the US and is responsible for almost 50% of deaths of hospitalized patients. With sepsis, one hour of delayed treatment increases mortality risk by nearly 8%, and T2's Bacteria panel enables detection in approximately five hours, a drastic improvement over the one to five days or more for existing diagnostic methods. Similarly, Alnylam used nanomedicine to help bring the first treatment to market for hATTR.

This trend of success arising from chasing the hardest problems in high impact areas without incumbent solutions is not exclusive to nanomedicine but has been critical for the success of a number of new modalities in biotechnology. Antisense, for example, had a long journey to market but ultimately found its place in enabling the first treatment for the fatal genetic disorder spinal muscular atrophy, in the drug Spinraza from Ionis and Biogen. Similarly, antibodies garnered excitement as an emerging treatment modality, and their initial strongest applications were in tackling liquid tumors (through Rituximab from Biogen and Genentech) and autoimmune conditions (also Rituximab for rheumatoid arthritis, as well as drugs like Novartis' Basiliximab for reversal of transplant rejection and J&J's Infliximab for Crohn's, colitis, RA, and more). Both disease categories were drastically underserved previously.

Like any emergent new biotechnology, nanomedicine has had a long journey, but the future looks bright, as the first products make their way to market and the industry matures around those. The recent waves of approved nanomedicines and nanotechnology-enabled products have brought more attention and confidence to the field, as well as increasingly more people experienced in the manufacturing, quality testing, and regulation of nanomedicines. There are many important unmet needs in medicine that can be solved through the application of nanotechnology, and as long as startups go after big problems with limited competing solutions, the opportunities for nanomedicine will be plentiful.

Innovation from the Industry to the Biomedical Field: Forged Composites® and CFRP 3D Printing for the Next Generation Medical Devices

Stefano Domenicali[1]

Author's Personal Note There are many years of friendship and there are many stories and anecdotes that bind me to Mauro. I could say so much but I prefer a "photograph" of a moment in time that connects us more than others.
We have both always considered, and continue to believe, that smiling is the extra gear in reaching great goals.
This was the case when we laid the foundation stone together, marking a long and fruitful collaboration between Houston Methodist Research Institute and Automobili Lamborghini.
It was a special day for us, a special day to smile!
Stefano Domenicali

Carbon fiber-reinforced polymer (CFRP), also carbon fiber-reinforced plastic, is a composite material made from polymer matrix reinforced with fibers. The polymer is usually an epoxy, vinylester, or polyester thermosetting plastic resin [*]. The reinforcement consists of filaments (fibers) of various types that can be also combined with each other: carbon fibers, glass fibers, Kevlar®, aluminum, ultrahigh-molecular-weight polyethylene (UHMWPE), or also organic fibers. These can have a length equal to that of the composite element (continuous fibers) that can be unidirectional or woven together as weft and warp, or they can be short (discontinuous fibers) typically randomly distributed.

The fundamental concept is that the matrix works to keep the fibers together and aligned and to transfer the load through the reinforcement that gives rigidity and high mechanical performance in its axial direction. For this reason, the orientation for the fibers drastically affects the mechanical and physical properties of the composite that is manufactured.

The main advantages of these materials are strictly linked to the high specific strength (strength/weight) and to the high specific modulus (stiffness/weight) combined with the possibility of varying the degree of anisotropy by acting on the concentration and the orientation of the reinforcement.

CFRPs have been in use since the 1970s for structural applications first in the aerospace field and then progressively in automotive, marine, construction, sport, and medical industries. The evolution of this technology created a wide range of resins and reinforcements dedicated to many different applications depending on required performance, costs, and production rate.

Different processes exist to produce CFRP parts: the choice is based on the geometries of the part, the required performance and costs: Pre-preg, resin transfer molding (RTM), braiding, filament winding, wrapping, press molding, and compression molding are some of them.

The most diverse one is the "wet layup methods" that consists by layering sheets of carbon fiber already impregnated with the resin (pre-preg) to a mold in the shape of the final product. The assembly is then placed in a vacuum bag to compact the layers and remove the air in between. The curing of the resin is performed with the use of an autoclave applying specific pressure and temperature for some hours. The whole process could take up to 10 days of work to get a middle complexity/middle dimension part.

Automobili Lamborghini has been significant in the composites world since 1983, expanding its experience over the years to apply the technologies and the engineering approach to its product inherited directly from the aerospace field. Automobili Lamborghini always faces the

S. Domenicali (✉)
Automobili Lamborghini, Sant'Agata Bolognese, Italy
e-mail: stefano.domenicali@lamborghini.com

A. Grattoni and T. Desai (eds.), *Biomedical Micro-Nanotechnologies Toward Translation*,
https://doi.org/10.1007/978-1-0716-0187-7_7

challenges of the future by resolutely marrying light engineering, focusing itself on the targeted use of carbon fiber composite materials in the creation of extremely dynamic super sports cars and at the same time maintaining sustainability in terms of consumption and emissions. One of its models, the Aventador LP 700-4, reflects exactly this carbon excellence, featuring an innovative monocoque made entirely of carbon fiber [*] that integrates various technologies and production processes to optimize the structural and weight performance and production costs.

In 2010 Automobili Lamborghini presented the "Sesto Elemento" (Sixth Element, or Carbon exactly), a car where intensive use of composite materials was made for the realization of most of the structural components from the monocoque, the entire front section, the bodywork, the bumpers, the rims, the arms suspension, to the transmission shaft. In this extraordinary "technological demonstrator" [*] the Forged Composites® process is found in the manufacturing of the full monocoque. This is a specific compression molding technology: a two-piece—male and female—mold, usually made out of aluminum or steel, where the material is pressed and heated between the two and for which Lamborghini owns the patent for A-class automotive applications.

Compression molding is one of the quicker methods of produce CFRP parts (we are talking about minutes) and, versus standard composite processes, it offers different technical advantages such as much more freedom in parts geometries and design, high integration, the possibility of using recycled CF fibers, and industrial advantages such as higher volume production, lower manufacturing costs, and innovative cosmetic appearance.

The material is based on pre-impregnated discontinued fibers that can flow and fill the cavity to create 3D shapes like ribs, directly out of the mold. The fibers' distribution on the part's surface is quite random, providing an element with quasi-isotropic characteristics and "notch un-sensitive", which means better fatigue resistance. Some of the mechanical characteristics are lower than the standard continuous fibers but they are anyway close to standard aluminum and the material can be hybridized with long fibers, with the final strength compensated via a specific design to reach the requirements.

The most recent and innovative production technologies are linked to the 3D printing technology for which the construction of a mold is not necessary, since the three-dimensional objects are realized through the additive production of successive layers of material, starting directly from a digital 3D model.

The additive approach to manufacturing has several big advantages over the conventional ones: it cuts costs by getting rid of production lines and some mechanical machining; it reduces waste enormously, requiring as little as one-tenth of the amount of material; it allows the creation of parts in shapes that conventional techniques cannot achieve, resulting in new and much more efficient designs [*].

An important concept is that a part design cannot be translated one-to-one from other technologies, but it must be peculiarly conceived and engineered to take advantage of all the specific production technology.

It has only been possible in recent years to produce parts via 3D printing in composite material reinforced with continuous fibers (carbon fiber, glass fiber, Kevlar®, etc.) in which the continuous filaments of resin and reinforcement are combined (fused) in the same structure with different distribution to obtain the maximum mechanical yield according to the geometry.

For many years, 3D printers were used just for prototyping, mainly in aerospace, but today the most recent 3D printing technologies are fast enough to satisfy a mass production rate with all the advantages to make slight modifications and customizations at almost no extra cost and no stop of production.

This can open a wide range of industrial applications for production, especially for composites where the classic manufacture methods need long times and high-level trained operators.

Carbon fiber can be also additivated by nanocomposite, which provides the original materials with new mechanical and physical properties.

Lamborghini is testing several different nanocomposite applications in its laboratories, in order to increase the fatigue limit of the original materials and to provide information about component damage.

Thanks of the use of carbon nanotubes (CNT) it will be possible in fact to monitor locally the level of stress in a given structure and to establish a proper health monitoring system for the whole vehicle by providing the driver with all information about potential damaged areas after low/high-speed crash.

In the near future, the next generation of cars will be hybridized or fully electric, but for sure with several disadvantages, such as the batteries weight.

With CNT it will be possible to get rid of batteries by obtaining a structural integration of the 3D batteries with laminated materials, including use of the battery as a sandwich core and fabrication of complete load-bearing panels using stacked CNT-based solid-state batteries. In this way, it will be possible to substitute current heavy lithium-ion batteries with light body panels with great advantages on the weight to power ratio.

In biomedical field, the CFRP materials are already used in many parts of medical machines mainly related to their characteristics of radio transparency, MRI compatibility, and lightness. For the same reasons, the CFRP can be used to substitute also the metal parts of many devices that are implanted on the patient or used in surgery room just for those applications where the characteristics of radio transparency, MRI compatibility, and lightness play an important role in the good success of the treatment and for the comfort of the patient.

For all the reasons above, Forged Composite® and 3D printing technologies as well as nanocomposites can offer many advantages for manufacturing biomedical devices with complex geometries and high volume production: the former is most valuable for mass production, the latter for dedicated and customized applications that can be obtained in a very short time.

There are already many known applications of CFRP for biomedical devices in the market but they are mainly for external applications and the high cost is still a limitation in their utilization.

The main challenge for the next years is to find the suitable composite materials that can be biocompatible for short-term or long-term applications inside the human body, with restrained costs and fast production cycles.

For 1 year, Automobili Lamborghini and Houston Methodist Research Institute has been collaborating in a self-funded research project to analyze the biocompatibility of different CFRP materials commonly used in automotive field.

Automobili Lamborghini is interested in investigate the compatibility of such materials with human skin that concerns just in vitro testing, in parallel, Houston Methodist Research Institute is interested in evaluating the biocompatibility of CFRP as a material for the development of innovative medical products. For these applications, the in vivo testing will be eventually performed autonomously by Methodist.

Methodist's laboratories are currently evaluating different CFRP materials manufactured with different technologies and different surface finishing and the performed in vitro tests already show promising results of biocompatibility for all the materials under investigation. This means a wide choice of materials potentially suitable also for biomedical application such as orthopaedic and neurosurgery devices that can be selected depending on proper mechanical characteristics, production technology, and costs.

References

Top Gear, *Lamborghini's Sesto Elemento*. https://www.topgear.com/car-news/lamborghinis-sesto-elemento

The Economist, *Print me a Stradivarius*. https://www.economist.com/leaders/2011/02/10/print-me-a-stradivarius?story_id=18114327

M. A. Masuelli, *Introduction of Fibre-Reinforced Polymers—Polymers and Composites: Concepts, Properties and Processes*, Submitted: October 24th 2012, Reviewed: October 24th 2012, Published: January 23rd 2013

Part II
Drug Delivery

Biomedical Microdevices (2019) 21: 45
https://doi.org/10.1007/s10544-019-0386-9

Polymers for extended-release administration

Marion S. Paolini[1] · Owen S. Fenton[1] · Chandrabali Bhattacharya[1] · Jason L. Andresen[1,2] · Robert Langer[1,3]

Published online: 8 April 2019

Author's Personal Note I first met Mauro when he was at the University of California at Berkeley. Since then it's been a pleasure to watch his career at Ohio state, the national cancer institute, and at Texas. In addition to his scientific achievements, I've always found Mauro to be a true gentleman. He has also trained many fine students. I want to wish Mauro a very very happy 60th birthday and I wish him many many more in the future.

Robert Langer

Abstract
Developing strategies to deliver the required dose of therapeutics into target tissues and cell populations within the body is a principal aim of controlled release and drug delivery. Specifically, there is an interest in developing formulations that can achieve drug concentrations within the therapeutic window, for extended periods of time, with tunable release profiles, and with minimal complication and distress for the patient. To date, drug delivery systems have been developed to serve as depots, triggers, and carriers for therapeutics including small molecules, biologics, and cell-based therapies. Notably, the efficacy of these systems is intricately tied to the manner in which they are administered. For example, systemic and oral routes of administration are common, but both can result in rapid clearance from the organism. Towards this end, what formulation and administration route strategies are available to prolong the bioavailability of therapeutics? Here, we discuss historical and modern drug delivery systems, with the intention of exploring how properties including formulation, administration route and chemical structure influence the ability to achieve extended-release drug release profiles within the body.

Keywords Controlled release system · Drug delivery · Extended release · Polymer

1 Introduction

Worldwide prescription drug sales are forecasted to grow at an annual compound rate of 6.5% in the next five years and expected to reach US$1.06 trillion in 2022 (Deloitte 2018). At the same time, and one probable cause of this growth, the population suffering from chronic diseases is also increasing. A recent study published by the RAND Corporation found that 60% of American adults live with at least one chronic condition, while 42% have more than one (Vogeli et al. 2007; Buttorff et al. 2017). To treat their chronic condition(s) patients often need an increased, prolonged, and repeated intake of medication. Drug delivery systems (DDS) have been designed to alleviate tedious administration schedules and improve patient compliance. Another issue arising from the repeated intake of medication is the difficulty in staying within the specific active therapeutic window of the treatment. Indeed, if the drug concentration is too low, no therapeutic effect is observed; if a drug's concentration is too high, issues associated with toxicity can occur. Controlled release systems (CRS) have been developed to enable the administration of a drug, a small molecule or even a biologic in a single dose, with a preset release rate in the body (Fig. 1), with optional triggered-release by physical, chemical or biological factors. Historically, there have been numerous attempts at creating materials capable of controlling the release of both small and large molecular weight drugs over a period of time. Early research was inspired by the diffusion of small molecular weight (< 300 g/mol) dyes through silicone tubing. Indeed, during these early studies, companies, including Alza, were formed and some of their early work focused on converting biocompatible and well-studied silicone tubing into materials that could deliver drugs including atropines, histamines, anesthetics, steroids, and antimalarial and antischistosomal agents. Following these research efforts and in tandem with the emergence of genetic engineering in the 1970s, interest in controlling the release of biologics developed. These efforts required preservation of the integrity and structure of proteins from degradation. Silicone and other traditional materials used for small molecule release, however, are impermeable to proteins given their hydrophobicity and, as a consequence, additional

✉ Robert Langer
rlanger@mit.edu

1 Koch Institute for Integrative Cancer Research, Massachusetts Institute of Technology, Cambridge, MA 02142, USA

2 Department of Chemistry, Massachusetts Institute of Technology, Cambridge, MA 02139, USA

3 Department of Chemical Engineering, Massachusetts Institute of Technology, Cambridge, MA 02142, USA

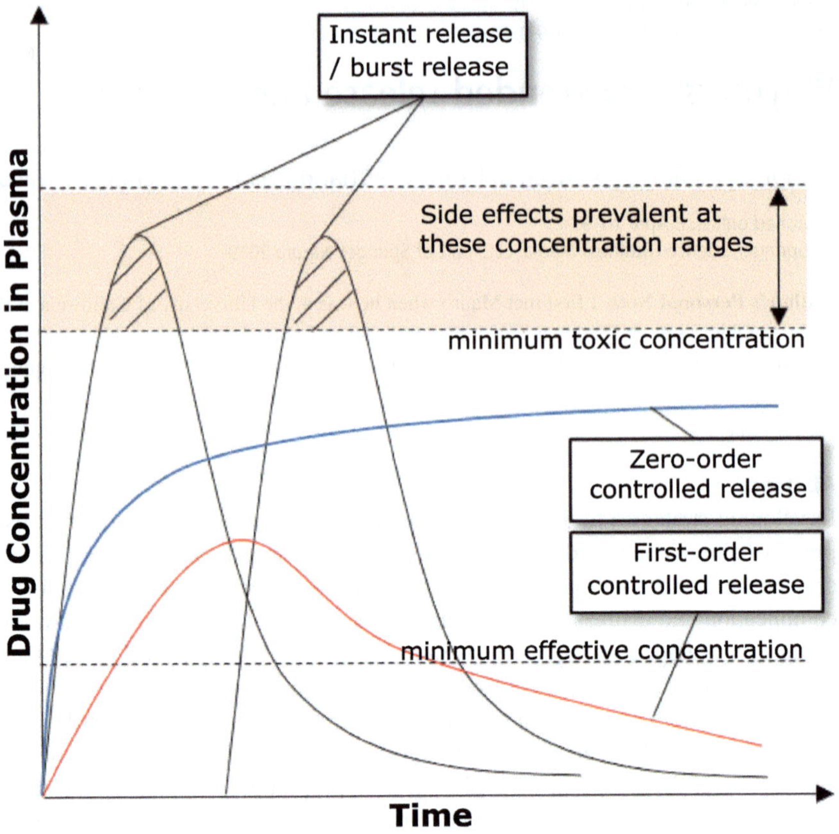

Fig. 1 Extended release systems allow to keep the drug concentration in plasma stable, between the therapeutic level, or the minimum effective concentration (MEC) and the toxic level, or the minimum toxic concentration (MTC). Adapted with permission from (Fenton et al. 2018)

materials and systems had to be designed (Hoffman 2008; Kleiner et al. 2014).

CRS therapies have significant clinical and economic benefits. Such systems usually allow for reduced treatment-related toxicity, fewer medical visits (providing socioeconomic savings and improved access to health care), and also reduced treatment burden and improvement in compliance. Such systems indeed help to solve compliance issues, or even improve the feasibility of treatments. In particular, patients with chronic conditions, such as diabetes or human immunodeficiency virus (HIV) infection, could benefit from extended release systems to avoid the burden that daily intake of pills or daily injections represent. Requiring such burdensome treatments to treat these diseases often causes a lack of adherence in the treatment regimen for patients. For example, 35% to 50% of medication for chronic diseases are not taken as prescribed (Osterberg and Blaschke 2005; Lauffenburger et al. 2017). This non-adherence is due to multiple factors, including: *i.* difficulty to renew prescriptions, *ii.* inaccessibility to prescribed drugs, *iii.* Inability to pay for treatment, *iv.* forgetting to take the treatment, and *iv.* psychological barriers against taking a treatment. The number of patients enduring chronic diseases is dramatically increasing, spurring the need for more convenient treatment options. Besides therapies for chronic diseases, some treatments such as contraceptive hormones or vaccines also require several repeated administrations. The need to inject vaccines at many different times cause feasibility issues in developing countries, where regular appointments with medical personnel cannot be easily guaranteed. On top of facilitating the access and the adherence to treatments for patients' population, CRS enable therapies targeted at difficult to reach areas of the body. Treatments in the joints, in the eye, or in the brain for instance can be either painful or technically difficult to realize. Implanting an CRS in these areas is a solution to prevent multiple interventions. CRS can therefore provide benefits to a vast number of patients, for many therapeutic indications.

Organic polymers in particular are an interesting option to design extended release systems because of their tunable mechanic properties, biocompatibility and biodegradability. Polymers, by definition, are molecules made of the repeat monomers. Polymers are the basis of some of the most fundamental materials of life. For example, proteins are polymers composed of chains of amino acids, and deoxyribonucleic acid (DNA) is also a polymer consisting of nucleotides. Organic polymers can also be synthetic, hydrophobic, hydrophilic, or amphiphilic, thermoplastics or thermosets. Polymers can also be tuned to have desired chemical and physical properties including molecular weight, elasticity, surface charge, and polarity can also be modified.

Here, we present (*i*) a brief summary of key theoretical rules to understand the mechanism of sustained delivery, as well as (*ii*) an overview of CRS, with triggered or passive release, and (*iii*) the limits of sustained delivery, in terms of difficulties of synthesis and biological responses.

2 Extended-release strategies – The chemistry behind it

2.1 Synthesis of sustained delivery devices and release mechanisms

Polymers as CRS can consist of hydrogels, i.e. hydrophilic matrices (such as hydroxypropyl methylcellulose (HPMC), sodium carboxymethylcellulose, alginates, hyaluronic acid, poly(hydroxyethyl methacrylate) or poly (HEMA), poly(ethylene glycol) (PEG), and poly(vinyl alcohol) (PVA)). Polymers can also form thermoplastics, thermosets, or hydrophobic matrices (such as polyethylene, polypropylene, ethylcellulose, polycaprolactone, poly(D, L-lactide), or poly(-ethylene-co-vinyl acetate) (PEVA)).

Hydrogels are cross-linked hydrophilic polymers, which swell in water. Thermoplastics are usually high molecular weight hydrophobic polymers which soften with temperature and can be remolded upon heating. The arrangements between the polymer chains of thermoplastics are based on intermolecular forces, not chemical bonds. As a consequence, the intermolecular forces can be reversible. Thermosets, by contrast, are high molecular weight polymers, which are chemically bound together. They cannot melt and reform (Hatefi and Amsden 2002; Sastri 2010). Such differences in chemical and mechanical properties give the possibility to adapt to different routes of administration, to different targeted sites, and to different durations of treatment.

Polymeric matrices loaded with a drug of interest can be prepared either by mixing, by compressing, or by solvent swelling (Grassi and Grassi 2005). Mixing or compressing are the simplest encapsulation methods: the drug is mixed with the pre-polymer or the polymer and added to the polymerization reactor, or compressed to afford a tablet. Solvent swelling techniques involve putting the pre-formed polymeric matrix in contact with a concentrated drug solution capable of swelling into the matrix. The delicate step is to remove the solvent without disrupting the drug within the matrix. The solvent may be a supercritical fluid, dense as a liquid but viscous as gas for easy removal from the matrix.

The principles underlying extended release systems are based on physical mechanisms; they depend on the design of the polymeric release system as well as on the drug's properties. Because extended release is by definition a long process, it is fundamental to model this process theoretically as a function of time, in order to be able to choose early on the most appropriate design for the desired application.

Models focusing on drug release from systems made of non-degradable, non-swellable polymers, where the release of the drug depends solely on the diffusion of the drug out of the polymer have been developed (Fu and Kao 2010; Siegel and Rathbone 2012). This type of release mechanism is known as Fickian diffusion. Fickian diffusion is described by Fick's first and second laws (Eq. 1). Solving Fick's second law with the relevant boundary limits is the simplest way to describe the Fickian diffusion of a drug from a polymer.

Equation 1 Fick's first and second Law

$$J = -D\nabla C \tag{1}$$

$$\frac{\partial C}{\partial t} = D\Delta C \tag{2}$$

where

- J *is the "diffusion flux vector" of which the dimension is amount of substance per unit area per unit time, so it is expressed in such units as* $mol.m^{-2}.s^{-1}$. *J measures the amount of substance that will flow through a unit area during a unit time interval;*
- D *is the diffusion coefficient or diffusivity. Its dimension is area per unit time, so typical units for expressing it would be* $m^2.s^{-1}$;
- C *is the concentration, of which the dimension is amount of substance per unit volume. It might be expressed in units of* $mol.m^{-3}$.

For such diffusion-driven drug delivery systems, the release rate of the drug will depend on the escape of the drug from the polymer, which relies on parameters such as the polymer's porosity and mesh size relative to the size of the drug, and the affinity of the drug to the matrix. Porosity depends on the nature of the polymer, the percentage of different polymers if it is a blend (e.g. L/G for PLGA), and also on the speed at which the solvent was removed (impacted by the solvent itself, and on the temperature used) (Pagels and Prud'Homme 2015; Kasyapi et al. 2017). Within the non-degradable category, a distinction is usually made between reservoirs and matrix-type systems (Fig. 2) (Fung and Saltzman 1997). Reservoirs are matrices coated with an inert material functioning as a rate-controlling membrane. As a consequence, the reservoir model is the simplest model of drug diffusion, viewed as solute released from a sphere. It assumes that drug is confined by a spherical shell of outer radius R and inner radius Ri; thus, the drug must diffuse through a layer of thickness (R − Ri) (Arifin et al. 2006). In this case, the solution of Fick's second law is the Zero order, or constant rate release model assuming a constant drug driving force (Eq. 2), the cumulative amount of drug release is linear in time (Table 1).

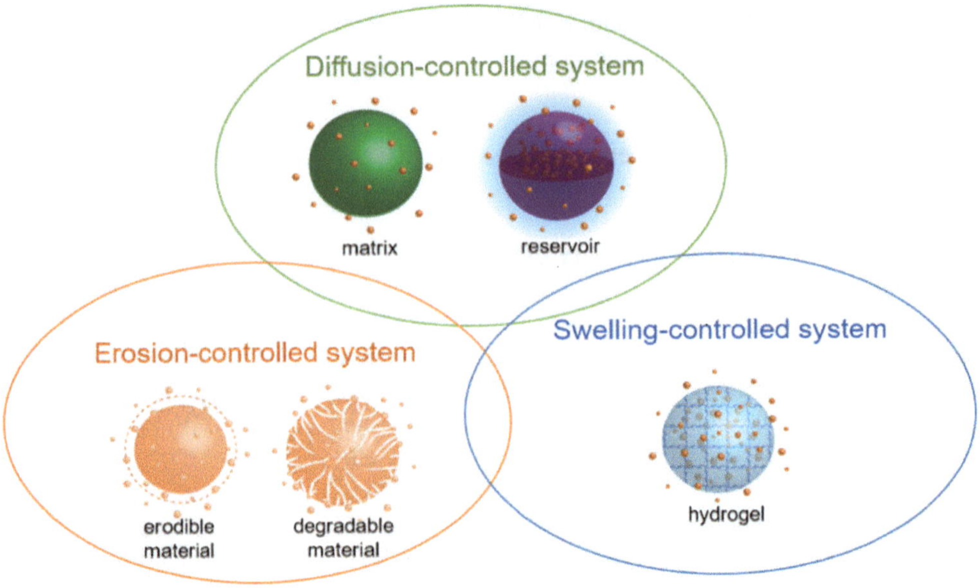

Fig. 2 Schematic representation of erodible, swellable or diffusion-controlled sustained drug delivery devices. Adapted with permission from (Tibbitt et al. 2016)

Equation 2 Zero order release model.

$$M_t = M_0 + K_0 t$$

where.

M_t	*is the amount of drug released until time t, in mol;*
M_0	*is the initial amount of drug, in mol;*
K_0	*is the zero order release constant, in mol.s^{-1};*
t	*is the time in seconds.*

In particular, for a spherical matrix, $K_0 = 4\pi \frac{RR_i}{R-R_i} DC_{i.}$ *where*

R	*is the outer radius R and $R_i$is the inner radius, both in m;*
D	*is the diffusion coefficient in the polymer matrix in m^2.s^{-1}*
C	*is the drug concentration inside the reservoir, in mol.m^{-3}.*

In the case of matrix-type devices, there is no diffusion rate-controlling membrane and the drug release is dependent on a non-constant drug concentration gradient (Fickian diffusion). Consequently, release is associated with concentration gradient, diffusion distance, and the degree of swelling of the polymer, which may be null in the simplest case (Eq. 1) (Fu and Kao 2010). Fickian diffusion refers to the solute transport process in which the polymer relaxation time (t_r) is much greater than the characteristic solvent diffusion time (t_d), and vice-versa when the characteristic solvent diffusion time (t_d) is much greater than the polymer relaxation time (t_r) (Peppas and Khare 1993; Grassi and Grassi 2005; Fu and Kao 2010). Mathematical models for matrix systems are often valid for DDS where the drug is assumed to be in high amount inside the non-biodegradable polymer matrix the system (constant source) and uniformly distributed. Here, we will consider dispersed drug system, meaning that the initial drug loading is higher than the solubility of the drug inside the polymer matrix ($C_0 > C_s$).

The diffusion-controlled mathematical model for dispersed drug system ($C_0 > C_s$) in a planar sheet and a sphere was initiated by Higuchi. Higuchi's model to describe the diffusion based on Fick's law, which is square root time dependent (Eq. 3) (Higuchi 1963).

Equation 3 Higuchi's equation.

$$M_t = A\sqrt{D(2C_0 - C_s)C_s t}$$

where

M_t	*is the amount of drug released until time t, in mol;*
A	*is the release area, in m^2;*
D	*is the drug diffusion coefficient, in m^2.s^{-1};*
C_0	*is the initial drug concentration in the matrix, it is expressed in mol.m^{-3};*
C_s	*is drug solubility, in mol.m^{-3}as well;*
t	*is the time in seconds.*

For reservoirs, the rate of diffusion mainly relies on the permeability of the polymeric membrane. Matrices or reservoirs may be made of polymers that are erodible or biodegradable, either by passive degradation or triggered by an external stimulus. The release of the drug they encapsulate will depend on erosion of the polymer itself (Fig. 2).

To model the release of a drug from a reservoir (that has a membrane) or from a degradable or erodible material, intrinsic parameters of the material must be considered. The Camera-Roda and Sarti equation (Eq. 4) models non-Fickian diffusion effects on macroscopic swelling kinetics (Camera-Roda and Sarti 1990). The authors assume that the swelling fluid flux J may be expressed as the sum of two terms: J_f, characterized by a zero relaxation time and representing the Fickian contribution to the global flux, and J_r, characterized by a non-zero relaxation time and representing the non-Fickian contribution to the global flux, respectively. Accordingly, the global flux can be expressed as:

Equation 4 Camera-Roda and Sarti equations

$$J = J_f + J_r$$
$$J_f = -D_f \times \nabla C$$
$$J_r = -D_r \times \nabla C - \tau \times \frac{\partial J_r}{\partial t}$$

where

J *is the total swelling fluid flux, in mol.m*$^{-2}$*.s*$^{-1}$*;*

J_f *is the Fickian contribution to the global flux, in mol.m*$^{-2}$*.s*$^{-1}$*;*

J_r *is the non-Fickian contribution to the global flux, in mol.m*$^{-2}$*.s*$^{-1}$*;*

C *is the swelling fluid concentration; it is expressed in such units as mol.m*$^{-3}$*;*

τ *is the relaxation time of the given polymer/swelling fluid system, in s;*

D_f *is the diffusion coefficient relative to the Fickian flux; it would be expressed in m*^{2}*.s*$^{-1}$*;*

D_r *is the diffusion coefficient relative to the non-Fickian flux; expressed in m*^{2}*.s*$^{-1}$ *as well;*

t *is time in s.*

From these equations are derived the power law equations for modeling release kinetics. Such equations are particularly interesting in studying extended release as they represent the cumulative drug release as a function of time and are straightforward. Several commonly used power law equations for modeling release kinetics are summarized (Table 1).

where

M_∞ *is the amount of drug released after an infinite time;*

K *is a constant;*

n *is the exponent characterizing the release process.*

In the case of Fickian diffusion, n is equal to 0.5, 0.45 and 0.43 for a thin film, a cylinder and a sphere, respectively (Siepmann and Peppas 2001). When n exceeds these thresholds, non-Fickian release occurs.

Release from most polymers follows Fickian diffusion and the release rate falls as the concentration of drug in the polymer decreases. However, a zero-order release rate (i.e. a release rate that is constant over time) is desired for drug delivery applications. This is the reason why a lot of sustained release delivery systems are coated with a non-permeable membrane. In the case of hydrogels, prolonged zero-order release kinetics may be obtained if the reduction in the driving force for diffusion matches the decrease in the resistance to diffusion due to the polymer degradation (Pagels and Prud'Homme 2015).

2.2 Triggered drug release

Both small and large molecules have been studied for sustained release application by exploiting the fact that the material properties of the delivery system influence drug diffusion (Fig. 2). Matrix and reservoir diffusion are united by the fact that the user commonly has only passive control over drug elution (Peppas et al. 1980; Kim et al. 1992; Peppas and Wright 1998; Freiberg and Zhu 2004). Indeed, upon impregnation of the material, the drug will release without a trigger. Accordingly, some challenges associated with these systems include burst release phenomena (wherein initial implantation would result in large amounts of the drug being released into the surroundings) and limited tunability of the sustained release profile (Lee 1985; Gupta et al. 2002; Mullarney et al. 2006). As an alternative, the concept behind triggered drug release is to still use a single administration of the therapeutic, but to control the drug's release using chemical, biological or physical cues (Jeong and Gutowska 2002; De Las Heras Alarcón et al. 2005; Hoare and Kohane 2008; Liu and Urban 2010; Aghabegi Moghanjoughi et al. 2016).

Table 1 Commonly used power law equations describing the release of drugs from polymeric matrices

Model	Expression	Application	Time dependence	Ref.
Ritger-Peppas	$\frac{M_t}{M_\infty} = Kt^n$	$n = 1/2$, Fickian diffusion (Higuchi model, see Eq. 3)	$f\left(t^{1/2}\right)$	(Ritger and Peppas 1987), (Higuchi 1963), (Serra et al. 2006)
		$n = 1$, swelling controlled, case II transport (Zero order model)	$f(t)$	
		$1/2 < n < 1$, Non-Fickian diffusion: dependent on diffusion and swelling (First order model).	$f(t^n)$	
Peppas-Sahlin	$\frac{M_t}{M_\infty} = k_1 t^m + k_2 t^{2m}$	Non-Fickian diffusion		(Peppas and Sahlin 1989), (Alfrey et al. 1966)

Toward this end, interest arose in using stimuli responsive materials to control and extend the release of drugs, while also avoiding issues associated with burst release (Okano et al. 1990; Huang and Brazel 2001; Min et al. 2010). Stimuli responsive materials represent promising platforms for controlled drug delivery and biomaterials application. By definition, stimuli responsive materials are those whose physical properties can be altered upon exposure to specific cues. These cues can include chemical signals (in the form of small molecules) (Bajpai et al. 2008; Dong et al. 2011; Lee et al. 2011; Yan et al. 2012; Schattling et al. 2014), pH (Peppas et al. 1999; Uhrich et al. 1999; Stuart et al. 2010; Gao et al. 2012), or alterations in chemical gradients including ion concentration (Sui et al. 2008)), biological signals (including proteins, enzymes, and antibodies) (Fischel-Ghodsian et al. 1988; Sui et al. 2008; Bawa et al. 2009; Maitz et al. 2013), and physical signals (including photonic (Lee et al. 2011; Azagarsamy et al. 2012; Han et al. 2012; Yan et al. 2012), electronic (Bawa et al. 2009), magnetic (Kost et al. 1987; Satarkar and Zach Hilt 2008; Bawa et al. 2009), and ultrasonic (Mitragotri et al. 1995; Bawa et al. 2009)), amongst others.

From a molecular design standpoint, stimuli responsive materials can be engineered from the bottom up to afford differential architectures of responsive polymers (Fig. 3). For example, some systems consist of block copolymers linked by degradable functional groups. Upon exposure to a specific cue, the degradable linkage can break to liberate the individual blocks. By contrast, other systems can have therapeutics appended directly to the main polymer chain, and these drugs can be release into the surroundings. Still other DDS consist of polymer chains synthesized from responsive monomers. Finally, cross-linking bonds can also be incorporated to join together with polymers to form responsive networks (Manouras and Vamvakaki 2017). Each of these architectural paradigms has been exploited in drug delivery research, highlighting the importance of unifying chemistry and materials design for biomaterials science. Overall release can therefore be extended through administration of a DDS with release occurring once or multiple times upon exposure to the designed signal. These responsive polymer architectures can be manipulated for triggered drug release ability and certain systems have been further engineered to utilize these cues for an extended release compared to a bolus administration of various drugs.

2.2.1 Chemical

To date, chemical cues are commonly employed to generate smart materials capable of extended release of drugs in vitro and in vivo. Chemically-responsive materials are those that can change in property (whether shape, stiffness, rheological property, chemical composition, molecular weight, etc.) upon interacting with a molecule in the environment (Culver et al. 2017). Chemical cues bear a number of strengths in creating responsive materials. For example, both nature and synthetic chemists have created materials replete with analyte-specific responsive functional groups (Bajpai et al. 2008; Bawa et al. 2009; Schattling et al. 2014). Towards that end, polymers that are either synthesized from responsive monomers, or are accessed using post-polymerization modification strategies, have been exploited for the development of extended release signals (Huang et al. 2000; Bawa et al. 2009; Wang et al. 2009; Mura et al. 2013; Yesilyurt et al. 2016).

From a strategic standpoint, chemically-responsive materials have a number of advantages. First, small molecules can be both implemented and studied with relative ease. Indeed, molecular characterization techniques including liquid chromatography mass spectrometry, nuclear magnetic resonance, infrared spectrometry, and gas chromatography ensure that we have chemical cues of precise and defined molecular structure. Nevertheless, chemical cues can be difficult to work with because of issues including diffusion, local concentration gradients, and imperfect chemical reactivity. For example, for an implanted material to be responsive to a small molecule, it is dependent both on the kinetics and thermodynamics of binding to the responsive functional group of interest (Hoffman and Stayton 2007). Therefore, many current materials were developed to target a naturally occurring cue that demonstrates elevated levels as a cause or symptom of a disease, such as blood glucose concentration in diabetes or glutathione levels both intracellularly and in certain cancers (Huang et al. 2000; Bawa et al. 2009; Mura et al. 2013). Many systems designed for insulin release have taken advantage of glucose-sensitive polymers containing terminal functional groups like concanavalin A (conA) or phenylboronic acid (PBA) that form reversible bonds with glucose, leading to release of insulin from the matrix (Wang et al. 2009; Lu et al. 2016; Yesilyurt et al. 2016).

2.2.2 pH

In pH-mediated release systems, a local change in pH can cause both chemical and mechanical changes in the polymeric material. Decreased pH can lead to the cleavage of acid-labile linkers connecting the polymer network or directly induce the cleavage of a covalently-bonded pendant drug molecule (Mura et al. 2013; Lu et al. 2016). Acid-degradable linkers or polymer chains are frequently utilized in hydrogel formation, where the cleavage of these bonds will lead to further swelling or degradation of the hydrogel, effectively triggering increased release kinetics of the encapsulated drug (Peppas et al. 1999; Schmaljohann 2006). These changes in pH can occur naturally in the surroundings of the polymeric system as they progress through the digestive system or encounter an acidic tumor microenvironment. Alternatively, pH changes can be produced internally through encapsulated enzymes

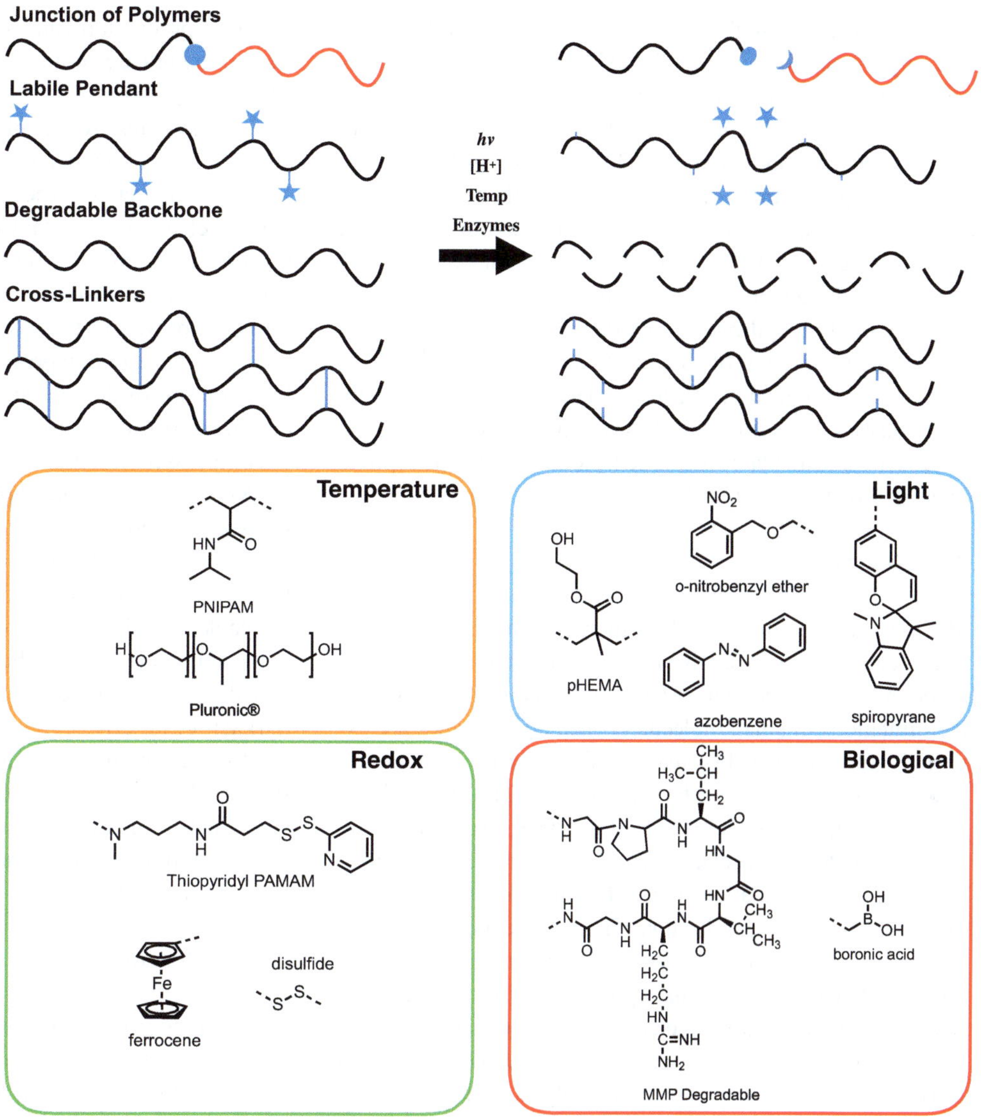

Fig. 3 Triggered extended release strategies. (A) Design of triggered release polymers: release mechanism after stimulus. (B) Selection of important polymers and functional groups exhibiting stimuli responsive behavior. Adapted with permission from (Schattling et al. 2014; Manouras and Vamvakaki 2017)

(Hasan et al. 2007; Dai et al. 2008; Min et al. 2010). One commonly used enzymatic trigger is glucose oxidase, catalyzing the conversion of glucose into gluconic acid, decreasing pH to enhance the solubility of lysine-modified insulin as well as to trigger the swelling or collapse of hydrogels (Fischel-Ghodsian et al. 1988; Podual et al. 2000; Gu et al. 2013a). A nanonetwork of chitosan and alginate coated dextran particles impregnated with glucose oxidase and catalase was able to achieve extended release of insulin to induce glycemic control for two weeks in diabetic mouse models (Gu et al. 2013b). One FDA approved polymer used as a pH-sensitive release trigger is Eudragit®, a taste-masking agent composed of methacrylic acid and cellulose esters (Hoy and Roche 1993). Eudragit® coatings of polymeric particles can be used for oral medicines to prevent burst release, as the polymer is unaffected by the neutral pH of saliva whereas degradation occurs upon entering the highly acidic gastric fluid (Moustafine et al. 2006; Hasan et al. 2007). Multiple delivery vehicles triggered by the slightly acidic conditions surrounding tumors have been explored to release a payload of doxorubicin (DOX) from polymeric particles (Car et al. 2014; Liu et al. 2016; Xu et al. 2016). A DOX-prodrug nanoparticle utilizing a pH-sensitive hydrazone linker to connect DOX to poly(L-glutamic acid) for targeted release to tumors was detectable up to one week after administration in mouse models (Xu et al. 2016).

2.2.3 Redox

Some DDS incorporate redox sensitive functional groups (ex. disulfides, ferrocenes, etc). These functional groups can respond to changes in redox potential which can alter the overarching polymeric structure. Naturally occurring reducing agents, such as glutathione, or the presence of reactive oxygen species, such as hydrogen peroxide, can be used to trigger the release of encapsulated or covalently modified drugs from polymeric carriers(Staff et al. 2012). One study explored the use of poly(amidoamine) (PAMAM) with thiopyridyl terminal groups to form disulfide crosslinks with an 8-armed thiol-terminated PEG subsequently loaded with antibiotics where the release was monitored over 72 h in vivo in the vaginal cavity of pregnant guinea pigs (Navath et al. 2011). Exposure to glutathione in the vaginal secretions caused a reductive cleavage of the disulfide crosslinks leading to degradation of the gel and release of amoxicillin. In another example, oxidative hydrogen peroxide was used as a trigger to initiate the production of carbon dioxide bubbles through an iron-mediated oxidation of ethanol. These bubbles ultimately could disrupt the barrier of PLGA microspheres, allowing the release of encapsulated dexamethasone sodium phosphate over 48 h (Chung et al. 2015).

2.2.4 Biological

Materials that are biologically responsive, that is, responsive to proteins or enzymes, have also been posited and explored given their high specificity. These systems overcome some of the limitations of small molecules – namely, enzymes are extremely specific, can be created using genetic engineering approaches, and can also be administered into the body or are found endogenously (Bajpai et al. 2008; Dai et al. 2008; Zhang et al. 2015a; Lu et al. 2016). Nevertheless, the stability, cost, and long circulation times of many antibodies and enzymes can limit their application. Chemical sequences degradable by esterases or peptides degradable by matrix metalloproteinases (MMPs) are often incorporated into polymeric materials for eventual biological cleavage and release (Dong et al. 2010; Zhu et al. 2012; Gu et al. 2013b; Amer and Bryant 2016). These enzymatic triggers can help to delay the release of therapeutics from hydrogels or polymeric vesicles compared to a bolus injection of the free drug. SDS may include multi-step triggers, such as the release of DOX-loaded quantum dots from a gelatin-based nanoparticle degraded by gelatinases (Wong et al. 2011). Certain SDS take advantage of overexpression or changes in activity levels of enzymes that occur in certain disease sates. Injection of a hyaluronic acid-dextran sulphate network containing MMP-cleavable peptide sequences was impregnated with an MMP-inhibitor TIMP3 thereby reducing the activity levels of overexpressed MMPs caused by myocardial infarction in a pig model. After 14 days, inhibitor levels were still detectable in plasma and the activity of the MMP in the myocardial infarction site was significantly decreased compared to control animals (Purcell et al. 2014). Other polymeric materials display a binding affinity for circulating biological molecules, including antibodies and glucose, thereby triggering a change in the binding preference or structure of the polymeric system. A poly(ethylene-co-vinyl acetate) chain functionalized with isocyanate groups to conjugate different haptens has been developed, initiating future antibody production causing competitive binding that could lead to the release of a narcotic antagonist from the polymer (Pitt et al. 1985).

Some systems are capable of interacting with receptors and antigens present on certain cell membranes. Such interactions allow both targeting a specific cell population as well as triggering release in a certain environment, enabling extended release of therapeutics through selective release conditions (Zhang et al. 2015b). A "virus-mimetic" nanoparticle comprised of poly(L-histidine-co-phenyalanine) and PEG-linked bovine serum albumin presenting folate ligands was shown to focus release of doxorubicin to the slightly acidic tumor microenvironment upon binding to folate receptors overexpressed in an ovarian carcinoma model (Lee et al. 2008). Certain targeting approaches that deliver to specific antigen-presenting cells are currently being explored in clinical trials. The safety and efficacy of a product composed of PEG-PLA polymeric nanoparticles modified with a small molecule ligand found to bind to the prostate specific membrane antigen (PSMA) is being investigated in order to target and deliver docetaxel to tumors that overexpress PSMA (Hrkach et al. 2012). Though the parent company recently went bankrupt and was purchased by Pfizer, a recent Phase II clinical trial indicated that BIND-014, a product utilizing this targeted release technology may be beneficial in treating patients with metastatic cancers that present circulating tumor cells (Autio et al. 2018).

2.2.5 Photodegradable

Light represents one of the most commonly employed cues used to trigger responsive materials. The advantages of using light as a trigger are myriad; for example, light can offer exquisite spatio-temporal control over reactive substrates – this is because the response of a polymer can be controlled using the wavelength, the intensity, and the duration of exposure to the light (Azagarsamy et al. 2012; Han et al. 2012; Yan et al. 2012). Moreover, light-based systems are often relatively affordable – indeed, lamps can be purchased with removable bulbs of different wavelength, and the versatility of these systems is one reason why light-based systems have become particularly pervasive in the scientific literature. Despite these advantages, light-based systems can also suffer from a variety of disadvantages. Some of these limitations include inability to penetrate deep enough into the system, uneven response across/throughout the surface, and inaccessibility of some wavelengths necessary for photoresponsivity (Tibbitt et al. 2012).

To create each of these architectures, specific chemical functional groups must be incorporated into each polymer to imbue photo-responsivity into the network. In choosing a photoresponsive functional group, several parameters must be considered. First, the wavelength of absorbance must be carefully selected – in many cases, it can be ideal to have functional groups that can selectively absorb the wavelength of absorbance. In this way, side processes including unwanted degradation of the polymer can be circumvented (Klinger and Landfester 2011). Additionally, the intensity of the light source must be considered which can affect the kinetics of degradation of the system. The intensity of the light source also has a delicate balance with the quantum yield of the functional group. Quantum yield is defined as the number of times a specific event occurs per photon absorbed by the system (Zhang et al. 2008). This property is functional group specific, and accordingly should be considered in deciding upon the polymer substrate design. Several photoactive functional group combinations have been incorporated into polymeric controlled release systems. These functional groups include, but are not limited to ortho-nitrobenzyl ethers, azobenzene, spiropyrans, spirooxazines, diarylethene, cinnamic esters, and coumarins, amongst others (Fig. 3) (Mura et al. 2013; Lee et al. 2014; Kamaly et al. 2016; Liu et al. 2016).

To date, several light responsive materials have been utilized for controlled release applications. These gels remain in place upon implantation or injection and can be triggered at a desired time to release their payload. Some of these systems are even capable of being triggered multiple times, extending the release of drugs compared to single administrations. Specifically, microparticles containing a photocleavable ortho-nitrobenzyl ether linker have been explored for the triggered release of proteins (Azagarsamy et al. 2012). These systems have been engineered to release growth factors from a PEG-diphotodegradable-diacrylate/PEG-thiol based network in vitro upon exposure to single photon UV light or multiphoton light at higher wavelengths (Tibbitt et al. 2012). Other DDS have used multiresponsive polymers sensitive to both pH and light triggers. One microgel composed of a poly(2-hydroxyethyl)methacrylate (pHEMA) and methacrylic acid demonstrated reversible swelling behavior that led to changes in release rates of myoglobin in vitro before irradiation with UV light led to complete release and degradation of the gel through the incorporation ortho-nitrobenzyl ether linkers (Klinger and Landfester 2011). Another study explored the conjugation of different moieties to an ortho-nitrobenzyl ether-PEG-methacrylate linker in order to conjugate growth factors, specific peptides or model proteins into a hydrogel scaffold for triggered UV initiated release (Griffin et al. 2013).

2.2.6 Magnetic

Magneto-responsive materials are an emerging material platform that is characterized by their ability to interact with externally applied magnetic fields. Magnetic responsive materials are becoming considerably more attractive with time for several reasons. First, magnetic responsive materials bear considerable amounts of orthogonality with respect to other more traditionally employed triggers. For example, unlike chemical or biological cues which are diffusion dependent, and unlike other physical cues, including light, which are exposure dependent and can have overlap with other UV-absorbing functionalities, magnetic responsive materials are quite specific (Li and Keller 2009; Esser-Kahn et al. 2011; Mura et al. 2013). What also makes magnetic responsive materials so attractive is that they can be multimodal in nature. For example, magnetic cues can be used for controlled release applications to release therapeutics with spatial and temporal control (Ling et al. 2014); additionally, these materials can be used for imaging purposes which provides an additional handle for achieving drug delivery goals. Despite these advantages, magnetic responsive materials can be quite challenging to implement within the drug delivery field. There are two predominant driving factors behind this difficulty which are in part tied to one another: the first is magnetic strength, and the second is cost. For example, to implement function, a magnetic field must be strong enough to interact with the magneto-responsive material. This can be challenging when devices are implanted into the body, or are located in difficult to reach areas that may be too far outside of the magnetic field. Although clinically we have access to strong magnets, particularly in the form of MR-based technologies, these instruments are extremely expensive and can only be operated by trained professionals, both of which limits their function for traditional materials.

From a synthesis and design standpoint, several classes of magnetic-responsive materials have been investigated. Magnetic gels are hydrogels that can respond to alterations in magnetic field. Recent efforts have demonstrated that hydrogels can be made magneto-responsive by either covalent modification of magnetic-responsive functionality into the polymer chains, or, more simply, by direct mixing of magnetic particles including iron or cobalt into the hydrogel system (Satarkar and Zach Hilt 2008). To date, these materials have found widespread use in drug delivery, including for the delivery of cancer therapeutics such as doxorubicin (Widder 1981; Sanson et al. 2011). Another study demonstrated magnetic control of insulin release from an ethylene-vinyl acetate copolymer depot containing a small magnet that led to triggered decreases in blood glucose over a 50 day study in diabetic mouse models (Kost et al. 1987). It is also important to note that from an atomic standpoint, the majority of these materials are imbued with their properties by combining iron, ferrocene, cobalt, modified PEGs, or folates – effectively, many of these processes are most commonly governed by free radical molecules with unequal spin quantum numbers and a large density of crosslinks, ultimately allowing for large net quantum values for either ferromagnetic or anti-ferromagnetic exchange coupling between the molecules. The administration of a constant low voltage as an electronic release trigger has also been previously explored (Yan et al. 2010; Kim et al. 2011).

2.2.7 Temperature

The solubility and mechanical properties of polymeric networks are affected by changes in temperature, allowing for systems to be developed with thermoresponsive release. Some polymers demonstrate a lower critical solution temperature (LCST), above which the polymers can undergo a phase transition from a solid gel to a liquid solution where the polymer will become miscible or the network will degrade (Zhang et al. 2008). The desired increase in temperature can be achieved through the placement of hot water bottles, a heating blanket, or a directed heat lamp for systems designed for subcutaneous delivery. The LCST of the polymer network can be manipulated through combinations of different polymer backbones as well as through the addition of hydrophobic groups. For example, a study varying the amount of acrylamide in a poly(N-isopropylacrylamide-co-acrylamide) (PNIPA-co-AAm) demonstrated different LCST properties resulting in varying levels of encapsulation and release of near-infrared dyes for in vivo imaging studies (Zhang et al. 2008).

Polymeric hydrogels can therefore be designed for drug delivery and extended release with the physiological temperature of 37 °C in mind, or systems can be tuned to promote release at slightly higher temperature as found in the tumor microenvironment. A polymeric hydrogel consisting of acrylamide-co-acrylic acid was designed with an upper critical solution temperature (UCST) slightly above 37 °C to promote the swelling of the gel and release of cisplatin (Shirakura et al. 2014). Another system utilized nanoparticles comprised of poly{γ-2-[2-(2-methoxyethoxy)-ethoxy]ethoxy-ε-caprolactone}-b-poly(γ-octyloxy-ε-caprolactone) (PMEEECL-b-POCTCL) that were engineered to achieve an LCST of 38 °C, resulting in the controlled in vitro release of doxorubicin and a fluorescent molecule with potential for release over 48 h (Cheng et al. 2012). A Pluronic® (poly(ethylene oxide)-poly(propylene oxide)-poly(ethylene oxide)) based gel loaded with vascular endothelial growth factor (VEGF)-laden nanoparticles demonstrated different LCST properties depending on the weight percentage of Pluronic® used. These gels were further encapsulated in a porcine bladder acellular matrix and the VEGF release was extended to 60 days in vitro (Geng et al. 2011).

Visudyne, a liposomal non-polymeric nanomedicine for photodynamic therapy, is the only nanomedicine with stimuli-responsive nanoplatform concept approved by Food and Drug Administration (FDA). Other stimuli-responsive drug delivery systems, including polymeric ones, are still at clinical stage. The potential of these promising clinical carriers for successful clinical trials and for market approval relies on the simplicity of their formulation. However, in order to obtain "smart" properties, most of stimuli-responsive drug delivery systems in development were designed with sophisticated features, difficult to scale up for industrial production. Therefore, the design simplicity is still one of the key properties for successful drug carrier translation (Liu et al. 2016).

3 Formulation of polymers as CRS: Implants, gels and particles

Polymers can be used to deliver therapeutics over a long period in different forms. Implants, either pure polymeric implants, or ones coated with polymers can be placed in the body for prolonged therapeutics release. Hydrogels, may be injected or implanted to sustain the administration of therapeutics. Lastly, the encapsulation of therapeutics within microparticles or nanoparticles, administered in gels or as depots, may prolong their release.

3.1 Achieving the sustained delivery of therapeutics with polymeric implants

Implants are man-made medical devices placed in the body to replace or support a biological structure. They may be hard implants, made of metal or ceramic, coated with polymers, entirely polymeric implants, or even cells artificially put in a polymeric envelope (Sabel et al. 1990; Elstad and Fowers 2009; Gershanik and Jenner 2012; Weaver et al. 2014; Cho and Kwon 2014; Appel et al. 2015; Blanco et al. 2015; Vegas et al. 2016; Bellinger et al. 2016; Kamaly et al. 2016; Ball et al. 2018; Tzeng et al. 2018; Fenton et al. 2018).

3.2 Routes of administration of sustained delivery implants

Implants can be placed close to the zone to be treated, or in a more accessible area for practical administration or explantation (Sussman et al. 2014). Implants can be subcutaneous, intraperitoneal, oral, intracranial, intraocular, mucosal, or transdermal with microneedles (Fig. 4). One issue with oral administration is that it encounters the hepatic first-pass metabolism. Hepatic first-pass metabolism reduces the effective dose of drug by metabolizing it rapidly and can have harmful effects on the liver (Pond and Tozer 1984; Lalka et al. 1993). Even intravenously injected drugs tend to be catabolized very rapidly, eliminated from circulation by the liver or the kidneys (Paolini et al. 2017; Germain et al. 2018). Placed subcutaneously, DDS may be injected as hard implants or in the form of depots. A depot consists in the deposition of a drug in a localized volume, from which the drug is gradually absorbed by surrounding tissue. Depot injections are usually either solid (hydrophobic drug aggregates) or oil-based. Although

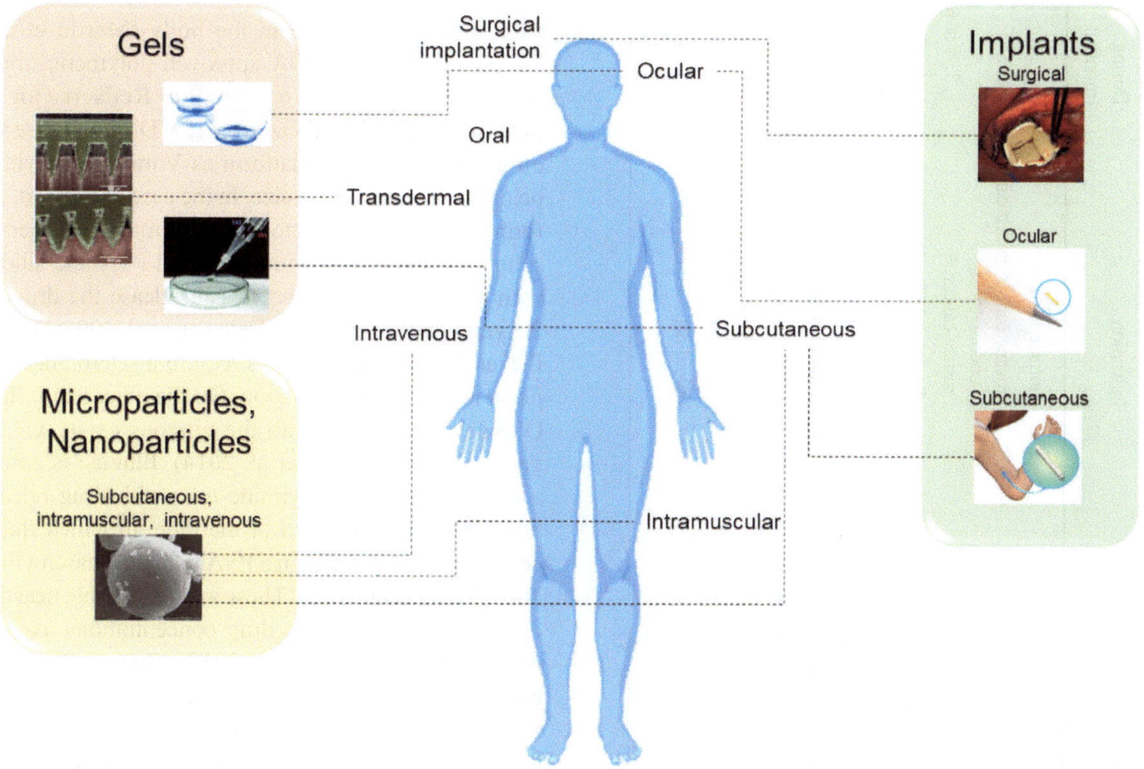

Fig. 4 Example of routes of administration by type of sustained release polymeric device. Adapted with permission from (Lesniak and Brem 2004; Park et al. 2011; Caffarel-Salvador et al. 2015; Lee et al. 2016; Fenton et al. 2018)

subcutaneous administration results in high dose variations from one injection to the other and may cause pain, it is still a very common route for most extended-release depots (Kim et al. 2017).

3.3 Approved implants for sustained delivery

Norplant® was the first extended-release polymeric implant approved by the Food and Drug Administration (FDA) in the US, in 1990. It was discontinued in the early 2000s and replaced by the Jadelle® (Norplant® II) implant (Table 2) consisting of two rods of a dimethylsiloxane / methylvinylsiloxane copolymer core enclosed in thin-walled silicone tubing (Diaz et al. 1982; Segal 1987). The original Norplant consisted of a set of six rods. Each rod is a cylindrical reservoir, placed subcutaneously, releasing levonorgestrel over 5 years. The outer membrane, as well as the cylindrical shape enable a zero-order release of the hormone. Other birth control implants were later approved by the FDA, such as the etonogestrel implants Implanon and Nexplanon (Blumenthal et al. 2008). Implanon and Nexplanon are made of ethylene vinyl acetate copolymer, a nonabsorbable material, and Nexplanon contains 15 mg of barium sulfate. Nexplanon was originally marketed under the brand name Implanon, but was subsequently modified and marketed as Nexplanon/ Implanon NXT. The presence of barium sulfate in Nexplanon makes it visible via X-ray, differentiating it from Implanon (Pedroso et al. 2015).

Contraceptive implants are placed subcutaneously for systemic diffusion, but other implants are designed to be implanted directly at their therapeutic sites. Implants such as ocular implants, designed to slowly release drugs in the vitreous chamber, present the opportunity to reach this organ protected by the so-called blood-ocular barrier (Bradbury and Lightman 1990). The eye is subject to conditions such as glaucoma, age-related macular degeneration (AMD), diabetic macular edema (DME), and retinal vein occlusions (RVOs). Since the commercialization of Vitrasert, the 1st generation of intravitreal implant, in 1996, several successive generations were designed and approved to treat different disorders (Fig. 5). Vitrasert, a ganciclovir implant (Bausch + Lomb), was approved by the FDA for the treatment of cytomegalovirus (CMV) retinitis in AIDS patients. It consisted of a pelleted form of ganciclovir (4 .5mg) coated with polyvinyl alcohol (PVA) and ethylene vinyl acetate (EVA) layers to control the release of the drug, for up to 8 months (Martin et al. 1994). This mode of delivery reduced morbidity from systemic use of ganciclovir and immunosuppressive drugs such as steroids, while alleviating the patients' eye disease. The marketing authorization in Europe was voluntarily withdrawn in 2002 due to a lack of demand. Furthermore, the implant only provides local protection against CMV, so oral ganciclovir must be taken by people with implants to prevent the disease

Table 2 Examples of approved polymeric implants allowing the sustained delivery of therapeutics

Brand	Company	FDA Approval	Active Pharmaceutical Ingredient (API)	Composition of the implant	Indication	Dosing (months)
Jadelle®	Population Council / Bayer	1996	levonorgestrel	silicone	contraception	60
Zoladex®	AstraZeneca	1998	goserelin	poly(lactic-co-glycolic acid)	Prostate cancer, breast cancer, endometriosis	1 or 3
Eligard	QLT Inc.	2002	leuprolide	poly(lactic-co-glycolic acid)	Prostate cancer	1, 3, 4 or 6
Gliadel®	Eisai inc.	2003	carmustine	polyanhydride	glioma and glioblastoma	N/A
Cypher™	Cordis, J&J	2003	sirolimus	poly(butyl methacrylate) (PBMA)	coronary stent	N/A
Taxus Express	Boston Scientific	2004	paclitaxel	poly (styrene-bisobutylene-b-styrene) (SIBS)	coronary stent	N/A
Implanon®, Nexplanon®	Organon / Merck	2006	etonogestrel	ethylene vinyl acetate	contraception	36
Ozurdex®	Allergan	2009	dexamethasone	poly(lactic-co-glycolic acid)	diabetic macular edema, uveitis	6
Propel™	Intersect ENT	2012	mometasone furoate	poly(lactic-co-glycolic acid)	Post-surgical inflammation	N/A
Iluvien®	pSivida Corp / Alimera Sciences	2014	fluocinole acetonide	Polyimide and polyvinyl alcohol	diabetic macular edema	36
Probuphine®	Titan Pharmaceuticals	2016	buprenorphine	ethylene vinyl acetate (EVA)	opioid dependence	6

from arising elsewhere in the body (Martin et al. 1999). Following Vitrasert, FDA approved polymeric implants for intravitreal drug delivery include Retisert (for uveitis), Iluvien (for DME) and Ozurdex (for DME). Retisert is based on the same delivery platform as Vitrasert but with Retisert being slightly smaller in size. In the case of Retisert, a pelleted form of fluocinolone acetonide (0.59 mg) is attached to a PVA suture tab and coated with additional PVA and silicon layers with a drug diffusion port. It can release the drug for up to 2.5 years (Jaffe et al. 2006; Callanan et al. 2008; Hazirolan and Pleyer 2013). Both implants require a sclerotomy at the pars plana region for implantation. On the other hand, Iluvien and Ozurdex are injected into the vitreous cavity via a 23–25-gauge needle (Cabrera et al. 2014). Illuvien is composed of a small cylindrical polyimide tube with drug release membrane caps on either end of the tube, in which fluocinolone acetonide is loaded within a PVA matrix (Ghasemi Falavarjani 2009; Wang et al. 2013). These systems enable near zero order drug release of effective drug concentrations over extended periods of time (Kaji et al. 2017). Since Vitrasert, Retisert and Iluvien are non-biodegradable, the drug-depleted devices need to be surgically removed or may accumulate in the vitreous cavity as in the case of Iluvien. However, the influence of the residual device left in the vitreous cavity is still unclear (Kaji et al. 2017; Mandal et al. 2017). Other than the eye, arteries are another site where implants are placed directly to release drugs. Since the approval of Cypher™, a sirolimus-eluting coronary stent in 2003, different generations of drug-eluting stents (DES), loaded with anticoagulant, immunosuppressant, anti-inflammatory and antiproliferative agents have been approved. As a whole, they demonstrated significant improvements compared to classical coronary stents in reducing in-stent restenosis, target lesion revascularization (TLR) and major adverse cardiac events (MACE) (Zilberman et al. 2010).

Eligard is a more classic implant: it is a subcutaneous depot of poly(lactic-co-glycolic acid) (PLGA) filled with leuprolide. Different formulations altering release durations are achieved notably by varying the ratio of of D,L-lactide to glycolide. In the 1-month formulation the PLGA has a 50:50 M ratio of D,L-lactide to glycolide with carboxyl end groups, the 3-month formulation has a molar ratio of 75:25 of D,L-lactide to glycolide with hexanediol, and the 6-month formulation has a molar ratio of 85:15 (Cox et al. 2005; Sartor 2006).

3.4 Sustained delivery implants in clinical or preclinical trials

Another type of intravitreal implant is in trials, the ECT cell capsule, Renexus® (NT-501), developed by Neurotech Pharmaceuticals (Fig. 6). It contains cells genetically engineered to produce ciliary neurotrophic factor and already completed a phase II clinical trial for Macular Telangiectasia

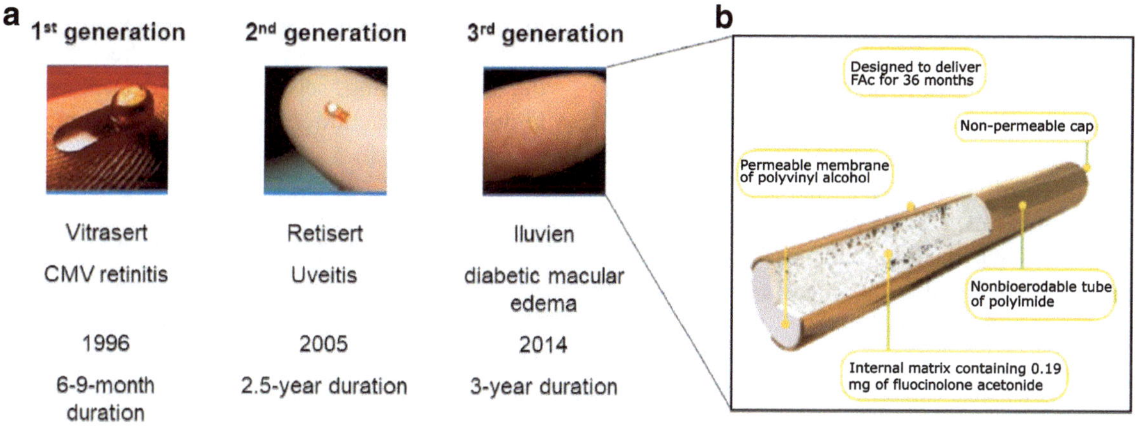

Fig. 5 (A) Evolution of hard implants for intravitreal delivery. (B) Details of the composition of Iluvien, 3rd generation of intravitreal implant

type 2 (MacTel) in addition to evaluating a current phase II clinical trial for glaucoma (Smith et al. 2015). In NT-501's phase I safety trial, researchers implanted the cells in 10 eyes of 10 patients with retinitis pigmentosa for six months. When the implants were removed at the six-month mark, all implants still contained viable cells, exhibited minimal cell loss, and were secreting CNTF at levels shown to be therapeutic towards retinal degeneration in dogs (Sieving et al. 2006; Chew et al. 2015; Birch et al. 2016).

This idea of implanting functional cells producing the therapeutics for an prolonged period of time has also been investigated to treat diabetes, this time encapsulating cells within a hydrogel (Fig. 6) (Köllmer et al. 2015; Vegas et al. 2016; An et al. 2018).

Another difficult area to access is the brain. The current gold standard to treat patients with Parkinson's disease is oral levodopa (L-dopa), the precursor of dopamine. However, the short half-life of L-dopa and its variable absorption through the gastro-intestinal tract and the blood-brain barrier limit the efficacy of this treatment. Moreover, the efficiency of L-dopa relies on the enzyme L-amino acid decarboxylase, which declines with the disease progression (Laloux et al. 2017). Solutions that are currently being investigated include a continuous administration of L-dopa, L-dopa implants in the duodenum, and neurostimulating electrodes for deep-brain stimulation in the brain (Sabel et al. 1990; Gershanik and Jenner 2012; Weaver et al. 2014; Laloux et al. 2017).

More classical subcutaneous or retinal implants are being redesigned in order to become refillable in order to extend their period of drug release. One example is a refillable microfabricated PDMS drug delivery device for the treatment of ocular diseases (Lo et al. 2008). This DDS consists of a reservoir and cannula with support structures to be secured to the sclera of the eye. Release is activated by applying mechanical pressure to the reservoir to force medication through the cannula. The device can be refilled by using a syringe and needle up to 24 times, with an expected refill frequency of 3 months at one delivery per day, meaning that the device will have a lifetime of approximately 6 years. Therefore, one surgery to implant the device and insert the cannula spares the patient from over 2000 injections into the eye (Koch et al. 2016). Refillable devices are also being developed for subcutaneous implants. To avoid the explantation and replacement of such devices, transcutaneously refillable implant are being developed. Notably, implants filled with antiretroviral drugs to achieve pre-exposure prophylaxis and avoid HIV transmission have been designed to be refilled after approximately 70 days of drug release (Chua et al. 2018).

Implants may also be designed to be administered in a less invasive manner: to be placed in the lumen of some organs, for

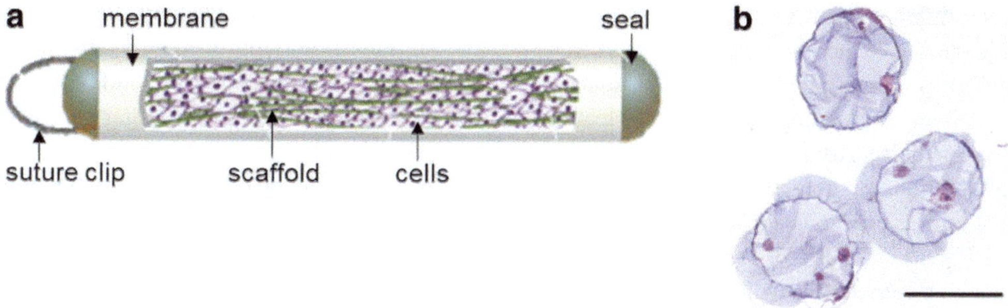

Fig. 6 Cell implants (A) The NT-501 implant puts genetically engineered cells into the retina (B) Synthetic pancreas: images of implants retrieved from the STZ-treated C57BL/6J mice presented in a–c that were implanted with SC-β cells encapsulated with TMTD alginate at a dose of 250 SC-β clusters/mouse. Scale bar, 400 μm. Adapted with permission from (Vegas et al. 2016)

instance, in the gastrointestinal (GI) tract or in the bladder. In the GI, star-shaped orally delivered capsules have been designed to unfold in the stomach and allow an extended drug delivery (Bellinger et al. 2016). After a single administration of the implant, a sustained delivery of mosquitocidal ivermectin has been achieved for up to 10 to 14 days. After this period, the polymers forming the capsule dissolve and allow the release of the remaining parts from the stomach to the intestine (Bellinger et al. 2016). To deliver lidocaine in the bladder for a similar period of time (about 14 days), the LiRIS (i.e., Lidocaine-Releasing Intravesical System, TARIS Biomedical, Lexington, MA, USA), a silicone tube with shape-memory was designed (Cima et al. 2014; Lee and Choy 2016). This tube is inserted through the ureter and coils once in the bladder cavity. It releases lidocaine by zero-order release for the first 24 h, followed by a first-order release for the rest of the 14-day period (Cima et al. 2014; Lee and Choy 2016).

Even less invasive implant options include microneedle patches. Microneedle patches have been developed for the release of flu vaccine or contraceptive hormones (Chu and Prausnitz 2011; Prausnitz 2017; Rouphael et al. 2017). Microneedles can be made of metal, silicon or polymers. The ideal shape for polymeric microneedles is a wide base to a sharp tip, which increases the mechanical strength. Nondissolving polymeric microneedles are often made of polycarbonate, polymethyl methacrylate, and silicon. Microneedles may also be made of swellable hydrogels that release the encapsulated drug upon gel hydration. Such designs have been realized with materials such as poly(methyl vinyl ether-comaleic acid) crosslinked with PEG. Work has also been done to develop microneedles as slow-release devices made out of biodegradable polymers such as PLGA and silk (Prausnitz 2017).

3.5 Using hydrogels as extended-release vehicles

Hydrogels are a macromolecular three-dimensional hydrophilic network of polymers formed by crosslinking through various mechanisms including hydrophobic association, electrostatic interactions, thermally induced entanglement, and covalent chemical crosslinking. Hydrogels are remarkably tunable, biocompatible, and have the ability to retain high water content which makes them extremely valuable biomaterial scaffolds for drug delivery, tissue engineering, soft electronics, and regenerative medicine. Additionally, hydrogels can be formed in aqueous solution, minimizing the denaturation and aggregation risks of biologic cargo (Pagels and Prud'Homme 2015; Tong et al. 2015; Li et al. 2016). Traditional hydrogels can break easily due to a lack of mechanical strength (Langer and Peppas 1981). However, engineered hydrogels with improved stability and structural complexities have provided enhanced spatial and temporal control over the extended release of various therapeutics (Li and Mooney 2016; Tibbitt et al. 2016). Various advancements have also been made to improve the shear-thinning, self-healing, and responsive nature of hydrogels (Guvendiren et al. 2012; Hozumi et al. 2018; Leach et al. 2018).

3.6 Routes of administration of hydrogels as CRS

Hydrogels may be administered through different routes. These include but are not limited to: oral, nasal, rectal, epidermal, vaginal, subcutaneous, intraperitoneal, intraocular, deposited on mucosa, etc. Gels present the advantage to be delivered through standard gauge needles, offering minimally invasive administration. Depending on the requirement of the delivery method, hydrogels can be developed into any size and shape, the major challenges being the efficiency of release and patient compliance. Hence, they are classified into macro, micro and nano gels (Fig. 7). Macrogels are typically millimeters to centimeters in size and are either transdermally administered (Tiwari et al. 2012) or surgically implanted, which is invasive and may result in patient discomfort (Yu and Ding 2008). To address this limitation,progress has been accomplished in the development of injectable alternatives designed to form the gel in the body (in situ hydrogels), outside the body but that transition to solution state under shear stress (shear-thinning hydrogels), or prepared externally but readily collapse and undergo shape recovery in the body (shape-memory hydrogels). In the case of in situ hydrogels, the sol-gel transition process is very slow and the solution solidifies after reaching the body. The various strategies of gelation applied in this case include click reactions, Michael addition, or charge complexation (Hiemstra et al. 2007; Silva and Mooney 2007; Silva et al. 2008; Jin et al. 2010). Another interesting class of hydrogels developed to bridge the gap between in situ sol-gel transition and biomaterial design is the exploration of temperature-responsive systems. Moreover, certain shear-thinning hydrogels transform into low-viscosity solutions under the mechanical stress exerted during injection and gelate quickly after removal of the force: self-healing polymers enable gels to resist the strain caused by an injection (Appel et al. 2015). This behavior is mainly due to hydrophobic and electrostatic interactions (Guvendiren et al. 2012). An alternative approach to using larger invasive hydrogels is to use micro- and nanogels.

Controlled release of a drug from the hydrogel is governed among other factors by the pore size of the polymeric network. The diffusion can be caused by degradation, swelling, mechanical deformation, or stimuli-responsive release. Hydrogels should also be biodegradable in a controllable manner after the drug is exhausted to avoid surgical removal (Table 3).

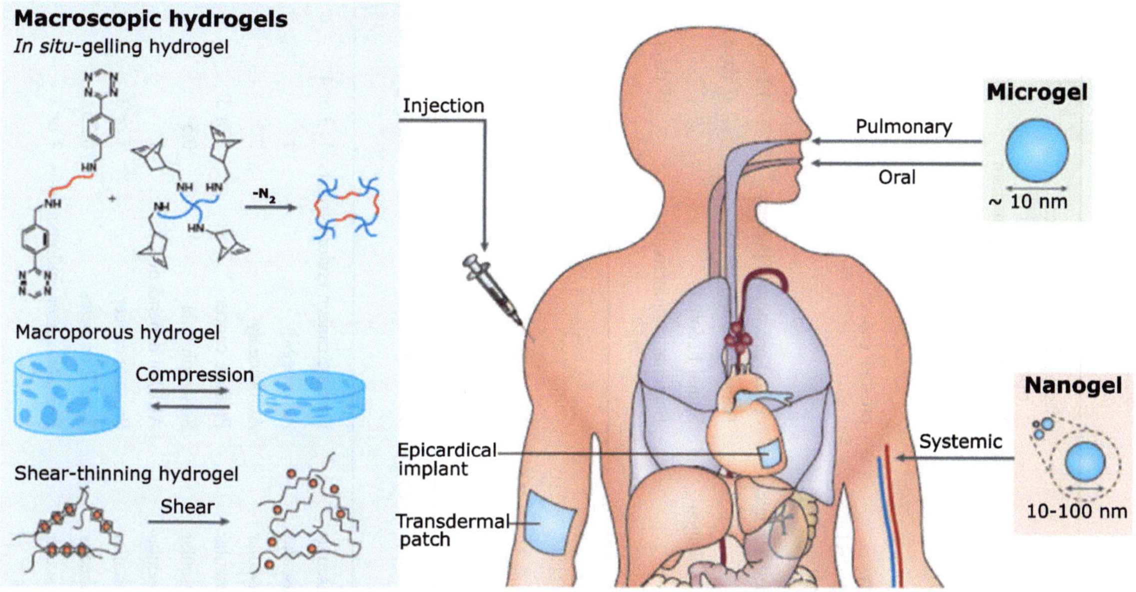

Fig. 7 Different hydrogel delivery systems (macroscopic hydrogels, microgels and nanogels) determines the route of administration. Adapted with permission from (Li and Mooney 2016)

3.7 Approved sustained release hydrogels

Medtronic's INFUSE, a collagen-based hydrogel, has been used to deliver recombinant human bone morphogenetic protein 2 (BMP2) for regeneration of bone and cartilage (Table 3). Excitingly, in a different application, the approval of Vantas hydrogel in the clinic for controlled and sustained release of histrelin in prostate cancer patients, opened the path to fuel further developments in cancer hydrogel therapy. A major drawback of the extended release formulation of hydrogels is the challenge of sterilization due to their high-water content. Hence, the cost of commercialization is often significantly high for hydrogel-based delivery systems.

3.8 In clinical trials / preclinical

In a clinical research study on the regeneration of bone and cartilage, a scaffold made of both hydrogel and porous hard material was used. In this scaffold, the atelocollagen gel promoted the production of cartilage matrix and PLGA or PGA served as porous support to build the bone on (Takato et al. 2014). In another example of hydrogels used as scaffolds, an alginate hydrogel was implanted as a prosthetic scaffold into LV heart muscle (Anker et al. 2015).

For ophthalmic conditions, drug-modified medicated contact lenses are being developed for use in prolonged and controlled release of drugs for the treatment of glaucoma (Carvalho et al. 2015). Contact lenses can be classified into two groups, namely rigid lenses consisting of poly (methylmethacrylate) (PMMA) and soft (gelatinous) lenses consisting mainly of polymers of hydroxyethyl methacrylate (pHEMA).

Hydrogels have also helped in suppressing immune responses. In this context, Vegas et al. developed a combinatorial library of alginate hydrogels having chemical modifications for encapsulation and protection of pancreatic islet cells from the host's immune system. In this study, triazole modified alginate showed anti-fibrotic and anti-inflammatory properties and helped in providing sufficient nutrient supply to the implanted insulin-producing cells up to six months, thereby maintaining normoglycemia in rodents and non-human primates (Vegas et al. 2016). Further work is still being conducted to improve the polymer encapsulating the cell islets (nature of the alginate, size and shape of the formulation), especially to mitigate the immune response. As An et al. pointed out, one major problem with alginate capsules is that it is almost impossible to remove or replace the implant due to the complicated organ structures and the large capsule number required (i.e., ~100,000 capsules needed for a human patient) (An et al. 2018). To address this hurdle, the authors designed an implant they call TRAFFIC (thread-reinforced alginate fiber for islets encapsulation), consisting of an alginate hydrogel with in situ cross-links to a nanoporous, wettable, $Ca2+$−releasing polymer thread.

More recently, Liu et al. engineered a novel set of materials that demonstrated prolonged stability of the hydrogel in the stomach and provided safe delivery of the active drug through gastrointestinal tract in a large animal model (Liu et al. 2017).

Oncogel, an injectable hydrogel for extended release of the chemotherapy drug paclitaxel, made it to phase II clinical trials (Elstad and Fowers 2009). In a similar fashion, the French company MedinCell is developing a platform for the extended release of drugs based on the biodegradable DB PEG-PLA and TB PLA-PEG-PLA copolymers dissolved in

Table 3 Examples of approved gels allowing the sustained delivery of therapeutics. Adapted with permission from (Li and Mooney 2016)

Brand	Company	FDA Approval	Active Pharmaceutical Ingredient (API)	Composition of the implant	Indication	Dosing
Cervidil	Ferring Pharmaceut-icals	1995	dinoprostone	PEG or urethane polymer	Vaginal insert for cervical ripening to induce labor	12 hours
AndroGel	AbbVie	2000	testosterone	Carbomer 980	Topical gel for hypogonadism treatment	N/A
INFUSE	Medtronic	2002	recombinant human bone morphogenetic protein 2 (BMP2)	Collagen	bone regeneration	N/A
Vantas®	Endo International	2004	histrelin acetate	acrylic co-polymer (Poly (2-hydroxyethyl methacrylate), poly(2-hydroxypropyl methacrylate))	prostate cancer	12 months
Supprelin LA®	Endo International	2007	histrelin acetate	Poly(2-hydroxyethyl methacrylate)	central precocious puberty	12 months
AzaSite	Akorn	2007	azithromycin	Poly(acrylic acid)	Bacterial conjunctivitis	7 days
MASTERGRAFT	Medtronic	2009	BMP2	Calcium phosphate and collagen	Spinal fusion	
Besivance	Bausch & Lomb	2009	Besifloxacin	Poly(acrylic acid)	Bacterial conjunctivitis	5 days

Table 4 Examples of approved depots and microparticles designed for sustained release of proteins and peptides

Brand	Company	FDA Approval	Active Pharmaceutical Ingredient (API)	Composition of the implant	Route of administration	Indication	Dosing (months)
Lupron Depot®	TAP	1989	leuprolide	PLGA microparticles	intramuscular	Prostate cancer, endometriosis	1, 3 or 4
Sandostatin® LAR®	Novartis	1998	octreotide	PLGA microparticles	intramuscular	Acromegaly	1
Somatuline® LA	Ipsen	1998	lantreotide	PLGA microparticles	intramuscular	Acromegaly	0.5
Trelstar Depot®	Debiopharm	2000	triptorelin	PLGA microparticles	intramuscular	Prostate cancer	1 or 3
Risperidal® Consta®	Alkermes/Janssen	2003	risperidone	PLGA microparticles	intramuscular	Schizophrenia	0.5
Vivitrex®/Vivitrol®	Alkermes Inc. / Cephalon Inc.	2006	naltrexone	PLGA microparticles	intramuscular	Alcohol dependence	1
Lutrate Depot®	G P Pharm	2010	leuprolide acetate	PLGA microparticles	intramuscular	Prostate cancer	1 or 3
Bydureon®	Amylin Pharmaceuticals Inc.	2012	exenatide	PLGA microparticles	subcutaneous	Type 2 diabetes	0.25
Lupaneta Pack	AbbVie	2012	leuprolide acetate / norethindrone acetate	polylactic acid microparticles	intramuscular	Prostate cancer, endometriosis	1 or 3

a bio-compatible solvent, tripropionin (Leconet et al. 2018). Their most advanced product, an extended treatment for schizophrenia, is currently in phase III clinical trials.

3.9 Microparticles and nanoparticles as CRS

The difficulty in slowing the release of therapeutics from hydrogels is the limiting factor in their success as a controlled delivery platform. Controlled release of small drugs from hydrogels requires correspondingly small mesh sizes. This has led to approaches which increase the effective size of small therapeutics by first encapsulating them in nanoparticles, microparticles, or charge complexes in order to prolong their release from hydrogels (Stenekes et al. 2000; Park et al. 2010; Seo et al. 2011; Pinkerton et al. 2014; Pagels and Prud'Homme 2015). Even without being embedded or bound into a gel, nanoparticles, microparticles or drug aggregates can be delivered as depots.

3.10 Routes of administration of microparticles and nanoparticles for extended release of therapeutics

It has been shown that microparticles and nanoparticles often undergo accelerated clearance and fail to accumulate in target tissues and deliver their payload for an extended period of time (Tsoi et al. 2016; Wilhelm et al. 2016; Zhang et al. 2016). Particles, of different sizes and shapes, combined to a gel formulation or not, may be administered through different routes: intravenous, subcutaneous, intraperitoneal, intraocular, mucosal, intra-tumoral, intra-cardiac, etc. The advantage of subcutaneous, intraperitoneal, or intraocular injection depots is an added distance from the blood circulation, resulting in a slower diffusion and elimination. Nanoparticle-based drug delivery to the heart has also been explored previously through intramyocardial administration for the treatment of infarcts, intracoronary administration for the treatment of in-stent restenosis, as well as through injection into the pericardial space (Segura-Ibarra et al. 2017). Nanoparticles injected into the pericardial space showed a prolonged presence in the heart, with a half-life of approximately 2.5 days (Segura-Ibarra et al. 2017).

The size and shape of particles influence their opsonisation (their mechanical exit from the bloodstream through noncontinuous endothelia with vascular fenestrations) as well as the diffusion kinetics of their content (Blanco et al. 2015). Their size and shape are fundamental parameters in their bioavailability and kinetics of elimination, especially when considering intravenously injected particles. To benefit from the so called and debated enhanced permeability and retention (EPR) effect and accumulate at the vicinity of solid tumors for instance, nanoparticles have to have defined mechanical properties such as size, shape, surface charge, and even elasticity (Blanco et al. 2015; Guo et al. 2018). To overcome barriers to achieve an optimal release in time and place after intravenous administration, new designs have been developed, such as the multistage vector (MSV) platform, combining objects on the micron scale and nanoparticles, or the injectable nanoparticle generators (Venuta et al. 2017).

Oral delivery of microparticles or nanoparticles is typically limited because the epithelium of the GI tract is not porous enough to let these objects through. However, oral administration of particles is still currently investigated, especially to deliver therapeutics to these epithelial GI cells (Ball et al. 2018).

3.11 Approved microparticles for sustained drug delivery

Some hormones have a very short half-life and may be used as chronic treatments, either to supplement a hormonal deficiency, or in the context of hormone therapy to treat hormone-dependent cancers, such as some prostate cancers (Table 4). Lupron Depot is a depot of leuprolide acetate, a hormone in the gonadotropin-releasing hormone (GnRH) analogue family of medications. GnRH agonists are hydrophilic small peptides with poor absorption and relatively short half-lives, so they were developed for daily injections (Periti et al. 2002). In the case of Lupron Depot, the hormone is formulated into biodegradable PLGA polymer microspheres, injected intramuscularly, which degrade and release drug over one, three or four months. Trelstar is another example of approved extended release of a GnRH agonist encapsulated in PLGA microparticles for treating prostate cancer. Yet another hormone based therapy, on the market in 1999 and later removed, Nutropin Depot was a growth hormone for children formulated in PLGA microparticles. It was withdrawn because of manufacturing issues, but also potentially because of adverse events such as pain at injection site (Yuen et al. 2018).

Vivitrex®/Vivitrol® contain naltrexone formulated into injectable microspheres of PLGA of approximately 100 μm in diameter, which contain other proprietary active moieties that lead to its extended-release properties lasting for several weeks (Anderson and Shive 1997; Johnson 2007). In this way, Vivitrex delivers 337 mg of Naltrexone per gram of microspheres in a monthly depot injection to control opioid abuse (Johnson 2007).

3.12 In clinical trials / preclinical

Innovative single injection vaccines are in development to facilitate the vaccination process in areas where access to medical care is difficult. Some of these single injection vaccines have been designed as spherical PLGA microparticles filled with inactivated polio vaccine (IPV) antigens. Formulations release two bursts of IPV 1 month apart, reproducing a typical vaccination schedule for those in developing countries (Tzeng et al. 2016, 2018). Fillable cubic microparticles are also being developed to deliver proteins during an extended period of time: for example at 0–16 weeks after injection (McHugh et al. 2017).

For ocular delivery, several nanosized micelles, for direct injection into the eye, or encapsulated into hydrogels, have been designed or are in development. As an example, cross-linked methoxy(polyethylene glycol)-block poly(ε-caprolactone) (mPEG-b-PCL) nanosized micelles, coated with silica and incorporated into a pHEMA hydrogel were developed and tested for sustained delivery of dexamethasone acetate (DMSA) into the eye. The in vitro drug release profile from DMSA-loaded SSCMs showed a biphasic distribution with a burst release over the first 8 h followed by a release of 6% per day over 6 days. In contrast, hydrogel containing DMSA-loaded SSCMs provided sustained release of the drug for periods up to 30 days (Lu et al. 2012; Mandal et al. 2017). This example of micelles embedded in a hydrogel, along with the multistage vector platform in development described previously, illustrate the nesting strategy to design optimized, long lasting drug delivery systems.

In addition to therapeutics, extended release can also be used to deliver probes, for continuous in vivo monitoring. For instance to monitor glucose concentration in blood, glucose-responsive fluorescent probes were designed to be slowly released: packed in microbeads, themselves formulated in an injectable hydrogel (Shibata et al. 2010).

4 Challenges in extended-release

4.1 Synthesis and formulation challenges

An important aspect in the design and development of CRS is avoiding burst release. Burst release leads to uncontrolled variations in dose, which can result in toxicity for the patient. Burst release also limits the amount of encapsulated therapeutics left available to be released over time.

Apart from burst release, some CRS present low or variable encapsulation rates and loading volumes (Tng et al. 2012; Markwalter and Prud'homme 2018), as well as issues regarding the stability of the encapsulated therapeutics. Difficulties in synthesis and development may indeed arise from the encapsulated therapeutics: their physical characteristics such as size, or their stability. For instance, cells cannot be encapsulated in any vehicle as they require specific circumstances to remain viable. Another example is that of proteins or peptides encapsulated in aliphatic polyesters, which may be degraded by the acidic products of these polymers (Park et al. 1995; Fu et al. 2000).

4.2 Body reaction to CRS

The biological reactions triggered by the implementation of extended-release systems pose a potential hurdle when designing these DDS. Such biological reactions may be local, such as an inflammation, and fibrosis around the foreign body, or result in a systemic inflammation or allergy. In a clinical trial for Vivitrex®, an intramuscular injection of PLGA microparticles filled with naltrexone, involving 25 individuals, one participant dropped out due to induration at the injection site, and another was discontinued because of an allergic reaction that resulted in angioedema, which resolved soon after the participant stopped taking the medication (Johnson 2007). After a local inflammation, a so-called foreign body reaction may occur: the end-stage response of the inflammatory and wound healing responses following implantation of a medical device or biomaterial (Anderson et al. 2008). On top of local and systemic reactions directly due to the implant, gel or depot, issues can arise from the mode of administration: complications after surgery or injection via large gauge needles. For instance, with Implanon and Nexplanon implants, implant site complications were reported by 3.6% of subjects during assessments in clinical trials. Pain during or after insertion was the most frequent implant site complication, occurring in 2.9% of subjects. Additionally, haematoma, redness and swelling were reported by 0.1%, 0.3% and 0.3% of patients, respectively (Pedroso et al. 2015). Local reactions, such as inflammation may also be sought because they are beneficial to the treatment: the recruitment of immune cells is useful for vaccines for instance. In such cases, inflammation caused by microneedle patches are turned into an advantage (Prausnitz 2017).

Systemic reactions consist mostly in systemic inflammation with allergy-like symptoms. For instance, the so-called CARPA syndrome due to the complement activation after the injection of liposomes (Nilsson et al. 2007; Szebeni et al. 2011). In the case of Nexplanon implants, which contain barium sulfate, although extremely rare, patients may present an allergic reaction to barium sulfate. To date, only two cases associated with Nexplanon have been reported in the literature (Sullivan 2012; Chaudhry 2013; Pedroso et al. 2015).

Because of their extended release purpose, the drug delivery systems may stay in the body for a prolonged period of time and cause not only a local or systemic inflammation, but also a chronic inflammation. Consequences of chronic inflammation include fibrosis (Kastellorizios et al. 2015) and even tumors, such as implant-associated anaplastic large cell lymphomas. Another long-term consequence of extended release drug delivery devices shared with any other chronic treatment is the induced tolerance to the extended released drug.

Solutions to decrease implant-site reactions and inflammations consist in using approved polymers known to minimize such reactions or adjusting other parameters such as shape, size, charge, and hydrophobicity of the biomaterials used (Veiseh et al. 2015; Vegas et al. 2016; Fenton et al. 2018). The use of PLGA-based devices may lead to an inflammatory response in the vitreous space due to the acidic degradation products of PLGA. Consequently, Graybug Vision, Inc. has developed a proprietary technology to reduce the inflammation related to PLGA degradation (Kaji et al. 2017). Another strategy involves encapsulation of inflammation resolution agonists, such as resolving D1 (RvD1), an ω-3 derived lipid mediator. Resolvin D1 (RvD1) loaded in Pluronic gels or PLGA films can significantly decrease arterial inflammation after sterile injury (Fenton et al. 2018).

Drug delivery devices may be explanted, either upon completion of the treatment or because of an unexpected reaction. The process of explantation may be difficult because of the material's degradation, movement inside the body, or because of the body's reaction around the device (Kleiner et al. 2014). Implanon rods can be located only through high-frequency ultrasound, or magnetic resonance imaging (MRI). Nexplanon can be located using traditional X-ray or CT-scan because of the inclusion of barium sulphate. There have been rare reports of Nexplanon implants having reached the lung via the pulmonary artery (D'Journo et al. 2015; Rowlands et al. 2017).

For the past 30 years, many CRS have been approved to treat chronic conditions and alleviate patients' treatment burden. Specific challenges to CRS such as burst release, stability, invasiveness and inflammation have been addressed in newer generations of approved CRS or in CRS in development. New materials, smaller objects, and different routes of administrations have also been explored. In order to control the release kinetics more precisely, triggered release features are being investigated, making more sophisticated, or "smart" CRS. With the increasing prevalence of chronic diseases and concern towards patients' quality of life, CRS development holds a promising future.

Acknowledgements We would like to congratulate Professor Mauro Ferrari on the occasion of his 60th birthday for his impactful scientific career and his contributions to the field of bioengineering. Some of us were fortunate enough to have the opportunity to attend a lecture he gave at MIT in February 2018. Professor Ferrari's talk was creative and exciting, and his approach to science, management, and culture was truly inspiring. Congratulations Professor Ferrari and we wish you the happiest of birthdays.

References

A. Aghabegi Moghanjoughi, D. Khoshnevis, A. Zarrabi, A concise review on smart polymers for controlled drug release. Drug Deliv. Transl. Res. **6**(3), 333–340 (2016). https://doi.org/10.1007/s13346-015-0274-7.

T. Alfrey, E.F. Gurnee, W.g. Lloyd, Diffusion in glassy polymers. J. Polym. Sci. C **12**, 249–261 (1966). https://doi.org/10.1002/pol.1969.160071204

Amer, L. D. and Bryant, S. J. (2016) 'The in vitro and in vivo response to MMP-sensitive poly(ethylene glycol) hydrogels', Ann. Biomed. Eng. Springer US, 44(6), pp. 1959–1969. doi: https://doi.org/10.1007/s10439-016-1608-4.

D. An et al., Designing a retrievable and scalable cell encapsulation device for potential treatment of type 1 diabetes. Proc. Natl. Acad. Sci. **115**(2), E263–E272 (2018). https://doi.org/10.1073/pnas.1708806115

J.M. Anderson, A. Rodriguez, D.T. Chang, Foreign body reaction to biomaterials. Semin. Immunol. **20**(2), 86–100 (2008). https://doi.org/10.1016/j.smim.2007.11.004

J.M. Anderson, M.S. Shive, Biodegradation and biocompatibility of PLA and PLGA microspheres. Adv. Drug Deliv. Rev. **28**, 5–24 (1997). https://doi.org/10.1016/j.addr.2012.09.004.

S.D. Anker et al., A prospective comparison of alginate-hydrogel with standard medical therapy to determine impact on functional capacity and clinical outcomes in patients with advanced heart failure (AUGMENT-HF trial). Eur. Heart J. **36**(34), 2297–2309 (2015). https://doi.org/10.1093/eurheartj/ehv259

E.A. Appel et al., 'Self-assembled hydrogels utilizing polymer-nanoparticle interactions', Nat. Commun. Nature Publishing Group, 6, pp. 1–9 (2015). doi: https://doi.org/10.1038/ncomms7295.

D.Y. Arifin, L.Y. Lee, C.H. Wang, Mathematical modeling and simulation of drug release from microspheres: Implications to drug delivery systems. Adv. Drug Deliv. Rev. **58**(12–13), 1274–1325 (2006). https://doi.org/10.1016/j.addr.2006.09.007

K.A. Autio et al., Safety and efficacy of BIND-014, a Docetaxel nanoparticle targeting prostate-specific membrane antigen for patients with metastatic castration-resistant prostate Cancer. JAMA Oncol. (2018). https://doi.org/10.1001/jamaoncol.2018.2168

M.A. Azagarsamy et al., Photocontrolled nanoparticles for on-demand release of proteins. Biomacromolecules **13**(8), 2219–2224 (2012). https://doi.org/10.1021/bm300646q.

A.K. Bajpai et al., 'Responsive polymers in controlled drug delivery', Prog. Polym. Sci. Oxford., pp. 1088–1118 (2008). doi: https://doi.org/10.1016/j.progpolymsci.2008.07.005.

R.L. Ball, P. Bajaj, K.A. Whitehead, Oral delivery of siRNA lipid nanoparticles: Fate in the GI tract. Sci. Rep. Springer US **8**(1), 1–12 (2018). https://doi.org/10.1038/s41598-018-20632-6

P. Bawa et al., 'Stimuli-responsive polymers and their applications in drug delivery', Biomed. Mater., p. 15 (2009). doi: https://doi.org/10.1088/1748-6041/4/2/022001.

A.M. Bellinger et al., Oral, ultra – Long-lasting drug delivery : Application toward malaria elimination goals. Sci. Transl. Med. **8**, 365ra157 (2016). https://doi.org/10.1126/scitranslmed.aag2374.

D.G. Birch et al., Long-term follow-up of patients with retinitis Pigmentosa receiving intraocular Ciliary Neurotrophic factor implants. Am J. Ophthalmol. Elsevier **170**, 10–14 (2016). https://doi.org/10.1016/J.AJO.2016.07.013

E. Blanco, H. Shen, M. Ferrari, Principles of nanoparticle design for overcoming biological barriers to drug delivery. Nat. Biotechnol. **33**(9), 941–951 (2015). https://doi.org/10.1038/nbt.3330

P.D. Blumenthal, K. Gemzell-Danielsson, M. Marintcheva-Petrova, Tolerability and clinical safety of Implanon®. Eur J Contracept Reprod Health Care **13**(supl), 29–36 (2008). https://doi.org/10.1080/13625180801960012

M.W.B. Bradbury, S.L. Lightman, The blood-brain interface. Eye (Basingstoke) **4**(2), 249–254 (1990). https://doi.org/10.1038/eye.1990.36

C. Buttorff, T. Ruder, M. Bauman, *Multiple Chronic Conditions in the United States*. (2017). doi: https://doi.org/10.7249/TL221.

M. Cabrera, S. Yeh, T.A. Albini, 'Sustained-release corticosteroid options', J. Ophthalmol. (2014). doi: https://doi.org/10.1155/2014/164692, 1.

E. Caffarel-Salvador et al., Hydrogel-forming microneedle arrays allow detection of drugs and glucose in vivo: Potential for use in diagnosis and therapeutic drug monitoring. PLoS One **10**(12), 1–21 (2015). https://doi.org/10.1371/journal.pone.0145644

D.G. Callanan et al., Treatment of posterior uveitis with a Fluocinolone Acetonide implant. Arch. Ophthalmol. **126**(9), 1191 (2008). https://doi.org/10.1001/archopht.126.9.1191.

G. Camera-Roda, G.C. Sarti, Mass transport with relaxation in polymers. AICHE J. **36**(6), 851–860 (1990). https://doi.org/10.1002/aic.690360606

A. Car et al., PH-responsive PDMS-b-PDMAEMA micelles for intracellular anticancer drug delivery. Biomacromolecules **15**(9), 3235–3245 (2014). https://doi.org/10.1021/bm500919z.

I.M. Carvalho et al., Sustained drug release by contact lenses for glaucoma treatment - a review. J. Control. Release Elsevier B.V. **202**, 76–82 (2015). https://doi.org/10.1016/j.jconrel.2015.01.023

F. Chaudhry, Adverse reaction to Nexplanon®. J. Fam. Plann. Reprod. Health Care **39**(4), 231–232 (2013). https://doi.org/10.1136/jfprhc-2013-100731.

Y. Cheng et al., Thermally controlled release of anticancer drug from self-assembled γ-substituted amphiphilic poly((ε-caprolactone) micellar nanoparticles. Biomacromolecules **13**(7), 2163–2173 (2012). https://doi.org/10.1021/bm300823y.

E.Y. Chew et al., 'Ciliary Neurotrophic factor for macular telangiectasia type 2: Results from a phase 1 safety trial', Am J. Ophthalmol. Elsevier, 159(4), p. 659–666.e1 (2015). doi: https://doi.org/10.1016/J.AJO.2014.12.013.

H. Cho, G.S. Kwon, Thermosensitive poly-(d,l-lactide-co-glycolide)-block-poly(ethylene glycol)-block-poly-(d,l-lactide-co-glycolide) hydrogels for multi-drug delivery. J. Drug Target. **22**(7), 669–677 (2014). https://doi.org/10.3109/1061186X.2014.931406

L.Y. Chu, M.R. Prausnitz, Separable arrowhead microneedles. J. Control. Release Elsevier B.V **149**(3), 242–249 (2011). https://doi.org/10.1016/j.jconrel.2010.10.033

C.Y.X. Chua et al., 'Transcutaneously refillable nanofluidic implant achieves sustained level of tenofovir diphosphate for HIV pre-exposure prophylaxis', J. Control. Release Elsevier, 286, pp. 315–325 (2018). doi: https://doi.org/10.1016/j.jconrel.2018.08.010.

M.F. Chung et al., Controlled release of an anti-inflammatory drug using an ultrasensitive ROS-Responsive gas-generating carrier for localized inflammation inhibition. J. Am. Chem. Soc. **137**(39), 12462–12465 (2015). https://doi.org/10.1021/jacs.5b08057

M.J. Cima et al., Single compartment drug delivery. J. Control. Release. Elsevier B.V. **190**, 157–171 (2014). https://doi.org/10.1016/j.jconrel.2014.04.049

M.C. Cox, C.D. Scripture, W.D. Figg, Leuprolide acetate given by a subcutaneous extended-release injection: Less of a pain? Expert. Rev. Anticancer. Ther. **5**(4), 605–611 (2005). https://doi.org/10.1586/14737140.5.4.605

H.R. Culver, J.R. Clegg, N.A. Peppas, Analyte-Responsive Hydrogels: Intelligent Materials for Biosensing and Drug Delivery. Acc. Chem. Res. **50**(2), 170–178 (2017). https://doi.org/10.1021/acs.accounts.6b00533

X.B. D'Journo, V. Vidal, A. Agostini, Intravascular pulmonary migration of a subdermal contraceptive implant. Ann. Thorac. Surg.. Elsevier **99**(5), 1828 (2015). https://doi.org/10.1016/j.athoracsur.2014.12.049

S. Dai, P. Ravi, K.C. Tam, pH-responsive polymers: Synthesis, properties and applications. Soft Matter **4**(3), 435–449 (2008). https://doi.org/10.1039/b714741d

Deloitte (2018) '2018 Global life sciences outlook Innovating life sciences in the fourth industrial revolution: Embrace, build, grow'.

S. Diaz et al., A five-year clinical trial of levonorgestrel silastic implants (Norplant TM). Contraception **25**(5), 447–456 (1982)

L. Dong et al., A pH/enzyme-responsive tumor-specific delivery system for doxorubicin. Biomaterials **31**(24), 6309–6316 (2010). https://doi.org/10.1016/j.biomaterials.2010.04.049

S. Dong et al., A dual-responsive Supramolecular polymer gel formed by crown ether based molecular recognition. Angew. Chem. Int. Ed. **50**(8), 1905–1909 (2011). https://doi.org/10.1002/anie.201006999

N.L. Elstad, K.D. Fowers, OncoGel (ReGel/paclitaxel) - clinical applications for a novel paclitaxel delivery system. Adv. Drug Deliv. Rev. Elsevier B.V. **61**(10), 785–794 (2009). https://doi.org/10.1016/j.addr.2009.04.010

A.P. Esser-Kahn et al., 'Triggered release from polymer capsules', Macromolecules, pp. 5539–5553 (2011). doi: https://doi.org/10.1021/ma201014n.

O.S. Fenton et al., Advances in biomaterials for drug delivery. *Adv. Mater.* **1705328**, 1–29 (2018). https://doi.org/10.1002/adma.201705328

F. Fischel-Ghodsian et al., Enzymatically controlled drug delivery. Proc. Natl. Acad. Sci. **85**(7), 2403–2406 (1988). https://doi.org/10.1073/pnas.85.7.2403

S. Freiberg, X.X. Zhu, Polymer microspheres for controlled drug release. Int. J. Pharm. **282**(1–2), 1–18 (2004). https://doi.org/10.1016/j.ijpharm.2004.04.013

K. Fu et al., Visual evidence of acidic environment within degrading poly(lactic-co- glycolic acid) (PLGA) microspheres. Pharm. Res. **17**(1), 100–106 (2000). https://doi.org/10.1023/A:1007582911958.

Y. Fu, W.J. Kao, Drug release kinetics and transport mechanisms of non-degradable and degradable polymeric delivery systems. Expert Opin. Drug Deliv. **7**(4), 429–444 (2010). https://doi.org/10.1517/17425241003602259

L.K. Fung, W.M. Saltzman, Polymeric implants for cancer chemotherapy. Adv. Drug Deliv. Rev. **26**(2–3), 209–230 (1997). https://doi.org/10.1016/S0169-409X(97)00036-7.

G.H. Gao et al., The use of pH-sensitive positively charged polymeric micelles for protein delivery. Biomaterials **33**(35), 9157–9164 (2012). https://doi.org/10.1016/j.biomaterials.2012.09.016

H. Geng et al., Sustained release of VEGF from PLGA nanoparticles embedded thermo-sensitive hydrogel in full-thickness porcine bladder acellular matrix. (2011). doi: https://doi.org/10.1186/1556-276X-6-312.

M. Germain et al., 'Priming the body to receive the therapeutic agent to redefine treatment benefit/risk profile', Sci. Rep., 8(1) (2018). doi: https://doi.org/10.1038/s41598-018-23140-9.

O. Gershanik, P. Jenner, Moving from continuous dopaminergic stimulation to continuous drug delivery in the treatment of Parkinson's disease. Eur. J. Neurol. **19**(12), 1502–1508 (2012). https://doi.org/10.1111/j.1468-1331.2011.03593.x

K. Ghasemi Falavarjani, Implantable posterior segment drug delivery devices; novel alternatives to currently available treatments. J. Ophthalmic Vis. Res.. Wolters Kluwer – Medknow Publications **4**(3), 191–193 (2009)

M. Grassi, G. Grassi, Mathematical modelling and controlled drug delivery: Matrix systems. Curr. Drug Deliv. **2**(1), 97–116 (2005). https://doi.org/10.2174/1567201052772906

D.R. Griffin et al., Synthesis of photodegradable macromers for conjugation and release of bioactive molecules. Biomacromolecules **14**(4), 1199–1207 (2013). https://doi.org/10.1021/bm400169d.

G. Gu et al., PEG-co-PCL nanoparticles modified with MMP-2/9 activatable low molecular weight protamine for enhanced targeted glioblastoma therapy. Biomaterials **34**(1), 196–208 (2013a). https://doi.org/10.1016/j.biomaterials.2012.09.044

Z. Gu et al., Injectable Nano-network for glucose-mediated insulin delivery. ACS Nano **7**(5), 4194–4201 (2013b). https://doi.org/10.1021/nn400630x

P. Guo et al., Nanoparticle elasticity directs tumor uptake. Nat. Commun.. Springer US **9**(1), 130 (2018). https://doi.org/10.1038/s41467-017-02588-9

P. Gupta, K. Vermani, S. Garg, Hydrogels: From controlled release to pH-responsive drug delivery. Drug Discov. Today **7**(10), 569–579 (2002). https://doi.org/10.1016/S1359-6446(02)02255-9

M. Guvendiren, H.D. Lu, J.A. Burdick, Shear-thinning hydrogels for biomedical applications *. Soft Matter **8**, 260–272 (2012). https://doi.org/10.1016/j.addr.2012.09.010.

D. Han, X. Tong, Y. Zhao, Block copolymer micelles with a dual-stimuli-responsive core for fast or slow degradation. Langmuir **28**(5), 2327–2331 (2012). https://doi.org/10.1021/la204930n

A.S. Hasan et al., Effect of the microencapsulation of nanoparticles on the reduction of burst release. Int. J. Pharm. **344**(1–2), 53–61 (2007). https://doi.org/10.1016/j.ijpharm.2007.05.066.

A. Hatefi, B. Amsden, Biodegradable injectable in situ forming drug delivery systems. J. Control. Release **80**(1–3), 9–28 (2002). https://doi.org/10.1016/S0168-3659(02)00008-1.

D. Hazirolan, U. Pleyer, Think global–act local: Intravitreal drug delivery systems in chronic noninfectious uveitis. Ophthalmic Res. Karger Publishers **49**(2), 59–65 (2013). https://doi.org/10.1159/000345477

C. Hiemstra et al., In vitro and in vivo protein delivery from in situ forming poly(ethylene glycol)-poly(lactide) hydrogels. J. Control. Release **119**(3), 320–327 (2007). https://doi.org/10.1016/j.jconrel.2007.03.014

T. Higuchi, Mechanism of sustained-action medication. Theoretical analysis of rate of release of solid drugs dispersed in solid matrices. J. Pharm. Sci. **52**(12), 1145–1149 (1963). https://doi.org/10.1002/jps.2600521210

T.R. Hoare, D.S. Kohane, Hydrogels in drug delivery: Progress and challenges. Polymer, 1993–2007 (2008). https://doi.org/10.1016/j.polymer.2008.01.027

A.S. Hoffman, The origins and evolution of "controlled" drug delivery systems. J. Control. Release Elsevier B.V. **132**(3), 153–163 (2008). https://doi.org/10.1016/j.jconrel.2008.08.012

A.S. Hoffman, P.S. Stayton, Conjugates of stimuli-responsive polymers and proteins. Prog. Polym. Sci. (Oxford) **32**, 922–932 (2007). https://doi.org/10.1016/j.progpolymsci.2007.05.005

M.R. Hoy and E.J. Roche (1993) Taste mask coatings for preparation of chewable pharmaceutical tablets.

T. Hozumi *et al.* (2018) 'Injectable hydrogel with slow degradability composed of gelatin and hyaluronic acid cross-linked by Schiff's base formation', Biomacromolecules, p. acs.biomac.7b01133. doi: https://doi.org/10.1021/acs.biomac.7b01133

J. Hrkach et al., Preclinical development and clinical translation of a PSMA-targeted Docetaxel nanoparticle with a differentiated pharmacological profile. Sci. Transl. Med. **4**(128), 128ra39–128ra39 (2012). https://doi.org/10.1126/scitranslmed.3003651

X. Huang, C.S. Brazel, On the importance and mechanisms of burst release in matrix-controlled drug delivery systems. J. Control. Release **73**(2-3), 121–136 (2001)

Z.-Z. Huang et al., Mechanism and significance of increased glutathione level in human hepatocellular carcinoma and liver regeneration. FASEB J. (2000). https://doi.org/10.1096/fj.00-0445fje

G.J. Jaffe et al., Fluocinolone Acetonide implant (Retisert) for noninfectious posterior uveitis: Thirty-Four–Week Results of a Multicenter Randomized Clinical Study. Ophthalmology. Elsevier **113**(6), 1020–1027 (2006). https://doi.org/10.1016/J.OPHTHA.2006.02.021

B. Jeong, A. Gutowska, 'Lessons from nature: Stimuli-responsive polymers and their biomedical applications', Trends Biotechnol., pp. 305–311 (2002). doi: https://doi.org/10.1016/S0167-7799(02)01962-5,20.

R. Jin et al., Synthesis and characterization of hyaluronic acid-poly(ethylene glycol) hydrogels via Michael addition: An injectable biomaterial for cartilage repair. Acta Biomater. Acta Materialia Inc. **6**(6), 1968–1977 (2010). https://doi.org/10.1016/j.actbio.2009.12.024

B.A. Johnson, Naltrexone long-acting formulation in the treatment of alcohol dependence. Ther. Clin. Risk Manag.. Dove Press **3**(5), 741–749 (2007)

H. Kaji et al., Drug delivery devices for retinal diseases. Adv. Drug Deliv. Rev.. Elsevier B.V. **128**, 148–157 (2017). https://doi.org/10.1016/j.addr.2017.07.002

N. Kamaly et al., Degradable controlled-release polymers and polymeric nanoparticles: Mechanisms of controlling drug release. Chem. Rev. **116**(4), 2602–2663 (2016). https://doi.org/10.1021/acs.chemrev.5b00346.

M. Kastellorizios, F. Papadimitrakopoulos, D.J. Burgess, Multiple tissue response modifiers to promote angiogenesis and prevent the foreign body reaction around subcutaneous implants. J. Control. Release Elsevier B.V. **214**, 103–111 (2015). https://doi.org/10.1016/j.jconrel.2015.07.021

N. Kasyapi, K. Dinesh Kumar, A.K. Bhowmick, Influence of microstructure of lactone-based triblock copolymers on drug release behavior of their microspheres. J. Appl. Polym. Sci. **134**(37), 1–14 (2017). https://doi.org/10.1002/app.45284

H. Kim, S.M. Jeong, J.W. Park, Electrical switching between vesicles and micelles via redox-responsive self-assembly of amphiphilic rod-coils. J. Am. Chem. Soc. **133**(14), 5206–5209 (2011). https://doi.org/10.1021/ja200297j

H. Kim, H. Park, S.J. Lee, Effective method for drug injection into subcutaneous tissue. Sci. Rep. Springer US **7**(1), 1–11 (2017). https://doi.org/10.1038/s41598-017-10110-w

S.W. Kim, Y.H. Bae, T. Okano, Hydrogels: Swelling, Drug Loading, and Release. Pharm. Res. **9**(3), 283–290 (1992). https://doi.org/10.1023/A:1015887213431.

L.W. Kleiner, J.C. Wright, Y. Wang, Evolution of implantable and insertable drug delivery systems. J. Control. Release Elsevier B.V. **181**(1), 1–10 (2014). https://doi.org/10.1016/j.jconrel.2014.02.006

D. Klinger, K. Landfester, Dual stimuli-responsive poly(2-hydroxyethyl methacrylate-co-methacrylic acid) microgels based on photo-cleavable cross-linkers: PH-dependent swelling and light-induced degradation. Macromolecules **44**(24), 9758–9772 (2011). https://doi.org/10.1021/ma201706r

B. Koch et al., Microfabrication for drug delivery. Materials **9**(8) (2016). https://doi.org/10.3390/ma9080646

M. Köllmer et al., Long-term function of alginate-encapsulated islets. Tissue Engineering Part B*: Reviews* **22**(1), ten.teb.2015.0140 (2015). https://doi.org/10.1089/ten.teb.2015.0140

J. Kost, J. Wolfrum, R. Langer, Magnetically enhanced insulin release in diabetic rats. J. Biomed. Mater. Res. **21**(12), 1367–1373 (1987). https://doi.org/10.1002/jbm.820211202

D. Lalka, R.K. Griffith, C.L. Cronenberger, The hepatic first-pass metabolism of problematic drugs. J. Clin. Pharmacol. **33**(7), 657–669 (1993). https://doi.org/10.1002/j.1552-4604.1993.tb04720.x

C. Laloux et al., Continuous cerebroventricular administration of dopamine: A new treatment for severe dyskinesia in Parkinson's disease? Neurobiol. Dis. The Authors **103**, 24–31 (2017). https://doi.org/10.1016/j.nbd.2017.03.013

R.S. Langer, N.A. Peppas, Present and future applications of biomaterials in controlled drug delivery systems. Biomaterials **2**(4), 201–214 (1981). https://doi.org/10.1016/0142-9612(81)90059-4

C. De Las Heras Alarcón, S. Pennadam, C. Alexander, 'Stimuli responsive polymers for biomedical applications', Chem. Soc. Rev. 276–285 (2005). doi: https://doi.org/10.1039/b406727d, 34.

J.C. Lauffenburger et al., Association between patient-centered medical homes and adherence to chronic disease medications: A cohort study. Ann. Intern. Med. **166**(2), 81–88 (2017). https://doi.org/10.7326/M15-2659.

D.G. Leach et al., STINGel: Controlled release of a cyclic dinucleotide for enhanced cancer immunotherapy. Biomaterials. Elsevier Ltd **163**, 67–75 (2018). https://doi.org/10.1016/j.biomaterials.2018.01.035

W. Leconet et al., 'Anti-PSMA/CD3 Bispecific antibody delivery and anti-tumor activity using a polymeric depot formulation', Mol. Cancer Ther. (2018). p. molcanther.1138.2017. doi: https://doi.org/10.1158/1535-7163.MCT-17-1138

B.K. Lee, Y. Yun, K. Park, PLA micro- and nano-particles. Adv. Drug Deliv. Rev. Elsevier B.V. **107**, 176–191 (2016). https://doi.org/10.1016/j.addr.2016.05.020

E.S. Lee et al., A virus-mimetic Nanogel vehicle. Angew. Chem. Int. Ed. **47**(13), 2418–2421 (2008). https://doi.org/10.1002/anie.200704121

H. Lee et al., A photo-degradable gene delivery system for enhanced nuclear gene transcription. Biomaterials **35**(3), 1040–1049 (2014). https://doi.org/10.1016/j.biomaterials.2013.10.030
P.I. Lee, Kinetics of drug release from hydrogel matrices. J. Control. Release **2**, 277–288 (1985). https://doi.org/10.1016/0168-3659(85)90051-3
S.-M. Lee et al., Multifunctional nanoparticles for targeted Chemophotothermal treatment of Cancer cells. Angew. Chem. Int. Ed. **50**(33), 7581–7586 (2011). https://doi.org/10.1002/anie.201101783
S.H. Lee, Y.B. Choy, Implantable devices for sustained, intravesical drug delivery. Int. Neurourol. J. **20**(2), 101–106 (2016). https://doi.org/10.5213/inj.1632664.332
M.S. Lesniak, H. Brem, Targeted therapy for brain Tumours. Nat. Rev. Drug Discov. **3**(June), 499–508 (2004). https://doi.org/10.1016/B978-0-12-397927-8.00005-1.
J. Li, D.J. Mooney, Designing hydrogels for controlled drug delivery. Nat. Rev. Mater. **1**(12) (2016). https://doi.org/10.1038/natrevmats.2016.71
M.H. Li, P. Keller, Stimuli-responsive polymer vesicles. Soft Matter **5**(5), 927–937 (2009). https://doi.org/10.1039/b815725a
Y. Li, F. Wang, H. Cui, Peptide-based supramolecular hydrogels for delivery of biologics. Bioeng. Translat. Med. **1**(3), 306–322 (2016). https://doi.org/10.1002/btm2.10041
D. Ling et al., Multifunctional tumor pH-sensitive self-assembled nanoparticles for bimodal imaging and treatment of resistant heterogeneous tumors. J. Am. Chem. Soc. **136**(15), 5647–5655 (2014). https://doi.org/10.1021/ja4108287
D. Liu et al., The smart drug delivery system and its clinical potential. Theranostics **6**(9), 1306–1323 (2016). https://doi.org/10.7150/thno.14858
F. Liu, M.W. Urban, Recent advances and challenges in designing stimuli-responsive polymers. Prog. Polym. Sci. **35**(1–2), 3–23 (2010). https://doi.org/10.1016/j.progpolymsci.2009.10.002.
J. Liu et al., Triggerable tough hydrogels for gastric resident dosage forms. Nat. Commun.. Springer US **8**(1), 1–9 (2017). https://doi.org/10.1038/s41467-017-00144-z
R. Lo et al., A refillable microfabricated drug delivery device for treatment of ocular diseases. Lab Chip **8**(7), 1027 (2008). https://doi.org/10.1039/b804690e
C. Lu et al., Hydrogel Containing Silica Shell Cross-linked Micelles for Ocular Drug Delivery. Int J Drug Dev Res **102**, 627–637 (2012). https://doi.org/10.1002/jps
Y. Lu et al., Bioresponsive materials. Nat. Rev. Mater. **2**(1) (2016). https://doi.org/10.1038/natrevmats.2016.75
M.F. Maitz et al., Bio-responsive polymer hydrogels homeostatically regulate blood coagulation. Nat. Commun. **4**, 2168 (2013). https://doi.org/10.1038/ncomms3168
A. Mandal et al., Polymeric micelles for ocular drug delivery: From structural frameworks to recent preclinical studies. J. Control. Release Elsevier B.V. **248**, 96–116 (2017). https://doi.org/10.1016/j.jconrel.2017.01.012
T. Manouras, M. Vamvakaki, 'Field responsive materials: Photo-, electro-, magnetic- and ultrasound-sensitive polymers', Polym. Chem. Royal Society of Chemistry, 8(1), pp. 74–96 (2017). doi: https://doi.org/10.1039/c6py01455k.
C.E. Markwalter, R.K. Prud'homme, Design of a Small-Scale Multi-Inlet Vortex Mixer for scalable nanoparticle production and application to the encapsulation of biologics by inverse flash NanoPrecipitation. J. Pharm. Sci.. Elsevier Ltd **107**, 1–7 (2018). https://doi.org/10.1016/j.xphs.2018.05.003
D.F. Martin et al., Treatment of cytomegalovirus retinitis with an intraocular sustained-release Ganciclovir implant. Arch. Ophthalmol. American Medical Association **112**(12), 1531 (1994). https://doi.org/10.1001/archopht.1994.01090240037023
D.F. Martin et al., Oral Ganciclovir for patients with cytomegalovirus retinitis treated with a Ganciclovir implant. N. Engl. J. Med.. Massachusetts Medical Society **340**(14), 1063–1070 (1999). https://doi.org/10.1056/NEJM199904083401402.
K.J. McHugh et al., Fabrication of fillable microparticles and other complex 3D microstructures. Science **1142**, 1138–1142 (2017). https://doi.org/10.1126/science.aaf7447.
K.H. Min et al., Tumoral acidic pH-responsive MPEG-poly(β-amino ester) polymeric micelles for cancer targeting therapy. J. Control. Release **144**(2), 259–266 (2010). https://doi.org/10.1016/j.jconrel.2010.02.024
S. Mitragotri, D. Blankschtein, R. Langer, Ultrasound-mediated transdermal protein delivery. Science **269**(5225), 850–853 (1995). https://doi.org/10.1126/science.7638603
R.I. Moustafine, I.M. Zaharov, V.A. Kemenova, Physicochemical characterization and drug release properties of Eudragit® E PO/Eudragit® L 100-55 interpolyelectrolyte complexes. Eur. J. Pharm. Biopharm. **63**(1), 26–36 (2006). https://doi.org/10.1016/j.ejpb.2005.10.005
M.P. Mullarney, T.A.P. Seery, R.A. Weiss, Drug diffusion in hydrophobically modified N,N-dimethylacrylamide hydrogels. Polymer **47**(11), 3845–3855 (2006). https://doi.org/10.1016/j.polymer.2006.03.096
S. Mura, J. Nicolas, P. Couvreur, Stimuli-responsive Nanocarriers for drug delivery. Nat. Mater.. Nature Publishing Group **12**, 991–1003 (2013). https://doi.org/10.1246/cl.151176.
R.S. Navath et al., Injectable PAMAM Dendrimer–PEG hydrogels for the treatment of genital infections: Formulation and in vitro and in vivo evaluation. Mol. Pharm. **8**(4), 1209–1223 (2011). https://doi.org/10.1021/mp200027z
B. Nilsson et al., The role of complement in biomaterial-induced inflammation. Mol. Immunol. **44**(1–3), 82–94 (2007). https://doi.org/10.1016/j.molimm.2006.06.020.
T. Okano et al., Thermally on-off switching polymers for drug permeation and release. J. Control. Release **11**(1–3), 255–265 (1990). https://doi.org/10.1016/0168-3659(90)90138-J.
L. Osterberg, T. Blaschke, Adherence to medication. N. Engl. J. Med. **353**(5), 487–497 (2005). https://doi.org/10.1056/NEJMra050100
R.F. Pagels, R.K. Prud'Homme, Polymeric nanoparticles and microparticles for the delivery of peptides, biologics, and soluble therapeutics. J. Control. Release Elsevier B.V. **219**, 519–535 (2015). https://doi.org/10.1016/j.jconrel.2015.09.001
M. Paolini et al., Nano-sized cytochrome P450 3A4 inhibitors to block hepatic metabolism of docetaxel. Int. J. Nanomedicine **12**, 5537–5556 (2017). https://doi.org/10.2147/IJN.S141145
K.M. Park et al., In situ cross-linkable gelatin-poly(ethylene glycol)-tyramine hydrogel via enzyme-mediated reaction for tissue regenerative medicine. J. Mater. Chem. **21**(35), 13180–13187 (2011). https://doi.org/10.1039/c1jm12527c
M.R. Park et al., Sustained delivery of human growth hormone using a polyelectrolyte complex-loaded thermosensitive polyphosphazene hydrogel. J. Control. Release Elsevier B.V. **147**(3), 359–367 (2010). https://doi.org/10.1016/j.jconrel.2010.07.126
T.G. Park, W. Lu, G. Crotts, Importance of in vitro experimental conditions on protein release kinetics, stability and polymer degradation in protein encapsulated poly (d,l-lactic acid-co-glycolic acid) microspheres. J. Control. Release **33**(2), 211–222 (1995). https://doi.org/10.1016/0168-3659(94)00084-8
C. Pedroso et al., Implant site Nexplanon reaction? *BMJ* Case Reports **2015**, 2014–2015 (2015). https://doi.org/10.1136/bcr-2014-206256
N.A. Peppas et al., Modelling of drug diffusion through swellable polymeric systems. J. Membr. Sci. **7**(3), 241–253 (1980). https://doi.org/10.1016/S0376-7388(00)80471-8
N.A. Peppas et al., Poly(ethylene glycol)-containing hydrogels in drug delivery. J. Control. Release **62**(1–2), 81–87 (1999). https://doi.org/10.1016/S0168-3659(99)00027-9.

N.A. Peppas, A.R. Khare, Preparation, structure and diffusional behavior of hydrogels in controlled release. Adv. Drug Deliv. Rev. **11**(1–2), 1–35 (1993). https://doi.org/10.1016/0169-409X(93)90025-Y

N.A. Peppas, J.J. Sahlin, A simple equation for the description of solute release. III. Coupling of diffusion and relaxation. Int. J. Pharm. **57**(2), 169–172 (1989). https://doi.org/10.1016/0378-5173(89)90306-2

N.A. Peppas, S.L. Wright, Drug diffusion and binding in ionizable interpenetrating networks from poly(vinyl alcohol) and poly(acrylic acid). Eur. J. Pharm. Biopharm. **46**(1), 15–29 (1998). https://doi.org/10.1016/S0939-6411(97)00113-6

P. Periti, T. Mazzei, E. Mini, Clinical pharmacokinetics of depot leuprorelin. Clin. Pharmacokinet. **41**(7), 485–504 (2002). https://doi.org/10.2165/00003088-200241070-00003.

N.M. Pinkerton et al., Gelation chemistries for the encapsulation of nanoparticles in composite gel microparticles for lung imaging and drug delivery. Biomacromolecules **15**(1), 252–261 (2014). https://doi.org/10.1021/bm4015232.

C.G. Pitt et al., Triggered drug delivery systems. J. Control. Release. Elsevier Science Publishers B.V **2**, 363–374 (1985)

K. Podual, F.J. Doyle, N.A. Peppas, Glucose-sensitivity of glucose oxidase-containing cationic copolymer hydrogels having poly(ethylene glycol) grafts. J. Control. Release. **67**(1), 19-17 (2000).

S.M. Pond, T.N. Tozer, First-pass elimination. Basic concepts and clinical consequences. Clin. Pharmacokinet. **9**(1), 1–25 (1984). https://doi.org/10.2165/00003088-198409010-00001

M.R. Prausnitz, Engineering microneedle patches for vaccination and drug delivery to skin. Annual Rev Chem Biomolecular Eng **8**(1), 177–200 (2017). https://doi.org/10.1146/annurev-chembioeng-060816-101514

B.P. Purcell et al., Injectable and bioresponsive hydrogels for on-demand matrix metalloproteinase inhibition. Nat. Mater. **13**(6), 653–661 (2014). https://doi.org/10.1038/nmat3922

P.L. Ritger, N.A. Peppas, A simple equation for description of solute release I. Fickian and non-fickian release from non-swellable devices in the form of slabs, spheres, cylinders or discs. J. Control. Release **5**(1), 23–36 (1987). https://doi.org/10.1016/0168-3659(87)90034-4

N.G. Rouphael et al., The safety, immunogenicity, and acceptability of inactivated influenza vaccine delivered by microneedle patch (TIV-MNP 2015): A randomised, partly blinded, placebo-controlled, phase 1 trial. Lancet **390**(10095), 649–658 (2017). https://doi.org/10.1016/S0140-6736(17)30575-5

S. Rowlands, D. Mansour, M. Walling, Intravascular migration of contraceptive implants: Two more cases. Contraception. Elsevier **95**(2), 211–214 (2017). https://doi.org/10.1016/j.contraception.2016.07.015

B.A. Sabel et al., Extended levodopa release from a subcutaneously implanted polymer matrix in rats. Ann. Neurol. **28**(5), 714–717 (1990). https://doi.org/10.1002/ana.410280519

C. Sanson et al., Doxorubicin loaded magnetic Polymersomes: Theranostic Nanocarriers for MR imaging and magneto-chemotherapy. ACS Nano **5**(2), 1122–1140 (2011). https://doi.org/10.1021/nn102762f

O. Sartor, Eligard®6: A new form of treatment for prostate Cancer. Eur. Urol. Suppl. **5**(18), 905–910 (2006). https://doi.org/10.1016/j.eursup.2006.08.006

V.R. Sastri, Plastics in medical devices : properties, requirements, and applications. Massachusetts: Elsevier Inc. (2010).

N.S. Satarkar, J. Zach Hilt, Hydrogel nanocomposites as remote-controlled biomaterials. Acta Biomater. **4**(1), 11–16 (2008). https://doi.org/10.1016/j.actbio.2007.07.009

P. Schattling, F.D. Jochum, P. Theato, 'Multi-stimuli responsive polymers-the all-in-one talents', Polym. Chem., pp. 25–36 (2014). doi: https://doi.org/10.1039/c3py00880k

D. Schmaljohann, Thermo- and pH-responsive polymers in drug delivery. Adv. Drug Deliv. Rev. **58**(15), 1655–1670 (2006). https://doi.org/10.1016/j.addr.2006.09.020

S. Segal, A new delivery system for contraceptive steroids. Am. J. Obstet. Gynecol. **157**(4 Pt 2), 1090–1092 (1987)

V. Segura-Ibarra et al., Nanoparticles administered intrapericardially enhance payload myocardial distribution and retention. J. Control. Release. Elsevier B.V. (2017). https://doi.org/10.1016/j.jconrel.2017.07.012

B.B. Seo et al., The biological efficiency and bioavailability of human growth hormone delivered using injectable, ionic, thermosensitive poly(organophosphazene)-polyethylenimine conjugate hydrogels. Biomaterials. Elsevier Ltd **32**(32), 8271–8280 (2011). https://doi.org/10.1016/j.biomaterials.2011.07.033

L. Serra, J. Doménech, N.A. Peppas, Drug transport mechanisms and release kinetics from molecularly designed poly(acrylic acid-g-ethylene glycol) hydrogels. Biomaterials **27**(31), 5440–5451 (2006). https://doi.org/10.1016/j.biomaterials.2006.06.011

H. Shibata et al., Injectable hydrogel microbeads for fluorescence-based in vivo continuous glucose monitoring. Proc. Natl. Acad. Sci. **107**(42), 17894–17898 (2010). https://doi.org/10.1073/pnas.1006911107

T. Shirakura et al., 'Hydrogel nanoparticles with thermally controlled drug release', ACS Macro Lett. UTC, 3(7), pp. 602–606 (2014). doi: https://doi.org/10.1021/mz500231e.

R.A. Siegel and M.J. Rathbone (2012) 'Fundamentals and Applications of Controlled Release Drug Delivery', pp. 19–44. doi: https://doi.org/10.1007/978-1-4614-0881-9.

J. Siepmann, N.A. Peppas, Modeling of drug release from delivery systems based on hydroxypropyl methylcellulose (HPMC). Adv. Drug Deliv. Rev. **48**(2–3), 139–157 (2001). https://doi.org/10.1016/S0169-409X(01)00112-0.

P.A. Sieving et al., Ciliary neurotrophic factor (CNTF) for human retinal degeneration: Phase I trial of CNTF delivered by encapsulated cell intraocular implants. Proc. Natl. Acad. Sci. **103**(10), 3896–3901 (2006). https://doi.org/10.1073/pnas.0600236103

E.A. Silva et al., Material-based deployment enhances efficacy of endothelial progenitor cells. Proc. Natl. Acad. Sci. **105**(38), 14347–14352 (2008). https://doi.org/10.1073/pnas.0803873105

E.A. Silva, D.J. Mooney, Spatiotemporal control of vascular endothelial growth factor delivery from injectable hydrogels enhances angiogenesis. J. Thromb. Haemost. **5**, 590–598 (2007)

J. Smith et al., New and emerging technologies for the treatment of inherited retinal diseases: A horizon scanning review. Eye. Nature Publishing Group **29**(9), 1131–1140 (2015). https://doi.org/10.1038/eye.2015.115.

R.H. Staff et al., Patchy nanocapsules of poly(vinylferrocene)-based block copolymers for redox-responsive release. ACS Nano **6**(10), 9042–9049 (2012). https://doi.org/10.1021/nn3031589

R.J. Stenekes et al., Controlled release of liposomes from biodegradable dextran microspheres: A novel delivery concept. Pharm. Res. **17**(6), 690–695 (2000). https://doi.org/10.1023/a:1007526114744.

M.A.C. Stuart et al., Emerging applications of stimuli-responsive polymer materials. Nat. Mater., 101–113 (2010). https://doi.org/10.1038/nmat2614.

Z. Sui, W.J. King, W.L. Murphy, Protein-based hydrogels with tunable dynamic responses. Adv. Funct. Mater. **18**(12), 1824–1831 (2008). https://doi.org/10.1002/adfm.200701288

M.J. Sullivan, Allergy to Nexplanon®. J. Family Plan. Reprod. Health Care **38**(4), 272.1–272.27272 (2012). https://doi.org/10.1136/jfprhc-2012-100366

E.M. Sussman et al., Porous implants modulate healing and induce shifts in local macrophage polarization in the foreign body reaction. Ann. Biomed. Eng. **42**(7), 1508–1516 (2014). https://doi.org/10.1007/s10439-013-0933-0

J. Szebeni et al., Activation of complement by therapeutic liposomes and other lipid excipient-based therapeutic products: Prediction and prevention. Adv. Drug Deliv. Rev. Elsevier B.V. **63**(12), 1020–1030 (2011). https://doi.org/10.1016/j.addr.2011.06.017

T. Takato et al., Preclinical and clinical research on bone and cartilage regenerative medicine in Oral and maxillofacial region. Oral Sci. Int.. Japanese Stomatological Society **11**(2), 45–51 (2014). https://doi.org/10.1016/S1348-8643(14)00008-1.

M.W. Tibbitt et al., Synthesis and application of photodegradable microspheres for spatiotemporal control of protein delivery. J. Biomed. Mater. Res. - Part A **100 A**(7), 1647–1654 (2012). https://doi.org/10.1002/jbm.a.34107

M.W. Tibbitt, J.E. Dahlman, R. Langer, Emerging Frontiers in Drug Delivery. J. Am. Chem. Soc. **138**(3), 704–717 (2016). https://doi.org/10.1021/jacs.5b09974

G. Tiwari et al., Drug delivery systems: An updated review. Int. J. Pharmaceut. Investig. **2**(1), 2 (2012). https://doi.org/10.4103/2230-973X.96920

D.J.H. Tng et al., Approaches and challenges of engineering implantable microelectromechanical systems (MEMS) drug delivery systems for in vitro and in vivo applications. Micromachines **3**(4), 615–631 (2012). https://doi.org/10.3390/mi3040615

X. Tong et al., Long-term controlled protein release from poly(ethylene glycol) hydrogels by modulating mesh size and degradation. Macromol. Biosci. **15**(12), 1679–1686 (2015). https://doi.org/10.1002/mabi.201500245

K.M. Tsoi et al., Mechanism of hard-nanomaterial clearance by the liver. Nat. Mater. **1**(August), 1–10 (2016). https://doi.org/10.1038/nmat4718.

S.Y. Tzeng et al., Thermostabilization of inactivated polio vaccine in PLGA-based microspheres for pulsatile release. J. Control. Release The Authors **233**, 101–113 (2016). https://doi.org/10.1016/j.jconrel.2016.05.012

S.Y. Tzeng et al., 'Stabilized single-injection inactivated polio vaccine elicits a strong neutralizing immune response', Proc. Natl. Acad. Sci., (21), p. 201720970 (2018). doi: https://doi.org/10.1073/pnas.1720970115, 115, E5269.

K.E. Uhrich et al., Polymeric systems for controlled drug release. Chem. Rev. **99**(11), 3181–3198 (1999). https://doi.org/10.1021/cr940351u

A.J. Vegas et al., Long-term glycemic control using polymer-encapsulated human stem cell-derived beta cells in immune-competent mice. Nat. Med. **22**(3), 306–311 (2016). https://doi.org/10.1038/nm.4030.

O. Veiseh et al., Size- and shape-dependent foreign body immune response to materials implanted in rodents and non-human primates. Nat. Mater. **14**(6), 643–651 (2015). https://doi.org/10.1038/nmat4290

A. Venuta et al., Post-nano strategies for drug delivery: Multistage porous silicon microvectors. J. Mater. Chem. B Royal Society of Chemistry **5**(2), 207–219 (2017). https://doi.org/10.1039/C6TB01978A

C. Vogeli et al., Multiple chronic conditions: Prevalence, health consequences, and implications for quality, care management, and costs. J. Gen. Intern. Med. **22**(SUPPL. 3), 391–395 (2007). https://doi.org/10.1007/s11606-007-0322-1.

B. Wang et al., Glucose-responsive micelles from self-assembly of poly(-ethylene glycol)-a-poly(acrylic acid-co-acrylamidophenylboronic acid) and the controlled release of insulin. Langmuir **25**(21), 12522–12528 (2009). https://doi.org/10.1021/la901776a

J. Wang et al., Drug delivery implants in the treatment of vitreous inflammation. Mediat. Inflamm. Hindawi **2013**, 780634 (2013). https://doi.org/10.1155/2013/780634

F.M. Weaver et al., Bilateral deep brain stimulation vs best medical therapy for patients. J. Am. Med. Assoc. **301**(1), 63–73 (2014). https://doi.org/10.1001/jama.2008.929.

K.J. Widder et al., Tumor remission in Yoshida sarcoma-bearing rats by selective targeting of magnetic albumin microspheres containing doxorubicin. Proc. Natl. Acad. Sci. USA. 78(1), 579–581 (1981)

S. Wilhelm et al., Analysis of nanoparticle delivery to tumours. Nat Rev Mater **1**, 1–12 (2016). https://doi.org/10.1038/natrevmats.2016.14

C. Wong et al., Multistage nanoparticle delivery system for deep penetration into tumor tissue. Proc. Natl. Acad. Sci. **108**(6), 2426–2431 (2011). https://doi.org/10.1073/pnas.1018382108

R. Xu et al., An injectable nanoparticle generator enhances delivery of cancer therapeutics. Nat. Biotechnol. **34**(4), 414–418 (2016). https://doi.org/10.1038/nbt.3506

Q. Yan et al., Voltage-responsive vesicles based on orthogonal assembly of two homopolymers. J. Am. Chem. Soc. UTC **132**(27), 9268–9270 (2010). https://doi.org/10.1021/ja1027502

X. Yan et al., Stimuli-responsive supramolecular polymeric materials. Chem. Soc. Rev., 6042–6065 (2012). https://doi.org/10.1039/c2cs35091b

V. Yesilyurt et al., Injectable self-healing glucose-responsive hydrogels with pH-regulated mechanical properties. Adv. Mater. **28**(1), 86–91 (2016). https://doi.org/10.1002/adma.201502902

L. Yu, J. Ding, Injectable hydrogels as unique biomedical materials. Chem. Soc. Rev. **37**(8), 1473–1481 (2008). https://doi.org/10.1039/b713009k

K.C.J. Yuen, B.S. Miller, B.M.K Biller, 'The current state of long-acting growth hormone preparations for growth hormone therapy.', Curr. Opin. Endocrinol. Diabetes Obes. p. 1 (2018). doi: https://doi.org/10.1097/MED.0000000000000416.

J. Zhang et al., The targeted behavior of thermally responsive nanohydrogel evaluated by NIR system in mouse model. J. Control. Release **131**(1), 34–40 (2008). https://doi.org/10.1016/j.jconrel.2008.07.019

S. Zhang et al., An inflammation-targeting hydrogel for local drug delivery in inflammatory bowel disease. Sci. Transl. Med. **7**(300), 300ra128 (2015a). https://doi.org/10.1126/scitranslmed.aaa5657

Y. Zhang et al., Active targeting of tumors through conformational epitope imprinting. Angewandte Chemie - International Edition. Wiley-Blackwell **54**(17), 5157–5160 (2015b). https://doi.org/10.1002/anie.201412114

Y.N. Zhang et al., Nanoparticle-liver interactions: Cellular uptake and hepatobiliary elimination. J. Control. Release. Elsevier B.V. **240**, 332–348 (2016). https://doi.org/10.1016/j.jconrel.2016.01.020

L. Zhu, P. Kate, V.P. Torchilin, Matrix metalloprotease 2-responsive multifunctional liposomal nanocarrier for enhanced tumor targeting. ACS Nano **6**(4), 3491–3498 (2012). https://doi.org/10.1021/nn300524f

M. Zilberman et al., in *Handbook of experimental pharmacology*, ed. by M. Schafer-Korting. Drug-eluting medical implants (Springer, Verlag, 2010), pp. 299–341. https://doi.org/10.1007/978-3-642-00477-3

Biomedical Microdevices (2019) 21: 46
https://doi.org/10.1007/s10544-019-0396-7

Organotropic drug delivery: Synthetic nanoparticles and extracellular vesicles

Sara Busatto[1,2,3] · **Anthony Pham**[1,2] · **Annie Suh**[1,2] · **Shane Shapiro**[4,5] · **Joy Wolfram**[1,2,6]

Published online: 15 April 2019

Author's Personal Note What do cowboy boots, Dante, marathons, and the saxophone have in common? Well, they are some of the features associated with one of the most visionary scientific leaders of our time. When the rest of the world is in slow motion, Dr. Mauro Ferrari is racing ahead, immersed in a reality distortion field that is soon birthed into existence. Between mentoring trainees, closing million-dollar deals, and flying terminally ill children to the United States for treatment, you can hear him muttering, "If I don't take on this burden, who will?" His message strikes a chord. Leaders are servers; we serve our patients, community, and colleagues. Working with Mauro comes with a curse. You are left with an underlying sense that no place is quite visionary, fast paced, or forward-thinking enough. You hear distant echoes. Where are the crazy ideas? How can we think bigger? What does physics have to do with cancer?

When you finally get a chance to catch up, Mauro announces that one day you will be President and he will move to Finland. You find yourself objecting that you want to stay in science, but are nonchalantly told you can do both.

Joy Wolfram

Abstract

Most clinically approved drugs (primarily small molecules or antibodies) are rapidly cleared from circulation and distribute throughout the body. As a consequence, only a small portion of the dose accumulates at the target site, leading to low efficacy and adverse side effects. Therefore, new delivery strategies are necessary to increase organ and tissue-specific delivery of therapeutic agents. Nanoparticles provide a promising approach for prolonging the circulation time and improving the biodistribution of drugs. However, nanoparticles display several limitations, such as clearance by the immune systems and impaired diffusion in the tissue microenvironment. To overcome common nanoparticle limitations various functionalization and targeting strategies have been proposed. This review will discuss synthetic nanoparticle and extracellular vesicle delivery strategies that exploit organ-specific features to enhance drug accumulation at the target site.

Keywords Extracellular vesicles · Exosomes · Nanoparticles · Nanomedicine · Organotropic

1 Introduction

A major goal of medicine is to develop treatment strategies that target specific organs, tissues, and cells (Kwon et al. 2012; Tiwari et al. 2012). Therapeutic agents that are effective *in vitro* often fail *in vivo* due to biological barriers that prevent site-specific accumulation (Barua and Mitragotri 2014; Blanco et al. 2015). In fact, conventional medications are usually subject to short circulation half-lives and rapid renal clearance (Podust et al. 2016). As a consequence, only around 0.001–0.01% of the administered therapeutic dose accumulates at the target site (Li et al. 2004; Wolfram et al. 2015a). Dose limitations due to side effects further restrict drug accumulation in diseased tissues (Tamargo et al. 2015). Thus, new strategies are essential to increase site-specific delivery of therapeutic agents. Nanomedicine, which consists on the application of synthetic or biological nanoparticles (NPs) for medical purposes, is a promising approach for overcoming barriers in the body (Salata 2004; Wagner et al. 2006; Wolfram and Ferrari 2019; Wolfram et al. 2015b). In particular, nanoparticles can act as carriers for the delivery of a wide variety of drugs to diseased tissues.

Nanosized particles can be synthesized from a plethora of materials, such as lipids (Shen et al. 2017a, b; Wolfram et al. 2014a, 2016), polymers (Molinaro et al. 2013; Yang et al. 2014), metals (Mu et al. 2018; Shen et al. 2014a), silica (Shen et al. 2014b, 2017b), and silicon (Shen et al. 2013), or isolated from biological sources. NPs exhibit unique properties that are lacking on the macro and molecular scale (Baer 2018; Navya and Daima 2016). For instance, NPs have high surface area to volume ratios (Baer 2018) that can be exploited for e.g. protein (Saha et al. 2014) and polymer (Makadia and Siegel 2011; Thamake et al. 2012) functionalization to prolong circulation and bestow specific targeting abilities (Dhar et al. 2008; Jokerst et al. 2011; Paolino et al. 2014; Scavo et al. 2015). In addition to the material properties, the shape

✉ Sara Busatto
busatto.sara@mayo.edu

✉ Joy Wolfram
wolfram.joy@mayo.edu

[1] Department of Transplantation Medicine, Mayo Clinic, Jacksonville, Florida 32224, USA

[2] Department of Physiology and Biomedical Engineering, Mayo Clinic, Jacksonville, Florida 32224, USA

[3] Department of Molecular and Translational Medicine, University of Brescia, Brescia, Italy

[4] Center for Regenerative Medicine, Mayo Clinic, Jacksonville, Florida 32224, USA

[5] Department of Orthopedic Surgery, Mayo Clinic, Jacksonville, Florida 32224, USA

[6] Department of Nanomedicine, Houston Methodist Research Institute, Houston, TX 77030, USA

(Caldorera-Moore et al. 2010; Kinnear et al. 2017; Longmire et al. 2011) and size (Betzer et al. 2017; Jiang et al. 2008; Prisner et al. 2017) of nanoparticles can confer specific transport abilities in the body. In particular, NP properties can be tailored to enhance interactions with tissue microenvironments that display distinct characteristics, including structural components, interstitial pressure (Torosean et al. 2013), pH (Du et al. 2015), and biomolecules (Friedman et al. 2013). Consequently, site-specific delivery of NPs is usually superior to that of small molecules (100 to 1000-fold improvement) (Wolfram et al. 2017).

Traditionally, NP drug delivery approaches have been divided into active and passive targeting strategies (Bertrand et al. 2014; Li et al. 2017). Active targeting involves utilization of NP surface molecules that preferentially bind to the tissue of interest (Bazak et al. 2015), while passive targeting involves exploitation of physical properties, such as NP shape and size (Bertrand et al. 2014; Gentile et al. 2013). An example of active targeting is the coupling of transferrin to NPs to target pathological cells that express high levels of transferrin receptors (Gan and Feng 2010). The most common example of passive targeting is exploitation of immature leaky vasculature and impaired lymphatic drainage for intratumoral accumulation of NPs, known as the enhanced permeability and retention (EPR) effect (Greish 2010). The EPR effect can be enhanced through NP pegylation, which attracts a protective water shell that hides NPs from macrophages and enhances circulation half-life (Jokerst et al. 2011; Pasut et al. 2015; Wolfram et al. 2014b). NP shapes can also be optimized for improved biodistribution and site-specific delivery (Kinnear et al. 2017). For example, discoidal particles pioneered by Dr. Mauro Ferrari mimic the dimensions of platelets to exploit fluid dynamics for adherence to inflamed vasculature (Decuzzi et al. 2010; Mi et al. 2016a, b; Shen et al. 2015; Venuta et al. 2017).

Despite major nanoparticle-mediated improvements in the biodistribution of drugs, analysis of more than 100 nanodelivery studies demonstrated that changing the material, shape, size, and charge of NPs results in minimal improvements in intratumoral accumulation (Wilhelm et al. 2016). Notably, tumor type, stage, and model had a greater impact on site-specific delivery compared to nanoparticle properties, indicating that the macro and microenvironment play a crucial role in promoting NP adhesion, extravasation, and tissue penetration (Wilhelm et al. 2016). Additionally, the success of nanomedicine is mainly hampered by the immune system (Boraschi et al. 2017) and anatomical obstacles, such as the endothelium (Blanco et al. 2015). One of the first barriers encountered upon intravenous injection is the mononuclear phagocyte system (MPS), which is composed of specialized phagocytic cells in the lymph nodes, spleen, and liver. The phagocytes are responsible for rapid clearance of nanosized endogenous and exogenous circulating material (Samuelsson et al. 2017). Additionally, the adaptive immune system produces antibodies against immunogenic molecules, leading to a potent immune response against subsequent exposure, which can also compromise NP therapy. For example, the formation of antibodies against molecules used for NP functionalization, such as polyethylene glycol (PEG), has been reported in animal models (Kierstead et al. 2015). Therefore, nanoparticle design strategies can be combined with other approaches that aid in overcoming biological barriers. For instance, optimization of the administration route can substantially improve biodistribution, as exemplary demonstrated by intranasal administration of nanoparticles to overcome the blood brain barrier (BBB) (Patel et al. 2018). Additionally, temporary inactivation of resident macrophages in the liver through pretreatment with pharmacological agents was shown to reduce immunological clearance of NPs, leading to improved site-specific delivery (Khalid et al. 2016; Pelt et al. 2018; Wolfram et al. 2017).

Another approach to overcome barriers in the body is exploitation of biological components that have been optimized through evolution for effective navigation within and between compartments in the body. One example is Abraxane, a United States Food and Drug Administration (FDA) approved NP, that contains albumin and paclitaxel (Peer et al. 2007). Albumin is the most abundant protein in the plasma and a carrier for endogenous hydrophobic molecules (Fanali et al. 2012). Cell-secreted nanoparticles, known as extracellular vesicles (EVs), are another example of natural carriers for short and long-distance transportation of biomolecules (Busatto et al. 2018; Théry et al. 2018; Vader et al. 2016). It has been shown that EVs exhibit organ-specific targeting abilities that are at least partially due to integrins, but most likely attributed to the interplay of several EV components (Hoshino et al. 2015). In particular, Hoshino et al. isolated EVs from tumor cell lines that colonized to specific organs during the metastatic process. Notably, each EV population mirrored the organ tropism of the secreting cell line promoting the formation of the pre-metastatic niche (Hoshino et al. 2015). EVs were also able to educate specific tumor cells to colonize to organs that were not preferred sites of metastasis (Hoshino et al. 2015; Nogues et al. 2018). Consequently, exploitation of EVs as drug carriers presents a promising opportunity for obtaining site-specific delivery. The vast majority of synthetic and EV-based nanodelivery strategies have focused on exploiting common characteristics of the tumor microenvironment, and several reviews have discussed this topic in detail (Chen et al. 2016a, b; Chulpanova et al. 2018; Wilhelm et al. 2016; Yang and Gao 2017).

Furthermore, innovative approaches for synthetic NP targeting and novel advances to deliver therapy in utero are emerging. Even though NP modifications augmenting site specific drug delivery has advanced, differences in macro environments make some organs (liver and lungs) more suitable for NP uptake than others (brain, eyes, lymphatics, and bone). For example, resident macrophages in the liver promote organotropic homing of NPs, making hepatic tissue an ideal target for NP-based therapeutics (Borrelli et al. 2018), while the brain and eyes are limited by specialized endothelial barriers (Del Amo and Urtti 2008; Hersh et al. 2016). The purpose of this review is to provide an overall perspective of NP-based drug delivery strategies designed to exploit organ-specific properties (Fig. 1).

2 Liver delivery

A large portion of systemically injected nanoparticles accumulate in the liver due to resident macrophages (Gustafson et al. 2015). Despite this hepatic tropism, the intra-organ biodistribution is often unfavorable for the treatment of liver diseases. Studies have focused on specifically delivering NPs to hepatocytes for improved therapeutic effects. For example, cationic solid lipid nanoparticles (CSLN) were designed to have a similar lipid composition as natural low-density lipoprotein (LDL), enabling hepatocyte recognition and intake of nanoparticles via lipoprotein receptors (Kim et al. 2008; Kong et al. 2013). The CSLN can form a complex with negatively charged siRNA that can silence pro-fibrogenic growth factors genes. This complex protects siRNA from enzymatic degradation during circulation, while increasing specificity for hepatocytes. *In vitro*, the CSLN/siRNA complex displayed higher uptake and improved gene silencing in HepG.2.2.15 cells compared to transfection with the commercially available agent, Lipofectamine. CSLN/siRNA treatment of rats with liver fibrosis reduced connective tissue growth factor (CTGF) expression and reversed the hepatic cell structure and lobular architecture to resemble that of non-fibrotic rats (Kong et al. 2013).

A study by Liang et al. explored the use of lyophilized conditioned media from mesenchymal stem cells (MSCs) in PLGA NPs encased by a red blood cell membrane for treatment of liver failure (Fig. 2a) (Liang et al. 2018). MSCs and the MSC-derived secretome have previously shown to be

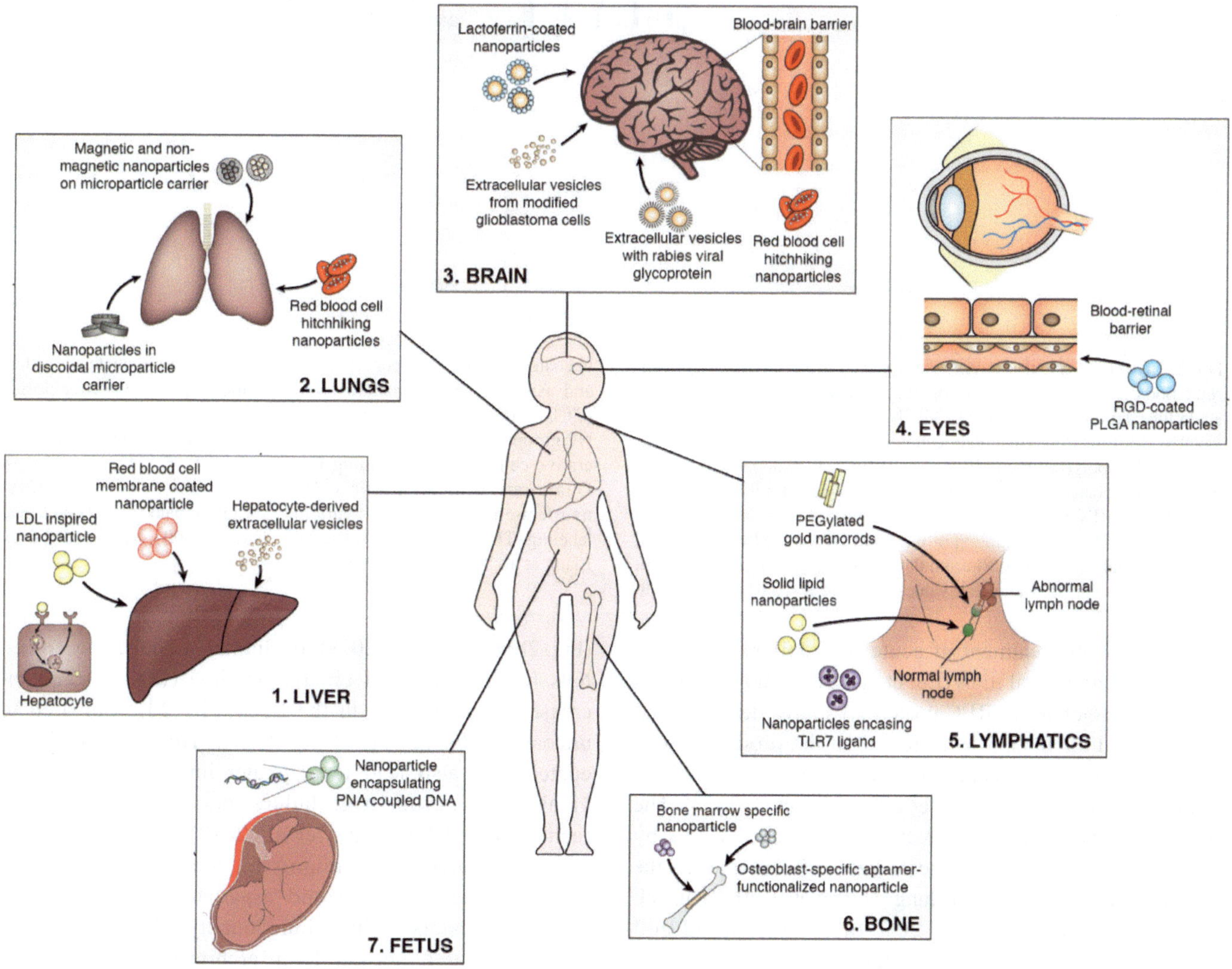

Fig. 1 Schematic of site-specific delivery strategies for nanotherapeutics to the brain, liver, bone, eye, lymphatic system, and fetus

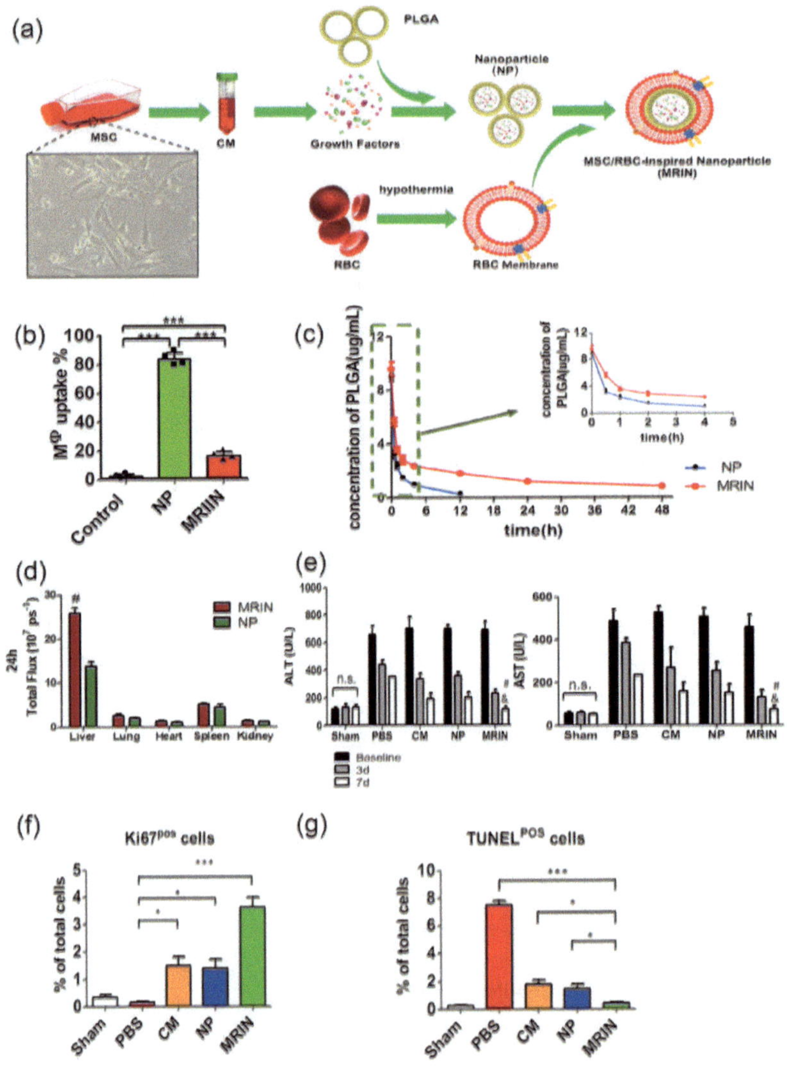

Fig. 2 Liver delivery of intravenously injected mesenchymal stem cell/red blood cell (RBC)-inspired nanoparticle (MRIN) in a mouse model of acute liver failure. **a** Schematic of MRIN fabrication: growth factors are isolated from mesenchymal stem cell-conditioned media (CM) and encapsulated in poly(lactic-co-glycolic acid) (PLGA) nanoparticles (NPs). NPs are then coated with cell membranes obtained from RBCs resulting in the final MRIN product. **b** Macrophage uptake of fluorescently labeled particles ($n = 4$). **c** Concentration of uncoated NPs and MRINs in blood ($n = 5$). **d** Analysis of particle accumulation in organs 24 h post treatment ($n = 3$). Data expressed as mean values ± s.d. #, $P < 0.01$ compared to uncoated NPs. **e** Serum concentrations of alanine aminotransferase (ALT) and aspartate aminotransferase (AST) following carbon tetrachloride (CCl_4)-induced liver injury ($n = 3$). #, $P < 0.05$ compared to uncoated NP group; &, $P < 0.05$ compared to CM group. **f** Quantitation of proliferative cells via Ki67 staining ($n = 3$). *. $P < 0.05$; ***, $P < 0.001$. **g** Quantitation of apoptotic cells via terminal deoxynucleotidyl transferase-mediated 2′-deoxyuridine, 5′-triphosphate (dUTP) nick end labeling (TUNEL) ($n = 3$). *, $P < 0.05$; ***; $P < 0.001$. Reproduced from Liang et al. 2018 with permission

effective in the treatment of acute liver failure in animal models and in clinical trials (Lin et al. 2017; Shi et al. 2017). Previous studies using RBC-coated NPs have demonstrated an increased circulation half-life due to the preservation of membrane proteins essential for long-term RBC blood circulation (Mohandas and Gallagher 2008; Piao et al. 2014). The RBC membrane prevents detection and uptake of NPs by macrophages, while the nanoscale size (200 nm) enables the NPs to bypass filtration in the lungs (Liang et al. 2018). In cell culture, it was shown that 20% of the red blood cell membrane-coated NPs were endocytosed by macrophages, while the corresponding value was 80% for non-coated ones (Fig. 2b) (Liang et al. 2018). In animal models, the coated NPs had a prolonged circulation time (Fig. 2c) and increased accumulation in the liver (Fig. 2d) compared to uncoated NPs (Liang et al. 2018). Treatment with the coated NPs also lead to a greater improvement in survival of mice with liver failure induced by carbon tetrachloride (CCl_4) (Liang et al. 2018). The coated NPs were also superior at improving the levels of alanine aminotransferase (ALT) and aspartate aminotransferase (AST), indicators of liver functionality (Fig. 2e), as well as promoting a normal liver architecture (Fig. 2f, g) (Liang et al. 2018).

EVs from hepatocytes have also been used to promote post-injury liver regeneration by delivery of endogenous biomolecules, such as sphingosine kinase 2 (SK2), which stimulates cell proliferation (Nojima et al. 2016). Treatment with these self-tropic EVs led to an increase in hepatocyte proliferation in mouse models of ischemia/reperfusion (I/R) injury and partial hepatectomy. In addition to utilizing targeting properties and endogenous therapeutic biomolecules in EVs, exogenous drugs have been loaded into EVs to further improve therapeutic efficacy (Borrelli et al. 2018). Taken together, various biomimicry approaches have shown promise for liver and hepatocyte-specific targeting.

3 Lung delivery

Lung cancer and other pulmonary disorders affect a large portion of the population and have high mortality rates. As is the case for most solid tumors, a small fraction of intravenously injected drugs reach lung cancer cells (Primeau et al. 2005). The pulmonary administration route is a promising alternative for delivery of NPs to lung tissue, as this method substantially increases site-specific exposure to therapeutic agents while reducing systemic side effects (De Backer et al. 2015; Rahhal et al. 2016; Tewes et al. 2016). In particular, NPs have shown promise as inhaled drug carriers for chemotherapeutic agents (Chen et al. 2016a, b; Nagesh et al. 2016; Nørgaard et al. 2010). A study by Abbas et al. sought to improve inhalation of NPs by incorporating them into larger microparticles (MPs), as previous research has demonstrated the failure of NPs to deposit deep into lung tissue due to size-dependent ease of exhalation. The use of aerodynamically favorable MPs as a carrier can aid in lung deposition (Ali and Lamprecht 2014; Rytting et al. 2008). Aerodynamics can be evaluated with a Next Generation Impactor (NGI), in which particles are dispersed in an air stream traveling through eight stages of NGI based on aerodynamic cutoff diameters. The results of this model suggest that approximately 10% of the MP dose would be retained in the upper respiratory track, while 90% would travel to the lower conducting and respiratory tracts, demonstrating the value of this model in aiding particle design for improved pulmonary delivery (Abbas et al. 2016). In addition to improving lung delivery, the authors sought to improve drug release through the use of magnetic and non-magnetic NPs, both containing epidermal growth factor receptor (EGFR) antibodies for targeting cancer cells. By administering both nanoparticles via MPs, a sustained basal drug release with intermittent increases in drug release via application of an alternating high frequency magnetic field (HFMF) could be established. This method of combined sustained and intermittent drug release has been shown improve therapeutic results in cancer treatment (Celikoglu et al. 2008; Howell 2001). Other studies have examined how to optimize NP design for lung targeting trough intravenous administration. A study by Decuzzi et al. examined how the size and shape of NPs and MPs affects their accumulation in specific organs. Mice were injected with particles with varying shapes and sizes. After two to six hours, the mice were euthanized, and organs (liver, spleen, heart, lung, kidneys, and brain) were harvested and examined. The authors noted that discoidal particles were found to be advantageous for lung treatment with the results showing discoidal particle accumulation in the lungs to be four times higher than that of spherical particles (Decuzzi et al. 2010). Mathematical modeling and experimental studies with discoidal particles have demonstrated improved adherence to the blood vessel wall when compared to other particle shapes (Decuzzi and Ferrari 2006; Gentile et al. 2008; Lee et al. 2009). Additionally, discoidal particles have been shown to avoid internalization by various cell types, possibly reducing uptake by liver macrophages (Champion and Mitragotri 2006; Gratton et al. 2008). A study by Mi et al. examined the use of discoidal MPs as carriers for therapeutic agents encapsulated into NPs for the treatment of melanoma lung metastases (Mi et al. 2016a). The discoidal MPs were selected in order to increase lung accumulation and enable simultaneous delivery of multiple types of NPs for combination therapy. Specifically, liposomes carrying small interfering RNA (siRNA) and docetaxel-encapsulated poly(lactide-co-glycolide) (PLGA) PEG NPs were incorporated in the same MP (Mi et al. 2016a). Mice studies revealed that liposome delivery resulted in siRNA retention solely in the kidneys, whereas MP-liposome delivery led to siRNA accumulation predominantly in the lungs after 24 h (Mi et al. 2016a). Similar results were observed when comparing PLGA-PEG NPs to NPs complexed with MPs; the NP only treatment resulted in significant accumulation in the liver, while the MP delivery system resulted in high lung retention. Analysis of cell-specific accumulation demonstrated that siRNA uptake in lung cancer cells was ten times higher with the MP platform compared to liposome delivery. Treatment with MPs resulted in longer survival times and significant reductions in metastatic nodules (Mi et al. 2016a). MPs were shown to be superior in treating mice when compared to other treatment groups, including combination therapy with siRNA-loaded liposomes and docetaxel-loaded PLGA-PEG NPs.

Another study explored the use of red blood cells (RBCs) as NP carriers for lung delivery. This RBC-hitchhiking (RH) strategy involved intravenous injection of NPs adhered to RBCs through mixing (RH) (Brenner et al. 2018). As the RBCs pass through the first capillary bed downstream of the injection, i.e. the pulmonary capillary bed, NPs transfer to the blood vessel wall (Brenner et al. 2018). Mice studies showed a significant improvement in lung accumulation when comparing RH NPs to NP alone. Several NPs were tested using the RH method, all of which showed a substantial increase in lung uptake when compared to free NPs. The fold improvement ranged from 4-fold for albumin-NPs and 40-fold for liposomes (Brenner et al. 2018). The efficacy of this treatment was demonstrated across multiple species: mice, rats, pigs, and humans. The experiment that determined the performance of RH treatment in humans was carried out ex vivo through injection into the pulmonary artery in human lungs that were unsuitable for transplantation. Analysis of the human lungs revealed that 41% of the injected RH NPs was deposited in the lung compared to 15–20% of the free NPs (Brenner et al. 2018).

Overall, these studies suggest that injection route and particle dimensions play a major role in lung delivery. In particular, inhaled MPs have been shown to increase drug bioavailability in the lungs, while discoidal MPs and RH improve intravenous treatment of lung pathologies.

4 Brain delivery

The BBB separates and protects the central nervous system (CNS) from the systemic blood circulation (Daneman and Prat 2015). This barrier is composed of endothelial cells with tight junctions that line the lumen of cerebral capillaries forming a physical obstacle that permits passage of a select few small molecules (Ballabh et al. 2004; Pandey et al. 2016). The BBB is an example of a biological barrier that substantially limits organotropic drug delivery (Hersh et al. 2016). Emphasis has been placed on designing drug delivery vehicles to enable therapeutic agents to reach the CNS. For example, engineered EVs have shown promise in delivering small interfering RNA (siRNA) to the brain following intravenous injection. Dendritic cells were transfected with a plasmid encoding for a fusion protein composed of the EV membrane protein, lysosome-associated membrane protein 2 (Lamp2b), the CNS-specific rabies viral glycoprotein (RVG), and a FLAG epitope (Fig. 3a,b) (Alvarez-Erviti et al. 2011). EVs with the RVG peptide were purified from cell media and loaded with siRNA through electroporation (Alvarez-Erviti et al. 2011). The siRNAs were against beta-amyloid cleaving enzyme 1 (BACE1) protease that is involved in the enzymatic cascade that causes ß-amyloid peptide aggregation in Alzheimer's pathogenesis. *In vitro* results showed that RVG-EVs had a delivery efficiency comparable to cationic liposomes, resulting in an equal and dose-dependent gene knock down without evident cytotoxic effects (Alvarez-Erviti et al. 2011). Notably, *in vivo* studies demonstrated that intravenously injected EVs efficiently targeted brain tissue delivering exogenous genetic material without inducing inflammatory responses (Alvarez-Erviti et al. 2011). Specifically, the EVs caused a substantial decrease in BACE1 mRNA (61%) (Fig. 3c) and protein (62%) (Fig. 3d) levels in cortical tissue (Alvarez-Erviti et al. 2011). RVG-EV therapy also decreased the concentration of the ß-amyloid 1–42 peptide (55%), which is one of the major components of amyloid plaques (Fig. 3e) (Alvarez-Erviti et al. 2011).

In addition to the use of viral peptides, NPs can be functionalized with endogenous ligands that recognize receptors that are overexpressed in the CNS. For example, neuronal cells of patients affected by age related neurodegenerative pathologies have increased levels of lactoferrin receptors (Qian and Wang 1998), which can be targeted with NPs. Specifically, lactoferrin-coated pegylated polymeric NPs have been used for delivery of neuroprotective molecules for the treatment of Alzheimer's disease in animal models (Liu et al. 2013). The intranasal administration route proved beneficial for bypassing the BBB, and NPs protected cargo against mucus and enzymes that compromise penetration and integrity in the nasal cavity. In mice studies, these intranasally administered lactoferrin-NPs loaded with a neuroprotective agent demonstrated a dose-dependent effect with an amelioration of cognitive defects (Liu et al. 2013). Notably, the lactoferrin-NPs were superior to the neuroprotective agent alone or untargeted NPs (Liu et al. 2013).

EV carriers for brain delivery have also been administered through the intranasal route (Sterzenbach et al. 2017). For example, human glioblastoma cells were transfected with a plasmid encoding Cre recombinase with a WW tag that promotes loading into EVs (Riling et al. 2015; Sterzenbach et al. 2017). EVs were then administered to transgenic mice carrying a stop cassette that inhibited expression of a gene encoding for a red fluorescent protein. Analysis of brain tissue demonstrated a high number of red fluorescent neurons (Sterzenbach et al. 2017), indicating that EV-based intranasal enzyme delivery was successful.

Delivery of NPs to the brain has also been enhanced utilizing RBCs (Brenner et al. 2018). Brenner et al. achieved improved NP-delivery to the brain after internal carotid artery injection of RH. Specifically, brain-specific uptake was 11.5% of the injected dose and the RH strategy displayed higher brain-to-blood ratios (27 fold) and brain to liver ratios (143 fold) compared to free NPs (Brenner et al. 2018). Notably, RBCs can efficiently shuttle NPs to several organs through administration of catheters placed into specific blood vessels that supply the target organ (Brenner et al. 2018).

5 Eye delivery

Macular degeneration is an age-related medical condition that promotes neovascularization of the subretinal space, leading to vision loss (Ambati et al. 2013). Presently, the only therapy available is in vitreous administration of anti-vascular endothelial growth factor (VEGF) molecules that block vascularization (Spaide et al. 2006). In the majority of patients, repeated intraocular injections result in retinal fibrosis or atrophy,

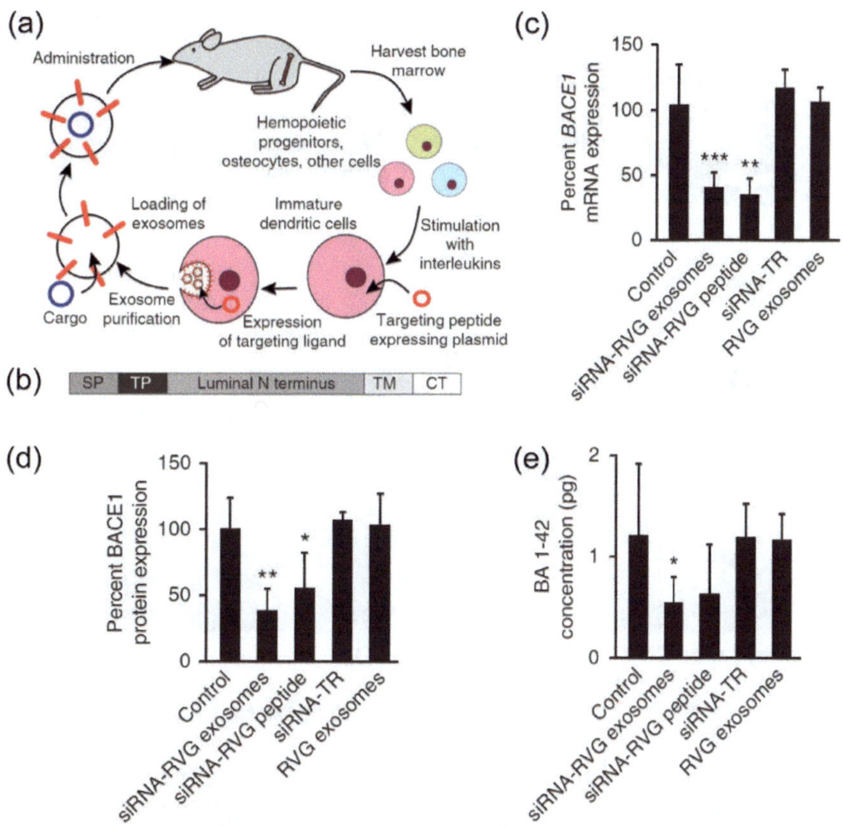

Fig. 3 Brain delivery of intravenously injected extracellular vesicles with small interfering RNA (siRNA) against beta-amyloid cleaving enzyme (BACE1) in a wild type mouse model. **a** Schematic of engineering, collection and administration of targeted extracellular vesicles for siRNA delivery. **b** Schematic representation on the fusion protein composed of: Lamp2b protein. SP, signal peptide; TP, targeting peptide; TM, transmembrane domain; CT, C terminus. **c** BACE1 mRNA levels in the brain cortical region (normalized to 18S RNA levels). **d** BACE1 protein levels in brain tissue. **e** Quantification of β-amyloid 1–42 (BA 1–42) in brain tissue samples. Data is presented as mean ± s.d. ($n = 5$). *, $P < 0.05$; **, $P < 0.01$; ***, $P < 0.001$ compared to untreated control. Reproduced from Alvarez-Erviti et al. 2011 with permission. RVG, rabies virus glycoprotein; TR, *in vivo* transfection reagent

consequently promoting eye damage. Intravenous injection of agents is also problematic, as the blood-retina barrier makes it challenging for therapeutic agents to reach the eye. To overcome these challenges, VEGF-targeting NPs have recently been developed (Luo et al. 2013). Intravenously administered arginylglycylaspartic acid (RGD)-coated PLGA NPs carrying a plasmid encoding for the VEGF intraceptor Flt23K were able to target the fundus of normal and degenerated eyes in three different induced choroid damage models: a gene-silenced and a laser-induced mouse model and a laser-induced monkey model (Luo et al. 2013). The plasmid encoding for Flt23K intraceptor efficiently targeted and seized VEGF into the endoplasmic reticulum, inhibiting paracrine and autocrine cell-stimulation, leading to decreased angiogenesis and fibrosis (Luo et al. 2013). Notably, in both mouse models, RGD-Flt23K-NPs localized preferentially to the extracellular matrix of lesion-containing eyes and were retained there for longer compared to Flt23K-NPs and Blank-NPs (Fig. 4a). Similarly, in the monkey model, the RGD-Flt23K-NPs were retained for one month after treatment, whereas the Flt23K-NPs and Blank-NPs were not detected (Fig. 4b) (Luo et al. 2013). Furthermore, RGD-Flt23K-NPs were more effective at reducing the fibrotic lesion volume and choroidal neovascularization volume (CNV) (Fig. 4c,d) compared to Flt23K-NPs, Blank-NPs, and sham groups (Luo et al. 2013). Both mouse models noted a significant reduction in CNV lesion volume when treated with RGD-Flt23K-NPs compared to the other groups (Fig. 4e, f) (Luo et al. 2013). Moreover, the RGD-Flt23K-NPs were superior at reducing CNV lesion volume compared to intravitreal administration of a VEGF antibody (Fig. 4g). Furthermore, six weeks after treatment, more than the 85% of the RGD-Flt23K-NPs treated mice showed an improvement in visual acuity greater than 10%, whereas the other groups displayed no significant changes (Fig. 4h) (Luo et al. 2013).

6 Lymphatic delivery

The lymphatic system is part of both the circulatory and immune system and is composed by vessels, primary organs (thymus and bone marrow), and secondary organs (lymph nodes and spleen). The function of lymph nodes is to defend against pathogens. Lymph nodes are the

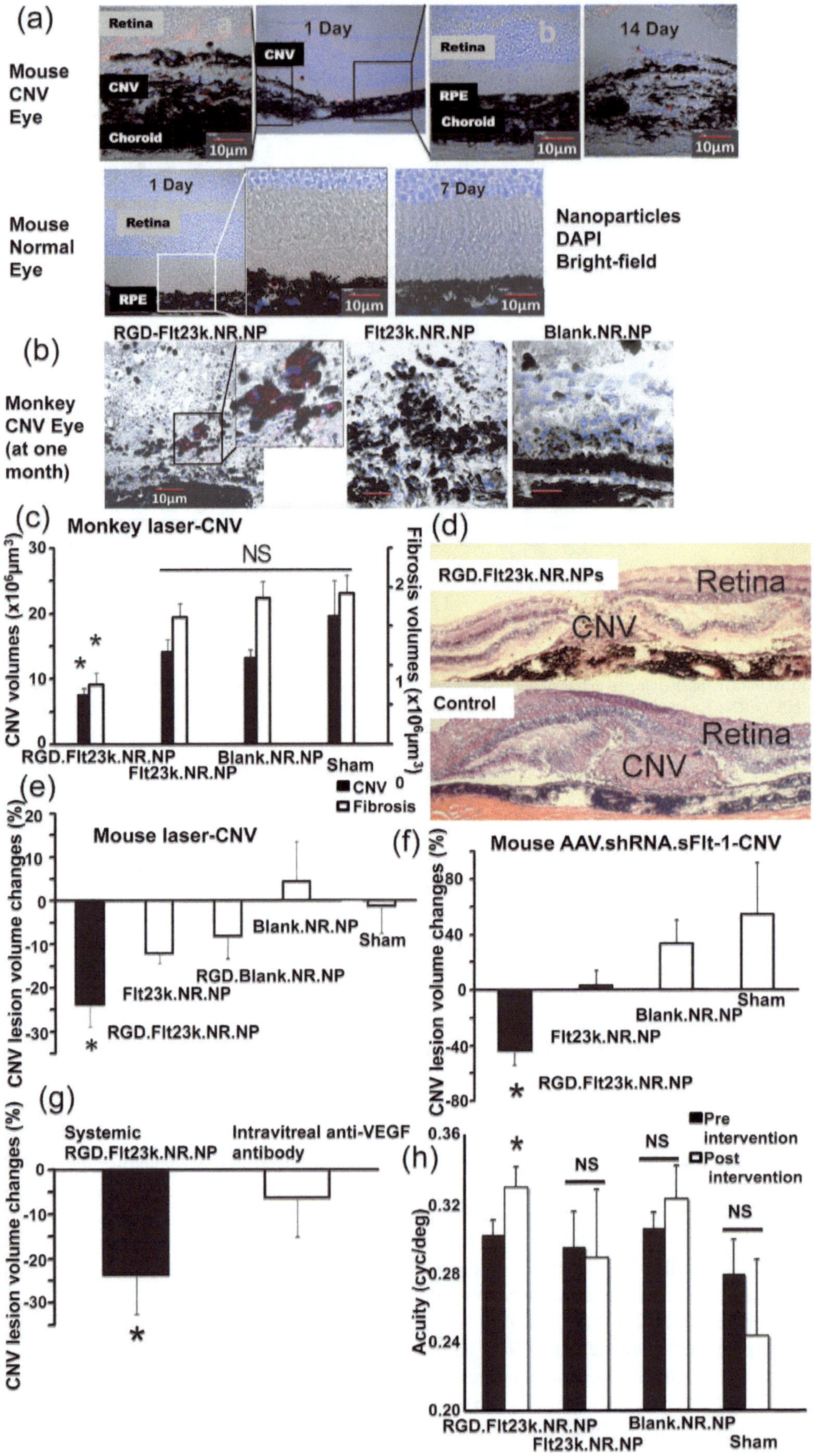
(a)
Mouse CNV Eye
Retina
CNV
Choroid
1 Day
14 Day
RPE
10μm
Mouse Normal Eye
7 Day
Nanoparticles
DAPI
Bright-field
(b)
Monkey CNV Eye (at one month)
RGD-Flt23k.NR.NP
Flt23k.NR.NP
Blank.NR.NP
(c) Monkey laser-CNV
CNV volumes (x10⁶μm³)
Fibrosis volumes (x10⁶μm³)
NS
RGD.Flt23k.NR.NP
Flt23k.NR.NP
Blank.NR.NP
Sham
CNV
Fibrosis
(d)
RGD.Flt23k.NR.NPs
Retina
CNV
Control
(e) Mouse laser-CNV
CNV lesion volume changes (%)
RGD.Flt23k.NR.NP
Flt23k.NR.NP
RGD.Blank.NR.NP
Blank.NR.NP
Sham
(f) Mouse AAV.shRNA.sFlt-1-CNV
CNV lesion volume changes (%)
RGD.Flt23k.NR.NP
Flt23k.NR.NP
Blank.NR.NP
Sham
(g)
CNV lesion volume changes (%)
Systemic RGD.Flt23k.NR.NP
Intravitreal anti-VEGF antibody
(h)
Acuity (cyc/deg)
NS
Pre intervention
Post intervention
RGD.Flt23k.NR.NP
Flt23k.NR.NP
Blank.NR.NP
Sham

◀ **Fig. 4** Choroidal neovascularization (CNV) eye lesion delivery of intravenously injected PLGA NPs in mice and monkeys. **a** Confocal analysis of red-labeled arginylglycylaspartic acid (RGD) functionalized NPs carrying the recombinant vascular endothelial growth factor intraceptor (Flt23k) intraceptor and Nile red (RGD.Flt23k.NR.NPs) in mouse ocular tissue cryosections. **b** RGD.Flt23k.NR.NPs in CNV eye lesions in a laser-induced monkey model. **c** Volumes of CNV lesions and fibrosis in a laser-induced CNV monkey model after treatment with RGD.Flt23k.NR.NPs ($n = 14$), red-labeled bare NPs carrying Flt23k intraceptor (Flt23k.NR.NPs) ($n = 16$ and $n = 8$ for CNV volume and fibrosis respectively), red-labeled NPs (Blank.NR.NPs) (n = 8 and $n = 19$ for CNV volume and fibrosis respectively), and sham control (n = 19 and n = 16 for CNV volume and fibrosis respectively). **d** Representative images of CNV in a laser-induced CNV monkey model treated with RGD.Flt23k.NR.NP or sham control. **e** CNV lesion regression in a laser-induced mouse models ($n = 11$). **f** CNV lesion volumes in a soluble fms-like tyrosine kinase-1 (sFlt-1) knockdown mouse model treated with RGD.Flt23k.NR.NPs ($n = 9$), Flt23k.NR.NPs ($n = 10$), red-labeled blank NPs functionalized with RGD (RGD.Blank.NR.NPs) ($n = 10$), Blank.NR.NPs (n = 10) and sham group (n = 10). **g** CNV lesion reduction after RGD.Flt23k.NR.NP treatment and intravitreal injection of anti-VEGF antibody (n = 10, 2 weeks). **h** Visual acuity in a soluble fms-like tyrosine kinase-1 (sFlt-1) knockdown mouse model treated with RGD.Flt23k.NR.NP, Flt23k.NR.NPs, Blank.NR.NPs, or sham (n = 11, 4 weeks). *, $P < 0.05$ versus sham control (c, e, f and h) and intravitreal injection (g). Data are presented as mean ± s.e.m. Reproduced from Luo et al. 2013 with permission. AAV, adeno-associated virus; DAPI, 4′,6-diamidino-2-phenylindole; NS, non-significant; RPE, retinal pigment epithelium; shRNA, short hairpin RNA; VEGF, vascular endothelial growth factor

location of lymphocytes and antigen-presenting cells that in healthy conditions can start and lead an efficient immune response (Ji 2016). Lymph nodes are also the primary site of metastasis from adjacent tumors due to their location and structure (Ji 2016). For these reasons, the lymphatic system is considered a good target for therapy, and several NP-based approaches have recently been proposed. In mice models of lymph node metastasis, the delivery efficiency of direct injection of pegylated gold nanorods (PAuNRs) into the lymph node adjacent to the site of metastasis has been studied (Oladipo et al. 2017). The NPs drained into the metastasized lymph node and remained there for an extended period of time, making photothermal therapy possible (Oladipo et al. 2017). Near-infrared irradiation caused a rise in temperature that destroyed surrounding cells and prevented metastatic spread (Oladipo et al. 2017).

Stimulation of the immune system with ligands that bind to toll-like receptors (TLRs) can drive antitumor responses mediated by a subset of T cells (cytotoxic and natural killer). However, systemic administration of TLR ligands is often inefficient due to dose limitations. To improve lymphatic delivery, polymeric NPs encasing TLR7 ligands have been injected subcutaneously close to the inguinal space (Widmer et al. 2018). The NPs protected ligands and delivered them to the draining lymph nodes, particularly to monocytes, macrophages and dendritic cells (Widmer et al. 2018).

In addition to cancer treatment, the lymphatic system is a promising target for NP-based therapies for viral infections. Among other viruses, the human immunodeficiency virus (HIV) mainly proliferates inside macrophages and attacks CD4+ T cells in the lymphatic system (Fevrier et al. 2011; Vidya Vijayan et al. 2017). It is challenging for systemically administered anti-retroviral drugs to these reach reservoir sites, causing continuous viral proliferation with poor outcomes for patients (Kumar et al. 2018). Therefore, there is a need to develop nanocarriers that are able to deliver antiviral drugs to lymphatic organs. SLNs are stable drug delivery vehicles that can be orally administered, as they resist gastric acidity (Kumar et al. 2018) and are taken up by the intestinal lymphatic system that transports lipids and nanosized particle from the gut into systemic blood circulation (Singh et al. 2014). In rats, it has been shown that orally administered poorly water-soluble antiretroviral drugs encapsulated into SLNs accumulate in lymphatic organs (spleen and thymus). In conclusion, there are several promising NP strategies for lymphatic delivery, which will become increasingly important with the rising number of immunotherapies in development. However, transport mechanisms within the lymphatic system are still poorly understood and thought to be characterized by unique and complex processes, hampering the development of NPs that exploit the anatomy and physical parameters of lymph vessels. For this reason, computational models of mass transport in the lymphatic system could be valuable for the design of NPs (Kojic et al. 2017).

7 Bone delivery

Bone tissue is composed of osteocytes that regulate osteoblast and osteoclast activity; orchestrating a homeostatic process that finely controls bone matrix deposition and adsorption (Raggatt and Partridge 2010; Schaffler and Kennedy 2012). Pathological conditions such as fractures or chronic bone damage can impair this homeostasis, resulting in tissue damage. For this reason, anabolic strategies are explored as therapies to improve bone regeneration (Roberts and Ke 2018). Osteoblast-specific oligonucleotide aptamer-functionalized lipid NPs shielding osteogenic siRNA have been evaluated for bone regeneration (Liang et al. 2015). In a rat model, the aptamer-targeted NPs displayed increased accumulation in the bone (Fig. 5a), specifically in osteoblasts (Fig. 5b), and enhanced gene knockdown (Fig. 5c) compared to NPs with a random sequence aptamer (Liang et al. 2015). Gene knockdown was considerably higher in cells expressing alkaline phosphatase (typically osteoblasts) compared to cells that were positive for an osteoclast-associated receptor (Oscar) (Fig. 5d). After repeated injections, osteopenic rats treated with aptamer-functionalized NPs displayed bone parameters comparable to healthy rats (C. Liang et al. 2015), indicating successful osteogenic treatment.

Bone marrow is a tissue encased by an inner bone structure responsible for the synthesis and secretion of functional circulating blood cells. Bone marrow can be affected by a spectrum of genetic disorders that cause medullary dysplasia or cytopenia, i.e. myelodysplastic syndromes (Wu et al. 2017). Potential effective therapies consist on chemotherapeutic drugs, which also exhibit systemic toxicity resulting in multiple collateral effects. For this reason, Wu et al. proposed the use of biodegradable NPs loaded with decitabine and arsenic trioxide, chemotherapeutic agents that when combined induce a favorable response in patients affected by myelodysplastic syndromes (Wu et al. 2017). NPs were functionalized with alendronate-PEG-lipid molecules to enhance targeting

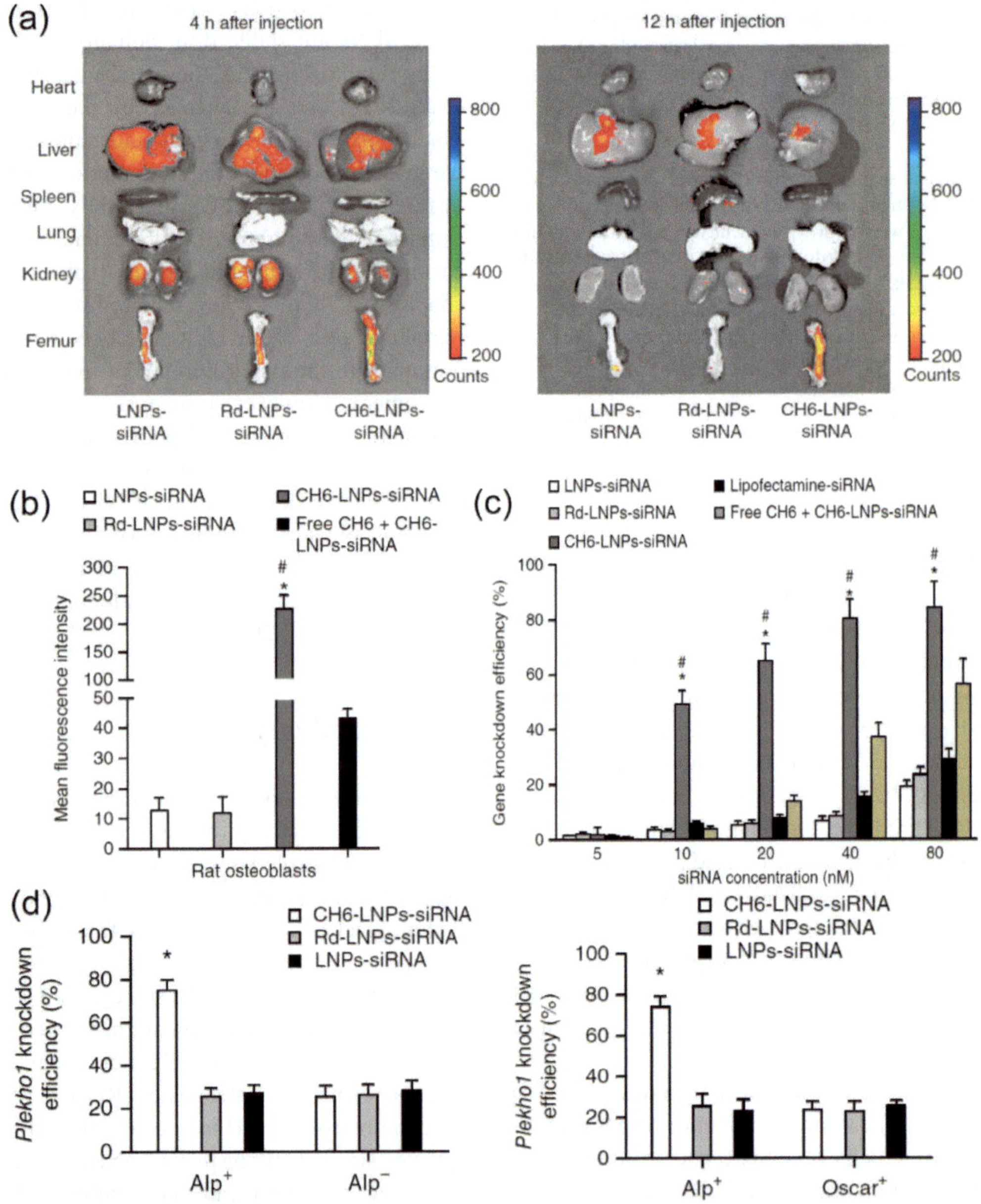

Fig. 5 Bone delivery of intravenously injected lipid-based nanoparticles (LNPs) with siRNA against pleckstrin homology domain containing O1 (Plekho1) in rats. **a** Fluorescent siRNA in organs of rats administered with LNPs-siRNA, random sequence (Rd) functionalized LNPs siRNA, or bone targeting aptamer (CH6) functionalized LNPs-siRNA. **b** siRNA uptake by primary osteoblasts. **c** Gene knockdown in primary rat osteoblasts. Data are presented as the mean ± s.d. ($n = 3$). **d** Plekho1 knockdown efficiency in alkaline phosphatase (Alp)+, Alp−, and osteoclast-associated receptor (Oscar) + cells following *in vivo* administration. Data are presented as means ± s.d. ($n = 6$). *, $P < 0.05$ for a comparison of CH6-LNPs-siRNA with LNPs-siRNA or Rd-LNPs-siRNA (b,c,d); #, $P < 0.05$ for a comparison of CH6-LNPs-siRNA with free CH6 + CH6- LNPs-siRNA (b,c). Reproduced from Liang et al. 2015 with permission

abilities. Alendronate is a bisphosphonate that mimics pyrophosphate and is used to treat pathological bone conditions, such as osteoporosis. In a mouse model, the bone-targeted NPs improved tissue-specific accumulation of the chemotherapeutic agents compared to untargeted NPs (up to 8-fold) (Wu et al. 2017). Four weeks after treatment, the white blood cell (WBC) and platelet counts were normal in 50% of the NP treated animals whereas the corresponding value was 25% in the free drug or untargeted NP groups. More than 200 days post-treatment mice that had received bone-targeted NPs displayed decreased mortality (~40%) compared to those that received the free drug (~70%), untargeted NPs (~75%) or phosphate buffered saline (~90%) (Wu et al. 2017). Ex vivo analysis of bone marrow cells demonstrated that bone-targeted NPs efficiently delivered chemotherapy to the target site, as evidenced by reduced transcription of target gene DNA methyltransferase (80% reduction) (Wu et al. 2017). Apoptosis was also increased in the bone marrow of mice injected with bone-targeted NPs (~70%) compared to free drugs (50%), untargeted NPs (~50%), or phosphate buffered saline (~10%). On the contrary, liver cell apoptosis was less prevalent in the bone-targeted NP group compared to the other treatment groups (Wu et al. 2017). These results indicate that the bone-targeted NPs efficiently deliver chemotherapeutic agents to bone tissue, while reducing chemotherapy-mediated hepatic toxicity.

8 In utero delivery

Pediatric patients affected by hemoglobinopathies, particularly β-thalassemia, are frequently subjected to blood transfusions or bone marrow transplants, which are medical procedures that can lead to life-threatening complications (Modell and Darlison 2008). Despite increasingly precocious pre-natal diagnoses, there is a lack of efficient in utero gene therapy able to edit the mutated gene before birth (Fan et al. 2012). Peptide nucleic acids (PNAs) are synthetic polymers that differ from DNA and RNA structures in that the backbone of the molecule is formed by amino acids instead of carbohydrates (Nielsen et al. 1994). PNAs bind to DNA through a stable bond resulting in an enzyme-resistant carrier for genetic material (McNeer et al. 2013). PLGA-NPs with PNA coupled with donor DNA were injected via the intravitelline vein or into the amniotic cavity of fetal mice carrying the human β-thalassemia mutation (Ricciardi et al. 2018). The NPs accumulated preferentially in the fetal liver, where hematopoietic stem cells undergo rapid expansion and transfer to bone marrow (Fig. 6a). The treatment resulted in successful gene editing with consequent improved postnatal outcomes compared to controls (Fig. 6b, c) (Ricciardi et al. 2018). Treated mice also had an increased hemoglobin concentration (Fig. 6d) along with an improved red blood cell (RBC) phenotype (Fig. 6e) and reduced reticulocyte count (Fig. 6f) (Ricciardi et al. 2018). After a single NP intravitelline injection, successful gene editing was detected in both fetal hematopoietic stem/progenitor cells and bone marrow cells of pups (Fig. 6g) (Ricciardi et al. 2018). The overall results confirm that in utero delivered NPs can mediate safe fetal gene editing resulting in normal fetal development and reduced post-natal β-thalassemia morbidity.

9 Protein corona

There has been extensive research into the modification of NPs following systemic administration to regulate organotropism, specifically the binding of biomolecules to the surface of NPs resulting in a protein-rich coating known as the protein corona. The protein corona significantly impacts the characteristics of NPs, causing variable changes in NP size, shape, charge, and aggregation capacity (Wolfram et al. 2014c). Consequently, these changes have implications in NP toxicity, targeting, biodistribution, and drug release, especially when transitioning from *in vitro* to *in vivo* environments (Wolfram et al. 2014c). With respect to biodistribution, the binding of certain plasma proteins to the protein corona can cause NPs to have affinities for specific tissues. A high concentration of opsonins, such as complement factors, fibrinogen, or immunoglobin G, in the protein corona promotes macrophage uptake of NPs (Nguyen and Lee 2017) and NP accumulation into the liver and spleen (Wolfram et al. 2014c). Enrichment with dysopsonins, including apolipoproteins (Apo) and albumin, was shown to increase NP circulation life and accumulation of NPs in organs other than the liver and spleen, such as the brain (Nguyen and Lee 2017; Wolfram et al. 2014c). Recent studies into the protein corona have explored methods to use this structure to improve therapeutic goals. One method involves the modification of NPs to create specific protein corona profiles from endogenous proteins. For instance, Schöttler et al. demonstrated that the stealth effect of pegylated NPs, referring to the reduction in non-specific cell uptake and increased blood circulation life of these nanoparticles, was due to the formation of protein coronas with high amounts of a clusterin, Apo J, from plasma (Schottler et al. 2016). Other nanomaterial compositions were shown to favorably bind to serum proteins that improved delivery to specific tissues, such as polysorbate-coated NPs binding to Apo E or Apo AI leading to an increased distribution in brain endothelium (Nguyen and Lee 2017). In regard to EVs, the characteristic and implications of the protein corona are largely unknown and are likely to be affected by the isolation method. Overall, to improve the efficiency of nanotherapeutics, studies must account for effects of the protein corona on NP functionality.

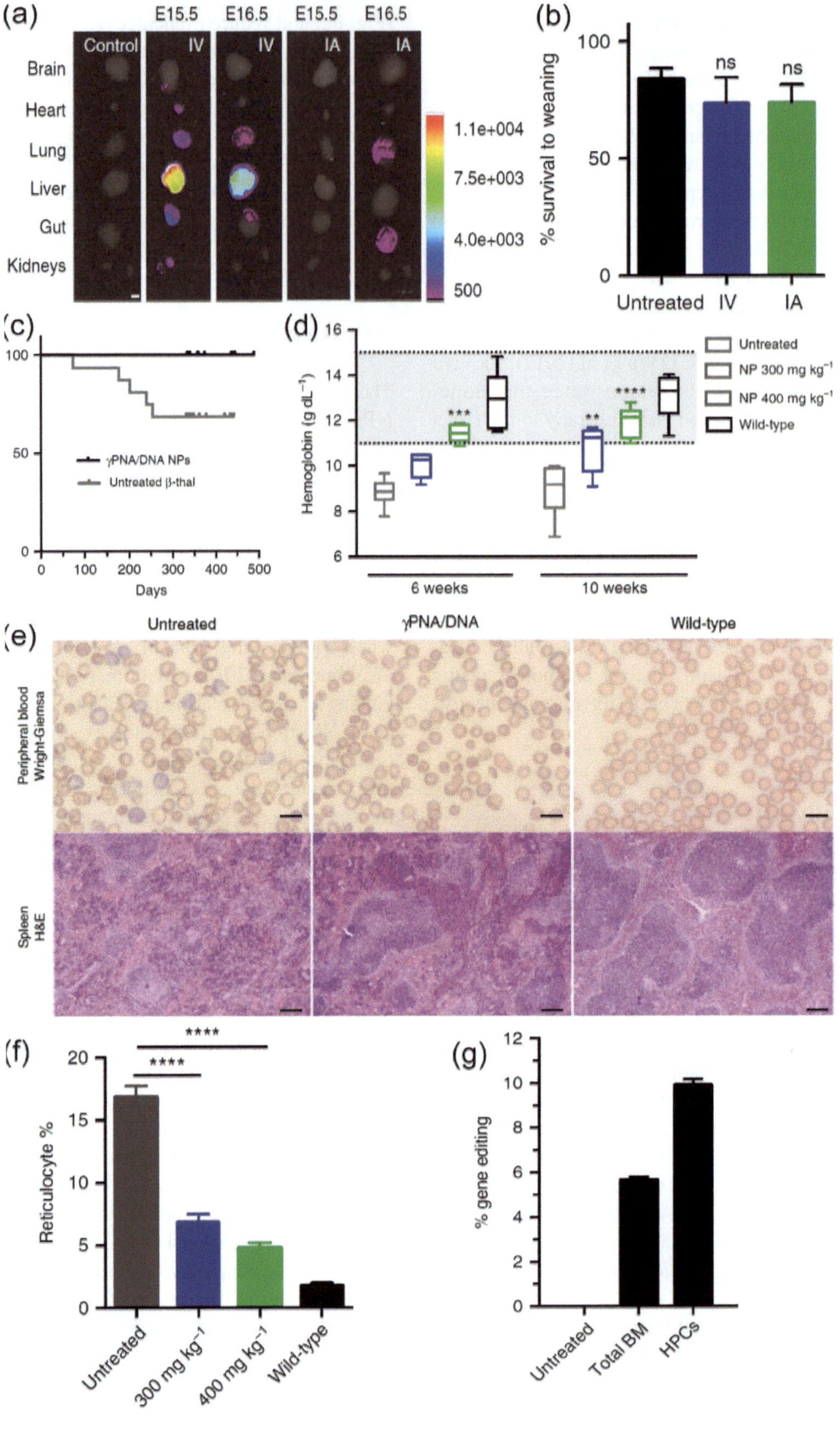
(a)
E15.5
E16.5
E15.5
E16.5
Control
IV
IV
IA
IA
Brain
Heart
Lung
Liver
Gut
Kidneys
1.1e+004
7.5e+003
4.0e+003
500
(b)
% survival to weaning
100
50
0
ns
ns
Untreated
IV
IA
(c)
100
50
0
0
100
200
300
400
500
Days
γPNA/DNA NPs
Untreated β-thal
(d)
Hemoglobin (g dL−1)
16
14
12
10
8
6
***
**
****
****
6 weeks
10 weeks
Untreated
NP 300 mg kg−1
NP 400 mg kg−1
Wild-type
(e)
Untreated
γPNA/DNA
Wild-type
Peripheral blood Wright-Giemsa
Spleen H&E
(f)
Reticulocyte %
20
15
10
5
0
****
****
Untreated
300 mg kg−1
400 mg kg−1
Wild-type
(g)
% gene editing
12
10
8
6
4
2
0
Untreated
Total BM
HPCs

◀ **Fig. 6** Fetus delivery of intravenously (IV) and intra-amniotic (IA) injected PLGA NPs with γ tail-clamp peptide nucleic acids (γtcPNA) in a mouse model of human β-thalassemia. **a** Fetus organ biodistribution of NPs 3 h post-injection at indicated embryonic (E) stages ($n = 9$ for control, $n = 10$ for E15.5 IV, $n = 7$ for E16.5 IV and E15.5 IA, $n = 8$ for E16.5 IA). Scale bars, 2 mm. **b** Survival to weaning (21 days) ($n = 7$ for untreated, $n = 4$ for IV E15.5, $n = 7$ for IA E15.5). **c** Survival of fetus at E15.5 (IV) ($n = 16$). **d** Blood hemoglobin levels (E15.5 IV) ($n = 6$), horizontal lines within the boxes indicate the median, the box indicates the first and third quartile, and the whiskers represent the range. **e** Blood cells (Wright–Giemsa staining) and spleen sections (hematoxylin and eosin staining) (E15.5 IV). Peripheral blood scale bars, 10 μm; spleen scale bars, 150 μm. **f** Reticulocyte count (E15.5 IV). Scale bar, 10 μm. **g** Gene edited total bone marrow and isolated hematopoietic progenitor cells (HPCs) (E15.5 IV) ($n = 3$). Data are presented as mean ± s.d. **, $P < 0.01$; ***, $P < 0.001$; ****, $P < 0.0001$. Reproduced from Ricciardi et al. 2018 with permission

10 Conclusion

Many therapeutic agents that show promising results in cell culture fail *in vivo* due to biological barriers that hinder site-specific accumulation. These barriers consist of the immune system, various anatomical obstacles, and pressure gradients. NPs have been developed as tools to aid in the transport of therapeutic agents within and across compartments in the body. Although some barriers are inherent to all tissues, others are organ-specific, indicating that preconditioning strategies, administration route, and NP properties should be tailored based on the target organ. Despite the development of several targeted nanodelivery strategies, there is still a pressing need to further understand organ-specific macro and microscale obstacles that impede drug delivery. Knowledge of the structure and function of the target organ is necessary to design delivery strategies that lead to favorable drug distributions and therapeutic outcomes.

Besides successfully delivering therapeutic agents to target organs, NPs must also ensure delivery to the appropriate subcellular compartments, as the intracellular environment poses additional obstacles, such as the lysosome that rapidly degrades biomolecules. Multifunctional delivery strategies are necessary to finely orchestrate drug transport across a diverse set of extracellular and subcellular barriers. Both synthetic NPs and extracellular vesicles provide promising multi-pronged strategies for site-specific delivery.

Acknowledgements This work was mainly supported by Mayo Clinic, the University of Brescia. Among various sources of intramural funding, the authors particularly acknowledge support from the Mayo Clinic in Florida Focused Research Team Program and the Center for Regenerative Medicine. This work was also partially supported by the National Cancer Institute Physical Sciences-Oncology Network of the National Institutes of Health, under award number U54CA210181. The content is solely the responsibility of the authors and does not necessarily represent the official views of the National Institutes of Health.

Compliance with ethical standards

Conflict of interest The authors declare that they have no conflict of interest.

References

Y. Abbas, H.M.E. Azzazy, S. Tammam, A. Lamprecht, M.E. Ali, A. Schmidt, S. Sollazzo, S. Mathur, Development of an inhalable, stimuli-responsive particulate system for delivery to deep lung tissue. Colloids Surf B Biointerfaces **146**, 19–30 (2016)

M.E. Ali, A. Lamprecht, Spray freeze drying for dry powder inhalation of nanoparticles. Eur J Pharm Biopharm **87**, 510–517 (2014)

L. Alvarez-Erviti, Y. Seow, H. Yin, C. Betts, S. Lakhal, M.J. Wood, Delivery of siRNA to the mouse brain by systemic injection of targeted exosomes. Nat Biotechnol **29**, 341–345 (2011)

J. Ambati, J.P. Atkinson, B.D. Gelfand, Immunology of age-related macular degeneration. Nat Rev Immunol **13**, 438–451 (2013)

D.R. Baer, The Chameleon Effect: Characterization Challenges Due to the Variability of Nanoparticles and Their Surfaces. Front Chem **6**, 145 (2018)

P. Ballabh, A. Braun, M. Nedergaard, The blood-brain barrier: an overview: structure, regulation, and clinical implications. Neurobiol Dis **16**, 1–13 (2004)

S. Barua, S. Mitragotri, Challenges associated with Penetration of Nanoparticles across Cell and Tissue Barriers: A Review of Current Status and Future Prospects. Nano Today **9**, 223–243 (2014)

R. Bazak, M. Houri, S. El Achy, S. Kamel, T. Refaat, Cancer active targeting by nanoparticles: a comprehensive review of literature. J Cancer Res Clin Oncol **141**, 769–784 (2015)

N. Bertrand, J. Wu, X. Xu, N. Kamaly, O.C. Farokhzad, Cancer nanotechnology: the impact of passive and active targeting in the era of modern cancer biology. Adv Drug Deliv Rev **66**, 2–25 (2014)

O. Betzer, M. Shilo, R. Opochinsky, E. Barnoy, M. Motiei, E. Okun, G. Yadid, R. Popovtzer, The effect of nanoparticle size on the ability to cross the blood-brain barrier: an *in vivo* study. Nanomedicine (Lond) **12**, 1533–1546 (2017)

E. Blanco, H. Shen, M. Ferrari, Principles of nanoparticle design for overcoming biological barriers to drug delivery. Nat Biotechnol **33**, 941–951 (2015)

D. Boraschi, L.R.C. Castellano, P. Italiani, Editorial: interaction of nanomaterials with the immune system: Role in nanosafety and nanomedicine. Front Immunol **8**, 1688 (2017)

D.A. Borrelli, K. Yankson, N. Shukla, G. Vilanilam, T. Ticer, J. Wolfram, Extracellular vesicle therapeutics for liver disease. J Control Release **273**, 86–98 (2018)

J.S. Brenner, D.C. Pan, J.W. Myerson, O.A. Marcos-Contreras, C.H. Villa, P. Patel, H. Hekierski, S. Chatterjee, J.Q. Tao, H. Parhiz, K. Bhamidipati, T.G. Uhler, E.D. Hood, R.Y. Kiseleva, V.S. Shuvaev, T. Shuvaeva, M. Khoshnejad, I. Johnston, J.V. Gregory, J. Lahann, T. Wang, E. Cantu, W.M. Armstead, S. Mitragotri, V. Muzykantov, Red blood cell-hitchhiking boosts delivery of nanocarriers to chosen organs by orders of magnitude. Nat Commun **9**, 2684 (2018)

S. Busatto, G. Vilanilam, T. Ticer, W.L. Lin, D.W. Dickson, S. Shapiro, P. Bergese, J. Wolfram, Tangential flow filtration for highly efficient concentration of extracellular vesicles from large volumes of fluid. Cells **7**, 273 (2018)

M. Caldorera-Moore, N. Guimard, L. Shi, K. Roy, Designer nanoparticles: incorporating size, shape and triggered release into nanoscale drug carriers. Expert Opin Drug Deliv **7**, 479–495 (2010)

F. Celikoglu, S.I. Celikoglu, E.P. Goldberg, Bronchoscopic intratumoral chemotherapy of lung cancer. Lung Cancer **61**, 1–12 (2008)

J.A. Champion, S. Mitragotri, Role of target geometry in phagocytosis. Proc Natl Acad Sci U S A **103**, 4930–4934 (2006)

S. Chen, K. Yang, R.G. Tuguntaev, A. Mozhi, J. Zhang, P.C. Wang, X.J. Liang, Targeting tumor microenvironment with PEG-based amphiphilic nanoparticles to overcome chemoresistance. Nanomedicine **12**, 269–286 (2016a)

Y. Chen, L. V. Minh, J. Liu, B. Angelov, M. Drechsler, V. M. Garamus, R. Willumeit-Römer, A. Zou, Baicalin loaded in folate-PEG modified liposomes for enhanced stability and tumor targeting. Colloids Surf B Biointerfaces **140**, 74–82 (2016b)

D. S. Chulpanova, K. V. Kitaeva, V. James, A. A. Rizvanov, V. V. Solovyeva, Therapeutic Prospects of Extracellular Vesicles in Cancer Treatment. Front Immunol **9**, 1534 (2018)

R. Daneman, A. Prat, The blood-brain barrier. Cold Spring Harb Perspect Biol **7**, a020412 (2015)

L. De Backer, T. Naessens, S. De Koker, E. Zagato, J. Demeester, J. Grooten, S.C. De Smedt, K. Raemdonck, Hybrid pulmonary surfactant-coated nanogels mediate efficient in vivo delivery of siRNA to murine alveolar macrophages. J Control Release **217**, 53–63 (2015)

P. Decuzzi, M. Ferrari, The adhesive strength of non-spherical particles mediated by specific interactions. Biomaterials **27**, 5307-5314 (2006)

P. Decuzzi, B. Godin, T. Tanaka, S.Y. Lee, C. Chiappini, X. Liu, M. Ferrari, Size and shape effects in the biodistribution of intravascularly injected particles. J Control Release **141**, 320–327 (2010)

E.M. Del Amo, A. Urtti, Current and future ophthalmic drug delivery systems. A shift to the posterior segment. Drug Discov Today **13**, 135–143 (2008)

S. Dhar, F.X. Gu, R. Langer, O.C. Farokhzad, S.J. Lippard, Targeted delivery of cisplatin to prostate cancer cells by aptamer functionalized Pt(IV) prodrug-PLGA-PEG nanoparticles. Proc Natl Acad Sci U S A **105**, 17356–17361 (2008)

J. Du, L.A. Lane, S. Nie, Stimuli-responsive nanoparticles for targeting the tumor microenvironment. J Control Release **219**, 205–214 (2015)

H.C. Fan, W. Gu, J. Wang, Y.J. Blumenfeld, Y.Y. El-Sayed, S.R. Quake, Non-invasive prenatal measurement of the fetal genome. Nature **487**, 320–324 (2012)

G. Fanali, A. di Masi, V. Trezza, M. Marino, M. Fasano, P. Ascenzi, Human serum albumin: from bench to bedside. Mol Asp Med **33**, 209–290 (2012)

M. Fevrier, K. Dorgham, A. Rebollo, CD4+ T cell depletion in human immunodeficiency virus (HIV) infection: role of apoptosis. Viruses **3**, 586–612 (2011)

A.D. Friedman, S.E. Claypool, R. Liu, The smart targeting of nanoparticles. Curr Pharm Des **19**, 6315–6329 (2013)

C.W. Gan, S.S. Feng, Transferrin-conjugated nanoparticles of poly(lactide)-D-alpha-tocopheryl polyethylene glycol succinate diblock copolymer for targeted drug delivery across the blood-brain barrier. Biomaterials **31**, 7748–7757 (2010)

F. Gentile, C. Chiappini, D. Fine, R.C. Bhavane, M.S. Peluccio, M.M.C. Cheng, X. Liu, M. Ferrari, P. Decuzzi, The effect of shape on the margination dynamics of non-neutrally buoyant particles in two-dimensional shear flows. J Biomech **41**, 2312–2318 (2008)

E. Gentile, F. Cilurzo, L. Di Marzio, M. Carafa, C.A. Ventura, J. Wolfram, D. Paolino, C. Celia, Liposomal chemotherapeutics. Future Oncol **9**, 1849–1859 (2013)

S.E.A. Gratton, P.A. Ropp, P.D. Pohlhaus, J.C. Luft, V.J. Madden, M.E. Napier, J.M. DeSimone, The effect of particle design on cellular internalization pathways. Proc Natl Acad Sci **105**, 11613–11618 (2008)

K. Greish, Enhanced permeability and retention (EPR) effect for anticancer nanomedicine drug targeting. Methods Mol Biol **624**, 25–37 (2010)

H.H. Gustafson, D. Holt-Casper, D.W. Grainger, H. Ghandehari, Nanoparticle Uptake: The Phagocyte Problem. Nano Today **10**, 487–510 (2015)

D.S. Hersh, A.S. Wadajkar, N. Roberts, J.G. Perez, N.P. Connolly, V. Frenkel, J.A. Winkles, G.F. Woodworth, A.J. Kim, Evolving Drug Delivery Strategies to Overcome the Blood Brain Barrier. Curr Pharm Des **22**, 1177–1193 (2016)

A. Hoshino, B. Costa-Silva, T.L. Shen, G. Rodrigues, A. Hashimoto, M. Tesic Mark, H. Molina, S. Kohsaka, A. Di Giannatale, S. Ceder, S. Singh, C. Williams, N. Soplop, K. Uryu, L. Pharmer, T. King, L. Bojmar, A.E. Davies, Y. Ararso, T. Zhang, H. Zhang, J. Hernandez, J.M. Weiss, V.D. Dumont-Cole, K. Kramer, L.H. Wexler, A. Narendran, G.K. Schwartz, J.H. Healey, P. Sandstrom, K.J. Labori, E.H. Kure, P.M. Grandgenett, M.A. Hollingsworth, M. de Sousa, S. Kaur, M. Jain, K. Mallya, S.K. Batra, W.R. Jarnagin, M.S. Brady, O. Fodstad, V. Muller, K. Pantel, A.J. Minn, M.J. Bissell, B.A. Garcia, Y. Kang, V.K. Rajasekhar, C.M. Ghajar, I. Matei, H. Peinado, J. Bromberg, D. Lyden, Tumour exosome integrins determine organotropic metastasis. Nature **527**, 329–335 (2015)

S.B. Howell, Clinical applications of a novel sustained-release injectable drug delivery system: DepoFoam technology. Cancer J **7**, 219–227 (2001)

R.C. Ji, Lymph Nodes and Cancer Metastasis: New Perspectives on the Role of Intranodal Lymphatic Sinuses. Int J Mol Sci **18** (2016)

W. Jiang, B.Y. Kim, J.T. Rutka, W.C. Chan, Nanoparticle-mediated cellular response is size-dependent. Nat Nanotechnol **3**, 145–150 (2008)

J.V. Jokerst, T. Lobovkina, R.N. Zare, S.S. Gambhir, Nanoparticle PEGylation for imaging and therapy. Nanomedicine (Lond) **6**, 715–728 (2011)

A. Khalid, S. Persano, H. Shen, Y. Zhao, E. Blanco, M. Ferrari, J. Wolfram, Strategies for improving drug delivery: nanocarriers and microenvironmental priming. Expert Opin Drug Deliv **14**, 865–877 (2016)

P.H. Kierstead, H. Okochi, V.J. Venditto, T.C. Chuong, S. Kivimae, J.M.J. Frechet, F.C. Szoka, The effect of polymer backbone chemistry on the induction of the accelerated blood clearance in polymer modified liposomes. J Control Release **213**, 1–9 (2015)

H.R. Kim, I.K. Kim, K.H. Bae, S.H. Lee, Y. Lee, T.G. Park, Cationic solid lipid nanoparticles reconstituted from low density lipoprotein components for delivery of siRNA. Mol Pharm **5**, 622–631 (2008)

C. Kinnear, T.L. Moore, L. Rodriguez-Lorenzo, B. Rothen-Rutishauser, A. Petri-Fink, Form follows function: nanoparticle shape and its implications for nanomedicine. Chem Rev **117**, 11476–11521 (2017)

M. Kojic, M. Milosevic, V. Simic, E.J. Koay, N. Kojic, A. Ziemys, M. Ferrari, Extension of the composite smeared finite element (CSFE) to include lymphatic system in modeling mass transport in capillary systems and biological tissue. J Serbian Soc Comput Mech **11**, 108–119 (2017)

W.H. Kong, K. Park, M.Y. Lee, H. Lee, D.K. Sung, S.K. Hahn, Cationic solid lipid nanoparticles derived from apolipoprotein-free LDLs for target specific systemic treatment of liver fibrosis. Biomaterials **34**, 542–551 (2013)

S. Kumar, R. Narayan, V. Ahammed, Y. Nayak, A. Naha, U.Y. Nayak, Development of ritonavir solid lipid nanoparticles by Box Behnken design for intestinal lymphatic targeting. J Drug Deliv Sci Technol **44**, 181–189 (2018)

I.K. Kwon, S.C. Lee, B. Han, K. Park, Analysis on the current status of targeted drug delivery to tumors. J Control Release **164**, 108–114 (2012)

S.Y. Lee, M. Ferrari, P. Decuzzi, Design of bio-mimetic particles with enhanced vascular interaction. J Biomech **42**, 1885–1890 (2009)

K.C.P. Li, S.D. Pandit, S. Guccione, M.D.J.B.M. Bednarski, Molecular imaging applications in Nanomedicine. Biomed Microdevices **6**, 113–116 (2004)

R. Li, K. Zheng, C. Yuan, Z. Chen, M. Huang, Be active or not: the relative contribution of active and passive tumor targeting of nanomaterials. Nanotheranostics **1**, 346–357 (2017)

C. Liang, B. Guo, H. Wu, N. Shao, D. Li, J. Liu, L. Dang, C. Wang, H. Li, S. Li, W.K. Lau, Y. Cao, Z. Yang, C. Lu, X. He, D.W. Au, X. Pan, B.T. Zhang, C. Lu, H. Zhang, K. Yue, A. Qian, P. Shang, J. Xu, L. Xiao, Z. Bian, W. Tan, Z. Liang, F. He, L. Zhang, A. Lu, G. Zhang, Aptamer-functionalized lipid nanoparticles targeting osteoblasts as a novel RNA interference-based bone anabolic strategy. Nat Med **21**, 288–294 (2015)

H. Liang, K. Huang, T. Su, Z. Li, S. Hu, P.U. Dinh, E.A. Wrona, C. Shao, L. Qiao, A.C. Vandergriff, M.T. Hensley, J. Cores, T. Allen, H. Zhang, Q. Zeng, J. Xing, D.O. Freytes, D. Shen, Z. Yu, K. Cheng, Mesenchymal stem cell/red blood cell-inspired nanoparticle therapy in mice with carbon tetrachloride-induced acute liver failure. ACS Nano **12**, 6536–6544 (2018)

B.L. Lin, J.F. Chen, W.H. Qiu, K.W. Wang, D.Y. Xie, X.Y. Chen, Q.L. Liu, L. Peng, J.G. Li, Y.Y. Mei, W.Z. Weng, Y.W. Peng, H.J. Cao, J.Q. Xie, S.B. Xie, A.P. Xiang, Z.L. Gao, Allogeneic bone marrow-derived mesenchymal stromal cells for hepatitis B virus-related acute-on-chronic liver failure: A randomized controlled trial. Hepatology **66**, 209–219 (2017)

Z. Liu, M. Jiang, T. Kang, D. Miao, G. Gu, Q. Song, L. Yao, Q. Hu, Y. Tu, Z. Pang, H. Chen, X. Jiang, X. Gao, J. Chen, Lactoferrin-modified PEG-co-PCL nanoparticles for enhanced brain delivery of NAP peptide following intranasal administration. Biomaterials **34**, 3870–3881 (2013)

M.R. Longmire, M. Ogawa, P.L. Choyke, H. Kobayashi, Biologically optimized nanosized molecules and particles: more than just size. Bioconjug Chem **22**, 993–1000 (2011)

L. Luo, X. Zhang, Y. Hirano, P. Tyagi, P. Barabas, H. Uehara, T.R. Miya, N. Singh, B. Archer, Y. Qazi, K. Jackman, S.K. Das, T. Olsen, S.R. Chennamaneni, B.C. Stagg, F. Ahmed, L. Emerson, K. Zygmunt, R. Whitaker, C. Mamalis, W. Huang, G. Gao, S.P. Srinivas, D. Krizaj, J. Baffi, J. Ambati, U.B. Kompella, B.K. Ambati, Targeted intraceptor nanoparticle therapy reduces angiogenesis and fibrosis in primate and murine macular degeneration. ACS Nano **7**, 3264–3275 (2013)

H.K. Makadia, S.J. Siegel, Poly Lactic-co-Glycolic Acid (PLGA) as biodegradable controlled drug delivery carrier. Polymers (Basel) **3**, 1377–1397 (2011)

N.A. McNeer, E.B. Schleifman, A. Cuthbert, M. Brehm, A. Jackson, C. Cheng, K. Anandalingam, P. Kumar, L.D. Shultz, D.L. Greiner, W. Mark Saltzman, P.M. Glazer, Systemic delivery of triplex-forming PNA and donor DNA by nanoparticles mediates site-specific genome editing of human hematopoietic cells *in vivo*. Gene Ther **20**, 658–669 (2013)

Y. Mi, C. Mu, J. Wolfram, Z. Deng, T.Y. Hu, X. Liu, E. Blanco, H. Shen, M. Ferrari, A micro/nano composite for combination treatment of melanoma lung metastasis. Adv Healthc Mater **5**, 936–946 (2016a)

Y. Mi, J. Wolfram, C. Mu, X. Liu, E. Blanco, H. Shen, M. Ferrari, Enzyme-responsive multistage vector for drug delivery to tumor tissue. Pharmacol Res **113(Pt A)**, 92–99 (2016b)

B. Modell, M. Darlison, Global epidemiology of haemoglobin disorders and derived service indicators. Bull World Health Organ **86**, 480–487 (2008)

N. Mohandas, P.G. Gallagher, Red cell membrane: past, present, and future. Blood **112**, 3939–3948 (2008)

R. Molinaro, J. Wolfram, C. Federico, F. Cilurzo, L. Di Marzio, C.A. Ventura, M. Carafa, C. Celia, M. Fresta, Polyethylenimine and chitosan carriers for the delivery of RNA interference effectors. Expert Opin Drug Deliv **10**, 1653–1668 (2013)

C. Mu, X. Wu, X. Zhou, J. Wolfram, J. Shen, D. Zhang, J. Mai, X. Xia, A.M. Holder, M. Ferrari, X. Liu, H. Shen, Chemotherapy sensitizes therapy-resistant cells to mild hyperthermia by suppressing heat shock protein 27 expression in triple-negative breast cancer. Clin Cancer Res **24**, 4900–4912 (2018)

P.K.B. Nagesh, N.R. Johnson, V.K.N Boya, P. Chowdhury, S.F. Othman, V. Khalilzad-Sharghi, B.B. Hafeez, A. Ganju, S. Khan, S.W. Behrman, N. Zafar, S.C. Chauhan, M. Jaggi, M.M. Yallapu, PSMA targeted docetaxel-loaded superparamagnetic iron oxide nanoparticles for prostate cancer. Colloid Surf B Biointerfaces **144**, 8-20 (2016)

P.N. Navya, H.K Daima, Rational engineering of physicochemical properties of nanomaterials for biomedical applications with nanotoxicological perspectives. Nano Converg **3**, 1 (2016)

V.H. Nguyen, B.J. Lee, Protein corona: a new approach for nanomedicine design. Int J Nanomedicine **12**, 3137–3151 (2017)

P.E. Nielsen, M. Egholm, O. Buchardt, Peptide nucleic acid (PNA). A DNA mimic with a peptide backbone. Bioconjug Chem **5**, 3–7 (1994)

L. Nogues, A. Benito-Martin, M. Hergueta-Redondo, H. Peinado, The influence of tumour-derived extracellular vesicles on local and distal metastatic dissemination. Mol Aspects Med **60**, 15–26 (2018)

H. Nojima, C.M. Freeman, R.M. Schuster, L. Japtok, B. Kleuser, M.J. Edwards, E. Gulbins, A.B. Lentsch, Hepatocyte exosomes mediate liver repair and regeneration via sphingosine-1-phosphate. J Hepatol **64**, 60–68 (2016)

A.W. Nørgaard, S. Larsen, M. Hammer, S.S. Poulsen, K.A. Jensen, G.D. Nielsen, P. Wolkoff, Lung damage in mice after inhalation of nanofilm spray products: The role of perfluorination and free hydroxyl groups. Toxicol Sci **116**, 216–224 (2010)

A.O. Oladipo, O.S. Oluwafemi, S.P. Songca, A. Sukhbaatar, S. Mori, J. Okajima, A. Komiya, S. Maruyama, T. Kodama, A novel treatment for metastatic lymph nodes using lymphatic delivery and photothermal therapy. Sci Rep **7**, 45459 (2017)

P.K. Pandey, A.K. Sharma, U. Gupta, Blood brain barrier: An overview on strategies in drug delivery, realistic *in vitro* modeling and *in vivo* live tracking. Tissue Barriers **4**, e1129476 (2016)

D. Paolino, D. Cosco, M. Gaspari, M. Celano, J. Wolfram, P. Voce, E. Puxeddu, S. Filetti, C. Celia, M. Ferrari, D. Russo, M. Fresta, Targeting the thyroid gland with thyroid stimulating hormone (TSH)-nanoliposomes. Biomaterials **35**, 7101–7109 (2014)

G. Pasut, D. Paolino, C. Celia, A. Mero, A.S. Joseph, J. Wolfram, D. Cosco, O. Schiavon, H. Shen, M. Fresta, Polyethylene glycol (PEG)-dendron phospholipids as innovative constructs for the preparation of super stealth liposomes for anticancer therapy. J Control Release **199**, 106–113 (2015)

A.A. Patel, R.J. Patel, S.R. Patel, Nanomedicine for intranasal delivery to improve brain uptake. Curr Drug Deliv **15**, 461–469 (2018)

D. Peer, J.M. Karp, S. Hong, O.C. Farokhzad, R. Margalit, R. Langer, Nanocarriers as an emerging platform for cancer therapy. Nat Nanotechnol **2**, 751–760 (2007)

J. Pelt, S. Busatto, M. Ferrari, E.A. Thompson, K. Mody, J. Wolfram, Chloroquine and nanoparticle drug delivery: A promising combination. Pharmacol Ther **191**, 43–49 (2018)

J.G. Piao, L. Wang, F. Gao, Y.Z. You, Y. Xiong, L. Yang, Erythrocyte membrane is an alternative coating to polyethylene glycol for prolonging the circulation lifetime of gold nanocages for photothermal therapy. ACS Nano **8**, 10414–10425 (2014)

V.N. Podust, S. Balan, B.C. Sim, M.P. Coyle, U. Ernst, R.T. Peters, V. Schellenberger, Extension of *in vivo* half-life of biologically active molecules by XTEN protein polymers. J Control Release **240**, 52–66 (2016)

A.J. Primeau, A. Rendon, D. Hedley, L. Lilge, I.F. Tannock, The distribution of the anticancer drug doxorubicin in relation to blood vessels in solid tumors. Clin. Cancer Res **11**, 8782–8788 (2005)

L. Prisner, N. Bohn, U. Hahn, A. Mews, Size dependent targeted delivery of gold nanoparticles modified with the IL-6R-specific aptamer AIR-3A to IL-6R-carrying cells. Nanoscale **9**, 14486–14498 (2017)

Z.M. Qian, Q. Wang, Expression of iron transport proteins and excessive iron accumulation in the brain in neurodegenerative disorders. Brain Res Brain Res Rev **27**, 257–267 (1998)

L.J. Raggatt, N.C. Partridge, Cellular and molecular mechanisms of bone remodeling. J Biol Chem **285**, 25103–25108 (2010)

T.B. Rahhal, C.A. Fromen, E.M. Wilson, M.P. Kai, T.W. Shen, J.C. Luft, J.M. DeSimone, Pulmonary delivery of butyrylcholinesterase as a model protein to the lung. Mol. Pharm **13**, 1626–1635 (2016)

A.S. Ricciardi, R. Bahal, J.S. Farrelly, E. Quijano, A.H. Bianchi, V.L. Luks, R. Putman, F. Lopez-Giraldez, S. Coskun, E. Song, Y. Liu, W.C. Hsieh, D.H. Ly, D.H. Stitelman, P.M. Glazer, W.M. Saltzman, In utero nanoparticle delivery for site-specific genome editing. Nat Commun **9**, 2481 (2018)

C. Riling, H. Kamadurai, S. Kumar, C.E. O'Leary, K.P. Wu, E.E. Manion, M. Ying, B.A. Schulman, P.M. Oliver, Itch WW domains inhibit its E3 ubiquitin ligase activity by blocking E2-E3 ligase trans-thiolation. J Biol Chem **290**, 23875–23887 (2015)

S.J. Roberts, H.Z. Ke, Anabolic strategies to augment bone fracture healing. Curr Osteoporos Rep **16**, 289–298 (2018)

E. Rytting, J. Nguyen, X. Wang, T. Kissel, Biodegradable polymeric nanocarriers for pulmonary drug delivery. Expert Opin Drug Deliv **5**, 629–639 (2008)

B. Saha, T.H. Evers, M.W. Prins, How antibody surface coverage on nanoparticles determines the activity and kinetics of antigen capturing for biosensing. Anal Chem **86**, 8158–8166 (2014)

O. Salata, Applications of nanoparticles in biology and medicine. J Nanotechnol **2**, 3 (2004)

E. Samuelsson, H. Shen, E. Blanco, M. Ferrari, J. Wolfram, Contribution of Kupffer cells to liposome accumulation in the liver. Colloids Surf B Biointerfaces **158**, 356–362 (2017)

M. P. Scavo, E. Gentile, J. Wolfram, J. Gu, M. Barone, M. Evangelopoulos, J. O. Martinez, X. Liu, C. Celia, E. Tasciotti, E. Vilar, H. Shen, Multistage vector delivery of sulindac and silymarin for prevention of colon cancer. Colloids Surf B Biointerfaces **136**, 694–703 (2015)

M.B. Schaffler, O.D. Kennedy, Osteocyte signaling in bone. Curr Osteoporos Rep **10**, 118–125 (2012)

S. Schottler, G. Becker, S. Winzen, T. Steinbach, K. Mohr, K. Landfester, V. Mailander, F.R. Wurm, Protein adsorption is required for stealth effect of poly(ethylene glycol)- and poly(phosphoester)-coated nanocarriers. Nat Nanotechnol **11**, 372–377 (2016)

J. Shen, R. Xu, J. Mai, H.C. Kim, X. Guo, G. Qin, Y. Yang, J. Wolfram, C. Mu, X. Xia, J. Gu, X. Liu, Z.W. Mao, M. Ferrari, H. Shen, High capacity nanoporous silicon carrier for systemic delivery of gene silencing therapeutics. ACS Nano **7**, 9867–9880 (2013)

J. Shen, H.C. Kim, C. Mu, E. Gentile, J. Mai, J. Wolfram, L. Ji, M. Ferrari, Z. Mao, H. Shen, Multifunctional gold nanorods for siRNA gene silencing and photothermal therapy. Adv Healthc Mater **3**, 1629–1637 (2014a)

J. Shen, H.C. Kim, H. Su, F. Wang, J. Wolfram, D. Kirui, J. Mai, C. Mu, Z. Mao, H. Shen, Cyclodextrin and polyethylenimine functionalized mesoporous silica nanoparticles for delivery of siRNA cancer therapeutics. Theranostics **4**, 487–497 (2014b)

J. Shen, X. Wu, Y. Lee, J. Wolfram, Z. Mao, M. Ferrari, H. Shen, Porous silicon microparticles for delivery of siRNA therapeutics. J Vis Exp **15**, 52075 (2015)

J. Shen, H.C. Kim, J. Wolfram, C. Mu, W. Zhang, H. Liu, Y. Xie, J. Mai, H. Zhang, Z. Li, M. Guevara, Z.W. Mao, H. Shen, A liposome encapsulated ruthenium polypyridine complex as a theranostic platform for triple-negative breast cancer. Nano Lett **17**, 2913–2920 (2017a)

J. Shen, H. Liu, C. Mu, J. Wolfram, W. Zhang, H.C. Kim, G. Zhu, Z. Hu, L.N. Ji, X. Liu, M. Ferrari, Z.W. Mao, H. Shen, Multi-step encapsulation of chemotherapy and gene silencing agents in functionalized mesoporous silica nanoparticles. Nanoscale **9**, 5329–5341 (2017b)

D. Shi, J. Zhang, Q. Zhou, J. Xin, J. Jiang, L. Jiang, T. Wu, J. Li, W. Ding, J. Li, S. Sun, J. Li, N. Zhou, L. Zhang, L. Jin, S. Hao, P. Chen, H. Cao, M. Li, L. Li, X. Chen, J. Li, Quantitative evaluation of human bone mesenchymal stem cells rescuing fulminant hepatic failure in pigs. Gut **66**, 955–964 (2017)

I. Singh, R. Swami, W. Khan, R. Sistla, Lymphatic system: a prospective area for advanced targeting of particulate drug carriers. Expert Opin Drug Deliv **11**, 211–229 (2014)

R.F. Spaide, K. Laud, H.F. Fine, J.M. Klancnik Jr., C.B. Meyerle, L.A. Yannuzzi, J. Sorenson, J. Slakter, Y.L. Fisher, M.J. Cooney, Intravitreal bevacizumab treatment of choroidal neovascularization secondary to age-related macular degeneration. Retina **26**, 383–390 (2006)

U. Sterzenbach, U. Putz, L.H. Low, J. Silke, S.S. Tan, J. Howitt, Engineered exosomes as vehicles for biologically active proteins. Mol Ther **25**, 1269–1278 (2017)

J. Tamargo, J.Y. Le Heuzey, P. Mabo, Narrow therapeutic index drugs: a clinical pharmacological consideration to flecainide. Eur J Clin Pharmacol **71**, 549–567 (2015)

F. Tewes, O.L. Gobbo, C. Ehrhardt, A.M. Healy, Amorphous Calcium Carbonate Based-Microparticles for Peptide Pulmonary Delivery. ACS Appl Mater Interfaces **8**, 1164–1175 (2016)

S.I. Thamake, S.L. Raut, Z. Gryczynski, A.P. Ranjan, J.K. Vishwanatha, Alendronate coated poly-lactic-co-glycolic acid (PLGA) nanoparticles for active targeting of metastatic breast cancer. Biomaterials **33**, 7164–7173 (2012)

C. Théry, K.W. Witwer, E. Aikawa, M.J. Alcaraz, J.D. Anderson, R. Andriantsitohaina, A. Antoniou, T. Arab, F. Archer, G.K. Atkin-Smith, D.C. Ayre, J.-M. Bach, D. Bachurski, H. Baharvand, L. Balaj, S. Baldacchino, N.N. Bauer, A.A. Baxter, M. Bebawy, C. Beckham, A. Bedina Zavec, A. Benmoussa, A.C. Berardi, P. Bergese, E. Bielska, C. Blenkiron, S. Bobis-Wozowicz, E. Boilard, W. Boireau, A. Bongiovanni, F.E. Borràs, S. Bosch, C.M. Boulanger, X. Breakefield, A.M. Breglio, M.Á. Brennan, D.R. Brigstock, A. Brisson, M.L.D. Broekman, J.F. Bromberg, P. Bryl-Górecka, S. Buch, A.H. Buck, D. Burger, S. Busatto, D. Buschmann, B. Bussolati, E.I. Buzás, J.B. Byrd, G. Camussi, D.R.F. Carter, S. Caruso, L.W. Chamley, Y.-T. Chang, C. Chen, S. Chen, L. Cheng, A.R. Chin, A. Clayton, S.P. Clerici, A. Cocks, E. Cocucci, R.J. Coffey, A. Cordeiro-da-Silva, Y. Couch, F.A.W. Coumans, B. Coyle, R. Crescitelli, M.F. Criado, C. D'Souza-Schorey, S. Das, A. Datta Chaudhuri, P. de Candia, E.F. De Santana, O. De Wever, H.A. del Portillo, T. Demaret, S. Deville, A. Devitt, B. Dhondt, D. Di Vizio, L.C. Dieterich, V. Dolo, A.P. Dominguez Rubio, M. Dominici, M.R. Dourado, T.A.P. Driedonks, F.V. Duarte, H.M. Duncan, R.M. Eichenberger, K. Ekström, S. El Andaloussi, C. Elie-Caille, U. Erdbrügger, J.M. Falcón-Pérez, F. Fatima, J.E. Fish, M. Flores-Bellver, A. Försönits, A. Frelet-Barrand, F. Fricke, G. Fuhrmann, S. Gabrielsson, A. Gámez-Valero, C. Gardiner, K. Gärtner, R. Gaudin, Y.S. Gho, B. Giebel, C. Gilbert, M. Gimona, I. Giusti, D.C.I. Goberdhan, A. Görgens, S.M. Gorski, D.W. Greening, J.C. Gross, A. Gualerzi, G.N. Gupta, D. Gustafson, A. Handberg, R.A. Haraszti, P. Harrison, H. Hegyesi, A. Hendrix, A.F. Hill, F.H. Hochberg, K.F. Hoffmann, B. Holder, H. Holthofer, B. Hosseinkhani, G. Hu, Y. Huang, V. Huber, S. Hunt, A.G.-E. Ibrahim, T. Ikezu, J.M. Inal, M. Isin, A. Ivanova, H.K. Jackson, S. Jacobsen, S.M. Jay, M. Jayachandran, G. Jenster, L. Jiang, S.M. Johnson, J.C. Jones, A. Jong, T. Jovanovic-Talisman, S. Jung, R. Kalluri, S.-I. Kano, S. Kaur, Y. Kawamura, E.T. Keller, D. Khamari, E. Khomyakova, A. Khvorova, P. Kierulf, K.P. Kim, T. Kislinger, M. Klingeborn, D.J. Klinke, M. Kornek, M.M. Kosanović, Á.F. Kovács, E.-M. Krämer-Albers, S. Krasemann, M. Krause, I.V. Kurochkin, G.D. Kusuma, S. Kuypers, S. Laitinen, S.M. Langevin, L.R. Languino, J. Lannigan,

C. Lässer, L.C. Laurent, G. Lavieu, E. Lázaro-Ibáñez, S. Le Lay, M.-S. Lee, Y.X.F. Lee, D.S. Lemos, M. Lenassi, A. Leszczynska, I.T.S. Li, K. Liao, S.F. Libregts, E. Ligeti, R. Lim, S.K. Lim, A. Linē, K. Linnemannstöns, A. Llorente, C.A. Lombard, M.J. Lorenowicz, Á.M. Lörincz, J. Lötvall, J. Lovett, M.C. Lowry, X. Loyer, Q. Lu, B. Lukomska, T.R. Lunavat, S.L.N. Maas, H. Malhi, A. Marcilla, J. Mariani, J. Mariscal, E.S. Martens-Uzunova, L. Martin-Jaular, M.C. Martinez, V.R. Martins, M. Mathieu, S. Mathivanan, M. Maugeri, L.K. McGinnis, M.J. McVey, D.G. Meckes, K.L. Meehan, I. Mertens, V.R. Minciacchi, A. Möller, M. Møller Jørgensen, A. Morales-Kastresana, J. Morhayim, F. Mullier, M. Muraca, L. Musante, V. Mussack, D.C. Muth, K.H. Myburgh, T. Najrana, M. Nawaz, I. Nazarenko, P. Nejsum, C. Neri, T. Neri, R. Nieuwland, L. Nimrichter, J.P. Nolan, E.N.M. Nolte-'t Hoen, N. Noren Hooten, L. O'Driscoll, T. O'Grady, A. O'Loghlen, T. Ochiya, M. Olivier, A. Ortiz, L.A. Ortiz, X. Osteikoetxea, O. Ostegaard, M. Ostrowski, J. Park, D.M. Pegtel, H. Peinado, F. Perut, M.W. Pfaffl, D.G. Phinney, B.C.H. Pieters, R.C. Pink, D.S. Pisetsky, E. Pogge von Strandmann, I. Polakovicova, I.K.H. Poon, B.H. Powell, I. Prada, L. Pulliam, P. Quesenberry, A. Radeghieri, R.L. Raffai, S. Raimondo, J. Rak, M.I. Ramirez, G. Raposo, M.S. Rayyan, N. Regev-Rudzki, F.L. Ricklefs, P.D. Robbins, D.D. Roberts, S.C. Rodrigues, E. Rohde, S. Rome, K.M.A. Rouschop, A. Rughetti, A.E. Russell, P. Saá, S. Sahoo, E. Salas-Huenuleo, C. Sánchez, J.A. Saugstad, M.J. Saul, R.M. Schiffelers, R. Schneider, T.H. Schøyen, A. Scott, E. Shahaj, S. Sharma, O. Shatnyeva, F. Shekari, G.V. Shelke, A.K. Shetty, K. Shiba, P.R.M. Siljander, A.M. Silva, A. Skowronek, O.L. Snyder, R.P. Soares, B.W. Sódar, C. Soekmadji, J. Sotillo, P.D. Stahl, W. Stoorvogel, S.L. Stott, E.F. Strasser, S. Swift, H. Tahara, M. Tewari, K. Timms, S. Tiwari, R. Tixeira, M. Tkach, W.S. Toh, R. Tomasini, A.C. Torrecilhas, J.P. Tosar, V. Toxavidis, L. Urbanelli, P. Vader, B.W.M. van Balkom, S.G. van der Grein, J. Van Deun, M.J.C. van Herwijnen, K. Van Keuren-Jensen, G. van Niel, M.E. van Royen, A.J. van Wijnen, M.H. Vasconcelos, I.J. Vechetti, T.D. Veit, L.J. Vella, É. Velot, F.J. Verweij, B. Vestad, J.L. Viñas, T. Visnovitz, K.V. Vukman, J. Wahlgren, D.C. Watson, M.H.M. Wauben, A. Weaver, J.P. Webber, V. Weber, A.M. Wehman, D.J. Weiss, J.A. Welsh, S. Wendt, A.M. Wheelock, Z. Wiener, L. Witte, J. Wolfram, A. Xagorari, P. Xander, J. Xu, X. Yan, M. Yáñez-Mó, H. Yin, Y. Yuana, V. Zappulli, J. Zarubova, V. Žėkas, J.-Y. Zhang, Z. Zhao, L. Zheng, A.R. Zheutlin, A.M. Zickler, P. Zimmermann, A.M. Zivkovic, D. Zocco, E.K. Zuba-Surma, Minimal information for studies of extracellular vesicles 2018 (MISEV2018): a position statement of the International Society for Extracellular Vesicles and update of the MISEV2014 guidelines. J Extracell Vesicles **7**, 1535750 (2018)

G. Tiwari, R. Tiwari, B. Sriwastawa, L. Bhati, S. Pandey, P. Pandey, S.K. Bannerjee, Drug delivery systems: An updated review. Int J Pharm Investig **2**, 2–11 (2012)

S. Torosean, B. Flynn, J. Axelsson, J. Gunn, K.S. Samkoe, T. Hasan, M.M. Doyley, B.W. Pogue, Nanoparticle uptake in tumors is mediated by the interplay of vascular and collagen density with interstitial pressure. Nanomedicine **9**, 151–158 (2013)

P. Vader, E.A. Mol, G. Pasterkamp, R.M. Schiffelers, Extracellular vesicles for drug delivery. Adv Drug Deliv Rev **106**, 148–156 (2016)

A. Venuta, J. Wolfram, H. Shen, M. Ferrari, Post-nano strategies for drug delivery: multistage porous silicon microvectors. J Mater Chem B **5**, 207–219 (2017)

K.K. Vidya Vijayan, K.P. Karthigeyan, S.P. Tripathi, L.E. Hanna, Pathophysiology of CD4+ T-Cell Depletion in HIV-1 and HIV-2 Infections. Front Immunol **8**, 580 (2017)

V. Wagner, A. Dullaart, A.K. Bock, A. Zweck, The emerging nanomedicine landscape. Nat Biotechnol **24**, 1211–1217 (2006)

J. Widmer, C. Thauvin, I. Mottas, V.N. Nguyen, F. Delie, E. Allemann, C. Bourquin, Polymer-based nanoparticles loaded with a TLR7 ligand to target the lymph node for immunostimulation. Int J Pharm **535**, 444–451 (2018)

S. Wilhelm, A.J. Tavares, Q. Dai, S. Ohta, J. Audet, H.F. Dvorak, W.C.W. Chan, Analysis of nanoparticle delivery to tumours. Nat Rev Mater **1**, 16014 (2016)

J. Wolfram, M. Ferrari, Clinical cancer nanomedicine. Nano Today, https://doi.org/10.1016/j.nantod.2019.02.005, (2019)

J. Wolfram, K. Suri, Y. Huang, R. Molinaro, C. Borsoi, B. Scott, K. Boom, D. Paolino, M. Fresta, J. Wang, M. Ferrari, C. Celia, H. Shen, Evaluation of anticancer activity of celastrol liposomes in prostate cancer cells. J Microencapsul **31**, 501–507 (2014a)

J. Wolfram, K. Suri, Y. Yang, J. Shen, C. Celia, M. Fresta, Y. Zhao, H. Shen, M. Ferrari, Shrinkage of pegylated and non-pegylated liposomes in serum. Colloids Surf B Biointerfaces **114C**, 294–300 (2014b)

J. Wolfram, Y. Yang, J. Shen, A. Moten, C. Chen, H. Shen, M. Ferrari, Y. Zhao, The nano-plasma interface: Implications of the protein corona. Colloids Surf B Biointerfaces **124**, 17–24 (2014c)

J. Wolfram, H. Shen, M. Ferrari, Multistage vector (MSV) therapeutics. J Control Release **219**, 406–415 (2015a)

J. Wolfram, M. Zhu, Y. Yang, J. Shen, E. Gentile, D. Paolino, M. Fresta, G. Nie, C. Chen, H. Shen, M. Ferrari, Y. Zhao, Safety of nanoparticles in medicine. Curr Drug Targets **16**, 1671–1681 (2015b)

J. Wolfram, B. Scott, K. Boom, J. Shen, C. Borsoi, K. Suri, R. Grande, M. Fresta, C. Celia, Y. Zhao, H. Shen, M. Ferrari, Hesperetin liposomes for cancer therapy. Curr Drug Deliv **13**, 711–719 (2016)

J. Wolfram, S. Nizzero, H. Liu, F. Li, G. Zhang, Z. Li, H. Shen, E. Blanco, M. Ferrari, A chloroquine-induced macrophage-preconditioning strategy for improved nanodelivery. Sci Rep **7**, 13738 (2017)

X. Wu, Z. Hu, S. Nizzero, G. Zhang, M.R. Ramirez, C. Shi, J. Zhou, M. Ferrari, H. Shen, Bone-targeting nanoparticle to co-deliver decitabine and arsenic trioxide for effective therapy of myelodysplastic syndrome with low systemic toxicity. J Control Release **268**, 92–101 (2017)

S. Yang, H. Gao, Nanoparticles for modulating tumor microenvironment to improve drug delivery and tumor therapy. Pharmacol Res **126**, 97–108 (2017)

Y. Yang, J. Wolfram, X. Fang, H. Shen, M. Ferrari, Polyarginine induces an antitumor immune response through binding to toll-like receptor 4. Small **10**, 1250–1254 (2014)

Publisher's note Springer Nature remains neutral with regard to jurisdictional claims in published maps and institutional affiliations.

Biomedical Microdevices (2018) 20: 59
https://doi.org/10.1007/s10544-018-0298-0

Micelle-like nanoparticles as siRNA and miRNA carriers for cancer therapy

Daniel F. Costa [1,2] · Vladimir P. Torchilin [1]

Published online: 12 July 2018

Abstract
Gene therapy has emerged as an alternative in the treatment of cancer, particularly in cases of resistance to chemo and radiotherapy. Different approaches to deliver genetic material to tumor tissues have been proposed, including the use of small non-coding RNAs due to their multiple mechanisms of action. However, such promise has shown limits in *in vivo* application related to RNA's biological instability and stimulation of immunity, urging the development of systems able to overcome those barriers. In this review, we discuss the use of RNA interference in cancer therapy with special attention to the role of siRNA and miRNA and to the challenges of their delivery *in vivo*. We introduce a promising class of drug delivery system known as micelle-like nanoparticles and explore their synthesis and advantages for gene therapy as well as the recent findings in *in vitro*, *in vivo* and clinical studies.

Keywords siRNA · miRNA · Cancer · Micelle-like nanoparticles · Polymers

1 Introduction

The genome is a complex machinery controlled by several pathways. Over the last decades, scientists have confirmed the influence of small RNA molecules, containing between 20 and 30 nucleotides in length, on post-transcriptional events (Carthew and Sontheimer 2009). This specific regulatory mechanism, so-called RNA interference (RNAi), has been shown to be essential in protecting germline and somatic lineages of eukaryotic cells against a variety of pathogens and diseases (Morrison 2018; Gandhi et al. 2014). The first record of RNAi predates 1984, when Nir and colaborators reported the induction of IFN-α1 gene expression and protein synthesis by human fibroblasts exposed to double-stranded RNA (Nir et al. 1984). Since then, the role of different types of nucleic acids in molecular biology has been investigated. In particular, small non-coding RNAs have emerged as one of the most promising categories for RNAi-based medicine. These RNAs comprise many subtypes of molecules, each one presenting a particular origin, structure, synthesis, and mechanism of action. However, two of them have been highlighted: short interfering RNA (siRNA) and microRNA (miRNA) (Carthew and Sontheimer 2009).

Studies in gene therapy application have expanded from simple to multifactorial diseases, such as cancer (Ginn et al. 2013; Wang et al. 2017). Current techniques, including surgical intervention, radio and chemotherapy have not always proven to be efficient strategies for successful treatment. One of the biggest challenges faced in clinics is the resistance of tumor cells to the most commonly used drugs (Tsouris et al. 2014). Among the possibilities for drug resistance, inhibition of efflux pumps and synthesis of antiapoptotic proteins are the most frequently affected by expression of a variety of genes, such as PLK1, Bcl2, survivin, MDR-1, and P-gp (Nakamura et al. 2011; Salzano et al. 2014; Abbasi et al. 2010; Liu et al. 2009; Yadav et al. 2009; Gibori et al. 2018). Knocking these genes down by RNAi has been seen as a way to transiently increase the sensitivity of cells to the action of antitumor drugs, overcoming resistance (Misso et al. 2014; Garofalo et al. 2012).

Despite the promising theory of a synergistic effect between RNAi and chemotherapeutic drugs, preclinical trials have shown many challenges when it comes to delivery of free nucleic acids to the tumor tissue *in vivo*. Poor serum

✉ Vladimir P. Torchilin
v.torchilin@northeastern.edu

Daniel F. Costa
costa.dan@husky.neu.edu

[1] Center for Pharmaceutical Biotechnology and Nanomedicine, Northeastern University, Boston, MA 02115, USA

[2] CAPES Foundation, Ministry of Education of Brazil, Brasília, DF 70040-020, Brazil

stability, off-target effects, and inefficient cellular entry are among the drawbacks (Aaldering et al. 2015; Conde et al. 2015). Thus, the development of drug delivery systems with the ability to carry and protect such payloads is required to improve the efficiency of RNAi-based medicine. Liposomes, dendrimers, lipid nanoparticles, polymeric micelles, inorganic nanoparticles, and some other materials have been used to deliver small RNAs in a safer and efficient manner. They have reduced dose frequency, facilitated targeting, improved biodistribution, and allowed in-time imaging (Navarro et al. 2015; Sarett et al. 2015). This review focuses specifically in one of these alternatives: the micelle-like nanoparticles, a new delivery system which combines the benefits from polymeric and hydrophobic moieties to create cationic amphiphilic conjugates able to form micellar structures by self-assembly in aqueous surroundings (Navarro et al. 2015). Here we discuss their production and characteristics as well as applications and recent findings in RNAi-based therapy for the treatment of cancer.

2 RNAi-based cancer therapy

GLOBOCAN, the latest worldwide project on cancer incidence, mortality, and prevalence conducted by the International Agency for Research on Cancer (IARC), estimated over 14 million new cases of cancer in 2012 (GLOBOCAN 2012). These are new and remissive cases of what has been considered one of the biggest public health challenges in the modern world. Although mortality has been decreasing in the past years, doctors still face reduction in efficacy of the treatments mainly caused by tumor type-dependent resistance and phenotypic variability between patients (Conde et al. 2015).

Several mechanisms of drug resistance have been reported, including pump-type resistance and apoptosis inhibition, the currently preferred targets for RNAi-based therapy (Tsouris et al. 2014). In the first case, cancer cells overexpress ATP-binding cassete (ABC) transporter proteins, such as P-glycoproteins (P-gp). As a result, increased amounts of drug are pumped out of the cells, reducing its concentration at the intracellular site of action. Overexpression of P-gp can be an intrinsic property of tumor tissues or induced by exposure to chemotherapy (Navarro et al. 2015). In the second case, failure of the treatment is caused by overproduction of antiapoptotic proteins, such as Bcl-2, that are responsible for inhibiting one of the major pathways of tumor cell death (Tsouris et al. 2014).

RNAi has been extensively studied among the strategies to overcome drug resistance. This is a natural well-conserved mechanism of most eukaryotic cells, where RNA components recognize a target by Watson-Crick base pairing and regulate post-transcriptional pathways (Carthew and Sontheimer 2009; Gandhi et al. 2014). Recent research has also shown regulation in chromatin structure levels, chromosome segregation, and transcription of genes. RNAi occurs in a programmable fashion, when endogenous patterns of gene expression are activated by metabolic changes in the diseased tissue. Similar events can happen through exogenous pathways, when gene sequences from foreign organisms are internalized and promote changes in the body response (Carthew and Sontheimer 2009; Gandhi et al. 2014; Li and Rana 2014; Ferracin et al. 2010).

This review focuses on two specific agents of RNAi, siRNA and miRNA, whose action depends on the same families of proteins. Briefly, double-stranded precursors are processed by a Dicer protein in the cytoplasm to form small fragments from 20 to 30 nucleotides long (siRNA or miRNA). One strand of this short sequence is then loaded into the RNA-induced silencing complex (RISC) by binding to its catalytic component, the Argonaute protein. This strand acts as a guide to a target mRNA. A high degree of complementarity to the target induces degradation by enzymatic cleavage of the molecule. However, the major pathway described in humans is imperfect binding with consequent repression of mRNA translation (Misso et al. 2014; Kumar and Mahato 2015; Chitkara et al. 2015; Pillai et al. 2007).

Although the mechanism of modulating mRNA expression is basically the same, variability at the end of the double-strand sequence seems to be a key factor differentiating siRNA and miRNA. It has become clear that siRNAs are fragments excised from fully complementary double-stranded RNAs of foreign nucleic acid and endogenous genomic loci (Tomari and Zamore 2005). In contrast, miRNAs are fragments from stem-loop precursors with an incomplete double-strand intimately related to cancer establishment and progression (Aaldering et al. 2015; Nicoloso et al. 2009). Compared to siRNAs, whose action relies on the degradation of a mRNA after its binding to RISC, miRNAs can also regulate genes involved in DNA damage and repair (Aaldering et al. 2015). Also, their lower sequence complementarity to the target reduces the chances of developing resistance to treatment, which would require mutations in multiple genes (Chitkara et al. 2015). On the other hand, it increases the induction of off-target effects (Kumar and Mahato 2015).

Some families of miRNA have a key role in promoting tumor suppression. Others, called oncomiRs, favor the development of cancer by targeting tumor suppressor genes (Table 1). These opposing classes divide miRNA-based therapy into two strategies: miRNA inhibition and replacement (Aaldering et al. 2015). The first is achieved by using specific molecules, such as anti-miRNA oligonucleotides (AMOs), able to neutralize and degrade oncomiRs (Conde et al. 2015). miR-519a and miR-32 are examples that have been considered relevant triggers to tumor development leading to decrease in patient survival (Shao et al. 2015; Jin et al. 2015;

Table 1 Examples of miRNAs and their action in miRNA-based cancer therapy

miRNA	Action	Type of cancer	REF
miR-519a	OncomiR	Hepatocellular carcinoma	(Shao et al. 2015)
miR-17-5p		Triple negative breast carcinoma	(Jin et al. 2015)
miR-32		Prostate carcinoma	(Liao et al. 2015)
miR-1290		Laryngeal carcinoma	(Janiszewska et al. 2015)
miR-21		Squamous cell carcinoma	(Ge et al. 2016)
miR-34a	Tumor suppressor	Osteosarcoma	(Wu et al. 2013)
let-7b		Lung carcinoma	(Stahlhut and Slack 2015)
miR-145		Colorectal carcinoma	(Drebber et al. 2011)
miR-494		Ovarian carcinoma	(Li et al. 2015a)
miR-212		Gastric carcinoma	(Li et al. 2015b)

Liao et al. 2015; Janiszewska et al. 2015; Ge et al. 2016). On the other hand, downregulation of tumor suppressor miRNAs, such as miR-34a and let-7b, are closely associated with poor prognosis, drug resistance, invasion and metastasis (Wu et al. 2013; Stahlhut and Slack 2015; Drebber et al. 2011; Li et al. 2015a, b). In these cases, exogenous miRNAs (or miRNA mimics) can be delivered to the tumor site to restore normal levels and promote cell homeostasis. However, miRNA mimics are known to be 100 to 1000-fold less potent than their endogenous counterparts (Zhang et al. 2013).

Such diversity in mechanisms of action suggests the use of miRNAs as prognostic, diagnostic, and therapeutic tools to improve treatment of cancer (Chitkara et al. 2015). The utility of these tools will depend on the design of an ideal vector, capable of delivering the payload specifically to the target tissue in sufficient amounts to modulate gene expression (Ferracin et al. 2010; de Planell-Saguer and Rodicio 2011).

3 Challenges for siRNA/miRNA delivery

A therapeutic effect of RNAi is the result of an efficient delivery of small non-coding RNAs into the cytoplasm of a target cell (Navarro et al. 2015). However, before reaching the site of action, many biological barriers must be overcome, including the blood milieu and tumor microenvironment (Chitkara et al. 2015). Technological advancement in transporting the payloads should greatly improve efficacy as well as reduce costs of the treatment (Aaldering et al. 2015).

Despite the promising results with RNAi in cell culture and animal studies, the majority of clinical trials have not advanced to late stages (Chen et al. 2018). Some of the non-conclusive results regarding its efficacy in humans rely partially on intrinsic characteristics of nucleotide-based molecules, such as: (1) serum instability due to rapid degradation by endo- and exonucleases in the bloodstream; (2) inefficient cell entry inherent in the negatively charged nature of siRNA and miRNA molecules; (3) poor pharmacokinetic profile associated with a half-life of about 5 min and rapid renal clearance due to their low molecular mass (≈ 13 kDa); (4) off-target effects, related mainly to non-specific binding of miRNA; (5) stimulation of the innate immune system after induction of interferon responses; (6) inefficient binding due to mutation in the sequence of the target mRNA and (7) short duration of the silencing effect, which requires high and sustained concentrations of payload in the target tissue (Gandhi et al. 2014; Aaldering et al. 2015; Conde et al. 2015; Navarro et al. 2015; Kumar and Mahato 2015).

In order to minimize most of these drawbacks and retain the bioactivity of the molecules, chemical modifications in the backbone or in the sugar moiety as well as the use of new delivery systems have been taken into consideration (Kumar and Mahato 2015). In the first case, *in vivo* instability and off-target effects are minimized by conjugating non-coding RNAs to lipids, peptides and carbohydrates through covalent linkages (Salzano et al. 2016; Moschos et al. 2007; Zimmermann et al. 2017). Conjugation with natural components of the cell membrane promotes miRNA and siRNA interaction with membrane receptors and thus facilitates their uptake (Broderick and Zamore 2011). In the second strategy, the uptake is increased due to characteristics of the designed carrier. Ideally, a successful delivery system should (1) promote long circulation by escaping immune system recognition, (2) maximize accumulation of the nucleic acid inside the target cell, (3) allow endosomal escape of siRNA and miRNA, (4) be easy to produce on a large scale and (5) be safe, biocompatible and cost-effective (Conde et al. 2015).

Many biomaterials have been investigated as candidates for an optimal delivery of nucleic acids, mostly non-viral lipids and protein carriers (Conde et al. 2015; Chen et al. 2018). Among the options, cationic liposomes and polymeric nanoparticles are the two major choices for delivery of oligonucleotides (Zhang et al. 2013; Gao et al. 2011). However, both systems have shown limitations in delivering chemotherapeutic drugs and small non-coding RNAs to the same cell at the same time. If the two delivery events are independent, the number of cells receiving both payloads is extremely reduced and the efficacy of a combination treatment is poor (Tsouris et al. 2014). Such challenge led to the development of multicomponent conjugates that combine the advantages of lipids and polymers in a single molecule. Polymeric chains able to form nanoparticles mainly by electronic interactions have been hydrophobized with lipid moieties similar to cellular membrane components. The amphiphilic product of this combination can self-assemble in a micellar structure, giving rise to what has been called micelle-like nanoparticles (MNPs) (Navarro et al. 2015).

4 Micelle-like nanoparticles: a new approach for RNAi-based therapy

Since the emerging of RNAi-based therapies, viral vectors have shown the ability to promote sustained gene silencing (Navarro et al. 2015). However, preclinical trials continue to reinforce safety concerns regarding the insertion of viral genome into human chromosomes. In addition, the production of viral vectors is costly and demands complex scaling up processes (Chen et al. 2018).

In face of these challenges, two different lines of research have emerged: one exploring lipid carriers and another focused on polymeric systems. Biocompatibility and similarity to the plasma membrane composition are two characteristics that made lipid carriers front runners in this field. The cationic surface charge of these particles facilitates their uptake by the anionic membrane of eukaryotic cells, which considerably improves transfection efficiency (Chitkara et al. 2015). Similar interaction is reported for the use of cationic polymers as main components of delivery systems. In comparison to lipids, they show particular advantages such as high stability, variety in architecture and molecular weights (MW), and multiple functional groups available for conjugation (Gandhi et al. 2014; Navarro et al. 2015).

Although both systems succeed in protecting the nucleic acid from degradation and metabolism to preserve bioactivity, the balance between efficacy and safety still needs to be considered carefully (Croy and Kwon 2006). Structure-function screenings have shown toxicity of cationic transfection reagents with high MW and charge, despite their higher transfection efficiency compared to low MW components (Grayson et al. 2006; Gebhart and Kabanov 2001). Additionally, they tend to form aggregates and lose activity in the presence of serum or salt-containing environments. A last concern about their behavior according to changes in pH has been addressed. These changes can significantly affect the release kinetics of the loaded siRNA or miRNA (Sarett et al. 2015).

To overcome some of the related drawbacks, the synthesis of environmentally-responsive copolymers has been suggested. The idea is to conjugate polymeric and lipid blocks with specific chemical groups to create amphiphilic molecules with multiple effects, such as: (1) release of the payload in reducible environments, (2) long circulation and escape from the reticuloendothelial system in the bloodstream, (3) incorporation of hydrophobic and hydrophilic drugs and (4) improvement of cell association and transfection (Navarro et al. 2015; Sarett et al. 2015; Salzano et al. 2016).

Amphiphilic multi-block conjugates can disperse as single units in water until they reach a limit called the critical micelle concentration (CMC), when the accumulation of free energy in the system favors the rearrange of the molecules towards a less energetic and more stable micellar structure (Prameela et al. 2015). In this way, hydrophobic moieties of the conjugates are grouped in the core of a micelle, while the hydrophilic moieties face the exterior and form a water-soluble shell in direct contact with water (Navarro et al. 2015). This shell is essential to protect miRNA and siRNA entrapped in the inner layers of the MNPs. The main challenge here is the complexity of a multi-block structure, which requires careful design and engineering to achieve reproducible formulations (Kumari et al. 2010). The characterization of the copolymers includes not only proving the proposed chemical structure, but also testing each functionalization in a biological model.

The incorporation of small non-coding oligonucleotides into MNPs can be done by electronic interaction or covalent binding (Fig. 1). In the first scenario, compact complexes are formed by interaction between positively charged amino groups of the carrier and negatively charged phosphate groups of the nucleic acids. The nanostructure decreases hydrophilicity and charge of the siRNA or miRNA, which facilitates its cellular uptake mainly by non-specific pinocytosis (Ruponen et al. 2001; Teo et al. 2013; Navarro et al. 2014). In the second case, components of the carrier are covalently bound to small non-coding RNAs mostly by disulfide bridge, which is highly sensitive to reductive intracellular environment (Salzano et al. 2015a).

Three synthetic polymers have been widely used for RNAi-based therapy: polyethylenimine (PEI), poly-L-lysine (PLL), and polyamidoamine (PAMAM). Among natural compounds, chitosan has been the first-choice in the majority of the studies (Navarro et al. 2015).

4.1 Polyethylenimine

The efficacy of PEI as a transfection agent has been proven in a broad range of cells, especially when polymeric branches over 25 kDa MW are the main component of the delivery systems (Chitkara et al. 2015). However, the cationic charge of high MW PEI is strong enough to also disrupt the plasma membrane, causing undesired toxicity in most clinical studies (Parmar et al. 2017). In contrast, low MW PEI (< 2 kDa) has shown minimal toxicity but up to 8-

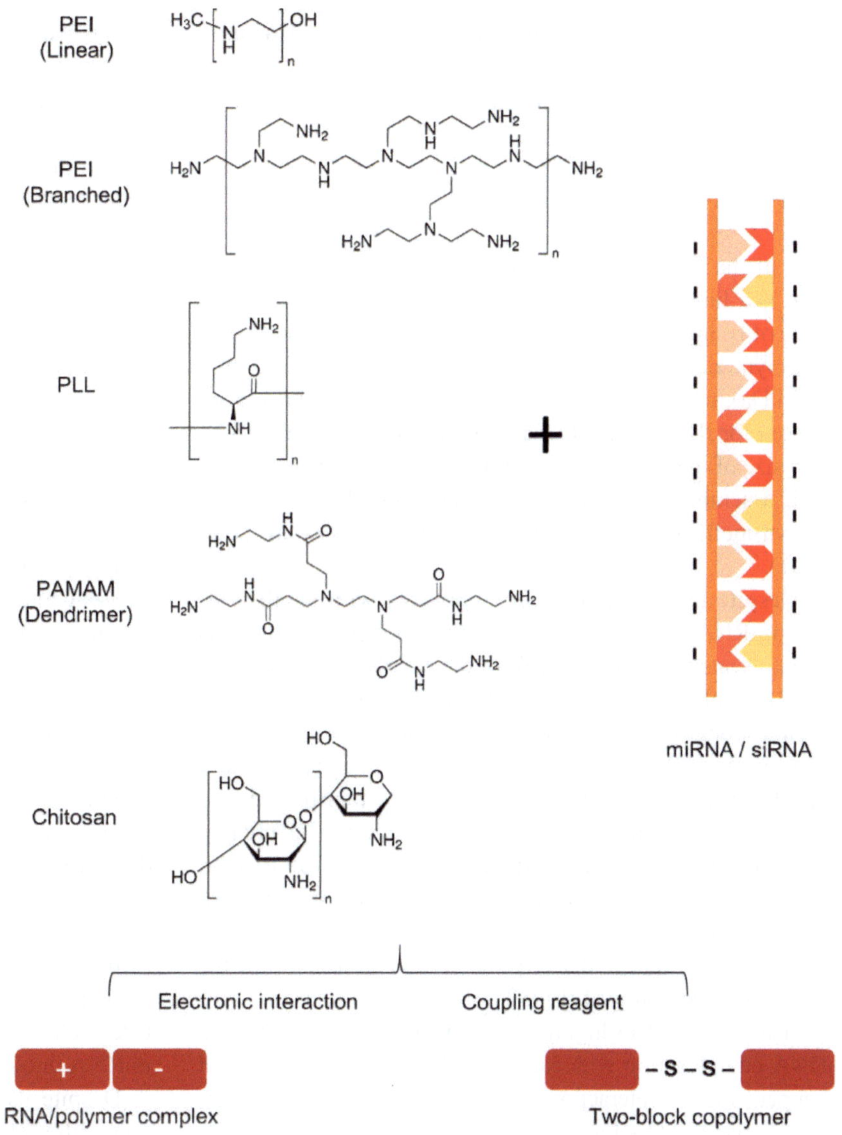

Fig. 1 Chemical structure of the main cationic polymers and schematic representation of their interaction with small non-coding oligonucleotides

fold decrease in gene transfection when compared to bigger branches (Grayson et al. 2006). To overcome such duality, low MW PEI has been attached to lipid moieties able to facilitate its uptake by eukaryotic cells. Copolymers of PEI and saturated fatty acids, such as 1,2-dipalmitoyl-sn-glycero-3-phosphoethanolamine (DPPE) and caprylic acid (CA), lowered the inherent cytotoxic effects of the delivery system and reduced the amount of polymer necessary to fully complex nucleic acid cargos (Navarro et al. 2014). The conjugation of PEI to unsaturated fatty acids, such as 1,2-dioleoyl-sn-glycero-3-phosphoethanolamine (DOPE) and linoleic acid (LA), could also improve gene silencing by facilitating endosomal escape of the carrier (Navarro et al. 2014; Aliabadi et al. 2011). Different degrees of substitution in the lipid chains causes variation in oligonucleotide binding affinity, surface charge, and cellular uptake (Aliabadi et al. 2011; Alshamsan et al. 2009).

The interaction of oligonucleotides and the majority of cationic polymers occurs through complexation with primary amine ends of the polymeric chains. PEI molecules, in particular, are also formed by secondary and tertiary amines in their chemical structure. These play a key role in the endosomal escape of delivery systems by causing the so-called proton sponge effect (Akinc et al. 2005). During endosomal saturation, ATPase pumps transfer protons from the cytosol into endosomes as an attempt to lower their internal pH. Unsaturated amino groups in the PEI molecule have the ability to sequester these protons, maintaining a continuous influx. The active pumping of protons is followed by passive entry of chloride ions and water. Consequently, there is an increase of osmotic pressure which leads to swelling, rupture of the endosomes and release of the delivery system into the cytoplasm (Nel et al. 2009; Salzano et al. 2015b). It is also suggested that protonated amino groups from the PEI can interact with the endosomal membrane, which would cause structural destabilization, formation of pores and consequent escape of the entrapped delivery system (Yue et al. 2011; Benjaminsen et al. 2013).

4.2 Poly-L-lysine

One of the first polymers used as carrier for gene delivery, PLL is a linear polypeptide composed of positively charged units of lysine (Toncheva et al. 1998). Although a wide range of polymeric chains is commercially available (0.50 to over 100 kDa), the molecules between 2.4 and 30 kDa have been chosen for RNAi-based therapy (Yogasundaram et al. 2012). The size of the chain is a determinant of the characteristics of PLL assemblies. Low MW ($\approx$ 5 kDa) peptide chains have shown inefficient transfection and high toxicity (Ward et al. 2001).

Despite its ability to form complexes with small non-coding RNAs, PLL shows high affinity for plasma proteins and presents rapid clearance from the circulation (Dash et al. 1999). Its pKa value of around 10.0 lowers the occurrence of proton sponge effect in comparison to other cationic polymers (Sonawane et al. 2003). Instead, particles containing PLL as a backbone are enzymatically digested once transported into lysosomes, which leads to a low transfection efficiency (Patil et al. 2011). Alternatively, conjugation of PLL to fusogenic peptides, fatty acids, histidine, and chloroquine improves endosomal escape. They destabilize the endosomal membrane and allow delivery of the nucleic acids into the cytoplasm (Pouton et al. 1998).

It is important to highlight that many findings related to DNA complexation with cationic polymers differ from what happens in their interaction with siRNA or miRNA. The molecular weight, topology, and charge density of oligonucleotides are some of the factors that directly affect the complexation rates (Zheng et al. 2012). Compared to DNA, flexible and shorter nucleic acid chains form more compact structures with PLL (Shukla et al. 2014).

4.3 Polyamidoamine

In contrast to linear polymers, PAMAM belongs to a class of spherical molecules with multiple branches, the so-called dendrimers. The layers forming a dendrimer (or generations, Gs) are synthesized one at a time, from the core to the periphery, which results in a structure that offers high functionality, defined molecular weight, and low polydispersity (Navarro et al. 2015; Yu et al. 2015). The repeating units of PAMAM provide flexibility to create spaces within the building blocks where small molecules, such as drugs and non-coding RNAs, can be entrapped. Also, the large number of peripheral amino groups allows covalent binding of numerous targets (Singh et al. 2008). Finally, their high stability, non-immunogenicity, and great capability to deliver nucleic acids into different types of cells make PAMAM a promising non-viral delivery system for RNAi-based therapies (Palmerston Mendes et al. 2017).

Increase in the number of generations have been associated with more potent transfection of the cargoes. However, dendrimers bigger than seven generations are known to cause significant toxicity (Boas and Heegaard 2004; Dufès et al. 2005). Alternatively, lipid-modification of non-toxic dendrimers has successfully improved transfection in comparison to their non-modified counterparts (Khopade et al. 2004; Movassaghian et al. 2011).

Since the first applications of PAMAM-based carriers as DNA transfection agents in the early 90s, many variations of this dendrimer have been released in the market (Yu et al. 2015; Haensler and Szoka Jr 1993). A common drawback is reported after the addition of functional groups to PAMAM molecules: it creates heterogeneity in the dendrimer scaffold. Consequently, controlling the number of ligands per particle becomes extremely difficult and could result in uncertain therapeutic effects (van Dongen et al. 2014).

4.4 Chitosan

Among the natural polymers used for drug delivery, chitosan has shown great potential for RNAi-based therapy. It is a derivative of chitin, composed of D-glucosamine and N-acetyl-D-glucosamine blocks, non-cytotoxic, non-immunogenic, biocompatible, and easy to adhere to mammalian cells (Modra et al. 2015). In an acidic environment, the amino groups of the D-glucosamine units (pKa $\approx$ 6.5) become positively charged and available to complex with siRNA or miRNA through electrostatic interactions (Layek et al. 2015).

Despite its advantages compared to synthetic polymers, unmodified chitosan has limitations in delivering nucleic acids at physiological pH. Experiments with pDNA revealed a very strong binding between the payload and the polymeric carrier, which resulted in inefficient unpacking in the cytoplasm. Furthermore, chitosan is associated with poor endosomal escape due to the lack of buffering amines (Wong et al. 2006; Lavertu et al. 2006). To overcome these drawbacks, the amino (C2 position) and hydroxyl groups of the molecule have been conjugated to hydrophobic moieties, such as lipid chains and bile acids (Hu et al. 2006; Layek and Singh 2012; Mandke and Singh 2012a; Chae et al. 2005). The modification weakens the polymer/nucleic acid interaction and facilitates unpacking inside the cell. Transfection efficiency is even greater when unsaturated lipid chains with one or two double bonds are bound to the polymeric chain (Mandke and Singh 2012b).

4.5 Functionalization of the polymeric blocks

Functionalization of nanoparticles by covering their surface with polyethylene glycol (PEG) was one of the first attempts used to optimize drug delivery systems. Apart from the advantages of each polymer already mentioned in this review, all of them present a common drawback when assembled in a nanostructure: fast clearance by the reticuloendothelial system (RES) (Brigger et al. 2002). Surface properties appear to be play a more important role in early interactions with components of the RES than size. Attachment of PEG in the surface of nanoparticles has efficiently reduced clearance from the bloodstream, helping to improve the delivery of small non-coding RNAs to sites of action (Yu et al. 2015; Jokerst et al. 2011; Dobrovolskaia and McNeil 2007).

Although the mechanisms of PEG effect are still unclear, recent findings drive attention to the presence of apolipoprotein E (ApoE) in the protein corona formed around nanoparticles after entry in the circulatory system. Lower levels of ApoE in the protein corona favor faster blood clearance. These levels are directly related to PEG density. Polymeric particles with less than 20 PEG chains per 100 nm^2 adsorb higher quantities of ApoE and are more likely to avoid early clearance (Bertrand et al. 2017).

Paradoxically, PEGylated MNPs can escape not only from RES uptake but also from internalization in the target tissue when the shielded cationic core is not available to interact with the negatively charged cell membrane (Choi et al. 1998; Oupicky et al. 2002). Introducing spacers between PEG and the cationic polymer is one of the strategies to overcome this challenge (Fig. 2) (Oumzil et al. 2011). Peptide chains sensitive to proteolytic activity are the linkers of choice when the targeted tissue shows upregulation of extracellular enzymes, such as matrix metalloproteinases (MMPs) and human leukocyte elastase (HLE). The increased enzymatic activity works as a trigger to de-shield the layer of PEG preferentially in the diseased tissue, optimizing the delivery of cargoes (Pak et al. 1999; Mansour et al. 2003).

Some cationic polymers, such as PEI, can be considered linkers themselves. In this case, the acidic environment of inflamed tissues and solid tumors destabilizes the structure of the particles by creating positive charges in the amino groups of the polymeric chain (Salzano et al. 2015b). A similar effect is induced by protonizable imidazole rings when poly-histidine is used as a linker between the core of MNPs and PEG (Midoux et al. 2009; Chang et al. 2010).

Another example of functionalization is the synthesis of bioconjugates sensitive to the reductive intracellular environment of tumor tissues. Elevated concentrations of glutathione (GSH) are commonly detected in the cytoplasm of cancer cells. GSH is capable of breaking disulfide bridges in lipid-modified oligonucleotides, which favors the release of free cargoes into the targeted cells (Salzano et al. 2016; Torchilin 2009). Breast, ovarian and lung tumors showed elevated GSH levels in comparison to healthy or peritumoral tissues, which helps to explain the stability of disulfide bonds in extracellular conditions (Gamcsik et al. 2012).

Functionalization of MNPs and the consequent increase in transfection efficiency of small non-coding RNAs can also be done by attaching cell-penetrating peptides (CPPs) to the surface of the particles (Kumari et al. 2010). Sometimes named "Trojan Horse" peptides, these amino acid sequences, such as the trans-activating transcriptional activator (TAT), improve internalization of the delivery systems through endocytic pathways or by direct translocation of components of the cell membrane (Layek et al. 2015).

Finally, a receptor-specific uptake can be achieved by coupling antibodies, growth factors, peptides, and other small molecules to the surface of the MNPs. The ligand must correspond to receptors overexpressed in the targeted cells, avoiding off-target effects and cytotoxicity to healthy tissues (Jing et al. 2013). Many studies have shown the benefits of attaching folic acid to MNPs. Folate receptor is overexpressed in most cancer cells, so that the ligand induces folate-assisted endocytosis and significantly increases transfection efficiency (Modra et al. 2015). Transferrin has also been used in RNAi-based therapy due to the upregulation of its receptor on the surface of many cancer cell-types (Grabowska et al. 2015; Daniels et al. 2012).

5 Recent achievements in cancer therapy

Many combinations of polymeric blocks and lipid chains have been used to create a broad variety of nanoparticles for the delivery of small non-coding RNAs (Wang et al. 2002; Incani et al. 2009; Abbasi et al. 2008). However, only few of them can be characterized as MNPs.

Low MW (1.8 kDa) and high MW (25 kDa) PEI were linked to DOPE to create amphiphilic conjugates. The resulting products were further functionalized with a peptide sequence sensitive to matrix metalloproteinases (MMPs) and PEG (2 kDa). The triblock conjugates were used to produce MNPs for the co-delivery of anti-GFP siRNA and paclitaxel to A549 cells (lung cancer). Enzymatic cleavage by human MMPs was efficiently proven *in vitro* as well as downregulation of GFP expression. After only a single treatment with low and high MW PEI conjugates, the levels of GFP decreased about 55% compared to untreated cells (Zhu et al. 2014). The same group functionalized low MW PEI-DOPE conjugate with azobenzene-modified PEG (2 kDa). Azobenzene is a linker sensitive to the lower pH of a tumor environment. Cancer cells treated with the formulation carrying anti-GFP siRNA internalized 3.2-fold more MNPs under hypoxia than under normoxia. *In vivo*, 32% of GFP downregulation was detected in A2780/GFP tumors (human ovarian carcinoma) (Perche et al. 2014). Aliabadi et al. (2011) analyzed downregulation of GAPDH and P-gp expression in MDA-MB-435 MDR cells (breast cancer) after treatment with MNPs of PEI attached to caprylic (C8), myristic (C14), palmitic (C16),

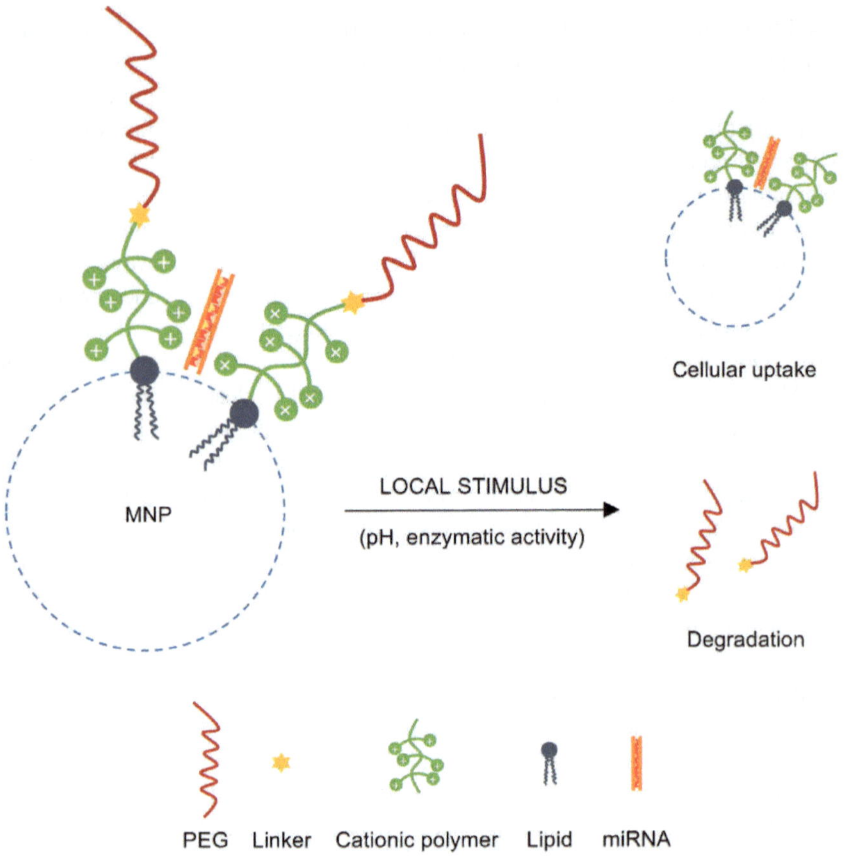

Fig. 2 Schematic representation of a functionalized MNP composed of three blocks: lipid, cationic polymer and PEG moieties. The spacer linking the cationic polymer and the PEG chain is cleaved by local stimuli in the diseased tissue. PEG is thus eliminated in the extracellular environment and the unprotected MNP becomes available to cellular uptake and miRNA transfection

stearic (C18), oleic (C18, one insaturation) and linoleic (C18, two insaturations) acids. The decrease in gene expression was dependent on the nature of the lipid and the level of substitution, but no clear tendency was found (Aliabadi et al. 2011).

PLL has also been lipid-modified to create MNPs. Guo et al. (2012) used cholic acid as modifier of PLL from 15 to 30 kDa. Later on, the amphiphilic chain was coupled to PEG (5 kDa) through a benzoic imine linker, known to be cleaved at the lower pH of extracellular tumor environments. The triblock conjugate formed MNPs able to complex with anti-VEGF siRNA (226–263 nm) and stable in serum for over 48 h. Studies in a mouse model of prostate carcinoma showed approximately 60% of reduction in the target mRNA levels and significant tumor suppression. Final tumor size of the treated group was approximately half of the size in the group control. Also, no marked toxicity or undesirable immune response were noticed (Guo et al. 2012). Zhang et al. (2016) created a triblock copolymer of N-succinyl chitosan, PLL, and palmitic acid (NSC–PLL–PA) to co-deliver Dox and anti-P-gp siRNA. The *in vitro* release of Dox was significantly higher in acidic pH compared to neutral environment. In subcutaneous model of HepG2/ADM tumors (multidrug-resistance hepatocellular carcinoma), the combination therapy with MNPs doubled the tumor growth inhibition reported after injections of free Dox. The antitumor effect was related to efficient downregulation of P-gp (Zhang et al. 2016).

Liu et al. (2015) conjugated the origin of PAMAM (G2) dendrimer to a hydrophobic alkyl chain (C18) and the terminals of the molecule to arginine, a cell-penetrating peptide. MNPs of about 50 nm in diameter protected anti-Hsp27 siRNA from enzymatic degradation *in vitro*. Also, the arginine-decorated particles caused downregulation of the target mRNA in PC-3 cells (prostate cancer) comparable to the commercial transfection vector (Liu et al. 2015). Corroborating with these findings, Márquez-Miranda et al. (2016) used coarse-grained molecular dynamics simulations to compare different combinations of PAMAM (G0, G1, G2) and lipid chains (18C, 15C, or 13C). The most effective siRNA transfection was given by the PAMAM (2G)-C18 alkyl chain conjugate (Márquez-Miranda et al. 2016).

Chitosan, the last example of polymer discussed in this review, was modified with Lipoïd® S75–3 (a soybean lecithin) to form MNPs of about 70 nm in diameter. The system was used to deliver anti-EGFR siRNA to U87MG cells (human glioblastoma). After 96 h of incubation with the treatment, EGFR expression was reduced by 52% (Messaoudi et al. 2014). The same group combined the delivery of anti-Galectin-1 and anti-EGFR siRNAs to decrease Temozolomide (TMZ) resistance in glioblastoma. Survival of the mice treated with combined siRNA therapy and 40 mg/kg of TMZ was significantly higher compared to groups treated with single RNA therapy and the chemotherapeutic drug. The effect was a result of efficient downregulation of EGFR and Galectin-1 analyzed by immunofluorescence in excised U87MG tumor tissues. Protein levels were, respectively, 24 and 28% lower than control group (Danhier et al. 2015). More recently, El-Sayed et al. (2018) successfully coupled three fatty acyl derivatives of CGKRK homing peptides to chitosan. The conjugates showed over 90% of binding affinity to siRNA targeting kinesin spindle protein and complete protection from early enzymatic degradation. The best cytotoxic effect (35%) on MDA-MB-231 cells (human breast cancer) was caused by the oleoyl derivative at polymer concentration of 10 μg/mL (El-Sayed et al. 2018).

The attachment of hydrophobic groups, such as palmitate, to PEG-PEI might cause false positive siRNA transfection results in luciferase assay. Rheiner et al. (2017) reported changes in the conformation of polymeric nanoassemblies when PEG-PEI was lipid-modified, which may have caused over-ubiquitination and degradation of luciferase in HT29 cells (human colon cancer) (Rheiner et al. 2017).

5.1 Clinical trials

According to the Clinical Trials Database, only eight studies in the United States investigated the use of non-viral drug delivery systems as carriers of siRNA or miRNA for cancer therapy (Table 2) (Clinical Trials Database 2018). These are phase I/IIa studies testing the safety, pharmacokinetics, and dose scaling of the treatments. Seven trials focus on the delivery of siRNA as payload. The only investigation of the therapeutic effect of miRNA was terminated due to occurrence of five serious immune-related adverse events. The majority of the studies on the action of miRNA in humans highlight the use of these oligonucleotides as markers for cancer diagnostic and prognostic, but not treatment.

None of the carriers is classified as MNPs. The most similar to this new class of delivery system is the cyclodextrin nanoparticle, which comprises the idea of linking different chemical blocks that assemble in a nanostructure (Zuckerman et al. 2014). In this case, a cationic cyclodextrin-based polymer forms the core of the particle, which surface is modified by attachment of PEG and human transferrin protein. The siRNA is entrapped in the core of the particle by electronic interaction with the cationic polymer. This was the first trial of RNAi-based cancer therapy in human patients.

6 Conclusion

The combination of chemo- and RNAi-based therapies has been considered a breakthrough in the treatment of cancer and a real alternative to overcome drug resistance. The understanding of the role of small non-coding RNAs in the disease led to the development of a wide range of prognostic, diagnostic, and therapeutic tools. Among them, non-viral carriers have emerged as a promising vector to deliver oligonucleotides specifically to the tumor tissue, which minimizes the off-target effects of gene therapy. In this review, we introduced a new class of non-viral drug delivery systems, the micelle-like nanoparticles. They represent the combination of polymeric and lipid systems in an attempt to have the advantages of both materials in a single carrier. The challenges of finding a perfect backbone for future drug carriers are followed by the idea of functionalization as a strategy to maximize the concentration of therapeutic agents in a targeted tissue. PEGylation, sensitivity to the action of enzymes and reductive microenvironments come along with the inclusion of CPPs and other ligands to the surface of delivery systems. These approaches have improved the therapeutic effect against a variety of cancer cells, including glioblastoma, prostate, lung, and breast carcinomas. However, the complexity of characterizing multi-component delivery systems and finding reproducible results still hold back the translation from preclinical studies to trials in humans.

Table 2 Registered clinical trials on RNAi-based cancer therapy using non-viral drug delivery systems (Clinical Trials Database 2018)

Payload	Carrier	Type of cancer	Study	Start date (status)
Anti-RMM2 siRNA	Cyclodextrin nanoparticle	Solid tumors	NCT00689065	2008 (terminated)
Anti-PKN3 siRNA	PEGylated lipoplex	Solid tumors	NCT00938574	2009 (completed)
Anti-PLK1 siRNA	Lipid nanoparticle	Solid tumors / Lymphoma	NCT01437007	2011 (completed)
Anti-PKN3 siRNA	PEGylated lipoplex	Pancreatic	NCT01808638	2013 (completed)
Anti-MYC siRNA	Lipid nanoparticle	Solid tumors / Myeloma / Lymphoma	NCT02110563	2014 (terminated)
Anti-MYC siRNA	Lipid nanoparticle	Hepatic	NCT02314052	2014 (terminated)
Anti-EphA2 siRNA	Liposome	Solid tumors	NCT01591356	2015 (recruiting)
miR-RX34	Liposome	Hepatic	NCT01829971	2013 (terminated)

Acknowledgements The authors thank Dr. William Hartner for helpful comments.

References

L.J. Aaldering, H. Tayeb, S. Krishnan, et al., RNA Biol. **12**, 412–425 (2015)

M. Abbasi, H. Uludag, V. Incani, et al., Biomacromolecules **9**, 1618–1630 (2008)

M. Abbasi, A. Lavasanifar, L.G. Berthiaume, et al., Cancer **116**, 5544–5554 (2010)

A. Akinc, M. Thomas, A.M. Klibanov, et al., J. Gene Med. **7**, 657–663 (2005)

H.M. Aliabadi, B. Landry, R.K. Bahadur, et al., Macromol. Biosci. **11**, 662–672 (2011)

A. Alshamsan, A. Haddadi, V. Incani, et al., Mol. Pharm. **6**, 121–133 (2009)

R.V. Benjaminsen, M.A. Mattebjerg, J.R. Henriksen, et al., Mol. Ther. **21**, 149–157 (2013)

N. Bertrand, P. Grenier, M. Mahmoudi, et al., Nat. Commun. **8**, 777 (2017)

U. Boas, P.M. Heegaard, Chem. Soc. Rev. **33**, 43–63 (2004)

I. Brigger, C. Dubernet, P. Couvreur, Adv. Drug Deliv. Rev. **54**, 631–651 (2002)

J.A. Broderick, P.D. Zamore, Gene Ther. **18**, 1104–1110 (2011)

R.W. Carthew, E.J. Sontheimer, Cell **136**, 642–655 (2009)

S.Y. Chae, S. Son, M. Lee, et al., J. Control. Release **109**, 330–344 (2005)

K.L. Chang, Y. Higuchi, S. Kawakami, et al., Bioconjug. Chem. **21**, 1087–1095 (2010)

X. Chen, L.S. Mangala, C. Rodriguez-Aguayo, et al., Cancer Metastasis Rev. **37**, 107–124 (2018)

D. Chitkara, A. Mittal, R.I. Mahato, Adv. Drug Deliv. Rev. **81**, 34–52 (2015)

Y.H. Choi, F. Liu, J.S. Kim, et al., J. Control. Release **54**, 39–48 (1998)

Clinical Trials Database (2018). https://clinicaltrials.gov. Accessed 5 Mar 2018

J. Conde, E.R. Edelman, N. Artzi, Adv. Drug Deliv. Rev. **81**, 169–183 (2015)

S.R. Croy, G.S. Kwon, Curr. Pharm. Des. **12**, 4669–4684 (2006)

F. Danhier, K. Messaoudi, L. Lemaire, et al., Int. J. Pharm. **481**, 154–161 (2015)

T.R. Daniels, E. Bernabeu, J.A. Rodríguez, et al., Biochim. Biophys. Acta **1820**, 291–317 (2012)

P.R. Dash, M.L. Read, L.B. Barrett, et al., Gene Ther. **6**, 643–650 (1999)

M. de Planell-Saguer, M.C. Rodicio, Anal. Chim. Acta **699**, 134–152 (2011)

M.A. Dobrovolskaia, S.E. McNeil, Nat. Nanotechnol. **2**, 469–478 (2007)

U. Drebber, M. Lay, I. Wedemeyer, et al., Int. J. Oncol. **39**, 409–415 (2011)

C. Dufès, I.F. Uchegbu, A.G. Schätzlein, Adv. Drug Deliv. Rev. **57**, 2177–2202 (2005)

N.S. El-Sayed, M. Sharma, H.M. Aliabadi, et al., Int. J. Biol. Macromol. **112**, 694–702 (2018)

M. Ferracin, A. Veronese, M. Negrini, Expert. Rev. Mol. Diagn. **10**, 297–308 (2010)

M.P. Gamcsik, M.S. Kasibhatla, S.D. Teeter, et al., Biomarkers **17**, 671–691 (2012)

N.S. Gandhi, R.K. Tekade, M.B. Chougule, J. Control. Release **194**, 238–256 (2014)

Y. Gao, X.L. Liu, X.R. Li, Int. J. Nanomedicine **6**, 1017–1025 (2011)

M. Garofalo, C. Quintavalle, G. Romano, et al., Curr. Mol. Med. **12**, 27–33 (2012)

Y. Ge, L. Zhang, M. Nikolova, et al., Nat. Cell Biol. **18**, 111–121 (2016)

C.L. Gebhart, A.V. Kabanov, J. Control. Release **73**, 401–416 (2001)

H. Gibori, S. Eliyahu, A. Krivitsky, et al., Nat. Commun. **9**, 16 (2018)

S.L. Ginn, I.E. Alexander, M.L. Edelstein, et al., J. Gene. Med. **15**, 65–77 (2013)

GLOBOCAN (2012). http://gco.iarc.fr. Accessed 13 Mar 2018

A.M. Grabowska, R. Kircheis, R. Kumari, et al., Biomater. Sci. **3**, 1439–1448 (2015)

A.C. Grayson, A.M. Doody, D. Putnam, Pharm. Res. **23**, 1868–1876 (2006)

J. Guo, W.P. Cheng, J. Gu, et al., Eur. J. Pharm. Sci. **45**, 521–532 (2012)

J. Haensler, F.C. Szoka Jr., Bioconjug. Chem. **4**, 372–379 (1993)

F.Q. Hu, M.D. Zhao, H. Yuan, et al., Int. J. Pharm. **315**, 158–166 (2006)

V. Incani, X. Lin, A. Lavasanifar, et al., ACS Appl. Mater. Interfaces **1**, 841–848 (2009)

J. Janiszewska, M. Szaumkessel, M. Kostrzewska-Poczekaj, et al., PLoS One **10**, e0144924 (2015)

Y.Y. Jin, J. Andrade, E. Wickstrom, PLoS One **10**, e0142574 (2015)

F. Jing, J. Li, D. Liu, et al., Pharm. Biol. **51**, 643–649 (2013)

J.V. Jokerst, T. Lobovkina, R.N. Zare, et al., Nanomedicine (London) **6**, 715–728 (2011)

A.J. Khopade, D.B. Shenoy, S.A. Khopade, et al., Langmuir **20**, 7368–7373 (2004)

V. Kumar, R.I. Mahato, Pharm. Res. **32**, 341–361 (2015)

A. Kumari, S.K. Yadav, S.C. Yadav, Colloids Surf. B: Biointerfaces **75**, 1–18 (2010)

M. Lavertu, S. Méthot, N. Tran-Khanh, et al., Biomaterials **27**, 4815–4824 (2006)

B. Layek, J. Singh, Carbohydr. Polym. **89**, 403–410 (2012)

B. Layek, L. Lipp, J. Singh, Int. J. Mol. Sci. **16**, 28912–28930 (2015)

Z. Li, T.M. Rana, Nat. Rev. Drug Discov. **13**, 622–638 (2014)

N. Li, X. Zhao, L. Wang, et al., Tumour Biol. **37**, 7767–7776 (2015a)

D. Li, Z. Li, J. Xiong, et al., Am. J. Cancer Res. **5**, 2980–2997 (2015b)

H. Liao, Y. Xiao, Y. Hu, et al., Oncol. Lett. **10**, 2055–2062 (2015)

C. Liu, G. Zhao, J. Liu, et al., J. Control. Release **140**, 277–283 (2009)

X. Liu, C. Liu, J. Zhou, et al., Nano **7**, 3867–3875 (2015)

R. Mandke, J. Singh, Pharm. Res. **29**, 883–897 (2012a)

R. Mandke, J. Singh, J. Pharm. Sci. **101**, 268–282 (2012b)

A.M. Mansour, J. Drevs, N. Esser, et al., Cancer Res. **63**, 4062–4066 (2003)

V. Márquez-Miranda, I. Araya-Durán, M.B. Camarada, et al., Sci. Rep. **6**, 29436 (2016)

K. Messaoudi, P. Saulnier, K. Boesen, et al., Int. J. Nanomedicine **9**, 1479–1490 (2014)

P. Midoux, C. Pichon, J.J. Yaouanc, et al., Br. J. Pharmacol. **157**, 166–178 (2009)

G. Misso, M.T. Di Martino, G. De Rosa, et al., Mol. Ther. Nucleic Acids. **e194**, 3 (2014)

K. Modra, S. Dai, H. Zhang, et al., Eng. Life Sci. **15**, 489–498 (2015)

C. Morrison, Nat. Rev. Drug Discov. **17**, 156–157 (2018)

S.A. Moschos, S.W. Jones, M.M. Perry, et al., Bioconjug. Chem. **18**, 1450–1459 (2007)

S. Movassaghian, H.R. Moghimi, F.H. Shirazi, J. Drug Target. **19**, 925–932 (2011)

K. Nakamura, A.S. Abu Lila, M. Matsunaga, et al., Mol. Ther. **19**, 2040–2047 (2011)

G. Navarro, S. Essex, R.R. Sawant, et al., Nanomedicine **10**, 411–419 (2014)

G. Navarro, J. Pan, V.P. Torchilin, Mol. Pharm. **12**, 301–313 (2015)

A.E. Nel, L. Mädler, D. Velegol, et al., Nat. Mater. **8**, 543–557 (2009)

M.S. Nicoloso, R. Spizzo, M. Shimizu, et al., Nat. Rev. Cancer **9**, 293–302 (2009)

U. Nir, B. Cohen, L. Chen, et al., Nucleic Acids Res. **12**, 6979–6993 (1984)

K. Oumzil, S. Khiati, M.W. Grinstaff, et al., J. Control. Release **151**, 123–130 (2011)

D. Oupicky, M. Ogris, K.A. Howard, et al., Mol. Ther. **5**, 463–472 (2002)

C.C. Pak, R.K. Erukulla, P.L. Ahl, et al., Biochim. Biophys. Acta **1419**, 111–126 (1999)
L. Palmerston Mendes, J. Pan, V.P. Torchilin, Molecules. **22**, 1401 (2017)
M.B. Parmar, D.N. Meenakshi Sundaram, R.B. K C, et al., Acta Biomater. **66**, 294–309 (2017)
M.L. Patil, M. Zhang, T. Minko, ACS Nano **5**, 1877–1887 (2011)
F. Perche, S. Biswas, T. Wang, et al., Angew. Chem. Int. Ed. Eng. **53**, 3362–3366 (2014)
R.S. Pillai, S.N. Bhattacharyya, W. Filipowicz, Trends Cell Biol. **17**, 118–126 (2007)
C.W. Pouton, P. Lucas, B.J. Thomas, et al., J. Control. Release **53**, 289–299 (1998)
G.K. Prameela, B.V. Phani Kumar, A. Pan, et al., Phys. Chem. Chem. Phys. **17**, 30560–30569 (2015)
S. Rheiner, D. Reichel, P. Rychahou, et al., Int. J. Pharm. **528**, 536–546 (2017)
M. Ruponen, S. Rönkkö, P. Honkakoski, et al., J. Biol. Chem. **276**, 33875–33880 (2001)
G. Salzano, R. Riehle, G. Navarro, et al., Cancer Lett. **343**, 224–231 (2014)
G. Salzano, G. Navarro, M.S. Trivedi, et al., Mol. Cancer Ther. **14**, 1075–1084 (2015a)
G. Salzano, D.F. Costa, V.P. Torchilin, Curr. Pharm. Des. **21**, 4566–4573 (2015b)
G. Salzano, D.F. Costa, C. Sarisozen, et al., Small **12**, 4837–4848 (2016)
S.M. Sarett, C.E. Nelson, C.L. Duvall, J. Control. Release **218**, 94–113 (2015)
J. Shao, J. Cao, Y. Liu, et al., FEBS Open Bio. **5**, 893–899 (2015)
R.S. Shukla, B. Qin, K. Cheng, Mol. Pharm. **11**, 3395–3408 (2014)
I. Singh, A.K. Rehni, R. Kalra, et al., Pharmazie **63**, 491–496 (2008)
N.D. Sonawane, F.C. Szoka Jr., A.S. Verkman, J. Biol. Chem. **278**, 44826–44831 (2003)
C. Stahlhut, F.J. Slack, Cell Cycle **14**, 2171–2180 (2015)
P.Y. Teo, C. Yang, J.L. Hedrick, et al., Biomaterials **34**, 7971–7979 (2013)
Y. Tomari, P.D. Zamore, Genes Dev. **19**, 517–529 (2005)
V. Toncheva, M.A. Wolfert, P.R. Dash, et al., Biochim. Biophys. Acta **1380**, 354–368 (1998)
V. Torchilin, Eur. J. Pharm. Biopharm. **71**, 431–444 (2009)
V. Tsouris, M.K. Joo, S.H. Kim, et al., Biotechnol. Adv. **32**, 1037–1050 (2014)
M.A. van Dongen, C.A. Dougherty, M.M. Banaszak Holl, Biomacromolecules **15**, 3215–3234 (2014)
D.A. Wang, A.S. Narang, M. Kotb, et al., Biomacromolecules **3**, 1197–1207 (2002)
T. Wang, G. Wang, X. Zhang, et al., Sci. Rep. **7**, 14736 (2017)
C.M. Ward, M.L. Read, L.W. Seymour, Blood **97**, 2221–2229 (2001)
K. Wong, G. Sun, X. Zhang, et al., Bioconjug. Chem. **17**, 152–158 (2006)
X. Wu, D. Zhong, Q. Gao, et al., Int. J. Med. Sci. **10**, 676–682 (2013)
S. Yadav, L.E. van Vlerken, S.R. Little, et al., Cancer Chemother. Pharmacol. **63**, 711–722 (2009)
H. Yogasundaram, M.S. Bahniuk, H.D. Singh, et al., Int. J. Biomater. **2012**, 584060 (2012)
M. Yu, X. Jie, L. Xu, et al., Biomacromolecules **16**, 2588–2598 (2015)
Y. Yue, F. Jin, R. Deng, et al., J. Control. Release **152**, 143–151 (2011)
Y. Zhang, Z. Wang, R.A. Gemeinhart, J. Control. Release **172**, 962–974 (2013)
C.G. Zhang, W.J. Zhu, Y. Liu, et al., Sci. Rep. **6**, 23859 (2016)
C. Zheng, L. Niu, J. Yan, et al., Phys. Chem. Chem. Phys. **14**, 7352–7359 (2012)
L. Zhu, F. Perche, T. Wang, et al., Biomaterials **35**, 4213–4222 (2014)
T.S. Zimmermann, V. Karsten, A. Chan, et al., Mol. Ther. **25**, 71–78 (2017)
J.E. Zuckerman, I. Gritli, A. Tolcher, et al., Proc. Natl. Acad. Sci. U. S. A. **111**, 11449–11454 (2014)

Biomedical Microdevices (2019) 21: 36
https://doi.org/10.1007/s10544-019-0385-x

Liquid crystal delivery of ciprofloxacin to treat infections of the female reproductive tract

Simone Pisano [1] · Matteo Giustiniani [2,3] · Lewis Francis [1] · Deyarina Gonzalez [1] · Lavinia Margarit [1,4] · I. Martin Sheldon [1] · Donatella Paolino [2,3] · Massimo Fresta [2,3] · R. Steven Conlan [1] · Gareth D. Healey [1]

Published online: 29 March 2019

Author's Personal Note

Two roads diverged in a yellow wood,
And sorry I could not travel both
And be one traveler, long I stood
And looked down one as far as I could
To where it bent in the undergrowth;
Then took the other, as just as fair,
And having perhaps the better claim,
Because it was grassy and wanted wear;
Though as for that the passing there
Had worn them really about the same,
And both that morning equally lay
In leaves no step had trodden black.
Oh, I kept the first for another day!
Yet knowing how way leads on to way,
I doubted if I should ever come back.
I shall be telling this with a sigh
Somewhere ages and ages hence:
Two roads diverged in a wood, and I—
I took the one less traveled by,
And that has made all the difference.
(Robert Frost, 1874-1963)

Mauro, the path you have taken, and the differences you have made, have been inspirational to us all.
Steve Conlan, Swansea, Wales, 2019.

Abstract

Infections of the female reproductive tract are a major cause of morbidity and mortality in humans, requiring significant investment to sustain treatment and representing a major challenge to health. The increasing prevalence of bacterial resistance, and an almost complete absence of new antibiotic therapies for the past five decades, mean there is a desperate need for novel approaches to the treatment of bacterial infections. Within the present study, we demonstrate the effective *ex vivo* treatment of bacterial infection of the female reproductive tract using a controlled-release, liquid crystal-based platform. Liquid crystal encapsulation of ciprofloxacin significantly enhanced its bactericidal efficacy and reduced cell toxicity. Liquid crystal structures are low-cost, simple to manufacture and provide a sustained-release profile of encapsulated ciprofloxacin. Treatment of *Escherichia coli* infected reproductive tract epithelial cells and whole organ cultures with liquid crystal encapsulated ciprofloxacin proved to be an effective strategy for reducing bacterial load and reproductive tract inflammatory responses to infection. These data suggest that such an approach could provide an efficacious treatment modality for enhancing the effectiveness of current antibiotic therapies.

Keywords Liquid crystal · Drug delivery · Female reproductive tract · Bacterial infection · Female reproductive tract infection · Ciprofloxacin

1 Introduction

Sexually transmitted infections and their associated diseases represent one of the main challenges in worldwide reproductive health (Horne et al. 2008). Despite sustained activity aimed at preventing such infections, they remain a significant socio-economic burden with the most prevalent causative agents being Herpes simplex virus type 2, group B streptococcus, *Treponema pallidum* (syphilis), bacterial vaginosis, Hepatitis B virus (hepatitis), *Neisseria gonorrhoeae* (gonorrhea), *Chlamydia trachomatis*, and human immunodeficiency viruses (HIVs) (Wira et al. 2011). According to World Health Organization (WHO) estimates, in 2008 there were more than 498 million new cases of over 30 different sexually transmitted infections. These included infection with *Trichomonas vaginalis* (276 million new cases), *Chlamydia trachomatis* (106 million new cases), *Treponema pallidum* (10 million new cases), HIV (2.7 million new cases) and *Neisseria gonorrhoeae* (106 million new cases) (WHO 2012).

The female reproductive tract has an abundant natural bacterial flora, dominated in most women by *Lactobacillus* species, which represent 90–95% of the total bacteria in the reproductive tract (Delaney and Onderdonk 2001; Larsen and Monif 2001; Srinivasan et al. 2010). *Lactobacillus*-dominance is not universal however, with several studies showing the

Electronic supplementary material The online version of this article (https://doi.org/10.1007/s10544-019-0385-x) contains supplementary material, which is available to authorized users.

✉ Gareth D. Healey
G.D.Healey@Swansea.ac.uk

[1] Institute of Life Science, Swansea University Medical School, Swansea University, Swansea SA2 8PP, UK

[2] Department of Clinical and Experimental Medicine, University of Catanzaro "Magna Graecia", Viale "S. Venuta", 88100 Catanzaro, Italy

[3] Inter-Regional Research Center for Food Safety & Health, University of Catanzaro "Magna Græcia", Viale "S. Venuta", 88100 Catanzaro, Italy

[4] Obstetrics & Gynecology Department Princess of Wales Hospital, Abertawe Bro Morannwg University Health Board, Coity Road, Bridgend CF31 1RQ, UK

presence of a diverse, heterogeneous microbiota that extends throughout the reproductive tract in women without reproductive tract disease (Chen et al. 2017; Lambert et al. 2013; Ravel et al. 2013; Srinivasan et al. 2010). *Lactobacillus* species also do not dominate throughout a woman's lifetime, with anaerobes and *Escherichia coli* most abundant during childhood and after menopause, when lower estrogen levels reduce the abundance of glycogen in the vaginal epithelium, which is essential for *Lactobacillus* growth and colonization (Brotman et al. 2014; Galhardo et al. 2006; Hillier and Lau 1997). The presence of *Lactobacillus* species is currently accepted as a biomarker of healthy status (Amabebe and Anumba 2018; das Neves et al. 2014). Through the production of lactic acid, which lowers vaginal pH, together with the production of hydrogen peroxide, bacteriocins and other antimicrobial compounds, *Lactobacilli* help to maintain vaginal homeostasis by creating an environment unsuitable for neutrophilic bacteria and by competing for nutrients, which deters the growth of pathogenic bacteria (Barbés and Boris 1999; Eschenbach et al. 1989; Tachedjian et al. 2018). Indeed, modification of these physiological conditions to a pH above 4.5 increases the risk of other bacterial infections including sexually transmitted infections and those associated with bacterial vaginosis (Aldunate et al. 2015; Daniel Johannes Rönnqvist et al. 2006; Mastromarino et al. 2014; McLaughlin et al. 2018).

Early and effective treatment of reproductive tract infections is essential to limit transmission of disease and to reduce and/or prevent associated morbidities. The development of antibiotic resistance by pathogens to many antibiotics has increased concerns about the effective management of these diseases (Alirol et al. 2017; Bala et al. 2007; P. Health England 2013; Laibl et al. 2005; Martin et al. 2006; McCarthy 2007; MMWR 2006; Nys et al. 2004; Wang et al. 2005). For example, a positive correlation between ciprofloxacin resistance and gonorrhea incident rates has been reported in several countries leading to the drug being removed as the primary treatment option (Bala et al. 2007; Chesson et al. 2014; McCarthy 2007; MMWR 2006). Therefore, finding a way to optimally identify patients more likely to respond positively to an antibiotic regimen can represent a competitive advantage for disease management. With this in mind, the UCLA Health Group introduced a genotypic assay to predict the susceptibility of *N. gonorrhoea* to ciprofloxacin, thus repositioning this drug as a valuable tool if the correct patient cohort is identified (Allan-Blitz et al. 2016). The same approach is true for other bacterial infections. In addition, a drive to reduce the amount of administered drug, which favors a more localized treatment approach, would reduce negative side effects such as diarrhea, nausea and vomiting, and possibly delay the onset of drug resistance.

Localised treatments of reproductive tract infections are favoured over oral approaches because they achieve higher local drug concentrations and avoid affecting the gastrointestinal tract microbiome (Alexander et al. 2004; Vermani and Garg 2000). Several different delivery modalities are currently marketed including tablets, capsules, pessaries and semisolids (Henderson 1958; Das Neves and Bahia 2006; Pavelić et al. 2005; Slack and Nichols 1982; Yao et al. 2016), but most are associated with poor distribution and retention due to the self-cleaning action of the vagina (Hussain and Ahsan 2005). A drive towards the use of semisolids (creams or gels), due to the ease of application and the symptomatic relief provided by the additional moisture provided, is evident (das Neves 2014; Palmeira-de-Oliveira et al. 2015).

The need to improve drug delivery beyond the classical pharmaceutical forms has driven the development of novel delivery systems designed to improve patient compliance, enhance cellular targeting, extend drug release profiles and reduce harmful side-effects. Several of these new pharmaceutical compositions such as liposomes, polymeric systems or cyclodextrins, are already marketed. Their advantage is to offer improved therapeutic profiles by encapsulating a drug, thereby protecting it from degradation, whilst also providing the opportunity for the attachment of targeting molecules on the outside of the carrier (Chen et al. 2014).

Liquid crystals are an emerging class of drug delivery systems made of polar lipids able to spontaneously reorganize themselves into three-dimensional structures (namely liquid crystalline phases) when in contact with water (Shah et al. 2001). They have a malleable structure, determined by the local physico-chemical properties, meaning their structure can be controlled to allow both ease of administration and sustained drug release (Chaudhary et al. 2016). Liquid crystals exhibit both optical and electrical anisotropy, as well as flow properties and molecular mobility. Whilst in practice liquid crystals do not transition directly from a liquid to a solid state, under certain conditions they exhibit molecularly organised intermediate phases (mesophases) with liquid and solid state properties (Zabara and Mezzenga 2014). Liquid crystals that form mesophases in appropriate solvents, named lyotropic liquid crystals, are based on lipids that spontaneously self-assemble in an aqueous environment. They are characterised by nanostructured hydrophilic and hydrophobic domains separated by lipid bilayers, and form following exposure to a polar (aqueous) environment (Zabara and Mezzenga 2012).

In the present study we evaluate controlled antibiotic delivery in cell and tissue models using ciprofloxacin, a second-generation fluoroquinolone, with broad-spectrum antibiotic activity against both Gram-positive and Gram-negative bacteria (LeBel 1988; Wiseman and Balfour 1994) encapsulated in liquid crystal. By demonstrating how liquid crystals can be used as a drug delivery system in the treatment of bacterial infections of the female reproductive tract, and that liquid crystal formulations of ciprofloxacin significantly enhance bacterial killing compared to the free drug formulation, we pave the way for tailored, controlled-release treatments of bacterial infections within the female reproductive tract.

2 Methods and materials

2.1 Preparation of lamellar phase ciprofloxacin emulsion (95% Monomuls® + 5% - 0.05% TFA)

A solution of 0.05% trifluoroacetic acid (TFA, Sigma-Aldrich, Dorset, UK) was prepared in deionised water, and ciprofloxacin (Sigma-Aldrich) was dissolved in it to the desired final concentration. The monoemulsion solution was obtained by adding 4.75 g of Monomuls® 90–018 (BASF, Cheadle, UK) to a glass beaker and heating it to 40 °C in a water bath. Once the monoemulsion became molten, 250 μl of the ciprofloxacin solution were added and the emulsion mixed. The ciprofloxacin emulsion could be maintained in the lamellar phase at 37 °C until needed.

2.2 Preparation of cubic phase ciprofloxacin emulsion (75% Monomuls® + 25% - 0.05% TFA)

A solution of 0.05% TFA was prepared in deionised water, and ciprofloxacin was dissolved in it to the desired final concentration. The monoemulsion solution was obtained by adding 3.25 g of Monomuls® 90–018 into a glass beaker and heating it to 40 °C in a water bath. Once the monoemulsion became molten, 1250 μl of the ciprofloxacin solution was added and mixed with a glass wand. The solution was cooled to ambient room temperature, undergoing a phase shift to the cubic structure in the process, and stored at room temperature until needed.

2.3 Characterization of liquid and cubic phases by scanning electron microscopy

Analysis was performed as previously described (Tan et al. 2008). Briefly, samples were deposited on a silicon surface and snap frozen in liquid nitrogen to ensure the conformational stability of the liquid crystals. The product was then coated with a 2 nm thick carbon coating and SEM images were acquired using an S-4800 scanning electron microscope (Hitachi) operated at accelerating voltages of 2 or 10 kV.

2.4 Drug release profiling by high-pressure liquid chromatography (HPLC)

The release of ciprofloxacin from cubic phase ciprofloxacin emulsion was measured using dialysis membrane and HPLC. Cubic phase ciprofloxacin emulsion was placed into dialysis tubing cellulose membrane (14,000 MWCO, Sigma-Aldrich, UK) and each end of the tubing sealed. The sealed membrane was then placed in a solution of 0.05% TFA in deionised water and samples extracted at various time points over a 72-h period and stored at 4 °C ready for HPLC analysis. HPLC analysis was performed as previously described by Locatelli and colleagues (Locatelli et al. 2015).

2.5 Primary cell and organ culture

Uteri with no gross evidence of reproductive disease or microbial infection, were collected from post-pubertal mixed-breed beef cattle ($n = 16$ over a 3-month period) within 15 min of slaughter as part of the routine operation of a commercial slaughterhouse. Postpartum cattle were not used because of the ubiquitous bacterial contamination and disruption of the epithelium that is typical of the puerperal endometrium (Herath et al. 2009; Wathes et al. 2009). The animals were 20–26 months old, reared on extensive grassland and had never been pregnant or inseminated. The stage of reproductive cycle was determined by examination of ovarian morphology and vasculature, as described previously, and animals on days 1 to 4 of the oestrus cycle were used because, similar to postpartum cows, peripheral plasma ovarian hormone concentrations are basal (Ireland et al. 1979). The uteri were kept on ice for approximately 1 h until further processing at the laboratory. External surfaces were washed with 70% ethanol and the uterine horn opened longitudinally with sterile scissors. Since innate immune responses are the same irrespective of the horn used, one horn was used for the isolation of purified endometrial cell populations, and the contralateral horn used for organ culture (Saut et al. 2014).

Endometrial cells were isolated as described previously (Cronin et al. 2012; Herath et al. 2006). Epithelial and stromal cell populations were distinguished by cell morphology, the presence of cytokeratin and vimentin respectively, and the absence of immune cell contamination was confirmed by the absence of CD45, as described previously (Herath et al. 2006; Turner et al. 2014). The epithelial and stromal cells were cultured in 1 ml complete medium per well, comprising: Phenol-red free Roswell Park Memorial Institute (RPMI) 1640 medium (Sigma-Aldrich, Dorset, UK) containing 10% heat-inactivated foetal bovine serum (FBS; Biosera, East Sussex, UK), and plated at 1×10^5 cells/ml in 24-well plates (TPP, Trasadingen, Switzerland) ready for treatment. *Ex vivo* organ cultures (EVOC) of the endometrium were collected using 8 mm diameter punch biopsies as previously described (Borges et al. 2012). Tissues were cultured in 24-well plates (TPP) containing 2 ml complete medium per well, and treatments initiated within 4 h of slaughter. During treatment, cells or tissues were maintained in a humidified, 5% CO_2 in air atmosphere incubator at 37 °C, with supernatants collected as indicated.

2.6 Immortalised cell-lines culture

The human HeLa cervical epithelial and human HEC1A endometrial epithelial cell lines were obtained from the European Collection of Cell Cultures (ECACC, Salisbury,

UK) and cultured in Dulbecco's Modified Eagle Medium/F-12 nutrient mix (Thermofisher, Gloucester, UK) supplemented with 10% foetal bovine serum (FBS). Sub-confluent cultures were split using trypsin/ethylenediaminetetraacetic acid (EDTA) and seeded into 24-well plates (TPP, Trasadingen, Switzerland) in 1 ml of medium at 1×10^5 cells/ml for bacterial infection experiments. Or were seeded onto 0.4 μm collagen-coated Transwell™ support membranes (Sigma-Aldrich, Dorset, UK) in 0.25 ml of medium at 1×10^5 cells/ml for cell viability experiments.

2.7 Cell viability

Cell viability was assessed by the mitochondria-dependent reduction of MTT to formazan, as described previously (Mosmann 1983). The correlation between MTT OD_{570} measurements and the number of live cells was confirmed using trypan blue exclusion and counting the number of live cells using a haemocytometer.

2.8 Bacterial culture

Cultures of *E. coli* (isolate MS499), collected from the uteri of postpartum cows with persistent uterine disease, were grown overnight in Luria-Bertani (LB) medium (Sigma-Aldrich) as described previously (Amos et al. 2014; Sheldon et al. 2010). The bacteria were diluted to 1×10^8 colony forming units (CFU)/mL in medium and centrifuged at 1500×g for 10 min at room temperature. The bacteria were then resuspended in sterile PBS (Life Technologies Ltd., Paisley, UK) followed by centrifugation at 6000×g for 10 min at 4 °C. After washing, bacteria were diluted to 1×10^3 CFU/ml in LB medium for bactericidal experiments, or complete cell culture medium for cell/organ infection experiments. Where required, the number of colony-forming units (CFU) was determined by plate counts on LB agar plates following serial dilution of bacterial cultures and/or cell/organ culture supernatants.

2.9 Enzyme-linked immuno sorbent assay (ELISA)

Concentrations IL-6 in cell and EVOC culture supernatants were measured by ELISA according to the manufacturer's instructions Bovine IL-6 Screening Set ESS0029, ThermoFisher Scientific). To account for differences between the weights of EVOC tissues, concentrations are reported as pg per mg tissue. The limit of detection for IL-6 was 35.6 pg/ml; the intra-assay coefficient of variance was 1.2% and the inter-assay coefficient of variance was 3.0%.

2.10 Statistical analysis

Statistical analyses were performed using IBM SPSS Statistics 20. Initially the data were tested for homogeneity, and log or square root transformed if appropriate. Data were analysed by analysis of variance (ANOVA) using Dunnett's pairwise multiple comparison t-test for individual group comparisons, or by student's t test. Data are presented as mean with standard deviation (SD), $P < 0.05$ was considered statistically significant, and n represents the number of independent experiments.

3 Results

3.1 The preparation of liquid crystal emulsions is simple and reproducible

The preparation of different liquid crystal encapsulated ciprofloxacin emulsions was achieved by altering the proportion of water used during synthesis (Fig. 1). Lamellar structures (Fig. 1a) were prepared using a lower proportion of water (5% compared to 25% in cubic phase preparations) and maintained in the lamellar phase by heating to 40 °C. Cubic structures (Fig. 1b) were maintained at ambient room temperature and readily phase shifted between a cubic and lamellar structure by the application of heat (40 °C). For both structures, ciprofloxacin was added whilst in the lamellar phase and becomes incorporated within the crystal structure once it is heated to 40 °C. Drug equivalencies were calculated based on the total amount of ciprofloxacin added and the total volume of the emulsion whilst the liquid crystal was in the lamellar phase.

A variety of techniques can be employed for liquid crystal characterization, including small-angle x-ray scattering (SAXS) (Freiberger and Glatter 2006; Goujon et al. 2015; Mendez and Hammouda 2013) and Fourier Transform Infrared Spectroscopy (FTIR) (Koenig et al. 1995; Pongali Sathya Prabu et al. 2011), amongst others, and for an exhaustive analysis of the different techniques used, the reader is referred to the review by An and colleagues (An et al. 2016). In light of the previously published methodology we used to prepare our liquid crystals, we chose scanning electron microscopy (SEM) as a simple and robust approach to liquid crystal characterization. Scanning electron microscopy (Fig. 2) corroborated the findings of other groups working with similar formulations (Nestor et al. 2013; Santamaría et al. 2013). Images obtained at 60×, 110×, 350×, 799×, 800× and 2500× (Fig. 2a–f, respectively) show a crystalline structure typical to cubic phase liquid crystal formulations.

3.2 Liquid crystal emulsions provide a sustained release delivery modality

After fabrication and characterization of the liquid crystal structures, the rate of release of encapsulated ciprofloxacin was analysed using high pressure liquid chromatography (HPLC). When plotted against a standard curve of

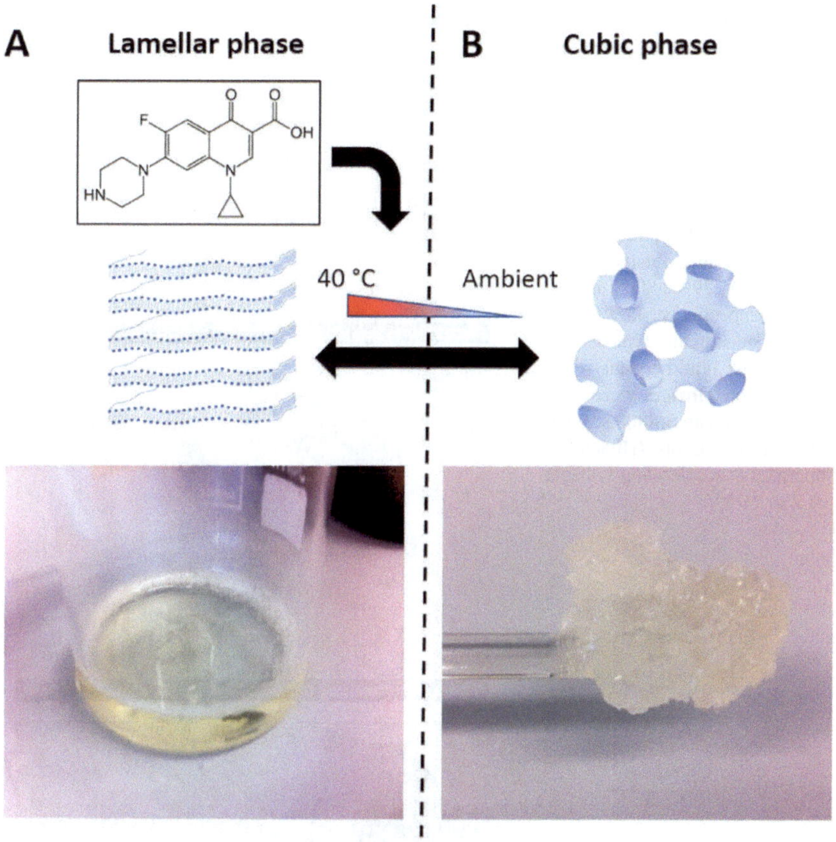

Fig. 1 Liquid crystal emulsions were prepared in different phases through controlled application of water content and heat: **a** Lamellar phase emulsions were obtained by mixing Monomuls® with a solution of 0.05% TFA (95% to 5%, *w/v*), heating the solution to 40 °C and then maintaining at 37 °C; **b** Cubic phase emulsions were obtained by mixing Monomuls® with a solution of 0.05% TFA (75% to 25%, *w/v*) and after heating to 40 °C, the solution was cooled to ambient room temperature, undergoing a phase shift to the cubic structure in the process, and stored at room temperature until needed. Ciprofloxacin was incorporated into the liquid crystal structures by addition whilst at 40 °C

ciprofloxacin release (Sup. Fig. 1), area under the curve (AUC) analysis revealed a sustained release profile from cubic phase liquid crystal, peaking at 4.07μg/ml after 72 h (Fig. 3), confirming the suitability of this formulation for sustained, slow-release treatment regimen.

3.3 Liquid crystal formulations have a limited impact on epithelial cell viability

Before evaluating the efficacy of liquid crystal formulations for the treatment of bacterial infections, we evaluated whether there were any cytotoxic effects of the various formulations on human cells (Fig. 4). HEC1A or HeLa epithelial cells were treated with ciprofloxacin (2 ng/ml to 1 μg/ml), lamellar phase liquid crystal emulsion (LP: 10 μl to 100 μl), LP encapsulated ciprofloxacin (LPC: 10 μl to 100 μl ≡ 50 ng/ml to 500 ng/ml of ciprofloxacin) or the control (TFA in deionised water at the same final concentration as for LPC) for 24 h. Initially, treatments were added directly onto cells. No observable effect was noted following addition of ciprofloxacin to the cell culture (data not shown), but following the addition of liquid crystal emulsions, a corona surrounding the liquid crystal, which was devoid of cells was noted in both HEC1A (Fig. 4a) and HeLa (Fig. 4b) cell cultures. Because of interference between the liquid crystal emulsions and the MTT assay reagents, cell cytotoxicity experiments were performed using Transwell™ support membranes with liquid crystal emulsions held in the Transwell™ and cells in the basal compartment of the plate. Treatment of HEC1A cells with ciprofloxacin resulted in reduced cell counts demonstrating a cytotoxic effect compared to treatment with the control (Fig. 4c). The effect, however, was not dose dependent and interestingly, was not observed in HeLa cells treated with ciprofloxacin (Fig. 4d) or in either HEC1A or HeLa cells once the ciprofloxacin was encapsulated in lamellar phase liquid crystal (LPC, Fig. 4g and h, respectively), suggesting that encapsulated ciprofloxacin is non-toxic to human cells. Treatment of HEC1A (Fig. 4e) or HeLa (Fig. 4f) cells with LP did not affect cell viability compared to control treated cells.

3.4 Encapsulation of ciprofloxacin within a liquid crystal emulsion significantly enhances bacterial killing

Effective and appropriate models for the evaluation of novel therapeutics are an essential tool for drug development. In considering an appropriate model for infection of the female reproductive tract we selected bovine endometrial tissue, which is readily available to us and a suitable model of the human female reproductive tract due to similarities of structure, function and response to bacterial infection (Sheldon

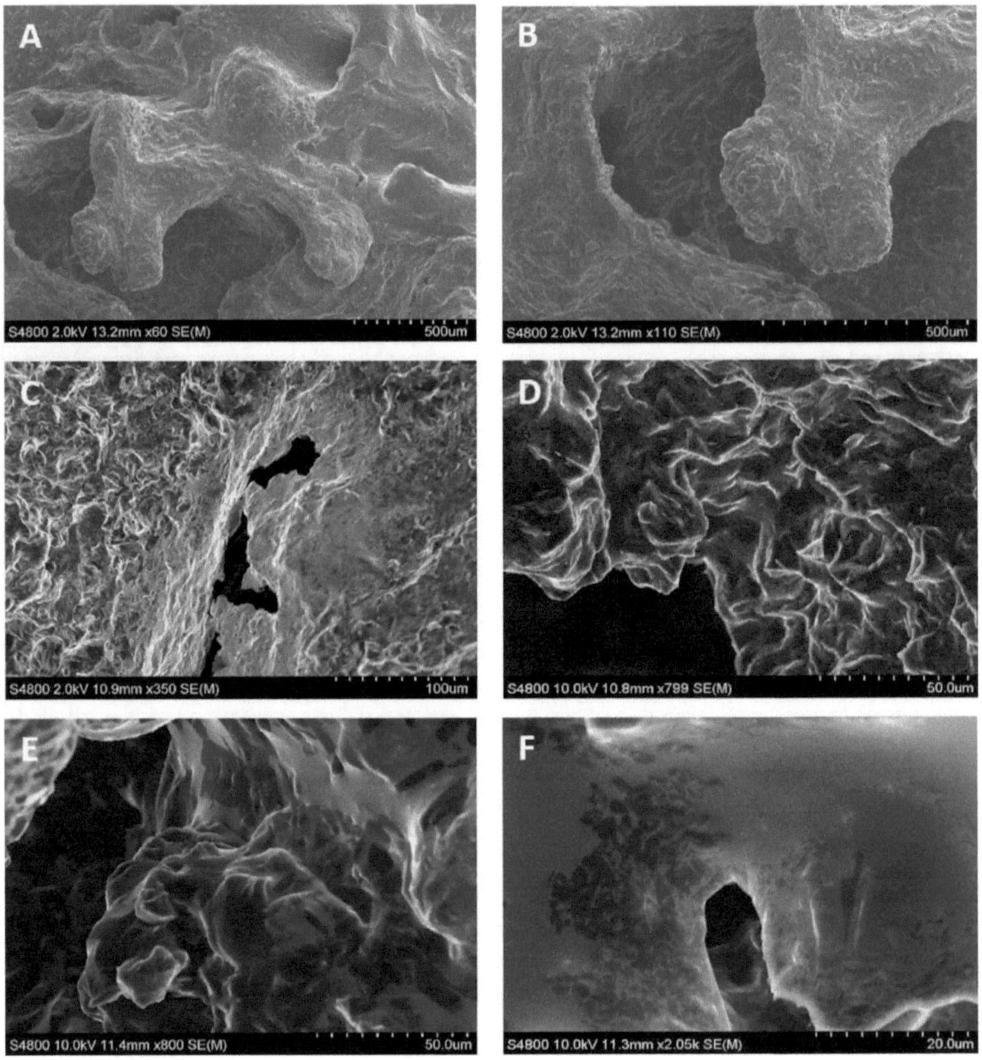

Fig. 2 Scanning electron microscope characterization of cubic phase liquid crystal formulations: samples were deposited on a silicon surface, snap frozen in liquid nitrogen to maintain the conformational stability of the liquid crystals and then coated with a 2 nm carbon layer. **a–f** images obtained at 60×, 110×, 350×, 799×, 800× and 2500× respectively, show a crystalline structure typical to cubic phase liquid crystal formulations. Images were acquired with an S-4800 scanning electron microscope (Hitachi) operated at accelerating voltages of 2 or 10 kV

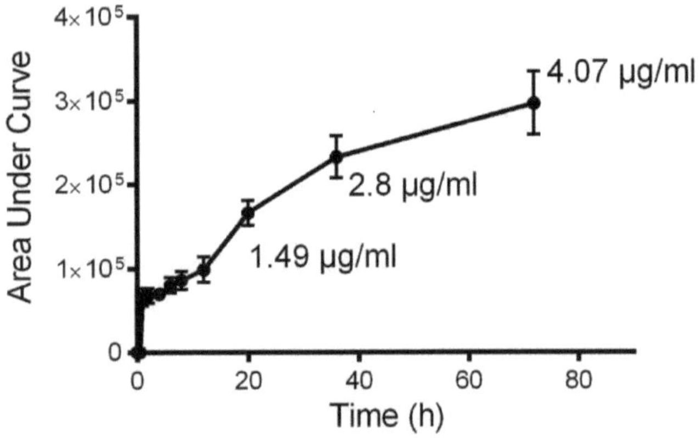

Fig. 3 Encapsulated ciprofloxacin displayed sustained release characteristics from cubic phase liquid crystal structures: cubic phase liquid crystal encapsulated ciprofloxacin was sealed in a semi-permeable membrane and placed in a 0.05%TFA solution. Samples were extracted at various time points over a 72-h period. Data shown is mean (SD), $n = 3$. Area under curve (AUC) analysis was performed using GraphPad PRISM 6

et al. 2017; Turner et al. 2012). A common bacterial infection within the bovine female reproductive tract, *E. coli* is one of the most prevalent bacterium associated with uterine disease in cattle. (Sheldon et al. 2009). Thus, we modelled infection of the female reproductive tract using primary bovine endometrial epithelial cells and EVOCs of the bovine endometrium, which have been previously used as effective models of infection, inflammation and for drug development (Borges et al. 2012; Healey et al. 2016; Saut et al. 2014).

Initial experiments investigated the effect of treatment on bacterial growth alone (Fig. 5). *E. coli* (MS499) was diluted to 1×10^3 colony forming units (CFU)/ml in LB medium for bactericidal experiments and treated with ciprofloxacin (Cipro: 10 ng/ml to 1 µg/ml), lamellar phase liquid crystal emulsion (LP: 10 µl to 100 µl) or LP encapsulated ciprofloxacin (LPC: 10 µl to 100 µl ≡ 50 ng/ml to 500 ng/ml of ciprofloxacin). Bacterial growth in the presence or absence of treatments was monitored for 24 h (Fig. 5a), 48 h (Fig. 5b) or 72 h (Fig. 5c), after which bacterial growth (number of CFU) was

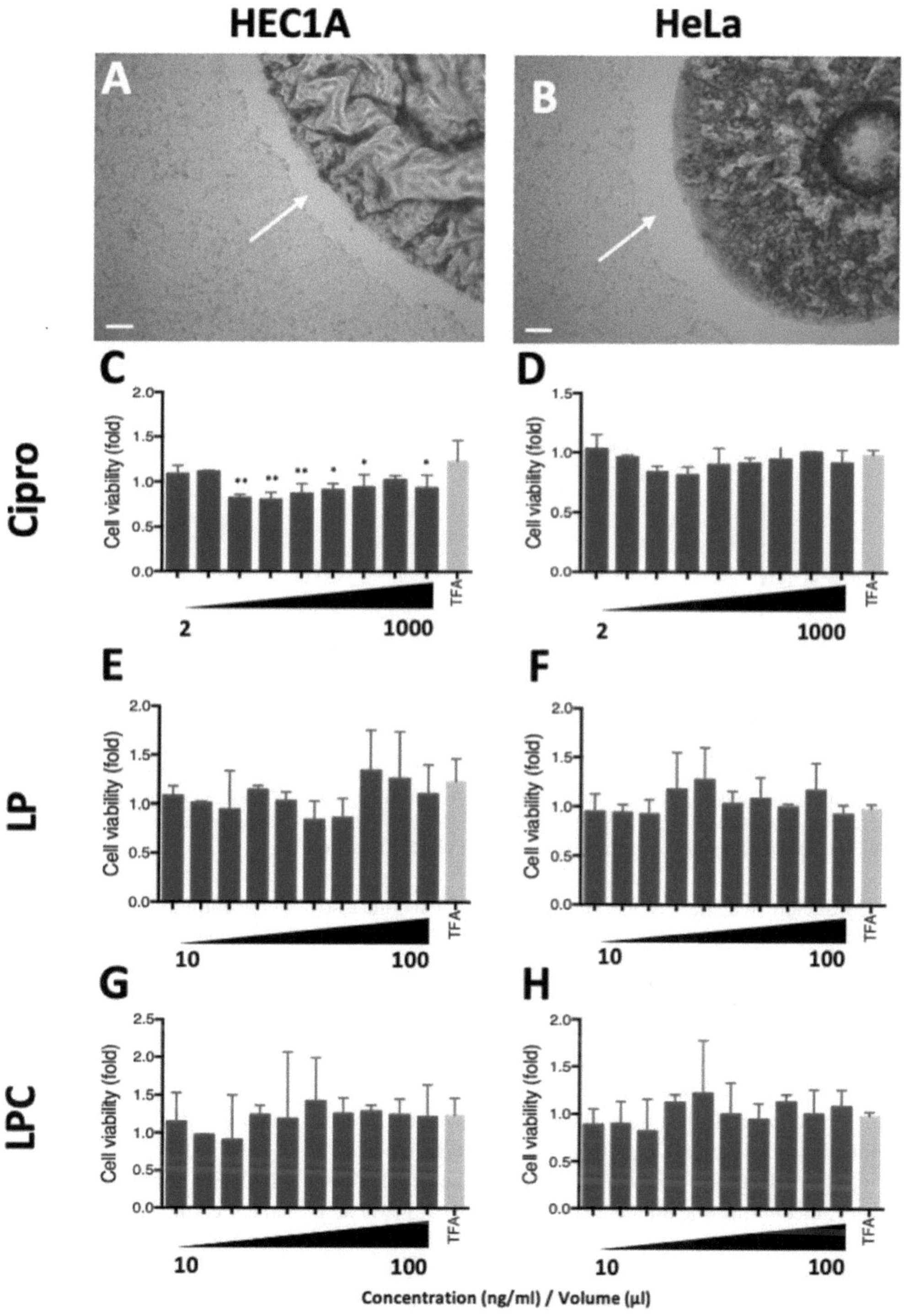

Fig. 4 Liquid crystal emulsions have a limited impact on epithelial cell viability: white arrows indicate a corona devoid of cells surrounding the liquid crystal in both cell lines; **a, b** Phase-contrast images of the interaction between cubic phase liquid crystal emulsion and HEC1A or HeLa epithelial cells. A corona devoid of cells can be seen around the liquid crystal, but no cell death was observed; **c–h** HEC1A or HeLa epithelial cells were treated with ciprofloxacin (Cipro: 2 ng/ml to 1 µg/ml; C-D), lamellar phase liquid crystal emulsion (LP) (10 µl to 100 µl; E-F) or LP encapsulated ciprofloxacin (LPC) (10 µl to 100 µl ≡ 50 ng/ml to 500 ng/ml of ciprofloxacin; G-H) for 24 h. Scale bars = 500 µm. Data were analysed by ANOVA using Dunnett's pairwise multiple comparison. Data are presented as mean (SD), $n = 4$. * $p < 0.05$, ** $p < 0.01$ compared to TFA

determined. At all three time points, treatment with LP had no effect on bacterial growth. Treatment with ciprofloxacin significantly reduced bacterial growth at all time points with no growth observed at ciprofloxacin concentrations above 350 ng/ml. A reduction in bacterial growth was also observed following treatment with LPC, which was significantly greater compared to ciprofloxacin treatment with no growth observed at LPC volumes above 12 µl ≡ 60 ng/ml. To determine the non-inhibitory concentration (NIC) and minimum inhibitory concentration (MIC) for ciprofloxacin (Sup. Fig. 2) and LPC (Sup. Fig. 3) treatments, data was also fitted to a Gompertz model using the method described by Lambert et al (Lambert and Pearson 2000). Table 1 shows the NICs and MICs for the two treatments at 24, 48 and 72 h, demonstrating the enhanced bactericidal effect (lower NIC and lower MIC) of treatment with LPC compared to ciprofloxacin. These data suggested that liquid crystals enhance the effectiveness of antibiotics in the treatment of bacterial diseases.

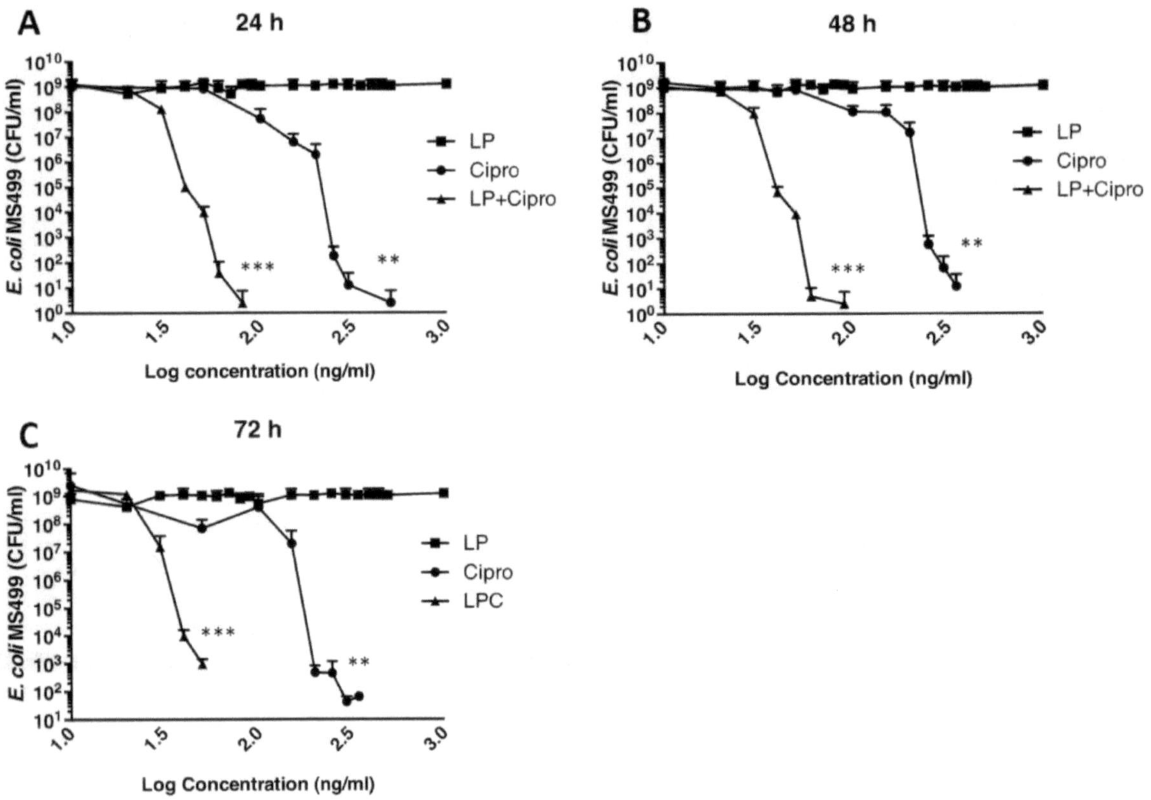

Fig. 5 Liquid crystal encapsulation enhances the bacterial toxicity of ciprofloxacin: cultures of *E. coli* (MS499) (at 1×10^3 CFU/ml) were prepared in LB medium and treated with ciprofloxacin (Cipro: 10 ng/ml to 1 μg/ml), lamellar phase liquid crystal emulsion (LP: 10 μl to 100 μl) or LP encapsulated ciprofloxacin (LPC: 10 μl to 100 μl ≡ 50 ng/ml to 500 ng/ml of ciprofloxacin) for 24 h (**a**), 48 h (**b**) or 72 h (**c**). Data were analysed by ANOVA using Dunnett's pairwise multiple comparison. Data are presented as mean (SD), $n = 4$. ** $p < 0.01$, *** $p < 0.001$ compared to LP

To explore whether reduced bacterial growth would also translate to reduced inflammation, we infected bovine epithelial cells, and EVOCs with *E. coli* (MS499) and monitored inflammatory responses by measuring interleukin (IL)-6 production, which is associated with epithelial driven innate immune responses in the endometrium (Cronin et al. 2012; Healey et al. 2016; Sheldon and Roberts 2010) and responsible for acute and chronic inflammation (Gabay 2006). Epithelial cells or EVOCs, previously treated with ciprofloxacin, LP or LPC for 4 h, were cultured in the presence of control medium or medium containing *E. coli* (MS499) at a concentration of 1×10^3 CFU/ml for 24 h (Fig. 6). Infection of epithelial cells with *E. coli* (MS499) significantly increased the secretion of IL-6 in primary epithelial cell cultures. Treatment with LP (Fig. 6a) did not reduce IL-6 secretion. Conversely, a significant reduction in IL-6 production was noted following treatment with ciprofloxacin concentrations above 200 ng/ml (Fig. 6b) and following treatment with LPC at a volume of 100 μl ≡ 500 ng/ml (Fig. 6c).

As with epithelial cells, infection of EVOCs with *E. coli* (MS499) significantly increased the secretion of IL-6 in primary epithelial cell cultures (Fig. 7). Treatment with LP (Fig. 7a) had no impact on IL-6 secretion and significant reduction in IL-6 production was noted following treatment with ciprofloxacin concentrations above 200 ng/ml (Fig. 7b). For EVOCs treated with LPC, a trend suggesting reduced IL-6 secretion was noted compared to *E. coli* (MS499) infected EVOCs that had not been treated with LPC (Fig. 7c).

Table 1 Non-inhibitory concentration (NIC) and minimum inhibitory concentration (MIC) of ciprofloxacin and LPC for the treatment of *E.coli* MS499

	NIC (ng/ml)		MIC (ng/ml)	
	Cipro	LPC	Cipro	LPC
24 h	48.96	17.48	97.58	31.26
48 h	52.03	16.59	156.0	36.34
72 h	15.6	19.59	125.1	33.48

Data was fitted to a Gompertz model using the method described by Lambert et al

4 Discussion

In humans, infections of the female reproductive tract are a major socio-economic burden that is associated with significant morbidity and mortality (Horne et al. 2008). Additionally,

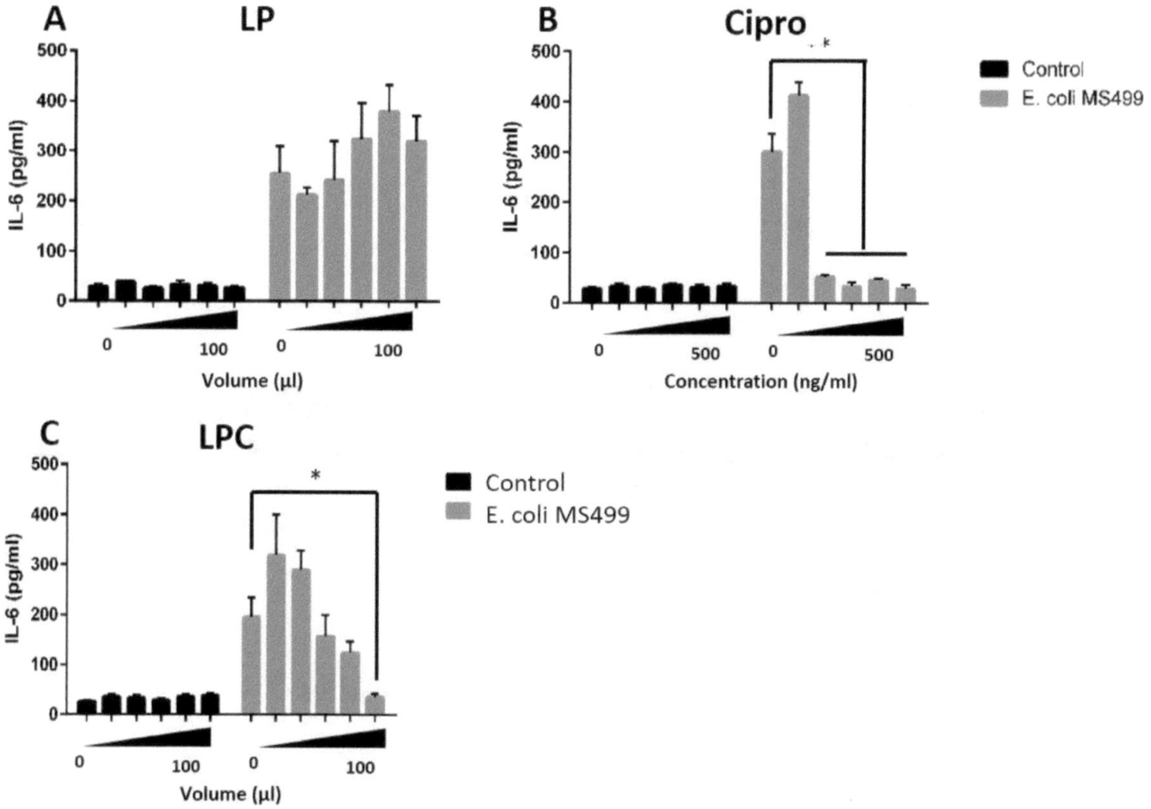

Fig. 6 IL-6 secretion by *E. coli* (MS499) infected epithelial cells is reduced following treatment with ciprofloxacin or liquid crystal encapsulated ciprofloxacin: primary bovine epithelial cells were pre-treated with control medium or medium containing ciprofloxacin (Cipro: 100 ng/ml to 500 ng/ml), lamellar phase liquid crystal emulsion (LP: 20 µl to 100 µl) or ciprofloxacin encapsulated liquid crystal emulsion (LPC: 20 µl to 100 µl ≡ 100 ng/ml to 500 ng/ml of ciprofloxacin) for 4 h. Cells were subsequently challenged with control medium or medium containing 1×10^3 CFU/ml *E. coli* (MS499) for 24 h. After challenge, supernatants were collected and stored at −20 °C prior to analysis for IL-6, by ELISA. Data were analysed by ANOVA using Dunnett's pairwise multiple comparison. Data are presented as mean (SD), $n = 4$. * $p < 0.05$, ** $p < 0.01$

an increasing prevalence of bacterial resistance over the past two decades means that previously effective antibiotic treatments are being abandoned except in cases where specific sensitivity can be demonstrated through additional testing (Allan-Blitz et al. 2016; Conly and Johnston 2005; So and Shah 2014). There is therefore an urgent need for novel approaches to the treatment of bacterial infections of the female reproductive tract.

Establishing new localised drug delivery vehicles could provide several benefits over conventional therapies that are crucial to microorganism eradication related to bacterial infections. Improved pharmacokinetic and pharmacodynamic profiles reduce the amount of administered drug needed, compared to systemic delivery, to obtain a clinically relevant effect. Improved half-lives and sustained, localised release at concentrations above the bacteria's minimum inhibitory concentration, when protecting against infection, or above the minimum bactericidal concentration, when treating current infection reduce the incidence of antibiotic resistance (Wu and Grainger 2006). Improved drug efficacy and a reduced need for repeated administrations also reduces costs and enhances patient compliance. Controlled-release platforms such as liquid crystals offer great potential due to their inherent structural characteristics that are ideally suited for encapsulation of antibiotics such as ciprofloxacin. Moreover, the liquid crystal platform presented here affords a slow-release profile, allowing for prolonged treatments avoiding multiple dosing regimens. Furthermore, the platform is minimally invasive and can be applied locally, removing the need for oral or intravenous administration.

Characterization of the liquid crystal formulations demonstrated that they were not toxic to cervical (HeLa) or endometrial (HEC1A) epithelial cells, corroborating previous works (Aldred et al. 2014; Card et al. 2015; Mason et al. 1995). Some cell toxicity was noted in HEC1A cells following treatment with ciprofloxacin, however the effect was not dose dependent suggesting the observed effect might be of limited biological significance. A variety of toxicities associated with ciprofloxacin therapy are known (Amorha et al. 2018; Oyebode et al. 2018; Stahlmann 1990; Stahlmann and Lode 2013) including arthropathogenic potential in animals, which led to the decision to not use ciprofloxacin or other quinolones in children or adolescents. Interestingly, cells treated with ciprofloxacin encapsulated within lamellar phase liquid crystal

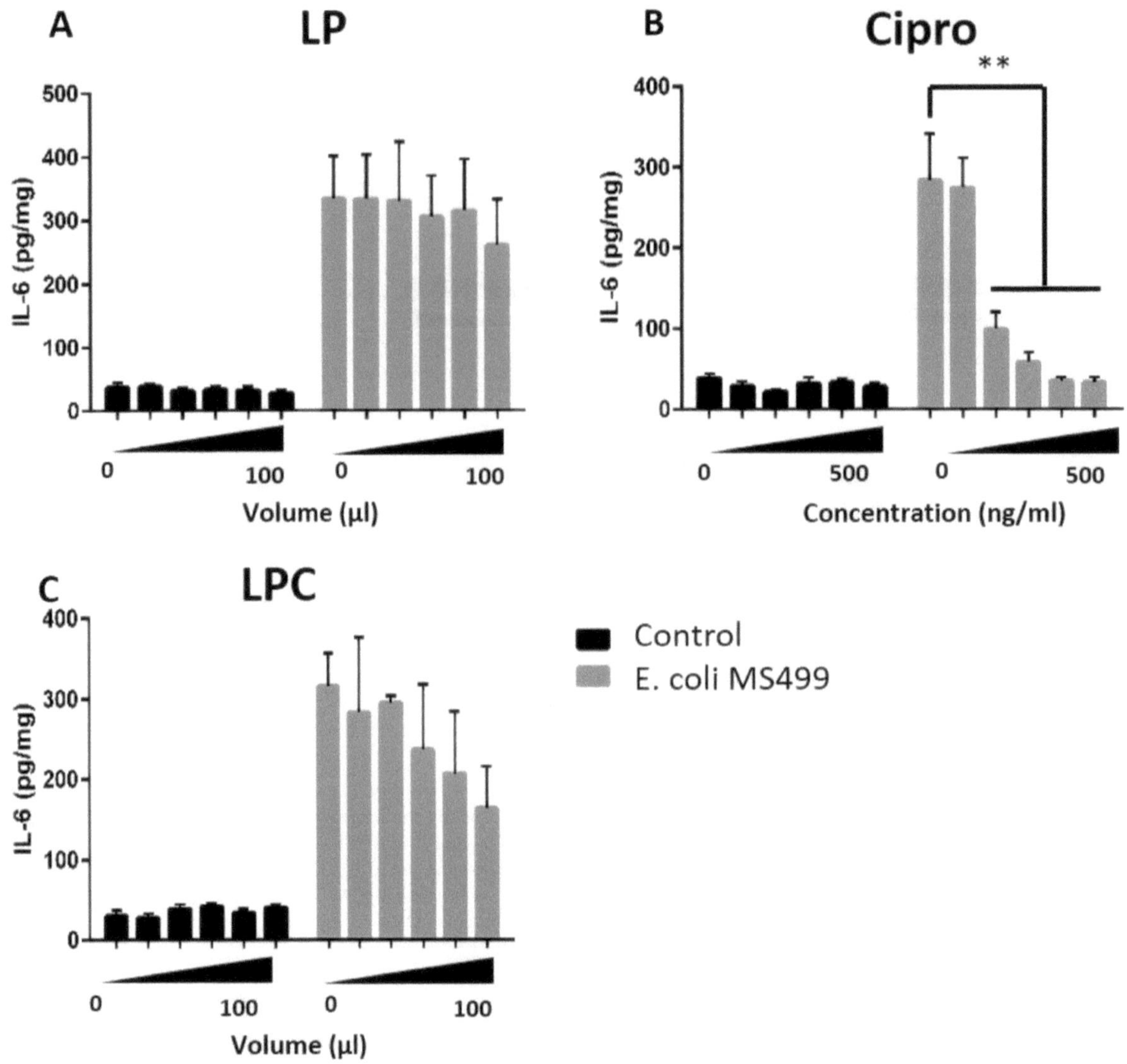

Fig. 7 IL-6 secretion by *E. coli* (MS499) infected EVOCs is reduced following treatment with ciprofloxacin or liquid crystal encapsulated ciprofloxacin: EVOCs were pre-treated with control medium or medium containing ciprofloxacin (Cipro: 100 ng/ml to 500 ng/ml), lamellar phase liquid crystal emulsion (LP: 20 μl to 100 μl) or ciprofloxacin encapsulated liquid crystal emulsion (LPC: 20 μl to 100 μl ≡ 100 ng/ml to 500 ng/ml of ciprofloxacin) for 4 h. EVOCs were subsequently challenged with control medium or medium containing 1×10^3 CFU/ml *E. coli* (MS499) for 24 h. After challenge, supernatants were collected and stored at −20 °C prior to analysis of IL-6, by ELISA. EVOC tissues were weighed to enable cytokine concentrations to be adjusted for tissue weight. Data were analysed by ANOVA using Dunnett's pairwise multiple comparison. Data are presented as mean (SD), $n = 4$. ** $p < 0.01$

did not display any toxicity, suggesting the notion that the use of such formulations may be a useful strategy to reduce toxicities associated with current treatment regimen.

In addition to reduced toxicity, ciprofloxacin encapsulation significantly enhanced bacterial cell killing. LPC treatment was five times more effective at killing *E. coli* than free-ciprofloxacin demonstrating the great potential of this formulation for enhancing the efficacy and sustained delivery of antibiotic agents.

Central to the response to bacterial challenge is the detection of pathogen-associated molecular patterns by toll-like receptors (Amjadi et al. 2014). Reproductive tract immune responses to *E. coli* infection require the binding of lipopolysaccharide from *E. coli* to toll-like receptor 4 on the surface of epithelial cells, which then produce IL-6 (Herath et al. 2006; Sheldon and Roberts 2010). The kinetics of IL-6 production reflecting its roles in the early response to infection such as leukocyte recruitment, B lymphocyte development, antibody secretion by plasma cells and the regulation of acute-phase proteins.

Interleukin-6 rapidly accumulates during the first 24 h following exposure to *E. coli* (Sheldon and Roberts 2010), and was attenuated by treatment with free- and encapsulated (LPC) ciprofloxacin. As expected, attenuation was more pronounced in epithelial cells than EVOCs. The purified cell model does not support inflammasome activation and IL1β secretion, which leads to the production of additional pro-inflammatory cytokines, such as more IL-6 (van de

Veerdonk et al. 2011), and chemokines such as IL-8 that recruit cells and promote phagocytosis and bacterial clearance (Pétrilli et al. 2007). Although treatment with free-ciprofloxacin appeared more efficacious than treatment with LPC, this is likely due to the concentration of ciprofloxacin present in the culture medium during this 24 h experiment. Whilst the full dose is immediately available following treatment with free-ciprofloxacin, the antibiotic is released slowly over the course of the experiment with LPC. Such a release profile will likely have significant advantages during longer-term infection, whereby the sustained presence of antibiotic will reduce IL-6 production and reduce bacterial load more effectively than the free drug. An expected effect corroborated by the application of liquid crystals for the slow release of drugs with a wide range of molecular weights and water solubilities (Esposito et al. 1996; Geraghty et al. 1996; Nesseem 2001; Nguyen et al. 2011; Nguyen TH et al. 2010; Omray 2013; Wyatt and Dorschel 1992). Myriad factors influence the dynamics of drug release from liquid crystal delivery platforms including temperature, pressure, pH, the molecular weight of the drug, hydrophobicity and salt concentration. Optimization based on the clinical application environment is therefore an important consideration of the drug formulation design process to ensure optimal efficacy.

These data demonstrate that the liquid crystal controlled-release platform presented is effective at delivering antibiotics to treat bacterial infections within the female reproductive tract. Such a platform has potential clinical application supporting a drive towards localised delivery of antibiotic therapies over systemic delivery. Liquid crystal systems have been extensively employed in the clinic as anticancer, anti-inflammatory, analgesic and immunosuppressive agents, amongst others (Chaudhary et al. 2016; Kim et al. 2015; Mei et al. 2018). Indeed, we envisage such a platform being used for topical delivery by the patient for the treatment of lower reproductive tract infections, and/or topical administration by a medical practitioner for upper reproductive tract infections. As well as addressing an important unmet need for efficacious therapies that can reduce negative side effects associated with systemic exposure and offer significant health and economic benefits. Such an approach also provides the potential to increase the efficacy and extend the effective therapeutic life-span of existing antibiotics by reducing the rate of emergence of antibiotic resistance.

Funding This work was supported by the ERASMUS+ student mobility programme

References

K.J. Aldred, R.J. Kerns, N. Osheroff, Biochemistry **53**, 1565 (2014)

M. Aldunate, D. Srbinovski, A.C. Hearps, C.F. Latham, P.A. Ramsland, R. Gugasyan, R.A. Cone, G. Tachedjian, Front. Physiol. **6**, 164 (2015)

N.J. Alexander, E. Baker, M. Kaptein, U. Karck, L. Miller, E. Zampaglione, Fertil. Steril. **82**, 1 (2004)

E. Alirol, T.E. Wi, M. Bala, M.L. Bazzo, X.-S. Chen, C. Deal, J.-A.R. Dillon, R. Kularatne, J. Heim, R. Hooft van Huijsduijnen, E.W. Hook, M.M. Lahra, D.A. Lewis, F. Ndowa, W.M. Shafer, L. Tayler, K. Workowski, M. Unemo, M. Balasegaram, PLoS Med. **14**, e1002366 (2017)

L.-T. Allan-Blitz, R.M. Humphries, P. Hemarajata, A. Bhatti, M.W. Pandori, M.J. Siedner, J.D. Klausner, Clin. Infect. Dis. **64**, ciw864 (2016)

E. Amabebe, D.O.C. Anumba, Front. Med. **5**, 181 (2018)

F. Amjadi, E. Salehi, M. Mehdizadeh, R. Aflatoonian, Adv. Biomed. Res. **3**, 1 (2014)

K.C. Amorha, O.B. Ugwuowo, E.E. Ayogu, S.O. Nduka, M.J. Okonta, Pak. J. Pharm. Sci. **31**, 1805 (2018)

M.R. Amos, G.D. Healey, R.J. Goldstone, S.M. Mahan, A. Düvel, H.-J. Schuberth, O. Sandra, P. Zieger, I. Dieuzy-Labaye, D.G.E. Smith, I.M. Sheldon, Biol. Reprod. **90**, 54 (2014)

J.-G. An, S. Hina, Y. Yang, M. Xue, Y. Liu, Characterization of liquid crystals: A literature review. Rev. Adv. Mater. Sci. **44**, 398 (2016)

M. Bala, K. Ray, S.M. Gupta, S. Muralidhar, R.K. Jain, J. Antimicrob. Chemother. **60**, 582 (2007)

C. Barbés, S. Boris, AIDS Patient Care STDs **13**, 747 (1999)

Á.M. Borges, G.D. Healey, I.M. Sheldon, Explants of intact endometrium to model bovine innate immunity and inflammation *Ex Vivo*. Am. J. Reprod. Immunol. **67**(6) (2012). https://doi.org/10.1111/j.1600-0897.2012.01106.x

R.M. Brotman, M.D. Shardell, P. Gajer, D. Fadrosh, K. Chang, M.I. Silver, R.P. Viscidi, A.E. Burke, J. Ravel, P.E. Gravitt, Menopause **21**, 450 (2014)

R.M. Card, M. Mafura, T. Hunt, M. Kirchner, J. Weile, M.-U. Rashid, A. Weintraub, C.E. Nord, M.F. Anjum, Antimicrob. Agents Chemother. **59**, 4410 (2015)

K. K. Chaudhary, P. Kannojia, N. Mishra, Liquid crystal systems in drug delivery. 2016. https://doi.org/10.4018/978-1-5225-0751-2.ch009

Y. Chen, P. Ma, S. Gui, Cubic and hexagonal liquid crystals as drug delivery systems. Biomed. Res. Int. (2014). https://doi.org/10.1155/2014/815981

C. Chen, X. Song, W. Wei, H. Zhong, J. Dai, Z. Lan, F. Li, X. Yu, Q. Feng, Z. Wang, H. Xie, X. Chen, C. Zeng, B. Wen, L. Zeng, H. Du, H. Tang, C. Xu, Y. Xia, H. Xia, H. Yang, J. Wang, J. Wang, L. Madsen, S. Brix, K. Kristiansen, X. Xu, J. Li, R. Wu, H. Jia, Nat. Commun. **8**, 875 (2017)

H.W. Chesson, R.D. Kirkcaldy, T.L. Gift, K. Owusu-Edusei, H.S. Weinstock, Emerg. Infect. Dis. **20**, 612 (2014)

J. Conly, B. Johnston, Can. J. Infect. Dis. Med. Microbiol. **16**, 159 (2005)

J.G. Cronin, M.L. Turner, L. Goetze, C.E. Bryant, I.M. Sheldon, Biol. Reprod. **86**, 51 (2012)

P. Daniel Johannes Rönnqvist, U.B. Forsgren-Brusk, E.E. Grahn-Håkansson, Acta Obstet. Gynecol. Scand. **85**, 726 (2006)

J. das Neves, *Mucosal Deliv. Biopharm* (Springer US, Boston, MA, 2014), pp. 261–280

J. das Neves, M.F. Bahia, Int. J. Pharm. **318**, 1 (2006)

J. das Neves, R. Palmeira-de-Oliveira, A. Palmeira-de-Oliveira, F. Rodrigues, B. Sarmento, *Mucoadhesive Mater. Drug Deliv. Syst.* (John Wiley & Sons, Ltd, Chichester, United Kingdom, 2014), pp. 99–132

M.L. Delaney, A.B. Onderdonk, Obstet. Gynecol. **98**, 79 (2001)

D.A. Eschenbach, P.R. Davick, B.L. Williams, S.J. Klebanoff, K. Young-Smith, C.M. Critchlow, K.K. Holmes, J. Clin. Microbiol. **27**, 251 (1989)

E. Esposito, V. Carotta, A. Scabbia, L. Trombelli, P. D'Antona, E. Menegatti, C. Nastruzzi, Int. J. Pharm. **142**, 9 (1996)
N. Freiberger, O. Glatter, (2006)
C. Gabay, Arthritis Res. Ther. **8**(Suppl 2), S3 (2006)
C.L. Galhardo, J.M. Soares, R.S. Simões, M.A. Haidar, G. Rodrigues de Lima, E.C. Baracat, Clin. Exp. Obstet. Gynecol. **33**, 85 (2006)
P.B. Geraghty, D. Attwood, J.H. Collett, Y. Dandiker, Pharm. Res. **13**, 1265 (1996)
N. Goujon, M. Forsyth, L.F. Dumée, G. Bryant, N. Byrne, Phys. Chem. Chem. Phys. **17**, 23059 (2015)
G.D. Healey, C. Collier, S. Griffin, H.-J. Schuberth, O. Sandra, D.G. Smith, S. Mahan, I. Dieuzy-Labaye, I.M. Sheldon, J. Immunol. **196**, 823 (2016)
W.H. Henderson, Antibiot. Med. Clin. Ther. (New York, NY) **5**, 470 (1958)
S. Herath, D.P. Fischer, D. Werling, E.J. Williams, S.T. Lilly, H. Dobson, C.E. Bryant, I.M. Sheldon, Endocrinology **147**, 562 (2006)
S. Herath, S.T. Lilly, N.R. Santos, R.O. Gilbert, L. Goetze, C.E. Bryant, J.O. White, J. Cronin, I.M. Sheldon, Reprod. Biol. Endocrinol. **7**, 55 (2009)
S.L. Hillier, R.J. Lau, Clin. Infect. Dis. **25**(Suppl 2), S123 (1997)
A.W. Horne, S.J. Stock, A.E. King, Reproduction **135**, 739 (2008)
A. Hussain, F. Ahsan, J. Control. Release **103**, 301 (2005)
J.J. Ireland, P.B. Coulson, R.L. Murphree, J. Anim. Sci. **49**, 1261 (1979)
D.-H. Kim, A. Jahn, S.-J. Cho, J.S. Kim, M.-H. Ki, D.-D. Kim, J. Pharm. Investig. **45**, 1 (2015)
J.L. Koenig, S.-Q. Wang, S.R. Challa, Appl. Spectrosc. **49**(3), 267–272 (1995)
V.R. Laibl, J.S. Sheffield, S. Roberts, D.D. McIntire, S. Trevino, G.D. Wendel, Obstet. Gynecol. **106**, 461 (2005)
R.J. Lambert, J. Pearson, J. Appl. Microbiol. **88**, 784 (2000)
J.A. Lambert, S. John, J.D. Sobel, R.A. Akins, PLoS One **8**, e82599 (2013)
B. Larsen, G.R.G. Monif, Clin. Infect. Dis. **32**, e69 (2001)
M. LeBel, Ciprofloxacin: Chemistry, Mechanism of Action, Resistance, Antimicrobial Spectrum, Pharmacokinetics, Clinical Trials, and Adverse Reactions. Pharmacotherapy: The Journal of Human Pharmacology and Drug Therapy. (1988). https://doi.org/10.1002/j.1875-9114.1988.tb04058.x
M. Locatelli, M.T. Ciavarella, D. Paolino, C. Celia, E. Fiscarelli, G. Ricciotti, A. Pompilio, G. Di Bonaventura, R. Grande, G. Zengin, L. Di Marzio, J. Chromatogr. A **1419**, 58 (2015)
I.M.C. Martin, S. Hoffmann, C.A. Ison, ESSTI Network, J. Antimicrob. Chemother. **58**, 587 (2006)
D.J. Mason, E.G. Power, H. Talsania, I. Phillips, V.A. Gant, Antimicrob. Agents Chemother. **39**, 2752 (1995)
P. Mastromarino, M. Di Pietro, G. Schiavoni, C. Nardis, M. Gentile, R. Sessa, Int. J. Med. Microbiol. **304**, 654 (2014)
M. McCarthy, Lancet (London, England) **369**, 1592 (2007)
S.E. McLaughlin, K.G. Ghanem, J.M. Zenilman, J. McLeod Griffiss, Sex. Transm. Dis. **1** (2018)
L. Mei, Y. Xie, Y. Huang, B. Wang, J. Chen, G. Quan, X. Pan, H. Liu, L. Wang, X. Liu, C. Wu, Acta Biomater. **67**, 99 (2018)
M.P. Mendez, B. Hammouda, J. Mater. Sci. Eng. B **3** (2013)
MMWR, CDC's Sex. Transm. Dis. Treat. Guidel. **54**,14: 332, (2006)
T. Mosmann, J. Immunol. Methods **65**, 55 (1983)
D.I. Nesseem, J. Pharm. Biomed. Anal. **26**, 387 (2001)
J. Nestor, A. Vílchez, C. Solans, J. Esquena, Langmuir **29**, 432 (2013)
B.B. Nguyen TH, T. Hanley, C.J. Porter, I. Larson, J. Pharm. Pharmacol. **62**, 696 (2010)
T.-H. Nguyen, T. Hanley, C.J.H. Porter, B.J. Boyd, J. Control. Release **153**, 180 (2011)
S. Nys, I.N. Okeke, S. Kariuki, G.J. Dinant, C. Driessen, E.E. Stobberingh, J. Antimicrob. Chemother. **54**, 952 (2004)
L. K. Omray, *Liquid Crystals as Novel Vesicular Delivery System: A Review* Current Trends in Technology and Science. **2** (2013)
O. T. Oyebode, O. R. Adebiyi, O. O. Olorunsogo, Toxicol. Mech. Methods **1**, 1 (2018)
P. Health England, *GRASP 2012 Report – Full: The Gonococcal Resistance to Antimicrobials Surveillance Programme* (2013). https://webarchive.nationalarchives.gov.uk/20140714081808/http://www.hpa.org.uk/webw/HPAweb&HPAwebStandard/HPAweb_C/1317140152044
R. Palmeira-de-Oliveira, P. Duarte, A. Palmeira-de-Oliveira, J. das Neves, M.H. Amaral, L. Breitenfeld, J. Martinez-de-Oliveira, Eur. J. Contracept. Reprod. Health Care **20**, 259 (2015)
Ž. Pavelić, N. Škalko-Basnet, I. Jalšenjak, Int. J. Pharm. **301**, 140 (2005)
V. Pétrilli, C. Dostert, D.A. Muruve, J. Tschopp, Curr. Opin. Immunol. **19**, 615 (2007)
N. Pongali Sathya Prabu, V.N. Vijayakumar, M.L.N.M. Mohan, J. Mol. Struct. **994**, 387 (2011)
J. Ravel, R.M. Brotman, P. Gajer, B. Ma, M. Nandy, D.W. Fadrosh, J. Sakamoto, S.S. Koenig, L. Fu, X. Zhou, R.J. Hickey, J.R. Schwebke, L.J. Forney, Microbiome **1**, 29 (2013)
E. Santamaría, M. Cortés, A. Maestro, M. Porras, J. M. Gutiérrez, C. González, J. Nepal Chem. Soc. **30**, (2013)
J.P.E. Saut, G.D. Healey, A.M. Borges, I.M. Sheldon, Reproduction **148**, 593 (2014)
J.C. Shah, Y. Sadhale, D.M. Chilukuri, Adv. Drug Deliv. Rev. **47**, 229 (2001)
I.M. Sheldon, M.H. Roberts, PLoS One **5**, e12906 (2010)
I.M. Sheldon, J. Cronin, L. Goetze, G. Donofrio, H.-J. Schuberth, Biol. Reprod. **81**, 1025 (2009)
I.M. Sheldon, A.N. Rycroft, B. Dogan, M. Craven, J.J. Bromfield, A. Chandler, M.H. Roberts, S.B. Price, R.O. Gilbert, K.W. Simpson, PLoS One **5**, e9192 (2010)
I.M. Sheldon, S.-E. Owens, M.L. Turner, J. Reprod. Immunol. **119**, 67 (2017)
M.P. Slack, W.W. Nichols, J. Antimicrob. Chemother. **10**, 368 (1982)
A.D. So, T.A. Shah, Ups. J. Med. Sci. **119**, 176 (2014)
S. Srinivasan, C. Liu, C.M. Mitchell, T.L. Fiedler, K.K. Thomas, K.J. Agnew, J.M. Marrazzo, D.N. Fredricks, PLoS One **5**, e10197 (2010)
R. Stahlmann, J. Antimicrob. Chemother. **26**(Suppl D), 31 (1990)
R. Stahlmann, H.M. Lode, Expert Opin. Drug Saf. **12**, 497 (2013)
G. Tachedjian, D.E. O'Hanlon, J. Ravel, Microbiome **6**, 29 (2018)
G. Tan, P. Xu, V.T. John, J. He, G.L. McPherson, V. Agarwal, A. Bose, Langmuir **24**, 10621 (2008)
M. Turner, G. Healey, I. Sheldon, Reprod. Domest. Anim. **47**, 402 (2012)
M.L. Turner, J.G. Cronin, G.D. Healey, I.M. Sheldon, Endocrinology **155**, 1453 (2014)
F.L. van de Veerdonk, M.G. Netea, C.A. Dinarello, L.A.B. Joosten, Trends Immunol. **32**, 110 (2011)
Vermani, Garg, Pharm. Sci. Technolo. Today **3**, 359 (2000)
S.A. Wang, J.R. Papp, W.E. Stamm, R.W. Peeling, D.H. Martin, K.K. Holmes, J. Infect. Dis. **191**, 917 (2005)
D.C. Wathes, Z. Cheng, W. Chowdhury, M.A. Fenwick, R. Fitzpatrick, D.G. Morris, J. Patton, J.J. Murphy, Physiol. Genomics **39**, 1 (2009)
WHO, Geneva, 2012. Available at: http://www.who.int/reproductivehealth/publications/rtis/stisestimates/en
C.R. Wira, M.V. Patel, M. Ghosh, L. Mukura, J.V. Fahey, Am. J. Reprod. Immunol. **65**, 196 (2011)
L.R. Wiseman, J.A. Balfour, Drugs Aging **4**, 145 (1994)
P. Wu, D.W. Grainger, Biomaterials **27**, 2450 (2006)
D. Wyatt, D. Dorschel, Pharm. Technol. **116** (1992)
Q. Yao, J. Wang, T. Cui, Z. Yang, M. Su, P. Zhao, H. Yan, Y. Zhan, H. Yang, Molecules **21**, 62 (2016)
A. Zabara, R. Mezzenga, Soft Matter **8**, 6535 (2012)
A. Zabara, R. Mezzenga, J. Control. Release **188**, 31 (2014)

Biomedical Microdevices (2019) 21: 31
https://doi.org/10.1007/s10544-019-0358-0

Engineered microscale hydrogels for drug delivery, cell therapy, and sequencing

Marissa E. Wechsler[1,2] · Regan E. Stephenson[3] · Andrew C. Murphy[2,4] · Heidi F. Oldenkamp[2,4] · Ankur Singh[3,5,6] · Nicholas A. Peppas[1,2,4,7,8]

Published online: 23 March 2019

Author's Personal Note

I have known Prof. Mauro Ferrari for 20 years and I have admired his vision, dedication to nanotechnology, and ability to come up with major medical projects. But I am also aware for his personal drama when in his 20s he lost his dear wife Maria Luisa to cancer and had to become the sole parent to his three pre-school children. This struggle and triumph are well presented in a truly magnificent talk (in Italian) entitled "Due parole sulla Nanomedicine: Simone di Cirene" (Two words about nanotechnology: Simon of Cyrene) https://www.youtube.com/watch?time_continue=3&v=K5NsHV5Hupk (starting at 29:40). Mauro uses his personal drama to conclude that his success in nanotechnology is the result of a call from God to do good to others. This is a powerful message that comes at the end of a wonderful scientific talk on nanotechnology and its promise to solving cancer problems. If you follow Mauro Ferrari's talk after 1:19:30 you will be mesmerized with his message, a message which reveals his deep faith. He finishes with the statement "Questi doni vengono dal Signore e devono tornare a Lui tramite i lavori che possiamo fare al servizio degli altri". And the audience breaks in a 5-minute standing ovation, recognizing how Mauro Ferrari has contributed to patients through his work in nanotechnology.

Happy 60th birthday Mauro

Nico Peppas

Abstract

Engineered microscale hydrogels have emerged as promising therapeutic approaches for the treatment of various diseases. These microgels find wide application in the biomedical field because of the ease of injectability, controlled release of therapeutics, flexible means of synthesis, associated tunability, and can be engineered as stimuli-responsive. While bulk hydrogels of several length-scale dimensions have been used for over two decades in drug delivery applications, their use as microscale carriers of drug and cell-based therapies is relatively new. Herein, we critically summarize the fundamentals of hydrogels based on their equilibrium and dynamics of their molecular structure, as well as solute diffusion as it relates to drug delivery. In addition, examples of common microgel synthesis techniques are provided. The ability to tune microscale hydrogels to obtain controlled release of therapeutics is discussed, along with microgel considerations for cell encapsulation as it relates to the development of cell-based therapies. We conclude with an outlook on the use of microgels for cell sequencing, and the convergence of the use of microscale hydrogels for drug delivery, cell therapy, and cell sequencing based systems.

Keywords Cell therapy · Drug delivery · Hydrogels · Sequencing

1 Introduction

Hydrogels are three-dimensional, hydrophilic networks, produced from crosslinked homopolymers, copolymers, or macromers (Peppas 1986; Slaughter et al. 2009). These hydrophilic polymer networks have a large affinity for water, but are insoluble due to the presence of either chemical or physical crosslinks which provide structure and physical integrity throughout the network (Peppas et al. 2000, 2006). Both synthetic and natural polymers have been used to synthesize hydrogels. They include polyacrylates (such as, poly(2-hydroxyethyl methacrylate)), poly(ethylene glycol), poly(vinyl alcohol), gelatin, alginate, chitosan, fibrin, collagen, in addition to many others (Suri and Schmidt 2009; Nikkhah et al. 2012; Morelli et al. 2016; Neves et al. 2017; Wang et al. 2018). Examples of common synthetic polymers are shown in Fig. 1. Hydrogels made from these materials are attractive for many biomedical applications due to their biocompatibility and water content that simulates natural tissues.

✉ Nicholas A. Peppas
peppas@che.utexas.edu

1 Department of Biomedical Engineering, The University of Texas at Austin, Austin, TX, USA

2 Institute for Biomaterials, Drug Delivery, and Regenerative Medicine, The University of Texas at Austin, Austin, TX, USA

3 Meinig School of Biomedical Engineering, Cornell University, Ithaca, NY, USA

4 McKetta Department of Chemical Engineering, The University of Texas at Austin, Austin, TX, USA

5 Sibley School of Mechanical and Aerospace Engineering, Cornell University, Ithaca, NY, USA

6 Englander Institute for Precision Medicine, Weill Cornell Medical College, New York, NY, USA

7 Division of Molecular Pharmaceutics and Drug Delivery, College of Pharmacy, The University of Texas at Austin, Austin, TX, USA

8 Department of Surgery and Perioperative Care, and Department of Pediatrics, Dell Medical School, The University of Texas at Austin, Austin, TX, USA

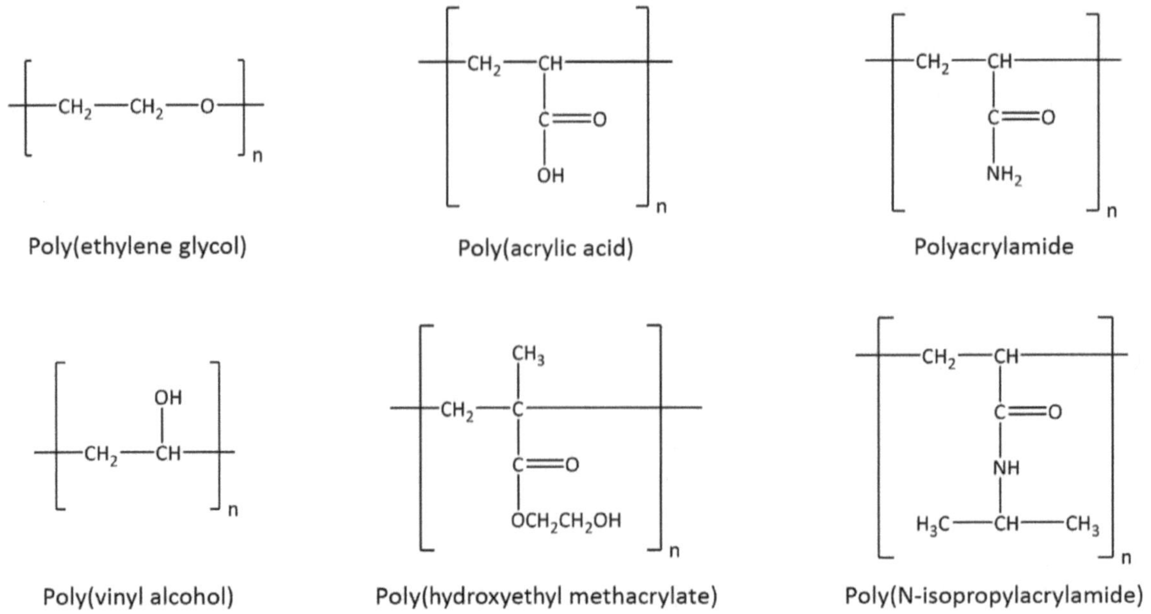

Fig. 1 Representative chemical structures of synthetic polymers used for biomedical applications

2 Hydrogel fundamentals

2.1 Network structure and characterization

Networks and the associated hydrogels can be neutral, cationic, anionic, or ampholytic as dictated by the pendant groups incorporated into the polymer backbone. Ionic gels (whether cationic, anionic, or amphoteric) in aqueous solvents are generally charged (to a certain extent) due to the protonation or deprotonation of charged pendant groups at the pH of the solvent. The extent to which charged pendant groups are either protonated or deprotonated can be controlled by adjusting the pH of the solvent. Hydrogels which deprotonate in the absorbed solvent are anionic, while hydrogels that protonate are cationic.

In anionic hydrogels, the quantity of deprotonated, negatively-charged pendant groups exceeds uncharged pendant groups when the pH of the solvent is greater than the pKa of the polyacid (commonly carboxylic acids). In contrast, cationic hydrogels have positively-charged, protonated pendant groups that are present at higher concentrations compared to neutral pendant groups when the pH is below the pKa of the conjugate polyacid (commonly protonated amines). As a result of varying pendant group incorporation into the hydrogel backbone, these networks can demonstrate environmentally responsive swelling behaviors. Factors which can affect hydrogel swelling in aqueous solutions include pH (Fig. 2a) and ionic strength (Fig. 2b); their influence is indicated by the general dependence shown in the Fig. 2.

Several quantitative parameters can be used to describe the network structure of hydrogels. The most important parameters used for hydrogel characterization include the polymer volume fraction in the swollen state ($\upsilon_{2,s}$), the number average molecular weight between crosslinks ($\overline{M_c}$), and the molecular distance between two consecutive crosslinks, otherwise known as mesh size (ξ). These related parameters can be predicted theoretically and determined experimentally. However, the two most prominent methods used to obtain these parameters which describe the hydrogel network structure are derived from equilibrium swelling theory and rubber elasticity theory (Lowman et al. 2004).

2.2 Equilibrium swelling theory

The Flory–Rehner theory (Flory and Rehner 1943), used for the analysis of neutral hydrogels, describes swelling by indicating that crosslinked polymer gels immersed in a fluidic environment will reach equilibrium as a subject of two opposing forces: (i) the thermodynamic force via mixing and (ii) the retractive force of the polymer chains.

Analytically, this is expressed in terms of the Gibbs free energy, shown in Eq. 1. Here, $\Delta G_{elastic}$ represents the contribution to the Gibbs free energy due to elastic, retractive forces inside the hydrogel, and ΔG_{mixing} is the Gibbs free energy change due to the compatibility of the polymer and the solvent.

$$\Delta G_{total} = \Delta G_{elastic} + \Delta G_{mixing} \tag{1}$$

Differentiation of Eq. 1 with respect to the moles of solvent, while maintaining constant pressure and temperature, results in Eq. 2. Here, μ_1 and $\mu_{1,o}$ are the chemical potential of the solvent in the hydrogel and pure solvent, respectively. At equilibrium, the difference between the two chemical potentials (solvent outside and inside of the hydrogel) must be

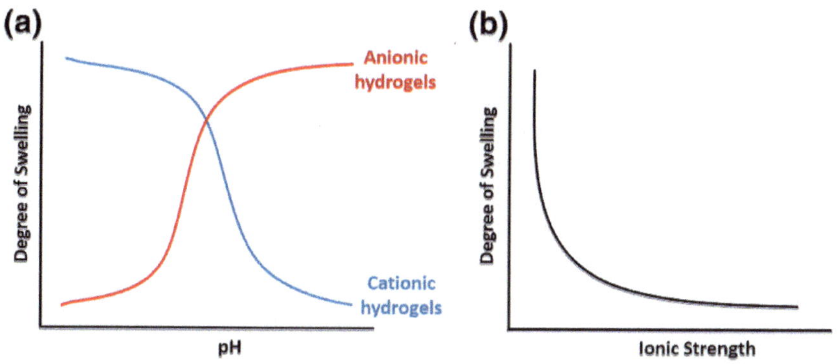

Fig. 2 Swelling behavior of ionizable hydrogels in aqueous solutions. **a** Cationic hydrogels remain in a swollen state at low pH and transition into a collapsed state at high pH, while anionic hydrogels are collapsed at low pH and swell at high pH. **b** Swelling of hydrogels decreases with increasing ionic strength. Figure adapted from (Culver and Peppas 2017)

equal to zero, and therefore, differences in chemical potential due to elastic forces and mixing must balance.

$$\mu_1 - \mu_{1,o} = \Delta\mu_{elastic} + \Delta\mu_{mixing} \quad (2)$$

The theory of rubber elasticity can be used to determine the change in chemical potential due to elastic forces, and moreover, lead to an expression (Eq. 3) to obtain the average molecular weight between crosslinks, $\overline{M_c}$, if the neutral hydrogel is prepared in the absence of solvent. In equation 3, $\overline{M_n}$ is the molecular weight of the polymer chains prepared under identical conditions but in the absence of the crosslinking agent, $\bar{\upsilon}$ is the specific volume of the polymer, V_1 is the molar volume of the swelling agent (water), and $\upsilon_{2,s}$ represents the polymer volume fraction in the swollen state. It must be noted that this equation was developed for crosslinked systems with tetrafunctional crosslinks.

$$\frac{1}{\overline{M_c}} = \frac{2}{\overline{M_n}} - \frac{\left(\frac{\bar{\upsilon}}{V_1}\right)\left[\ln(1-\upsilon_{2,s}) + \upsilon_{2,s} + \chi_1\upsilon_{2,s}^2\right]}{\left(\upsilon_{2,s}^{1/3} - \frac{\upsilon_{2,s}}{2}\right)} \quad (3)$$

This theory was further modified by Peppas and Merrill (Peppas and Merrill 1977) for hydrogels prepared in the presence of water (Equation 4). This adaption modifies the change in chemical potential due to elastic forces to take into account the volume fraction density of the chains during crosslinking. In Equation 4, $\upsilon_{2,r}$ denotes the polymer volume fraction in the relaxed state (the state of the polymer immediately after crosslinking), while $\upsilon_{2,s}$ represents the polymer volume fraction in the swollen state in water.

$$\frac{1}{\overline{M_c}} = \frac{2}{\overline{M_n}} - \frac{\left(\frac{\bar{\upsilon}}{V_1}\right)\left[\ln(1-\upsilon_{2,s}) + \upsilon_{2,s} + \chi_1\upsilon_{2,s}^2\right]}{\upsilon_{2,r}\left[\left(\frac{\upsilon_{2,s}}{\upsilon_{2,r}}\right)^{1/3} - \left(\frac{\upsilon_{2,s}}{2\upsilon_{2,r}}\right)\right]} \quad (4)$$

The presence of ionic moieties within hydrogels adds complexity to the swelling analysis by incorporating a term to account for the total change in Gibbs free energy due to the ionic properties of the network, ΔG_{ionic}. This modification is shown in Equation 5.

$$\Delta G_{total} = \Delta G_{elastic} + \Delta G_{mixing} + \Delta G_{ionic} \quad (5)$$

Again, upon differentiating Equation 5 with respect to the moles of solvent molecules, while maintaining constant pressure and temperature, Equation 6 can be obtained. Here, while the expression is similar to that of Equation 2, the term $\Delta\mu_{ionic}$ represents the change in chemical potential due to the ionic nature of the hydrogel.

$$\mu_1 - \mu_{1,o} = \Delta\mu_{elastic} + \Delta\mu_{mixing} + \Delta\mu_{ionic} \quad (6)$$

A more complex equation was derived by Brannon-Peppas and Peppas (Brannon-Peppas and Peppas 1990) for the swelling behavior of ionic hydrogels. For anionic (Equation 7) and cationic (Equation 8) hydrogels, two separate expressions were obtained. Use of these expressions for analysis of the average molecular weight between crosslinks now requires information and numerical values of the ionic strength (*I*) and the equivalent dissociation constants (K_a and K_b for acids and bases, respectively).

$$\frac{V_1}{4I}\left(\frac{\upsilon_{2,s}^2}{\bar{\upsilon}^2}\right)\left(\frac{K_a}{10^{-pH} + K_a}\right)^2 = \left[\ln(1-\upsilon_{2,s}) + \upsilon_{2,s} + \chi_1\upsilon_{2,s}^2\right] + \left(\frac{V_1}{\bar{\upsilon}\overline{M_c}}\right)\left(1 - \frac{2\overline{M_c}}{\overline{M_n}}\right)\upsilon_{2,r}\left[\left(\frac{\upsilon_{2,s}}{\upsilon_{2,r}}\right)^{1/3} - \left(\frac{\upsilon_{2,s}}{2\upsilon_{2,r}}\right)\right] \quad (7)$$

$$\frac{V_1}{4I}\left(\frac{\upsilon_{2,s}^2}{\bar{\upsilon}^2}\right)\left(\frac{K_b}{10^{pH-14} + K_a}\right)^2 = \left[\ln(1-\upsilon_{2,s}) + \upsilon_{2,s} + \chi_1\upsilon_{2,s}^2\right] + \left(\frac{V_1}{\bar{\upsilon}\overline{M_c}}\right)\left(1 - \frac{2\overline{M_c}}{\overline{M_n}}\right)\upsilon_{2,r}\left[\left(\frac{\upsilon_{2,s}}{\upsilon_{2,r}}\right)^{1/3} - \left(\frac{\upsilon_{2,s}}{2\upsilon_{2,r}}\right)\right] \quad (8)$$

For solute (drug, peptide, protein, therapeutic agent) transport, the primary mechanism of transport or release is diffusion through the molecular space between polymer chains, often referred to as "molecular pores" or mesh. Depending on the size of the pores, hydrogels can be classified either as nonporous, microporous, or macroporous. In addition, the type and structure of crosslinking throughout the hydrogel can greatly affect its properties (such as porosity) through factors such as entanglements, hydrogen bonding, ionic bonding, crystallites, and covalent bonding.

The porosity of hydrogels can be defined by the structural parameter (ξ), the mesh size. The mesh size, or correlation length, determines the average linear distance between crosslinks as shown in Equation 9. In this expression, the unperturbed end-to-end distance between neighboring crosslinks is $(\bar{r}_o^2)^{1/2}$ (a known value) and the extension ratio (α) can be determined using Equation 10 from the polymer volume fraction in the swollen state ($\upsilon_{2,s}$). However, mesh size can be predicted without the unperturbed distance using Equation 11 if the values of certain parameters are available or can be determined: the bond length of the polymer backbone (l), the characteristic ratio (C_n), the average molecular weight between crosslinks ($\overline{M_c}$), the polymer volume fraction in the swollen state ($\upsilon_{2,s}$), and the molecular weight of repeating units (M_r).

$$\xi = \alpha\left(\bar{r}_o^2\right)^{1/2} \tag{9}$$

$$\alpha = \upsilon_{2,s}^{1/3} \tag{10}$$

$$\xi = \upsilon_{2,s}^{-1/3}\left(\frac{2C_n\overline{M_c}}{M_r}\right)^{1/2} l \tag{11}$$

The ability to tailor the molecular structure of networks and the thermodynamic/swelling behavior of hydrogels using several key parameters enables tuning of their diffusive, responsive, and mechanical properties (Peppas et al. 2006). These parameters are the polymer volume fraction in the swollen state ($\upsilon_{2,s}$), the average molecular weight between crosslinks ($\overline{M_c}$), and the mesh size (ξ).

2.3 Solute diffusion from hydrogels

Understanding solute transport is likely the single most important design parameter for hydrogels used as drug carriers. Since solute transport within hydrogels is primarily based on diffusion, Fick's law of diffusion can be used to describe this phenomenon shown in Equations 12 and 13. Here, c_i represents the concentration, j_i indicates the mass flux of solute i, and $D_{i,p}$ is the diffusion coefficient of solute i in the polymer matrix, p.

$$j_i = -D_{i,p}\frac{dc_i}{dx} \tag{12}$$

$$\frac{\partial c_i}{\partial t} = D_{i,p}\frac{\partial^2 c_i}{\partial x^2} \tag{13}$$

It is important to note that several assumptions have been incorporated into the equations above. First, these equations describe solute release from a carrier of a thin, planar geometry; although equivalent equations for solute release from carries with a variety of other geometries have been derived (Crank 1975).

In addition, in the form of Fick's law shown above, the diffusion coefficient is assumed to be independent of concentration. Thus, to improve the predictive power of Fickian diffusion, a concentration-dependent diffusion coefficient is used and can be written as Equation 14. In Equation 14, $D_{i,p}(c_i)$ is the concentration-dependent diffusion coefficient, and its form is dependent upon characteristics of the polymer matrix (Peppas and Narasimhan 2014).

$$\frac{\partial c_i}{\partial t} = \frac{\partial}{\partial z}\left(D_{i,p}(c_i)\frac{\partial c_i}{\partial x}\right) \tag{14}$$

In typical solute eluting systems, diffusion commonly dictates distribution of solutes throughout hydrogels and delivery to its surrounding environment. Drug release from these swelling-controlled systems (where solute release is mediated by the inward flux of solvent molecules and subsequent swelling of the polymer matrix) can be predicted via a power law relation derived by Korsmeyer and Peppas (Korsmeyer et al. 1983), shown in Equation 15. Here, M_t and M_∞ are the cumulative amounts of drug released at time t and equilibrium, respectively, k is a constant dictated by the structural and geometric information about the matrix, and n is indicative of the drug release mechanism. Release by Fickian diffusion using Equation 15 is approximated with $n = 0.5$.

$$\frac{M_t}{M_\infty} = kt^n \tag{15}$$

3 Methods of synthesizing microscale hydrogels

Polymeric microgels have emerged as a popular alternatives to bulk hydrogels because of their ease of injectability, size control, immunoisolation, and enhanced nutrient transport. Microscale hydrogels can be tuned to release encapsulated drugs and proteins at controlled rates based on diffusion through the hydrogel network, hydrolytic degradation of polymer network, or stimuli-responsive swelling of the network. In contrast to bulk hydrogels, the microscale diameter of microgels also maximizes transport of oxygen and nutrients into and out of hydrogels, which is specifically beneficial to encapsulated cells. For applications in drug delivery and cell encapsulation, approaches for the synthesis of controlled hydrogel architectures at the microscale level have been developed. Figure 3 summarizes some of the most prevalent techniques used to synthesize polymeric microscale hydrogels for biomedical applications. Methods like microfluidics allow for the production of microdroplet carriers that are well-suited for drug delivery applications (Foster et al. 2017). In contrast, lithography offers improved control over the three-dimensional nanoscale structure of the hydrogel which has been exploited for drug delivery applications (Gratton et al. 2007; Caldorera-Moore et al. 2011) and microscale patterning of cell-laden hydrogels for tissue engineering applications (Suri and Schmidt 2009; Mironi-Harpaz et al. 2015). In each case, the hydrogel precursor is chosen to produce the desired network properties and can be a monomer solution or uncrosslinked polymer combined with a crosslinking agent. Cells or therapeutic agents can then be included in the pre-polymer solution, allowing for simultaneous encapsulation, or in the case of drug delivery applications, the drug can be loaded into the carrier post-synthesis through complexation with functional groups such as carboxylate anions.

3.1 Microfluidics

Droplet-producing microfluidic devices are emerging as the method of choice for microgel fabrication because they are easily customizable for the design requirements of the microgel application. These devices work by sectioning off pre-gel materials in an oil emulsion phase and typically achieved using a T-junction (Tan and Takeuchi 2007) or flow-focusing design (Headen et al. 2014, 2018a; Chen et al. 2016b; Foster et al. 2017; Lienemann et al. 2017; Zhang et al. 2018). By adjusting the flow rate ratios of input fluids, microgels of varying sizes and dispersity can be produced at a rapid rate with the fastest production of monodisperse microgels occurring during the jetting phase (Headen et al. 2014; Zhang et al. 2018). Microgel production rate has been reported between 250 and 1200 Hz depending on particle diameter (Griffin et al. 2015). Through adopting a device design with parallel flow focusers, this rate has been increased by up to 6-fold (Headen et al. 2018a), thereby reducing the time cells interact with pre-gel components and increasing throughput capacity. Headen et al. demonstrated that parallel microfluidic devices can increase microgel production rate by 600% with minimal impact on the size or polydispersity compared to single nozzle microfluidic devices.

Successful microfluidic-directed gelation of PEG-based macromers specifically, PEG-maleimide (Headen et al.

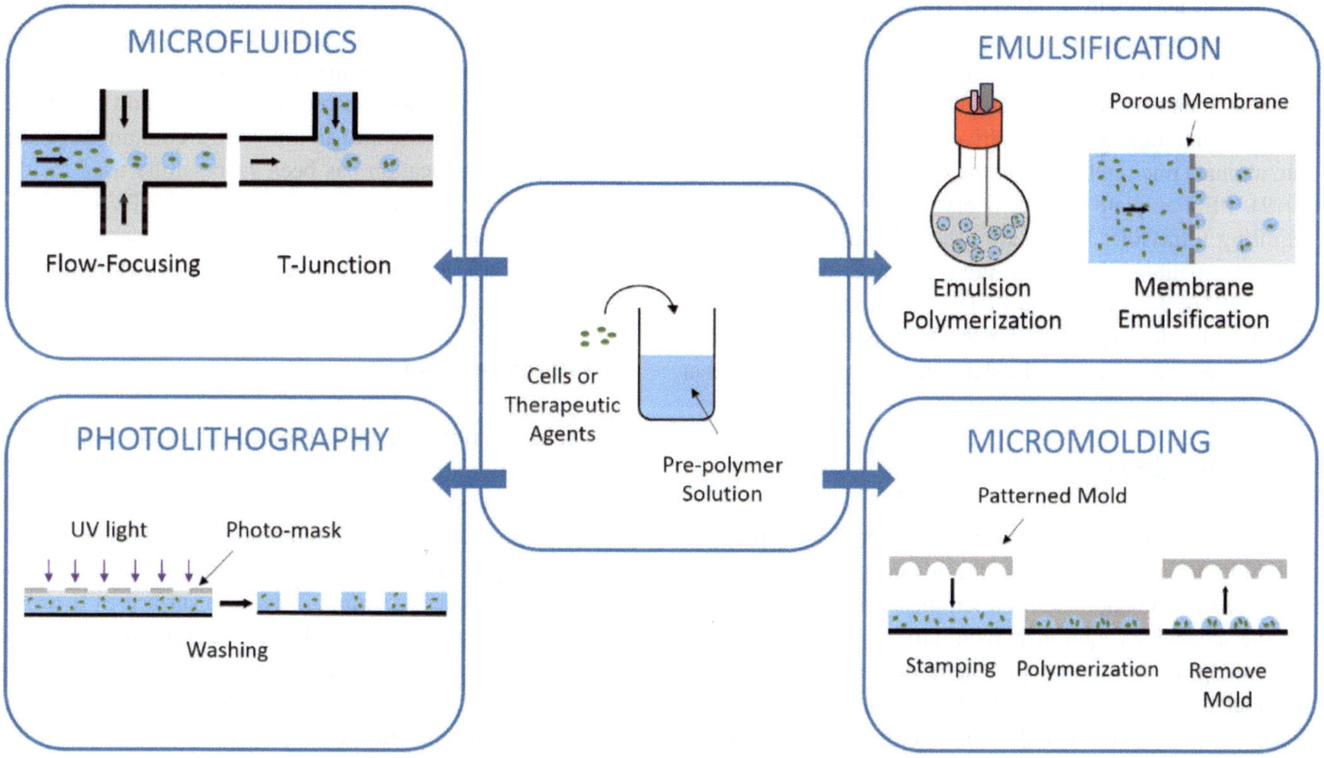

Fig. 3 Schematic of microscale hydrogel synthesis techniques. A pre-polymer solution containing the desired components of the hydrogel network is combined with cells or therapeutic agents for simultaneous encapsulation. Microstructures are formed using techniques such as microfluidics, emulsification, photolithography, or micromolding, and the network is crosslinked to form microgels

2014, 2018a, 2018b; Foster et al. 2017), PEG-vinylsulfone (Allazetta et al. 2013; Lienemann et al. 2017), PEG-acrylates (Choi et al. 2016), alginate (Tan and Takeuchi 2007; Hirama et al. 2012; Kim et al. 2014; Chen et al. 2016a; Mao et al. 2017; Zhang et al. 2018), and methacrylated gelatin (Rossow et al. 2012; Lee and Cha 2018) have been established to date. These biomaterial systems vary in crosslinking chemistry, pH requirements, and number of precursor solutions required before gelation, all of which drive design requirements of the microfluidic device for the microgel on-chip gelation process.

Combination of a macromer and a crosslinking agent, for example in the case of PEG-maleimide, requires fast on-chip combination to avoid premature crosslinking (Foster et al. 2017), or a diffusive design when the crosslinking agent is small enough to allow diffusion through macromer on-chip (Headen et al. 2014, 2018a, 2018b). Due to the slower speed of gelation for PEG-vinylsulfone, fast on-chip gelation does not impose a significant limitation. The ion-based requirements for crosslinking of alginate have synergized well with microfluidic device fabrication of alginate microgels.

One approach is to use calcium- ethylenediaminetracetic acid or calcium carbonate as the source for calcium ions, made available for crosslinking alginate by co-flowing acid-laden oil (Tan and Takeuchi 2007; Chen et al. 2016a; Zhang et al. 2018). Alternative methods involve off-chip pairing of microgels with an agarose slab containing calcium chloride. After the alginate pre-gel is sectioned into droplets within a decane and surfactant continuous phase by a flow focusing microfluidic device, the pre-microgel droplets are collected over a macro-scale agarose gel containing solubilized calcium chloride. Innate osmotic pressure differences between the agarose slab and alginate droplet drive diffusion of calcium ions into alginate microgels without the use of acids (Hirama et al. 2012). PEG-acrylate and methacrylated gelatin chemistries requiring ultraviolet light and a photo initiator have shown to be compatible with microfluidic microgel droplets (Choi et al. 2016; Lee and Cha 2018). Optically transparent microfluidic devices allow for the brief penetration of UV facilitating gelation as the droplet proceeds to be collected at the output.

3.2 Emulsification

Emulsification has been extensively studied for the preparation of hydrogel carriers due to the advantage of synthesizing hydrogels from nanometer to micrometer length scales without the specialized equipment requirements of other methods. Depending on the solubility of the desired monomers, a prepolymer solution is chosen to form either an oil-in-water or a water-in-oil emulsion with the continuous phase upon sonication or mechanical stirring, although other systems such as water-in-water emulsions (Aydın and Kızılel 2017) have been explored. Surfactants are commonly used to improve the emulsion stability, and nonionic, cationic, anionic, zwitterionic, or a combination of surfactants can be chosen to tune the desired microdroplet size. The polymer network is then formed by either polymerization of the monomer and crosslinking agent (Yan and Gemeinhart 2005; Tokuyama and Kato 2008; González-Sánchez et al. 2011; Aydın and Kızılel 2017; Wang et al. 2018), or by physical or chemical crosslinking of the polymer (Wang et al. 2013; Morelli et al. 2016; Piacentini et al. 2017).

A wide range of homopolymer and copolymer systems have been explored with emulsion techniques including polyacrylamide (González-Sánchez et al. 2011), poly(acrylic acid) (Yan and Gemeinhart 2005), poly(ethylene glycol) (Aydın and Kızılel 2017; Wang et al. 2018), chitosan (Wang et al. 2013; Morelli et al. 2016), and poly(vinyl alcohol) (Morelli et al. 2016; Piacentini et al. 2017). For drug delivery applications, emulsion polymerization is frequently used to produce nanoparticle carriers (Sahana et al. 2008; You and Auguste 2008), which offer improved surface area-to-volume ratios; however, microparticle systems have also been studied (Yan and Gemeinhart 2005). With emulsification approaches for cell encapsulation, microparticles can be prepared by the physical or chemical crosslinking of the polymer solution to avoid cytotoxicity of many monomers and surfactants. This approach has successfully been used for the encapsulation of human mesenchymal stem cells in chitosan and collagen microbeads (Wang et al. 2013).

The major limitations of emulsification are the lack of control over particle morphology, size, and polydispersity. One strategy to produce monodisperse carriers is membrane emulsification, which forces the pre-polymer solution through a microporous membrane into an immiscible continuous phase. Membrane emulsification has been successfully used for the entrapment of lipase in poly(vinyl alcohol) (PVA) microspheres (Piacentini et al. 2017), as well as pH dependent release from chitosan and PVA microparticles (Morelli et al. 2016).

3.3 Photolithography

The use of photolithography in biomedical applications for the synthesis of microgels has grown in interest due to improved control over the size and shape of the hydrogel, in addition to greater uniformity. Photolithographic preparation of microstructures begins with coating the pre-polymer solution containing a light reactive initiator onto the surface of a wafer, which can be functionalized with reactive groups to anchor the resulting hydrogel to the surface. The pre-polymer solution is then selectively exposed to light in a desired pattern, resulting in polymerization of only the exposed regions, and then the unreacted solution is removed.

Photolithographic techniques for cell encapsulation and drug delivery have successfully used poly(ethylene glycol) acrylates (Revzin et al. 2001; Hoffmann and West 2010), chitosan (Li et al. 2015), gelatin methacrylate (Nikkhah et al. 2012), fibrinogen (Mironi-Harpaz et al. 2015), and interpenetrating networks of collagen and hyaluronic acid (Suri and Schmidt 2009) for the creation of patterned microstructures. Compared to other synthesis methods, photolithography offers the advantage of precisely controlling particle size and shape through selective initiation without the need for the additional continuous phase required by emulsification or microfluidics for the creation of pre-patterned mold. Additionally, the ability to sequentially coat multiple layers of pre-polymer solution has been exploited to prepare multilayer microparticles of poly(N-vinyl-2-pyrrolidone) and PEG triacrylate that selectively incorporate PEG-biotin into a single layer (Li et al. 2016).

The major challenges of photolithographic microstructure synthesis come from the reliance on photo-masks for selective exposure, which are limited to two-dimensional patterns and suffer from diffraction limitations on the resolution of the resulting structures. One alternative is two photon photolithography, which can be performed by the absorption of two photons of the same wavelength (Hoffmann and West 2010), or with a combination of photoinitiator and photoinhibitor that each absorb different wavelengths (Scott et al. 2009). Because of the need to absorb two photons simultaneously, two photon photolithography offers the advantage of smaller initiation volumes, which allow for nanoscale structures in addition to three-dimensional patterns (Kawata et al. 2001; Deforest et al. 2009). For cell encapsulation, there is an additional challenge of obtaining successful crosslinking of patterned structures in the dilute pre-polymer solutions necessary for cell compatibility. In order to address this issue, microscopic laser photolithography was successfully used to pattern cell-compatible hydrogels with fibrinogen and Pluronic F127 to exploit temperature-dependent phase transitions as a way to minimize diffusion during crosslinking (Mironi-Harpaz et al. 2015).

3.4 Micromolding

Micromolding uses a patterned master template that contains a negative of the desired microstructure. The pre-polymer solution is then either filled into the mold, or pressed between the mold and a solid substrate prior to initiation, and removed once polymerization has completed, leaving structured hydrogels with the shape of the pattern's indentation. For biomedical applications, micromolding has been successfully employed for the creation of patterned microgels of natural polymers including hyaluronic acid (Khademhosseini et al. 2006), collagen (McGuigan et al. 2008), and chitosan (Fukuda et al. 2006; Jung and Yi 2015), as well as synthetic polymers including poly(ethylene-glycol) (Subramani and Birch 2009; Heath et al. 2015; Jung and Yi 2015) and poly(dimethyl siloxane) (PDMS) (Fu et al. 2010). Moreover, the Singh group previously reported a microarray of composite bioadhesive microgels, specifically, maleimide functionalized polyethylene glycol with an interpenetrating network of gelatin ionically cross-linked with silicate nanoparticles, engineered by integrating microfabrication with Michael-type addition chemistry and ionic gelation (Patel et al. 2014). These composite bioadhesive hydrogels represent a platform that could be used to study the independent effect of stiffness and adhesive ligand density on cell survival and function, as was demonstrated with select cancer cells.

Micromolding can be used to synthesize hydrogels with a variety of two-dimensional shapes, including clovers or crosses (McGuigan et al. 2008), as well three-dimensional structures, such as micropillars (Fu et al. 2010), but it is difficult to prepare patterns with a high aspect ratio or spherical structures. Recent techniques have improved the ability to generate complex microstructures. Chitosan-poly(ethylene glycol) microspheres with a core-shell structure were prepared by exploiting surface tension induced droplet formation in PDMS molds (Jung and Yi 2015). Additionally, microparticles, rather than pre-polymer solution, can be used in order to create microstructures from multiple materials (Park et al. 2007).

The ability of micromolding to simultaneously pattern multiple three-dimensional structures without the need for spin coating multiple pre-polymer layers or limiting exposure area is a significant advantage over photolithography. Despite this advantage, the need for a mold presents several challenges for micromolding approaches. For micromolding methacrylated hyperbranched polyglycerol microparticles, as many as 50% of the particles prepared using a rigid SU-8 photoresist mold were found to break apart or remain attached to the mold after the synthesis, and although switching to a flexible PDMS mold improved yield, the difficulty completely filling the mold with the pre-polymer solution led to greater defects than photolithography (Oudshoorn et al. 2007). Additionally, the use of PDMS for the patterned template has been found to increase protein adsorption and cell adhesion on PEG microstructures due to residual PDMS left behind on the surface (Heath et al. 2015).

4 Microscale hydrogels for drug delivery

Microscale hydrogels are excellent candidates for use as drug delivery carriers, especially for obtaining controlled release of therapeutic agents. The polymer carriers are primarily used for several reasons: to protect therapeutic molecules from undesired degradation, increase the efficacy of delivery, and decrease potential side effects. To obtain controlled release, the carriers help maintain the concentration of drug within the therapeutic window over an extended time, which allows for a prolonged time period between administrations (Liechty et al. 2010; Vermonden et al. 2012). Controlled release systems are preferred compared to traditional drug delivery platforms because there is no sharp increase in drug concentration shortly following administration, preventing a potentially toxic response. In addition, the need for less frequent administration could potentially increase patient compliance.

Physical and chemical parameters of the microgels can be altered to tune their delivery properties. The structure and density of the crosslinking agent used in the formulation will affect the network mesh size, and thus the rate of the diffusion of water-soluble therapeutic drug out of the microgel (Kim and Bae 1992; Peppas et al. 2000). The diameter of the hydrogels will also affect controlled release, as a larger diameter will lead to longer release if the microgels are completely loaded with the molecule of interest. Different tissues of interest can also be targeted by altering the diameter of the microgels. For instance, microgels 1–5 μm in diameter would be best suited to passively target antigen-presenting cells (Wattendorf et al. 2008), while those 10–20 μm in diameter would be better suited to target capillary beds, especially in tumors (Salem et al. 2005).

Precision particle formation is commonly used to produce uniform particles made of various materials. This technique, which utilizes special solvent extraction and evaporation methods, has been adapted to synthesize microparticles for controlled release drug delivery platforms (Xia and Pack 2015). Precision particle formation can control the particle diameter to within one micron (Berkland et al. 2003). This work demonstrated that properties of both the drugs and the particles impact the distribution of the drug molecules within the particles, as well as the degradation and drug release rates. Furthermore, Berkland and colleagues were able to alter the particle size and shell thickness to obtain particles with zero-order, pulsatile, or tandem drug release (Berkland et al. 2003).

Moreover, the kinetics of drug release from hydrogel microcarriers can be tailored by adjusting properties of the carrier, such as monomers utilized, crosslinking density, use of a coating (to create a core-shell structure), as well as several other parameters. An example of this is the recently developed hydrogel microparticle drug delivery system comprised of poly(lactic-co-glycolic) acid (synthesized via a water-in-oil emulsion) for the release of platelet derived growth factor-AA (Pinezich et al. 2018). This specific growth factor has been studied as a signaling molecule to initiate a response in oligodendrocyte precursor cells, which can lead to proliferation or differentiation of central nervous system tissue after injury or disease. When oligodendrocyte precursor cells were exposed to drug loaded microgels, it was found that a burst release, followed by withdrawal of the growth factor, stimulated survival, proliferation, and differentiation of the cells *in vitro*. This type of system could allow the development of improved strategies for regeneration of neural tissues, which are intrinsically limited.

Understanding the conditions that affect hydrogel properties can also be used to cultivate targeted delivery strategies. For example, anionic pH-responsive materials are commonly used in oral drug delivery platforms due to their ability to protect the therapeutic molecule in the harsh gastric conditions of the stomach and then release it in the small intestine for absorption into the bloodstream (Torres-Lugo et al. 2002; Carr and Peppas 2009; Durán-Lobato et al. 2014; Koetting et al. 2016; Steichen et al. 2016; Sharpe et al. 2018). Other responsive hydrogel systems have been used in additional drug delivery platforms, such as an enzyme-responsive microgel system studied for pulmonary drug delivery (Secret et al. 2014). An inverse emulsion polymerization method was used to synthesize poly(ethylene glycol) diacrylate microparticles containing specific peptides that degrade in the presence of metalloproteinases that are overexpressed in certain pulmonary diseases such as lung cancer, tuberculosis, and chronic obstructive pulmonary disease (COPD) (Chang et al. 1996; Demedts et al. 2006; Leinonen et al. 2006; Ullah and Aatif 2009). Hydrogels are especially useful for pulmonary drug delivery because their size can be precisely engineered for bronchial delivery, but their swelling behavior prevents the microgels from uptake and clearance by macrophages in the alveoli.

5 Applications of microscale hydrogels for cell therapy

Microscale hydrogels are appealing for cell encapsulation and therapy because they are highly permeable to water and easily customizable for the cell type and application. For example, incorporation of short, cell-recognizable polypeptide sequences enhances bioactivity of the microgel, as well as allows for countless permutations of possible microgel qualities. Many PEG-based hydrogels have been developed to promote enzymatic degradation through the incorporation of protease-degradable peptides (Raeber et al. 2005; Lee et al. 2005; Phelps et al. 2012; Gould et al. 2012; Schultz and Anseth 2013). These functionalized PEG hydrogels have been used to regulate cellular responses (Cayrol et al. 2015; Lee et al. 2015), and have been engineered for the development of tissue engineering constructs (Lutolf and Hubbell 2005; Sridhar et al. 2015; Schultz et al. 2015), tumor organoids (Tian et al. 2015), and synthetic ex vivo immune tissues (Purwada et al. 2018). The diverse spectrum of cell types used for such applications each require preferred culturing conditions that can be supplied through a personalized microgel with major applications in cell therapy and *in vitro* tissue and disease models. In order to develop a microgel strategy for these uses, special considerations and validations must be implemented.

Many pre-gel chemicals, as well as the oil phase in microfluidic approaches, can threaten cell viability during the formation and preparation of microgels. In fact, for an acid-laden oil double emulsion set-up, Zhang et al. showed that viability was best at high oil flow rates because cell contact with the oil was therefore shorter (Zhang et al. 2018). Oil cytotoxicity concerns were further addressed by introducing perfluorooctanol on-chip to destabilize oil coating on microgels and allow for fast oil removal upon collection of microgels downstream (Zhang et al. 2018). Along similar lines, a double emulsion approach minimizes the required amount of oil phase, thereby increasing viability over a single emulsion device format of aqueous gel droplets in oil (Choi et al. 2016). Microgels fabricated using the double-emulsion approach offer the added ease of spontaneous "dewetting" of oil in Dulbecco's Modified Eagle Medium (DMEM) culture media, providing for easy oil removal immediately following gelation.

Confirmation of microgel encapsulated cell viability and biological activity is critical in the validation of any microgel platform for cell therapy or *in vitro* models. Survival from the microgel fabrication process and extended culture is easily confirmed using live/dead fluorescent stains adopted by many microgel groups (Steinhilber et al. 2013; Headen et al. 2014; Chen et al. 2016a; Choi et al. 2016; Zhang et al. 2018). Following proof of cell survival in microgels, biological activity and proliferation of cells should be assessed. This second step is arguably more important because it determines the biological relevance of the biomaterial and microgel parameters. For example, in encapsulating pancreatic islet cells for targeted cell delivery, insulin secretion should remain undisrupted by the surrounding microgel (Headen et al. 2014).

By adopting non-cytotoxic crosslinking chemistries, cell viability is easily sustained through microgel fabrication and beyond. One main driver for this approach is the development of modular, high-throughput three-dimensional cell culture models. Siltanen and colleagues demonstrated the efficacy of heparin-PEG microgels dosed with the growth factors Nodal and FGF-2 in directing the differentiation of mouse embryonic stem cells toward an endoderm lineage (Siltanen et al. 2016). This approach showed differentiation potential similar to hanging droplet cultures but at a far higher throughput by fabricating hydrogel droplets in a microfluidic device. Other cell microgel culture models include a bilayered alginate model of the liver by encapsulating hepatocytes surrounded by fibroblasts in the outer alginate layer (Chen et al. 2016b). Three-dimensional cancer models are also increasingly desired as platforms for drug discovery and dose finding. In a tumor model of prostate cancer, hyaluronic acid microgel models drove overexpression of two multi-drug resistant genes greater than two-dimensional culture (Xu et al. 2011). Additionally, there are models devoted to development of a mesenchymal stem cell niche (Lienemann et al. 2017), some with applications for immunoprotected cell therapy (Headen et al. 2018a). While there have been several microgel approaches proposed for cell therapeutic delivery, many simply propose this application, and stop short of validating utility using *in vivo* studies.

Microgel delivery of cells has been a promising solution to immune rejection of donor cells (Headen et al. 2014) which causes cell death and limits the therapeutic window of cell therapy. Immunoisolation of cells using microgels is achieved in two ways: most commonly, by physical isolation from immune cells by the biomaterial (Headen et al. 2014; Mao et al. 2017; Bochenek et al. 2018), but also by inducing localized immune suppression through apoptosis of adaptive immune cells (Headen et al. 2018b). The latter approach utilized T cell sensitivity to Fas ligand (Fas-L) following activation. By presenting biotin-streptavidin conjugated Fas-L to microgel surfaces, immune response to co-delivered unmodified donor islet cells was diminished through depletion of effector T cells while regulatory T cells remain (Fig. 4a). This approach is notable because it demonstrates the use of microgels as drivers of immune tolerance for cell therapy when the physical encapsulation of cells in microgels is not required.

Since the inception of microgel technology, it has been proposed as possible platform for cell-based insulin delivery for a long-term solution to insulin-resistant diabetes (Lim and Sun 1980). In line with disappointing clinical trials for islet encapsulated alginate microgels, recent work in the hydrogel-mediated cell delivery field may suggest that gels larger than the microscale are a better conduit than microgels for promoting immune tolerance to encapsulated cell therapy. Comparison of the fibrotic response to alginate gels in mouse peritoneum showed that hydrogel spheres up to 1.9 mm in diameter demonstrated markedly less cellular deposition of macrophages and myofibroblasts compared to microscale gels (Veiseh et al. 2015). Size-dependent foreign body response was conserved in several of the other biomaterials compared (Veiseh et al. 2015). This size theory was later confirmed in non-human primates, when delivery of allogeneic pancreatic islet cells survived 4 months of implantation in modified alginate spheres 1.5 mm in size and further exhibited minimal signs of a foreign-body response (Bochenek et al. 2018) (Fig. 4b). This study highlighted the importance of hydrogel placement, alginate chemistry, and gel size on the foreign body response, and provided strong evidence for the validity of cell encapsulation in hydrogel beads as an effective method of immunoisolation for long-term cell therapy.

Alternatively, it should be considered that encapsulated cells for cell therapy may not require long-term survival or microgel maintenance in every application. In an ischemic model, arginine-glycine-aspartic acid (RGD)-modified alginate microgels were used to deliver human outgrowth endothelial cells and blood vessel sprouting growth factors VEGF and HGF in mice (Kim et al. 2014). In this study, long-term cell viability and microgel biocompatibility was not confirmed, but rather the appearance of new blood vessel in a

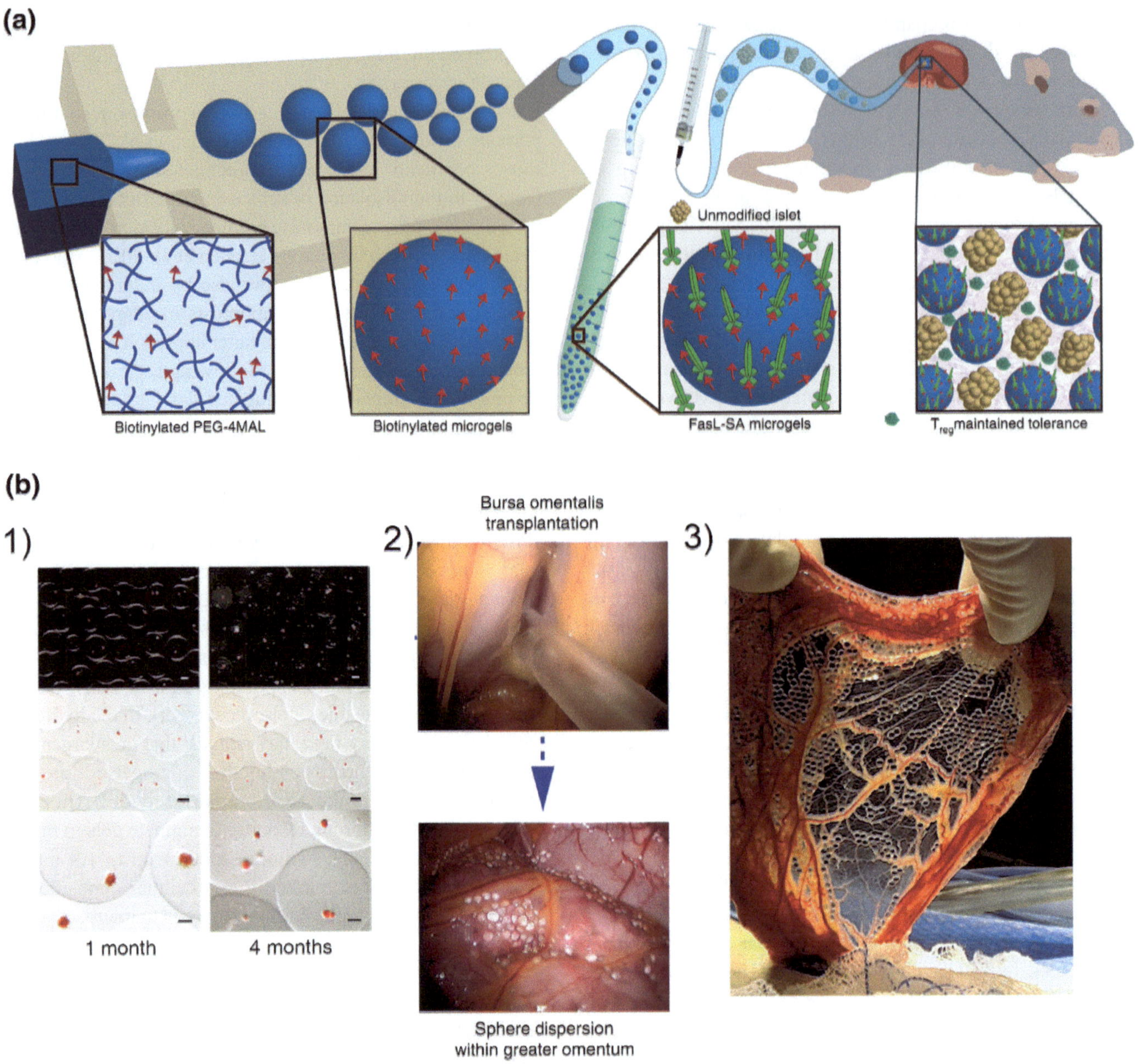

Fig. 4 **a** Fabrication process of immune modulating microgels with biotin-streptavidin surface modification to present Fas ligand. These microgels drive effector T cell apoptosis while maintaining immune-tolerant regulatory T cells in a cell therapy implantation site to sustain insulin delivery from donor islet cells. Reprinted by permission from Springer Nature: Springer, Nature Materials (Headen et al. 2018b), Copyright (2018). **b** Explanted and implanted allogeneic pancreatic islet cells in modified alginate spheres. 1) Explanted modified alginate Z1-Y15 beads with encapsulated islet cells after 1 month and 4 months in a non-human primate. Cells were stained with dithizone (red) to demonstrate sustained insulin production 2) Gels showed minimal clumping in experimental implantation site, the bursa omentalis in the abdomen. 3) Explanted gels nested within the greater omentum remained visibly clear and loosely lodged in connective tissue, demonstrating minimal immune response to donor cells and modified alginate gels. Figure adapted and reprinted by permission from Springer Nature: Springer, Nature Biomedical Engineering (Bochenek et al. 2018), Copyright (2018)

mouse hindlimb ischemia model proved its success as a cell therapy approach. This application provides an alternate look at cell therapy in microgels and widens their application to more than just the long-term, cell-based production of bioproducts.

Additionally, microgel approaches to tissue engineering and cell therapy have looked at how microgel-formed macropores mitigate immune response to biomaterial implantation. The latter was achieved by engineering a microgel system with dual crosslinking mechanisms whereby initial microgel formation occurs on chip, and secondary microgel crosslinking establishes connections between microgels (Griffin et al. 2015). Cells can be incorporated during secondary crosslinking to form a cell-laden hydrogel with macropores. This two-pronged approached was also evaluated for use in a wound healing model without the addition of cells and found that the presence of these pores increased immune cell penetration into the material rather than accumulating at the surface and initiating the foreign body response. Evidence of reduced immune response was maintained out to 5 days post-implantation. This approach provides the framework for an additional microgel application where implanted cells or infiltrating cells from the wound healing process inhabit the negative space surrounding microgels rather than from within microgels. Additionally, this promising achievement successfully bridges advances in microgel technology with macro-scale biomaterials applications and expands the microgel application realm from drug delivery and cell therapy to tissue engineering and regenerative medicine.

6 Future outlook

Independent of the progress of biomedical microgel research, the scientific community has entered into "the genomic era" (Guttmacher and Collins 2003) as the cost of sequencing genomes, transcriptomes, proteomes, and the ever expanding other "omes" has diminished as the technologies have improved. Many of the recent advances in the sequencing field involve scaling size down to the single-cell sampling level. This approach is necessary because sequencing in bulk samples produces information on the population level, losing key information about biological subsets within the sample (Trapnell 2015). By partitioning single cells into separate wells or droplets, individualized sequencing can be performed producing 'omic' data on a cell by cell level. In this way, outlier populations are not buried by the population average and sample heterogeneity is brought to light. Single-cell sequencing, similar to microgel fabrication, has benefited from microfluidic water-in-oil emulsions to separate cells into individual droplets that serve as personalized cell bioreactors for nucleic acid barcoding and other pre-sequencing steps. This approach has successfully allowed for high-throughput single-cell sequencing of DNA (Lan et al. 2017; Demaree et al. 2018), mRNA (Klein et al. 2015; Macosko et al. 2015; Zilionis et al. 2017), chromatin states (CHIP-seq) (Wu et al. 2009; Rotem et al. 2015; Murphy et al. 2018), immune repertoires (Vollmers et al. 2013), and multiplexed mRNA and protein expression (Stoeckius et al. 2017).

Simultaneously, microgels and microbeads are permeating the sequencing sphere. Many sequencing schemes are based off of the technique Dropseq (Macosko et al. 2015), combining single cells with oligonucleotide barcodes extending from a hydrogel bead. These beads work by collecting nucleic acids and serve as an easy method for transcript barcoding. Each bead contains oligonucleotide barcode sequences that are unique to the bead, and incorporation of this barcode into the collected material tags it by each source cell. Bead pairing on a chip is used in Dropseq and InDrops single cell mRNA sequencing (Klein et al. 2015; Macosko et al. 2015; Zilionis et al. 2017), CITE-seq mRNA sequencing paired with epitope expression (Stoeckius et al. 2017), single cell ChIP-seq (Rotem et al. 2015) and LIFE-ChIP-seq (Murphy et al. 2018) chromatin mapping, DroNc-seq intranuclear RNA sequencing (Murphy et al. 2018), and SiC-seq single cell genome sequencing(Lan et al. 2017). In addition to microgels, immobile hydrogels are synergizing with digital microfluidic devices as microscale bioreactors (Fiddes et al. 2012). Luk et al. proposed combining digital microfluidics with agarose gels as microscale liquid handling system for multiplexing enzyme digestion of protein samples for proteomic studies (Luk et al. 2012). In this work, solid agarose discs were chemically modified and conjugated to the enzymes of interest, thereby fixing them in space, simplifying the task of sample and enzyme after digestion.

These examples demonstrate the early stages of the convergence of microgels and sequencing technology development. The authors envision this marriage to continue to develop as the symbiotic utility of these fields is further explored. For example, oiligonucleotide barcoded beads could be combined with cell-laden microgels on-chip to track cell-fate and population heterogeneity following challenge with time, growth factors, environmental stimuli, and drug exposure. Such a model would provide clues to developmental biology, infectious disease, cancer research and the heterogeneity of cell therapy.

Acknowledgements This work is submitted in honor of Professor Mauro Ferrari's 60th birthday. Mauro has been an inspirational force in the fields of nanotechnology and bionanotechnology. His writings have been extremely important in defining these fields, and the senior authors who have known him and worked with him appreciate all the contributions he has made to their research, directly and indirectly.

We acknowledge financial support from the National Institutes of Health (R01-AI132738-01A1 and 5R33CA212968-02 awarded to AS; R01-EB022025 awarded to NAP), the National Science Foundation CAREER award (DMR-1554275 awarded to AS), Department of Defense CDMRP and Cancer Career Development Award (W81XWH-17-1-0215 awarded to AS). In addition, NAP acknowledges support from the Cockrell Family Chair Foundation and the office of the Dean of the Cockrell School of Engineering at the University of Texas at Austin for the Institute for Biomaterials, Drug Delivery, and Regenerative Medicine. We acknowledge financial support from the National Science Foundation Graduate Research Fellowship Program (DGE-1610403 awarded to MEW; DGE-1650441 awarded to RES). Any opinions, findings, and conclusions or recommendations expressed in this material are those of the author(s) and do not necessarily reflect the views of the funding agencies.

References

S. Allazetta, T.C. Hausherr, M.P. Lutolf, Microfluidic synthesis of cell-type-specific artificial extracellular matrix hydrogels. Biomacromolecules **14**, 1122–1131 (2013). https://doi.org/10.1021/bm4000162

D. Aydın, S. Kızılel, Water-in-water emulsion based synthesis of hydrogel Nanospheres with tunable release kinetics. JOM **69**, 1185 (2017). https://doi.org/10.1007/s11837-016-1969-z

C. Berkland, K. Kim, D.W. Pack, PLG Microsphere Size Controls Drug Release Rate Through Several Competing Factors. Pharm. Res. **20**, 1055–1062 (2003). https://doi.org/10.1023/A:1024466407849

M.A. Bochenek, O. Veiseh, A.J. Vegas, et al., Alginate encapsulation as long-term immune protection of allogeneic pancreatic islet cells transplanted into the omental bursa of macaques. Nat Biomed Eng **2**, 810 (2018)

L. Brannon-Peppas, N.A. Peppas, Dynamic and equilibrium swelling behaviour of pH-sensitive hydrogels containing 2-hydroxyethyl methacrylate. Biomaterials **11**, 635–644 (1990)

M. Caldorera-Moore, M.K. Kang, Z. Moore, et al., Swelling behavior of nanoscale, shape- and size-specific, hydrogel particles fabricated using imprint lithography. Soft Matter **7**, 2879 (2011). https://doi.org/10.1039/c0sm01185a

D.A. Carr, N.A. Peppas, Assessment of poly(methacrylic acid- *co* - *N* -vinyl pyrrolidone) as a carrier for the oral delivery of therapeutic proteins using Caco-2 and HT29-MTX cell lines. J. Biomed. Mater. Res. Part A 9999A, NA-NA (2009). https://doi.org/10.1002/jbm.a.32395

F. Cayrol, M.C. Diaz Flaque, T. Fernando, et al., Integrin v 3 acting as membrane receptor for thyroid hormones mediates angiogenesis in malignant T cells. Blood **125**, 841–851 (2015). https://doi.org/10.1182/blood-2014-07-587337

J.C. Chang, A. Wysocki, K.M. Tchou-Wong, et al., Effect of Mycobacterium tuberculosis and its components on macrophages and the release of matrix metalloproteinases. Thorax **51**, 306–311 (1996)

Q. Chen, D. Chen, J. Wu, J.M. Lin, Flexible control of cellular encapsulation, permeability, and release in a droplet-templated bifunctional copolymer scaffold. Biomicrofluidics **10**, 1–9 (2016a). https://doi.org/10.1063/1.4972107

Q. Chen, S. Utech, D. Chen, et al., Controlled assembly of heterotypic cells in a core-shell scaffold: Organ in a droplet. Lab Chip **16**, 1346 (2016b). https://doi.org/10.1039/c6lc00231e

C.-H. Choi, H. Wang, H. Lee, et al., One-step generation of cell-laden microgels using double emulsion drops with a sacrificial ultra-thin oil shell. Lab Chip **16**, 1549–1555 (2016). https://doi.org/10.1039/C6LC00261G

J. Crank, *The Mathematics of Diffusion*, 2nd edn. (Oxford University Press, New York, 1975)

H.R. Culver, N.A. Peppas, Protein-imprinted polymers: The shape of things to come? Chem. Mater. **29**, 5753–5761 (2017). https://doi.org/10.1021/acs.chemmater.7b01936

C.A. Deforest, B.D. Polizzotti, K.S. Anseth, Sequential click reactions for synthesizing and patterning three-dimensional cell microenvironments. Nat. Mater. **8**, 659 (2009). https://doi.org/10.1038/nmat2473

B. Demaree, D. Weisgerber, F. Lan, A.R. Abate, An ultrahigh-throughput microfluidic platform for single-cell genome sequencing. J. Vis. Exp. **57598** (2018). https://doi.org/10.3791/57598

I.K. Demedts, A. Morel-Montero, S. Lebecque, et al., Elevated MMP-12 protein levels in induced sputum from patients with COPD. Thorax **61**, 196–201 (2006). https://doi.org/10.1136/thx.2005.042432

M. Durán-Lobato, B. Carrillo-Conde, Y. Khairandish, N.A. Peppas, Surface-modified P(HEMA- *co* -MAA) Nanogel carriers for Oral vaccine delivery: Design, characterization, and *in vitro* targeting evaluation. Biomacromolecules **15**, 2725–2734 (2014). https://doi.org/10.1021/bm500588x

L.K. Fiddes, V.N. Luk, S.H. Au, et al., Hydrogel discs for digital microfluidics. Biomicrofluidics **6**, 14112 (2012). https://doi.org/10.1063/1.3687381

P.J. Flory, J. Rehner, Statistical mechanics of cross-linked polymer networks I. rubberlike elasticity. J. Chem. Phys. **11**, 512–520 (1943). https://doi.org/10.1063/1.1723791

G.A. Foster, D.M. Headen, C. González-García, et al., Protease-degradable microgels for protein delivery for vascularization. Biomaterials **113**, 170–175 (2017). https://doi.org/10.1016/j.biomaterials.2016.10.044

J. Fu, Y.K. Wang, M.T. Yang, et al., Mechanical regulation of cell function with geometrically modulated elastomeric substrates. Nat. Methods **7**, 733 (2010). https://doi.org/10.1038/nmeth.1487

J. Fukuda, A. Khademhosseini, Y. Yeo, et al., Micromolding of photocrosslinkable chitosan hydrogel for spheroid microarray and co-cultures. Biomaterials **27**, 5259 (2006). https://doi.org/10.1016/j.biomaterials.2006.05.044

M.I. González-Sánchez, J. Rubio-Retama, E. López-Cabarcos, E. Valero, Development of an acetaminophen amperometric biosensor based on peroxidase entrapped in polyacrylamide microgels. Biosens. Bioelectron. **26**, 1883 (2011). https://doi.org/10.1016/j.bios.2010.03.024

S.T. Gould, N.J. Darling, K.S. Anseth, Small peptide functionalized thiol–ene hydrogels as culture substrates for understanding valvular interstitial cell activation and *de novo* tissue deposition. Acta Biomater. **8**, 3201–3209 (2012). https://doi.org/10.1016/j.actbio.2012.05.009

S.E.A. Gratton, P.D. Pohlhaus, J. Lee, et al., Nanofabricated particles for engineered drug therapies: A preliminary biodistribution study of PRINT™ nanoparticles. J. Control. Release **121**, 10 (2007). https://doi.org/10.1016/j.jconrel.2007.05.027

D.R. Griffin, W.M. Weaver, P.O. Scumpia, et al., Accelerated wound healing by injectable microporous gel scaffolds assembled from annealed building blocks. Nat. Mater. **14**, 737–744 (2015). https://doi.org/10.1038/nmat4294

A.E. Guttmacher, F.S. Collins, Welcome to the genomic era. N. Engl. J. Med. **349**, 996 (2003). https://doi.org/10.1056/NEJMe038132

D.M. Headen, G. Aubry, H. Lu, A.J. García, Microfluidic-based generation of size-controlled, biofunctionalized synthetic polymer microgels for cell encapsulation. Adv. Mater. **26**, 3003–3008 (2014). https://doi.org/10.1002/adma.201304880

D.M. Headen, J.R. García, A.J. García, Parallel droplet microfluidics for high throughput cell encapsulation and synthetic microgel generation. Microsyst. Nanoeng. **4**, 17076 (2018a). https://doi.org/10.1038/micronano.2017.76

D.M. Headen, K.B. Woodward, M.M. Coronel, et al., Local immunomodulation with Fas ligand-engineered biomaterials achieves allogeneic islet graft acceptance. Nat. Mater. **17**, 732–739 (2018b). https://doi.org/10.1038/s41563-018-0099-0

D.E. Heath, A.R.M. Sharif, C.P. Ng, et al., Regenerating the cell resistance of micromolded PEG hydrogels. Lab Chip **15**, 2073 (2015). https://doi.org/10.1039/c4lc01416b

H. Hirama, T. Kambe, K. Aketagawa, et al., Hyper alginate gel microbead formation by molecular diffusion at the hydrogel/droplet Interface. Langmuir **29**, 519–524 (2012). https://doi.org/10.1021/la303827u

J.C. Hoffmann, J.L. West, Three-dimensional photolithographic patterning of multiple bioactive ligands in poly(ethylene glycol) hydrogels. Soft Matter **6**, 5056 (2010). https://doi.org/10.1039/c0sm00140f

S. Jung, H. Yi, Facile micromolding-based fabrication of biopolymeric - synthetic hydrogel microspheres with controlled structures for improved protein conjugation. Chem. Mater. **27**, 3988 (2015). https://doi.org/10.1021/acs.chemmater.5b00920

S. Kawata, H.-B. Sun, T. Tanaka, K. Takada, Finer features for functional microdevices. Nature **412**, 697–698 (2001). https://doi.org/10.1038/35089130

A. Khademhosseini, G. Eng, J. Yeh, et al., Micromolding of photocrosslinkable hyaluronic acid for cell encapsulation and entrapment. J. Biomed. Mater. Res. Part A. **79A**, 522 (2006). https://doi.org/10.1002/jbm.a.30821

S.W. Kim, Y.H.O.T. Bae, Hydrogels: swelling, drug loading, and release. Pharm. Res. **9**, 283–290 (1992)

P.-H. Kim, H.-G. Yim, Y.-J. Choi, et al., Injectable multifunctional microgel encapsulating outgrowth endothelial cells and growth factors for enhanced neovascularization. J. Control. Release **187**, 1–13 (2014). https://doi.org/10.1016/j.jconrel.2014.05.010

A.M. Klein, D.A. Weitz, M.W. Kirschner, et al., Droplet barcoding for single-cell transcriptomics applied to embryonic stem cells. Cell **161**, 1187–1201 (2015). https://doi.org/10.1016/j.cell.2015.04.044

M.C. Koetting, J.F. Guido, M. Gupta, et al., pH-responsive and enzymatically-responsive hydrogel microparticles for the oral delivery of therapeutic proteins: Effects of protein size, crosslinking density, and hydrogel degradation on protein delivery. J. Control. Release **221**, 18–25 (2016). https://doi.org/10.1016/j.jconrel.2015.11.023

R.W. Korsmeyer, R. Gurny, E. Doelker, et al., Mechanisms of solute release from porous hydrophilic polymers. Int. J. Pharm. **15**, 25–35 (1983). https://doi.org/10.1016/0378-5173(83)90064-9

F. Lan, B. Demaree, N. Ahmed, A.R. Abate, Single-cell genome sequencing at ultra-high-throughput with microfluidic droplet barcoding. Nat. Biotechnol. **35**, 640–646 (2017). https://doi.org/10.1038/nbt.3880

D. Lee, C. Cha, Combined effects of co-culture and substrate mechanics on 3D tumor spheroid formation within microgels prepared via flow-focusing microfluidic fabrication. Pharmaceutics **10**, 1–14 (2018). https://doi.org/10.3390/pharmaceutics10040229

S.-H. Lee, J.S. Miller, J.J. Moon, J.L. West, Proteolytically degradable hydrogels with a Fluorogenic substrate for studies of cellular proteolytic activity and migration. Biotechnol. Prog. **21**, 1736–1741 (2005). https://doi.org/10.1021/bp0502429

T.T. Lee, J.R. García, J.I. Paez, et al., Light-triggered *in vivo* activation of adhesive peptides regulates cell adhesion, inflammation and vascularization of biomaterials. Nat. Mater. **14**, 352–360 (2015). https://doi.org/10.1038/nmat4157

T. Leinonen, R. Pirinen, J. Böhm, et al., Expression of matrix metalloproteinases 7 and 9 in non-small cell lung cancer: Relation to clinicopathological factors, β-catenin and prognosis. Lung Cancer **51**, 313–321 (2006). https://doi.org/10.1016/J.LUNGCAN.2005.11.002

B. Li, L. Wang, F. Xu, et al., Hydrosoluble, UV-crosslinkable and injectable chitosan for patterned cell-laden microgel and rapid transdermal curing hydrogel *in vivo*. Acta Biomater. **22**, 59 (2015). https://doi.org/10.1016/j.actbio.2015.04.026

B. Li, M. He, L. Ramirez, et al., Multifunctional hydrogel microparticles by polymer-assisted photolithography. ACS Appl. Mater. Interfaces **8**, 4158 (2016). https://doi.org/10.1021/acsami.5b11883

W.B. Liechty, D.R. Kryscio, B.V. Slaughter, N.A. Peppas, Polymers for drug delivery systems. Annu. Rev. Chem. Biomol. Eng. **1**, 149–173 (2010). https://doi.org/10.1146/annurev-chembioeng-073009-100847

P.S. Lienemann, T. Rossow, A.S. Mao, et al., Single cell-laden protease-sensitive microniches for long-term culture in 3D. Lab Chip **17**, 727–737 (2017). https://doi.org/10.1039/C6LC01444E

F. Lim, A.M. Sun, Microencapsulated islets as bioartificial endocrine pancreas. Science **210**, 908–910 (1980)

A.M. Lowman, T.D. Dziubla, P. Bures, N.A. Peppas, Structural and dynamic response of neutral and intelligent networks in biomedical environments. Adv. Chem. Eng. **29**, 75–130 (2004). https://doi.org/10.1016/S0065-2377(03)29004-9

V.N. Luk, L.K. Fiddes, V.M. Luk, et al., Digital microfluidic hydrogel microreactors for proteomics. Proteomics **12**, 1310–1318 (2012). https://doi.org/10.1002/pmic.201100608

M.P. Lutolf, J.A. Hubbell, Synthetic biomaterials as instructive extracellular microenvironments for morphogenesis in tissue engineering. Nat. Biotechnol. **23**, 47–55 (2005). https://doi.org/10.1038/nbt1055

E.Z. Macosko, A. Basu, A. Regev, et al., Highly parallel genome-wide expression profiling of individual cells using Nanoliter droplets Resource Highly Parallel Genome-wide Expression Profiling of Individual Cells Using Nanoliter Droplets. Cell **161**, 1202–1214 (2015). https://doi.org/10.1016/j.cell.2015.05.002

A.S. Mao, J.-W. Shin, S. Utech, et al., Deterministic encapsulation of single cells in thin tunable microgels for niche modelling and therapeutic delivery. Nat. Mater. **16**, 236–243 (2017). https://doi.org/10.1038/nmat4781

A.P. McGuigan, D.A. Bruzewicz, A. Glavan, et al., Cell encapsulation in sub-mm sized gel modules using replica molding. PLoS One **3**, e2258 (2008). https://doi.org/10.1371/journal.pone.0002258

I. Mironi-Harpaz, L. Hazanov, G. Engel, et al., In-situ architectures designed in 3D cell-laden hydrogels using microscopic laser photolithography. Adv. Mater. **27**, 1933 (2015). https://doi.org/10.1002/adma.201404185

S. Morelli, R.G. Holdich, M.M. Dragosavac, Chitosan and poly (vinyl alcohol) microparticles produced by membrane emulsification for encapsulation and pH controlled release. Chem. Eng. J. **288**, 451 (2016). https://doi.org/10.1016/j.cej.2015.12.024

T.W. Murphy, Y.-P. Hsieh, S.S. Ma, et al., Microfluidic low-input fluidized-bed enabled ChIP-seq device for automated and parallel analysis of histone modifications. Anal. Chem. **90**, 7666–7674 (2018). https://doi.org/10.1021/acs.analchem.8b01541

M.I. Neves, M.E. Wechsler, M.E. Gomes, et al., Molecularly imprinted intelligent scaffolds for tissue engineering applications. Tissue Eng. B Rev. **23**, 27–43 (2017). https://doi.org/10.1089/ten.TEB.2016.0202

M. Nikkhah, N. Eshak, P. Zorlutuna, et al., Directed endothelial cell morphogenesis in micropatterned gelatin methacrylate hydrogels. Biomaterials **33**, 9009 (2012). https://doi.org/10.1016/j.biomaterials.2012.08.068

M.H.M. Oudshoorn, R. Penterman, R. Rissmann, et al., Preparation and characterization of structured hydrogel microparticles based on cross-linked hyperbranched polyglycerol. Langmuir **23**, 11819 (2007). https://doi.org/10.1021/la701910d

J.H. Park, S.O. Choi, R. Kamath, et al., Polymer particle-based micromolding to fabricate novel microstructures. Biomed. Microdevices **9**, 223 (2007). https://doi.org/10.1007/s10544-006-9024-4

R.G. Patel, A. Purwada, L. Cerchietti, et al., Microscale bioadhesive hydrogel arrays for cell engineering applications. Cell. Mol. Bioeng. **7**, 394 (2014). https://doi.org/10.1007/s12195-014-0353-8

N.A. Peppas, *Hydrogels in Medicine and Pharmacy* (CRC Press, Boca Raton, 1986)

N.A. Peppas, E.W. Merrill, Crosslinked poly(vinyl alcohol) hydrogels as swollen elastic networks. J. Appl. Polym. Sci. **21**, 1763–1770 (1977). https://doi.org/10.1002/app.1977.070210704

N.A. Peppas, B. Narasimhan, Mathematical models in drug delivery: How modeling has shaped the way we design new drug delivery systems. J. Control. Release **190**, 75–81 (2014). https://doi.org/10.1016/j.jconrel.2014.06.041

N.A. Peppas, P. Bures, W. Leobandung, H. Ichikawa, Hydrogels in pharmaceutical formulations. Eur. J. Pharm. Biopharm. **50**, 27–46 (2000). https://doi.org/10.1016/S0939-6411(00)00090-4

N.A. Peppas, J.Z. Hilt, A. Khademhosseini, R. Langer, Hydrogels in biology and medicine: From molecular principles to bionanotechnology. Adv. Mater. **18**, 1345–1360 (2006). https://doi.org/10.1002/adma.200501612

E.A. Phelps, N.O. Enemchukwu, V.F. Fiore, et al., Maleimide cross-linked bioactive PEG hydrogel exhibits improved reaction kinetics and cross-linking for cell encapsulation and *in situ* delivery. Adv. Mater. **24**(64–70), 2 (2012). https://doi.org/10.1002/adma.201103574

E. Piacentini, M. Yan, L. Giorno, Development of enzyme-loaded PVA microspheres by membrane emulsification. J. Membr. Sci. **524**, 79 (2017). https://doi.org/10.1016/j.memsci.2016.11.008

M.R. Pinezich, L.N. Russell, N.P. Murphy, K.J. Lampe, Encapsulated oligodendrocyte precursor cell fate is dependent on PDGF-AA release kinetics in a 3D microparticle-hydrogel drug delivery system. J. Biomed. Mater. Res. A **106**, 2402–2411 (2018). https://doi.org/10.1002/jbm.a.36432

A. Purwada, S.B. Shah, W. Béguelin, et al., Ex vivo synthetic immune tissues with T cell signals for differentiating antigen-specific, high affinity germinal center B cells. Biomaterials (2018). https://doi.org/10.1016/j.biomaterials.2018.06.034

G.P. Raeber, M.P. Lutolf, J.A. Hubbell, Molecularly engineered PEG hydrogels: A novel model system for Proteolytically mediated cell migration. Biophys. J. **89**, 1374–1388 (2005). https://doi.org/10.1529/biophysj.104.050682

A. Revzin, R.J. Russell, V.K. Yadavalli, et al., Fabrication of poly(ethylene glycol) hydrogel microstructures using photolithography. Langmuir **17**, 5440 (2001). https://doi.org/10.1021/la010075w

T. Rossow, J.A. Heyman, A.J. Ehrlicher, et al., Controlled synthesis of cell-laden microgels by radical-free gelation in droplet microfluidics. J. Am. Chem. Soc. (2012). https://doi.org/10.1021/ja300460p

A. Rotem, O. Ram, N. Shoresh, et al., Single-cell ChIP-seq reveals cell subpopulations defined by chromatin state HHS public access author manuscript. Nat. Biotechnol. **33**, 1165–1172 (2015). https://doi.org/10.1038/nbt.3383

D.K. Sahana, G. Mittal, V. Bhardwaj, M.N.V.R. Kumar, PLGA nanoparticles for oral delivery of hydrophobic drugs: Influence of organic solvent on nanoparticle formation and release behavior *in vitro* and *in vivo* using estradiol as a model drug. J. Pharm. Sci. **97**, 1530 (2008). https://doi.org/10.1002/jps.21158

R. Salem, R.J. Lewandowski, B. Atassi, et al., Treatment of Unresectable hepatocellular carcinoma with use of 90Y microspheres (TheraSphere): Safety, tumor response, and survival. J. Vasc. Interv. Radiol. **16**, 1627–1639 (2005). https://doi.org/10.1097/01.RVI.0000184594.01661.81

K.M. Schultz, K.S. Anseth, Monitoring degradation of matrix metalloproteinases-cleavable PEG hydrogels via multiple particle tracking microrheology. Soft Matter **9**, 1570–1579 (2013). https://doi.org/10.1039/C2SM27303A

K.M. Schultz, K.A. Kyburz, K.S. Anseth, Measuring dynamic cell–material interactions and remodeling during 3D human mesenchymal stem cell migration in hydrogels. Proc. Natl. Acad. Sci. **112**, E3757–E3764 (2015). https://doi.org/10.1073/pnas.1511304112

T.F. Scott, B.A. Kowalski, A.C. Sullivan, et al., Two-color single photon Photoinitiation and Photoinhibition for subdiffraction photolithography. Science **324**, 913–917 (2009). https://doi.org/10.1126/science.116761

E. Secret, S.J. Kelly, K.E. Crannell, J.S. Andrew, Enzyme-responsive hydrogel microparticles for pulmonary drug delivery. ACS Appl. Mater. Interfaces **6**, 10313–10321 (2014). https://doi.org/10.1021/am501754s

L.A. Sharpe, J.E. Vela Ramirez, O.M. Haddadin, et al., pH-responsive microencapsulation Systems for the Oral Delivery of Polyanhydride nanoparticles. Biomacromolecules **19**, 793–802 (2018). https://doi.org/10.1021/acs.biomac.7b01590

C. Siltanen, M. Yaghoobi, A. Haque, et al., Microfluidic fabrication of bioactive microgels for rapid formation and enhanced differentiation of stem cell spheroids. Acta Biomater. **34**, 125–132 (2016). https://doi.org/10.1016/J.ACTBIO.2016.01.012

B.V. Slaughter, S.S. Khurshid, O.Z. Fisher, et al., Hydrogels in regenerative medicine. Adv. Mater. **21**, 3307–3329 (2009). https://doi.org/10.1002/adma.200802106

B.V. Sridhar, J.L. Brock, J.S. Silver, et al., Development of a cellularly degradable PEG hydrogel to promote articular cartilage extracellular matrix deposition. Adv. Healthc. Mater. **4**, 702–713 (2015). https://doi.org/10.1002/adhm.201400695

S. Steichen, C. O'Connor, N.A. Peppas, Development of a P((MAA-co-NVP)-g-EG) hydrogel platform for Oral protein delivery: Effects of hydrogel composition on environmental response and protein partitioning. Macromol. Biosci. **17**, 1600266 (2016). https://doi.org/10.1002/mabi.201600266

D. Steinhilber, T. Rossow, S. Wedepohl, et al., Angewandte Chemie stimuli-responsive microgels a microgel construction kit for bioorthogonal encapsulation and pH-controlled release of living cells. Angew. Chem. Int. Ed. **52**, 13538 (2013). https://doi.org/10.1002/anie.201308005

M. Stoeckius, C. Hafemeister, W. Stephenson, et al., Simultaneous epitope and transcriptome measurement in single cells. Nat. Methods **14**, 865–868 (2017). https://doi.org/10.1038/nmeth.4380

K. Subramani, M.A. Birch, Micropatterning of Poly (Ethylene Glycol)-Diacrylate (PEG-DA) Hydrogel by Soft-Photolithography for Analysis of Cell-Biomaterial interactions. J. Biomimetics Biomater. Tissue Eng. (2009). https://doi.org/10.4028/www.scientific.net/JBBTE.2.3

S. Suri, C.E. Schmidt, Photopatterned collagen-hyaluronic acid interpenetrating polymer network hydrogels. Acta Biomater. **5**, 2385 (2009). https://doi.org/10.1016/j.actbio.2009.05.004

W.-H. Tan, S. Takeuchi, Monodisperse alginate hydrogel microbeads for cell encapsulation. Adv. Mater. **19**, 2696–2701 (2007). https://doi.org/10.1002/adma.200700433

Y.F. Tian, H. Ahn, R.S. Schneider, et al., Integrin-specific hydrogels as adaptable tumor organoids for malignant B and T cells. Biomaterials **73**, 110–119 (2015). https://doi.org/10.1016/j.biomaterials.2015.09.007

H. Tokuyama, Y. Kato, Preparation of poly(N-isopropylacrylamide) emulsion gels and their drug release behaviors. Colloids Surf. B: Biointerfaces **67**, 92 (2008). https://doi.org/10.1016/j.colsurfb.2008.08.003

M. Torres-Lugo, M. García, R. Record, N.A. Peppas, Physicochemical behavior and cytotoxic effects of p(methacrylic acid–g-ethylene glycol) nanospheres for oral delivery of proteins. J. Control. Release **80**, 197–205 (2002). https://doi.org/10.1016/S0168-3659(02)00027-5

C. Trapnell, Defining cell types and states with single-cell genomics. Genome Res. **25**, 1491 (2015)

M.F. Ullah, M. Aatif, The footprints of cancer development: Cancer biomarkers. Cancer Treat. Rev. **35**, 193–200 (2009). https://doi.org/10.1016/J.CTRV.2008.10.004

O. Veiseh, J.C. Doloff, M. Ma, et al., Size- and shape-dependent foreign body immune response to materials implanted in rodents and non-human primates. Nat. Mater. **14**, 643–651 (2015). https://doi.org/10.1038/nmat4290

T. Vermonden, R. Censi, W.E. Hennink, Hydrogels for Protein Delivery. Chem. Rev. **112**, 2853–2888 (2012). https://doi.org/10.1021/cr200157d

C. Vollmers, R.V. Sit, J.A. Weinstein, et al., Genetic measurement of memory B-cell recall using antibody repertoire sequencing. Proc. Natl. Acad. Sci. **110**, 13463 (2013). https://doi.org/10.1073/pnas.1312146110

L. Wang, R.R. Rao, J.P. Stegemann, Delivery of mesenchymal stem cells in chitosan/collagen microbeads for orthopedic tissue repair. Cells Tissues Organs **197**, 333 (2013). https://doi.org/10.1159/000348359

G. Wang, D. Chen, L. Zhang, et al., A mild route to entrap papain into cross-linked PEG microparticles via visible light-induced inverse emulsion polymerization. J. Mater. Sci. **53**, 880 (2018). https://doi.org/10.1007/s10853-017-1484-9

U. Wattendorf, G. Coullerez, J. Vörös, et al., Mannose-based molecular patterns on stealth microspheres for receptor-specific targeting of human antigen-presenting cells. Langmuir **24**, 11790–11802 (2008). https://doi.org/10.1021/la801085d

A.R. Wu, J.B. Hiatt, R. Lu, et al., Automated microfluidic chromatin immunoprecipitation from 2,000 cells. Lab Chip **9**, 1365–1370 (2009). https://doi.org/10.1039/b819648f

Y. Xia, D.W. Pack, Uniform biodegradable microparticle systems for controlled release. Chem. Eng. Sci. **125**, 129–143 (2015). https://doi.org/10.1016/J.CES.2014.06.049

F. Xu, C.M. Wu, V. Rengarajan, et al., Three-dimensional magnetic assembly of microscale hydrogels. Adv. Mater. **23**, 4254–4260 (2011). https://doi.org/10.1002/adma.201101962

X. Yan, R.A. Gemeinhart, Cisplatin delivery from poly(acrylic acid-co-methyl methacrylate) microparticles. J. Control. Release **106**, 198 (2005). https://doi.org/10.1016/j.jconrel.2005.05.005

J.O. You, D.T. Auguste, Feedback-regulated paclitaxel delivery based on poly(N,N-dimethylaminoethyl methacrylate-co-2-hydroxyethyl methacrylate) nanoparticles. Biomaterials **29**, 1950 (2008). https://doi.org/10.1016/j.biomaterials.2007.12.041

L. Zhang, K. Chen, H. Zhang, et al., Microfluidic templated multicompartment microgels for 3D encapsulation and pairing of single cells. Small **14**, 1–8 (2018). https://doi.org/10.1002/smll.201702955

R. Zilionis, J. Nainys, A. Veres, et al., Single-cell barcoding and sequencing using droplet microfluidics. Nat. Protoc. **12**, 44–73 (2017). https://doi.org/10.1038/nprot.2016.154

Biomedical Microdevices (2019) 21: 39
https://doi.org/10.1007/s10544-019-0397-6

Investigation of parameters that determine Nano-DC vaccine transport

Jingxin Zhang [1,2] · Junhua Mai [1] · Feng Li [1,3] · Jianliang Shen [1,4] · Guodong Zhang [1] · Jun Li [1,2] · Louis E. Hinkle [1] · Daniel Lin [1] · Xuewu Liu [1] · Zheng Li [3] · Rong-fu Wang [5] · Elizabeth A Mittendorf [6,7] · Mauro Ferrari [1,8] · Haifa Shen [1,9]

Published online: 4 April 2019

Author's Personal Note I have had the pleasure of collaborating with Mauro for almost 10 years. I first met him when I joined the University of Texas Health Science Center at Houston in 2010, and moved with him to Houston Methodist Research Institute in the same year. One of my earliest memories with Mauro was making it sharp to his 8 am daily meetings, a killer when I had to commute a 30-mile stretch through morning traffic from The Woodlands. Mauro has been genuinely and extremely passionate for science and translational research, especially as it relates to finding a cure for patients with metastatic cancers. His multidisciplinary approach to understand the cancer problem lead to the Transport Oncophysics program which provides guidance on effective delivery of therapeutic agents to the cancer tissue by overcoming sequential physical and biological barriers. He has been the driving force to push our projects from research to development, and eventually to clinic. I have also seen how he has been very passionate and dedicated to serving the underprivileged; one more quality of admiration. Mauro's great leadership and knowledge not only in science but in all the experiences he would share with us make him one of the most unique and admirable individuals I have met.

Haifa Shen

Abstract

Effective migration of dendritic cells into the lymphatic system organs is the prerequisite for a functional dendritic cell vaccine. We have previously developed a porous silicon microparticle (PSM)-based therapeutic dendritic cell vaccine (Nano-DC vaccine) where PSM serves both as the vehicle for antigen peptides and an adjuvant. Here, we analyzed parameters that determined dendritic cell uptake of PSM particles and Nano-DC vaccine accumulation in lymphatic tissues in a murine model of HER2-positive breast cancer. Our study revealed a positive correlation between sphericity of the PSM particles and their cellular uptake by circulating dendritic cells. In addition, the intravenously administered vaccines accumulated more in the spleens and inguinal lymph nodes, while the intradermally inoculated vaccines got enriched in the popliteal lymph nodes. Furthermore, mice with large tumors received more vaccines in the lymph nodes than those with small to medium size tumors. Information from this study will provide guidance on design and optimization of future therapeutic cancer vaccines.

Keywords Silicon · Microparticle · Nano-DC vaccine · Biodistribution · Transport

1 Introduction

Dendritic cell (DC) is the most powerful antigen-presenting cell (APC) in the body. DCs pick up tumor antigens and migrate to lymphatic tissues where they process the tumor antigen and present antigen epitopes to T cells, resulting in T cell activation. Based on the essential role of DCs in mediating anti-tumor immunity, dendritic cell vaccines have been developed to fight against human cancers. One peptide DC vaccine, sipuleucel-T, was approved by the US Food and Drug Administrations in 2010 for treatment of castration-resistant metastatic prostate cancers (Kantoff et al. 2010; Cheever and Higano 2011), and many dendritic cell vaccines are currently in various stages of clinical trials (Gelao et al. 2014; Carreno et al. 2015; Tanyi et al. 2018).

Since tumor antigen-pulsed DCs need to be properly activated in order for the DC vaccine to function properly, adjuvants are usually added into the vaccines to stimulate DC maturation. The most commonly used adjuvant in peptide DC vaccines is granulocyte-macrophage colony-stimulating

✉ Mauro Ferrari
mferrari@houstonmethodist.org

✉ Haifa Shen
hshen@houstonmethodist.org

1 Department of Nanomedicine, Houston Methodist Research Institute, Houston, TX 77030, USA

2 Xiangya Hospital of Central South University, Changsha 410000, Hunan, China

3 Center for Bioenergetics, Houston Methodist Research Institute, Houston, TX 77030, USA

4 State key laboratory of ophthalmology, School of Biomedical Engineering, Wenzhou Medical University, Wenzhou 325035, China

5 Center for Inflammation and Epigenetics, Houston Methodist Research Institute, Houston, TX 77030, USA

6 Department of Surgery, Brigham and Women's Hospital, Dana-Farber/ Brigham and Women's Cancer Center, Boston, MA 02115, USA

7 Breast Oncology Program, Dana-Farber/ Brigham and Women's Cancer Center, Boston, MA 02115, USA

8 Department of Medicine, Weill Cornell Medical College, 1300 York Avenue, New York, NY 10065, USA

9 Department of Cell and Developmental Biology, Weill Cornell Medical College, 1300 York Avenue, New York, NY 10065, USA

factor (GM-CSF) (Knutson et al. 2001; Kantoff et al. 2010; Mittendorf et al. 2012; Tanyi et al. 2018). The Toll-like receptor ligands/agonists are also good candidates for vaccine development (Napolitani et al. 2005; Duthie et al. 2011). Various forms of nanoparticles have also been incorporated into different types vaccine forms to boost vaccine activity in recent years (Goldberg 2015). However, most DC vaccines have not met the expectation to potently fight against human cancers in clinic. Among the many factors that could negatively impact DC vaccine activity is the existence of physical and biological barriers that block effective migration of the cells into lymphatic organs (Shen et al. 2017). It has been estimated that less than 5% of total injected DCs can migrate out of the inoculation site, and the majority of the cells die *in situ* (Turnis et al. 2010). Thus, identification of parameters that dictate DC migration and cell entry into lymphatic tissues holds the key to generate effective therapeutic DC vaccines.

In an effort to develop effective treatment for human cancers, we have developed a porous silicon microparticle (PSM)-based DC vaccine (Nano-DC vaccine) (Xia et al. 2015). PSMs are biocompatible and water-soluble microparticles that have been successfully applied to deliver chemotherapy drugs and therapeutic small inhibitory RNA and microRNA molecules in our laboratories (Shen et al. 2013a, b; Xu et al. 2013; Mai et al. 2014; Xu et al. 2016; Zong et al. 2016; Ban et al. 2017). In the Nano-DC vaccine, PSM serves both as an adjuvant to stimulate TRIF/MAVS-dependent signaling and as a vehicle for packaging and sustained release of antigen peptides. A HER2 peptide-pulsed Nano-DC vaccine is effective in mediating anti-tumor immunity in murine models of HER2-positive breast cancer (Xia et al. 2015). In this study, we have examined correlation between physical parameters of the PSM and DC uptake of the particles. We have also analyzed impact from different routes of vaccination on Nano-DC vaccine accumulation in lymphatic tissues. Finally, we have compared Nano-DC vaccine enrichment in lymphatic tissues in mice with small, medium, and big size tumors. Our findings will be important for designing next-generation therapeutic DC vaccines.

2 Materials and methods

2.1 Cells and cell culture

TUBO cells were originally provided by Dr. Zen Wei (Wayne State University, Detroit, MI) and Guido Forni (University of Turin, Italy). Cells were cultured in DMEM medium supplemented with 10% fetal bovine serum (FBS) and 1% penicillin-streptomycin at 37 °C with 5% CO_2 and 95% humidity. Bone marrow cells were collected from the femurs and tibias of Balb/c mice and suspended in RPMI 1640 medium supplemented with 10% FBS and 1% penicillin-streptomycin. After cells were washed once with the medium, bone marrow cells were suspended in 1 ml ACK lysing buffer for 1 min. The suspension was then filtered using a sterile cell strainer 70 μm Nylon mesh and centrifuged. Cells were resuspended in growth medium supplemented with β-mercaptoethanol, granulocyte-macrophage colony-stimulating factor (GM-CSF, 200 μg/ml) and interleukin 4 (IL4, 200 μg/ml) at a final concentration of 1×10^6 cells/ml, and seeded into 6 well plates (4 ml cell suspension per well). Cells were incubated at 37 °C with 5% CO_2 and 95% humidity for 7 days, and growth medium was changed every other day. On day 6, cells were stimulated with100 ng/ml lipopolysaccharide (LPS). They were harvested on day 7, washed once, and resuspended in Opti-MEM medium at a concentration of 1×10^6 cells/ml.

2.2 Nano-DC vaccine preparation

Nano-DC vaccine was prepared following the same procedure as we have previously described (Xia et al. 2015). Briefly, the p66 HER2 peptide (TYVPANASL) was packaged into 1,2-dioleoyl-sn-glycero-3-phosphocholine (DOPC) liposomes and the liposomes were subsequently loaded into the nanopores of PSM particles. The liposomal peptide-loaded PSM particles were then incubated with LPS-activated BMDCs.

2.3 PSM for positron emission tomography/computed tomography (PET-CT) study

PSM particles were produced by a combination of photolithography and electrochemical etching, and conjugated with APTES on surface to enhance stability in aqueous solution as previously described (Godin et al. 2012; Shen et al. 2012, 2013a). PSM images were taken with a JEOL JSM-7100F microscope (TouchScope). To prepare zirconium-89 labeled PSM (^{89}Zr-PSM) for PET-CT study, PSM particles were first conjugated with deferoxamine (DFO) and then labeled with ^{89}Zr. To prepare PSM particles for ^{89}Zr labeling, PSM suspension in dimethylformamide (DMF) were mixed with disuccinimidyl suberate (DSS, Thermo Fisher Scientific) and triethylene amine (TEA, Sigma-Aldrich), and the reaction was left for 30 min at room temperature. After washing twice with DMF, PSM particles were suspended in 1 ml dimethyl sulfoxide (DMSO, Sigma-Aldrich) containing 20 μl TEA and 20 μl DFO mesylate salt for 4 h at room temperature. Particles were washed twice and then vacuum dried. To label PSMs with ^{89}Zr, 1 to

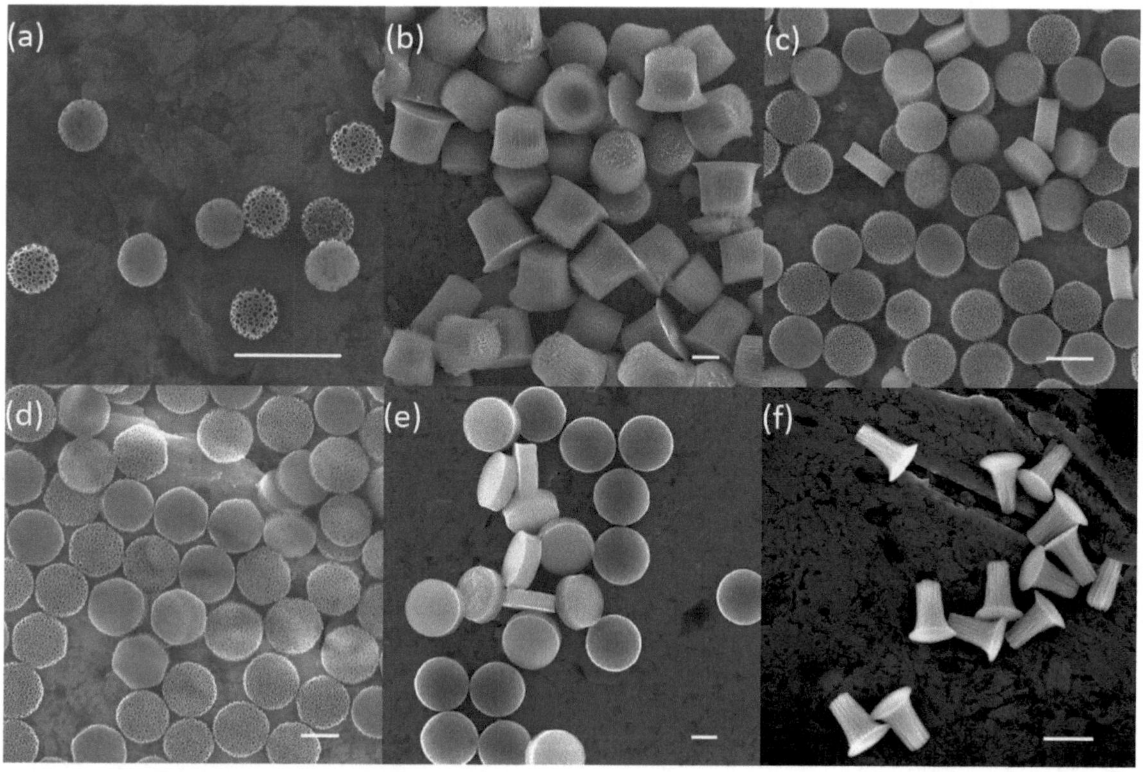

Fig. 1 SEM images of PSM particles with different size and shape. (**a**) 0.38 μm × 0.24 μm; (**b**) 1.64 μm × 1.76 μm; (**c**) 1 μm × 0.4 μm; (**d**) 1.5 μm × 0.4 μm; (**e**) 2.6 μm × 0.7 μm; (**f**) 0.88 μm × 1 μm. Scale bars: 1 μm

4 mCi ^{89}Zr-oxalate (Washington University, School of Medicine, Mallinckrodt Institute of Radiology) was added to the DFO-modified silicon particles. The suspension was mixed and sonicated for 1 min, and then incubated at room temperature for 2 h. The ^{89}Zr labeled PSMs were spinned down, and verified with a radio thin-layer chromatography TLC scanner (Bioscan AR2000).

2.4 Flow cytometry analysis on particle uptake by circulating cells

The following antibodies were used to stain single cells: antibodies for DC staining - anti-mouse CD45 (30-F11, Invitrogen), anti-mouse CD11b (M1/70, TONBO), anti-mouse CD11c (N418, TONBO), and anti-mouse MHC II (AF6–120.1, eBioscience), anti-mouse CD115 (T38–320, BD), and anti-mouse Ly-6G (RB6-8C5, eBioscience). After incubation, the antibody/cell mixtures were left on ice for 30 min. The cells were then washed 3 times with 1 ml 2% FBS in PBS, and centrifuged at 300×g for 3 min. Cells were resuspended in 300 μl 2% FBS in PBS, and separated with flow cytometry using the LSR II (BD Biosciences) for the $CD45^+CD11b^+CD11c^+MHCII^+$ DCs, $CD45^+CD11b^+CD115^+$ macrophages, and $CD45^+CD11b^+Ly\text{-}6G^+$ neutrophils. Results were analyzed with FACSDiva (BD Biosciences).

Table 1 Zeta potential and sphericity of PSMs with different shapes and sizes

Size (μm × μm)	Sphericity	Zeta potential
1.5 × 0.4	0.71	+ 6 mV
2.6 × e0.7	0.71	+ 6 mV
1 × 0.4	0.79	+ 6 mV
0.4 × 0.2	0.85	+ 6 mV
0.9 × 1	0.86	+ 5 mV
1.6 × 1.7	0.87	+ 6 mV

2.5 Animal studies

All animal procedures were approved by Houston Methodist Research Institute IACUC. Balb/c mice were purchased from the Jackson Laboratory or Charles River. To test particle uptake by circulating cells, mice were treated in the foot pads (*i.d.*) with PSM particles of different size and shape, and blood samples were collected 30 min later for flow cytometry. To examine biodistribution of Nano-DC vaccines, 6 to 8-week-old female Balb/c mice were inoculated with TUBO cells (5 × 10^5 cells/mouse in phosphate buffer saline (PBS)/Matrigel) in the mammary fat pads. The tumor-bearing mice were treated *i.d.* or intravenously (*i.v.*) with vaccines, and imaged using PET-CT at different time points. Mice were sacrificed after

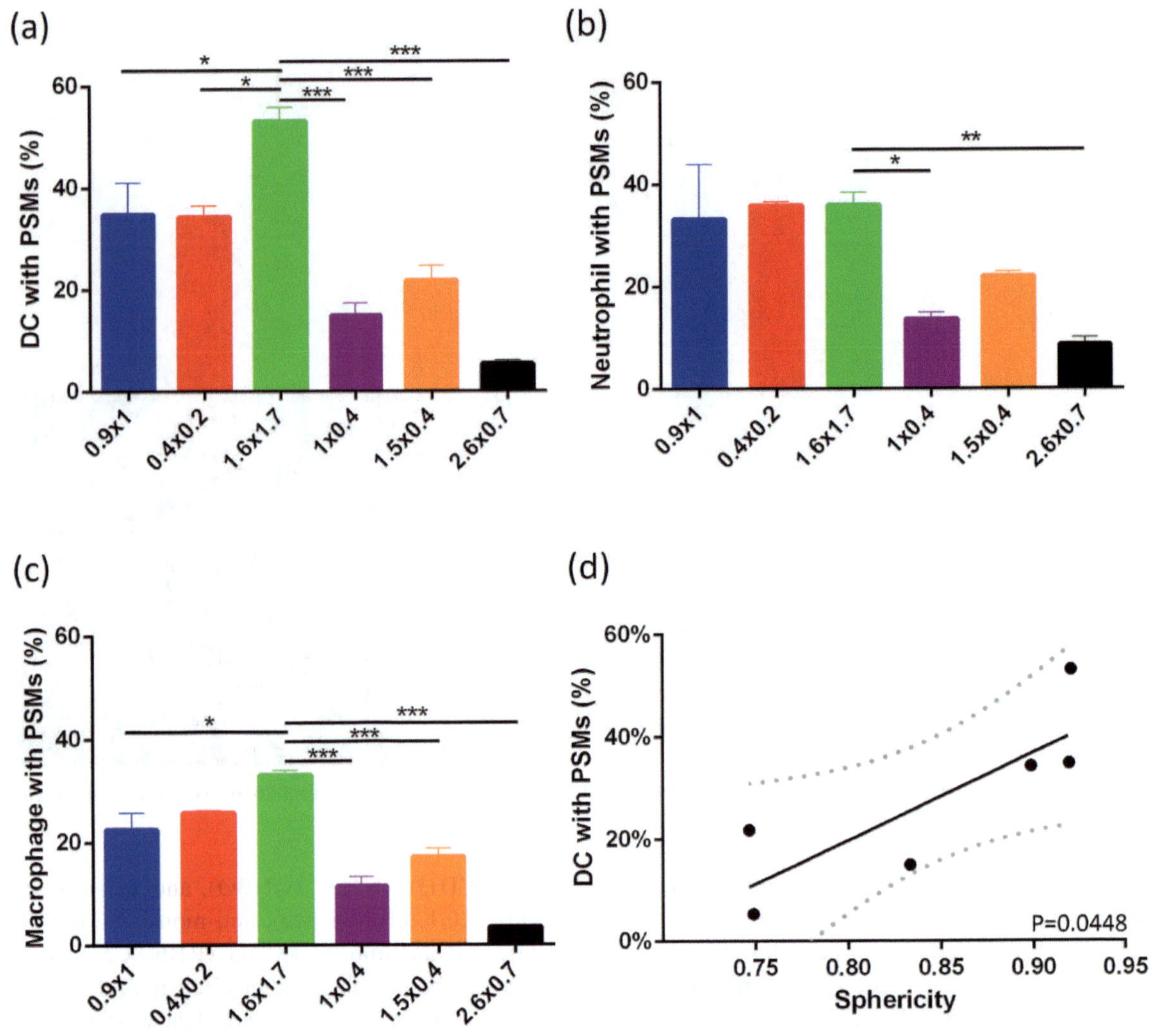

Fig. 2 Internalization of PSMs by dendritic cells, neutrophils, and monocytes. (**a**-**c**) Flow cytometry analysis on percentage of DCs, neutrophils, and monocytes internalized with PSMs in popliteal lymph nodes from vaccinated mice ($n = 5$) after i.d. injection. Data is presented as mean ± SEM. $*p < 0.05$, $**p < 0.01$, $***p < 0.001$. (**d**) Correlation between percentage of PSM internalization by DCs and particle sphericity. Sphericity is calculated by $\Psi = \frac{\pi^{\frac{1}{3}}(6V)^{\frac{2}{3}}}{A}$, where V is the volume and A is surface area of the particle

PET-CT imaging, and major organs, and tumors and lymph nodes were collected and processed for flow cytometry or radioactivity measurement.

2.6 Statistical analyses

Statistical significance between 2 test groups was analyzed using Student's t test. One-way ANOVA was applied to compare statistical significance among multiple groups.

3 Results

3.1 Characterization of PSMs

Six groups of PSMs were applied to evaluate particle uptake by phagocytic cells in circulation including DCs, monocytes and neutrophils *in vivo*. Although they differed by size and shape (Fig. 1a-f), all particles contained nanometer-size pores that could be loaded with antigen peptides. In addition, PSMs were chemically modified on surface with APTES and subsequently conjugated with the cyanine 5 (Cy5) fluorescent dye for imaging and detection. Detailed parameters of the particles are summarized in Table 1.

3.2 Physical properties of PSM on particle uptake efficiency by circulating phagocytic cells

Since Nano-DC vaccine was prepared by incubating antigen peptide-loaded PSM particles with BMDCs and there was a possibility that the particles could be passed to other cells *in vivo* (Xia et al. 2015), we evaluated uptake by DCs and other cell types after the same number of Cy5-labeled PSMs were administered in the foot pads of BALB/c immuno-

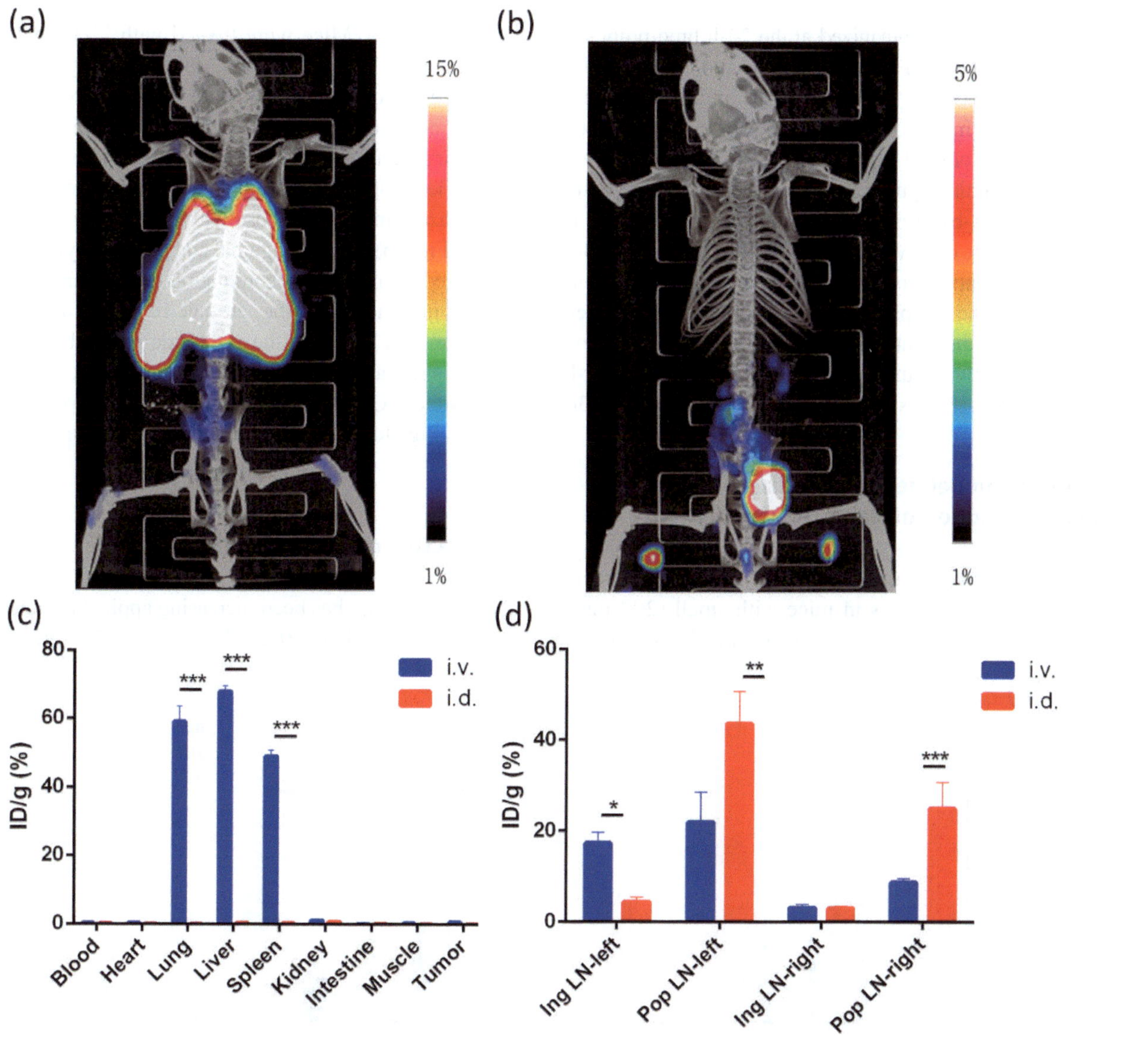

Fig. 3 Biodistribution of Nano-DC vaccine in major organs and lymph nodes after *i.v.* and *i.d.* inoculation. (**a-b**) PET-CT images of mice 24 h after vaccination by *i.v.* and *i.d.*, respectively. (**c-d**) Accumulation of radioactive Nano-DC vaccine in major organs and lymph nodes 24 h after vaccination. Y axis is presented by counts per injection dosage per gram ($n = 3$)

competent mice. Flow cytometry was applied to determine percentage of Cy5-positive cells. Interestingly, the six types of particles could be assigned into two groups based on their uptake by DCs, macrophages and neutrophils. Specifically, mice treated with 0.9×1 μm, 0.4×0.2 μm, and 1.6×1.7 μm particles had higher levels of particle uptake by all three cell types than those treated with 1×0.4 μm, 1.5×0.4 μm, and 2.6×0.7 μm particles (Fig. 2a-c). The highest DC uptake rate was detected in mice treated with the 1.6×1.7 μm particles. A positive correlation was established between sphericity of the particles and cellular uptake (Fig. 2d). Based on this result, we concluded that particle sphericity, but not particle size, that determined particle uptake by the phagocytic cells *in vivo*. Thus, the 1.6×1.7 μm particles were selected for further analysis in the study.

3.3 Route of vaccination-dependent biodistribution of Nano-DC vaccines

In a previous study, we developed a Nano-DC vaccine contained a HER2-specific antigen peptide, and applied the vaccine to treat mice with primary HER2-positive TUBO tumors (Xia et al. 2015). The TUBO tumor line was originally derived from a mammary gland tumor in a BLAB/c mouse carrying an MMTV-HER2 transgenic gene (Lucchini et al. 1992). In order to track Nano-DC vaccine migration inside the body, we labeled PSMs with ^{89}Zr before they were used to assemble the vaccine. Mice with primary TUBO tumors were treated with an equal amount of Nano-DC vaccine (normalized by radioactivity) either by tail vein injection (*i.v.*) or intra-footpad injection (*i.d.*). PET-CT was performed 3, 6, and 24 h after vaccination to monitor time-dependent vaccine migration. Mice were euthanized at the 24-h time point, and major organs and lymph nodes were collected to quantitatively measure radioactivity. As expected, the intravenously inoculated vaccines mainly accumulated in the major organs, while the intra-footpad injected vaccines slowly migrated through the draining lymph nodes, i.e., the popliteal lymph nodes (Fig. 3a-b), resulting in major differences on biodistribution in the liver, lung and spleen (Fig. 3c). Both routes of vaccination could lead to accumulation of Nano-DC vaccine in the lymph nodes, although there was a trend of asymmetrical distribution pattern, with the left side lymph nodes consistently accumulating higher number of ^{89}Zr-labeled Nano-DC vaccines than those on the right side (Fig. 3d).

3.4 Correlation between tumor size and biodistribution of Nano-DC vaccine

Correlation between tumor size and Nano-DC vaccine biodistribution was studies in mice with small (200 mm^3), medium (500 mm^3), and large (1000 mm^3) size primary TUBO tumors. Mice were treated with ^{89}Zr-labeled Nano-DC vaccines by i.v. injection, and major organs and lymph nodes were collected 24 h later to measure radioactivity. While no statistical significance was obtained in the liver, lung, or spleen (Fig. 4a), more Nano-DC vaccine accumulated in the left popliteal and inguinal lymph nodes in mice with large-size tumors over those with small and medium size tumors (Fig. 4b). In both inguinal and popliteal lymph nodes, we observed asymmetric accumulation of the Nano-DC vaccine, with lymph nodes on the left side receiving more Nano-DC vaccine than those on the right side (Fig. 4b). Quantitative analysis revealed a positive correlation between tumor weight and Nano-DC vaccine accumulation in the popliteal lymph nodes (Fig. 4c), but not in the inguinal lymph nodes (Fig. 4d).

4 Discussion and conclusion

Nanotechnology has been increasing applied in cancer immunotherapy (Goldberg 2015; Shen et al. 2017). Various types of

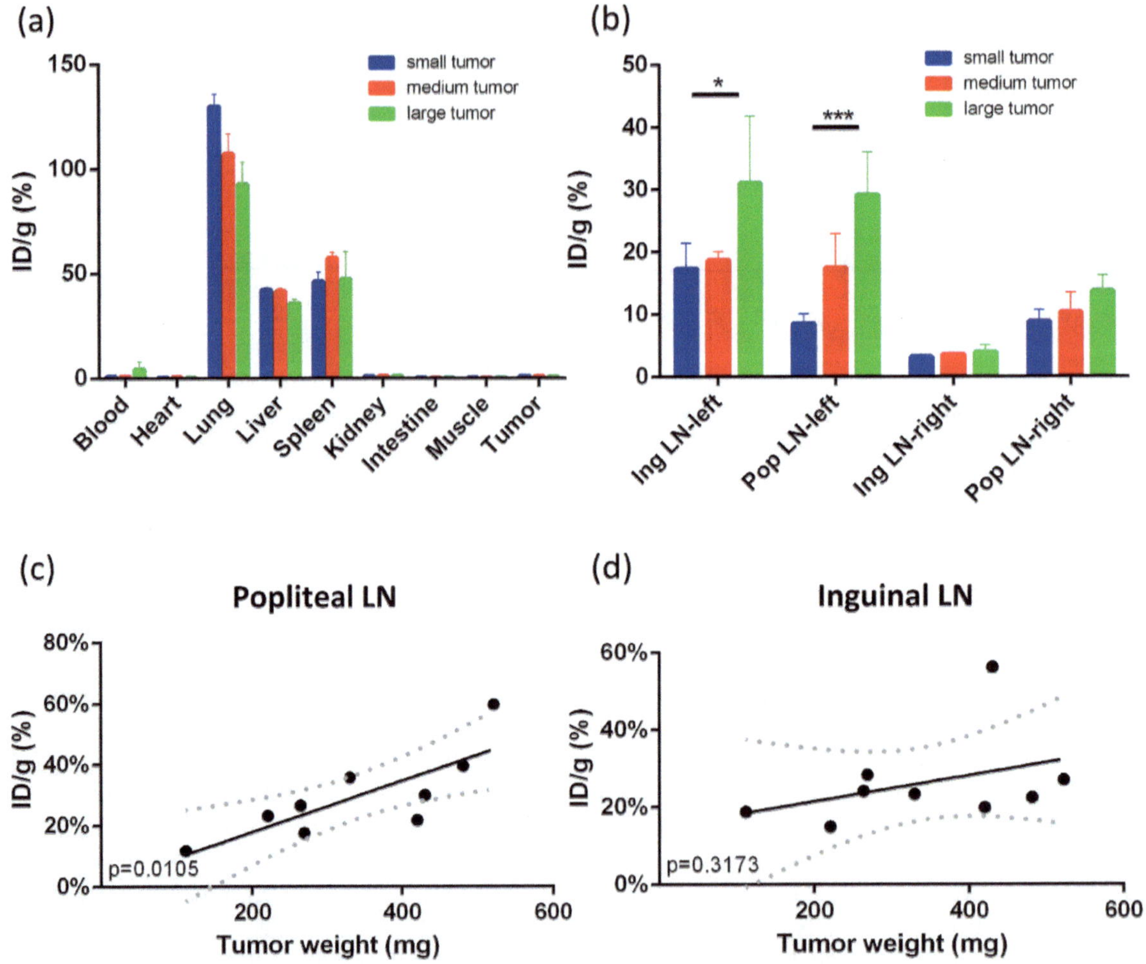

Fig. 4 Biodistribution of intravenously inoculated Nano-DC vaccine in mice with different size tumors. (**a**) Accumulation of Nano-DC vaccine in major organs in mice with different size tumors 24 h after vaccination (n = 3). (**b-d**) Correlation between tumor weight (mg) and Nano-DC vaccine accumulation in tumors and lymph node. Y axis is presented by count cper injection dosage per gram

nanoparticles with different physicochemical properties have been used ranging from inorganic mesoporous silica (Li et al. 2018) to polymeric nanoparticles (Luo et al. 2017) to hybrid nanodiscs (Kuai et al. 2017). Regardless the type, all nanometer-scale particles need to interact with the immune cells in order to exert their biological activities. It is thus important to understand the parameters that are required for effective particle uptake by the APCs.

It has been reported that physical properties of the antigen-carrying particles can dictate the type of immune responses (Kumar et al. 2015). In our study, we have applied particles in the nanometer to micrometer range to carry tumor antigens and demonstrated that the overall shape of the PSM particles is a key parameter to determine uptake by DCs and macrophages, two of the major APC populations. We have also found that different routes of vaccination can drastically alter the distribution pattern of Nano-DC vaccine in the spleen and major lymph nodes. Interestingly, we detected an asymmetric distribution pattern of Nano-DC vaccines in the tumor-bearing mice. There have been reports on asymmetric distribution of lymph nodes (Ghezzi et al. 2006; Yoruk et al. 2014). It remains to be determined whether asymmetric distribution of Nano-DC vaccine is a result of the asymmetric pattern of lymph nodes.

Overall, information generated from our study will provide guidance on design of future particulate cancer vaccines for effective uptake by the APCs in order to achieve optimal antitumor immunity.

Acknowledgements This work was partially supported by National Institutes of Health grants R01CA193880, R01CA222959 and U54CA210181, US Department of Defense grant W81XWH-17-1-0389, and METAvivor.

References

Y. Ban, J. Mai, X. Li, M. Mitchell-Flack, T. Zhang, L. Zhang, L. Chouchane, M. Ferrari, H. Shen, X. Ma, Targeting autocrine CCL5-CCR5 Axis reprograms immunosuppressive myeloid cells and reinvigorates antitumor immunity. Cancer Res. **77**(11), 2857–2868 (2017)

B.M. Carreno, V. Magrini, M. Becker-Hapak, S. Kaabinejadian, J. Hundal, A.A. Petti, A. Ly, W.R. Lie, W.H. Hildebrand, E.R. Mardis, G.P. Linette, Cancer immunotherapy. A dendritic cell vaccine increases the breadth and diversity of melanoma neoantigen-specific T cells. Science **348**(6236), 803–808 (2015)

M.A. Cheever, C.S. Higano, PROVENGE (Sipuleucel-T) in prostate cancer: The first FDA-approved therapeutic cancer vaccine. Clin. Cancer Res. **17**(11), 3520–3526 (2011)

M.S. Duthie, H.P. Windish, C.B. Fox, S.G. Reed, Use of defined TLR ligands as adjuvants within human vaccines. Immunol. Rev. **239**, 178–196 (2011)

L. Gelao, C. Criscitiello, A. Esposito, M. De Laurentiis, L. Fumagalli, M.A. Locatelli, I. Minchella, M. Santangelo, S. De Placido, A. Goldhirsch, G. Curigliano, Dendritic cell-based vaccines: Clinical applications in breast cancer. Immunotherapy **6**(3), 349–360 (2014)

F. Ghezzi, A. Cromi, S. Uccella, S. Giudici, M. Franchi, P. Bolis, Left-right asymmetry in pelvic lymph nodes distribution: Is there a right-side prevalence? Eur. J. Obstet. Gynecol. Reprod. Biol. **127**(2), 236–239 (2006)

B. Godin, C. Chiappini, S. Srinivasan, J.F. Alexander, K. Yokoi, M. Ferrari, P. Decuzzi, X. Liu, Discoidal porous silicon particles: Fabrication and biodistribution in breast Cancer bearing mice. Adv. Funct. Mater. **22**(20), 4225–4235 (2012)

M.S. Goldberg, Immunoengineering: How nanotechnology can enhance cancer immunotherapy. Cell **161**(2), 201–204 (2015)

P.W. Kantoff, C.S. Higano, N.D. Shore, E.R. Berger, E.J. Small, D.F. Penson, C.H. Redfern, A.C. Ferrari, R. Dreicer, R.B. Sims, Y. Xu, M.W. Frohlich, P.F. Schellhammer, Sipuleucel-T immunotherapy for castration-resistant prostate cancer. N. Engl. J. Med. **363**(5), 411–422 (2010)

K.L. Knutson, K. Schiffman, M.L. Disis, Immunization with a HER-2/neu helper peptide vaccine generates HER-2/neu CD8 T-cell immunity in cancer patients. J. Clin. Invest. **107**(4), 477–484 (2001)

R. Kuai, L.J. Ochyl, K.S. Bahjat, A. Schwendeman, J.J. Moon, Designer vaccine nanodiscs for personalized cancer immunotherapy. Nat. Mater. **16**(4), 489–496 (2017)

S. Kumar, A.C. Anselmo, A. Banerjee, M. Zakrewsky, S. Mitragotri, Shape and size-dependent immune response to antigen-carrying nanoparticles. J. Control. Release **220**(Pt A, 141–148 (2015)

A.W. Li, M.C. Sobral, S. Badrinath, Y. Choi, A. Graveline, A.G. Stafford, J.C. Weaver, M.O. Dellacherie, T.Y. Shih, O.A. Ali, J. Kim, K.W. Wucherpfennig, D.J. Mooney, A facile approach to enhance antigen response for personalized cancer vaccination. Nat. Mater. **17**(6), 528–52+ (2018)

F. Lucchini, M.G. Sacco, N. Hu, A. Villa, J. Brown, L. Cesano, L. Mangiarini, G. Rindi, S. Kindl, F. Sessa, et al., Early and multifocal tumors in breast, salivary, harderian and epididymal tissues developed in MMTY-Neu transgenic mice. Cancer Lett. **64**(3), 203–209 (1992)

M. Luo, H. Wang, Z. Wang, H. Cai, Z. Lu, Y. Li, M. Du, G. Huang, C. Wang, X. Chen, M.R. Porembka, J. Lea, A.E. Frankel, Y.X. Fu, Z.J. Chen, J. Gao, A STING-activating nanovaccine for cancer immunotherapy. Nat. Nanotechnol. **12**(7), 648–654 (2017)

J. Mai, Y. Huang, C. Mu, G. Zhang, R. Xu, X. Guo, X. Xia, D.E. Volk, G.L. Lokesh, V. Thiviyanathan, D.G. Gorenstein, X. Liu, M. Ferrari, H. Shen, Bone marrow endothelium-targeted therapeutics for metastatic breast cancer. J. Control. Release **187**, 22–29 (2014)

E.A. Mittendorf, G.T. Clifton, J.P. Holmes, K.S. Clive, R. Patil, L.C. Benavides, J.D. Gates, A.K. Sears, A. Stojadinovic, S. Ponniah, G.E. Peoples, Clinical trial results of the HER-2/neu (E75) vaccine to prevent breast cancer recurrence in high-risk patients: From US military Cancer institute clinical trials group study I-01 and I-02. Cancer **118**(10), 2594–2602 (2012)

G. Napolitani, A. Rinaldi, F. Bertoni, F. Sallusto, A. Lanzavecchia, Selected toll-like receptor agonist combinations synergistically trigger a T helper type 1-polarizing program in dendritic cells. Nat. Immunol. **6**(8), 769–776 (2005)

H. Shen, J. You, G. Zhang, A. Ziemys, Q. Li, L. Bai, X. Deng, D.R. Erm, X. Liu, C. Li, M. Ferrari, Cooperative, nanoparticle-enabled thermal therapy of breast cancer. Adv Healthc Mater **1**(1), 84–89 (2012)

H. Shen, C. Rodriguez-Aguayo, R. Xu, V. Gonzalez-Villasana, J. Mai, Y. Huang, G. Zhang, X. Guo, L. Bai, G. Qin, X. Deng, Q. Li, D.R. Erm, B. Aslan, X. Liu, J. Sakamoto, A. Chavez-Reyes, H.D. Han, A.K. Sood, M. Ferrari, G. Lopez-Berestein, Enhancing chemotherapy response with sustained EphA2 silencing using multistage vector delivery. Clin. Cancer Res. **19**(7), 1806–1815 (2013a)

J. Shen, R. Xu, J. Mai, H.C. Kim, X. Guo, G. Qin, Y. Yang, J. Wolfram, C. Mu, X. Xia, J. Gu, X. Liu, Z.W. Mao, M. Ferrari, H. Shen, High capacity nanoporous silicon carrier for systemic delivery of gene silencing therapeutics. ACS Nano **7**(11), 9867–9880 (2013b)

H. Shen, T. Sun, H.H. Hoang, J.S. Burchfield, G.F. Hamilton, E.A. Mittendorf, M. Ferrari, Enhancing cancer immunotherapy through nanotechnology-mediated tumor infiltration and activation of immune cells. Semin. Immunol **34**, 114 (2017)

J.L. Tanyi, S. Bobisse, E. Ophir, S. Tuyaerts, A. Roberti, R. Genolet, P. Baumgartner, B.J. Stevenson, C. Iseli, D. Dangaj, B. Czerniecki, A. Semilietof, J. Racle, A. Michel, I. Xenarios, C. Chiang, D.S. Monos, D.A. Torigian, H.L. Nisenbaum, O. Michielin, C.H. June, B.L. Levine, D.J. Powell Jr., D. Gfeller, R. Mick, U. Dafni, V. Zoete, A. Harari, G. Coukos, L.E. Kandalaft, Personalized cancer vaccine effectively mobilizes antitumor T cell immunity in ovarian cancer. Sci. Transl. Med. **10**(436), eaao5931 (2018)

M.E. Turnis, X.T. Song, A. Bear, A.E. Foster, S. Gottschalk, M.K. Brenner, S.Y. Chen, C.M. Rooney, IRAK-M removal counteracts dendritic cell vaccine deficits in migration and longevity. J. Immunol. **185**(7), 4223–4232 (2010)

X. Xia, J. Mai, R. Xu, J.E. Perez, M.L. Guevara, Q. Shen, C. Mu, H.Y. Tung, D.B. Corry, S.E. Evans, X. Liu, M. Ferrari, Z. Zhang, X.C. Li, R.F. Wang, H. Shen, Porous silicon microparticle potentiates anti-tumor immunity by enhancing cross-presentation and inducing type I interferon response. Cell Rep. **11**, 957–966 (2015)

R. Xu, Y. Huang, J. Mai, G. Zhang, X. Guo, X. Xia, E.J. Koay, G. Qin, D.R. Erm, Q. Li, X. Liu, M. Ferrari, H. Shen, Multistage vectored siRNA targeting ataxia-telangiectasia mutated for breast Cancer therapy. Small **9**(9–10), 1799–1808 (2013)

R. Xu, G. Zhang, J. Mai, X. Deng, V. Segura-Ibarra, S. Wu, J. Shen, H. Liu, Z. Hu, L. Chen, Y. Huang, E. Koay, Y. Huang, J. Liu, J.E. Ensor, E. Blanco, X. Liu, M. Ferrari, H. Shen, An injectable nanoparticle generator enhances delivery of cancer therapeutics. Nat. Biotechnol. **34**(4), 414–418 (2016)

O. Yoruk, R. Yuksel, Y. Yuksel, S. Dane, Left-right asymmetry in neck lymph nodes distribution in patients with bilateral laryngeal cancer. Surg. Radiol. Anat. **36**(3), 239–242 (2014)

H. Zong, S. Sen, G. Zhang, C. Mu, Z.F. Albayati, D.G. Gorenstein, X. Liu, M. Ferrari, P.A. Crooks, G.J. Roboz, H. Shen, M.L. Guzman, In vivo targeting of leukemia stem cells by directing parthenolide-loaded nanoparticles to the bone marrow niche. Leukemia **30**(7), 1582–1586 (2016)

Part III
Biomedical Devices

Biomedical Microdevices (2019) 21: 47
https://doi.org/10.1007/s10544-019-0389-6

Advanced implantable drug delivery technologies: transforming the clinical landscape of therapeutics for chronic diseases

Fernanda P. Pons-Faudoa[1,2] · Andrea Ballerini[1,3] · Jason Sakamoto[1] · Alessandro Grattoni[1,4,5]

Published online: 18 May 2019

Author's Personal Note I first met Mauro in person on the train when he was travelling from Venice to give a talk in Milan. I had to jump on the train to have enough time to present the results of my work to him. At that time, I was a Master's student in Mechanical Engineering, working on experimental evaluation of Mauro's theory on osmosis. Despite my green outlook and casual attire, Mauro showed great interest in my results and invited me for a follow up meeting in Catanzaro. A few months later, I met Mauro at Lamezia Terme airport. He kindly offered to drive me to campus. My mother had just passed away three days earlier. To my surprise, Mauro stopped at a chapel to say a prayer for her. I am not a religious person, but that gesture had a profound and everlasting impact on me. That day, Mauro asked me to join his laboratory in Texas. "What about California and UC Berkeley? Oh well…"
Shortly thereafter, I arrived in Houston, Texas. Mauro, Paola and their family showed great hospitality. My experience in Mauro's laboratory turned out to be extraordinary! Mauro gave me a lot of independence, offered invaluable opportunities to grow scientifically and to meet inspirational people across numerous scientific and artistic disciplines. He entrusted me with the responsibility to lead one of his pillar projects. In the past 12 years, I have witnessed Mauro's successes, visionary scientific quests and incredible determination. I saw his passion for science, his selfless dedication to patients even in the most desperate situations, and experienced first-hand his commitment to his team, which he often refers to as a family.
Mauro, I want to personally thank you. You have my utmost respect. I acknowledge that most of what I have accomplished stemmed from you, the trust that you put in me and the invaluable opportunities that you have offered.
Alessandro Grattoni

Abstract

Chronic diseases account for the majority of all deaths worldwide, and their prevalence is expected to escalate in the next 10 years. Because chronic disorders require long-term therapy, the healthcare system must address the needs of an increasing number of patients. The use of new drug administration routes, specifically implantable drug delivery devices, has the potential to reduce treatment-monitoring clinical visits and follow-ups with healthcare providers. Also, implantable drug delivery devices can be designed to maintain drug concentrations in the therapeutic window to achieve controlled, continuous release of therapeutics over extended periods, eliminating the risk of patient non-compliance to oral treatment. A higher local drug concentration can be achieved if the device is implanted in the affected tissue, reducing systemic adverse side effects and decreasing the challenges and discomfort of parenteral treatment. Although implantable drug delivery devices have existed for some time, interest in their therapeutic potential is growing, with a global market expected to reach over $12 billion USD by 2018. This review discusses implantable drug delivery technologies in an advanced stage of development or in clinical use and focuses on the state-of-the-art of reservoir-based implants including pumps, electromechanical systems, and polymers, sites of implantation and side effects, and deployment in developing countries.

Keywords MEMS · NEMS · Non-biodegradable polymers · Long-acting formulations · Implants

1 Introduction

Chronic disease can be defined as a disease that continues or reoccurs over a long period of time. Chronic diseases encompass cardiovascular diseases, diabetes, respiratory diseases, and other disorders that affect a large number of people, are costly to manage, and increase disability-adjusted life years (DALYs) (Bernell and Howard 2016). With one DALY equaling a year loss of disease-free life, DALYs are a measure of the burden of disease across the population (World Health Organization n.d.). Some chronic health disorders are manageable with appropriate treatment. However, the prevalence of chronic diseases such as cardiovascular diseases, diabetes, cancers, respiratory diseases, and human immunodeficiency virus/acquired immunodeficiency syndrome (HIV/AIDS) is steadily increasing and expected to affect 157 million people in the US by 2020 (Comlossy 2013). The global burden of chronic conditions also continues to rise (Fig. 1) and is projected to account for 69% of all deaths worldwide, of which 80% will be in developing countries, by 2030 (Alwan et al. 2010; Samb et al. 2010). As the global economic impact of chronic diseases is estimated to reach $47 trillion in the next two decades, concerted efforts are focused on relieving this burden (World Economic Forum and Harvard School of Public Health 2011).

Traditional intervention via oral or intravenous administration of therapeutics has several limitations. Some drugs have poor bioavailability and require multiple doses, augmenting the risk of resistance and side effects as well as the potential for drug abuse. Additionally, poor patient adherence has direct effects on medication efficacy. Non-adherence is a major concern, 30 to 50% of adults with chronic conditions in the US do not take their medications as prescribed and this has been correlated with 125,000 deaths and 10% of hospitalizations annually. This results in an annual economic burden of $100

✉ Alessandro Grattoni
agrattoni@houstonmethodist.org

[1] Department of Nanomedicine, Houston Methodist Research Institute, 6670 Bertner Avenue, Houston, TX 77030, USA

[2] School of Medicine and Health Sciences, Tecnologico de Monterrey, Avenida Eugenio Garza Sada 2501, 64849 Monterrey, NL, Mexico

[3] Department of Oncology and Onco-Hematology, University of Milan, Via Festa del Perdono 7, 20122 Milan, Italy

[4] Department of Surgery, Houston Methodist Hospital, 6550 Fannin Street, Houston, TX 77030, USA

[5] Department of Radiation Oncology, Houston Methodist Hospital, 6550 Fannin Street, Houston, TX 77030, USA

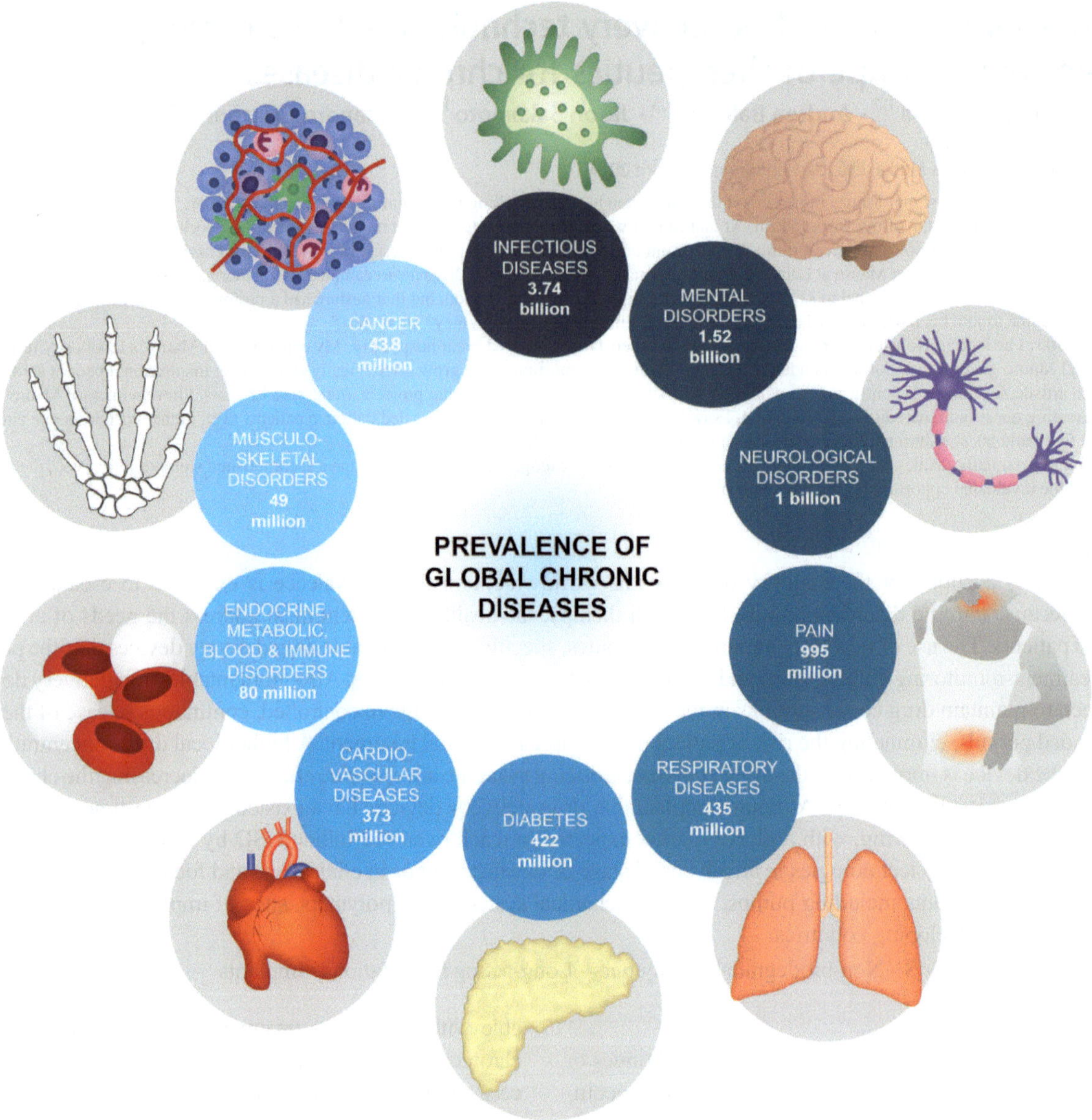

Fig. 1 Top ten global chronic diseases by prevalence (Bertolote 2005; Ferkol and Schraufnagel 2014; Goldberg and McGee 2011; Steel et al. 2014; The Global Cancer Observatory 2018; Vos et al. 2015; World Health Organization 2016)

billion USD in health care services (Cutler et al. 2018; Kini and Ho 2018; Pagès-Puigdemont et al. 2016; Oung et al. 2017). Social and technological efforts such as patient education services, health care provider interventions, reminder tools, and electronic monitoring devices have tried to tackle with medication non-adherence with no significant success (Kini and Ho 2018; Pagès-Puigdemont et al. 2016; Oung et al. 2017; World Health Organization 2003). Because of the correlation between increased non-adherence and higher illness prevalence there is an obvious need to find a solution for medication non-adherence (Atinga et al. 2018; Pagès-Puigdemont et al. 2016).

Compared with traditional systemic delivery, implantable drug delivery devices offer many advantages. Site-specific implantation can bypass the absorption and distribution phase of oral and peripheral regimens, resulting in higher drug concentrations in targeted areas (Danckwerts and Fassihi 1991). Thus, drug levels can be maintained in the therapeutic window by virtue of controlled, continuous release of therapeutics. Importantly, as this technology can be used over extended periods, it eliminates the possibility of poor patient compliance and decreases the discomfort of parenteral treatment (Park 2014). Therefore implantable drug delivery technologies provide site-specificity and deal with medication non-adherence, transforming the clinical landscape of therapeutics for chronic diseases.

Controlled drug delivery technologies have progressed over the last six decades to third-generation modulated delivery systems, with increasing interest in long-term delivery systems (Farina et al. 2017; Meng and Hoang 2012a; Park

2014; Yun et al. 2015). Accordingly, the global market for implantable drug delivery is growing—valued at $9.05 billion USD in 2013 and expected to be $12.42 billion by the end of 2018. Newer, more easily applicable machineries improve the scalability of implantable drug delivery devices. Companies and small start-ups find implantable devices profitable because they are cost-effective and lower overall treatment cost (Kumar and Pillai 2018), and there is high demand to file patents on versatile implantable drug delivery devices that can be tailored for multiple drugs (Coherent Market Insights 2017; Yang and Pierstorff 2012). Another potential benefit is the opportunity for pharmaceutical companies to exploit medications coming off patent, as patent expirations can be extended by creating new products that combine patented medications and implantable devices (Beall et al. 2016). Implantable drug delivery devices can also be advantageous for less prevalent chronic diseases such as drug abuse, pain management, and neurological disorders. Furthermore, telemedicine can allow physicians to remotely control drug release rate from the implant or maximize treatment effectiveness through the use of artificial intelligence and machine learning algorithms (Ross et al. 2017).

In this article, we highlight current technologies for long-term drug delivery in advanced stages of development or in clinical use, with a brief discussion of the use, mechanism of function, advantages, and limitations of each system. This review will demonstrate how advanced implantable drug delivery technologies can transform the clinical landscape of therapeutics for chronic illnesses. The drug delivery systems covered include reservoir-based polymer systems, pumps, and electromechanical systems, excluding polymeric fully degradable systems and long-term delivery devices that are not completely implanted, which are thoroughly revised elsewhere (Kamaly et al. 2016; Majeed and Thabit 2018). We further present a clinical perspective on sites of implantation and potential strategies to improve device development associated with patient acceptance, and device deployment in the developing world.

2 Reservoir-based polymer systems

Reservoir-based polymer systems are passive implants with a simple design consisting of a drug core surrounded by a non-degradable polymeric film (Fig. 2a). Drug release rate is controlled by polymeric coating properties, such as polymer configuration, molecular weight, and coating thickness, as well as physicochemical properties of the drug, such as solubility, particle size, and molecular weight. Historically, polymeric systems have been employed for site-specific mid-/long-term systemic drug administration after subcutaneous implantation. However, most polymeric systems suffer from an initial drug release burst, which can potentially reach toxic levels and endanger the patient. After this burst, drug core concentrations decrease, possibly to below their therapeutic window (Kumar and Pillai 2018; Yang and Pierstorff 2012).

Polymer systems require constant drug concentration within the drug core to achieve zero-order kinetic drug release. These implants often employ the polymers silicone, polyvinyl alcohol (PVA), and ethylene vinyl acetate (EVA). By contrast, biodegradable implants use naturally occurring polymers (e.g., human serum albumin, collagen, gelatin) or synthetic polymers (e.g., polylactic acid, polyglycolic acid, polylactic-co-glycolic acid copolymer) (Kumar and Pillai 2018; Yang and Pierstorff 2012).

Son et al. (2017) developed a 3D-printed porous cylindrical device called Biocage that can be filled with a drug. The Biocage is small enough to fit inside a 22-gauge needle for direct delivery and robust enough to be implanted directly into the target issue. The Biocage has the following dimensions: 300-μm hollow inner diameter, 20-μm outer wall, 40-μm solid base, 900-μm height, and 5-μm-diameter pores (Fig. 2b). The creators demonstrated fluorescent microsphere release from the implant but did not determine the drug release rate. They also confirmed that the Biocage can be used for local drug delivery within the brain and explain how 3D printing offers structural and material versatility to the device. However, although the materials are biocompatible and biodegradable, they are not yet approved by the US Food and Drug Administration (FDA). Nevertheless, this technological platform shows much promise, as it offers drug versatility, has high drug loading efficiency, and can be implanted within the target organ.

In 2006, the FDA approved a similar reservoir-based polymer system, Implanon® (Merck, Kenilworth, NJ, USA), a 4 cm × 22 mm non-biodegradable implant, as a female hormone-based contraceptive. An EVA copolymer rod encompasses 68 mg etonogestrel, which controls the daily release of progestin for up to 3 years. However, the release rate decreases over time, from 60 to 70 μg/day in first couple of weeks to 35–45, 30–40, and 25–30 μg/day at the end of the first, second, and third year, respectively (FDA Reference IDs: 3080389, 4,100,681) (Allen et al. 2016; Huber 1998). However, another study extending its use to 5 years indicated an efficacy of 100% (Ali et al. 2016), suggesting that if the device is still effective after 5 years, patients have likely received supraoptimal doses. Therefore, this implant should be further improved to deliver at a constant rate for 3–5 years. In some cases, the Implanon® was incorrectly inserted, making its localization for removal difficult for healthcare professionals. This led to the design of Nexplanon®, a second-generation device with the addition of the radiopaque ingredient barium sulfate, which entered the US market in 2011 (FDA Reference IDs: 3080389, 4,100,681) (Allen et al. 2016; Huber 1998).

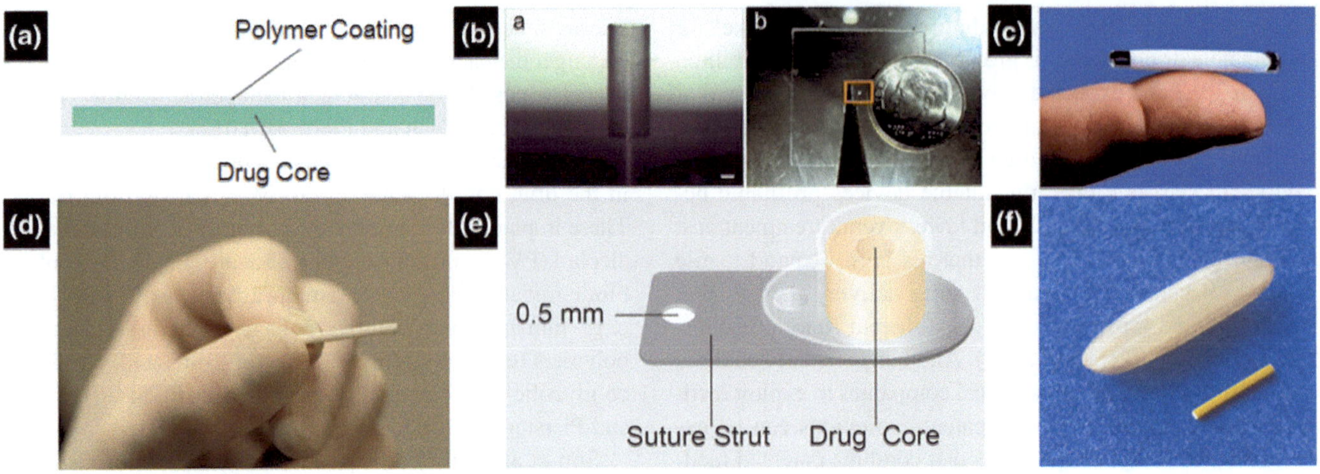

Fig. 2 FDA-approved and experimental non-biodegradable reservoir-based polymer systems. **a** Non-biodegradable polymer schematic depicting an outer polymer coating encompassing an inner drug core. **b** Drug-versatile 3D-printed Biocage device (a) Magnified light microscopy image to detect porosity with 100-μm scale bar. (b) Biocage device boxed in orange in relation to pencil tip and dime to appreciate its minute size and how it can be inserted using a 22-gauge needle. (Image 2B adapted from (Son et al. 2017) licensed under CC BY 4.0). **c** Vantas® and SUPPRELIN® LA 50 mg histrelin acetate implants for prostate cancer symptom relief and childhood central precocious puberty treatment, respectively (Image reproduced from (Rudlang and Brasso 2016). **d** Probuphine® 80 mg buprenorphine hydrochloride implant for opioid dependence treatment (Image used with permission from Titan Pharmaceuticals Inc.). **e** Retisert® implant design consists on a platform for suturing device and drug core with 0.59 mg fluocinolone acetonide enclosed in silicone elastomer cup with a PVA membrane outlet for treating chronic noninfectious uveitis. **f** Intraocular ILUVIEN® 0.19 mg fluocinolone acetonide device for diabetic macular edema treatment in relation to a grain of rice to demonstrate its size (Image is courtesy of Alimera Sciences Inc.)

The Hydron® implant (Endo Pharmaceuticals Solutions Inc., Malvern, PA, USA) consists of a hydrogel polymeric reservoir called MedLaunch™ that is spun-cast into a 3.5 cm × 3 mm tube (Stevenson et al. 2012). Two of these non-biodegradable reservoir-based polymeric system implants are already on the market: Vantas® and SUPPRELIN® LA. The drug core contains 50 mg histrelin acetate in both implants, but the drug delivery rate is modified for the treatment of two different diseases (Fig. 2c). The Vantas® implant delivers 50 μg/day for 12 months to relieve symptoms of prostate cancer, whereas the SUPPRELIN® LA implant releases 65 μg/day for 12 months to treat children with central precocious puberty (FDA Reference IDs: 4099967, 2,887,911). Currently, there are no reports of decreasing drug release rates from these implants, which could be attributed to their shorter treatment periods. The Hydron® implant technology was also adapted to deliver 84 mg octreotride, a somatostatin analog, for up to 6 months to treat acromegaly. However, the phase 3 clinical trial was terminated for business reasons (NCT01295060) (Endo Pharmaceuticals n.d.; Stevenson et al. 2012).

ProNeura™ (Titan Pharmaceuticals Inc., San Francisco, CA, USA) is a non-biodegradable rod composed of an EVA matrix and a drug formulation. The Probuphine® implant, ProNeura™ with buprenorphine, was FDA-approved in 2016 for the maintenance treatment of opioid dependence. Four 26 × 2.5 mm implants are needed to maintain therapeutic drug levels (Fig. 2d). Each device contains 80 mg buprenorphine hydrochloride, a partial opioid agonist, delivered at a controlled rate for up to 6 months (FDA Reference ID: 4215185). Probuphine® has proved more cost-effective than sublingual buprenorphine, as it minimizes fluctuations in plasma concentrations and reduces clinic and pharmacy visits by eliminating the need for daily supervision (Barnwal et al. 2017; Carter et al. 2017). Currently, preclinical studies are testing the use of ProNeura™ to deliver a dopamine agonist (ropinirole) and T3 for the treatment of Parkinson's disease and hypothyroidism, respectively (Titan Pharmaceuticals n.d.).

All above-mentioned reservoir-based polymer systems are subcutaneously implanted in the inner arm, as they require systemic therapeutic levels. However, two non-biodegradable implants are FDA-approved for intravitreal management of ophthalmology-related diseases: Retisert® (Bausch & Lomb, Rochester, NY, USA) and ILUVIEN® (Alimera Sciences Inc., Alpharetta, GA, USA). Because ocular diseases affecting the posterior chamber require constant drug exposure, both devices take advantage of the higher viscosity in the vitreous humor, which increases drug half-life. Retisert® treats chronic noninfectious uveitis and can achieve drug release for 30 months but must then be removed (FDA Reference ID: 2955048) (Borkar et al. 2017; Haghjou et al. 2011; Logan et al. 2016; Yasin et al. 2014). ILUVIEN® can maintain therapeutic levels in the vitreous humor for up to 36 months for the treatment of diabetic macular edema (DME) in vitrectomized and non-vitrectomized eyes (Carle et al. 2014; Hawrami et al. 2016; Kumar et al. 2016; Meireles et al. 2017; Pessoa et al. 2018). After 36 months, a new implant can be inserted without removing the previous implant, as no side effects have been reported from having multiple implants in the eye (FDA Reference ID: 3635981)

(Borkar et al. 2017; Hawrami et al. 2016; Logan et al. 2016; Wang et al. 2013; Yasin et al. 2014).

Retisert® consists of a drug formulation tablet enclosed in a silicone elastomer cup with an outlet consisting of a PVA membrane (Fig. 2e). The tablet contains 0.59 mg fluocinolone acetonide (FA), a corticosteroid, and the following inactive ingredients: microcrystalline cellulose, PVA, and magnesium stearate. Retisert® passively delivers FA into the vitreous humor for the treatment of chronic noninfectious uveitis affecting the posterior chamber (FDA Reference ID: 2955048) (Borkar et al. 2017; Haghjou et al. 2011; Logan et al. 2016; Yasin et al. 2014). ILUVIEN® is a 3.5 × 0.37 mm rod made of polyimide with a non-permeable cap on one end and a permeable PVA membrane on the other end. The inside of the rod is composed of a PVA matrix with 0.19 mg FA (Fig. 2f). This implant is not the first line of therapy but is only approved for DME eyes that did not respond to laser therapy and anti-VEGF therapy (Elaraoud et al. 2016a; Figueira et al. 2017; Massin et al. 2016). Real-world results indicate the efficacy of ILUVIEN®, demonstrating improved best corrected visual acuity and central foveal thickness (Alfaqawi et al. 2017; Amoaku et al. 2015; Bailey et al. 2017; Bertelmann et al. 2013; Bertelmann and Schulze 2015; Cunha-Vaz et al. 2014; Elaraoud et al. 2016b, c; El-Ghrably et al. 2017; Fusi-Rubiano et al. 2018; Gonçalves et al. 2017; Mourtzoukos 2017; Quhill and Quhill 2016; Saedon et al. 2017; Schmit-Eilenberger 2015; Syed 2017; Veritti et al. 2017; Yang et al. 2015). Another implant, Vetrisert®, was FDA-approved for the treatment of cytomegalovirus retinitis but was later discontinued. A pellet of 4.5 mg ganciclovir was enclosed between PVA and EVA and was found to relieve symptoms for up to 8 months (Yasin et al. 2014). Vetrisert® was also effective in treating cytomegalovirus retinitis in AIDS patients, extending the progression of retinitis from 15 to 226 days (Martin 1994).

In summary, reservoir-based polymer systems are the type of implant that has received the most FDA approval and has been on the market the longest. All use the same mechanism of release: drug diffusion through non-biodegradable polymer film. A comparison of their advantages and limitations is shown in Table 1.

3 Pumps

3.1 Osmotic pumps

Osmotic pumps were developed in the 1950s by Rose and Nelson for drug delivery in animals. Since then, numerous designs have found clinical use for the treatment of human diseases (Keraliya et al. 2012; Santus and Baker 1995). Implantable osmotic pumps are drug delivery devices developed for the sustained administration of therapeutics over extended periods of time ranging from months to years. Osmotic pumps are conventionally composed of a hollow cylinder containing a drug reservoir and an osmotic engine separated by a movable piston. The drug reservoir is directly connected to the outside through micro-holes, and the osmotic engine is separated from the outside by means of a semipermeable membrane.

The mechanism of osmotic pump-driven drug release occurs after the pump is implanted. The osmotic engine, which contains high concentration of osmolytes (i.e., salts), drives an osmotic flow of interstitial fluid through the semipermeable membrane. The inward H_2O flow increases hydrostatic pressure in the osmotic reservoir, which exerts force on the piston (Fig. 3a). The piston is pushed toward the drug reservoir and causes injection of the drug solution in an equivalent amount to the volume of drug solution displaced. Ideally, this process is continuous and terminates when the piston has displaced the entire amount of drug solution and has reached the extreme end of the drug reservoir. Other osmotic pumps have a different design in which the osmotic engine surrounds the drug reservoir (Fig. 3b). In these pumps, a high salt concentration in the osmotic engine displaces the drug out through the micro-orifice at a controlled rate by compressing the drug reservoir (Cobo et al. 2015; Herrlich et al. 2012; Kumar and Pillai 2018; McConville 2011).

Viadur® (Bayer Healthcare Pharmaceuticals, Berlin, Germany) was a non-biodegradable titanium osmotic implant that utilized a DUROS® controlled release pump to administer leuprolide acetate, a gonadotropin-releasing hormone analog, for 12 months for the palliative treatment of advanced prostate cancer (FDA Reference ID: 2888026) (Rohloff et al. 2008). Despite successful clinical trials and FDA approval, Viadur® was removed from the market in 2007 due to its lack of cost-effectiveness and limited long-term market viability. In general, the fabrication and assembly procedures as well as the quality control of osmotic implants may ultimately be too expensive to justify their clinical use as an alternative to conventional drug administration approaches.

The Medici Drug Delivery System™ (Intarcia Therapeutics Inc., Boston, MA, USA) is an osmotic mini-pump tailored to hold a certain drug volume over different dosing intervals (Intarcia Therapeutics n.d.-a). ITCA 650 utilizes the Medici Drug Delivery System™ to achieve continuous delivery of exenatide, a glucagon-like peptide-1 receptor agonist, for the treatment of type 2 diabetes. The pump maintains exenatide release for 6 months and is undergoing further development for a 1-year dose (Intarcia Therapeutics n.d.-c). A challenge to delivering a 1-year dose is the necessity of maintaining a constant concentration of osmolyte in the osmotic engine over the entire duration of the treatment to achieve constant drug elution. As such, the osmolyte must be included in a supersaturated form to maintain its constant concentration despite the inward flow of H_2O. When a

Table 1 Comparison among non-biodegradable reservoir-based polymer systems

Implant	Development status	Advantages	Limitations
Biocage	Experimental	• Small size • Drug versatility • Site-specific drug release • Easy insertion procedure	• Materials not FDA-approved
Hydron® implant technology	FDA-approved	• 1-year drug release in comparison to conventional drug administration • Small size • Easy insertion and removal procedures • Alternative implantation sites	• Implant must be exchanged after 1 year • Only one drug formulation (histrelin acetate)
Implanon®/Nexplanon®	FDA-approved	• 3-year drug release in comparison to conventional drug administration • Small size • Easy insertion and removal procedures • Radiopaque • Alternative implantation sites • Soft and flexible shape	• Implant must be exchanged after 3 years • Patients likely initially receiving supraoptimal doses • Decline in drug release rate over time
Probuphine®	FDA-approved	• Small size • Easy insertion and removal procedures	• Four implants needed • Implants must be removed after 6 months • Not radiopaque
Retisert®	FDA-approved	• 30-month drug release • Site-specific drug release	• Requires invasive surgery • Implant must be exchanged • Adverse side effects • Not cost-effective
ILUVIEN®	FDA-approved	• 36-month drug release • Cost-effective • Site-specific drug release	• Non-biodegradable implant not removed from vitreous humor after treatment

substantial amount of drug has been released, release rate may decline as a result of reduced osmotic flow.

ITCA 650 has completed its phase 3 clinical trial, called FREEDOM. However, the FDA issued a Complete Response Letter regarding manufacturing aspects, and the device is currently on an FDA clinical hold (Genetic Engineering and Biotechnology News 2018). Titanium osmotic pump manufacturing can be very expensive, as these pumps require extremely tight dimensional and geometrical tolerances as well as lathe machining for minimal surface roughness in the inner implant cavity. Intarcia is currently resolving these issues, and the Medici Drug Delivery System™ will be adapted for the continuous delivery of HIV pre-exposure prophylaxis (PrEP) (Intarcia Therapeutics n.d.-b).

Osmotic pumps have been further developed to improve intravesical drug delivery using osmotic flow of H_2O from urine instead of interstitial fluid. GemRIS™ and lidocaine-releasing intravesical system (LiRIS®) (TARIS Biomedical®, Lexington, MA, USA), which utilize the TARIS® System, are elastomeric tubular osmotic intravesical implants that deliver gemcitabine and lidocaine, respectively, to treat bladder diseases. The TARIS® System is a dual-lumen silicone tube containing an osmotic engine encompassing the solid drug core in one lumen and nitinol wireform in the other (Fig. 3c). The permeability of silicone permits H_2O from the urine to diffuse through the osmotic engine into the drug core and dissolve the drug. This creates an osmotic pressure in the osmotic engine that forces drug solution out through the orifice (Fig. 3b).

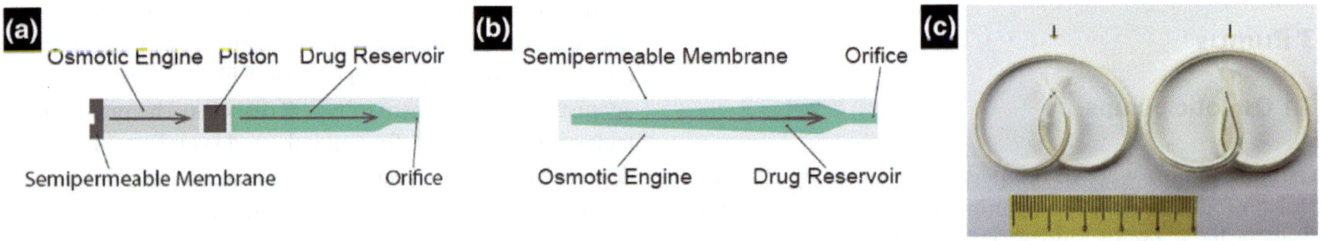

Fig. 3 Osmotic pump drug release schematics and FDA-approved osmotic pump implant. **a** Osmotic pump drug release mechanism for liquid drug formulations use a high salt concentration osmotic engine driven by osmotic flow through semipermeable membrane to move piston and displace drug through orifice. **b** Osmotic pumps with an inner solid drug reservoir encompassed by a high osmolyte concentration osmotic engine surrounded by semipermeable membrane osmotically displace solubilized drug through orifice. **c** Intravesical GemRIS™ implant loaded with solid gemcitabine for bladder cancer treatment (Image adapted from (Cima et al. 2014))

Intravesical osmotic pumps are currently undergoing clinical trials. GemRIS™ completed a phase 1b clinical trial to assess its safety and tolerability in muscle-invasive bladder cancer patients (NCT02722538) (Taris Biomedical LLC n.d.-a). GemRIS™ will also undergo a phase 1b clinical trial with Opdivo® (nivolumab) in the same patient population as well as two other clinical trials for non-muscle-invasive bladder cancer (NCT02720367) and muscle-invasive bladder cancer unfit for radical cystectomy (NCT03404791) (Taris Biomedical LLC n.d.-b; Taris Biomedical LLC n.d.-c).

An advantage of the TARIS® System is that the drug is loaded in solid form, which augments its loading efficiency. Also, the implant does not have moving components, decreasing the risk of potential failure and reducing fabrication costs. The device can achieve local sustained release of drug, minimizing side effects and frequent drug catheter injections to the bladder. These implants have received positive feedback from people who suffer from bladder diseases seeking a new drug administration strategy (Cima et al. 2014; Herrlich et al. 2012; Matheson 2014; Nickel et al. 2012; Taris Biomedical LLC n.d.-d). Nonetheless, accidental rupture of the implant can cause drug overdose from the dissolving solid drug core, and the device may be difficult to efficiently remove from the body without cystoscopy.

3.2 Peristaltic pumps

Peristaltic pumps have been used clinically for many years. In 1881, Eugene Allen was the first to patent the peristaltic pump in the US for blood transfusions (*US249285A*) (Allen 1881; INTEGRA Biosciences n.d.). Years later, cardiothoracic surgeon Dr. Michael DeBakey created the DeBakey pump that was used in the Gibbon heart-lung machine in 1953 (Winters 2015). Positive displacement is the driving force for pumping fluids contained in a tube inside the peristaltic pump. Rollers attached to the external circumference of a rotor compress the flexible tube, trapping liquid drug doses between rollers. As the rotor rotates, the rollers displace the drug in the tube and the tube returns to its natural state after passage of the drug, a process known as peristalsis (Fig. 4a). This peristalsis transports the drug toward the pump outlet and into a catheter for delivery to the target site.

This technology has been applied to create an implantable peristaltic pump capable of chronically administering therapeutics at the target site (Berg and Dallas 2013). However, the implant is relatively large to accommodate the mechanical components, battery, and drug. As such, the volumetric loading efficiency, defined as the ratio of drug reservoir volume to implant volume, is greatly limited to 22–30% (Medtronic 2011). A disadvantage of this pump is that its size restricts the implantation site, requiring a catheter to administer the drug at the target site. This technological platform is already on the market as the SynchroMed™ II pump (Medtronic, Fridley, MN, USA), an implantable FDA-approved system composed of a pump reservoir, reservoir fill port, reservoir valve, pump tubing, check valve, catheter port, and implanted catheter (Fig. 4b) (Kosturakis and Gebhardt 2012; Pope and Deer 2015). Drug is percutaneously loaded in the reservoir fill port and passes through the reservoir valve into the pump reservoir. The design of the pump reservoir involves pressurized gas stored below the reservoir. Thus, at normal body temperature, the gas expands and displaces the drug in the pump reservoir into the pump tubing. The SynchroMed™ II pump then transports the drug in a peristaltic motion through the pump tube, check valve, catheter port, and implanted catheter, where it is released at the target site (Bolash et al. 2015; Christo and Bottros 2014; Meng and Hoang 2012b; Pope and Deer 2015).

This pump is FDA-approved for the chronic delivery of treprostinil, morphine sulfate, and ziconotide. Intravenous treprostinil, epidural/intrathecal morphine sulfate, and intrathecal ziconotide are delivered for the treatment of pulmonary arterial hypertension, chronic intractable pain, and severe chronic pain management, respectively (Bourge et al. 2016; Medtronic 2017). In Sweden, the SynchroMed™ II pump is administering intracerebroventricular PDFG-BB in Parkinson's disease patients in a phase 1/2a study evaluating its safety and tolerability (NCT00866502) (Newron Sweden AB n.d.; Paul et al. 2015).

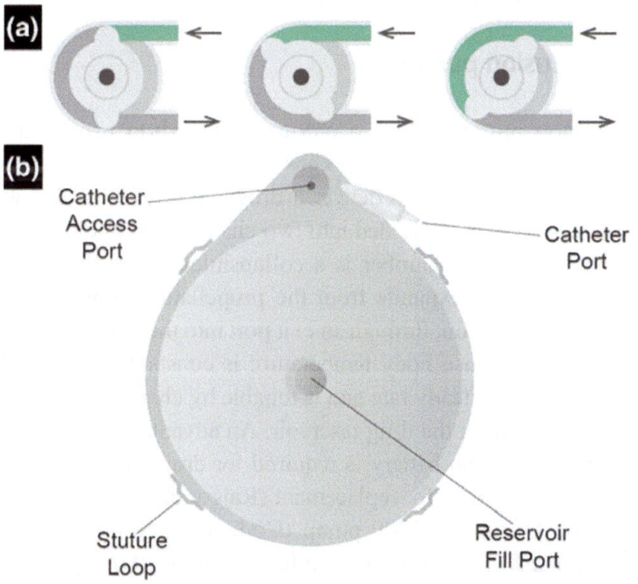

Fig. 4 Peristaltic pump drug release mechanism and design. **a** Peristaltic pump drug release mechanism: a central rotor with rollers attached to its circumference rotates, compressing the flexible tube, trapping liquid drug doses between rollers and displacing it through the catheter. **b** General outer schematic of implantable pumps: a discoidal-shaped implant with a central reservoir fill port that can be accessed percutaneously, a catheter port that connects the catheter and implant, and suture loops to securely anchor the implant in the abdominal pump pocket

Chronic drug delivery requires careful dose monitoring by a healthcare professional to maintain adequate therapeutic levels. The SynchroMed™ II pump can be programmed by telemetry to deliver a wide range of therapeutic flow rates, thus personalizing the drug dose for each patient (Li et al. 2012; Medtronic 2017). Likewise, pain tolerance differs between patients, so an advantage of this pump is that the patient can self-administer an additional dose through a personal therapy manager (PTM), a handheld accessory with a lockout system ensuring that the patient does not administer more drug than is approved by the doctor (Bhatia et al. 2014). Currently, the PTM is undergoing a phase 4 clinical trial for patient-controlled intrathecal analgesia with bupivacaine for chronic low back pain (NCT02886286) (Ilias et al. 2008; Salim M Hayek and University Hospitals Cleveland Medical Center n.d.).

Because these pumps require a battery, their lifespan is limited to 4–7 years. Also, their low volumetric loading efficiency of 22–30% requires a large pump and limits the size of the reservoir compared with the volume of the device. Consequently, patients must go to a healthcare professional to refill the pump every 3–4 months, which affects patient acceptability (Bolash et al. 2015; Christo and Bottros 2014; Meng and Hoang 2012b; Pope and Deer 2015). Another setback is reports that magnetic resonance imaging (MRI) temporarily stops the pump motor rotor. As a result, all patients must undergo assessment of pump motor function after an MRI (Kosturakis and Gebhardt 2012; Pope and Deer 2015). Peristaltic systems are also costly to manufacture (Rajgor et al. 2011).

3.3 Infusion pumps

Infusion pumps utilize a chlorofluorocarbon propellant, whose change from liquid to gas at body temperature serves as the driving force to deliver a drug. This implantable mechanical system is divided into two chambers: propellant and drug. The drug chamber is a collapsible bellow that compresses as gas expands from the propellant chamber. This forces the drug out through an exit port into the pump catheter (Fig. 5). Because body temperature is constant, the drug is delivered at a steady rate and is tunable by changing the drug concentration in the drug reservoir. An advantage of infusion pumps is that no battery is required for drug administration, avoiding the need for replacement (Rajgor et al. 2011).

The Codman® 3000 pump (Codman & Shurtleff, Inc., Raynham, MA, USA) is an infusion pump FDA-approved for intrathecal delivery of morphine sulfate for pain management and hepatic arterial infusion of chemotherapy to the tumor site. The pump achieves a constant flow rate by maintaining a pump drive pressure of approximately 0.6 bar at body temperature. There are different titanium Codman® 3000 pump drug reservoir sizes: 16, 30, or 50 ml. Thus, the Codman® 3000 pump size depends on the model and can measure 6.12–8.64 × 3.20–3.74 cm and weigh 98–173 g. As a result, a disadvantage of the pump is its low volumetric loading efficiency of 14–29%. However, the pump can be transcutaneously refilled every 4–8 weeks through a self-sealing silicone central port (Baert et al. 2008; Codman and Shurtleff 2003; Codman & Shurtleff n.d.).

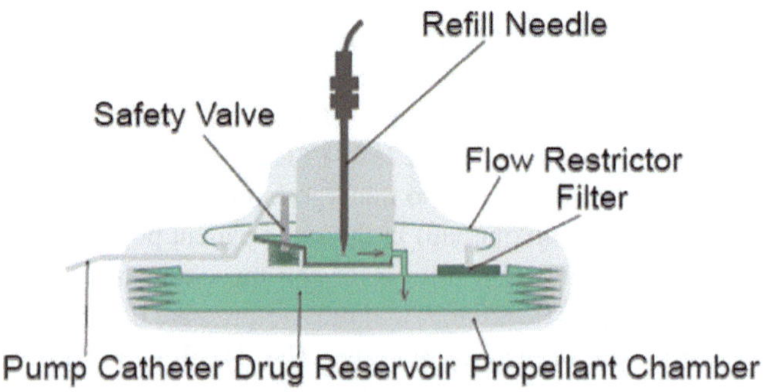

Fig. 5 Infusion pump drug release schematic. The infusion pump is divided into two chambers: a collapsible drug reservoir and a propellant chamber. At body temperature the propellant changes from liquid to gas compressing the drug reservoir thus forcing the drug out through the restrictor filter into the pump catheter. The drug reservoir is refilled with a designated needle that closes the safety valve avoiding drug release while refilling

In a study evaluating baclofen delivery for severe spasticity treatment, the Codman® 3000 pump demonstrated an accuracy higher than 90% (Ethans et al. 2005). Although this accuracy is similar to that of peristaltic pumps, the infusion pump has a lifetime warranty advantage as it omits the battery. Furthermore, a pilot study of the delivery of darunavir via the caudal vena cava by the Codman® 3000 pump for HIV PrEP confirmed a steady-state plasma drug concentration with an average of 40 ng/ml in two dogs. This study also highlights the versatility of the pump and catheter through its adaption to deliver viscous solutions (Baert et al. 2008). Although the Codman® 3000 pump is highly acceptable by patients, especially for hepatic arterial infusion for chemotherapy, its production stopped in April 2018. This halt was likely due to low profitability, with pumps costing from $7000 to $11,000 USD, and low demand, with only 300 sales per year in the US (Grady and Kaplan 2018).

Another dynamic implant that relies on a positive driving force to modulate drug dosing is the Prometra® pump (Flowonix Medical Inc., Mt. Olive, NJ, USA). This FDA-approved chronic pain management pump delivers morphine intrathecally and uses the same positive pressure gas expansion actuation design as the Codman® 3000 pump but with battery-powered valves for flow regulation (Fig. 4) (Christo and Bottros 2014; Cobo et al. 2015; Kumar and Pillai 2018; Wilkes 2014). The titanium device is relatively large to accommodate the electrical components that permit remotely controlled drug release, measuring 7.1 × 2 cm with an unfilled weight of 150 g and drug reservoir volume of 20 ml. Programmable dose changes are a big advance for implants, as they give patients the ability to self-administer drug from an implant as they would with oral pills. The FDA-approved patient therapy controller (PTC™) offers patients flexibility to manage their pain (Deer and Pope 2015; Flowonix Medical n.d.). Also, external control of dosing is a requirement for pain management because dosing throughout the day is variable (Kumar and Pillai 2018).

In a study of 110 patients with chronic pain, Prometra® pumps had higher dosing accuracy when administering morphine sulfate compared with SynchroMed™ II pumps (Christo and Bottros 2014; Rauck et al. 2010). This could be attributed to the Prometra® pump valves delivering more precise drug doses due to their employment of simple open-and-close mechanisms. By contrast, SynchroMed™ II pumps have a fixed drug dose between rollers that cannot be finely tuned. Furthermore, the accuracy, efficacy, and safety of Prometra® pumps were demonstrated in patients for up to 12 months (Kalyvas et al. 2014; Rauck et al. 2010, 2013).

A major disadvantage of Prometra® pumps is the need to completely remove medication prior to MRI, as magnetic fields may open the valves and empty the drug reservoir, causing drug overdose (Christo and Bottros 2014; Pope and Deer 2015). To avoid this procedure and achieve an MRI-compatible implant, a flow-activated safety valve (FAV™) was incorporated in the new pump model, Prometra® II. However, the pump was recalled in 2017 due to a failure of the FAV™ during an MRI scan, resulting in a patient receiving a fatal dose (U.S. Food and Drug Administration 2018). Although Prometra® II was designed to prevent the need for pre-MRI medication removal, the recall mandates emptying the drug reservoirs in Prometra® and Prometra® II pumps before an MRI scan (Flowonix Medical 2018). However, as physicians and healthcare workers are aware of this necessity, this is not a restrictive problem with careful monitoring.

In summary, pumps with different mechanisms of action can be chosen depending on the patient's disease, drug release longevity, and site of implantation. Larger peristaltic and infusion pumps can possibly be used to treat chronic diseases due to their larger drug reservoir and refill feature, whereas smaller osmotic pumps maintain constant drug release for systemic or site-specific effects (Kumar and Pillai 2018). A comparison of advantages and limitations of peristaltic, osmotic, and infusion pumps is shown in Table 2.

4 Microfabricated systems

In the biomedical field, electromechanical systems offer distinctive solutions for drug release related to precision dosing. There is much interest in implants that incorporate this technology and are fabricated in the micro- and nanometer range (Kumar and Pillai 2018). Microscopic and nanoscopic devices with features in the microscale and nanoscale array are termed microelectromechanical systems (MEMS) and nanoelectromechanical systems (NEMS), respectively.

When these implants are scaled down, the driving forces of drug release change with respect to the decrease in area and volume; forces such as adhesion and surface tension have a greater effect on molecules, which is convenient for controlled drug delivery (Bhushan 2007). Given the incredible variety of technologies proposed, in this section we will review some representative MEMS and NEMS undergoing preclinical research or clinical development.

4.1 Mems

Fluidic MEMS show potential for drug delivery applications and can be integrated with electronic components to allow remote control over drug administration. Santini et al. (1999) developed one of the first microfluidic devices capable of pulsatile release. It consists of a microfabricated silicon wafer containing an array of drug reservoirs capped by gold membranes (Fig. 6a, b) (Maloney et al. 2005). The device allows the selective opening of single reservoirs by applying an electrical potential to the gold membranes. An electrochemical reaction causes the complete dissolution of the membrane, allowing drug release (Fig. 6c). This technology was adapted for leuprolide release by Microchips Biotech, Inc. (Bedford, MA, USA).

Micro-CHIP allows the remote control of drug delivery. This system utilizes electronic circuitry for radio-frequency communication with the remote control unit for triggering dissolution of capping reservoir membranes. This system is complex and requires a power source consisting of a battery (Fig. 6d), which occupies ~40% of the implant volume due to the significant power consumption of the device. Micro-CHIP has arrays of drug reservoirs of 300–600 nl each that can be individually opened over time, creating a pulsatile delivery profile (Farra et al. 2012; Grayson et al. 2004). To mimic constant delivery, the reservoirs should be opened at frequent time points. The rate of release from each reservoir can be controlled by modification of the dissolving capping layer (Santini et al. 2000).

Micro-CHIP devices have been developed for different experimental applications. One of the most relevant applications is the delivery of leuprolide in a canine model (Prescott et al. 2006). For this purpose, the implant contains 100 drug reservoirs providing a total reservoir volume of 30 μl, corresponding to 2.5 mg leuprolide acetate powder (Grayson et al. 2004). The device, with approximate dimensions of $4.5 \times 5.5 \times 1$ cm^3, has a volume of approximately 30 ml, meaning that its nominal loading efficiency is 0.1%. More recently, a different version of Micro-CHIP was tested in a clinical trial (Farra et al.

Table 2 Comparison among peristaltic, osmotic, and infusion pumps

Implant	Development Status	Advantages	Limitations
SynchroMed™ II pump	• FDA-approved • Clinical trials: NCT00866502, NCT02886286	• Transcutaneous refilling • Telemetry dosing • PTM patient self-administration • Site-specific drug release with catheter	• Large size • Needs battery • Movable mechanical components • Low volumetric loading efficiency • Requires invasive surgery • Pump and catheter malfunctions • Requires specific drug formulation • Drug instability requires refill every 3–4 months
Medici Drug Delivery System™	• FDA clinical hold	• Small size and optimal shape • Drug versatility	• Holds less than 1 year dose • Implant must be exchanged after 6 months • Movable mechanical components
TARIS® System	• Clinical trials: NCT02722538, NCT02720367, NCT03404791	• Small size and optimal shape • Site-specific (intravesical) • Drug released for weeks to months for optimal treatment	• Difficult insertion and removal procedures
Codman® 3000 pump	• FDA-approved • Discontinued	• Transcutaneous refilling • Overfill safety features • Drug reservoir versatility • Nominal flow rate versatility • No battery • Site-specific drug release with catheter	• Large size • Requires invasive surgery • Requires refill every 4–8 weeks to expand bellow • Low volumetric loading efficiency • Requires specific drug formulation
Prometra® and Prometra® II pumps	• FDA-approved • Recalled by Flowonix Medical Inc.	• Dosing accuracy • PTC™ patient self-administration • Programmable drug doses • Transcutaneous refilling • Site-specific drug release with catheter	• Large size • Needs battery • Movable mechanical components • Low volumetric loading efficiency • Requires invasive surgery • Pump and catheter malfunctions • Requires specific drug formulation • Drug instability requires refill every 3 months • Drug reservoir must be emptied prior to MRI scan

2012). This implant delivers teriparatide, synthetic human parathyroid hormone fragments [hPTH(1–34)], which is the only treatment approved for anabolic osteoporosis and requires daily injection (Watson 2012). The device, with approximate dimensions of $5.4 \times 3.1 \times 1.1$ cm^3, contains 20 reservoirs containing 40 μg teriparatide each, providing a total reservoir volume of 12 μl. This Micro-CHIP has a volume of approximately 15 ml and a nominal loading efficiency of 0.08%. However, Farra et al. (2012) report that the drug loading procedure does not allow complete yield of all drug reservoirs, which reduces the effective loading efficiency to below 0.08%.

These implants have the benefit of being made of components that can be microfabricated with conventional semiconductor technologies. The shell can be machined or injection-molded for the low-cost parallel fabrication of a large number of parts. Nonetheless, the assembly, loading, and sterilization of the device is expensive. Additionally, the extremely low loading efficiency significantly limits its applicability for long-term sustained delivery of therapeutics. However, the Micro-CHIP will be tested with a variety of chronic drug therapies since Teva Pharmaceuticals partnered with Microchips Biotech Inc. in 2015 (Microchips Biotech 2015).

Humayun et al. developed prototypes of one of the first ocular MEMS pumps for the treatment of DME and noninfectious uveitis. The Posterior MicroPump Delivery System (PMP) is implanted on the sclera beneath the conjunctiva and delivers micro- and nanodoses intravitreally. The PMP can be wirelessly programmed with The Eye™. The device is $13 \times 16 \times 5$ mm in size and is custom-contoured for a reduced front height to fit on the outer surface of the eye. The PMP has a drug reservoir with a refill port, battery, electronics, electrolysis chamber, and cannula (Fig. 7a). When the device is turned on, an electrical potential electrolyzes H_2O into H_2 and O_2, which returns to H_2O when turned off. The gases generate pressure on the drug reservoir and force the drug into the cannula at a desired dose (Fig. 7b) (Cobo et al. 2015; Gutiérrez-Hernández et al. 2014; Humayun et al. 2014; Yasin et al. 2014). Use of the PMP for delivering ranibizumab, an angiogenic inhibitor, for 90 days was demonstrated to be safe. However, four of the eleven patients received a lower than target dose (Humayun et al. 2014). PMP safety was

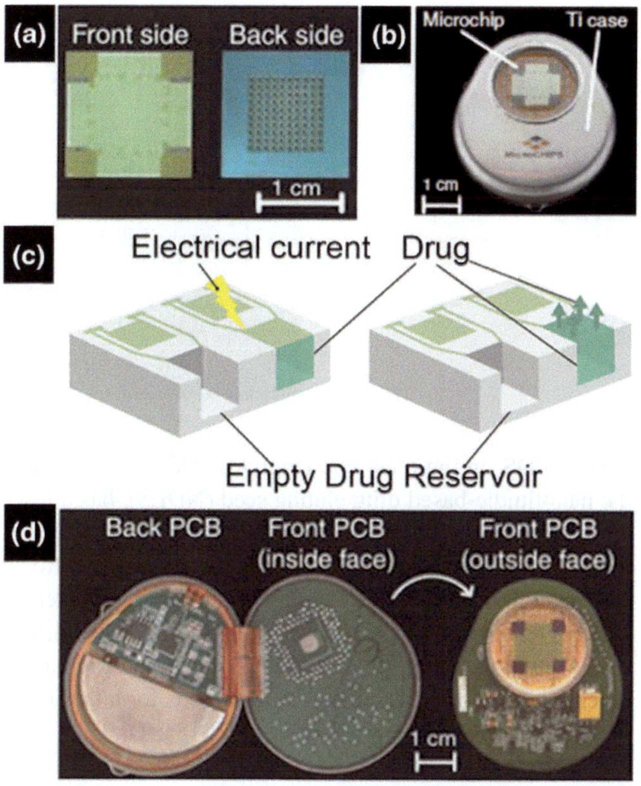

Fig. 6 30-ml Micro-CHIP device used for leuprolide release in dogs. **a** Silicon wafer with 100 30 μl drug reservoirs capped by gold membranes. **b** Assembled 30-ml Micro CHIP device. **c** Micro-CHIP drug release schematic: an electrical potential to the gold membrane permits selective opening of specific reservoir for drug release. **d** Internal circuitry of the implant. (Image 6A, 6B, 6D adapted from (Prescott et al. 2006))

previously assessed in a 1-year canine study (Gutiérrez-Hernández et al. 2014). Humayun et al. patented the technology and created the company Replenish Inc. (Pasadena, CA, USA), which produces Replenish MicroPumps.

4.2 NEMS

4.2.1 NEMS for constant delivery

At the nanoscale, the properties of fluids under confinement can be beneficially leveraged. Nanochannels constitute highly precise and accurate delivery vehicles for the delivery of therapeutics in a controllable manner. When the size of the channels shrinks to the size of the diffusing analytes, wall-to-molecule interactions play a dominant role in molecular release, causing constrained and saturated diffusion (Ziemys et al. 2011; Ziemys et al. 2010). Therefore, nanochannels can passively control the release of molecules through concentration-driven transport as long as the drug reservoir is supersaturated (Bruno et al. 2018). Taking advantage of these nanoscale effects, constant, sustained release of drugs can be achieved by judiciously tailoring the size and surface chemistry of nanochannels. This nanochannel approach was developed by various groups, with pioneering studies of silicon nanochannels conducted by Ferrari et al. and Desai et al. in the 1990s (Chu et al. 1997; Desai et al. 1999; Ferrari et al. 1995; Grattoni et al. 2009; Chu et al. 1999).

Nanochannel membranes can easily be mounted on a drug reservoir to achieve constant rate, zero-order kinetic drug release from the reservoir. These implants are still in the preclinical phase but have a high market acceptability due to low manufacturing costs. The titanium oxide nanotube membrane, NanoPortal Membrane (Nano Precision Medical, Emeryville, CA, USA), is attached to a small, rice grain-sized cylindrical implant (Fig. 8a). This implant was designed to be subcutaneously implanted through an in-office procedure. Currently, this technology is in preclinical development to release glucagon-like peptide-1 agonists for 3 months to up to 1 year (Nano Precision Medical n.d.-a; Nano Precision Medical n.d.-b; Nano Precision Medical n.d.-c). A larger cylindrical titanium device measuring 4 cm × 4 mm has two membranes with NANOPOR™ (Delpor Inc., San Francisco, CA, USA) technology fixed at each end (Fig. 8b) (Delpor n.d.-a; Delpor n.d.-f). DLP-202 and DLP-414 can release hGH for 3 months and exenatide for 3–6 months, respectively (Delpor n.d.-d; Delpor n.d.-e). To maintain constant release, drugs must be soluble to form supersaturated solutions within the drug reservoir to saturate the nanochannels. Insoluble drugs cannot saturate nanochannels and thus do not have zero-order kinetics. Prozor™ (Delpor Inc., San Francisco, CA, USA) technology enables release of insoluble drugs by maintaining an acidic pH in the drug reservoir (Delpor n.d.-g). DLP-114 and DLP-119 are a 6–12-month formulation of risperidone and 3-month

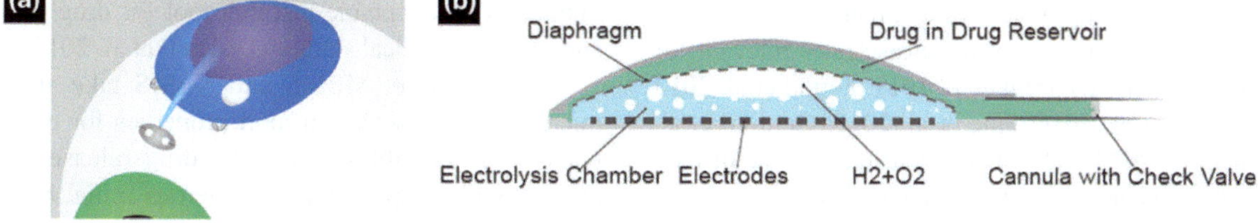

Fig. 7 Replenish MicroPump schematic and drug release mechanism. **a** Replenish MicroPump implanted on the sclera beneath the conjunctiva. **b** Replenish MicroPump drug release mechanism: electrodes in the electrolysis chamber generate an electric potential electrolyzing H_2O into H_2 and O_2 when the device is turned on. This creates pressure on the diaphragm that shifts drug in drug reservoir and displaces it through the cannula

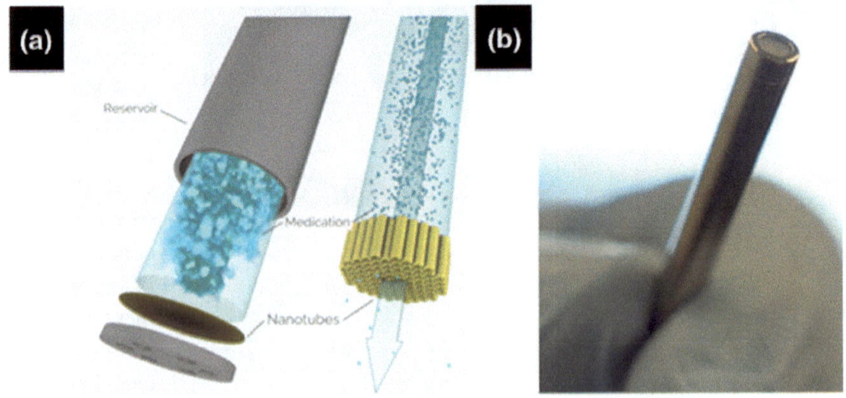

Fig. 8 NEMS translational research devices. **a** Schematic of nanoportal membrane from Nano Precision Medical showing how the nanotubes are the rate-limiting step for drug release from the reservoir (Image used with permission from Nano Precision Medical). **b** Drug-versatile Delpor Inc. implant with two membranes with NANOPOR™ technology fixed at each end (Image used with permission from Delpor Inc.)

formulation of olanzapine, respectively. Both are antipsychotic drugs, with risperidone used to treat schizophrenia and olanzapine used to treat bipolar disorder (Delpor n.d.-b; Delpor n.d.-c).

The nanochannel Delivery System (nDS) has nanochannels as small as 2.5 nm with tight tolerances on size, geometry, and surface properties. It was further developed by Grattoni et al. and is currently in clinical translation (Grattoni et al. 2009, 2011a c). The nDS membrane, a 5 × 20 × 12.3 mm or 43 × 28.5 × 8.7 mm silicon chip, represents the core of the nDS implantable technology (Fig. 9a). The membrane exploits the previously mentioned nanoscale phenomena to passively control the constant release of drugs, biological molecules, and nanoparticles without requiring movable components or actuation (Fig. 9b) (Fine et al. 2010; Grattoni et al. 2011a). The implant contains the nDS membrane, a mechanically robust shell, and loading ports with sealing components. This simple architecture allows for high effective loading efficiency, which, depending on the size and shape of the implant, may range from 60 to 90% (Fig. 9c). This technology is suitable for the use of drugs in liquid, suspension, solid, and powder forms in water, organic solvent, or lipid-based formulations. This offers flexibility in terms of its employment for a broad spectrum of therapeutic applications, enabling the delivery of drugs in their most stable formulation for long-term treatment. Transcutaneous refilling allows the treatment of chronic pathologies over several years without need for explantation and replacement.

The nDS technology was validated *in vitro* and *in vivo* in rodents, dogs, pigs, and non-human primates with a constant, sustained release of drug molecules and nanoparticles over a broad range of molecular sizes at release rates relevant for medical applications (Di Trani et al. 2019; Ferrati et al. 2013 2015; Filgueira et al. 2016; Fine et al. 2010; Grattoni et al. 2011c; Sih et al. 2013). The nDS can sustain release of HIV PrEP antiretroviral drugs, tenofovir alafenamide fumarate, and emtricitabine for 83 days in non-human primates and allow transcutaneous drug refilling (Chua et al. 2018b). Grattoni et al. developed a cylindrical intratumoral device approximately 3.5 mm long with a silicone cap at one end and a smaller nDS on the other end (Fig. 9d). This device, termed the nanofluidic-based drug eluting seed (NDES), has a reservoir capacity of 3.3–5 μl and is percutaneously delivered intratumorally via minimally invasive insertion with a trocar. The authors demonstrated that the intratumoral sustained release of CD40 and OX40 from the NDES increases immune cell infiltration. Thus, the nDS nanochannel platform has the potential to expand available clinical options for intratumoral immunotherapy delivery (Chua et al. 2018a; Di Trani et al. 2017; Hood et al. 2016).

4.2.2 NEMS for tunable delivery

Some diseases require a variable rather than a constant drug delivery dose. The passively controlled nDS membrane can be accompanied by electrodes to adjust the delivery rate of drugs, allowing for programmable dose modulation, remote titration, and responses to sensor feedback (Fine et al. 2011; Grattoni et al. 2011b). After applying a 1 .5V direct current electrical field across the membrane, ionic species redistribute across the nanochannels, causing ionic concentration polarization that can be modified to tune drug release rate (Bruno et al. 2015, 2016; Di Trani et al. 2017; Grattoni et al. 2011a). The dynamically controlled nDS membrane can be mounted on a drug reservoir remotely controlled via Bluetooth Low Energy communication (Fig. 9e). This technology has been validated *in vitro,* demonstrating changes in methotrexate release when transmembrane potential is applied. A disadvantage of this device is that its volumetric loading efficiency is low (22%) due to the volume of the circuitry chamber. However, this implant is an adaptable research tool for drug development and pharmacological studies (Di Trani et al. 2017).

To summarize, MEMS and NEMS take advantage of micro- and nanoscale transport properties for drug delivery. Nanofluidics enables zero-order drug release kinetics for months with no potentially dangerous initial burst release. Although most are not yet FDA-approved, there is great potential to make small implants that can treat a wide variety of diseases. Table 3 shows a comparison of advantages and limitations of MEMS and NEMS.

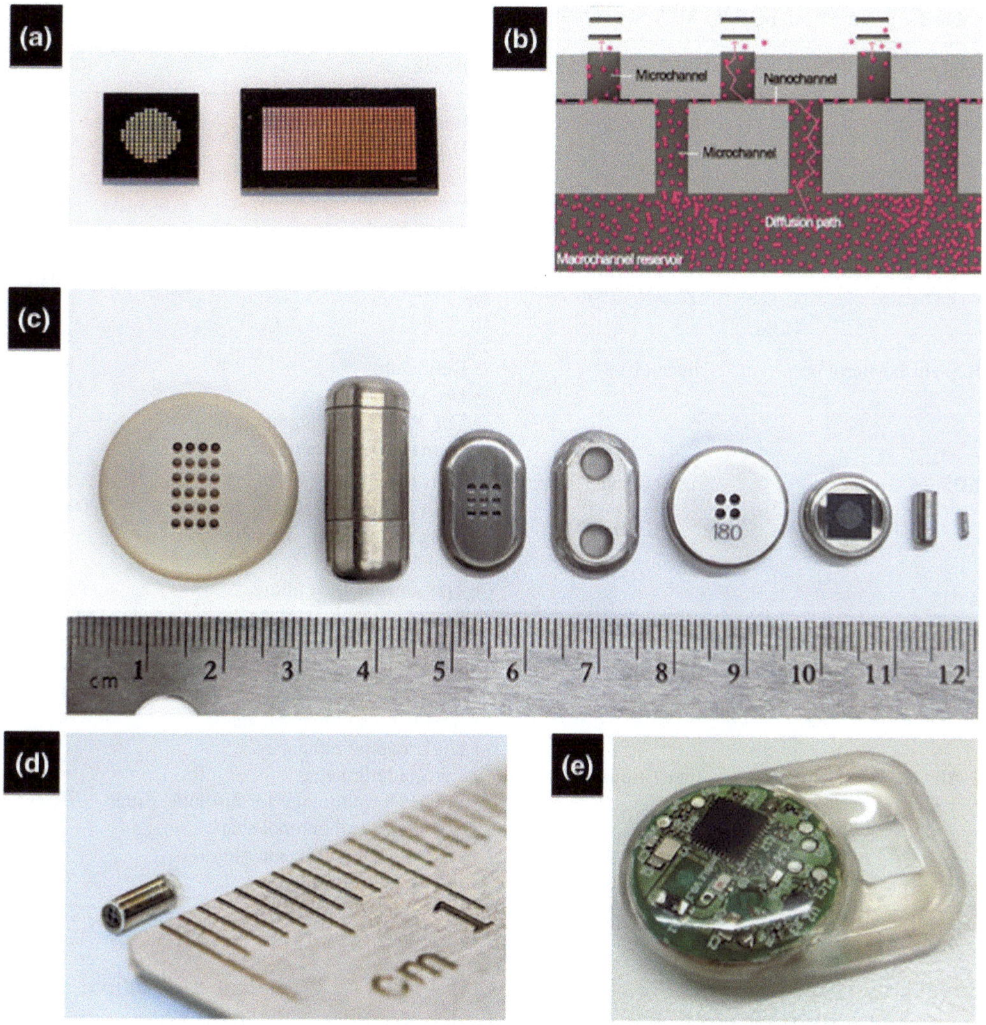

Fig. 9 nDS. **a** Differently sized mechanically robust silicon microfabricated nDS membranes, which house a defined number of densely packed slit-nanochannels to achieve constant and sustained delivery of therapeutics over extended periods of time. **b** Drug release diffusion path through nDS membrane: first from drug reservoir to perpendicular microchannels, then rate-limiting horizontal nanochannels, and then out through perpendicular microchannels. **c** The membrane is conveniently mounted on a drug delivery reservoir with a size and shape that can be optimized for the therapeutic application, drug, duration of treatment, and site of implantation. **d** Image of NDES with nDS membrane next to ruler to illustrate its small size. **e** Dynamically controlled nDS membrane mounted on polyether ether ketone, sized 24 × 34 × 4.5 mm^3, with an 800-μl drug reservoir chamber and a circuitry chamber with the electronics and battery

5 Sites of implantation

Although implants offer refined and efficacious means for controlled drug delivery, they all require placement by a healthcare professional. Surgical procedures vary based on the site of implantation and are associated with potential challenges and adverse effects. Although side effects are usually mild, they can be significant in some cases. Here, we describe various insertion procedures and provide an overview of their most common challenges to provide insight that can aid in the development of the next generation of drug delivery implants. Figure 10 illustrates the different sites of implantation for implants.

5.1 Intraocular placement

In addition to drug loading efficiency, the implantation procedure of a device is also of great importance when designing an implant. The end goal is the use of a minimally invasive in-office procedure by a trained healthcare professional that does not require post-operative care. Both ILUVIEN® and Retisert® posterior chamber implantations are performed in a doctor's office, as they require aseptic conditions and anesthesia. ILUVIEN® is minimally invasive, whereas Retisert® is invasive due to its shape. The small cylindrical shape of ILUVIEN® fits inside a needle and permits intravitreal injection (Fig. 11a). A benefit is that there is no need for stitches as the sclera can self-heal from the needle wound, reducing complications (Borkar et al. 2017; Logan et al. 2016). By contrast, Retisert® requires sclerotomy along with blood vessel cauterization to insert the irregularly shaped device and sutures to anchor it within the posterior chamber (Fig. 11b). The sclerotomy incision also requires subconjunctival antibiotics and a steroid injection (Bausch & Lomb n.d.; Yasin et al. 2014). The incision must be re-opened for removal, but the implant can be replaced using the same anchoring suture in the sclera. However, some ophthalmologists prefer to insert a new Retisert® at another incision site, leaving the old implant in place. If the patient requires a third implant, ophthalmologists will replace the first implant (Nicholson et al. 2012). Retisert® limitations could be addressed by changing the implant shape to a cylinder to allow a non-invasive procedure. The

Table 3 Comparison between MEMS and NEMS

Implant	Development status	Advantages	Limitations
Microchip	Human trial	• Reservoir-specific trigger • Drug versatility • Remote control of drug administration	• Large size • Extremely small drug reservoir • Very low loading efficiency • Complex technology • Requires battery • Difficult insertion and removal procedures • High fabrication and assembly costs • Pulsatile drug release • Rigid implant
Replenish MicroPump	Human trial	• Refillability • Drug versatility • The Eye™ programmable doses • Human safety evaluated	• Requires battery • Difficult insertion procedure • High manufacturing cost • Rigid implant
nDS	Translational research	• Zero-order kinetics • High loading efficiency • Drug and reservoir versatility • Scalability • Transcutaneous refilling • Remote tunable release • Systemic or site-specific drug release	• High rates of drug delivery (mg/day) require large membrane surface area • Difficult insertion and removal procedures • Rigid implant
NanoPortal Membrane	Translational research	• Drug versatility • Zero-order kinetics • Small size and optimal shape • High loading efficiency	• Drug release for 3 months to 1 year • Implant must be exchanged • Rigid implant
NANOPOR™ technology	Translational research	• Zero-order kinetics • Prozor™ technology for insoluble drugs • Small size and optimal shape • High Loading efficiency • Low manufacturing cost	• Implant must be exchanged Rigid implant

irregularly shaped Retisert® device contains 0.59 mg FA compared with 0.19 mg in the smaller, cylinder-shaped ILUVIEN®. Accordingly, multiple smaller cylinder implants could be injected into the vitreous humor to maintain the therapeutic dose for DME treatment.

Even though implantable devices offer the obvious and needed advantages of site-specific therapeutic delivery, there are some limitations and challenges directly related to the physical presence of an object in the eye. Patients must be monitored for the most common complications, intraocular pressure elevation and endophthalmitis, after implantation, because a foreign device is introduced into a pressure-regulated chamber (FDA Reference IDs: 2955048; 3635981) (Alfaqawi et al. 2017; Bausch & Lomb n.d.; Borkar et al. 2017; Logan et al. 2016; Parrish et al. 2016; Wright and Hall 2016; Yasin et al. 2014). Although ILUVIEN® has the convenience of a sutureless procedure, there are reports of implant migration into the anterior chamber, blocking the visual axis, and dislodgement into the infusion cannula during vitrectomy (Andreatta et al. 2017; El-Ghrably et al. 2015; Moisseiev and Morse 2016; Papastavrou et al. 2017). Reported Retisert® adverse effects are mostly sclerotomy-related: hypotony, temporary decrease in visual acuity, cataract formation, choroidal detachment, retinal detachment, vitreous hemorrhage, wound dehiscence, implant dislocation, and scleral melt (FDA Reference ID: 2955048) (Almeida et al. 2015; Chang et al. 2015; Freitas-Neto et al. 2015; Petrou et al. 2014; Yasin et al. 2014). For these reasons, the doctor performs indirect ophthalmoscopy to verify correct placement of the implant, adequate central retinal artery perfusion, and absence of complications (FDA Reference IDs: 2955048; 3635981) (Bausch & Lomb n.d.; Yasin et al. 2014). Nonetheless, as previously mentioned, the benefits of both of these implants outweigh the few complication reports.

An approach to avoiding placement of foreign devices in the vitreous humor is to anchor an implant episclerally and deliver drugs into the posterior chamber through a cannula, like the Replenish MicroPump. The PMP reservoir and intraocular cannula are sutured episclerally between the superior and lateral rectus muscles to impede movement (Fig. 11c). A sclerotomy is required, but the incision is small as only the cannula must be inserted into the posterior chamber, followed by suturing of the conjunctiva (Humayun et al. 2014). An advantage of the PMP is its ease of access and the possibility of refilling the drug reservoir, eliminating future device removal or replacement (Yasin et al. 2014). A major limitation of the PMP is that it has not undergone clinical trials, so although it has a refill feature, its safety and efficacy have yet to be established.

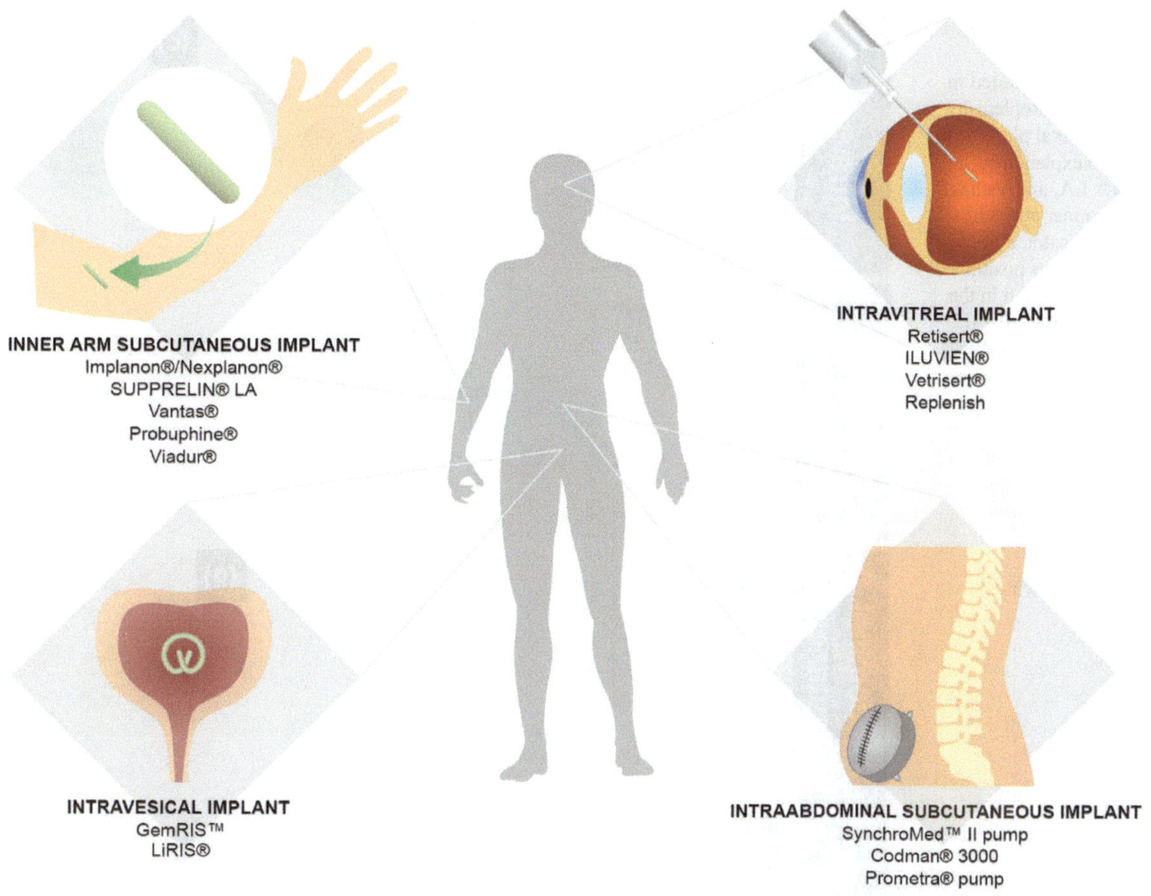

Fig. 10 Sites of implantation for FDA-approved implants and devices in clinical trials

5.2 Upper inner arm subcutaneous placement

The subcutaneous tissue is advantageous for drug delivery because the gastrointestinal tract is bypassed, thus improving drug bioavailability for systemic administration (Kumar and Pillai 2018). Also, the implantation site should be discrete but readily accessible for a quick in-office procedure and not uncomfortable to the patient. Thus, the most widely used implantation site in the clinic is the upper inner arm, for which implants are cylindrical and have a personalized applicator device that facilitates their insertion. Implant placement is an in-office procedure performed by a trained healthcare provider that takes approximately 10 min. Implanon®, Nexplanon®, SUPPRELIN® LA, and Vantas® are inserted in the inner side of the non-dominant upper arm approximately 8–10 cm above the medial epicondyle of the humerus after injection of local anesthesia (Fig. 11d) (FDA Reference IDs: 2887911; 3080389; 4099967; 4100681).Probuphine® implants are inserted at the same site but require a minor incision followed by insertion of four implants. These are positioned in a close fan-shaped distribution 4–6 mm apart with the fan opening toward the shoulder (Fig. 11e) (FDA Reference ID: 4215185) (Itzoe and Guarnieri 2017; Smith et al. 2017). Interestingly, different subcutaneous implantation sites have been adapted to meet the patient's needs. Nexplanon® was inserted in the scapular region in patients at risk of self-removal of the implant (Pragout et al. 2018). In elderly patients, Vantas® is subcutaneously inserted in the abdominal region due to patient-limited arm mobility (Woolen et al. 2014).

When inserting implants into the upper arm, it is important to avoid the sulcus between the biceps and triceps muscles and the neurovascular bundle that lies deeper in the subcutaneous tissue to avoid complications such as peripheral nerve injury and paresthesia (Laumonerie et al. 2018). Equally important, the presence of the device must always be verified immediately after insertion to circumvent implant migration. There are reports of difficulty removing Implanon®/Nexplanon® devices associated with peripheral nerve injury and implant migration (Barlow-Evans et al. 2017; Chevreau et al. 2018; Chung et al. 2017; Diego et al. 2017; Guiahi et al. 2014; Laumonerie et al. 2018; Odom et al. 2017). Regardless, the benefits are indisputable given the high efficacy rate and that most common adverse effects reported for these implants include erythema, hematoma, application site pain, and edema are quick to resolve (FDA Reference IDs: 2887911; 3080389; 4099967; 4100681; 4215185) (Davis et al. 2014; Donnelly et al. 2015; Eugster 2015; Fisher et al. 2014; Itzoe and

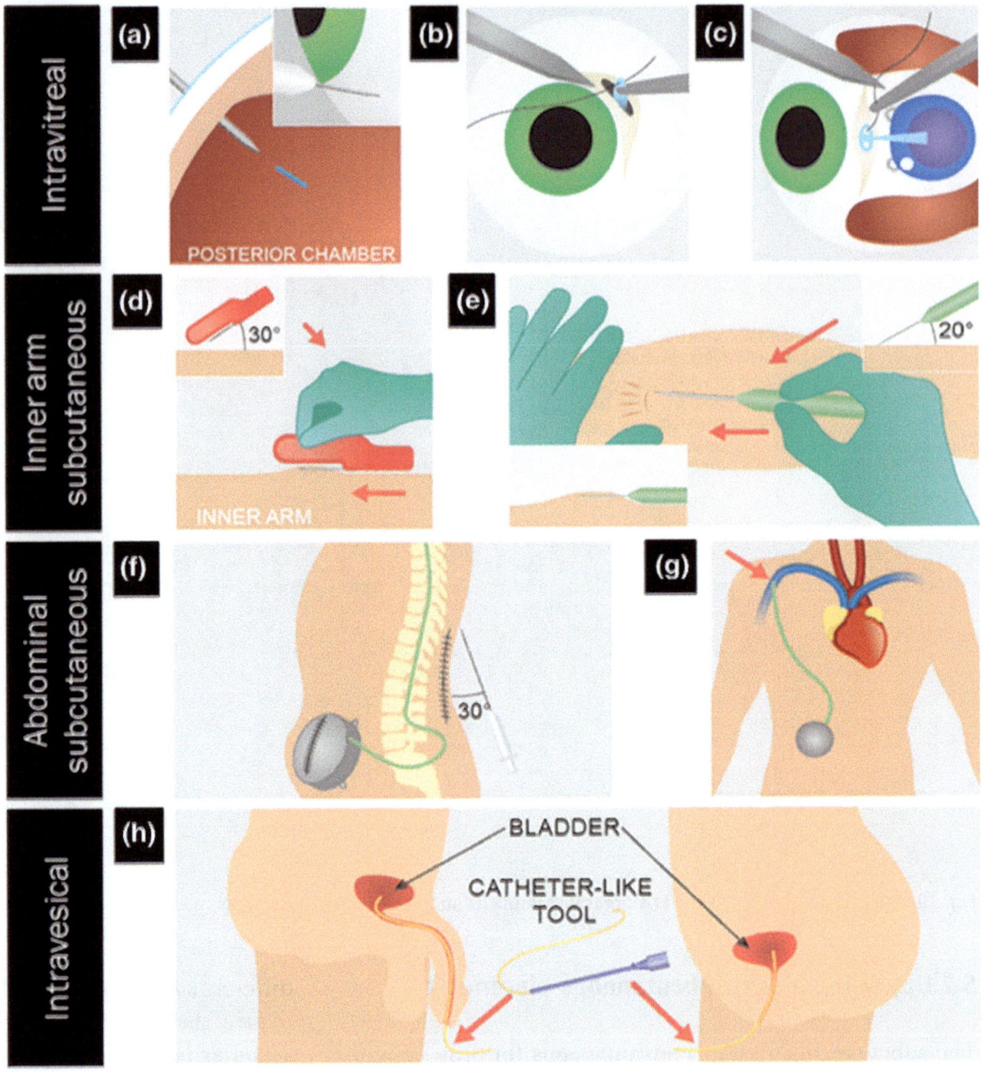

Fig. 11 Implantation procedures. **a** ILUVIEN® intravitreal insertion. **b** Retisert® sutured in posterior chamber. **c** Replenish MicroPump episcleral placement. **d** Implanon®, Nexplanon®, SUPPRELIN® LA, and Vantas® insertion in the inner arm with personalized applicator. **e** Four Probuphine® implants positioned in fan-shaped distribution in the inner arm. **f** SynchroMed™ II, Codman® 3000, and Prometra® pump surgery for abdominal subcutaneous placement of the pump and intrathecal catheter. **g** SynchroMed™ II pump surgery for abdominal subcutaneous placement of the pump and intravenous catheter. **h** GemRIS™ and LiRIS intravesical insertion with a catheter-like tool

Guarnieri 2017; Pedroso et al. 2015; Serati et al. 2015; Shumer et al. 2016; Silverman et al. 2015; Simon et al. 2016; Smith et al. 2017).

5.3 Abdominal subcutaneous placement

The size of peristaltic and infusion pumps necessitates catheters for site-specific drug delivery and restricts them to a subcutaneous abdominal implantation site. SynchroMed™ II, Codman® 3000, and Prometra® pumps require surgery under general anesthesia, resulting in a significantly higher costs, as this procedure take 1–3 h in an operating room due to the requirement for fluoroscopy for intrathecal catheter placement and verification (Flowonix Medical Inc. 2017; Medtronic 2017). At this implantation site, 27% of reported complications of intrathecal delivery are related to surgical procedures (Stetkarova et al. 2010). A surgeon implants a filled pump subcutaneously in the abdomen in a pump pocket no more than 2.5 cm from the surface of the skin and connects the intracatheter (Fig. 11f). The pump pocket and the spinal incision site are irrigated, sutured, and covered in dressing to avoid infection (Flowonix Medical Inc. 2017; Medtronic 2017). For intravenous treprostinil administration, a catheter is inserted into the superior vena cava via a subclavian, cephalic, jugular, or axillary puncture, anchored to the venotomy site, and connected to the abdominal pump pocket (Fig. 11g) (Bourge et al. 2016).

A benefit of these pumps is transcutaneous drug refill, allowing longer treatment durations. However, potential severe complications from erroneous subcutaneous injection of drug during device refilling have been reported (Maino et al. 2014; Perruchoud et al. 2012; Ruan et al. 2010). With this in mind, ultrasound-guided pump refill is a feasible and simple technique that reduces the probability of refill-related complications (Gofeld and McQueen 2011; Saulino and Gofeld 2014). Common minor adverse effects reported for SynchroMed™ II, Codman® 3000, and Prometra® pumps are implant site pain, edema, and hematoma (Codman & Shurtleff; Ethans et al. 2005; Pope and Deer 2017; Rauck et al. 2013). As previously mentioned, these pumps have movable components that control drug delivery, increasing the risk of device malfunctions. There are reports of cases in which the pump had to be explanted due to pump failures (Kalyvas et al. 2014; Riordan and Murphy 2015; Sgouros et al. 2010). These discoidal pumps are limited by their size and therefore necessitate catheters to deliver the drug to the site of interest. Given that most complications are attributable to catheter malfunctions, implants should be redesigned to omit the need for catheters (Ethans et al. 2005; Kalyvas et al. 2014; Miracle et al. 2011; Stetkarova et al. 2010). Even though abdominal subcutaneous placement of pumps clearly has limitations and challenges, the devices have had a positive impact on improving patient health and living conditions, outweighing the complications and reported adverse effects.

5.4 Intravesical placement

Other implants that have a drug release rate dependent on the targeted organ are GemRIS and LiRIS. Both of these site-specific devices are placed in the bladder as an in-office procedure that does not require an operating room. Insertion of the implant was a priority in their design, as it changes shape after it is implanted in the bladder. At first, it is shaped as a long tube positioned in a catheter-like tool that enables easy insertion into the bladder (Fig. 11h). After delivery, the implant wire restructures the implant into a pretzel-like shape that impedes expulsion of the device through the urethra. After the treatment period, the implant is removed via cystoscopy (Matheson 2014; Nickel et al. 2012). Possible complications are yet to be reported, as these implants are currently in clinical trials.

5.5 Next generation of implantable drug delivery systems

The complexity and limitations of surgical procedures for implantation and explantation have significant effects on patient acceptability of the technology. As such, future device designs should employ minimally invasive approaches, smaller implant volumes, and fewer insertion-explantation procedures to fully leverage their potential. Cylindrical devices, such as the polymeric implants Implanon®/Nexplanon®, SUPPRELIN® LA, Vantas, and ILUVIEN® and the osmotic pumps GemRIS, LiRIS, and Viadur®, have minimally invasive insertion procedures. Yet, their major disadvantage is the need for implant replacement if the patient wishes to continue the medication. For this reason, implants intended to be inserted subcutaneously in the inner arm need to incorporate a drug refill feature, like pumps. As previously mentioned, the discoidal subcutaneous abdominal pumps SynchroMed™ II, Codman® 3000, and Prometra® II have this feature, but the pump size is a key limitation, as it requires surgery for implantation-explantation. If implants are to become a mainstream drug administration route, implants should be carefully designed with a small volume for minimally invasive implantation and chronically sustained drug delivery eluding insertion-explantation with drug refillability.

Even if new implants are designed with these considerations, patient counseling will be crucial to increase acceptability. Implantation procedures lower patient acceptance because all procedures guarantee pain and discomfort, even if only from the local anesthetic. Therefore, patients will require counseling to show the potential cumulative benefits of prolonged compliance-free therapy, with optimized drug delivery far outweighing potential risks and immediate discomfort (Danckwerts and Fassihi 1991; Kumar and Pillai 2018; Rajgor et al. 2011). Also, data demonstrating the value of using implants over conventional treatment could drive insurance companies to cover the costs. Insurance companies may be more willing to pay for less expensive conventional therapy than to reimburse an outpatient procedure to insert the implant as well as cover the costs associated with the implant (i.e. refilling, removal, etc.).

6 Deployment in the developing world

The burden of chronic diseases in developing countries is rapidly increasing and has unfavorable social, economic, and health consequences (Alwan et al. 2010). Often, these countries have unreliable healthcare services that incite poor health practices, medication non-adherence and subsequently increase mortality rates. An important cause of non-adherence in developing countries is the high cost of therapeutics and paucity of health resources, which results in waste and underutilization of already limited resources. Also, healthcare center visits are linked to patient compliance since these trips are time-consuming and expensive. To ensure that all countries receive quality healthcare, national and international agencies must invest in developing countries (Atinga et al. 2018; Fullman et al. 2017; Pagès-Puigdemont et al. 2016; World Health Organization 2003). A proposal to resolve medication non-adherence due to lack of health resources and healthcare centers are long-term drug delivery devices. The Bill and Melinda Gates Foundation, among others, strive to resolve this issue by supporting different companies for the development of long-term sustained release implants that can be administered in developing countries.

Pregnancy is not a chronic disease but does require sustained prenatal care to ensure that both the baby and mother are healthy. Sustainable Development Goals call for a reduction in maternal deaths to fewer than 70 per 100,000 by 2030 (Kassebaum et al. 2016). Moreover,

Millennium Development Goals call for universal access to reproductive healthcare, specifically contraceptives. Providing free contraception implants to girls and women in developing countries could benefit 120 million women and help prevent approximately 30 million unintended pregnancies, which in turn would reduce infant and maternal mortality by 280,000 and 30,000 deaths, respectively (Bill & Melinda Gates Foundation n.d.).

The Bill and Melinda Gates Foundation invested in Microchips Biotech to develop a microchip that releases levonorgestrel, a progestin, for 16 years and can be stopped at any time with a wireless controller (Lee 2014). Jadelle® (Bayer, Leverkusen, Germany) is a polymer-based levonorgestrel-releasing implant similar to Implanon®. Jadelle® was prequalified by the World Health Organization in 2009, and the Jadelle Access Program was launched by the Bill and Melinda Gates Foundation and Bayer in 2013. The goal of this program is to deliver 27 million implants in 6 years. Bayer will supply the Jadelle® implants, and the Foundation will cover default risk. Although Jadelle® is approved in the US, it was not sold as of 2015 (Bayer AG Pharmaceuticals n.d.).

The Bill and Melinda Gates Foundation has also taken interest in the prevention of HIV and supported the research of Intarcia Therapeutics, Inc. to develop a pump that can store enough drug doses for 6–12 months, enabling people in developing countries to have HIV protection (Intarcia Therapeutics Inc n.d.-b). Although the Bill and Melinda Gates Foundation has taken a big step toward investing in drug delivery implants in developing countries, more efforts are needed.

The feasibility of deploying implants in developing countries largely depends on cost, simplicity in device implantation, and patient acceptance. The cost of devices can certainly represent a barrier for their deployment. In principle, the implant cost should be close to the cost of the drug itself. Complex electromechanical systems and pumps may be too expensive to achieve ample utilization. Due to the limited health care resources, deployment in developing countries may be exclusively limited to implants which require minimally invasive procedures such as subcutaneous placement in the inner arm. In this case, the surgical technique and expertise needed is nominal and can be easily taught ad hoc. As previously discussed, larger systems for either subcutaneous or deeper implantation would not be as attainable because of the invasive surgery required and the needs for follow-ups and longer recovery periods. Refillable implants could be very relevant in the context of preventive therapies for infectious diseases or for chronic treatments. Transcutaneous refilling can extend the life span of an implant while avoiding repeated surgical removal and replacement procedures. However, as it has been shown for SynchroMed™ pump-like systems, refilling requires specilized training to avoid failures that could be catastrophic with drug leakage in the surrounding tissues. Patient acceptability is key for the success of implantable systems. In cases such as the treatment and prevention of HIV, the use of implants is faced with the issue of 'stigma' associated with the disease. In these circumstances, the implantation site is determinant as patients may be concerned about the visibility of the device underneath the skin. Other factors, such as religion and spiritual beliefs also play an important role: studies revealed one of the reasons for medication non-adherence is that patients usually prefer spiritual or divine healing causations beyond medical treatment, as a result of low trust in new medication technologies, their efficacy and fear of side effects (Atinga et al. 2018; Pagès-Puigdemont et al. 2016).

On a different note, implants have the potential of addressing one key problem in deliverying medical treatments. Implants can minimize the frequency with which patients have to visit the health care centers, avoiding multiple lengthy travels and thus improving adherence to treatment. Ideally, tunable implants where clinicians can remotely adjust medication regimes and monitor patients through telemedicine could be advantageous. However, these systems may in turn be more expensive and therefore not implementable.

7 Conclusions

The treatment of chronic diseases will shift from oral dosing to implantable drug delivery devices as they obviate patient non-adherence and potentially limit side effects. Different passive and dynamic drug delivery technologies have received FDA approval, are in clinical trials, or are in an advanced stage of development. Consideration of current implantation procedures can further improve implant designs and thereby increase patient acceptability. Long-term delivery devices for chronic illness treatment with minimally invasive approaches for implantation should be employed for use in developing countries and to reduce DALYs. In conclusion, advanced implantable drug delivery devices hold promise as more effective treatment tools, transforming the clinical landscape of therapeutics given the growing incidence of chronic diseases.

Acknowledgements We thank Virginia Facciotto (virginia.facciotto@gmail.com) for the design of graphics and figures. Funding support from The Nancy Owens Breast Cancer Foundation, Center for the Advancement of Science in Space (CASIS) GA-2013-118, CASIS GA-2014-145, NIH NIAID R01 AI120749-01A1, NIH NIGMS R01 GM 127558.

Compliance with ethical standards J.S and A.G. disclose a financial interest in NanoMedical systems, Inc. F.P.P. and A.B. disclose no competing financial interest.

References

F. Alfaqawi, P.L. Lip, S. Elsherbiny, R. Chavan, A. Mitra, B. Mushtaq, Eye **31**, 650 (2017)

M. Ali, A. Akin, L. Bahamondes, V. Brache, N. Habib, S. Landoulsi, D. Hubacher, Hum. Reprod. **31**, 2491 (2016)

E. E. Allen, Instrument for transfusion of blood with patent 249,285. 8 November 1881.

R.H. Allen, A.M. Kaunitz, M. Hickey, in *Williams Textb. Endocrinol*, 13th edn., ed. by S. Melmed, K. S. Polonsky, P. R. Larsen, H. M. Kronenberg. (Elsevier, Philadelphia, 2016), pp. 664–693

D.R.P. Almeida, E.K. Chin, K. Mears, S.R. Russell, V.B. Mahajan, Retin. Cases Brief Rep. **9**, 142 (2015)

A. Alwan, D.R. MacLean, L.M. Riley, E.T. d'Espaignet, C.D. Mathers, G.A. Stevens, D. Bettcher, Lancet **376**, 1861 (2010)

W.M.K. Amoaku, S. Saker, E.A. Stewart, Eye **29**, 1115 (2015)

W. Andreatta, I. Elaraoud, A. Mitra, Retin. Cases Brief Rep., 1 (2017)

R.A. Atinga, L. Yarney, N.M. Gavu, PLoS One **13**, e0193995 (2018)

L. Baert, L. Schueller, Y. Tardy, D. Macbride, G. van't Klooster, H. Borghys, E. Clessens, G. Van Den Mooter, E. Van Gyseghem, P. Van Remoortere, Int. J. Pharm. **355**, 38 (2008)

C. Bailey, U. Chakravarthy, A. Lotery, G. Menon, J. Talks, C. Bailey, A. Kamal, F. Ghanchi, C. Khan, R. Johnston, M. McKibbin, A. Varma, B. Mustaq, C. Brand, J. Talks, N. Glover, Eye **31**, 1707 (2017)

R. Barlow-Evans, K. Jaffer, M. Balogun, BMJ Case Rep. bcr (2017)

P. Barnwal, S. Das, S. Mondal, A. Ramasamy, T. Maiti, A. Saha, Ther. Adv. Psychopharmacol. **7**, 119 (2017)

Bausch & Lomb, Retisert®: Surgical Procedure For Implantation (n.d.), http://www.bauschretinarx.com/Portals/50/files/baus3920-retisert-surgical-Instructions-rgb-p6.pdf. Accessed 5 Mar 2018

Bayer AG Pharmaceuticals, Bayer AG Pharmaceuticals: Implant Initiative Family Planning Choices in secure long-term contraception (n.d.), http://pharma.bayer.com/en/commitment-responsibility/family-planning/implant-initiative/. Accessed 25 Feb 2018

R.F. Beall, J.W. Nickerson, W.A. Kaplan, A. Attaran, PLoS One **11**, e0148939 (2016)

J.M. Berg, T. Dallas, in *Encycl. Microfluid. Nanofluidics*, ed. by D. Li. (Springer US, Boston, 2013), pp. 1–12

S. Bernell, S. W. Howard, Front. Public Health 4, (2016)

T. Bertelmann, S. Schulze, Ophthalmol. Ther. **4**, 51 (2015)

T. Bertelmann, Senior Physician, A. Messerschmidt-Roth, and Study Coordinator, Department of Ophthalmology, Philipps University, Marburg, Germany, Eur. Ophthalmic Rev. 07, 122 (2013)

J. M. Bertolote, World Health Organization: Neurological Disorders Affect Millions Globally WHO Report (2005). http://www.who.int/mediacentre/news/releases/2007/pr04/en/. Accessed 18 June 2018

G. Bhatia, M. E. Lau, K. M. Koury, P. Gulur, F1000Research (2014)

B. Bhushan, Microelectron. Eng. **84**, 387 (2007)

Bill & Melinda Gates Foundation, Bill & Melinda Gates Foundation: What We Do Family Planning Strategy Overview (n.d.). https://www.gatesfoundation.org/What-We-Do/Global-Development/Family-Planning. Accessed 25 Feb 2018

R. Bolash, B. Udeh, Y. Saweris, M. Guirguis, J.E. Dalton, N. Makarova, N. Mekhail, Neuromodulation Technol. Neural Interface. **18**, 150(2015)

D.S. Borkar, C. Ung, L. Sobrin, Int. Ophthalmol. Clin. **57**, 193 (2017)

R.C. Bourge, A.B. Waxman, M. Gomberg-Maitland, S.M. Shapiro, J.H. Tarver, D.L. Zwicke, J.P. Feldman, M.M. Chakinala, R.P. Frantz, F. Torres, J. Cerkvenik, M. Morris, M. Thalin, L. Peterson, L.J. Rubin, Chest **150**, 27 (2016)

G. Bruno, T. Geninatti, R.L. Hood, D. Fine, G. Scorrano, J. Schmulen, S. Hosali, M. Ferrari, A. Grattoni, Nanoscale **7**, 5240 (2015)

G. Bruno, G. Canavese, X. Liu, C.S. Filgueira, A. Sacco, D. Demarchi, M. Ferrari, A. Grattoni, Nanoscale **8**, 18718 (2016)

G. Bruno, N. Di Trani, R.L. Hood, E. Zabre, C.S. Filgueira, G. Canavese, P. Jain, Z. Smith, D. Demarchi, S. Hosali, A. Pimpinelli, M. Ferrari, A. Grattoni, Nat. Commun. **9**, 1682 (2018)

M.V. Carle, T.G. Chu, D.S. Boyer, Expert Rev. Ophthalmol **9**, 81 (2014)

J.A. Carter, R. Dammerman, M. Frost, J. Med. Econ. **20**, 893 (2017)

P.Y. Chang, Z. Kresch, C.M. Samson, R.C. Gentile, Ocul. Immunol. Inflamm. **23**, 453 (2015)

J. Chevreau, D. Krief, O. Abou Arab, M. Zitoun, A. Foulon, F. Sergent, J. Gondry, Eur. J. Obstet. Gynecol. Reprod. Biol. **224**, 81 (2018)

P. Christo, M. Bottros, J. Pain Res., 615 (2014)

W. H. Chu, T. Huen, J. K. Tu, M. Ferrari, in edited by P. L. Gourley (1997), pp. 111–122

W.-H. Chu, R. Chin, T. Huen, M. Ferrari, J. Microelectromechanical Syst. **8**, 34 (1999)

C. Y. X. Chua, P. Jain, A. Susnjar, J. Rhudy, M. Folci, A. Ballerini, A. Gilbert, S. Singh, G. Bruno, C. S. F., C. Yee, E. B. Butler, A. Grattoni, (2018a) Nanofluidic drug-eluting seed for sustained intratumoral immunotherapy in triple negative breast cancer. Journal of Controlled Release 285:23-34

C.Y.X. Chua, P. Jain, A. Ballerini, G. Bruno, R.L. Hood, Manas Gupte, S. Gao, N. D. Trani, A. Susnjar, K. Shelton, L. R. Bushman, M. Folci, C. S. Filgueira, M. A. Marzinke, P. L. Anderson, M. Hu, P. Nehete, R. C. Arduino, J. K. Sastry, A.o Grattoni, (2018b) Transcutaneously refillable nanofluidic implant achieves sustained level of tenofovir diphosphate for HIV pre-exposure prophylaxis. Journal of Controlled Release **286**315-325

M. Chung, C. Loudill, M. Wieler, J. Farnam, S.H. Ahn, J. Vasc. Interv. Radiol. **28**, 466 (2017)

M.J. Cima, H. Lee, K. Daniel, L.M. Tanenbaum, A. Mantzavinou, K.C. Spencer, Q. Ong, J.C. Sy, J. Santini, C.M. Schoellhammer, D. Blankschtein, R.S. Langer, J. Controlled Release **190**, 157 (2014)

A. Cobo, R. Sheybani, E. Meng, Adv. Healthc. Mater. **4**, 969 (2015)

Codman & Shurtleff, *Patient Information Booklet for Pain Management Therapy CODMAN 3000 Implantable Constant-Flow Infusion Pump* (2003).

Codman & Shurtleff, CODMAN 3000 Implantable Constant-Flow Infusion Pump: Prescribing Information (n.d.). http://www.codmanpumps.com/Products_pumps_epi.asp. Accessed 23 Apr 2018.

Coherent Market Insights, Global Implantable Drug Delivery Devices Market, by Product Type (Drug Infusion Pumps, Intraocular Drug Delivery Devices, Contraceptive Drug Delivery Devices and Stents), by Technology (Biodegradable Implants and Non Bio Degradable Implants), by Application (Ophthalmology, Cardiovascular, Birth Control/Contraception and Others) and by Region - Trends and Forecast till 2025 (Coherent Market Insights, 2017), p. 144.

M. Comlossy, Chronic Disease Prevention and Management (2013).

J. Cunha-Vaz, P. Ashton, R. Iezzi, P. Campochiaro, P.U. Dugel, F.G. Holz, M. Weber, R.P. Danis, B.D. Kuppermann, C. Bailey, K. Billman, B. Kapik, F. Kane, K. Green, Ophthalmology **121**, 1892 (2014)

R.L. Cutler, F. Fernandez-Llimos, M. Frommer, C. Benrimoj, V. Garcia-Cardenas, BMJ Open **8**, e016982 (2018)

M. Danckwerts, A. Fassihi, Drug Dev. Ind. Pharm. **17**, 1465 (1991)

J.S. Davis, F. Alkhoury, C. Burnweit, J. Pediatr. Surg. **49**, 807 (2014)

T.R. Deer, J.E. Pope, Expert Rev. Clin. Pharmacol. **8**, 507 (2015)

Delpor, Delpor Delivering the Future of Medicine: Delpor's Device Device (n.d.-a). https://www.delpor.com/delpors-device/ . Accessed 24 Feb 2018.

Delpor, Delpor Delivering the Future of Medicine: Products DLP-114 (n.d.-b), https://www.delpor.com/products/dlp-114/. Accessed 24 Feb 2018.

Delpor, Delpor Delivering the Future of Medicine: Products DLP-119 (n.d.-c), https://www.delpor.com/products/dlp-119/. Accessed 24 February 2018

Delpor, Delpor Delivering the Future of Medicine: Products DLP-202 (n.d.-d). https://www.delpor.com/products/dlp-202/. Accessed 24 February 2018

Delpor, Delpor Delivering the Future of Medicine: Products DLP-414 (n.d.-e), https://www.delpor.com/products/dlp-414/. Accessed 24 February 2018

Delpor, Delpor Delivering the Future of Medicine: Technologies NANOPOR (n.d.-f). https://www.delpor.com/technologies/nanopor/ Accessed 24 February 2018

Delpor, Delpor Delivering the Future of Medicine: Technologies PROZOR (n.d.-g), https://www.delpor.com/technologies/prozor/. Accessed 24 February 2018

T.A. Desai, D.J. Hansford, L. Kulinsky, A.H. Nashat, G. Rasi, J. Tu, Y. Wang, M. Zhang, M. Ferrari, Biomed Microdevices **2**, 11 (1999)

N. Di Trani, A. Grattoni, M. Ferrari, in (IEEE, 2017), pp. 40.6.1–40.6.4.

N. Di Trani, P. Jain, C.Y.X. Chua, J.S. Ho, G. Bruno, A. Susnjar, F.P. Pons-Faudoa, A. Sizovs, R.L. Hood, Z.W. Smith, A. Ballerini, C.S. Filgueira, A. Grattoni, Nanomedicine **16**, 1 (2019)

D. Diego, E. Tappy, J. Carugno, Contraception **95**, 218 (2017)

S. Donnelly, D. Shuriff, S. Stylianos, J. Zitsman, J. Pediatr. Surg. Case Rep. **3**, 523 (2015)

I. Elaraoud, W. Andreatta, A. Kidess, A. Bhatnagar, M. Tsaloumas, F. Quhill, Y. Yang, BMC Ophthalmol. **16** (2016a)

I. Elaraoud, A. Attawan, F. Quhill, Ophthalmol. Ther. **5**, 95 (2016b)

I. Elaraoud, H. Quhill, F. Quhill, Ophthalmol. Ther. **5**, 105 (2016c)

I.A. El-Ghrably, A. Saad, C. Dinah, Ophthalmol. Ther. **4**, 129 (2015)

I. El-Ghrably, D.H.W. Steel, M. Habib, D. Vaideanu-Collins, S. Manvikar, R.J. Hillier, Eur. J. Ophthalmol. **27**, 357 (2017)

Endo Pharmaceuticals, ClinincalTrials.gov Identifier NCT01295060 (n.d.). https://clinicaltrials.gov/ct2/show/NCT01295060. Accessed 21 Sept 2018.

K.D. Ethans, O.I. Schryvers, P.W. Nance, A.R. Casey, Spinal Cord **43**, 214 (2005)

E. A. Eugster, in Endocr. Dev., edited by M. Cappa, S. Cianfarani, L. Ghizzoni, S. Loche, and M. Maghnie (S. Karger AG, 2015), pp. 54–59.

M. Farina, A. Ballerini, G. Torchio, G. Rizzo, D. Demarchi, U. Thekkedath, A. Grattoni, Biomed. Microdevices **19**, 42 (2017)

R. Farra, N.F. Sheppard, L. McCabe, R.M. Neer, J.M. Anderson, J.T. Santini, M.J. Cima, R. Langer, Sci. Transl. Med. **4**, 122ra21 (2012)

T. Ferkol, D. Schraufnagel, Ann. Am. Thorac. Soc. **11**, 404 (2014)

M. Ferrari, I. Chu, I. Desai, D. Hansford, G. Mazzoni, T. Huen, M. Zhang, MRS Proc. **414** (1995)

S. Ferrati, D. Fine, J. You, E. De Rosa, L. Hudson, E. Zabre, S. Hosali, L. Zhang, C. Hickman, S.S. Bansal, et al., J. Control. Release **172**, 1011 (2013)

S. Ferrati, E. Nicolov, E. Zabre, T. Geninatti, B.A. Shirkey, L. Hudson, S. Hosali, M. Crawley, M. Khera, G. Palapattu, A. Grattoni, J. Sex. Med. **12**, 1375 (2015)

J. Figueira, J. Henriques, M. Amaro, V. Rosas, D. Alves, J. Cunha-Vaz, Ophthalmic Res. **57**, 166 (2017)

C.S. Filgueira, E. Nicolov, R.L. Hood, A. Ballerini, J. Garcia-Huidobro, J.Z. Lin, D. Fraga, P. Webb, O.M. Sabek, A.O. Gaber, K.J. Phillips, A. Grattoni, Int. J. Obes. **40**, 1776 (2016)

D. Fine, A. Grattoni, S. Hosali, A. Ziemys, E.D. Rosa, J. Gill, R. Medema, L. Hudson, M. Kojic, M. Milosevic, L.B. Iii, R. Goodall, M. Ferrari, X. Liu, Lab. Chip **10**, 3074 (2010)

D. Fine, A. Grattoni, E. Zabre, F. Hussein, M. Ferrari, X. Liu, Lab. Chip **11**, 2526 (2011)

M.M. Fisher, D. Lemay, E.A. Eugster, J. Pediatr. **164**, 912 (2014)

Flowonix Medical, Flowonix: Healthcare provider MRI information (2018), http://www.flowonix.com/healthcare-provider/mri-information Accessed 18 Sept 2018

Flowonix Medical Inc., Prometra® II Programmable Pump (REF 13827) For Use with Intrathecal Catheter (2017)

Flowonix Medical, PR Newswire: Flowonix Medical, Inc. Announces First Use of Its Patient Therapy Controller (PTC™) (n.d.). https://www.prnewswire.com/news-releases/flowonix-medical-inc-announces-first-use-of-its-patient-therapy-controller-ptc-300134985.html. Accessed 23 Apr 2018

C.A. Freitas-Neto, A. Maghsoudlou, S. Dhanireddy, A. Payal, S. Boonsopon, M.D. Ratwatte, C.S. Foster, Ocul. Immunol. Inflamm. **23**, 425 (2015)

N. Fullman et al., The Lancet **390**, 1423 (2017)

W. Fusi-Rubiano, C. Mukherjee, M. Lane, M.D. Tsaloumas, N. Glover, A. Kidess, A.K. Denniston, H.E. Palmer, A. Manna, R. Morjaria, BMC Ophthalmol. **18** (2018)

Genetic Engineering & Biotechnology News, Genetic Engineering & Biotechnology News: Intarcia Axes 60, Halts Studies after FDA Clinical Hold (2018), https://www.genengnews.com/gen-news-highlights/intarcia-axes-60-halts-studies-after-fda-clinical-hold/81255457. Accessed 23 Apr 2018

M. Gofeld, C.K. McQueen, Pain Med. **12**, 607 (2011)

D.S. Goldberg, S.J. McGee, BMC Public Health **11** (2011)

R.S. Gonçalves, C. Teixeira, P. Coelho, Case Rep. Ophthalmol. **8**, 465 (2017)

D. Grady, S. Kaplan, New York Times: A lifesaving pump for cancer patients is being phased out (2018), https://www.nytimes.com/2018/04/25/health/liver-pump-cancer-medical-device.html. 13 Sept 2018

A. Grattoni, E. De Rosa, S. Ferrati, Z. Wang, A. Gianesini, X. Liu, F. Hussain, R. Goodall, M. Ferrari, J. Micromech. Microeng. **19**(115), 018 (2009)

A. Grattoni, D. Fine, E. Zabre, A. Ziemys, J. Gill, Y. Mackeyev, M.A. Cheney, D.C. Danila, S. Hosali, L.J. Wilson, et al., ACS Nano **5**, 9382 (2011a)

A. Grattoni, S. Parazynski, F. Hussain, Mech. Eng. Mag. Sel. Artic. **133**, 23 (2011b)

A. Grattoni, H. Shen, D. Fine, A. Ziemys, J.S. Gill, L. Hudson, S. Hosali, R. Goodall, X. Liu, M. Ferrari, Pharm. Res. **28**, 292 (2011c)

A.C.R. Grayson, R.S. Shawgo, A.M. Johnson, N.T. Flynn, Y. Li, M.J. Cima, R. Langer, Proc. IEEE **92**, 6 (2004)

M. Guiahi, K. Tocce, S. Teal, T. Green, P. Rochon, Contraception **90**, 606 (2014)

J.-C. Gutiérrez-Hernández, S. Caffey, W. Abdallah, P. Calvillo, R. González, J. Shih, J. Brennan, J. Zimmerman, J.-C. Martínez-Camarillo, A.R. Rodriguez, R. Varma, A. Santos, G. Sánchez, M. Humayun, Transl. Vis. Sci. Technol. **3**, 8 (2014)

N. Haghjou, M. Soheilian, M.J. Abdekhodaie, J Ophthalmic Vis Res **6**, 317 (2011)

A. Hawrami, H. Laviers, S. Patra, H. Zambarakji, Acta Ophthalmol. (Copenh.) **94** (2016)

Salim M Hayek and University Hospitals Cleveland Medical Center, ClinicalTrials.gov Identifier NCT02886286 (n.d.). https://clinicaltrials.gov/ct2/show/NCT02886286 Accessed 21 September 2018.

S. Herrlich, S. Spieth, S. Messner, R. Zengerle, Adv. Drug Deliv. Rev. **64**, 1617 (2012)

R.L. Hood, G. Bruno, P. Jain, J. Anderson, T. Wolfe, C. Quini, J. Schmulen, X.C. Li, E.B. Butler, S. Krishnan, A. Grattoni, J. Biomed. Nanotechnol. **12**, 1907 (2016)

J. Huber, Contraception **58**, 85S (1998)

M. Humayun, A. Santos, J.C. Altamirano, R. Ribeiro, R. Gonzalez, A. de la Rosa, J. Shih, C. Pang, F. Jiang, P. Calvillo, J. Huculak, J. Zimmerman, S. Caffey, Transl. Vis. Sci. Technol. **3**, 5 (2014)

W. Ilias, B. le Polain, E. Buchser, L. Demartini, the oPTiMa study group, Pain Pract. **8**, 164 (2008)

Intarcia Therapeutics Inc., Intarcia Therapeutics: Investors & Media Press Releases (n.d.). http://www.intarcia.com/media/press-releases/2016-dec-29-intarcia-equity-financing-gates-foundation.html Accessed 25 February 2018.

Intarcia Therapeutics, Intarcia Therapeutics: Pipeline & Technology (n.d.-a). http://www.intarcia.com/pipeline-technology/ Accessed 24 February 2018

Intarcia Therapeutics, Intarcia Therapeutics: Pipeline & Technology HIV Prevention (n.d.-b). http://www.intarcia.com/pipeline-technology/hiv-prevention.html Accessed 24 February 2018
Intarcia Therapeutics, Intarcia Therapeutics: Pipeline & Technology Pipeline (n.d.-c). http://www.intarcia.com/pipeline-technology/pipeline.html Accessed 24 February 2018
INTEGRA Biosciences, INTEGRA: Peristaltic Pump - How It Works (n.d.). https://www.integra-biosciences.com/united-states/en/news/peristaltic-pump-how-it-works Accessed 18 June 2018.
M. Itzoe, M. Guarnieri, Drug Des. Devel. Ther. **11**, 1429 (2017)
A. V. Kalyvas, M. S. Themistocleous, N. Mpoutos, K. Themistoklis, S. Korfias, and D. Sakas, in Conf. Congr. Eur. Soc. Ster. Funct. Neurosurg. (2014).
N. Kamaly et al., The Lancet **388**, 1775 (2016)
N. J. Kassebaum et al., The Lancet **388**, 1775(2016)
R.A. Keraliya, C. Patel, P. Patel, V. Keraliya, T.G. Soni, R.C. Patel, M.M. Patel, ISRN Pharm. **2012**, 1 (2012)
V. Kini, P.M. Ho, JAMA **320**, 2461 (2018)
A. Kosturakis, R. Gebhardt, Pain Physician **15**, 475 (2012)
A. Kumar and J. Pillai, In Nanostructures Eng. Cells Tissues Organs (Elsevier, 2018), pp. 473–511.
A. Kumar, Q. Alfahad, A. Mitra, S. Elsherbiny, P.L. Lip, Eye **30**, 763 (2016)
P. Laumonerie, L. Blasco, M.E. Tibbo, O. Leclair, P. Kerezoudis, E. Chantalat, P. Mansat, World Neurosurg. **111**, 317 (2018)
D. Lee, BBC News: 'Remote control' contraceptive chip available 'by 2018' (2014). http://www.bbc.com/news/technology-28193720 Accessed 22 February 2018.
T. Li, A.T. Evans, S. Chiravuri, R.Y. Gianchandani, Y.B. Gianchandani, Adv. Drug Deliv. Rev. **64**, 1639 (2012)
S.A. Logan, C.Y. Weng, P.E. Carvounis, Int. Ophthalmol. Clin. **56**, 127 (2016)
P. Maino, E. Koetsier, R.S.G.M. Perez, R. Anesth, Pain Med. **39**, 434 (2014)
W. Majeed, H. Thabit, Expert Rev. Med. Devices **15**, 579 (2018)
J.M. Maloney, S.A. Uhland, B.F. Polito, N.F. Sheppard, C.M. Pelta, J.T. Santini, J. Controlled Release **109**, 244 (2005)
D.F. Martin, Arch. Ophthalmol. **112**, 1531 (1994)
P. Massin, A. Erginay, B. Dupas, A. Couturier, R. Tadayoni, Clin. Ophthalmol. **10**, 1257 (2016)
R. Matheson, MIT News: Big step in battling bladder disease (2014). http://news.mit.edu/2014/startups-bladder-disease-medical-device-acquired-1016 Accessed 18 June 2018.
C. McConville, **30**, 764 (2011).
Medtronic, Medtronic SYNCHROMED® II Programmable Pumps Implant Manual (2011).
Medtronic, Medtronic: Healthcare Professionals SynchroMed II Drug Infusion Pump (2017). http://www.medtronic.com/us-en/healthcare-professionals/products/neurological/intrathecal-baclofen-therapy-systems/synchromed-ii.html Accessed 24 February 2018.
A. Meireles, C. Goldsmith, I. El-Ghrably, A. Erginay, M. Habib, B. Pessoa, J. Coelho, T. Patel, R. Tadayoni, P. Massin, J. Atorf, A.J. Augustin, Eye **31**, 684 (2017)
E. Meng, T. Hoang, Ther. Deliv. **3**, 1457 (2012a)
E. Meng, T. Hoang, Adv. Drug Deliv. Rev. **64**, 1628 (2012b)
Microchips Biotech, Microchips Biotech: Teva and Microchips Biotech Announce Partnership to Enhance Patient Outcomes through Digital Drug Delivery Technology (2015). http://microchipsbiotech.com/news-pr-item.php?news=10 Accessed 24 February 2018.
A.C. Miracle, M.A. Fox, R.N. Ayyangar, A. Vyas, S.K. Mukherji, D.J. Quint, Am. J. Neuroradiol. **32**, 1158 (2011)
E. Moisseiev, L.S. Morse, JAMA Ophthalmol. **134**, 1067 (2016)
S. Mourtzoukos, Expert Rev. Ophthalmol. **12**, 95 (2017)
Nano Precision Medical, Nano Precision Medical Technology: The Goldilocks Effect (n.d.-a). http://www.nanoprecisionmedical.com/technology/the-goldilocks-effect Accessed 24 February 2018.
Nano Precision Medical, Nano Precision Medical Technology: NanoPortal Technology (n.d.-b). http://www.nanoprecisionmedical.com/technology/nanoportal-technology Accessed 24 February 2018.
Nano Precision Medical, Nano Precision Medical Technology: NanoPortal Capsule (n.d.-c). http://www.nanoprecisionmedical.com/technology/nanoportal-capsule Accessed 24 February 2018.
Newron Sweden AB, ClinicalTrials.gov Identifier NCT00866502 (n.d.). https://clinicaltrials.gov/ct2/show/NCT00866502 Accessed 21 September 2018.
B.P. Nicholson, R.P. Singh, J.E. Sears, C.Y. Lowder, P.K. Kaiser, Am. J. Ophthalmol. **154**, 969 (2012)
J.C. Nickel, P. Jain, N. Shore, J. Anderson, D. Giesing, H. Lee, G. Kim, K. Daniel, S. White, C. Larrivee-Elkins, J. Lekstrom-Himes, M. Cima, Sci. Transl. Med. **4**, 143ra100 (2012)
E.B. Odom, D.L. Eisenberg, I.K. Fox, Contraception **96**, 89 (2017)
A.B. Oung, E. Kosirog, B. Chavez, J. Brunner, J.J. Saseen, Ther. Adv. Chronic. Dis. **8**, 113 (2017)
N. Pagès-Puigdemont, M.A. Mangues, M. Masip, et al., Adv. Ther. **33**, 1740 (2016)
V.T. Papastavrou, H. Zambarakji, I. Dooley, H. Eleftheriadis, T.L. Jackson, Retin. Cases Brief Rep. **11**, 44 (2017)
K. Park, J. Controlled Release **190**, 3 (2014)
R.K. Parrish, P.A. Campochiaro, P.A. Pearson, K. Green, C.E. Traverso, FAME Study Group, Ophthalmic Surg. Lasers Imaging Retina **47**, 426(2016)
G. Paul, O. Zachrisson, A. Varrone, P. Almqvist, M. Jerling, G. Lind, S. Rehncrona, B. Linderoth, H. Bjartmarz, L.L. Shafer, R. Coffey, M. Svensson, K.J. Mercer, A. Forsberg, C. Halldin, P. Svenningsson, H. Widner, J. Frisén, S. Pålhagen, A. Haegerstrand, J. Clin. Invest. **125**, 1339 (2015)
C. Pedroso, I. Martins, F. Palma, and A. I. Machado, Case Rep. **2015**, bcr2014206256 (2015).
C. Perruchoud, M. Bovy, A. Durrer, M. Rosato, B. Rutschmann, J.-P. Mustaki, E. Buchser, Neuromodulation Technol. Neural Interface **15**, 31 (2012)
B. Pessoa, J. Coelho, N. Correia, N. Ferreira, M. Beirão, A. Meireles, Ophthalmic Res. **59**, 68 (2018)
P. Petrou, C. Koutsandrea, D. Papaconstantinou, D. Mpouritis, I. Georgalas, Drug Des. Devel. Ther., 2373 (2014)
J.E. Pope, T.R. Deer, Pain Manag. **5**, 175 (2015)
J. E. Pope and T. R. Deer, in Treat. Chronic Pain Cond., ed. By J. E. Pope and T. R. Deer (Springer New York, New York, NY, 2017), pp. 191–193.
D. Pragout, F. Darrouzain, H. Marret, Eur. J. Obstet. Gynecol. Reprod. Biol. (2018)
J.H. Prescott, S. Lipka, S. Baldwin, N.F. Sheppard, J.M. Maloney, J. Coppeta, B. Yomtov, M.A. Staples, J.T. Santini, Nat. Biotechnol. **24**, 437 (2006)
H. Quhill, F. Quhill, Case Rep. Ophthalmol. **7**, 301 (2016)
N. Rajgor, V. Bhaskar, M. Patel, Syst. Rev. Pharm. **2**, 91 (2011)
R. Rauck, T. Deer, S. Rosen, G. Padda, J. Barsa, E. Dunbar, G. Dwarakanath, Neuromodulation Technol. Neural Interface **13**, 102 (2010)
R. Rauck, T. Deer, S. Rosen, G. Padda, J. Barsa, E. Dunbar, G. Dwarakanath, Neuromodulation Technol. Neural Interface **16**, 163 (2013)
J. Riordan, P. Murphy, Neuromodulation Technol. Neural Interface **18**, 433 (2015)
C.M. Rohloff, T.R. Alessi, B. Yang, J. Dahms, J.P. Carr, S.D. Lautenbach, J. Diabetes Sci. Technol. **2**, 461 (2008)
J.S. Ross, J. Bates, C. Parzynski, J. Akar, J. Curtis, N. Desai, J. Freeman, G. Gamble, R. Kuntz, S.-X. Li, D. Marinac-Dabic, F. Masoudi, S.-L. Normand, I. Ranasinghe, R. Shaw, H. Krumholz, Med. Devices Evid. Res. **10**, 165 (2017)
X. Ruan, P. Couch, H. Liu, R.V. Shah, F. Wang, S. Chiravuri, Pain Physician **13**, 337 (2010)

T. Rudlang, K. Brasso, Austin J Urol **3**, 1045 (2016)

H. Saedon, A. Anand, Y. Yang, Clin. Ophthalmol. **11**, 583 (2017)

B. Samb, N. Desai, S. Nishtar, S. Mendis, H. Bekedam, A. Wright, J. Hsu, A. Martiniuk, F. Celletti, K. Patel, F. Adshead, M. McKee, T. Evans, A. Alwan, C. Etienne, The Lancet **376**, 1785 (2010)

J.T. Santini, M.J. Cima, R. Langer, Nature **397**, 335 (1999)

J.J. Santini, A. Richards, R. Scheidt, M. Cima, R. Langer, Angewandte Chemie International Edition **39**, 2396 (2000)

G. Santus, R.W. Baker, J. Controlled Release **35**, 1 (1995)

M. Saulino, M. Gofeld, Neuromodulation Technol. Neural Interface **17**, 696 (2014)

V. Schmit-Eilenberger, Clin. Ophthalmol., 801 (2015)

M. Serati, G. Bogani, S. Kumar, A. Cromi, F. Ghezzi, Contraception **91**, 91 (2015)

S. Sgouros, C. Charalambides, P. Matsota, I. Tsangaris, G. Kostopanagiotou, Pediatr. Neurosurg. **46**, 62 (2010)

D.E. Shumer, N.J. Nokoff, N.P. Spack, Adv. Pediatr. **63**, 79 (2016)

J. Sih, S.S. Bansal, S. Filipini, S. Ferrati, K. Raghuwansi, E. Zabre, E. Nicolov, D. Fine, M. Ferrari, G. Palapattu, A. Grattoni, Anal. Bioanal. Chem. **405**, 1547 (2013)

L.A. Silverman, E.K. Neely, G.B. Kletter, K. Lewis, S. Chitra, O. Terleckyj, E.A. Eugster, J. Clin. Endocrinol. Metab. **100**, 2354 (2015)

C. Simon, M.S. Agier, J. Béné, C. Muller, L. Vrignaud, H. Marret, A.P. Jonville-Bera, J. Gynécologie Obstétrique, Biol. Reprod. **45**, 1074 (2016)

L. Smith, J. Mosley, J. Johnson, and M. Nasri, **42**, 505 (2017).

A.I. Son, J.D. Opfermann, C. McCue, J. Ziobro, J.H. Abrahams, K. Jones, P.D. Morton, S. Ishii, C. Oluigbo, A. Krieger, J.S. Liu, K. Hashimoto-Torii, M. Torii, Sci. Rep. **7** (2017)

Z. Steel, C. Marnane, C. Iranpour, T. Chey, J.W. Jackson, V. Patel, D. Silove, Int. J. Epidemiol. **43**, 476 (2014)

I. Stetkarova, S.A. Yablon, M. Kofler, D.S. Stokic, Neurorehabil. Neural Repair **24**, 609 (2010)

C.L. Stevenson, J.T. Santini, R. Langer, Adv. Drug Deliv. Rev. **64**, 1590 (2012)

Y.Y. Syed, Drugs **77**, 575 (2017)

Taris Biomedical LLC, ClinicalTrials.gov Identifier NCT02722538 (n.d.-a). https://clinicaltrials.gov/ct2/show/NCT02722538 Accessed 21 September 2018.

Taris Biomedical LLC, ClinicalTrials.gov Identifier NCT02720367 (n.d.-b). https://clinicaltrials.gov/ct2/show/NCT02720367 Accessed 21 September 2018.

Taris Biomedical LLC, ClinicalTrials.gov Identifier NCT03404791 (n.d.-c). https://clinicaltrials.gov/ct2/show/NCT03404791 Accessed 21 September 2018.

Taris Biomedical LLC, Taris Biomedical: Taris Advantage (n.d.-d). http://www.tarisbiomedical.com/taris-advantage.php Accessed 24 February 2018.

The Global Cancer Observatory, International Agency for Research on Cancer: All cancers fact sheet (2018). http://gco.iarc.fr/today/data/factsheets/cancers/39-All-cancers-fact-sheet.pdf Accessed 26 October 2018.

Titan Pharmaceuticals, Titan Pharmaceuticals: Pipeline (n.d.). https://www.titanpharm.com/pipeline/overview Accessed 25 February 2018.

U.S. Food and Drug Administration, U.S. Food and Drug Administration: Class 2 Device Recall Prometra II Programmable Infusion PumpPump (2018). https://www.accessdata.fda.gov/scripts/cdrh/cfdocs/cfRES/res.cfm?id=156065 Accessed 23 April 2018.

D. Veritti, V. Sarao, L. Diplotti, F. Samassa, P. Lanzetta, Expert Opin. Pharmacother. **18**, 1507 (2017)

T. Vos et al., The Lancet **386**, 743 (2015)

J. Wang, A. Jiang, M. Joshi, J. Christoforidis, Hindawi Publ. Corp. **2013**, 1 (2013)

J.T. Watson, Sci. Transl. Med. **4**, 122ed1 (2012)

D. Wilkes, J. Pain Res., 571 (2014)

W.L. Winters, Methodist DeBakey Cardiovasc. J. **11**, 5 (2015)

S. Woolen, C. Holzmeyer, E. Nesbitt, P.F. Siami, Prostate Cancer **2014**, 1 (2014)

World Economic Forum and Harvard School of Public Health, Methodological Appendix: The Global Economic Burden of Non-Communicable Diseases (2011).

World Health Organization, Adherence to Long-Term Therapies: Evidence for Action (2003).

World Health Organization, *Global Report on Diabetes* (WHO Press, World Health Organization, Geneva, 2016)

World Health Organization, World Health Organization Metrics: Disability-Adjusted Life Year (DALY) (n.d.). http://www.who.int/healthinfo/global_burden_disease/metrics_daly/en/ Accessed 18 September 2018.

C. Wright, J. Hall, Drug Des. Devel. Ther., 1543 (2016)

W.-W. Yang, E. Pierstorff, J. Lab. Autom. **17**, 50 (2012)

Y. Yang, C. Bailey, F.G. Holz, N. Eter, M. Weber, C. Baker, S. Kiss, U. Menchini, J.M. Ruiz Moreno, P. Dugel, A. Lotery, Eye **29**, 1173 (2015)

M.N. Yasin, D. Svirskis, A. Seyfoddin, I.D. Rupenthal, J. Controlled Release **196**, 208 (2014)

Y.H. Yun, B.K. Lee, K. Park, J. Controlled Release **219**, 2 (2015)

A. Ziemys, A. Grattoni, D. Fine, F. Hussain, M. Ferrari, J. Phys. Chem. B **114**(11), 117 (2010)

A. Ziemys, M. Kojic, M. Milosevic, N. Kojic, F. Hussain, M. Ferrari, A. Grattoni, J. Comput. Phys. **230**, 5722 (2011)

Biomedical Microdevices (2019) 21: 37
https://doi.org/10.1007/s10544-019-0392-y

Project honeybee: Clinical applications for wearable biosensors

Leland Hartwell[1] · **Heather M. Ross**[2,3] · **Jeffrey T. La Belle**[4,5]

Published online: 1 April 2019

Author's Personal Note Dear Mauro: My strongest link to nanomedicine is as a fortunate 2014 NanoGagliate participant. The unlikely bridge from the elite of Houston to underserved of Gagliato remains an inspiration.

Leland Hartwell

Abstract

The Project Honeybee Observational Clinical Trials were 12-month studies designed to validate the use of commercially available ambulatory medical devices costing $50–$300 for clinical applications. Each trial had a patient population of about 15 15–30 subjects with a broad range of disease types including heart failure, diabetes, sepsis, and Parkinson's disease. Over 30 supported proposals were funded in the 4-year period, as well as the creation of a database of all commercially available devices. Each year a call for proposals was published within ASU and Mayo Clinic Arizona. Proposals were selected for funding by a committee of ASU faculty from engineering, nursing, and exercise physiology departments. The progress of each research trial was monitored through monthly colloquia with the nursing, biomedical engineering, computer science, and nutrition graduate research assistants, to discuss the challenges and opportunities arising with each research trial. PIs were required to report on study progress 6 months into the trial period and 3 months following the conclusion of the 12-month project. The project was very successful in meeting our goals of testing consumer wearable devices on patients for a variety of conditions across a variety of clinical settings in the greater Phoenix community. The following clinical sites participated in one or more of these clinical trials: Adelante Healthcare, Arizona Arrhythmia Consultants, Arizona Cardiology Group, Banner University Medical Center, Barrow Neurological Institute, Honor Health, Mayo Clinic, and St Joseph's Hospital. A total of 12 ASU faculty and 39 clinicians participated.

Keywords Project honeybee · Wearables · Sensors · Clinical medicine

1 Introduction

Modern airline travel is widely recognized as one of the safest modes of transportation worldwide. This is due, in part, to the advent of continuous sensory monitoring (i.e. sound, vibration, temperature, and pressure) of mechanical aircraft components (Reliability Centered Maintenance) in order to detect parts needing replacement prior to mechanical failure. Similarly, improving health outcomes while reducing healthcare system costs will require increasing emphasis on prevention, personalized medicine and prediction. Drawing on the airline experience, the recent proliferation of wearable devices to continuously monitor physiological parameters offers promise of a technological solution to improve health outcomes and the healthcare system.

In 2013, ASU's Center for Sustainable Health (now the Biodesign Pathfinder Center) launched a research initiative, Project Honeybee, to test the utility of wearable physiological sensors (devices) in the clinical management of patients through a series of small observational clinical trials (OCT). Each Honeybee OCT began with a clinical question and embedded a highly experienced doctoral level Graduate Research Assistant in order to facilitate workflow as well as train the health workforce of the future. We named the effort after the honeybee, which is thought to be nature's best collector of information. Just as honeybees gather data from nature and return collected knowledge to the hive, we sought to investigate the potential of wearable devices to gather information

✉ Jeffrey T. La Belle
JEFFREY.LABELLE@asu.edu

1 Biodesign Pathfinder Center, Biodesign Institute at Arizona State University, Tempe, AZ 85281, USA

2 School for the Future of Innovation in Society, Tempe, AZ 85287, USA

3 College of Nursing and Health Innovation, Tempe, AZ 85004, USA

4 School of Biological and Health Systems Engineering, Arizona State University, Tempe, AZ 85287, USA

5 School of Medicine, Mayo Clinic Arizona, Scottsdale, AZ 85259, USA

from a patient's daily living and return it to the clinical "hive" to better inform the healthcare experience.

Funded by ASU for a series of four years, Project Honeybee OCTs incorporated ACO quality measures and performance standards in each of our yearly trials as a value-added metric and service. A key feature of our effort has been the interdisciplinary support from ASU's innovative nursing, engineering, business, and design leaders engaging in research.

We had several goals for these trials, including expanding ASU engagement with clinicians in the greater Phoenix area, initiating collaborations between ASU faculty and area clinical institutions, and testing the challenges that implementing sensors in clinical medicine would face. We hoped to gain insights into the functioning of commercial devices, the needs of clinical teams implementing devices, patient compliance with device monitoring, physician engagement, the value of data obtained from wearable devices, and the range of applications devices could be utilized to monitor.

2 Methods

2.1 Clinical trials

Each Spring we issued a Call for Proposals (CFP) to ASU investigators and area clinicians. Our CFP specified that we supported observational clinical trials of wearable devices available at a cost of 50–300 USD with anticipated study enrollment of 20–50 participants during the 1-year funded period of each study. The CFP clarified that we would not support interventional trials for devices that did not have FDA clearance. The CFP further specified the requirement for qualified clinician leadership by a healthcare provider in Maricopa county. The expected outcome for each study was pilot data suitable for publication and/or further external funding applications.

We sought to maximize experiences while minimizing costs. Each trial was allotted 27,000 USD to support a doctoral nursing student half-time for one year to manage patient enrollment and data collection, and 3000 USD for procuring devices. We did not support Principal Investigator (PI) salaries, clinical time, or patient incentives. Consequently, each trial was limited both in the number of patients and the duration of the trial. However, the many short trials approach provided very diverse experiences across the range of clinical conditions monitored.

In spite of the limited support, our call for proposals was nearly fully subscribed each year. The range of devices ad study approaches was broad. In addition to observational clinical trials, the advisory committee elected to fund some studies each year that did not precisely conform to the CFP requirements for a clinical observational trial including some preclinical validation studies, algorithm development studies, and novel device development. Tables 1, 2, 3, and 4 below provide an overview of the funded projects for the four years of Project Honeybee.

2.2 Database

In addition to the clinical studies, Project Honeybee also funded the development of a database of commercially available devices. The database was constructed by conducting regular searches of consumer and trade publications, as well as weekly Google searches to identify consumer wearables in development and/or available for purchase. The database captured device name, manufacturer information, cost, published technical specifications, and target biosignals. The intention was for the database to be a searchable resource for clinicians to identify options for wearable monitors to recommend to patients.

We employed one graduate engineering student for 20 h per week for database construction and maintenance. Over the period of two years, we found that rapid changes in the field of commercial wearables exceeded the work capacity of a single 0.5 fulltime equivalent (FTE) worker and would require at least 1 fulltime staff to manage the constant changes. We explored the possibility of commoditizing the database to support the resources necessary to fund the development and ongoing maintenance of a sophisticated and clinically-relevant searchable platform. After several months of early business model development, we determined that the ongoing database project did not fit with the long-term goals of Project Honeybee at this time, and we discontinued the database project.

3 Results

As Project HoneyBee consisted of 32 unique studies, for the sake of brevity we will highlight two very different ones. An early study on the evaluation of the wearables to support the device database to aid the other studies will be highlighted as well as one of the clinical evaluations of placing devices on patients.

3.1 Project honeybee preclinical trial

3.1.1 Goals

We sought to evaluate the performance of commercially available wearable devices against a gold standard in a laboratory-based rest and exercise protocol. In addition, we evaluated each device with innovation, quality, cost, and comfort scores derived by our research team.

Table 1 Observational clinical trials 2014–2015

Funded Trials - Year 1

Project	Principal investigator(s)	Clinical partner(s)	Device(s)	Endpoints	Subjects enrolled
Testing patient mobility as a predictor of 30-day hospital readmission of heart failure patients	Dr. Matthew Buman GRA: Erin Krzywicki, RN, BSN	Scottsdale Healthcare John C. Lincoln Dignity Health	activPAL3c (PAL Technologies, Ltd., Glasgow, Scotland, UK) sensor (in-hospital), Tractivity (Kineteks Corp., Warwick, RI, USA) monitor (in-hospital), Jawbone (Jawbone, San Francisco, CA, USA) activity monitor (post-hospital)	Correlate inpatient mobility to post-discharge activity	29
Testing arrhythmia self-monitoring of atrial fibrillation to reduce hospitalization	Dr. Heather Ross GRA: Alyssa Smith, RN, BSN	Arizona Arrhythmia Consultants	AliveCor monitor (AliveCor, Inc., Mountain View, CA, USA	Self-assessment accuracy of heart rhythm	18
Validating self-monitoring devices against intrathoracic impedance measurements from implantable cardiac devices for self-monitoring of heart failure	Dr. Heather Ross GRA: Lexy Cook, RN, BSN	Arizona Arrhythmia Consultants	Wello monitor (Azoi, Inc., San Francisco, CA, USA) [note the Wello monitor was never shipped] Jawbone Up3 (Jawbone, San Francisco, CA, USA)	Validation of resting heart rate Correlation of resting heart rate data to changes in fluid status	4
Long-term gait monitoring to improve patient selection for shunt surgery in normal pressure hydrocephalus	Dr. Narayanan Krishnamurthi Dr. James Abbas et al. GRA: Debra Corley, RN, BSN	St. Joseph's Hospital & Medical Center Barrow Neurological Institute	Inertial Measurement Units (IMUs) (proprietary device, Krishnamurthi Lab, Phoenix, AZ, USA)	Gait assessment	0
Heart Failure	Dr. Steve Lester GRA: Mary Kourieh, RN, BSN	Mayo Clinic	Withings Smart Scale, Blood Pressure Cuff and Activity Tracker (Withings, Issy-les-Moulineaux, Île-de-France, France)	Quality of life, Self-efficacy	Not available
Feasibility study for physical activity monitoring devices for diabetes patients	Dr. Jim Levine Dr. Curtiss Cook GRA: EJ Milas, RN, BSN	Mayo Clinic	KAM (Kersh Health, Richardson, TX, USA)	Acquisition of physical activity data	17
Daily energy and neurocognitive function	Dr. Jim Levine	Mayo Clinic	Apple Watch (Apple, Inc., Cupertino, CA, USA)	Activity	0
Pre-clinical trial to validate patch and wearable devices against state of the art (SOTA) seeking gold standard.	Dr. Jeffrey La Belle Dr. Mark Spano GRA: Kevin Uchimura	Mayo Clinic	Multiple patch and ambulatory devices: • Braun forehead thermometer (Braun, Kronberg, Germany) • Braun thermoscan thermometer (Braun, Kronberg, Germany) • Mabis digital thermometer (Briggs Healthcare, Waukegan, IL, USA) • Omron 5 series (Omron Healthcare, Kyoto, Japan) • Omron 7 series (Omron Healthcare, Kyoto, Japan) • Omron HEM-18 (Omron Healthcare, Kyoto, Japan) • KAM (Kersh Health, Richardson, TX, USA) • Withings wireless BP monitor (Withings, Issy-les-Moulineaux, Île-de-France, France) • Fitbit One (Fitbit, San Francisco, CA, USA) • AliveCor (AliveCor, Inc., Mountain View, CA, USA) • Tanita BF 679 W (Tanita Corporation, Tokyo, Japan) • VitalConnect Healthpatch Biosensor (VitalConnect, San Jose, CA, USA) • Fitbit Surge (Fitbit, San Francisco, CA, USA) • Withings body composition monitor (Withings, Issy-les-Moulineaux, Île-de-France, France) • Tinke (Zensorium, Singapore) • Finger Pulse OX SM-110 (Santamedical, Tustin, CA, USA) • Withings Pulse O2 (Withings, Issy-les-Moulineaux, Île-de-France, France) • FeverSmart (Fever Smart, Philadelphia, PA, USA) • GENEActiv Original (Activinsights, Kimbolton, Cambridgeshire, UK) • Jawbone Up3 (Jawbone, San Francisco, CA, USA) • Apple Watch (Apple, Cupertino, CA, USA)	Lab validation	38 participants, generating ≥20 data sets per device

Table 2 Observational clinical trials 2015–6

Funded Trials - Year 2

Project	Principal investigator(s)	Clinical partner(s)	Device	Endpoints	Subjects enrolled
Develop a machine-learning based algorithm to quantify type, context, and intensity of wearable device data collected at wrist and hip. Make algorithm available to other PHB teams and the research community at large. ACO-5, ACO-7	Matthew Buman, PhD Pavan Turaga, PhD CO-PI: Brian Ashcroft, PhD	ASU, SNHP ASU, SECEE ASU, Biodesign, CSMB	GENEactiv Actigraph GT3x+ (Activinsights, Kimbolton, Cambridgeshire, UK)	Machine learning algorithm development	0, secondary data analysis only)
Measure the impact of outpatient Red Blood Cell (RBC) transfusions on at-home functional status through physiological parameters and physical activity metrics. ACO-7	Daryl Kor, MD Jyotishman Pathak, MD CO-PI: Karl Poterack, MD	Mayo Clinic Rochester Mayo Clinic Rochester Mayo Clinic Scottsdale	Fitbit Charge HR (Fitbit, San Francisco, CA, USA)	Post-op activity assessment	4
Long-term continuous monitoring to characterize the effects of deep brain stimulation on gait and balance dysfunction in people with Parkinson's disease ACO-6, ACO-7	Narayanan Krishnamurthi, PhD CO-PI: James Abbas, PhD PARTNERS: Padma Mahant, MD Francisco Ponce, MD	ASU, CONHI ASU, SBHSE Banner Good Samaritan Medical Center St. Joseph's Hospital and Medical Center	Inertial Measurement Units (IMU) (proprietary device, Krishnamurthi lab, Phoenix, AZ, USA)	Gait and balance assessment	3
Verification and validation of customized device sensors to reduce hospital readmission to demonstrate equivalency between custom array and gold standard in the lab. ACO-1, ACO-2, ACO-5, ACO-6, ACO-7	Jeffrey La Belle, PhD Mark Spano, PhD CO-PI: Eric Steidley, MD	ASU, SBHSE ASU, SBHSE Mayo Arizona	Customized lab-built sensor array/ device monitoring activity, temperature, ECG and pulse oximetry (La Belle lab, Tempe, AZ, USA)	Sensor validation	10
To create a scientifically validated weight loss program to reach patients utilizing wearable technologies. ACO-5, ACO-7	James Levine, MD CO-PI: Gabriel Koepp	Mayo Clinic Arizona / ASU Apple, Inc. SP Health	Apple Watch (Apple, Cupertino, CA, USA)	Activity, Weight change	0
Defining the healing trajectory after elective surgery. Use wearable activity monitoring devices pre and post elective inguinal hernia surgery (2 types) to ascertain quality/duration of return to baseline activity and identify patients at risk for opioid dependence. ACO-2, ACO-5, ACO-6, ACO-7	Karl Poterack, MD CO-PI: Richard T. Schlinkert, MD CO-PI: Matthew Buman, PhD	Mayo Clinic Arizona Mayo Clinic Arizona ASU	GENEactiv (Activinsights, Kimbolton, Cambridgeshire, UK)	Activity	7
Assess the feasibility and effectiveness of patient-directed deployment of an individualized medication management plan to treat recurrences of Atrial Fibrillation (AF) utilizing wearable monitor. ACO-1, ACO-4, ACO-5, ACO-6, ACO-7	Heather Ross, DNP Partners: Mitchell J. Ross, MD John Beshai, MD	Arizona Arrhythmia Consultants / ASU Arizona Cardiology Group Mayo Clinic Arizona	AliveCor ECG monitor (AliveCor, Mountain View, CA, USA)	Quality of life, Healthcare utilization	8
Assess the feasibility and effectiveness of patient-directed Congestive Heart Failure (CHF) assessment and management using an ambulatory health tracking device. ACO-1, ACO-4, ACO-5, ACO-6, ACO-7	Heather Ross, DNP Partners: Mitchell J. Ross, MD Eric Steidley, MD John Beshai, MD Mitchell J. Cohen, MD Edward Rhee, MD	Arizona Arrhythmia Consultants / ASU Arizona Cardiology Group Mayo Clinic Arizona Mayo Clinic Arizona Phoenix Children's Hospital Phoenix Children's Hospital	Jawbone Up3 (Jawbone, San Francisco, CA, USA)	Resting heart rate validation	11
Early recognition of sepsis to reduce mortality, organ failure, hospital length of stay, and cost of care. Monitor relevant physiological trends with high signal-noise ratio and determine wearable device feasibility of use in a high-risk Hematopoietic stem cell transplantation (HSCT) cohort. ACO-1, ACO-7	Ayan Sen, MD Pierre Noel, MD Charles Bruce, MD CO-PIs: Michelle Alore, RN, MSN, DNP Bhavesh Patel, MD Jane Olsen, NP Brittany Williams, RN	Mayo Clinic Arizona Mayo Clinic Arizona Mayo Clinic Arizona Mayo Clinic Arizona Mayo Clinic Arizona Mayo Clinic Arizona	Body Guardian (Preventice Technologies, Rochester, MN, USA)	Heart rate validation	11

Table 3 Observational clinical trials 2016–7

Funded Trials - Year 3

Project	Principal investigator(s)	Clinical partner(s)	Device	Endpoints	Subjects enrolled
Assessing the feasibility and effectiveness of using a wearable health tracking device to identify Military Veterans at risk for service-related conditions	Dr. Tamiko Azuma Karen Gallagher, MS Dimple Patel GRA	ASU Pat Tillman Center ASU Speech and Language Clinic	Vivofit HR (Garmin, Olathe, KS, USA)	Correlation of activity and sleep with self-rated stress, anxiety, sleep, and cognition	21
Toward a portable wearable system for detection and diagnosis of obstructive sleep apnea	Dr. Matt Buman Dr. Pavan Turaga Maynak Gupta GRA	None	Zephyr BioHarness (Medtronic, Minneapolis, MN, USA) GeneActiv (Activinsights, Kimbolton, Cambridgeshire, UK)	Algorithm development	0 (secondary data analysis only)
Mayo 360 Personalized Support System (Mayo 360PSP). Patient-Driven Lifestyle Solutions for Weight and Wellness.	Amy Foxx-Orenstein, DO Dr. George Runger Tanya Carroccio GRA	Mayo Clinic Arizona Weight and Wellness Solutions Program	Fitbit Charge HR (Fitbit, San Francisco, CA, USA)	Activity measurement	27
Monitoring to Investigate the Effects of Sleep Behavior Disorder on Gait Dysfunction in People with Parkinson's Disease	Dr. Narayanan Krishnamurthi Stephanie De Santiago GRA	St. Joseph's Hospital and Medical Center	Actiwatch Spectrum (Koninklijke Philips N.V., Amsterdam, The Netherlands) APDM Inertial Measurement Units (proprietary device, Krishnamurthi lab, Phoenix, AZ, USA)	Sleep amount and gait correlation	0
Expand development on ambulatory medical wearable device captures three biosignals for patients	Dr. Jeffrey La Belle Heather Ross, DNP Alejo Chavez/Ashwin Sundar GRA's	None	Design and fabricate a device in-house to measure outcomes	Device fabrication	3
Measurement of medical staff and trainees' sleep and physical activity during various call and shift schedules	Dr. Karl Poterack Dr. Andrew Gorlin Dr. Matt Buman Ran McBryde GRA	Anesthesiology Department at Mayo Clinic in Phoenix, AZ	GENEactiv (Activinsights, Kimbolton, Cambridgeshire, UK)	Sleep amount and quality	11
ACT NOW: Use of ACT9X activity and sleep tracker to help with early detection of Alzheimer's disease in Older adults with Down syndrome	Dr. Shannon Ringenbach Nathaniel Arnold GRA	Alzheimer's and Cognitive Disorders Program, Barrow Neurological Institute	Actigraph GT9X Link (Actigraph LLC, Pensacola, FL, USA)	Physical activity and sleep correlation Determination of effects of activity training on physical activity and sleep	15
Validating Force-Sensing Mouthguards Against Expert Concussion Analysis to Guide Removal from Play in Collegiate Athletics	Heather Ross, DNP Lexy Richards GRA Alyssa Smith GRA	ASU Athletics	FITGuard mouthguard (Force Impact Technologies, Los Angeles, CA, USA)	Head trauma detection	5
Wearable wireless vital signs monitoring in predicting clinical decompensation	Dr. Ayan Sen Dr. Holly Geyer Dr. Brian Hull Dr. Christopher Farmer Madeleine Goetz GRA	Mayo Clinic Arizona	Zephyr BioHarness (Medtronic, Minneapolis, MN, USA)	Heart rate, Respiratory rate, Heart rate variability, Clinical decompensation	11

Table 4 Observational clinical trials 2017–8

Funded Trials - Year 4

Project	Principal investigator(s)	Clinical partner(s)	Device	Endpoints	Subjects enrolled
Patient-Driven Lifestyle Solutions for Weight and Wellness – Part 2	Amy Foxx-Orenstein D.O. George Runger, PhD, Julie Liss, PhD, Matthew Buman, PhD (Co-PIs) Tanya Carroccio GRA Miriam Zerio-Levering	Ken and Linda Morris Weight and Wellness Solutions Program at Mayo Clinic, Arizona	Fitbit Charge HR (Fitbit, San Francisco, CA, USA)	Activity measurement	27 (continued from Year 3)
Patient-Driven Lifestyle Solutions for Weight and Wellness – Development of a Software Platform	Amy Foxx-Orenstein D.O. George Runger, PhD, Julie Liss, PhD, Matthew Buman, PhD (Co-PIs) Tanya Carroccio GRA Miriam Zerio-Levering	Ken and Linda Morris Weight and Wellness Solutions Program at Mayo Clinic, Arizona	Data from the Fitbit Charge HR (Fitbit, San Francisco, CA, USA) will be utilized in the creation of an online dashboard	Dashboard software development	0 (secondary data analysis)
Feasibility and Explorative Study to test the superiority of Connected Health Care using wearable devices to improve body composition and fitness levels	Regis Fernandes, MD Robert Scales, MS, Minako Katayama, MD, Hari Chaliki, MD, Sophie Bersoux, MD, Temkit Hamy, PhD (CO-PIs) GRA - TBD	Robert Scales, MS Hari Chaliki, MD Sophie Bersoux, MD	24alife (Mikropis Holding, Zalec, Slovenia) FitBit Zip (Fitbit, San Francisco, CA, USA) Heart-rate chest sensor (TBD)	Activity, Weight Change	17
Metabolic Tracking During Pregnancy to Promote Adequate Gestational Weight Gain	Elizabeth Reifsnider, PhD, RN Corrie Whisner, PhD, Kiley Vander Wyst, MPH (Co-PIs) Kiley Vander Wyst - GRA	Adelante Healthcare	Breezing Device (Breezing Metabolism, Tempe, AZ, USA)	Pregnancy weight gain	23
Real-World Validation of a Novel Multi-sensor Wearable for Heart Failure	Jeffrey La Belle, PhD Matthew Buman, PhD Heather Ross, PhD, DNP Pavan Turaga, PhD (Co-PI) Raja Nagappan - GRA Alexan Gomez - GRA Robin Faulkner - GRA	Honor Health Heart & Vascular Institute	Design and fabricate a multi-sensor CHF device in-house	Device validation	0
Validation of Force and Vector in Athletic Head Contact	Heather Ross, PhD, DNP Lexy Richards, RN, Carter Pruitt, Joshua Beaumont (Co-PIs) Austin Tielke - GRA Dimple Patel - GRA	ASU Athletics ASU Wrestling	IM Mouthguard (Prevent Biometrics, Edina, MN, USA)	Head trauma validation	4

3.1.2 Methods

We recruited participants from the ASU student community to perform multiple "typical use" protocols with 22 different devices across 7 categories (Table 5) with 15 unique sensor types (Table 6). Table 5 illustrates the total data collection that took place during the 12-month funding period. We achieved our goal to collect a minimum 20 basal and post-exercise readings (data sets) from unique participants for all devices when possible. Only two devices were tested less frequently: the VitalConnect Healthpatch Biosensor, a device now not available to the consumer ($n = 9$), and the FeverSmart wearable thermometer (n = 9), which had various issues during use, preventing accurate data collection.

Testing Protocols:

Activity Testing: During activity testing, the participant in the study was filmed from the waist down for the duration of the protocol. The participant was instructed to walk naturally at five different speeds. Then, they walked at each speed for one minute before changing to the next. The speeds were 2.5, 3.0, 3.4, and 4.0 mph. The participant wore several devices on their wrists, including the GENEActiv, Apple Watch, Jawbone UP3, and Fitbit Surge. The Fitbit One, KAM, and Withings Pulse O2 were worn on a fitness belt that included a pocket for the iPhone 6 (necessary for use with the Apple Watch). The total steps collected by each device was recorded before and after exercise; total steps walked during the five-minute session were compared to an average manually counted by three investigators separately.

Blood Pressure Testing: Blood pressure testing began with participant seated, at rest while the BPMs were used. First, the participant's manual blood pressure was taken on their right or left arm. The rest of the devices were randomized before testing began to ensure no bias in time effect after exercise. According to specific device use, the Omron 7 was always used on the left wrist and the other devices, Omron 5 and Withings Wireless, were used in succession on alternating arms. These devices were the Omron 5 and the Withings Wireless. Once each device was tested, the participant was asked to run for five minutes on a treadmill, beginning at a speed of 2 mph and increasing or decreasing at the participant's preference. Immediately after running, the devices were tested in the same order the specific participant used during baseline testing. If this protocol followed in succession from a different protocol, the participant rested for 10 min to ensure baseline reading accuracy. Time stamps were recorded for every measurement taken.

GSR/Total Body Fat Testing: While the participant was at rest, an investigator applied iWorx GSR leads to their index and middle finger and information was collected from the participant to ensure the settings of the devices were personalized. With shoes and socks removed, data were continuously recorded using the iWorx. The participant first was instructed to use the Tanita scale followed by the Withings Smart Body Analyzer. Each scale was used three times, each in alternating succession. With their shoes back on, the participant was asked to run for five minutes on a treadmill, beginning at a speed of 2 mph and increasing or decreasing at the

Table 5 Devices tested in preclinical trial

Devices to be evaluated	Data sets collected	Device usage category
Braun Forehead Thermometer	32/20	Thermometer
Braun Thermoscan Thermometer	32/20	Thermometer
Mabis Digital Thermometer	32/20	Thermometer
Omron 5 series	31.5/20	Blood Pressure Monitor
Omron 7 series	31.5/20	Blood Pressure Monitor
Omron HEM-18	30/20	Blood Pressure Monitor
KAM	34/20	Activity Monitor
Withings Wireless BP Monitor	23/20	Blood Pressure Monitor
Fitbit One	28/20	Activity Monitor
AliveCor	22/20	EKG/Heart Rate Monitor
Tanita BF 679 W	23/20	Bioelectrical Impedance Analysis
VitalConnect Healthpatch Biosensor*	9/9	Thermometer, Activity Monitor, Respiratory Rate Monitor
Fitbit Surge	29/20	Activity Monitor, Heart Rate Monitor
Withings Body Composition Monitor	22/20	Heart Rate Monitor, Bioelectrical Impedance Analysis
Tinke	22/20	Pulse Oximeter, Respiratory Rate Monitor
Finger Pulse OX SM-110	21/20	Pulse Oximeter
Withings Pulse O2	30/20	Activity Monitor
Withings Pulse O2	21/20	Pulse Oximeter
FeverSmart*	9/9	Thermometer
GeneActiv Original, Sleep, Action, Wireless	22/20	Activity Monitor
Jawbone UP3	20/20	Activity Monitor
Apple Watch	20/20	Activity Monitor, Heart Rate Monitor
Braun Forehead Thermometer	32/20	Thermometer

Table 6 Sensor types and references for biosignals of interest

Biosignal	Sensor Type 1	Sensor Type 2	Sensor Type 3	Sensor Type 4	Gold Standard Method/Reference
Temperature	Oral thermometer	Tympanic thermometer	Infrared thermometer	Auxiliary thermometer	Interdevice comparison
Activity	3-axis accelerometer	3-axis accelerometer, 3-axis gyroscope, & compass	N/A	N/A	Video movement observation
Blood Pressure	Inflatable arm cuff	Inflatable wrist cuff	Wireless inflatable arm cuff	N/A	Manual blood pressure
Respiratory Rate	3-axis accelerometer	LED	N/A	N/A	Research grade spirometers
EKG	Single-lead EKG	N/A	N/A	N/A	Research grade 3-lead EKG and self-comparison
GSR	Single-lead GSR	N/A	N/A	N/A	Research grade 2-lead GSR
SpO_2	Infrared LED	Spectro-photometer & reflective films	Research grade finger pulse oximeter	N/A	Research grade finger pulse oximeter

participants' preference. Immediately after exercise, with shoes and socks quickly removed, the same steps were repeated with continuous iWorx data collection and alternating scale measurements. If this protocol followed in succession from a different protocol, the participant rested for 10 min to ensure baseline reading accuracy. Time stamps were recorded for every measurement taken.

EKG Testing: While the participant was at rest, an investigator applied the iWorx 3-lead EKG system to the participant's wrists and right ankle. Due to possible noise, the participant was instructed to sit at about six feet away from iWorx module and desktop computer. Then, the AliveCor was given to the participant and the iWorx was started. The AliveCor collected a total of three readings, 30 s each, in succession. The AliveCor was used until three successful files were written. The time the readout occurred was recorded for later iWorx processing. When all three readings were completed, the participant was asked to run for five minutes on a treadmill, beginning at a speed of 2 mph and increasing or decreasing at the participant's preference. Immediately after exercise, the same steps were repeated with continuous iWorx data collection and AliveCor measurements. If this protocol followed in succession from a different protocol, the participant rested for 10 min to ensure baseline reading accuracy. Time stamps were recorded for every measurement taken.

Pulse Oximetry Testing: While the participant was at rest, the SM-110 was applied to their right index finger and the iWorx pulse oximeter module on their right middle finger. With the iWorx recording and SM-110 turned on, their left-hand alternated between the Tinke and the Withings Pulse O2. When either the Tinke or Withings Pulse O2 finished a reading, its value and the concurrent SM-110 value were recorded, meaning a total of 3 values for the Tinke and Withings Pulse O2, and a total of 6 values for the SM-110 were recorded. The participant was asked to run for five minutes on a treadmill, beginning at a speed of 2 mph and increasing or decreasing at the participant's preference. Immediately after exercise, the same steps were repeated with continuous iWorx data collection and device measurements. If this protocol followed in succession from a different protocol, the participant rested for 10 min to ensure baseline reading accuracy. Time stamps were recorded for every measurement taken.

Respiratory Rate Testing: The participant was asked to rest for ten minutes before testing began. This allowed time for the breathing rate to normalize and the iWorx spirometer to calibrate. The spirometer was set up on a ring stand for the participant's comfort and to reduce noise from movement. The participant was instructed to wear a nose clip during the data collection periods. After the 10-min rest and initialization period, the participant was asked to breathe entirely through the spirometer while the Tinke was used on their left hand three times in succession. The participant was asked to run for five minutes on a treadmill, beginning at a speed of 2 mph and increasing or decreasing at the participant's preference. Immediately after exercise, with the nose clip reapplied by the participant, the same steps were repeated with continuous iWorx data collection and Tinke measurements. If this protocol followed in succession from a different protocol, the participant rested for 10 min to ensure baseline reading accuracy. Time stamps were recorded for every measurement taken.

Temperature Testing: While the participant was at rest, the FeverSmart was applied to their underarm with sticky gauze. The Mabis Digital Oral thermometer was first placed in the participant's mouth while an investigator used the Braun Forehead thermometer on the participant. Then, the Braun Thermoscan was handed to the participant for them to insert in their ear. Each device was used three times in the order described previously. The participant was asked to run for five minutes on a treadmill, beginning at a speed of 2 mph and increasing or decreasing at the participant's preference. Immediately after exercise, the devices were used again in the same order. When all three values were recorded for each device, the data from the FeverSmart were collected. If this protocol followed in succession from a different protocol, the participant rested for 10 min to ensure baseline reading accuracy. Time stamps were recorded for every measurement taken.

VitalConnect Testing: For the first part of the VitalConnect Protocol, the VitalConnect patch was applied to their chest, just below the heart in the orientation indicated in the manual. The FeverSmart patch from Temperature Testing was also applied in the same way as before. While the participant was at rest, the iWorx spirometer was set up to calibrate and allowed the participant's breathing rate to normalize. The spirometer was set up on a ring stand for the participant's comfort and to reduce noise from movement. The participant was instructed to wear a nose clip during the data collection periods. After the 10-min rest and initialization period, the participant was asked to breathe entirely through the spirometer while the Tinke was used on their left hand three times in succession. The participant's temperature was also recorded using the Braun Thermoscan and the Braun Forehead thermometer in alternation. The participant was asked to run for five minutes on a treadmill, beginning at a speed of 2 mph and increasing or decreasing at the participant's preference. Immediately after exercise, with the nose clip reapplied by the participant, the same steps were repeated with continuous iWorx data collection, Tinke, and thermometer measurements. If this protocol followed in succession from a different protocol, the participant rested for 10 min to ensure baseline reading accuracy. Time stamps were recorded for every measurement taken.

Immediately following the aforementioned VitalConnect Testing, a second part was completed. During activity testing for the VitalConnect, the VitalConnect patch was still continuously worn, and the steps described previously for Activity Testing were followed.

Innovation Score: Device innovation was rated on a scale of 0 to 5 and represents an average of equally weighted scores that were given to the device that include ease of use, design aesthetics, invasiveness, wireless capabilities, operating constraints, and size. Invasiveness, wireless capabilities, operating constraints, and size had a defined range based on manufacturer metrics provided for each device. Ease of use and design aesthetics were the only categories that were subjective to the study participants, as they were the device operators. Participants were asked a series of questions to determine how intuitive the device was to use, ease of operation, clarity of formal instructions, clarity of the interface and menus, and ease of interpreting the data. Design aesthetics averaged the opinions of all participants who used the device with respect to their idea of the device's overall attractiveness, and how attractive the device is in its class. The specific questions the participants were asked included the following, each with its own 0 to 5 rating.

1. How intuitive is the device to use?
2. How easy is it to hold and successfully and safely operate it at the same time?
3. How clear are the formal instructions?
4. How clear is the device interface or menus?
5. How clearly are the results communicated?
6. How attractive is the device design in general?
7. How attractive is the device compared to other devices in its class?

Comfort Score: Like the innovation score, comfort was quantitatively evaluated by our participants. After a participant completed a test, they were given a form that includes the following scale. The participant was able to view the scale before testing so that they could think about each device in the following terms:

9. Unnoticeable
8. Occasionally Noticeable
7. Constantly Noticeable
6. Occasionally Annoying
5. Constantly Annoying
4. itchy Irritant
3. Concerning Pressure
2. Minor Pain
1. Major Pain

This score of 0 to 9 was then converted to a scale of 0 to 5 and weighted equally in the invasiveness category within the innovation score.

Quality Score: Device quality was rated on a scale of 0 to 5 and represents an average of scores that were given to the device that include data accessibility and performance. Data accessibility represents the average value assigned for accessibility to raw data, memory of results, and exportability of data. Performance was designated as a function of reliability and reusability, which was again assigned a score of 0 to 5.

Reliability scores were based on the p values for hypotheses tested, and the slope and the corresponding R^2 of the gold standard graphs (GSG). Reusability was defined as how quickly the device could repeat a measurement as well as the relative standard deviation (RSD) found in baseline testing where applicable.

Cost Score: The cost score for each device was updated based on market prices for the devices on 08/13/2015. The actual cost score relates the cost of the device overall as well as its relative cost in its class wherein the overall cost is weighted more heavily than the relative cost. In this way, the most important aspect is general cost, but the cost of similar devices is still considered.

3.1.3 Data analysis

Basal variance and relative standard deviation For all devices where replicate measurements were taken at rest (basal readings of at least $n = 3$), we determined population averages, standard deviation, and RSD. RSD is a comparison of the average and standard deviation that measures how precise the average is and measures random error.

Gold standard graphical comparisons Similar to the Yellow Springs Instrument for blood glucose, clinical GS methods exist for other physiological measurements. With respect to the 7 biosignals of interest to this OCT (Table 6), both GS devices and reference methods were used to evaluate commercial device performance. Manual blood pressure, core body thermometers, multi-lead EKG and galvanic skin response (GSR) systems, finger pulse oximeters as well as blood gas analyzers, and spirometers, are widely used as clinical GSs for assessment in the hospital. The reference used for quantifying physical activity was direct observation and monitoring. A simple test system, using versions of the clinical GS devices listed above, allowed for gold standard graphs (GSG) to be created for each device tested.

Only three devices were not able to be calibrated directly against a true GS (i.e. AliveCor ECG monitor, Tanita scale [using GSR for calculating body fat percentage (BF%) and water percentages], and Withings Smart Body Analyzer [using GSR for calculating BF% and water percentages].

In the case of the AliveCor, raw signal data were not available. The GS method, a three lead EKG system, was taken alongside the AliveCor measurements, and side-by-side images were evaluated for the clear detection and counting of QRS complexes (represented by tick marks) and associated waves. In order to evaluate the device's accuracy in the presence of noise or other interference, the total tick marks for the 30 s period are counted and recorded. The total number of QRS complexes is also counted within the same document. Any QRS complex that does not have a corresponding tick mark is recorded as well as any tick mark that does not have a corresponding QRS complex. These particular tick marks are recorded as *miscounted ticks* and this method was used as the reference for comparison of EKG.

For Tanita and Withings scales, the GSR recordings collected via the iWorx unit did not provide the expected data. The data collected from the iWorx was categorized as the skin impedance change in relation to emotional stress. We found that physical stress (as a result of exercise via the scales) and emotional stress (via iWorx) data are not directly comparable. As a result, we developed an alternative reference method to use with the Withings and Tanita body fat percentage readings. The participant's demographic information was used to predict a body fat percentage using information gained from their pre-testing survey and equations below, taken from "Body mass index as a measure of body fatness: age-and-sex specific prediction formulas" (Deurenberg et al., 1991):

$$BMI = \frac{weight(kg)}{height(cm)^2} \tag{1}$$

$$\begin{aligned}&(1.2^{*}BMI) + (0.23^{*}age) - (10.8^{*}gender) - 5.4\\&wherein, 1\ is\ coded\ for\ \text{``}male\text{''}\ and\ 0\ is\ coded\ for\ \text{``}female\text{''}\end{aligned} \tag{2}$$

GSGs were created using the references listed above, and error zones were created by estimating between 5% and 20% error at the lowest and highest possible physiological levels This standard can determine the device's relationship to its reference thereby elucidating the dynamic range, percent data off the calibration line, upper and lower limits of detection, accuracy, and precision of the device.

Testing for statistical significance In order to measure the significance of exercise, t-tests were conducted on all devices for their readings before and after exercise. We analyzed data from all devices to test the hypothesis that exercise will cause a significant change in blood pressure ($p < .05$ indicates statistical significance). A second set of t-tests were conducted to test the hypothesis that a specific device differs significantly from its reference method ($p < 0.05$ indicates statistical significance). The associated p values for each device with respect to each hypothesis and reference used can be seen in Table 7.

3.1.4 Results

Best-in-class Using the innovation, cost, and comfort scores as seen in Table 7, the best-in-class devices were identified in each category: Braun Thermoscan (temperature), Apple Watch (activity), Omron 5 arm meter (blood pressure), VitalConnect Healthpatch Biosensor (respiratory rate),

Table 7 Summary of statistical results and analysis for all devices

Name	Category	Innovation	Quality	Cost	Comfort	Device Name	Baseline v. Reference	Exercise v. Reference	Baseline v. Exercise	N
Braun Forehead †	Thermometer	3	2	4	8	Mabis Digital Temp	5.64E-04	0.19	1.71E-07	32
Braun Thermoscan †	Thermometer	2	2	3	8	Mabis Digital Temp	2.79E-05	3.46E-05	0.16	32
Mabis Digital †	Thermometer	2	1	5	6	Braun Thermoscan	2.79E-05	3.46E-05	0.06	32
FeverSmart †	Thermometer	4	3	1	8	Mabis Digital Temp	0.23	0.19	0.57	8
Vital Connect Healthpatch Biosensor †	Thermometer	4	4	0	8	Braun Thermoscan	1.39E-06	3.21E-06	0.45	8
Withings Wireless BP Monitor ‡	Blood Pressure Monitor	4	3	1	6	Omron HEM-18	0.33/0.69	6.16E-05/0.19	0.02/0.005	21
Omron 5 series	Blood Pressure Monitor	3	3	4	7	Omron HEM-18	0.01/0.62	0.31/0.13	0.0004/0.06	30
Omron 7 series ‡	Blood Pressure Monitor	3	3	3	7	Omron HEM-18	0.18/0.16	0.003/0.73	0.002/0.08	30
KAM ¶	Activity Monitor	2	3	4	9	Steps from Manual step count	–		–	38
Fitibit One	Activity Monitor	4	3	3	9	Steps from Manual step count	–	0.34	–	28
Withings Pulse O2 †	Activity Monitor	4	3	4	9	Steps from Manual step count	–	0.01	–	33
Fitbit Surge †	Activity Monitor	4	3	0	8	Steps from Manual step count	–	0.002	–	30
Vital Connect Healthpatch Biosensor	Activity Monitor	4	4	0	8	Steps from Manual step count	–	0.002	–	8
Alivecor	EKG Monitor	3	4	3	7	iWorx.	0.001	0.001	7.83E-06	21
Tanita BF 679 W †	BIA Monitor	2	3	4	8	Calculated Body Fat	0.002	0.003	0.04	22
Withings Smart Body Analyzer †	BIA Monitor	4	3	1	8	Calculated Body Fat	0.33	0.33	0.58	21
Tinke *	Pulse Oximeter	3	1	2	8	iWorx RR SpO2	2.28E-08	7.60E-07	0.88	20
Finger Pulse OX SM- 110 *	Pulse Oximeter	3	2	5	8	iWorx RR SpO2	6.03E-37	1.70E-40	0.78	20
Withings Pulse O2 *	Pulse Oximeter	4	3	4	8	iWorx RR SpO2	2.60E-05	1.36E-05	0.88	20
Vital Connect Healthpatch Biosensor ¶	Respiration Monitor	4	4	0	8	iWorx RR (BPM)	0.99	0.07	4.05E-05	8
Tinke	Respiration Monitor	3	2	2	8	iWorx RR (BPM)	0.001	5.99E-12	0.021	20
Jawbone UP3	Activity Monitor	4	3	1	8	Steps from Manual step count	–	0.001	–	25
Apple Watch	Activity Monitor	4	3	0	8	Steps from Manual step count	–	0.27	–	25

Device labels		Gold standard graphs						Error		
Name	Category	All Data Slope	All Data R2	N	Device Error Slope	Device Error R2	N	Total Number of Successful Collections	Number of device failures or user errors (data collected)	Number of device or user failures (data not collected)
Braun Forehead †	Thermometer	0.10	0.05	32	0.10	0.02	32	191	0	1
Braun Thermoscan †	Thermometer	0.09	0.04	32	0.24	0.12	32	192	0	0
Mabis Digital †	Thermometer	0.83	0.12	32	0.48	0.12	32	192	19	2
FeverSmart †	Thermometer	0.58	0.08	8	0.58	0.08	8	48	30	4
Vital Connect Healthpatch Biosensor †	Thermometer	0.06	0.02	8	0.06	0.02	8	138	0	0
Withings Wireless BP Monitor ‡	Blood Pressure Monitor	0.63	0.75	21	0.68	0.78	19	42	0	5
Omron 5 series	Blood Pressure Monitor	0.82	0.76	30	0.87	0.77	28	61	0	2
Omron 7 series ‡	Blood Pressure Monitor	0.64	0.65	30	0.68	0.67	28	63	0	5
KAM ¶	Activity Monitor	0.04	0.35	38	0.04	0.35	38	38	0	2
Fitibit One	Activity Monitor	2.51	0.13	30	2.51	0.13	28	30	2	1
Withings Pulse O2 †	Activity Monitor	0.78	0.01	33	0.78	0.01	33	33	1	0
Fitbit Surge †	Activity Monitor	1.4	0.61	30	−0.18	0.01	30	30	0	2
	Activity Monitor	0.78	0.01	8	0.78	0.01	8	8	0	0

Table 7 (continued)

Vital Connect Healthpatch Biosensor										
Alivecor	EKG Monitor	0.94	0.90	21	0.94	0.90	21	126	8	3
Tanita BF 679 W †	BIA Monitor	1.06	0.31	22	1.08	0.70	22	132	1	1
Withings Smart Body Analyzer †	BIA Monitor	0.98	0.52	21	0.91	0.35	21	114	1	5
Tinke*	Pulse Oximeter	−0.09	0.002	21	−0.12	0.003	20	123	1	6
Finger Pulse OX SM-110 *	Pulse Oximeter	0.42	0.44	21	0.42	0.44	20	246	1	0
Withings Pulse O2 *	Pulse Oximeter	−0.09	0.004	21	−0.11	0.01	20	110	1	7
Vital Connect Healthpatch Biosensor ¶	Respiration Monitor	0.87	0.54	8	0.87	0.54	8	8	0	0
Tinke	Respiration Monitor	0.12	0.14	21	0.12	0.16	20	123	1	3
Jawbone UP3	Activity Monitor	0.16	0.01	25	0.34	0.10	25	25	0	1
Apple Watch	Activity Monitor	1.37	0.66	25	1.19	0.73	25	25	0	0

* Quantization in data caused speculation in reliability of slope and R^2 values for this device

† Best fit line given by data shown to be highly speculative

‡ Unknown possible fitting issues seen in gold standard graphs

¶ Major outliers caused fitting issues with best fit line

AliveCor (heart rate/ECG), Tanita BF 679 W (GSR), and SM-110 (pulse oximetry).

3.1.5 Device failures and usability issues

Activity devices The most common failure associated with activity devices was that the devices were not charged, or the devices were not activated by the investigator for collection, which largely falls under user error. With respect to actual device failures, there were three possible errors related to dramatic step undercounting, though it was not clear in some instances if it was user error or actual device malfunction.

Blood pressure devices The BPMs accrued the most failures of all the device types, most often due to user error or user body habitus (i.e. cuff to small for user's arm). The most significant issue that was observed during blood pressure testing was impossibly low or abnormal manual blood pressure readings. In these instances, all of the other BPMs recorded significantly higher blood pressure for the individual participant, and likely represents user error with the manual blood pressure measurement. This observation highlights a challenge in providing gold standard comparison testing for a physiological measurement in a lab where staff may not be trained healthcare providers.

GSR/Total body fat devices GSR devices had errors related to incorrect participant settings resulting in invalid findings. These errors were device-related in failure to connect to the internet, as well as user-related in failure to manually input correct settings.

EKG devices Failure associated with the AliveCor device mainly stemmed from user error, as it was not always easy for the participant to position their hands in the proper orientation and apply the correct amount of pressure to the electrodes. We also observed noise from external conditions that interfered with the high frequency waves that the AliveCor used to transmit the EKG signal.

Pulse oximetry devices We observed numerous device failures with the Tinke, with multiple software crashes and failure to connect to the mobile device. We observed user failures with the Withings Pulse O2, with a participant squeezing the device too firmly.

Respiratory rate devices As the Tinke was used for pulse oximetry testing as well as respiratory rate testing, the software issues associated with this device are listed previously.

Temperature devices We experienced numerous failures with the FeverSmart device, notably related to early battery depletion. Even though the device would go through proper power down procedures, it would exhaust between 50% and 60% of its battery per test. Another failure with the FeverSmart was due to the sweat buildup on the participant. The adhesive gauze lost its ability to stick to the participant and would commonly lift or fall off completely. These were device failures rather than user error.

The Mabis Digital Oral thermometer demonstrated user error when the participant would open their mouth and talk or not close their moth completely while using the oral thermometer, resulting in impossible hypothermic values. We also observed device error with the Mabis device from an auto-shutoff during testing.

3.1.6 Study summary

Thirty-eight subjects were tested providing a total of 576 data sets that represent 7 different device categories and 22 unique devices. This study provided many opportunities for engineering student investigators to increase their experience in their field, but also for professional development. By working with new technologies, the investigators gain valuable hands-on experience with and exposure to successful wearable devices that can be translated into robust medical devices. Additionally, by working with human subjects the investigators are able to become comfortable when working with a study participant, to understand the need for professional behavior, and to better comprehend the needs of the participant as well as observing for user error. Student investigators also sharpened their critical thinking and problem-solving skills through their thorough analysis of the study data.

This study led to the development of an integrated wireless on-body sensor currently being investigated by the La Belle Labs. Results of this study with best-in-class identifications also provided guidance to other studies related to Project Honey Bee in terms of wearable device selection.

3.2 Heart failure monitoring study

3.2.1 Goals

Building on the notion of lab-based wearable device validation studies, we sought to perform field-based validation of consumer wearable devices using implanted FDA-cleared medical devices (i.e. pacemakers, implantable defibrillators) as the gold standard comparator. We focused on congestive heart failure, using changing cardiopulmonary measures as an indicator for worsening heart failure status. Researchers identified the Wello health tracking device (Azoi, Inc.) as a target ambulatory health tracker with capacity to track heart rate, blood pressure, and lung capacity. Despite broadly advertised pre-orders and product launch promises, by December 2014 the Wello device failed to ship for consumer purchase. Researchers then turned to a true wearable device that incorporated resting heart rate measurements using a bioimpedance sensor, the Jawbone Up3 (Jawbone, Inc.). After a several months' long delay in delivery, researchers were able to obtain Jawbone Up3 devices with bioimpedance-based resting heart rate measures for the clinical validation study.

3.2.2 Methods

Before recruiting study participants, researchers wore the Jawbone Up3 activity monitors for several weeks to identify challenges with the hardware, web interface, and smartphone app in order to better manage the research participant experience. Researchers identified several issues, including hardware malfunction that caused the device to smoke on charging, challenges with the bracelet clasp design, and battery life shorter than advertised. Researchers used these experiences to design an instruction guide for study participants including instructions for band placement and a prescribed charging schedule.

In addition to hardware and software function, we were concerned about protecting subject privacy in the context of using commercially-available wearable devices that may not employ privacy protections consistent with United States Health Insurance Portability and Accountability Act (HIPAA) laws. We trained patients in the use of the Jawbone Up3 smartphone app and desktop interface, using password protection and data sharing capabilities as available from the manufacturer. Language on the participant consent document included, "The researchers will review your Jawbone Up3 data daily, and will report any abnormal findings to your healthcare provider," and "Efforts will be made to limit the use and disclosure of your personal information, including research study and medical records, to people who have a need to review this information. We cannot promise complete secrecy. Organizations that may inspect and copy your information include the IRB and other representatives of this organization."

Researchers recruited 11 subjects over a 10-month period January 2016 to December 2016. Subjects included patients with an established diagnosis of heart failure who had an implanted cardiac monitor as the gold-standard comparator to validate wearable device findings related to resting heart rate and heart failure status. Subjects uploaded study data to the Jawbone website and shared device data with researchers. Researchers monitored subjects' data to ensure that they were wearing the monitor on a daily basis and particularly at night in order to capture resting heart rate data.

3.2.3 Data analysis

Subject data from the Jawbone were compared to data from the implanted cardiac monitor to assess correlation in resting heart rate and indicators predicting worsening heart failure. Eight subjects completed the 4-month monitoring period. No subjects reported adverse outcomes. We sought to compare tabular Jawbone data to raw data from implanted devices. However, our clinician partners were unable to obtain raw data from implanted device manufacturers in all but one case. Therefore, we were limited to performing visual correlation of graphic data available from implanted devices to the tabular data from the Jawbone.

In addition to collecting device data, we measured quality of life using validated psychometric scales (Minnesota Living with Heart Failure [MLHF], University of Minnesota; SF-36, RAND Corporation).

3.2.4 Results

Two subjects had extremely poor data correlation with the wearable device failing to generate actionable signals to indicate impending worsening of heart failure status. Six subjects had adequate data capture from the wearable device with reasonable correlation to implantable monitors. However, the majority of the measures clustered around 60 beats per minute, correlating to the programmed lower pacing rate for most study participants.

Six of 8 subjects had clinically significant score improvement on MLHF (>5-point increase). Paired t-test revealed with statistically significant improvement in physical subscore ($p = .03$) and nonsignificant improvement on emotional subscore and overall score. SF-36 demonstrated statistically significant improvement in emotional wellbeing subscore ($p = .03$) with nonsignificant changes in other subscores including increases in physical functional, role physical, and pain subscores, as well as general health score. Despite significant improvements in emotional wellbeing subscores, we noted decline in role emotional, energy, and social subscores (Tables 8 and 9).

3.2.5 Study summary

Conclusions from this pilot study suggested that patients with chronic heart failure can use a wearable monitor without adverse effects, though the utility of the monitor for precise tracking of resting heart rate to predict heart failure status is limited. However, there may be some improvement in quality of life for patients using wearable monitors to measure chronic health status. The validation component of the study failed to demonstrate that the Jawbone Up3 consumer-grade wearable device could be a reliable resting heart rate monitor for patients with heart failure. However, the findings of this study were limited by the fact that most of the study participants had a basement heart rate of 60 beats per minute with chronic pacing, limiting useful resting heart rate data for many participants.

Additional observations related to study outcomes noted challenges related to the lack of funding directed to the clinical site. Without funding incentives for the clinical site, clinical staff were modestly helpful with subject recruitment but otherwise minimally involved with the study, including no engagement in device management and data acquisition. Device-related responsibilities fell to the graduate research assistant (employed at 20 h per week) and highlights the challenge with integrating new technology into clinical practice in terms of workforce sustainability.

Regarding data privacy, we noted that no study participants expressed concerns about privacy protections at the time of recruitment, study consent, or at any point during the study period. We noted at around the same time we were conducting this study in 2014–2015, the Connected & Open Research Ethics (CORE) Project (https://thecore.ucsd.edu) was taking shape to address ethical research with sensor-based devices.

Table 8 Minnesota living with heart failure results

	Mean difference pre- to post-test (range)	p value (* = significant)
Total score	−10.625 (−46 to +14)	.07
Physical subscore	−6.875 (−23 to +4)	.03*
Emotional subscore	−2.25 (−9 to +4)	.12

Table 9 SF-36 results

	Mean difference pre- to post-test (range)	p value
General Health	.625 (−25 to +30)	.46
Physical Functional	8.75 (−20 to +35)	.09
Role Physical	5.21 (−25 to +67)	.32
Role Emotional	−16.67 (−67 to 0)	.10
Energy	−1.88 (−20 to +20)	.35
Emotional Wellbeing	7.17 (−8 to +24)	.03*
Social	−9.06 (−48 to +13)	.14
Pain	9.69 (−30 to +55)	.16

4 Discussion

The Project Honeybee observational clinical trial experience yielded several successes and highlighted some significant challenges. Some issues were specific to our experience, while others may be extrapolatable to other project settings.

4.1 Successes

4.1.1 Nursing students

The ASU doctoral nursing students were an inspiration for us all. They made things work under the most trying of circumstances. They are highly dedicated, entrepreneurial and problem solvers. They universally reported the experience they had with Project Honeybee was one of their most important learning opportunities of their careers. They explained that while the trials were often compromised through low patient numbers, faulty devices, poor patient compliance, etc. the experience gained in overcoming unanticipated challenges would allow them to do larger and more successful studies with wearable devices in the future.

4.1.2 Publications and presentations

As of this writing, investigators had 9 manuscripts submitted or accepted to peer-reviewed journals, and 10 conference posters or papers delivered.

4.1.3 Faculty and clinician engagement

One of our overarching goals for the Project Honeybee experience was to engage ASU research faculty with area clinicians to establish or build on existing relationships for clinical-based studies. Over the four years of Project Honeybee funded studies, 12 ASU faculty and 39 area clinicians participated in one or more projects.

4.1.4 Clinical institutional engagement

Two of the eight design aspirations of Arizona State University's charter as a New American University are to conduct use-inspired research with purpose and impact and be socially embedded by connecting with communities through mutually beneficial partnerships. In line with these principles, Project Honeybee studies engaged participation with nine external healthcare institutions, including private practice groups, large tertiary medical centers, and a federally qualified healthcare center.

4.2 Challenges

4.2.1 Devices

Although hundreds of commercial ambulatory and wearable devices are on the market, device performance was the major limitation in nearly all trials. First, they are nearly all meant for the health enthusiast consumer market and have not been validated for the intended application or patient population. Device validation is the most important deficiency in the field and any team intending to obtain reliable data must go through a validation process with their own patient population. Second, algorithms for analyzing data are proprietary and hence it is usually impossible to reanalyze the data for key signals. Third, device memory is often a limitation and many devices cannot store the amount of data needed. Fourth, data downloads for many devices require that the device be brought in by the patient. Fifth, many devices advertised are not yet available and plans built on their availability are often frustrated. Sixth, many devices are not comfortable to wear, and patient compliance can be compromised.

4.2.2 IRB approval

ASU's Institutional Review Board does not provide oversight for clinical trials, even if only observational. Therefore, IRB approval was typically obtained through the clinical site of the trial with ASU's IRB providing secondary oversight for ASU-based investigators. Although the ASU IRB has developed relationships with other area IRBs, this process proved quite cumbersome for investigators. Typically, teams underestimated the time required for IRB approval and this often delayed the onset of trials by several months. We began withholding awarded funds until IRB approval was obtained.

4.2.3 Institutional affiliations

Permission for ASU doctoral nursing students to enroll and monitor patients at clinical sites required a formal approval process at the clinical site. Some sites, especially Mayo Clinic, were slow to provide authorization. This factor delayed the onset of some trials.

4.2.4 Patient enrollment

Teams frequently overestimated the availability of patients at a clinical site for the trial. Not all patients are willing to participate, satisfy exclusion criteria and are compliant with study needs. Moreover, on-site recruitment support from clinicians and staff, particularly at non-research institutions, was often lacking. In future trials a comprehensive assessment of patient numbers and on-site subject recruitment support is essential before a trial is launched.

4.2.5 Leadership

Clinicians are very busy and were not reimbursed in our study. Nursing students frequently found it difficult to get guidance from the clinician. In addition, many clinicians were unwilling to devote scarce time to talk with patients about study participation. Coupled with an ASU IRB prohibition for study staff to conduct chart review for potential subject screening, this limitation proved a significant barrier to study subject recruitment. What seemed to work best was to establish a brief weekly meeting between student and clinician. In addition, it would undoubtedly help to reimburse clinicians for their time by incorporating a per-subject enrollment fee as is common in clinical drug and device trials.

5 Conclusion

The Project Honeybee studies have identified many challenges to the deployment of consumer wearable devices in clinical medicine. The field is still immature, and progress will depend significantly on improvements in devices, their validation, wearability, data storage and access to raw data as well as in their ability to monitor several physiological parameters at the same time.

In addition, our experience highlighted the need for adequate time and funding in order to successfully complete a clinical project. Although doctoral nursing students provided outstanding research support interfacing with clinical investigators, clinical support staff, and patient subjects, the 12-month timeframe often posed a challenge for recruitment and longitudinal follow-up even for a small-sized study population of 15–30 subjects. Similarly, the funding limitation that excluded covering tuition costs for graduate research assistants, clinical site incentives, or subject incentives may have contributed to the challenges this project experienced.

Looking ahead, we hope that our early experiences with wearable research may inform continued work to understand the applications, potential, and practice implications of incorporating wearable devices into clinical care.

Acknowledgments The authors and all participants wish to acknowledge the generous support from the Strategic Investment Fund, Arizona State University.

Compliance with ethical standards

Ethical approval All procedures performed in studies involving human participants were in accordance with the ethical standards of the institutional and/or national research committee and with the 1964 Helsinki declaration and its later amendments or comparable ethical standards.

Reference

Deurenberg P, Weststrate JA, Seidell JC. Body mass index as a measure of body fatness: age- and sex-specific prediction formulas. Br. J. Nutr. **65**(2):105–114 (1991)

Biomedical Microdevices (2019) 21: 38
https://doi.org/10.1007/s10544-019-0393-x

Functionalization of endovascular devices with superparamagnetic iron oxide nanoparticles for interventional cardiovascular magnetic resonance imaging

Elvin Blanco[1,2] · **Victor Segura-Ibarra**[1,3] · **Danish Bawa**[2] · **Md Nafiujjaman**[1] · **Suhong Wu**[1] · **Haoran Liu**[1] · **Mauro Ferrari**[1,4] · **Alan B. Lumsden**[5] · **Dipan J. Shah**[2] · **C. Huie Lin**[2]

Published online: 1 April 2019

Author's Personal Note I first joined Mauro Ferrari's team in Houston, Texas 10 years ago, a few days shy of his 50th birthday. I had first met him several months prior. At our first encounter, as we were walking outside, he signaled towards the Texas Medical Center in the background and pointed out its long-standing claim as the largest medical center anywhere in the world. It was fitting that I should find Mauro in this setting; a larger-than-life man on a quest to make a big difference in people's lives by solving humanity's toughest medical challenges. When I was formally offered the chance to join his team, I accepted. Since then, I've been fortunate to join Mauro on several of his adventures, from those in Houston to those in the south of Italy. Always very gracious and generous, I've seen the tremendous impact Mauro has had on the careers of several of his protégés, many of whom have contributed their work to this book. It has been inspiring to hear him talk of his frustration with the time it takes to translate medical breakthroughs to the clinic, and his passion for wanting to make a difference in patients' lives. Our research article tries its best to celebrate micro-nanotechnologies toward translation, which is not only the focal theme of this book, but a vision and fundamental driver of Mauro's career. On behalf of all of the authors of our small research contribution, I congratulate Mauro on the occasion of his 60th birthday.

Elvin Blanco

Abstract

Presently, cardiovascular interventions such as stent deployment and balloon angioplasty are performed under x-ray guidance. However, x-ray fluoroscopy has poor soft tissue contrast and is limited by imaging in a single plane, resulting in imprecise navigation of endovascular instruments. Moreover, x-ray fluoroscopy exposes patients to ionizing radiation and iodinated contrast agents. Magnetic resonance imaging (MRI) is a safe and enabling modality for cardiovascular interventions. Interventional cardiovascular MR (iCMR) is a promising approach that is in stark contrast with x-ray fluoroscopy, offering high-resolution anatomic and physiologic information and imaging in multiple planes for enhanced navigational accuracy of catheter-based devices, all in an environment free of radiation and its deleterious effects. While iCMR has immense potential, its translation into the clinical arena is hindered by the limited availability of MRI-visible catheters, wire guides, angioplasty balloons, and stents. Herein, we aimed to create application-specific, devices suitable for iCMR, and demonstrate the potential of iCMR by performing cardiovascular catheterization procedures using these devices. Tools, including catheters, wire guides, stents, and angioplasty balloons, for endovascular interventions were functionalized with a polymer coating consisting of poly(lactide-co-glycolide) (PLGA) and superparamagnetic iron oxide (SPIO) nanoparticles, followed by endovascular deployment in the pig. Findings from this study highlight the ability to image and properly navigate SPIO-functionalized devices, enabling interventions such as successful stent deployment under MRI guidance. This study demonstrates proof-of-concept for rapid prototyping of iCMR-specific endovascular interventional devices that can take advantage of the capabilities of iCMR.

Keywords Magnetic resonance imaging (MRI) · Interventional cardiovascular magnetic resonance (iCMR) · Superparamagnetic iron oxide (SPIO) nanoparticles · Catheters · Polymer coatings

Elvin Blanco and C. Huie Lin share senior authorship.

Electronic supplementary material The online version of this article (https://doi.org/10.1007/s10544-019-0393-x) contains supplementary material, which is available to authorized users.

✉ Elvin Blanco
eblanco@HoustonMethodist.org

✉ C. Huie Lin
clin@HoustonMethodist.org

1 Department of Nanomedicine, Houston Methodist Research Institute, 6670 Bertner Ave., Houston, TX 77030, USA

2 Department of Cardiology, Houston Methodist DeBakey Heart and Vascular Center, Houston Methodist Hospital, 6550 Fannin St., Suite 1901, Houston, TX 77030, USA

3 Escuela de Ingeniería y Ciencias, Tecnológico de Monterrey, 64710 Monterrey, NL, Mexico

4 Department of Medicine, Weill Cornell Medical College, New York, NY 10065, USA

5 Department of Cardiovascular Surgery, Houston Methodist DeBakey Heart and Vascular Center, Houston Methodist Hospital, Houston, TX 77030, USA

1 Introduction

Cardiovascular disease remains the leading cause of hospitalization and death in the US (Benjamin et al. 2018). With over one million procedures performed annually in the US (Slicker et al. 2016), cardiac catheterization remains a vital treatment modality. X-ray fluoroscopy-based cardiovascular catheterization procedures have enabled interventions that reduce the need for open surgery, including coronary angioplasty with stenting, aortic valve replacement, and atherectomy, the latter involving removal of atherosclerotic tissue from vessels (Bhat et al. 2017). However, x-ray fluoroscopy is unable to properly delineate soft tissue, relying on pre-procedural x-ray angiography, computed tomography (CT), or magnetic resonance imaging (MRI) for anatomical information (Ma et al. 2013). Poor soft tissue contrast and imaging in a single plane limits accurate navigation and positioning of endovascular tools such as catheters. Indeed, 2D fluoroscopy integrates 3D structures into a single plane, creating challenges when there are three degrees of freedom of motion in aligning an endovascular device (depicted as a screwdriver in Fig. 1) with a cardiovascular structure (depicted as a screw in Fig. 1). This loss of 3D information causes continuous misregistration errors that either prolong the procedure time or result in complications. Importantly, x-ray fluoroscopy exposes patients to ionizing radiation (Berrington de Gonzalez and Darby 2004) and iodinated contrast agents (Pucelikova et al. 2008), which can prove particularly detrimental to children born with congenital heart disease and pregnant women (Chambers et al. 2011), as well as patients with renal insufficiency. Taken together, novel approaches for imaging of cardiovascular procedures are needed to improve catheter steering and intervention, as well as patient and operator safety.

MRI-guided catheterization has tremendous transformative potential in the realm of cardiac interventional procedures by bringing together both form and function through a combination of high-resolution anatomical imaging and physiologic flow data. Cardiac MR (CMR) has superb soft tissue contrast and has the ability to obtain three-dimensional structural information and continuous imaging across multiple planes (Mazal et al. 2016). Simultaneous CMR allows the empiric measurement of blood flow through cardiovascular structures without the need for mathematical assumptions required in standard cardiac catheterization, providing superior physiological information. Consequently, interventional CMR (iCMR) enables more accurate navigation and positioning of devices, due largely to the advantage of imaging in up to three planes simultaneously. Importantly, MR imaging can highlight cardiovascular structures without the use of ionizing radiation (Bock et al. 2005), potentially reducing the risk for malignancies in patients, particularly children. Lastly, iCMR can promote the development of novel, minimally-invasive procedures that extend beyond vascular lumens, including extra-anatomic bypass, structural heart interventions, and local delivery of drugs, genes, or cells (Saikus and Lederman 2009).

While iCMR stands to bring about a true paradigmatic shift in cardiac catheterization, the modality has yet to gain widespread clinical use. A challenge exists in the form of integrating an iCMR environment in the clinical setting, including the adjacent proximity of an MRI laboratory and a fluoroscopy suite (Mazal et al. 2016). However, a significant hurdle that limits the commonplace clinical applicability of iCMR is the lack of endovascular devices suitable for CMR (Lotz 2013). Catheters currently used clinically have ferrous materials that impart the mechanical properties necessary for proper navigation and steering of the device (Ratnayaka et al. 2008). Unfortunately, these create large magnetic susceptibility artifacts due to distortion of the main static magnetic field (B_0 field) and linear nature of the frequency encoding gradient proximal to ferromagnetic objects (Stadler et al. 2007), which in turn impede proper imaging of cardiovascular structures. Of note, catheters containing materials such as nylon and polyester have been shown to possess similar mechanical properties to conventional devices containing ferrous materials (Clogenson and Dobbelsteen 2016). Research efforts have been devoted to the development of both passive and active catheters for use in iCMR. Passive catheters relying on incorporation of MRI-visible coatings or rings in the device for visualization have been explored over the last few decades (Rubin et al. 1990; Bakker et al. 1996; Omary et al. 2000; Zhang et al. 2010). Other notable examples consist of balloon-tipped devices that employ CO_2 (Miquel et al. 2004) or a T2-shortening agent (Krueger et al. 2006) to fill the balloon. Active catheters have micro-coils and circuitry that generate an MRI-visible echo upon excitation (Saeed et al. 2012), including devices containing inductively coupled radiofrequency (RF) coils (Bock et al. 2004). Balloon-tipped catheters, however, can only be used for flow-directed movement, with minimal specificity, and cannot be used for interventions. Similarly, active devices require continuous connection to an RF source, precluding this strategy for visualization of implantable devices.

Herein, our objective was to functionalize clinically-used interventional tools, including catheters, wire guides, stents, and angioplasty balloons, with a polymer coating consisting of poly(lactide-co-glycolide) (PLGA) and superparamagnetic iron oxide (SPIO) nanoparticles (NPs), in an attempt to make these devices suitable for iCMR. Following examination of commercially-available catheter-based instruments in the iCMR setting in pigs, SPIO-functionalized devices were examined and used to perform cardiovascular interventions.

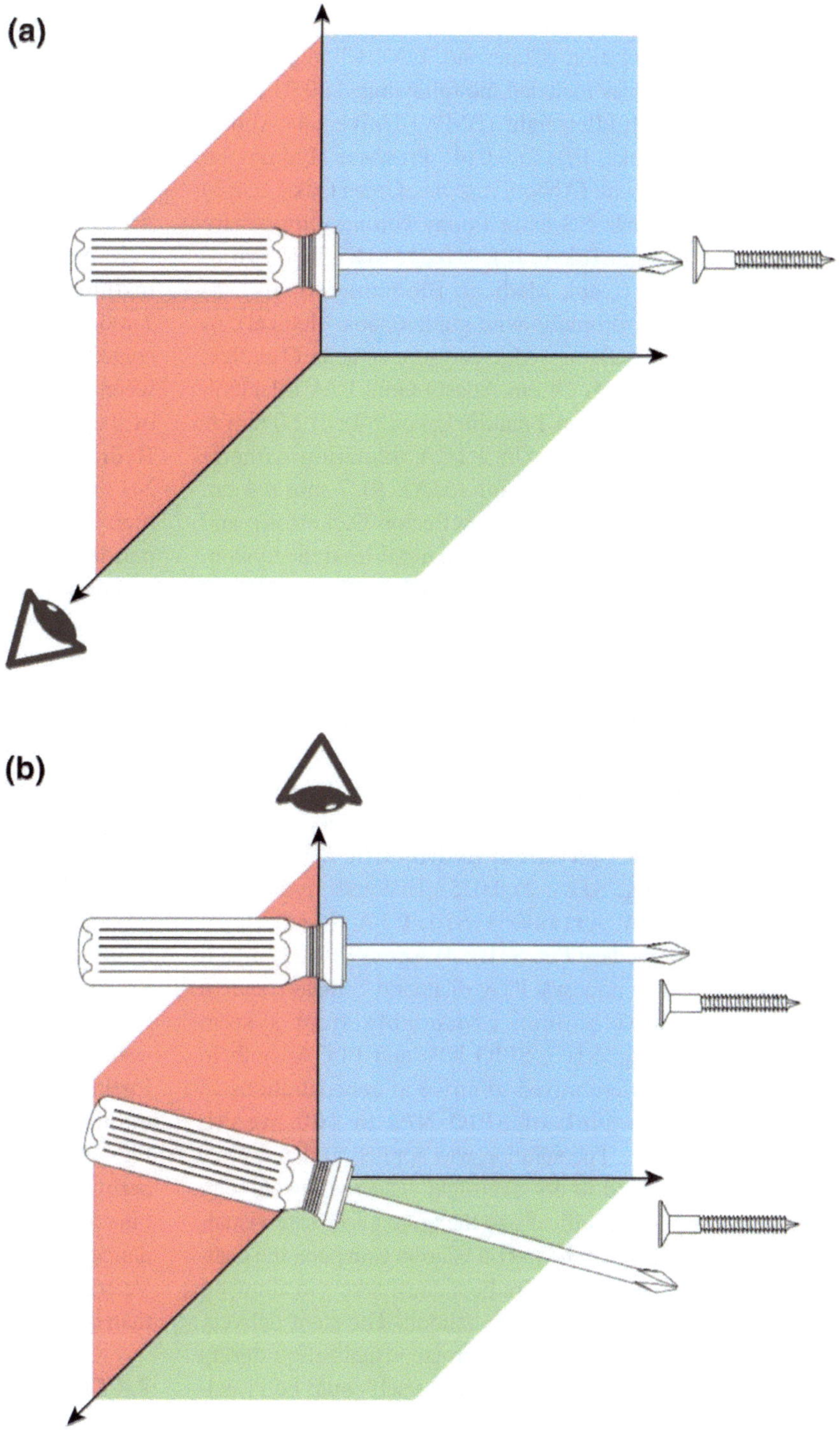

Fig. 1 Limitations of 2-D fluoroscopy for cardiovascular catheterization procedures. Due to the flattening effect of x-ray fluoroscopy, a screwdriver (endovascular device) appears aligned with a screw (cardiovascular structure) from one angle (**a**). However, an orthogonal view reveals that the screwdriver may be out of plane with the screw, or rotated off-axis from the screw (**b**). These imaging errors drive increased x-ray dose, contrast use, intervention times, and/or may require additional imaging modalities such as ultrasound or the use of preoperative computed tomography (CT) or MRI

Findings highlight the ability to properly visualize and navigate a series of SPIO functionalized endovascular instruments in cardiovascular structures, as well as successful placement and imaging of stents, with the capability to image at all stages of deployment. This study demonstrates the advantages of iCMR for endovascular procedures and the need for catheter-based tools with MRI-visible design considerations.

2 Materials and methods

2.1 Materials

Superparamagnetic iron oxide (SPIO) nanoparticles (5 nm, Lot No. 051012) were purchased from Ocean Nanotech (San Diego, CA). Ester-terminated poly(lactide-co-glycolide)

(PLGA, 50:50, 0.55–0.75 inherent viscosity) was obtained from Durect Corporation (Cupertino, CA). Clinically-used catheterization devices included the following: 1) 0.014" Hi-Torque Balance Middleweight (BMW) Universal (Abbott Vascular, Green Oaks, IL); 2) 0.014" Prowater Hydrophilic Asahi PTCA guidewire (Abbott Vascular, Green Oaks, IL); 3) 0.014" Runthrough® NS extra floppy coronary guidewire (Terumo Runthrough, Tokyo, JP); 4) 0.018" HiWire® hydrophilic wire guide (Cook Medical, Bloomington, IN); 5) 0.035" HiWire® hydrophilic wire guide (Cook Medical); 6) 7-French Arrow® balloon wedge-pressure catheter (Teleflex, Wayne, PA); 7) 14 mm × 4 cm Atlas® Gold PTA dilatation catheter (Becton Dickinson, Franklin Lakes, NJ); 8) 2.0 mm × 12 mm NC Quantum Apex™ PTCA dilatation catheter (Boston Scientific, Marlborough, MA); 9) 7 mm × 4 cm Conquest® PTA dilatation catheter (Becton Dickinson); and 10) Assurant Cobalt® balloon expandable stent system (Medtronic, Minneapolis, MN). GORE® PRECLUDE® Pericardial Membrane was obtained from Gore Medical (Flagstaff, AZ).

2.2 Polymer coating for MRI visualization

Clinically-used catheterization devices were coated with an SPIO-containing polymer solution. These included the following: 1) 0.018" HiWire® hydrophilic wire guide (Supplementary Fig. S1A); 2) 0.035" HiWire® hydrophilic wire guide; 3) Atlas® Gold PTA dilatation catheter (Supplementary Fig. S1B); 4) Apex™ PTCA dilatation catheter; 5) Conquest® PTA dilatation catheter; and 6) Assurant Cobalt® balloon expandable stent system (Supplementary Fig. S1C). SPIO NPs and PLGA, both in dichloromethane, were mixed to arrive at concentrations of 2.5, 5, and 10 mg/mL of SPIO NPs in 100 mg/mL of polymer solution. The solution was applied to the devices by brush and allowed to dry overnight. The coating on wire guides was applied specifically to the tip to allow safe visualization and navigation, whereas on balloon catheters, the coating was applied to the areas of radiopaque markers to allow for visualization of the length of the balloon. The stent delivery system was coated at the tip to optimize visualization during movement and positioning, as the stent itself could be visualized under iCMR when not in motion.

There are currently no endovascular stent grafts suitable for iCMR due to the presence of ferromagnetic metals in the delivery systems. Thus, we fabricated an endovascular stent graft system suitable for MRI use (Supplementary Fig. S2). An SPIO polymer coating was applied to a GORE® PRECLUDE® Pericardial Membrane consisting of expanded polytetrafluoroethylene (ePTFE). The Assurant Cobalt® stent was deployed and the SPIO-coated ePTFE membrane was then sutured to the exterior of the stent with 4–0 Prolene (Ethicon Inc., Somerville, NJ). Additional reinforcement was achieved using cyanoacrylate. The covered stent was then crimped on an SPIO-functionalized 14 mm × 4 cm Atlas® Gold PTA dilatation catheter.

2.3 Animal model

All animal studies were approved by the Institutional Animal Care and Use Committee of the Houston Methodist Research Institute. Male and female Domestic Yorkshire pigs (3–7 months, 35–50 kg) were used for all in vivo studies. All coated and non-coated wire guides, catheters, and balloons were each examined in two pigs ($N = 2$), with the exception of the 0.014" Hi-Torque BMW Universal, 0.014" Prowater Hydrophilic Asahi PTCA guidewire, 0.014" Runthrough® NS extra floppy coronary guidewire, 7-French Arrow® balloon wedge-pressure catheter, Assurant Cobalt® balloon expandable stent system, and the aforementioned custom endovascular stent graft system, which were each examined in one pig ($N = 1$). In each experiment, the animal was sedated with intramuscular injections of ketamine, acepromazine, and atropine. An endotracheal tube was inserted into the trachea and anesthesia maintained by administration of 1–5% isoflurane. An intravenous catheter was inserted into a marginal ear vein for fluid maintenance, and an intravenous (IV) saline solution started and continued throughout the procedure. The animal was transported to a hybrid OR/CMR suite, consisting of an adjoining catheterization lab and MRI suite present in the Houston Methodist Institute for Technology, Innovation, & Education (MITIE[SM], Supplementary Fig. S3), and placed in a supine position. Surgical areas were sterilized with alcohol and chlorhexidine. Vital signs were continuously monitored by the Houston Methodist Research Institute Comparative Medicine Program (CMP) vet staff throughout the procedure and documented every 15 min. Pigs were humanely euthanized upon completion of the experiment with an IV injection of phenytoin/pentobarbital combination in conjunction with an isoflorane overdose, in accordance with acceptable practices as outlined by American Veterinary Medical Association (AVMA) guidelines for euthanasia of animals.

2.4 Catheterization and MR image acquisition

Percutaneous access was performed under fluoroscopy using the Seldinger technique (Higgs et al. 2005), with introducer sheaths placed in the left and right femoral arteries and veins. Heparin (150 units/kg) was administered IV as an anticoagulant, with subsequent hourly bolus IV injections (2000–3000 units) to maintain the anticoagulation status of the animal. Animals were then transferred to the MRI suite.

Image acquisition and interventions were performed using a 1.5 T Siemens Magnetom Aera MRI Scanner (Siemens Healthineers, Erlangen, DE). Images were obtained in a series of ECG gated 2D images in the axial, coronal, and saggital planes using a balanced steady state free precession sequence (SSFP) with typical parameters of: 1) spatial resolution 1.5–2.0 cm in plane; 2) slice thickness of 4–5 mm; and 3) flip angle of 80°, gated to mid-diastole. This was followed by an ECG gated 2D gradient echo cine with typical parameters: 1) spatial resolution 1.0–1.5 cm in plane; 2) slice thickness of 4.5 mm; 3) flip angle of 15°, 4) bandwidth of 250 Hz/pixel; and 5) temporal resolution of 30 msec per frame. Approximately 10 min following administration of the gadolinium-based contrast agent Gadobutrol (Gadavist®, Bayer Corporation, Whippany, NJ), an equilibrium phase 3-D T1 weighted gradient echo volume was acquired with typical parameters: 1) spatial resolution 1.0 mm isotropic: 2) flip angle of 10°; and 3) bandwidth of 300 Hz/pixel. Catheter manipulations were done using a real-time SSFP cine sequence with typical parameters: 1) spatial resolution of 2.0 × 2.0 mm in plane; 2) slice thickness of 8 mm; and 3) temporal resolution of 50 msec.

3 Results

Standard fluoroscopically-active catheter-based instruments are simply not suitable for use under MRI, preventing a facile translation of present-day devices into the realm of iCMR. In an attempt to highlight this limitation and demonstrate the need for MRI-visible catheters, commercially-available "off-the-shelf" catheterization devices were examined in the iCMR setting. A 0.014″ coronary wire guide with a ferromagnetic core (RUNTHROUGH® NS extra floppy coronary guidewire) produced a large artifact that obscured cardiovascular and non-cardiovascular structures in the MR image when placed in the ascending thoracic aorta (Fig. 2a), brought about by ferromagnetic metals in the wire construction. Similar massive MRI artifacts were obtained with Hi-Torque BMW and Prowater Asahi PTCA guidewires (Supplementary Fig. S4). A 0.018″ peripheral wire guide (HiWire® hydrophilic wire guide) composed of Nitinol with a tungsten-impregnated polyurethane jacket without a ferromagnetic core produced only a limited artifact in the shaft in the form of a bright signal (Fig. 2b). However, the tip of the wire guide could not be visualized in standard MRI sequences. Lastly, the Apex™ PTCA dilatation catheter, a coronary angioplasty balloon, generated a very limited signal from the radiopaque non-ferromagnetic metal ring markers (Fig. 2c).

CMR-safe balloon-tipped catheters are currently used clinically for purposes of right heart hemodynamic catheterization (Rogers et al. 2017). We examined an Arrow® balloon wedge-pressure catheter positioned at the level of the pulmonary valve, and upon inflation with a gadolinium-based contrast agent, the balloon tip was visualized in standard real-time CMR (Fig. 3a). However, the catheter shaft, and thus, the position, torque, or potential entanglement of the catheter, could not be evaluated. An adjusted saturation CMR sequence provided for enhanced visualization of the gadolinium-filled balloon, but at the expense of loss of anatomical data (Fig. 3b).

In iCMR, visualization of catheter-based devices is of paramount importance, with either limited signal or substantial artifacts proving detrimental to proper steering and positioning of catheters. We hypothesized that successful, real-time navigation of endovascular instruments is possible by applying an SPIO polymer coating to these devices. Having identified a basic toolbox of existing cardiovascular instruments that generated minimal to no MRI artifacts, these were functionalized with the SPIO polymer coating and evaluated in an iCMR setting. The dose of SPIO coating was varied in range from low concentration to a high concentration on wire guides and catheters to identify an iCMR signal that optimized both

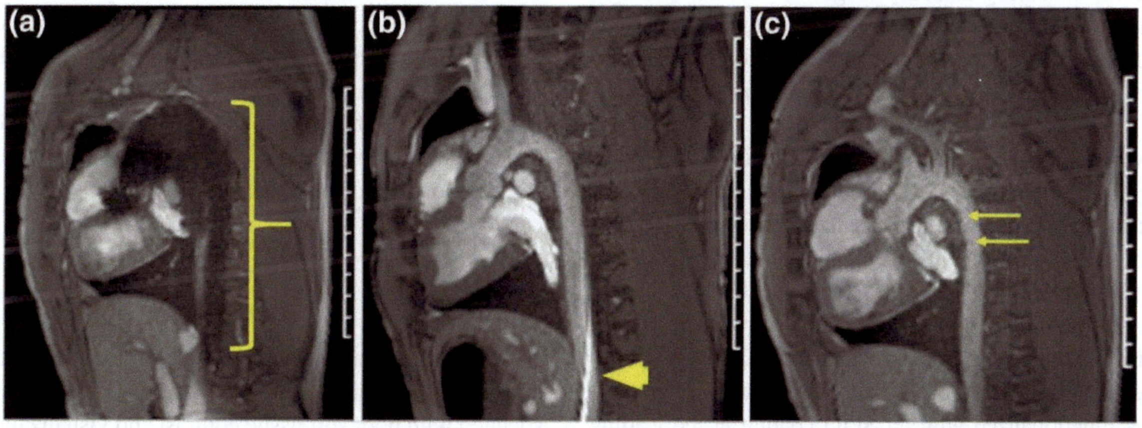

Fig. 2 CMR of "off-the-shelf" cardiovascular catheterization devices. **a** A 0.014" Runthrough® NS extra floppy coronary guidewire with a ferromagnetic core was advanced into the ascending aorta. The bracket highlights the artifact generated by the ferromagnetic core. **b** 0.018" HiWire® hydrophilic wire guide without ferromagnetic metallic components was positioned in the descending thoracic aorta. The arrowhead points to bright signal emanating from the shaft of the wire guide in the abdominal aorta. **c** An Apex™ PTCA dilatation catheter with metallic marker rings located in the descending thoracic aorta. The arrows signal the position of the marker rings. Images were obtained in gradient echo sequence

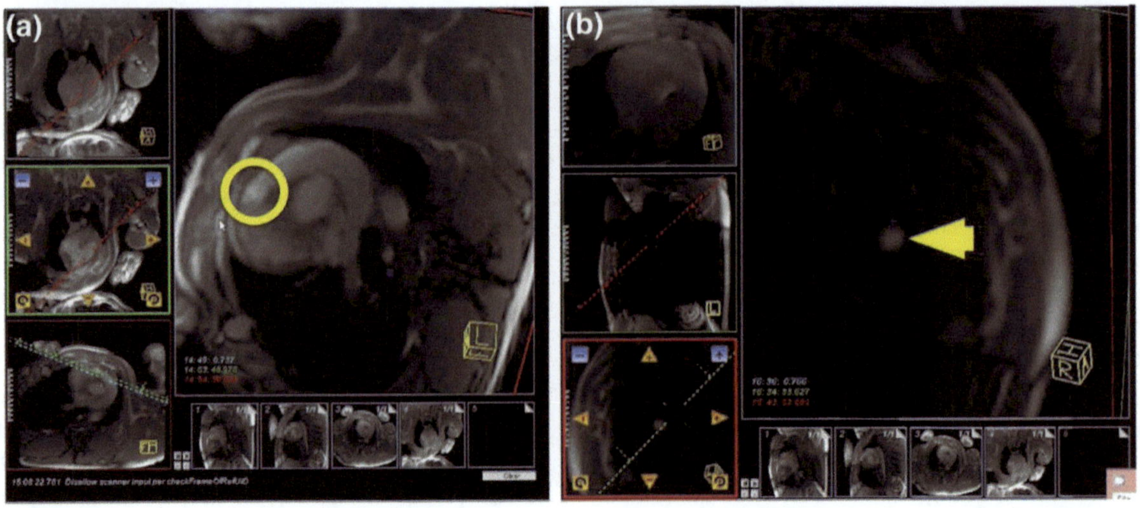

Fig. 3 iCMR of 7-French Arrow® balloon wedge-pressure catheter. **a** Balloon tipped catheter positioned at the level of the pulmonary valve and filled with diluted gadolinium-based contrast agent visualized in standard, real-time CMR. The circle indicates the tip of the catheter. Figure insets indicate multiplanar images representing 3D images acquired in real-time sequence via Interactive Front End software (Siemens Healthineers, Erlangen, DE). **b** Adjusted saturation CMR sequence highlighting enhanced visualization of gadolinium-filled balloon (arrowhead)

visibility and spatial resolution (Supplementary Fig. S5). As can be seen in Fig. 4a, the uncoated 0.018" HiWire® wire guide (the same as Fig. 2b, shown here for purposes of comparison) generated no signal at the tip of the wire guide under iCMR when positioned in the descending thoracic aorta. Upon functionalization of the wire guide, signal can be seen emanating from the SPIO-coated tip (Fig. 4b). Similarly, an ATLAS® balloon was functionalized with an SPIO polymer coating and positioned in the abdominal aorta. While the tip of uncoated controls produced a subtle signal due to the presence of metallic marker rings (Fig. 4c), the coated Atlas balloon with SPIO coating produced a very strong signal under MRI that made the balloon easily discernible without obscuring anatomy (Fig. 4d). Similar findings, wherein the catheter was easily discernible under MRI compared to uncoated catheters, were observed following SPIO functionalization of Apex™ PTCA and Conquest® PTA dilatation catheters (Supplementary Fig. S6).

Unlike steel and ferromagnetic metal containing stent platforms which create large artifact obscuring the anatomy, cobalt-based stent platforms could be visualized under iCMR when not in motion and did not obscure the anatomy (data not shown). However, visualization of the stent delivery system while in motion proved more challenging. To optimize visibility of stent delivery and deployment under iCMR, a cobalt-based device (Assurant Cobalt® balloon expandable stent system) was functionalized with an SPIO coating at the leading edge of the stent delivery system, and stent delivery was examined under MRI. This allowed visualization of the stent delivery system through the abdominal aorta (Fig. 5a), thoracic aorta, and selective positioning in the subclavian artery. The stent was then deployed in real-time via inflation of the balloon with diluted gadolinium-based contrast agent. As can be seen in Fig. 5b, the cobalt stent was successfully deployed in the subclavian artery, while the shaft of the 0.035" HiWire® wire guide was also visible in the descending thoracic aorta.

As endovascular aortic repair is an ideal target for iCMR application, our goal was then to attempt to translate endovascular stent graft technology to the iCMR setting. However, we were unable to identify any endovascular stent graft systems suitable for MRI use. To overcome this challenge, the cobalt-based balloon-expandable stent platform was modified with an SPIO-coated ePTFE membrane, giving rise to a customized iCMR-functionalized stent graft. In an attempt to highlight the ability to apply iCMR-enabled devices to cardiac interventions, the customized iCMR stent system was deployed in the pig abdominal aorta. As can be seen in the simultaneous real time MR images obtained before (Fig. 5c) and after (Fig. 5d) deployment of the stent under iCMR-guidance, the stent graft was successfully deployed in vivo in the pig abdominal aorta, best demonstrated by significant foreshortening of the device upon expansion to its final diameter.

4 Discussion

Cardiovascular catheter-based interventional procedures are currently performed under x-ray fluoroscopy, a modality associated with poor soft tissue contrast and visualization in two dimensions, limiting accurate navigation of endovascular tools (Fig. 1a). Importantly, present-day standard fluoroscopy typically requires repeated injection of iodinated contrast in order to visualize cardiovascular structures. The risk of acute renal injury related to iodinated contrast exposure frequently

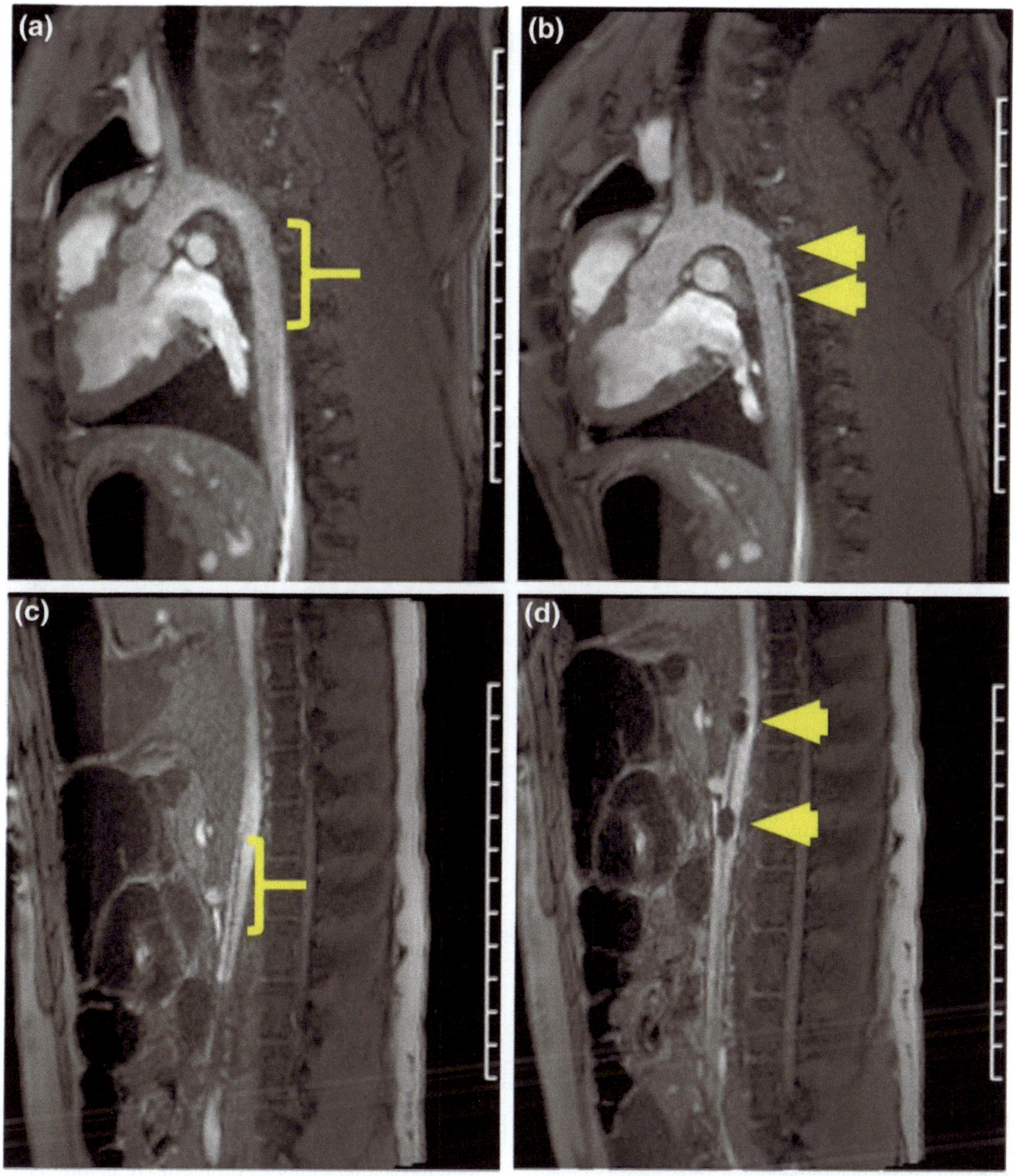

Fig. 4 SPIO functionalization of wire guide and angioplasty balloon for iCMR. **a** Control (uncoated) 0.018" HiWire® hydrophilic wire guide positioned in the descending thoracic aorta. The bracket highlights the position of the tip of the wire guide. **b** SPIO functionalized 0.018" HiWire® hydrophilic wire guide positioned in the descending thoracic aorta. Arrowheads denote the location of SPIO markers. **c** Control (uncoated) Atlas® Gold PTA dilatation catheter in the abdominal aorta. The bracket highlights signal generated by the metallic marker rings as seen on gradient echo imaging. **d** SPIO functionalized Atlas® Gold PTA dilatation catheter in the abdominal aorta. Arrowheads indicate positioning of SPIO markers

limits the use of catheterization and transcatheter intervention in patients with renal insufficiency (Andreucci et al. 2017). The inability to visualize cardiovascular structures or devices with depth lengthens the duration of complex catheterization cases, further increases x-ray and contrast exposure, reduces the likelihood of success, and may even augment the risk of complications.

MRI-guided catheterization combines high-resolution anatomical imaging with superior soft tissue contrast and physiologic flow data. Three-dimensional structural information and imaging across multiple planes enables more accurate positioning of endovascular devices in iCMR (Fig. 1b), in a setting free of radiation and harmful side effects to patients. Thus, there exists a monumental potential for iCMR to shift

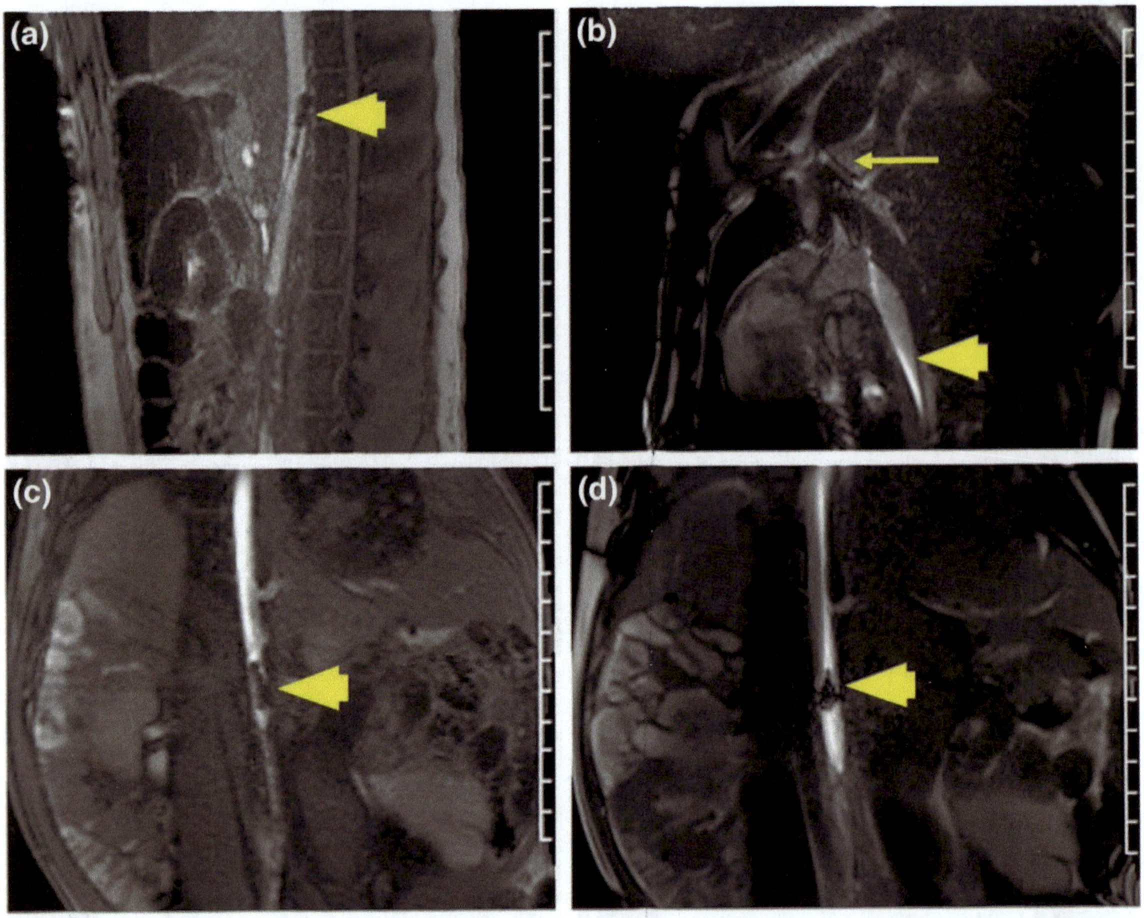

Fig. 5 Stent deployment in iCMR. **a** Assurant Cobalt® balloon expandable stent system advanced to the abdominal aorta. The tip of the catheter, which has been coated with SPIO, is denoted by the arrowhead. **b** Cobalt-based stent (arrow) deployed in the subclavian artery. The shaft of the wire guide can be seen in the descending thoracic aorta (arrowhead). Pre- (**c**) and post-deployment (**d**) of a custom endovascular stent graft system suitable for MRI use in the abdominal aorta under iCMR. The system was composed of a SPIO-coated ePTFE membrane sutured onto a cobalt balloon-expandable stent, which in turn was crimped on an Atlas® Gold PTA dilatation catheter. After deployment of the stent in the abdominal aorta, a significant foreshortening of the stent material (arrowhead) was observed

the paradigm of transcatheter interventions. It is important to note that while currently, diagnostic right heart catheterization in CMR could potentially be performed in any MRI environment that can be rendered a sterile procedure room with an invasive hemodynamic monitoring system, at present, the ability to begin to explore interventional applications clinically is extremely limited due to the non-existence of a basic toolkit suitable for iCMR. As such, the objective of this study was to determine the limitations of existing cardiac catheterization devices in the iCMR environment, and then functionalize a basic set of existing endovascular instruments using an SPIO coating, so as to realize the potential of iCMR. Specifically, we aimed to functionalize wire guides, angioplasty balloon catheters, a guiding catheter, and a stent delivery system for use in iCMR.

Herein, we demonstrated that standard wire guides used for percutaneous coronary intervention, as well as coronary diagnostic and guiding catheters, were not suitable for iCMR due to the profound artifact created by the ferromagnetic metallic braiding in these devices that obstruct visualization of anatomical structures (Fig. 2a, Supplementary Fig. S4). Moreover, ferrous material in catheters may even generate heat in the MRI environment (Klemm et al. 2000). These metallic braidings are crucial design considerations for properties (e.g. torqueability) necessary for proper device steering, and exclusion of these comes at the cost of mechanical performance (Saikus and Lederman 2009). Thus, design of catheters suitable for MRI use not only have to account for the incorporation of MRI-visible components, but also overcome the challenge of ensuring that novel catheter designs maintain specific mechanical properties essential to proper device navigation and positioning. Krueger et al. developed a non-metallic, fiber-reinforced material for use in guidewires, with mechanical properties that were similar to a Nitinol-based guidewire (Krueger et al. 2008). Of note, incorporation of iron powder particles, ranging in diameter from 6 to 8 μm, allowed for visualization under MRI.

In this study, we also examined the iCMR performance of a balloon-tipped catheter filled with diluted gadolinium-based contrast agent, devices that are currently used clinically for purposes of right heart catheterization (Ratnayaka et al. 2013). However, limitations include the inability to visualize the catheter shaft (Fig. 3a). Moreover, while adjusting the saturation resulted in an increase in signal emanating from the balloon, there was a loss in visualization of anatomical structures (Fig. 3b). In another study, we examined an FDA approved guidewire specifically dedicated for use in MRI, the 0.035" MRWire EmeryGlide (Nano4Imaging GmbH, Aachen, DE) with discrete markers consisting of carboxydextran-coated SPIO nanoparticles (Resovist®) at 0, 2, 4 cm from the tip. The device generated a very robust signal under MRI from the three markers (data not shown). While these examples highlight the ongoing pursuit of MRI-suitable devices, commercially-available catheters and interventional instruments for use in iCMR remain scarce. At the present, the design and development of iCMR-suitable passive and active catheter-based platforms remains largely within the academic arena, with several notable examples from research groups. Kocaturk et al. developed an active, clinical-grade guidewire with a loop and dipole antenna to visualize the tip and shaft of the guidewire, respectively (Kocaturk et al. 2009). The resulting guidewire maintained the mechanical properties of commercially-available x-ray guidewires and was visible under MRI. Magnusson et al. designed a catheter containing a hyperpolarized ^{13}C contrast agent under flow with the ability to visualize the entire catheter length (Magnusson et al. 2007). Lastly, Unal et al. applied gadolinium-based coatings to commercially available catheters and guidewires and were able to image these devices in the canine aorta (Unal et al. 2006).

Herein, our goal was to functionalize several commercially-available endovascular instruments with an SPIO polymer coating for purposes of iCMR. Given that the polymer-jacketed wire guides did not generate significant artifact, we successfully functionalized these such that an SPIO-coated tip could be visualized in arterial anatomy, allowing for safe navigation of these wire guides in vivo in iCMR (Fig. 4b). Balloon catheters of a wide spectrum of sizes were also functionalized and visualized under MRI (Fig. 4d). Previous studies have shown the potential that iCMR has for stent deployment. Raval et al. demonstrated the use of active catheters for accurate stent deployment under real-time MR imaging in a porcine model of aortic coarctation (Raval et al. 2005). We sought to develop a basic stent delivery system based on cobalt stents that: 1) did not cause CMR image distortion (in contrast to stainless steel), which was demonstrated in Fig. 5a and b; and 2) could be visualized during and after deployment. Incorporating a layer of SPIO-functonalized ePTFE into the cobalt stent platform resulted in an iCMR-visible endovascular stent graft, which we successfully deployed in the abdominal aorta of a pig (Fig. 5c-d).

During the course of the study, development of devices suitable for MRI use highlighted important advantages for iCMR. Selective contrast injection for angiography is standard before stent deployment in fluoroscopy. This often requires larger-sized catheters for purposes of injecting contrast around the stent delivery system, or for more vascular access points to inject contrast from a separate catheter. This was wholly unnecessary in the iCMR environment, potentially reducing the risk of vascular complications. For example, while transcatheter aortic valve replacement (TAVR) is typically guided by repeated contrast angiography through a second arterial catheter placed via second arterial access site, TAVR could be performed under real-time CMR, with positioning check before deployment and functional assessment after deployment using high-resolution cine sequences, all without the need for simultaneous contrast injection or additional arterial access. Additionally, iCMR allows for depth perception, as well as measurement of physiologic flow data during the intervention.

While promising, limitations to development of iCMR-suitable devices exist. A major limitation is the use of a non-braided catheter, which results in loss of torqueability, trackability in tortuous vessels, and the potential to kink the catheter. The development of non-ferrous braiding materials that can be incorporated within devices and still provide the requisite mechanical properties can surmount this challenge. Likewise, catheter shape selection to optimize cannulation and back-up support is extremely limited as most high-performance catheters utilize a metal-braid construction. The present study did not evaluate heating of functionalized wire guides, catheters, angioplasty balloons, or stents, but will be an important consideration upon clinical translation. Moreover, current real-time iCMR has significant time and spatial resolution limitations. As a result, the potential utility of iCMR is unlikely to extend to small vessel interventions such as coronary angioplasty. Regardless of these challenges, it is our belief that iCMR represents a safe and accurate modality for cardiovascular interventions, and this study contributes to the burgeoning field by highlighting the feasibility of performing iCMR through incorporation of MRI-visible design considerations to endovascular devices.

5 Conclusions

iCMR has tremendous potential to revolutionize catheter-based interventional cardiovascular procedures, with significant advantages over present-day x-ray fluoroscopy, which remains the clinical gold standard. However, translating iCMR into a clinical reality is principally limited by the lack of MRI-visible endovascular tools and devices. Herein, we demonstrated that SPIO functionalization of a basic set of endovascular instruments, including wire guides and catheters, allowed for active visualization under real-time MRI.

The ability to image these devices in an MRI setting resulted in superior navigation in cardiovascular structures, as well as the ability to perform interventions such as stent deployment. The spatial and temporal resolution provided by MRI, and the potential to visualize these instruments across multiple planes, resulted in enhanced awareness of catheter and stent placement. The capability of performing interventions without the use of x-ray fluoroscopy is expected to eliminate radiation exposure, allowing for procedures to be performed in a wider cohort of patients (e.g. pregnant women, children) and also encourage the pursuit of novel, minimally-invasive cardiovascular interventions. Although MR acquisition is typically slower than fluoroscopy due to factors such as lower frame rate and the need for breath-hold, future experiments will consist of improving image quality and speed of acquisition, and will aim to take advantage of the power of real-time 3D imaging and physiologic flow data during interventions. It is our belief that incorporation of MRI-visible components in endovascular tools will contribute immensely towards translating iCMR into routine clinical practice. Our findings demonstrate the feasibility of iCMR if endovascular tools include design considerations for the MRI environment, and we believe this study will contribute towards continued research in the field of MRI-visible endovascular device development.

Acknowledgements The authors thank Matthew G. Landry for assistance with schematics. This work was supported by the George and Angelina Kostas Research Center for Cardiovascular Nanomedicine. CHL acknowledges support from the Houston Methodist Specialty Physician Group Grant Program. VS-I is grateful for support from the Instituto Tecnológico y de Estudios Superiores de Monterrey and the Consejo Nacional de Ciencia y Tecnología (CONACyT, 490202/278979). MF gratefully acknowledges support from the Ernest Cockrell Jr. Presidential Distinguished Chair at the Houston Methodist Research Institute. MF serves on the Board of Directors of Arrowhead Pharmaceuticals. The authors declare that they have no conflict of interest.

References

M. Andreucci, T. Faga, R. Serra, G. De Sarro, A. Michael, Update on the renal toxicity of iodinated contrast drugs used in clinical medicine. Drug Healthc. Patient Saf. **9**, 25–37 (2017)

C.J. Bakker, R.M. Hoogeveen, J. Weber, J.J. vanVaals, M.A. Viergever, W.P. Mali, Visualization of dedicated catheters using fast scanning techniques with potential for MR-guided vascular interventions. Magn. Reson. Med. **36**(6), 816–820 (1996)

E.J. Benjamin, S.S. Virani, C.W. Callaway, A.M. Chamberlain, A.R. Chang, S. Cheng, et al., Heart disease and stroke Statistics-2018 update: A report from the American Heart Association. Circulation **137**(12), e67–e492 (2018)

A. Berrington de Gonzalez, S. Darby, Risk of cancer from diagnostic X-rays: Estimates for the UK and 14 other countries. Lancet **363**(9406), 345–351 (2004)

T.M. Bhat, M.E. Afari, L.A. Garcia, Atherectomy in peripheral artery disease: A review. J Invasive Cardiol **29**(4), 135–144 (2017)

M. Bock, S. Volz, S. Zuhlsdorff, R. Umathum, C. Fink, P. Hallscheidt, et al., MR-guided intravascular procedures: Real-time parameter control and automated slice positioning with active tracking coils. J. Magn. Reson. Imaging **19**(5), 580–589 (2004)

M. Bock, R. Umathum, S. Zuehlsdorff, S. Volz, C. Fink, P. Hallscheidt, et al., Interventional magnetic resonance imaging: An alternative to image guidance with ionising radiation. Radiat. Prot. Dosim. **117**(1–3), 74–78 (2005)

C.E. Chambers, K.A. Fetterly, R. Holzer, P.J. Lin, J.C. Blankenship, S. Balter, et al., Radiation safety program for the cardiac catheterization laboratory. Catheter. Cardiovasc. Interv. **77**(4), 546–556 (2011)

H. Clogenson, J. Dobbelsteen, Catheters and guidewires for interventional MRI: Are we there yet? J Imaging Intervent Radiol **2**, 28 (2016)

Z.C. Higgs, D.A. Macafee, B.D. Braithwaite, C.A. Maxwell-Armstrong, The Seldinger technique: 50 years on. Lancet **366**(9494), 1407–1409 (2005)

T. Klemm, S. Duda, J. Machann, K. Seekamp-Rahn, L. Schnieder, C.D. Claussen, et al., MR imaging in the presence of vascular stents: A systematic assessment of artifacts for various stent orientations, sequence types, and field strengths. J. Magn. Reson. Imaging **12**(4), 606–615 (2000)

O. Kocaturk, A.H. Kim, C.E. Saikus, M.A. Guttman, A.Z. Faranesh, C. Ozturk, et al., Active two-channel 0.035″ guidewire for interventional cardiovascular MRI. J. Magn. Reson. Imaging **30**(2), 461–465 (2009)

J.J. Krueger, P. Ewert, S. Yilmaz, D. Gelernter, B. Peters, K. Pietzner, et al., Magnetic resonance imaging-guided balloon angioplasty of coarctation of the aorta: A pilot study. Circulation **113**(8), 1093–1100 (2006)

S. Krueger, S. Schmitz, S. Weiss, D. Wirtz, M. Linssen, H. Schade, et al., An MR guidewire based on micropultruded fiber-reinforced material. Magn. Reson. Med. **60**(5), 1190–1196 (2008)

J. Lotz, Interventional vascular MRI: Moving forward. Eur. Heart J. **34**(5), 327–329 (2013)

Y. Ma, N. Gogin, P. Cathier, R.J. Housden, G. Gijsbers, M. Cooklin, et al., Real-time x-ray fluoroscopy-based catheter detection and tracking for cardiac electrophysiology interventions. Med. Phys. **40**(7), 071902 (2013)

P. Magnusson, E. Johansson, S. Mansson, J.S. Petersson, C.M. Chai, G. Hansson, et al., Passive catheter tracking during interventional MRI using hyperpolarized 13C. Magn. Reson. Med. **57**(6), 1140–1147 (2007)

J.R. Mazal, T. Rogers, W.H. Schenke, A.Z. Faranesh, M. Hansen, K. O'Brien, et al., Interventional-cardiovascular MR: Role of the interventional MR technologist. Radiol. Technol. **87**(3), 261–270 (2016)

M.E. Miquel, S. Hegde, V. Muthurangu, B.J. Corcoran, S.F. Keevil, D.L.G. Hill, et al., Visualization and tracking of an inflatable balloon catheter using SSFP in a flow phantom and in the heart and great vessels of patients. Magn. Reson. Med. **51**(5), 988–995 (2004)

R.A. Omary, O. Unal, D.S. Koscielski, R. Frayne, F.R. Korosec, C.A. Mistretta, et al., Real-time MR imaging-guided passive catheter tracking with use of gadolinium-filled catheters. J. Vasc. Interv. Radiol. **11**(8), 1079–1085 (2000)

T. Pucelikova, G. Dangas, R. Mehran, Contrast-induced nephropathy. Catheter. Cardiovasc. Interv. **71**(1), 62–72 (2008)

K. Ratnayaka, A.Z. Faranesh, M.A. Guttman, O. Kocaturk, C.E. Saikus, R.J. Lederman, Interventional cardiovascular magnetic resonance: still tantalizing. J. Cardiovasc. Magn. Reson. **10**, 62 (2008)

K. Ratnayaka, A.Z. Faranesh, M.S. Hansen, A.M. Stine, M. Halabi, I.M. Barbash, et al., Real-time MRI-guided right heart catheterization in adults using passive catheters. Eur. Heart J. **34**(5), 380–389 (2013)

A.N. Raval, J.D. Telep, M.A. Guttman, C. Ozturk, M. Jones, R.B. Thompson, et al., Real-time magnetic resonance imaging-guided stenting of aortic coarctation with commercially available catheter devices in Swine. Circulation **112**(5), 699–706 (2005)

T. Rogers, K. Ratnayaka, J.M. Khan, A. Stine, W.H. Schenke, L.P. Grant, et al., CMR fluoroscopy right heart catheterization for cardiac output and pulmonary vascular resistance: Results in 102 patients. J. Cardiovasc. Magn. Reson. **19**(1), 54 (2017)

D.L. Rubin, A.V. Ratner, S.W. Young, Magnetic-susceptibility effects and their application in the development of new ferromagnetic catheters for magnetic-resonance-imaging. Investig. Radiol. **25**(12), 1325–1332 (1990)

M. Saeed, S.W. Hetts, J. English, M. Wilson, MR fluoroscopy in vascular and cardiac interventions (review). Int. J. Card. Imaging **28**(1), 117–137 (2012)

C.E. Saikus, R.J. Lederman, Interventional cardiovascular magnetic resonance imaging: A new opportunity for image-guided interventions. JACC. Cardiovasc. Imaging **2**(11), 1321–1331 (2009)

K. Slicker, W.G. Lane, O.O. Oyetayo, L.A. Copeland, E.M. Stock, J.B. Michel, et al., Daily cardiac catheterization procedural volume and complications at an academic medical center. Cardiovasc. Diagn. Ther. **6**(5), 446–452 (2016)

A. Stadler, W. Schima, A. Ba-Ssalamah, J. Kettenbach, E. Eisenhuber, Artifacts in body MR imaging: Their appearance and how to eliminate them. Eur. Radiol. **17**(5), 1242–1255 (2007)

O. Unal, J. Li, W. Cheng, H. Yu, C.M. Strother, MR-visible coatings for endovascular device visualization. J. Magn. Reson. Imaging **23**(5), 763–769 (2006)

K. Zhang, A.J. Krafft, R. Umathum, F. Maier, W. Semmler, M. Bock, Real-time MR navigation and localization of an intravascular catheter with ferromagnetic components. Magn Reson Mater Phy **23**(3), 153–163 (2010)

Publisher's note Springer Nature remains neutral with regard to jurisdictional claims in published maps and institutional affiliations.

Biomedical Microdevices (2018) 20: 65
https://doi.org/10.1007/s10544-018-0301-9

3D printing for preoperative planning and surgical training: a review

Anurup Ganguli [1,2] · Gelson J. Pagan-Diaz [1,2] · Lauren Grant [1,2] · Caroline Cvetkovic [1,2,3] · Mathew Bramlet [4,5] · John Vozenilek [1,4,5,6] · Thenkurussi Kesavadas [5,6] · Rashid Bashir [1,2,7,8]

Published online: 4 August 2018

Abstract
Surgeons typically rely on their past training and experiences as well as visual aids from medical imaging techniques such as magnetic resonance imaging (MRI) or computed tomography (CT) for the planning of surgical processes. Often, due to the anatomical complexity of the surgery site, two dimensional or virtual images are not sufficient to successfully convey the structural details. For such scenarios, a 3D printed model of the patient's anatomy enables personalized preoperative planning. This paper reviews critical aspects of 3D printing for preoperative planning and surgical training, starting with an overview of the process-flow and 3D printing techniques, followed by their applications spanning across multiple organ systems in the human body. State of the art in these technologies are described along with a discussion of current limitations and future opportunities.

Keywords Rapid prototyping · Organ models · Surgical training · Preoperative planning · 3D printing

1 Introduction

Surgeons have always faced the challenging task of carrying out surgeries on complex anatomical structures of the human body. With advances in medical imaging, tools such as multi-detector computed tomography (MDCT) and magnetic resonance imaging (MRI) have made radiological diagnosis less invasive and more informative. These tools provide high resolution two-dimensional images, but are limited in their ability to accurately depict complex 3D structures. Virtual 3D reconstruction methods, such as volume or surface rendering, offer a better understanding of structural complexity by allowing for rotation or sectioning of layers of the virtual 3D model. This ability has proven useful for the visualization of complex structures, such as congenital heart defects and aneurysms (Ou et al. 2007; Spevak et al. 2008; Kellenberger et al. 2007). However, differences between real anatomical structures and the interpretation of virtual 3D images still remain (Shiraishi et al. 2010). As a result, the success of surgical operations relies heavily on the surgeon's previous training and past experiences. To address the limitations in virtual image analysis, 3D printing is now being used to create physical models of patient anatomies for surgical simulation and preoperative planning.

Surgical simulations allow surgeons to learn new techniques or prepare for specific operations without the use of patients, cadavers, or animal models. Through physical recreation of a patient's organs or tissues, a realistic surgical procedure can be simulated. With recent advancements in the field of 3D printing, MRI or CT images can now be used to fabricate 3D anatomical models. 3D printing is a process by which 3D digital (CAD) models are converted into physical models through a layer-by-layer fabrication process. Patient-specific printed models of the anatomical region in need of surgical repair can provide the surgeon with a better visualization of the defected tissue, thus helping him to prepare for the surgery. These models can also be used to train new surgeons, given the short supply of donated human tissues and organs. A number of advantages of 3D printed models have been noted, such as better tactile understanding of an anatomy, rehearsal of the exact surgical route, implant and prosthesis

✉ Rashid Bashir
rbashir@illinois.edu

[1] Department of Bioengineering, University of Illinois at Urbana-Champaign, Urbana, IL, USA

[2] Micro and Nanotechnology Laboratory, University of Illinois at Urbana-Champaign, Urbana, IL, USA

[3] Center for Neuroregeneration, Houston Methodist Research Institute, Houston, TX, USA

[4] OSF Saint Francis Medical Center, University of Illinois College of Medicine at Peoria, Urbana, IL, USA

[5] Jump Trading Simulation & Education Center, Peoria, IL, USA

[6] Healthcare Engineering Systems Center, College of Engineering, University of Illinois at Urbana-Champaign, Urbana, IL, USA

[7] Department of Industrial and Enterprise Systems Engineering, University of Illinois at Urbana-Champaign, Urbana, IL, USA

[8] Carle Illinois College of Medicine, Urbana, IL, USA

planning and tailoring, easy intraoperative access for surgical guidance, and better communication between surgeon and patient (Müller et al. 2003; Kalejs and von Segesser 2009).

Given the growing interest in 3D printing for surgical simulation over the past decade, we feel there is a need for a comprehensive, systematic review of the subject. This review aims to summarize the use of 3D printing for surgical simulation in a variety of clinical areas, as well as the developed techniques to optimize the implementation of rapid prototyping to such areas. Beyond summarizing techniques, this review also aims to identify potential shortcomings in current surgical applications of 3D printing, so that future researchers can streamline their efforts towards these goals.

2 3D printing technique overview

3D printing can be broadly described as an additive fabrication process by which free-form 3D structures are formed in a layer-by-layer fashion from 3D computer models. The fabrication of physical 3D models from anatomical imaging data (MRI/CT) can be described as a 3-step process: (1) image acquisition, (2) image processing, and (3) 3D printing (Fig. 1).

The accuracy of the 3D printed object depends on the accuracy of the acquired image. Different imaging modalities used include MRI, MDCT, cone beam computed tomography (CBCT), positron emission tomography (PET), and ultrasonography (US) among others. MDCT is the most widely used modality owing to the ease of post-processing its data. Furthermore, images can be acquired with a spatial resolution of 400-600 microns, showing how MDCT offers a 3D dataset with better resolution, but can still result in procedural failure owing to the complex morphology (Rengier et al. 2010). The final images are saved in DICOM (digital imaging and communications in medicine) format.

Several software packages, such as Mimics (Materialise NV, USA) and MeVislab (Mevismedical Solutions, Germany), have been used for the post processing of images. This software segments the images to produce a final STL file. Segmentation is the process of isolating an area of interest in the data files, where the dataset is a series of DICOM images which when stacked together form the complete structure. Advanced image processing algorithms have also been proposed for special cases like non-enhanced or low resolution images (Frakes et al. 2005). Triangular meshing is usually performed on the contours of segmented region of interest (Hahn et al. 2004). This data can be further processed using various CAD software for modification of geometry or mesh optimization. The final modified data is converted to STL format, which can then be sent to a 3D printer for fabrication.

The 3D printing machine reads data from a CAD file and adds successive layers of material, thus building the model from a series of cross-sections. Compared to alternative fabrication tools such as CNC machining, which consists of CAD-guided material removal, 3D printing is more appealing due to its geometric flexibility, lower fabrication time, lower cost and minimal technical expertise requirement. Its geometric flexibility is the main reason that 3D printing is best suited for the fabrication of complex anatomical structures. The downtime for 3D printing (a few hours) is usually lower than that for contemporary techniques (several hours to days) but actual fabrication time depends on the particular 3D printing method, as well as the complexity of the structure.

The STL file obtained from the DICOM images of the part of interest is sent to a 3D printer and the final 3D part is obtained as the output. This technique truly allows medical practitioners to employ "patient specific" surgical planning and problem understanding. Figure 1 describes this process of file preparation for the 3D printing of an anatomical model.

A common approach for converting this computational data into 3D solid models is rapid prototyping (Ngan et al. 2006; Sodian et al. 2007). Rapid prototyping is a method for 3D model fabrication from computer-generated virtual 3D surface models, usually derived from CT or MRI imaging data set. The virtual model is first broken down into thin slices or layers. A rapid prototyping machine then builds the solid

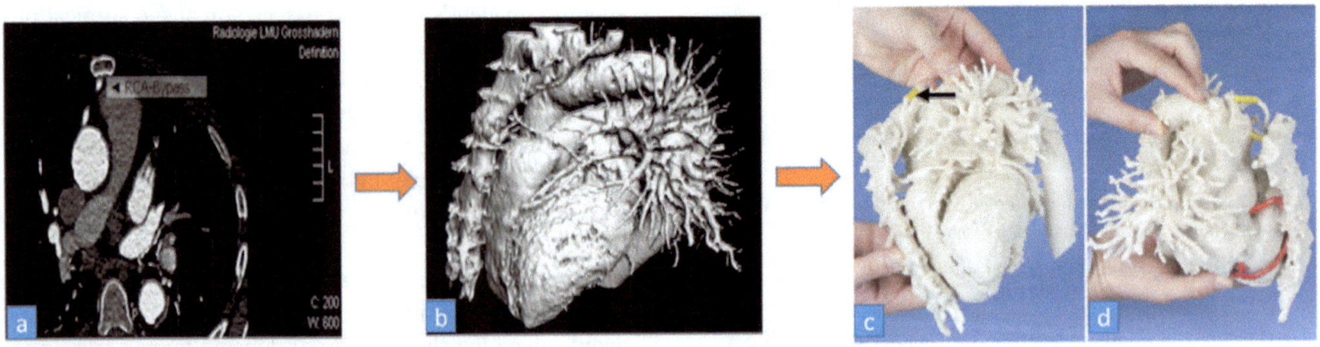

Fig. 1 Process flowchart for 3D printing models for clinical simulations. **a** Step1: *CT scan* of a patient with a patent left anterior descending artery and right coronary artery bypass close to sternum (arrow) (**b**) Step2: *Image processing-* 3D segmentation of the CT data (**c**) Step 3: *3D printing-* SLA model of the patient's cerebrovascular anatomy (Sodian et al. 2008b). (Copyright © 2008 The Society of Thoracic Surgeons. Published by Elsevier Inc.)

model layer upon layer, resulting in a physical replica of the virtual model. Rapid prototyping encompasses several different techniques each with its own benefits and drawbacks (Fig. 2) (Park et al. 2016). Stereolithography (SLA) is one of the most common techniques, and uses an ultraviolet laser to polymerize a photo-curable resin (Winder and Bibb 2005). Fused deposition modelling (FDM) is similar to SLA with the main difference being that the layers are formed by the extrusion of small beads of fused thermoplastic material from a fine tip (Kruth 1991). Selective laser sintering (SLS) uses an infrared laser to cure thermoplastic powder (Berry et al. 1997). Laminated object manufacturing is yet another technique, which uses layers of paper or plastic cut by a laser and bonded by a heating process (Mueller and Kochan 1999).

3 3D printing technique development by surgery area

3.1 Cardiac surgery

Since their invention, surgeons have used preoperative diagnostic imaging techniques such as MRI, CT, echocardiography, and angiography to prepare for cardiovascular surgery and interventional cardiology. However, even with high resolution imaging capabilities it remains difficult to predict the actual anatomical structures for patients with congenital or other intra-cardiac defects, owing to great morphological diversity. Surgical preparation has a high associated risk for patients with a history of previous operations. Vessels can tear open and cavities can burst while scars are dissected. A realistic setting with lifelike heart dimensions provides an ideal platform for training surgeons.

MRI angiography and CT data have been successfully used to fabricate 3D models that are helpful in strategizing treatment paths, intraoperative orientation, and patient demonstrative guides since early 2000 (Sodian et al. 2007). By implementing SLA as a rapid prototyping method, researchers were able to model complex features in congenital heart disease (CHD) such as damaged pulmonary vascular tree (Ngan et al. 2006), complex vascular pathology, and intracardiac lesions (Sodian et al. 2007). Furthermore, FDM has also been used as a successful technique to model cardiovascular anatomy for various cases (Valverde et al. 2017). These studies served as proof of concepts for the use of rapid prototyping for cardiac surgery applications, achieving +95% accurate representation by the models (Noecker et al. 2006; Mottl-Link et al. 2008; Jacobs et al. 2008).

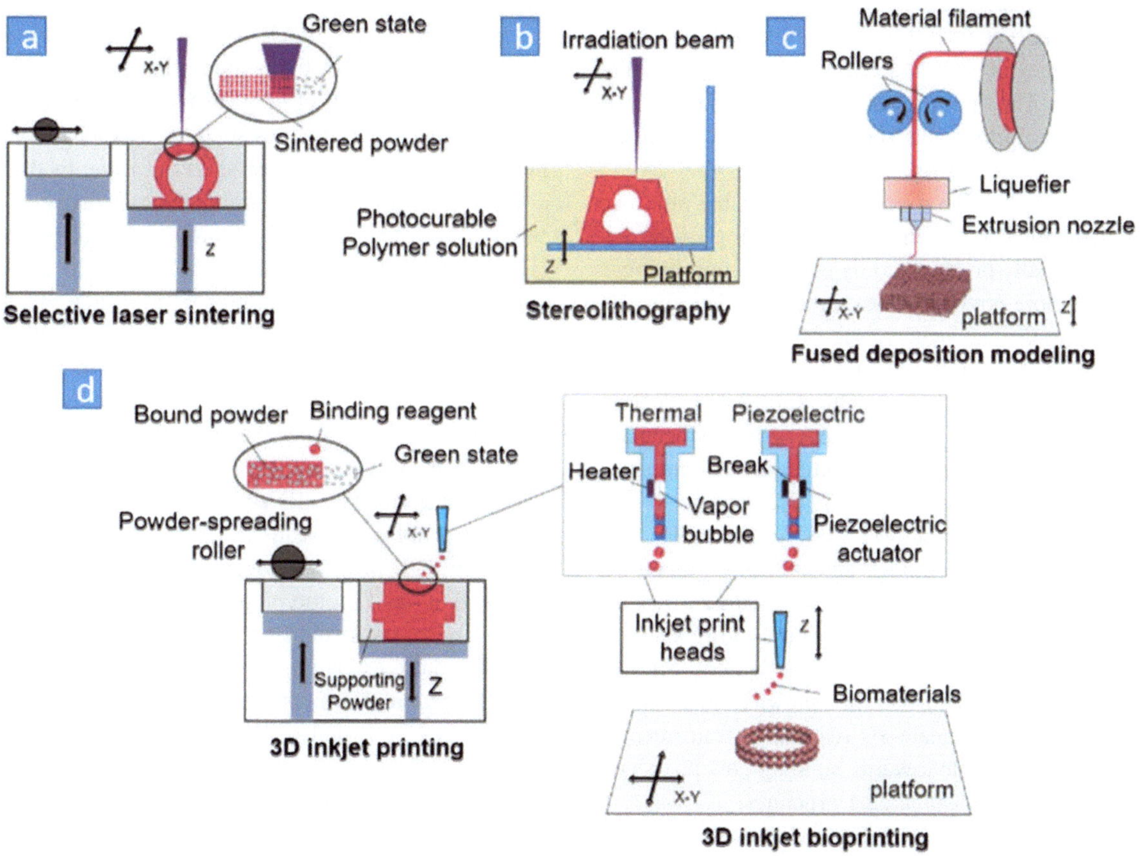

Fig. 2 Schematic representation of rapid prototyping techniques for surgery applications: (**a**) Selective laser sintering (**b**) Stereolithography (**c**) Fused deposition modeling (**d**) 3D inkjet printing (Park et al. 2016). (Copyright © 2016 Royal Society of Chemistry)

Several successes have achieved recreation of case specific 3D heart models from different imaging techniques. While most image sources are MRI or CT, these methods may involve intravenous contrast, sedation, and ionizing radiation. Contrastingly, 3D echocardiography has improved in resolution substantially with the development of new transducers and advancement of software and hardware; achieving higher resolution imaging without the drawbacks. For this reason, 3D echocardiography data has been tested as a suitable image source for 3D modeling (Olivieri et al. 2015). By converting 3D echocardiography image data into compatible files and 3D printing through rapid prototyping, models displaying ventricle septal defects and perivalvular leaks were successfully developed and were shown to be highly correlated to the 2D raw data.

Alterations to standard rapid prototyping techniques have enabled unique approaches to 3D modeling of cardiac structures. A novel rapid prototyping technique was developed, for example, using a powder and binder-based technique to fabricate lifelike models of exact aortic anatomy (Schmauss et al. 2012). This method was used as a postoperative approach to effectively assess decisions made in a diseased patient that had undergone a failed transcatheter aortic valve replacement. Another approach showed the use of the "drop on demand" 3D printing technique for evaluation of appropriate transcatheter percutaneous pulmonary valve implantation (PPVI) approaches (Schievano et al. 2007). While PPVI can overcome many disadvantages of surgery, it is limited to patients with a specific anatomy. The development of accurate 3D representations of such anatomies was shown to be effective in the cardiologists' decision making.

Exploring methods to improve representation has also been of interest to clinical researchers. 3D modeling of cardiac structures can usually be hindered by complex morphologies which become more erratic in diseased tissue. Therefore, the use of techniques that can achieve sub-millimeter reconstruction is key for proper diagnosis and preoperative planning with the use of 3D printed models. One approach to generating solid 3D models with sub-millimeter resolution (accurately depicting complex 3D cardiac geometries) is combining high spatial resolution imaging (MDCT) with iterative closest point algorithms (ICP) (Greil et al. 2007). This enabled the reconstruction of STL heart models and cardiac segments utilizing a laser sintering technique with a mean surface distance of approximately 250 μm, which were successfully analyzed in comparison to standard techniques of assessment (echocardiography, angiography, etc.). This has shown how the development of algorithms to create 3D renderings from medical images is a critical variable towards building patient-specific reconstructions of individual cardiac structures.

Some approaches towards high fidelity replicas have taken advantage of different material properties to achieve soft or clear models of cardiac structures. By combining multi-slice computed tomography data and SLA with *rubber-like* urethane and solid epoxy, flexible models of CHD cases were accurately fabricated. These models served for both planning innovative cardiac surgical procedures (Shiraishi et al. 2010), as well as effectively teaching multiple seminars on targeted surgical simulation of VSDs in the pediatric heart (Costello et al. 2015, 2014). The high-fidelity replicas were made by mixing a transparent plastic and a rubber-like material, which provided the freedom to cut with a scalpel blade while preserving structural integrity of the model. Towards similar training efforts, (trans-apical aortic valve replacement training), translucent heart models were effectively fabricated out of CT imaging data by using silicone, which allowed the trainees to see the stent valve throughout the exercise (Abdel-Sayed et al. 2009; Abdel-Sayed and Von Segesser 2011). This improvement was necessary as this procedure requires trainees to be able to see through the heart to visualize their performance. Current artificial heart models are usually opaque, as they are made of polyurethane, and thus are not ideal.

3.2 Neurosurgery

Use of rapid prototyping technology for neurosurgery simulation is still new and vastly unexplored due to the complexity of structures in the brain. In the case of preoperative planning in neurosurgery the aim is to find the safest approach that minimizes permanent brain tissue damage (Spottiswoode et al. 2013; Kettenbach et al. 1999). Since their development, medical imaging techniques such as MRI have been used by neurosurgeons to plan brain surgeries. Due to the excellent contrast of soft tissues, these images can display a clear distinction between the healthy tissue and that to be resected (Hirsch et al. 2000; Geerts et al. 2007). Up until recently surgical approaches were simulated in virtual reality. Recent advances in virtual surgical simulation for the brain using MRI imaging data and haptic devices have improved surgical outcomes by allowing surgeons to simulate the exact surgery on a computer and finalize the optimal approach (Lemole et al. 2007). But even with these simulation techniques, correct depth perception on a 2D screen is difficult, and these simulations do not allow realistic interaction with the anatomy, which only 3D printed models offer.

The first rapid prototyping proof of concept approaches to neurosurgery consisted of virtual simulations derived from MRI/CT data, followed by 3D printing of the anatomy for the rehearsal and refinement of the final surgical plan. In a recent study, a virtual reality pre-surgical simulation called interactive virtual simulation (IVS) in combination with 3D printed models demonstrated the use of rapid prototyping for preoperative planning and rehearsal in cases of skull base and deep tumors (Oishi et al. 2013). The many advantages of the use of this combination for pre-surgical planning included an increased anatomical understanding for the complicated bone dissection necessary to achieve canals in skull base without any damage to surrounding tissue. Various novel approaches to the 3D printing of nervous tissue focus on developing methods to fabricate brain tissue models capable of incorporating cerebral circulation details.

Even though virtual surgical simulation has greatly advanced preoperative planning, it fails to provide the complete experience of rehearsing the surgeries or planning a surgical route with actual instruments. The need for considerable computer skills and hardware have also been limiting factors in widespread application. As presented above, initial efforts combined standard CT and MR angiography data to demonstrate the first application of 3D printing (SLA) for cerebrovascular biomodeling of patients with cerebral aneurysms and cerebral arteriovenous malformations (AVM) (Urso et al. 1999). These replicated vasculature models were solid in structure and displayed only the internal lumen structure, highlighted by contrast in CTA or by blood flow in MRA.

In a later study, 3D printed SLA models of cerebral aneurysms with parent and surrounding vessels were built from 3D CT scans and 3D rotational angiography data (Wurm et al. 2004). This data acquisition method enabled the fabrication small vessels, with a resolution of 0.4 mm. However, because the models were fabricated of a rigid synthetic resin, they were of very limited value in evaluating neck geometry, performing clipping simulation, and practicing dissection exercises. Therefore, while achieving higher resolution than previous works, these models did not aid in preoperative planning such as determining optimal patient head position.

Other works have focused on demonstrating the first 3D printed hollow models of cerebral aneurysms which could be practically used for preoperative planning and surgical training (Kimura et al. 2009). The 3D elastic vasculature model was fabricated using a rubber-like polymer from CTA data and MagicsRP software. UV light was used to cure the vessel walls while the uncured center part of the vessel cross-section was curetted away manually to make the vessels hollow. For surgical simulation, the models were immobilized with plastic clay in the direction of the surgical window, and the clipping procedure was performed under the operative microscope. Post-clipping, the lumen of the vessels was checked using a flexible vascular endoscope for patency of the parent vessel. To simulate the surgery in cases with complex, deep-seated vertebrobasilar aneurysms, the authors used a hard, 3D printed model with aneurysms, vessels, and a cranial base bone, also performing a craniotomy simulation to gain access to the aneurysms. This method allowed for the optimization of the surgical window location for clipping placement. Overall, this approach proved useful in preoperative planning and could also be an ideal tool for new surgeon training. However, the model fabrication took several days, and the lumen formation technique was not ideal.

Later in 2013, another approach to fabricating a hollow, clipable cerebral elastic model for surgical simulation demonstrated the first solid 3D model of an aneurysm and the parent artery using a FDM-based 3D printer (Mashiko et al. 2015). In this instance, the surfaces of the fabricated models were manually smoothed and dip coated with a layer of elastic silicone and hardener mixture. The silicone was allowed to harden for 6 hours, followed by dissolution and removal of ABS, resulting in hollow elastic models of the aneurysm. The advantages of this technique included lower cost and reduced fabrication time (14-24 hours). Also, because the hollowing was not performed by curettage, branched and complicated parent artery systems could be fabricated. The hollow elastic models also simulate the artery deformation that occurs during surgery. This simulation was performed preoperatively and aided the surgeon in choosing the optimal surgical approach and correct clip design (Fig. 3i–h).

Other efforts on developing standard techniques of rapid prototyping for neurosurgery have explored the implementation of new materials to achieve more realistic 3D models. Treatment of some cerebral aneurysms and tumors requires drilling of the temporal bone during skull base surgery, and although cadaver dissection has been the primary form of training in such cases, it has very limited accessibility. To address this, one case modified 3D skull base model using selective laser sintering (SLS), designing it to mimic surface details, inner bony structures such as the inner ear and air cells, artificial dura matter, cranial nerves, the internal carotid artery, and venous sinuses for surgical training (Mori et al. 2008). Materials such as silicone, sponge, and rubber fiber were used to model intricate and soft tissue structures. Two simulation surgeries were successfully performed on the model using ultrasonic bone curette or a high-speed drill.

3.3 Craniomaxillofacial surgery

The inherent complexity and variations in the craniomaxillofacial anatomy make reconstructive surgeries a challenging task. Software simulations in conjunction with high resolution 3D images have significantly improved preoperative planning by allowing the manipulation of images to simulate surgery. However, tasks such as realistic simulations of osteotomies, positioning of bone fragments, and assessment of intraoperative position and symmetry of deformed anatomy are difficult to accomplish with virtual reconstructions. Complications including improper assessment of deep skeletal contours of orbit and skull base, variability in the positioning of different facial components such as sagittal jaw and skull base relative to each other, and variations in craniomaxillofacial development, also demand the use of 3D printed biomodels.

There have been very few alterations to the basic techniques to rapid prototyping for craniomaxillofacial surgery.

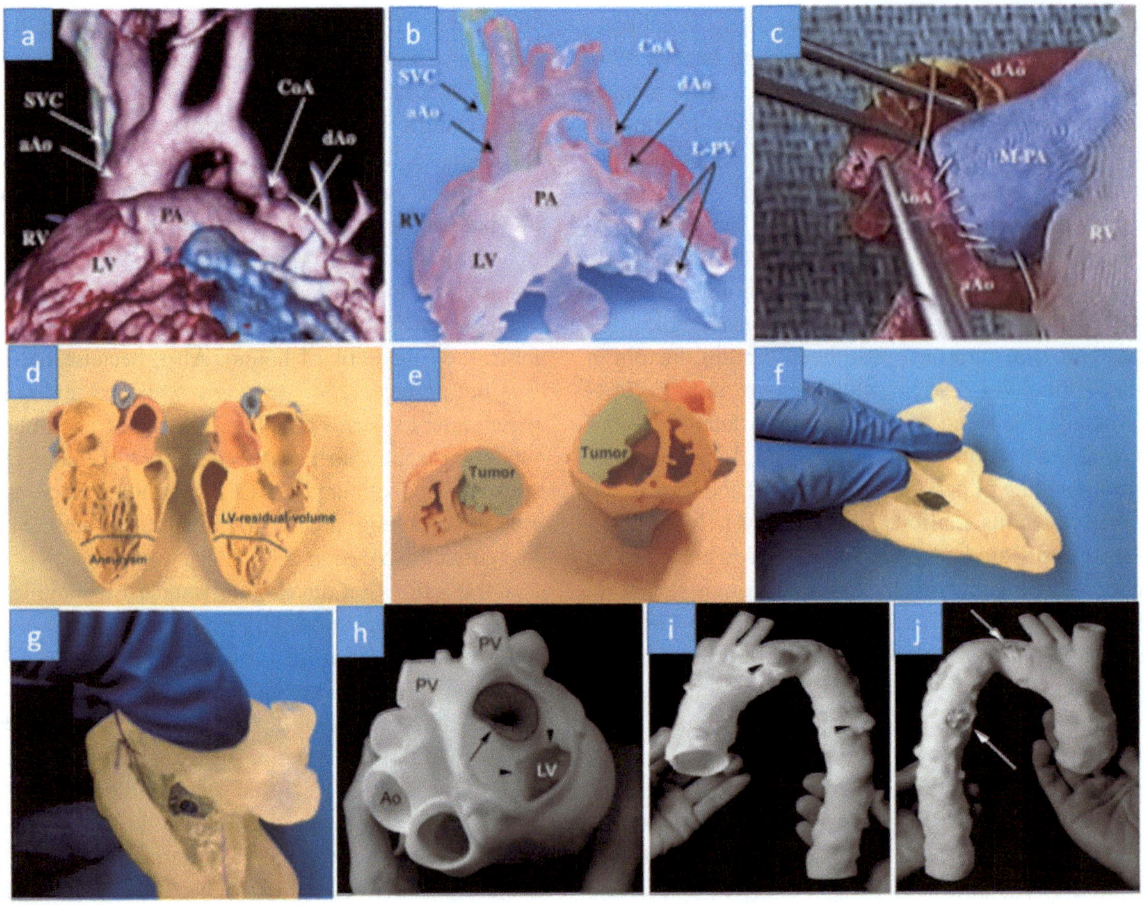

Fig. 3 3D printing for cardiac surgery. **a** and **b** CT data and 3D printed solid resin model of a 2 month old with isolated coarctation of aorta (Shiraishi et al. 2010) (**c**) Simulative surgery with rubber-like urethane model of 4-day old neonate with hypoplastic left heart syndrome (Shiraishi et al. 2010) (Copyright © 2010, Oxford University Press) (**d**) Model of a patient with aneurysm and LV-residual volume (Jacobs et al. 2008) (**e**) Massive cardiac tumor with infiltration to the right ventricular and tricuspid valve (Jacobs et al. 2008) (Copyright © 2008, Oxford University Press) (**f**) and (**g**) surgeon training exercise with flexible 3D printed model – surgical right atriotomy revealing VSD location and suture placement around VSD lesion (Costello et al. 2014) (Copyright © 2014, SAGE Publications) (**h**) close up on left atrial disc of an amplatzer 35 mm ASD occlude (arrow) demonstrating no obstruction of the mitral valve annulus (arrowheads) (Kim et al. 2008) (**i**) and (**j**) anterior view of multiple aortic ulcerations and posterior view of thoracic endoprosthesis placement (arrows) (Kim et al. 2008) (Copyright © 2008, American Heart Association, Inc.)

Minor deviation from gold standards has been implemented to achieve more accurate models. A distinct approach used cone beam CT (CBCT) instead of the standard multi slice CT (MSCT) to fabricate 3D models using an SLS technique for two patients, both yielding favorable results (Adolphs et al. 2013). For one patient with severe skeletal class III malocclusion in an edentulous upper jaw, the 3D model clearly demonstrated the need for a two-jaw correction to achieve alignment, whereas for the other patient with a severe class III malocclusion due to a bilateral cleft lip and palate, maxillary distraction was performed after planning the procedure on the model. In yet another demonstration, 3D printing was used for implant pre-shaping in the cases of bilateral craniomaxillofacial post-traumatic deformities, instead of simply serving as a model reference (Cui et al. 2014). In this study, 3D printed SLS models of the skull were fabricated and a rehearsal surgery was performed to discover the location, reduction direction, and shift distance of the osteotomy. Pre-shaping of the titanium plates or mesh was done on the models before implantation. Satisfactory aesthetic and functional results were achieved in all the cases with no post-operative complications (Fig. 5a). This approach has also been implemented to pre-bend titanium mesh implants for orbital floor fractures and to intraoperatively guide the correct placement of the implant (Kozakiewicz et al. 2009; Mustafa et al. 2011). Complex orbital anatomy makes intraoperative titanium mesh shaping and cutting very difficult, and results in less than satisfactory results in terms of final shape. The models helped in achieving the desired shape of the implant preoperatively and resulted in improved surgical outcomes in a shorter operating time. Improvement of vision was demonstrated by the elimination of diplopia and enophthalmos.

The testing of rapid prototyping techniques has also been implemented for ear surgery. Tympanoplasty is especially difficult because of the anatomical complexity of the ear. Safe surgery to restore good hearing in cases such as atresia is difficult even after using CT imaging and virtual reconstruction. To address these challenges, Suzuki et al (2005) demonstrated the use of selective laser sintering (SLS) to fabricate a 3D model for surgical simulation of a malformed ear. Temporal bone models of congenital aural atresia were prototyped, and conventional surgical instruments were used to dissect the models as a part of surgical planning and rehearsal. The SLS models performed better than stereolithographic models as they permitted the use of a surgical drill, such as that used on real bone, and could simulate other surgical procedures such as dissection. This finding was further confirmed by various reports on building 3D printed models of middle and inner ear anatomy to improve anatomical understanding of trainees (Suzuki et al. 2004a, b). In the enlarged models, the mastoid air cells had already been dissected, exposing sinodural angle, surgical dome, endolymphatic duct, and the sigmoid sinus. An enlarged model of the osseous labyrinth was also shown. The in-depth three-dimensional anatomical knowledge obtained from these models could greatly assist trainees in understanding the complexity of tympanoplasty.

4 Case studies

While there have been gradual improvements in the techniques used for rapid prototyping of physiological structures, most of the medical studies performed on the use of 3D printed models for preoperative planning and surgical training have focused on providing a proof of concept for morphologies of specific diseased tissues or organs of interest. The following is a summary of case studies that have focused on showing the feasibility of the rapid prototyping technique as a method for preoperative planning or training and whose novelty lies on the organ structure and not the technique itself.

4.1 Cardiac surgery

4.1.1 Congenital heart disease (CHD)

Congenital heart disease (CHD) is a common cardiac pathology and occurs in 0.8-1.2% of all live births worldwide, 50% of which are ventricular septal defects (VSD) (Costello et al. 2015; Hoffman and Kaplan 2002; Reller et al. 2008). Surgical planning and intraoperative orientation is difficult in CHD reoperation cases due to excessive scar tissue and fibrous strands which make it challenging to identify anatomical structures and dimensions. Congenital heart disease (CHD) is associated with complex intra-cardiac anatomy, due to the variability of defects which may be present. The complexity of anatomical structures in CHD cases warrants the need for 3D printing to understand the spatial organization, plan the surgery preoperatively, carry out surgical simulations, and guide the surgeons during operation. Current approaches for surgeon training, including pathological specimens, pictorial 2D images, and animal models, fail to provide a sufficient understanding of specific scenarios. The accurate visualization of intra-cardiac structure is essential for the diagnosis and treatment of such conditions. Greil et.al. (2007) demonstrated the ability to fabricate accurate 3D models for CHD cases, utilizing high spatial resolution imaging (MRI/MDCT) of patient anatomy, accurately depicting complex 3D geometries such as connections of the atria, ventricles, and great arteries of a criss-cross heart, as well as fine ventricular trabeculations and the proximal and distal course of the main coronary artery. However, small cardiac structures such as papillary muscles, semilunar valves, and atrioventricular valves were not captured due to limited image resolution caused by blood pooling and motion artefacts. Surgeons performed a simulative Norwood procedure on a 4 day old neonate with hypoplastic left heart syndrome (Fig. 3a–c). Noecker et al. (2006) fabricated rigid and flexible models of 11 patients with and without CHD, capturing the complex arch anatomy that is present in CHD cases. These powder and starch-based inkjet printed models provided tactile and spatial representation and could be used for medical device development and preoperative testing.

Mottl-Link et al (2008) demonstrated the clinical application of a powder-based 3D printed (inkjet) model of a CHD case by using it intraoperatively during a high-risk procedure. With the aid of the replica, pathological localization of highly abnormal structures and coronary arteries was easily achieved, allowing the closure of a large ventricular septal defect. In another instance of preoperative planning, Vranicar et al. (2008) fabricated hand-held models of CHD cases to evaluate the abnormality of the great arteries. 3D printed models of cardiovascular anatomy of 9 patients with aortic coarctation and 3 patients with vascular rings were constructed. The clear depictions of pathologies in the fabricated replicas aided in surgical and transcatheter intervention.

Pulmonary atresia with ventricular septal defect (VSD) and major aortopulmonary collateral arteries (MAPCAs) is an uncommon form of cyanotic congenital heart disease in which the native pulmonary arteries are highly hypoplastic or even absent. In this case, the corrective surgery aims to repair intracardiac defects, connect the MAPCAs to the pulmonary arterial system, and form continuity of right ventricle to pulmonary artery. In this process, identification and isolation of MAPCAs is one of the most challenging tasks.

In some cases of CHD in which native pulmonary arteries are highly deficient, identification and isolation of major aortopulmonary collateral arteries can prove to be highly difficult without 3D models. Using CT angiography and 3D printing, Ngan et. al. (2006) built accurate replicas of vascular anatomy that proved very helpful for preoperative planning and intraoperative success. The authors found that 96% and 93% of the major aortopulmonary collateral arteries identified during surgery and conventional angiography, respectively, were accurately represented by the models. Moreover, surgeons found the models to be highly useful in visualizing the vascular anatomy.

Sodian et al. (2007) used MRI angiography and CT data to fabricate stereolithographic models of cardiovascular anatomy of 2 pediatric patients, one with a left retro-esophageal abnormal subclavian artery and a right descending aorta and the other with a ventricular septal defect. These replicas were used for surgical planning and intraoperative orientation and were found to greatly facilitate both. The authors report that such models may not be necessary in every case, but when incorporated can aid in anticipating intraoperative problems by providing better anatomical information. Selection of the correct segmentation values during MRI and CT image processing was identified as a bottleneck of this technique. In yet another study, Sodian et al. (2008a) evaluated 3 cases of pediatric cardiac transplantation in patients with univentricular heart by using stereolithography to fabricate 3D printed models from MRI imaging data. The patients had undergone multiple corrective procedures and had relatively unpredictable anatomy. The 3D replicas assisted in preoperatively planning the heart transplantation by providing specific dimensions, spatial organization, and special requirements for the donor heart. In a similar work by Oliveri et al. (2015), fabrication using a polyjet printer from 3D echocardiography data (instead of the much established MRI or CT) provided accurate 3D printed models of congenital cardiac defects such as ventricular septal defects and periprosthetic aortic valve leaks. Comparisons between the 3D digital model and the physical model showed the congruency between the two, thus establishing that echocardiography imaging data can be reliably used for bio-modelling.

4.1.2 Coronary artery bypass grafting (CABG)

Coronary artery bypass grafting (CABG) is a process by which a healthy artery or vein from the body is grafted to the patient's blocked coronary artery, bypassing the blocked portion of the coronary artery and creating a new path for oxygen-rich blood to flow to the heart muscle. A large patient population with previous CABG have reported significant aortic stenosis (Odell et al. 1996). Reoperation on such patients for aortic valve replacement has a high mortality rate owing to the unpredictable anatomic structure (Fighali et al. 1995; Collins and Aranki 1994). Repeat sternotomy, an incision on the sternum or breastbone, entails the risk of dissecting the patient's vein or internal mammary graft from previous surgery. Sodian et. al. (2008b) fabricated a 3D printed SLA model of cardiovascular anatomy of a patient with symptomatic aortic valve stenosis and previous coronary bypass grafting. The model helped locate the bypass grafts and other structures intraoperatively and helped in opening the sternum without any damage.

4.1.3 Ventricular aneurysmectomy and reshaping

Alterations in ventricular volume and shape, along with reduced contraction of anterior wall and septum, are common conditions post anterior myocardial infarction (Zardini et al. 1993). Such ventricular dysfunction leads to aneurysm formation, which may finally lead to congestive heart failure (CHF). The corrective methods for such cases involves surgical resection of the aneurysm and ventricular reshaping. There are several techniques for ventricular reshaping, such as volume reduction (only if a dyskinetic scar is present), or surgical repair of the postinfarction dyskinetic aneurysm (Jatene 1985). Ventricular rest volume and geometry are important parameters for correct reshaping of the ventricle, and any mistake directly affects ejection fraction and stroke volume. For this particular application, Jacobs et al. (2008) fabricated powder-based 3D printed (inkjet) models of cardiovascular anatomy for 2 patients with ventricular aneurysms and used them to determine the ideal resection lines for aneurysmectomy as well as the shape of the residual ventricle post-reconstruction. Using the left ventricle (LV) aneurysm 3D model, reshaping of the left ventricle with a sufficient LV volume was easily accomplished (Fig. 3d).

4.1.4 Cardiac tumors

Primary cardiac tumors are extremely rare but have the potential to be fatal. Complete surgical resection is required for improved survival (Murphy et al. 1990). Echocardiography, MRI, and CT are all useful in determining the extent of infiltration of the tumor (Kaminaga et al. 2003), but because of the beating heart, radical complete resection is rarely possible. Additionally, due to the rarity of cardiac tumors, many surgeons have limited experience in their treatment. A 3D model of the target area could help in identifying the optimal resection volume, thus enabling maximal organ preservation. Jacobs et al. (Jacobs et al. 2008) used an inkjet powder-based technique to 3D print a cardiovascular model of a patient with cardiac fibroma and used it to plan the tumor resection procedure. Owing to the massive tumor infiltration to the tricuspid valve, the authors decided to access the tumor through the right ventricle instead of the right atrium (Fig. 3e). A complete tumor mass reduction along with tricuspid valve replacement was achieved. In a similar work, Schmauss et al (2013) built a 3D printed SLA model of a primary cardiac tumor on the right ventricular wall and successfully used it to define the infiltration of the tumor and plan for its resection.

4.1.5 Cardiovascular

The past few years have seen a tremendous rise in cases of catheter based treatment for congenital as well as structural heart diseases. Both structural modifications and novel device implantation have been shown using catheter based treatment. Cardiac valve interventions can be very complex and intra-cardiac navigation of delivery systems requires an in-depth understanding of spatial relationship of the anatomy, which varies from patient to patient. Ablation procedures in electro-physiology, which are done as intra-cardiac therapeutic procedures, also require thorough understanding of the patient's cardiovascular anatomy.

Using CT angiography data and fused deposition modelling, Kim et al. (2008) fabricated 3D printed models of cardiovascular anatomy for 4 cases: congenital muscular VSD, fenestrated ASD with large atrial septal aneurysm, prosthetic mitral valve perivalvular leak, and thoracic aortic pseudoaneurysm and ulcerations. For the VSD case, the 3D replica was used to simulate different approaches to crossing the VSD using a catheter, including superior vena cava, inferior vena cava and retrograde aortic approaches. The retrograde approach was found to be optimal in minimizing bending and maximizing primary and secondary bends of the catheter. A 12-mm Amplatzer muscular VSD occluder device was chosen for closing the defect. For the second case of ASD and aneurysm, the 3D replica was used to choose a 35 mm Amplatzer device to cover all fenestrations. Various catheters were placed in the model to finalize the approach of crossing the fenestration to place the device and to select the make of the catheter. For the prosthetic mitral valve perivalvular leak, the 3D replica was used to discover the distance needed to be travelled by the transseptally placed delivery catheter, the angular orientation required to reach the defect, and the size of the occluder device. In the last case, the 3D replica helped in identifying the number and precise locations of pseudoaneurysms and ulcers. A 31 mm diameter by 15 mm length gore TAG thoracic endoprosthesis was placed that covered both the aneurysm and ulcerations (Fig. 3h–j).

After the replacement of the ascending aorta and aortic arch, and re-implantation of the supra-aortic vessels; development of a false aneurysm is regarded as a severe complication (Mohammadi et al. 2005). The standard treatment for this condition has been surgical, with a high mortality rate (Mohammadi et al. 2005). The other option uses endovascular interventional techniques but is often complicated (Angeli et al. 2004). Decision-making is difficult in this scenario, as important information about aneurysm dimensions, its relationship to supra-aortic arteries, and the shape of the opening between the aneurysm and aortic lumen are difficult to gauge. Sodian et al (2009) used SLA to print a model of aortic anatomy of a patient with aortic pseudoaneurysm after replacement of the ascending aorta. The model helped in fabricating a custom-made occluding device and in simulating the intervention procedure. Using the 3D replica, surgeons could identify the dimensions of the pseudoaneurysm as well as the shape of the opening between the aortic lumen and false aneurysm. Due to the slit shaped opening, the occluder device had to be custom built, as the Amplatzer commercially available devices are round in shape. This conclusion likely would not have been made without the printed model.

4.1.6 Transcatheter AVR

In recent years, transcatheter aortic valve replacement (TAVR) has become a promising alternative to open surgical methods. However, the standard outline for preoperative planning for such procedures is largely missing. Schmauss et al (2012) report a case of a 70 year old patient with severe aortic stenosis and a porcelain aorta who was referred for transcatheter aortic valve replacement but died during operation due to ischemia. The authors, using powder and binder-based technique (inkjet), did a postoperative 3D fabrication of the aortic anatomy of the patient and used it to re-evaluate the course of surgery. The model assisted in the discovery that the heart valve should be placed deep into the aortic annulus in cases with a small and inelastic sinus of Valsalva, preventing the occlusion of the coronary ostia and improving the outcome of such surgeries.

4.1.7 Transcatheter percutaneous pulmonary valve implantation (PPVI)

The number of patients requiring surgical replacement of the pulmonary valve or conduit continues to increase. Pulmonary valve abnormalities are most common in patients with repaired congenital heart disease as pulmonary incompetence can occur in such cases. Scheivano et al (2007) showed the use of the "drop on demand" 3D printing technique for transcatheter percutaneous pulmonary valve implantation (PPVI), which can overcome many disadvantages of surgery but is limited to patients with a specific anatomy. This procedure requires an exact size, shape, and physical properties of the implantation site in order to safely anchor the valved stent. Preoperative planning and patient screening is even more crucial due to large variation in 3D geometry and the relationship of the right ventricular outflow tract (RVOT) and pulmonary trunk among patients. The authors used 3D printed rigid models of RVOT and pulmonary trunk for the patient selection process and showed that use of printed models resulted in a more accurate selection as compared to the MRI alone.

4.1.8 Left atrial appendage closure

Left atrial appendage closure, a treatment to reduce the risk of blood clots entering the bloodstream and causing stroke in non-valvular atrial fibrillation patients, is currently done using the Watchman device under fluoroscopic and transesophageal echocardiographic guidance. MDCT offers a 3D dataset with better resolution but can still result in procedural failure owing to the complex morphology. Otton et al (2015) fabricated a 3D printed (polyjet) patient-specific model to guide a left atrial occlusion surgery using the Watchman device. The model was fabricated using a rubber-like material to simulate atrial mechanical properties. Watchman devices of varying sizes were placed in the model and anatomic deformation tests were performed. This exercise helped in selecting the correct dimension of the device to avoid any post-operative complications.

Ensuring patient safety and improving surgical success rates are the ultimate goals for any surgeon. Fueled by these motives, the field of surgical training has seen widespread acceptance of simulation-based training curriculum. Surgical simulation creates a safe environment for practice and error that maximizes learning (Akaike et al. 2012). In spite of the widespread acceptance of the concept, its application is still scant in critical areas like congenital heart disease surgeries. Below we discuss various specific applications of 3D printing for surgeon training in the field of cardiac surgery.

4.1.9 Ventricular septal defect - pediatric heart

Pictorial images and 2D echocardiography data of CHD prove insufficient for users who lack the real-world 3D perspective. Animal models with structural defects are limited, and do not provide an accurate representation of CHDs in humans. Costello et al. (2015, 2014) conducted multiple teaching seminars on targeted surgical simulation of VSDs in the pediatric heart with 3D printed (polyjet) replicas of 5 common VSD subtypes. The models were made of a plastic and a rubber-like material mixture, and could be cut with a scalpel blade without damaging the structural integrity. Although limited MRI image resolution prevented accurate depiction of the heart valves in the models, all participants in the seminar reported significant improvement in knowledge acquisition, knowledge reporting, and structural conceptualization (Fig. 3f–g).

4.1.10 Trans-apical aortic valve replacement

Trans-apical aortic valve replacement (AVR) is a growing technique that requires a high level of surgical skill. Existing training approaches have been similar to the aforementioned techniques used in conventional CHD training. For trans-apical AVR training, the trainees must be able to see through the heart to visualize their performance. Current artificial heart models are usually opaque, as they are made of polyurethane, and thus are not ideal. Abdel-Sayed et al. 2009; Abdel-Sayed and Von Segesser 2011) found a solution to this problem by using CT imaging data to fabricate translucent heart models made of silicone which allow the trainee to see the stent valve throughout the exercise.

4.2 Neurosurgery

4.2.1 Brachytherapy

Poulson et al. (1999) described the use of 3D printed models for neurosurgery planning as early as 1999. A fabricated stereolithographic model for a patient diagnosed with an orbital tumor was used to assist an orbital brachytherapy. Brachytherapy is an advanced cancer treatment where, using hollow plastic or metal needles, radioactive seeds or sources are placed in or near the tumor, giving a high radiation dose to the tumor while reducing the radiation exposure to the surrounding healthy tissues (Tyl et al. 1997). This is usually done under direct visualization in cases where the tumor is palpable. Correct needle placement is critical as the radiation dose from the seeds falls off rapidly. Incorrect needle placement will result in "hot" or "cold" spots causing complications in healthy tissues, such as necrosis. Application of brachytherapy to orbital tumors poses two specific challenges: first, the tumor is not visually obvious so accurate needle placement is difficult; second, the needles may damage the optic nerve. Using the 3D printed replica, the authors determined important pre-surgical questions, including the number and length of the needles needed to cover the tumor, the trajectory of needle insertion, and the separation between the needles being used (Fig. 4a–b).

4.2.2 Skull base surgeries

Muller et al (2003) fabricated SLA and SLS models of patients to assist in the planning of skull base procedures. Exposure to the lesion through precise planning is critical for surgery, and many standard approaches to different skull base regions have been demonstrated (Fisch 1985; Al-Mefty et al. 1987; Bertalanffy 1991). Lesions which cannot be reached or exposed through standard approaches are often encountered, and dealing with such cases requires further understanding of the patient's anatomical variations. In this study, a standardized approach using only imaging (MRI or MDCT) was found to be ineffective. The skull base lesion and surrounding structures were reconstructed in 3D on a computer screen, using Materialize 3D software. These models also aided in determining the surgical route and provided perspective for deep tumors. Skull base tumors require tailoring of procedures to individual patients with respect to craniotomy, bone resection, spatial composition of the target, and the surgical trajectory. This combination approach proved very useful for these processes (Fig. 4d–g).

4.2.3 Lesion near motor cortex

Spottiswoode et al. (2013), using powder based inkjet printing, fabricated 3D printed brain models for two patients with lesions in proximity to the motor cortex. They also attached the fMRI

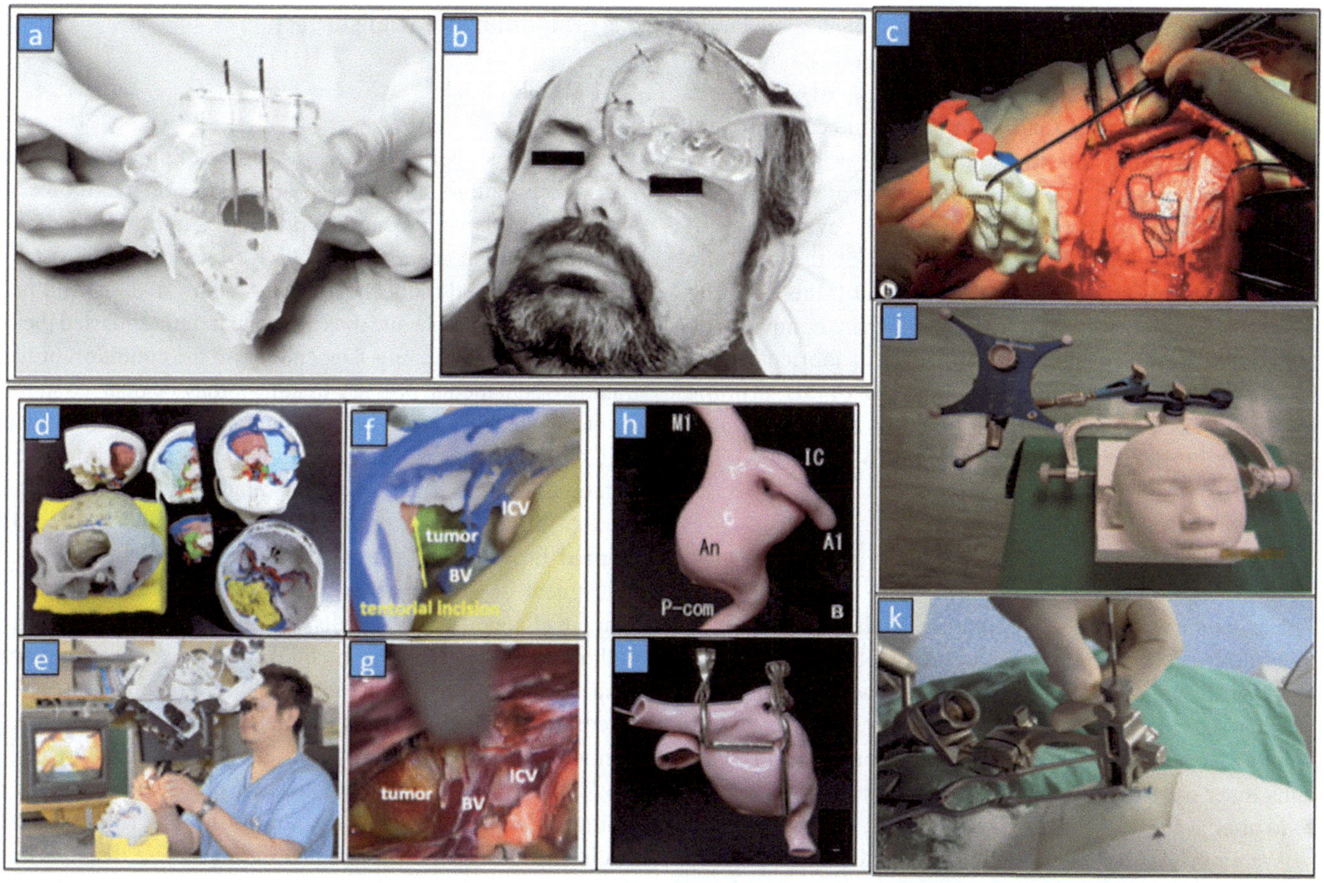

Fig. 4 3D printing applications in neurosurgery. **a–b** SLA biomodel for brachytherapy with template and needles and the patient with the template (on the right) (Poulsen et al. 1999) (Copyright © 1999 Elsevier Science Inc.) (**c**) 3D printed model showing lesion in blue and fMRI region in red and the corresponding features shown intraoperatively (Spottiswoode et al. 2013) (Copyright © 2013, Karger Publishers) (**d–g**) 3D printed color models of cerebrovascular anatomy; surgical simulation using model; tumor seen in the model and corresponding view intraoperatively (Oishi et al. 2013) (Copyright © 2013, American Association of Neurological Surgeons) (**h–i**) Flexible model of left internal cerebral artery aneurysm and simulation of angioplastic neck clipping with 2 fenestrated clips (Mashiko et al. 2015) (Copyright © 2015 Elsevier Inc.) (**j–k**) 3D printed model of a patient with deep seated tumor and with image guidance frame in place for surgical simulation; tissue biopsy in progress during surgeon training (Waran et al. 2014a) (Copyright © 2014 Association of Program Directors in Surgery. Published by Elsevier B.V.)

volumes to the 3D printed models to get a clear indication of proximity of functionally eloquent areas to the tumor. This additional information can help surgeons plan the safest surgery route. The model showed detailed grey matter surface features, extent of the tumor infiltration, and gyri for intuitive navigation compared to actual brain morphology (Fig. 4c). The model fabricated in this study was found to have the appropriate dimensional accuracy (x,y,z spatial resolution of 0.042mm, 0.047mm, 0.089mm, respectively) to be useful in neurosurgical planning as well as training purposes. These models can also be used to facilitate communication with patients, in the explanation of the surgical procedure and associated risks.

4.2.4 Cerebrovascular

Cerebral aneurysm surgery is another major class of neurosurgery that requires thorough understanding of both the shape of the aneurysm and its relationship with the parent artery, cranial nerves, brain, and bones. The standard technique for open cerebral aneurysm surgery is microsurgical clipping, wherein a small metal clip is used to stop the blood flow into the aneurysm. However, the use of endovascular embolization has also recently increased (Gnanalingham et al. 2006). Cases unsuitable for endovascular intervention are referred for clipping surgery and are usually complex and require in-depth understanding of the anatomical structure, effective planning, and practice. The diagnostic techniques to achieve the thorough understanding required for clipping surgery have evolved from cerebral angiography, wherein dyes and X-rays were used to visualize the blood flow patterns in brain, to high resolution techniques like CT angiography and MR angiography that also allow virtual 3D reconstruction of the images. Many researchers have exploited this advantage to demonstrate applications of virtual surgical simulation in neurosurgery (Futami et al. 2004; Satava 2001; Stadie et al. 2008). The cerebral aneurysms were found to be captured accurately by the 3D printed models and demonstrated the relationship of the vasculature to the skull base. Clipping was simulated on the printed model, and surgeons found that they could optimize the patient positioning for surgery by manipulating the model to find the best angle of approach. However, vessels less than 1mm in diameter could not be fabricated using this method. The 3D printed models assisted in surgical planning, selection of the correct treatment (i.e clip or coil application), evaluation of the neck of the aneurysm, and selection of the best clip positions during surgery. The accuracy of the vascular models was found to be satisfactory when compared to the intraoperative video. Others have also fabricated rigid anatomical models validating the use of 3D printing for intracranial aneurysm biomodelling and confirming its feasibility for this application (Erbano et al. 2013; Xu et al. 2014).

4.2.5 Cerebral tumors

The cavernous sinus (CS) is comprised of the venous plexus, multiple cranial nerves, and the internal carotid artery. Thus, it is a key location for skull base surgery owing to the many intracranial pathologies that can occur in this region. Another such interesting region is the temporal bone that contains inner ear structures, cranial nerves, and the internal carotid artery.

Waran et al. (2014a, b; 2015) published multiple reports using 3D printing for simulating neurosurgical procedures. In their first report, patient head models with deep-seated small thalamic lesions were prepared using MRI and CT data and used for educating trainees on patient positioning in relation to pathology, utilization of navigation systems for cranial surgery, target selection, and surgical planning. The simulation included creating a burr hole in the skull followed by a biopsy on the underlying tumor. Medtronic S7 and BrainLAB Kolibri navigation platforms were used for navigation related steps. It was possible to load all imaging information onto both platforms and subsequently register the models as if the trainees were operating on a real patient (Fig. 4j–k). In a second report, the authors used similar 3D printed models for surgical training in navigation and planning of skin flap, to performing initial steps in a craniotomy and simple tumor excision. Most recently, the authors used a similar model for simulating third ventriculostomies and pineal biopsies for hydrocephalus secondary to a pineal tumor. In each of these cases, the model was fabricated using patient MRI and CT data. These models allowed endoscopic intraventricular biopsy training in addition to basic craniotomy procedures. The authors concluded that 3D printed models provide standardized and reproducible procedures, allowing candidates to repeat steps until they are confident, and thus improving the surgical training experience.

4.3 Craniomaxillofacial surgery

4.3.1 Maxillo-facial

Maxillofacial surgeries involve the region of upper jaw, nose, eye sockets, and the face. The earliest works on maxillofacial preoperative planning using 3D printing can be traced back to the late twentieth century. In 1998, Kermer et al. (1998) fabricated 3D printed models using SLA with CT data from patients suffering from acute maxillofacial trauma and used them for preoperative planning. Depending on the condition of the traumatized patients, primary treatment may be delayed in the favor of intensive care, and this time can be utilized to plan and rehearse surgery on 3D printed models for improved outcomes. In all of the patients treated in this study, the intraoperative situation was adequately captured by the 3D printed models and simulation surgery helped in better adjustment of mini or micro plates, resulting in satisfactory functional and aesthetic results. As an additional remark, the authors advised against the use of 3D printing for cases of very small fragments in the midface fracture due to the disconnected nature of the anatomy. Around the same time, Santler et al. (1998) published a cumulative report of their work with 3D printed SLA models for preoperative planning from 1988 to 1997. They found that in the case of congenital malformations the models helped in the precise diagnosis of the deformity whereas simulation surgery allowed prediction of intraoperative problems beforehand. In the cases of trauma, osteoradionecrosis, and osteomyelitis, the size, shape, and localization of the defect determined the transplant donor site and 3D printed models greatly facilitated this process. The models helped in exact jaw positioning in patients suffering from dysgnathia and allowed better planning for asymmetrical malformations. For patients suffering from severe alveolar crest atrophy, exact bone measurements on the 3D printed models helped in deciding whether bone transplants, dental implants, or a combination would be ideal. Also, rehearsing the surgery on the models allowed perfect implant positioning in relation to the opposite jaw. Overall, the authors believed that 3D printing facilitated precise diagnosis and correct surgical approach selection, and reduced operation time. In another work, Sinn et al. (2006) demonstrated the application of 3D printing (SLA) to fabricate custom adaptation of internal distracter device to be placed at the level of the zygomatic body and arch of a patient suffering from Cruzon's disease with severe hypoplastic maxilla. Custom-made devices are most useful where the already complex anatomy is further distorted as they can help avoid some of the problems like mechanism failure or loosening of the device due to inadequate bone stock. The 3D printed SLA models helped identify the area of greatest bone stock prior to surgery so that the device could be placed in the area of maximum bone.

The surgical training reports in craniomaxillofacial surgery have been confined to surgeries related to the ear. The complex three-dimensional anatomy of the temporal bone, along with its close association with neurovascular structures, requires extensive skill and experience. The established training module utilizes cadaveric temporal bones, but such specimens for trainee dissection have limited accessibility with significant costs of procurement (Longfield et al. 2015). 3D printed temporal bone models are a lucrative substitute to this conventional training approach.

4.3.2 Mandibular

Mandible reconstruction is a very complex procedure owing to the intricate geometry, presence of connected muscles which act in different directions, position of condyles in the glenoid fossa, and occlusion. Reconstruction is needed after partial resection and continuity defect, and is usually done with a titanium bone plate followed by an autogenous bone graft at a later stage. The bone graft allows dental rehabilitation and also provides aesthetic and functional restoration to the face. 3D printing is useful in providing the anatomical detail and the flexibility to physically carry out surgical procedures on models before the actual surgery. Multiple authors have demonstrated its relevance to mandibular reconstruction post tumor resection (Cunningham et al. 2005; Robiony et al. 2008). The focus of the treatment was to resect the tumor and preserve the form of the mandible. The 3D printed SLA models of the mandibular anatomy based on patient CT data allowed prefabrication of contour reconstruction plates, which in turn provided a more conservative surgical approach (Fig. 5b). The biggest disadvantage of 3D printing models was the added time and cost. In addition, Robiony et al. (2008) presented another mandibular application of 3D printing for class II dentofacial deformities where the 3D model was used to plan the elongation of the hypoplastic mandible in the desired occlusion. Previously, Gateno et al. (2000) had developed a protocol using virtual reality to simulate distraction osteogenesis of the mandible, and the authors used this to plan the correct elongation vector axis. After virtual evaluation of the functional and aesthetic aspects, a stereolithographic model of the surgical solution was prepared and used for intraoperative guidance. Good aesthetic and occlusal results were seen (Fig. 5c–e). In another demonstration, Sinn et al. (2006) used 3D printing to plan and create a custom HTR implant for the left mandibular body of a patient suffering from hemifacial microsomia, a congenital disorder which affects the development of the lower half of the face, usually including the ear, mouth, and mandible. In this case, the 3D printed model helped in constructing the implant to exactly match the contour of the deformed left mandibular body. Satisfactory aesthetic and functional results were observed and no radiographic evidence of implant migration or bony resorption beneath the implant were found in the 5 year follow-up. Zhou et al. (2009) reported similar use of 3D printing in another case of unilateral hemifacial macrosomia. The left jaw of the patient was less developed than the right. CT data showed the hypoplastic left mandible with wasting soft tissue and thinning skin and fat of the left cheek. As a standard procedure in unilateral deformities, the image of the right mandible was mirrored to the left using computer aided design (CAD) to restore the facial symmetry. The 3D printed model of the patient's mandibular anatomy helped in understanding the deformity and in deciding the best surgical treatment path. Satisfactory results were achieved. Oliveira et al. (2008) showed the use of 3D printed (SLA) biomodels in the surgical planning for two cases. One patient, suffering from partial mandible destruction due to a gunshot wound, had previously been implanted with a titanium plate to restore facial symmetry. In this surgery, the 3D printed models were used to determine the volume, length, and angulation required of the free flap obtained from autologous peroneal bone. A second patient, with ameloblastoma in the mandible, underwent tumor resection followed by autologous peroneal bone graft implantation. The 3D printed model helped in understanding the exact defect dimensions and positioning to determine the general morphology of the implant. Complete restoration of the defect and biofunctionality was seen for both the patients. Similar work on amelobastoma cases in the mandible has been shown by other authors (Sannomiya et al. 2008; Cohen et al. 2009). In addition, Cohen et al. (2009) demonstrated the use of a 3D printed model as a template for iliac crest bone graft in second stage reconstruction where a comparison between the template and the bone graft revealed that the bone graft was short. This allowed the surgeons to plan appropriately, and a few bony fragments were fixed to the bridging plate instead of a monoblock.

4.3.3 Craniofacial

The earliest reports of 3D printing in craniofacial surgery date back to the late twentieth century. Bill et al. (1995) published a report in 1995 showing application of 3D printing for reconstruction of severe skull defects. With fabricated models of calavarial defect and the original skull with the defect, the SLA model was used to shape the donor bone for implant and the sterilized skull model was used to simulate the actual transplant surgery. The authors concluded that 3D printing helped in reducing surgical time and improving results for complex craniofacial surgeries (Fig. 5j–l). Sailer et al. (1998) used 3D printing (SLA) in multiple cases of craniofacial surgery including plagiocephaly, hypertelorism, Apert's syndrome, Crouzon's syndrome, Pfeiffer's syndrome, Treacher-Collins syndrome, and post-traumatic skull base and periorbital lesions, among others. The stereolithographic models of the various anatomies served diverse functions such

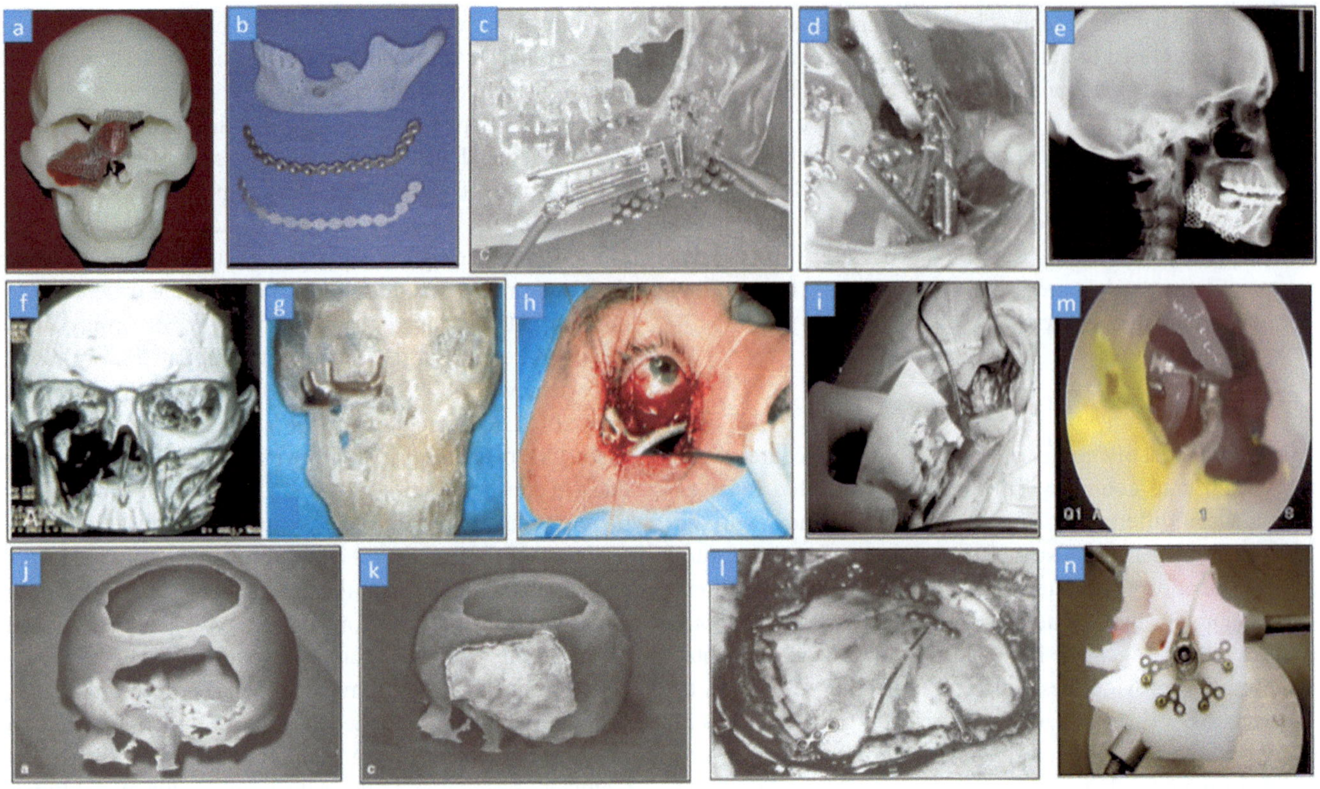

Fig. 5 3D printing applications in cranio-maxillo-facial surgery (**a**) Preshaping of titanium mesh implant on 3D printed model of a patient suffering from post-traumatic deformity (Cui et al. 2014) (Copyright © 2014 American Association of Oral and Maxillofacial Surgeons. Published by Elsevier Inc.) (**b**) SLA model of mandible and prebent reconstruction plate (Cunningham et al. 2005) (Copyright © 2005 American Association of Oral and Maxillofacial Surgeons. Published by Elsevier Inc.) (**c–e**) Distraction osteogenesis simulated on SLA model; intraoperative distractor position and radiographic image of the patient post operation (Robiony et al. 2008) (Copyright © 2008, 2008 by Mutaz B. Habal, MD.) (**f–h**) CT scan of right orbital bony defect; SLA model of the same with custom fit titanium implant; intraoperative view of the implant (Holck et al. 1999) (Copyright © 1999 American Academy of Ophthalmology, Inc. Published by Elsevier Inc.) (**i**) 3D model of patient suffering from atresia of right ear used for surgical planning and its intraoperative view (Suzuki et al. 2005) (Copyright © 2005 Taylor & Francis) (**j–l**) 3D printed model of a cranial defect; shaping of the calavarial bone implant; the implant fit perfectly intraoperatively (Bill et al. 1995) (Copyright © 1995 Published by Elsevier Ltd.) (**m–n**) Surgeon training- posterior of the tympanic cavity and with a prosthesis carina screw (Bakhos et al. 2009) (Copyright © 2010, 2010 Otology)

as planning of cranial, skull base, orbital, or nasoethmoidal osteotomies, 3D-planning of orbital reassembly, assessment of periorbital deficiency, and planning of contouring and implant location. Satisfactory aesthetic and functional results were seen for all the cases. D'Urso et al. (1998) also used SLA in many similar cases. Two prominent cases included one patient with bicoronal craniosynostosis, a facial cleft, and hypertelorism; and the other patient with a very large right sided facial encephalocoele. The fabricated model was used to perform a surgical rehearsal by doing the exact osteotomy and reconstruction on the biomodels of both patients. These models helped predict possible intraoperative complications. The surgeons concluded that the ability to shape bone grafts intraoperatively using a biomodel template and to physically reconstruct the biomodels greatly improved the results of surgery. Later, similar use of 3D printed models specifically for hypertelorism surgical planning was demonstrated by Hidalgo et al. (2009). In another report, D'Urso et al. (2000) showed the use of 3D printing (SLA) to fabricate customized cranioplastic implants for patients with craniotomy defects. The ideology and process route was similar to that of Bill et al. (1995), described previously. D'Urso et al. (2000) presented a 3D printing (inkjet and FDM) based cranial reconstruction approach for a patient with a large post-trauma defect in the left-frontal bone. The use of the 3D printed model, in addition to software simulation, allowed for a prosthesis creation that perfectly fit the defect and restored facial symmetry while reducing the operation time.

4.3.4 Orbital

Orbital reconstruction is a prominent craniofacial sub-section where 3D printing has shown a large presence in clinical simulation. Perry et al. (1998) published one of the first accounts in 1998, using stereolithography to produce models for patients suffering from various orbital deformities such as post-traumatic, fibrous dysplasia, radiation hypoplasia, high facial cleft, and partial Romberg syndrome. The models were crucial in the planning of various procedures, including orbital volume restoration by resection of bone or grafting, and augmentation of a hypoplastic orbit using a lyo-cartilage graft to restore symmetry. For planning osteotomies of the orbital margins and the zygomatic complex, the models were useful for planned repositioning of the superior orbital rim and recontouring of the frontal bone. Satisfactory aesthetic and functional results were seen in all cases. Holck et al. (1999) published another report where 3D printed models were used for preoperative planning in two cases. For one patient, who was suffering from binocular vertical diplopia and globe compression due to displaced left orbital roof fracture, the SLA model was used to measure the dimensions of the displaced bone fragment along with its relationship to the anterior cranial fossa and the left frontal sinus. It was also used to plan and practice osteotomy. For a second patient suffering from binocular oblique diplopia and globe ptosis due to a previous excision surgery for cylindrical cell papilloma of the paranasal sinuses, the SLA model served as a template to fabricate a temporary custom-fit prosthesis to repair the defect of the orbital walls (Fig. 5f–h). The 3D printing-based procedure improved post-surgical outcomes with no complications. In a later report, Fan et al. (2007) demonstrated the application of SLA models for cases involving late reconstruction of complex orbital fractures, which can lead to aesthetic deformities as well as functional impairment, and are usually treated by pulling back herniated tissue into orbit and repairing the defect using an alloplastic implant. Reconstructing complex orbital fractures becomes more difficult if the bone healing has already occurred. Such late reconstruction procedures usually involve osteotomy, fractured bones, and globe reposition and fixation (Patel and Hoffmann 1998; Carr and Mathog 1997). The authors found that the 3D models in conjunction with software simulations helped in planning osteotomy, movement, reposition, and fixation of fractured bones and globe. Better pre-surgical planning resulted in improved surgical outcomes in all cases. Rohner et al. (2013) fabricated a 3D printed model for a patient requiring radical resection of the left orbit due to recurrence of an adenoid cystic carcinoma of the palate. The reconstruction was done with cranial bone grafts and a free latissimus dorsi flap. The model was used for planning of the resection, pre-bending of the reconstruction plate, and planning grafts from parietal bones apart from analyzing the anatomy. On a similar note, Park et al. (2015) used 3D printed skull-orbital anatomies to pre-mold synthetic scaffolds for optimal orbital cavity reconstruction. Postoperative complications were analyzed, and volumetric measurement of the orbital cavity showed insignificant volume differences between the damaged and reconstructed orbits. Moreover, successful reconstruction using this method was shown for patients who had lost more than 40% of their orbital wall in addition to soft tissue loss.

4.3.5 Ear

Surgery on a malformed ear demands a thorough anatomical understanding of the patient in addition to surgical experience. Tympanoplasty is especially difficult because of the anatomical complexity of the ear. Safe surgery to restore good hearing in cases such as atresia is difficult even after using CT imaging and virtual reconstruction. To address these challenges, Suzuki et al. (2005) demonstrated the use of selective laser sintering (SLS) to fabricate a 3D model for surgical simulation of a malformed ear. Abnormal structures of the middle ear such as the labyrinth, ossicles, and facial nerves were easily visualized in the models to aid in preoperative surgical planning (Fig. 5i). The authors showed similar surgical training work for ear surgeries. In another report, Bakhos et al. (2009) used stereolithography to prepare temporal bone models in which the surgical simulation of the posterior cavities, including a mastoidectomy followed by a trans-labyrinthine approach was possible. As a part of their surgical simulation exercise, two types of implantable middle ear prosthesis were introduced with satisfactory results (Fig. 5m–n). In a similar work, Longfield et al. (2015) showed the use of 3D printing for producing pediatric temporal bone for surgical training.

Cadaveric pediatric temporal bones are extremely rare, and no such training models existed before this. The fabricated temporal bone models were drilled for mastoid and facial recess as well as cochleostomy. Both models were assessed to be anatomically accurate with good mastoid pseudopneumatization and intact landmarks. The drilling efficacy for the models was found to be slightly softer than cadaveric temporal bones, and without the accompanied changes in density associated with the labyrinth, but was overall satisfactory. These models would greatly improve trainee learning experience, although at the current stage they cannot replace cadaveric models. Further work needs to address this limitation and this is discussed in the later sections.

4.4 Orthopedic surgery

This section discusses cases with applications of 3D printing in the general area of bone surgery, apart from the cranio-maxillo-facial cases discussed above. As a general orthopedic surgeon would usually see patients with fractures in different parts of the body, we found the same trend in the reports published. The reports in orthopedic surgery frequently involve work by a single author overlapping various parts of the body and are usually not limited to a particular anatomical location. Thus, we describe the review of these articles in their chronological order instead of dividing them into anatomy-based subsections.

Orthopedic surgery cases involving extensive primary injuries with multiple bone fragmentation or bone deformities pose significant challenges for the surgeon. In such cases, custom implants and tools might be required to correctly shape the fixation plates and plan the surgical route, which requires thorough understanding of local shape variations and bone dimensions (McGurk et al. 1997). The revision of an orthopedic procedure is another example requiring a complex reconstructive process (John et al. 1995; Lachiewicz and Hussamy 1994; Kavanagh 1991). Changes in patient bone geometry, such as variation in local residue bone thickness after the last implant operation, can further complicate the case. It has been shown that joint replacement revisions have higher failure rates than primary cases (Kavanagh 1991; Bryan and Rand 1982). CT imaging data and 3D virtual reconstructions have been used to assist orthopedic surgical planning but they lack in providing the intuitive three-dimensional extent of bone defects, understanding of deformed bone anatomy, visualization of integration of bone fragments, or rehearsal of complex manipulations. 3D printed physical models from CT data can solve all the above problems, while also serving as a medium for explanation of the procedure to patients and obtaining informed consent. These physical models can be used, for example, to show the extent of tumor invasion into a bone, to study bone reassembly for cases of bone fragmentation, to build grooves to mark drilling procedures, and to test implants before the actual surgery.

Potamianos et al. (2015) described the use of 3D printing (SLA) for a case involving a right shoulder injury with the possibility of a clavicle and scapula double fracture. Such a fracture, which would completely disconnect the bone between shoulder and trunk, would require extensive surgery with plate and screw stabilization and hence would need a confirmed diagnosis before proceeding. CT imaging data and 3D virtual reconstruction did not allow for a definitive diagnosis given the complexity of fracture lines in the case, so a 3D printed SLA model was fabricated to analyze the fracture. The model clearly showed that the fracture was not a floating shoulder type and that the clavicle sustained a complete transverse fracture, while the scapula remained partially attached. Once the diagnosis was confirmed, the patient was spared the unnecessary surgery.

In a later study, Brown et al. (2003) used 3D printing as a template guide to understand and correct several cases of severe acetabulum fractures, spine pathology, and fractures of the tibia, ankle, foot, pelvic ring, shoulder, arm, femur, knee, and wrist. 3D printed models of an unfractured hemipelvis were used to understand the anatomy, contour the reconstruction plates, and establish the drilling trajectories. Plaster models of the fractured hemipelvis were then used to test the implant positioning template prior to application in the patient. For cases of the spine, midline split spine models helped to elucidate the position of fracture fragments, the compromise of the spinal canal, and the complete pedicle visualization for pedicle screw placement including anatomic starting point, screw trajectory, and length. 3D printed models with trajectories of pedicle screw placement were split into right and left halves and taken to the operating room with the patient for guidance. The models were also used to fix articular fractures, precisely position the plates, and direct osteotomies. For acetabular fractures, the contoured plate was positioned and locked with screws onto reduced fracture anatomy (Fig. 6). Similar works were published by Frame and Huntley (2012) and Guarino et al. (2007) showing applicability of rapid prototyping in forearm and spine-pelvis surgeries, respectively. In a different demonstration, Ahn et al. (2006) published 3D printing (polyjet) application in pre-surgical planning for distal tibia comminuted fracture and iliac wing fracture of the pelvis. In the first case, the physical model was used for proper implant selection and preforming of the implant to obtain satisfactory geometrical conformity to the contour of the physical model. For the second case, because the iliac wing of the pelvis had an excessive flexion, the pelvic reconstruction plate was chosen as the implant, as it could bend to conform to the bone shape. The implant fixation position, incision position, and its size were decided by using the 3D printed model. Satisfactory results and reduction in operation time were seen post-surgery in both situations. Hurson et al. (2007) published another report on the use of selective laser sintering (SLS) 3D printed models for the preoperative planning of acetabular fractures. The authors concluded that the models allowed better appreciation of the degree of anterior displacement of the fracture fragments and significantly reduced inter-observer variability during injury classification; additionally, SLS provided a full sized pelvic model in a relatively faster time (12 h). (Dhakshyani et al. 2010, 2012) published multiple papers on the use of rapid prototyping models for dysplastic hip orthopedic surgery. Dysplastic hip is a developmental hip disorder in which the acetabular roof is shallow, leading to early degeneration of the bone (Su et al. 2008; Mavčič et al. 2002; Umer et al. 2006). The authors showed that the FDM models allowed better surgical planning towards securing hip stability by proper implant size selection, bone grafting, femoral shortening, or proper consideration of high hip center and leg length discrepancy. The models also allowed for the selection of the best surgical route in advance.

Congenital scoliosis is a lateral curvature in the normally straight vertical line of the spine where early and aggressive deformity treatment can minimize risk to patient health. Posterior hemivertebra resection has proven to be an effective method for treatment of congenital scoliosis (Winter et al. 1984; Thompson and Marks 1995; Ruf and Harms 2002; Lazar and Hall 1999; Andrew and Piggott 1985). The surgical goal involves restoration of a straight spine and a physiological sagittal profile with the shortest fusion segment. Wu et al. (2011) showed that rapid prototyped models (such as FDM) can reduce the operation time, the risk

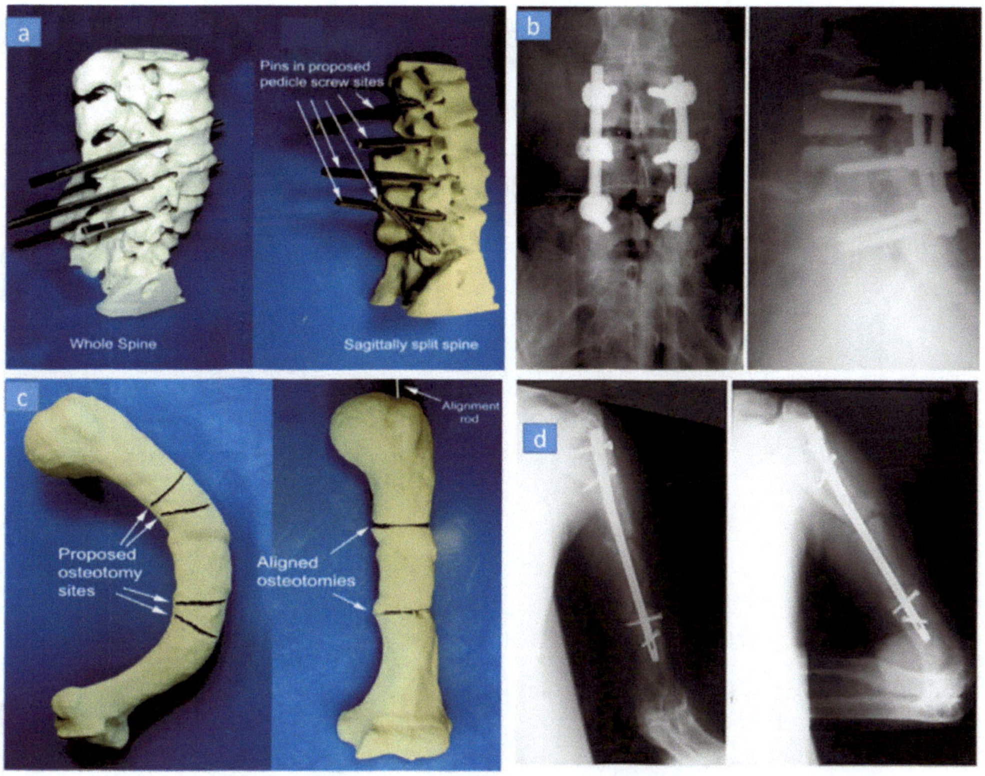

Fig. 6 3D printing applications in orthopedic surgery. **a–b** 3D printed model of lumbar spine of a fifty-year old woman with instability of L3-L4 and L4-L5 and with pins placed in proposed trajectories for pedicle screw insertions; anteroposterior and lateral radiographs of lumbar spine post-surgery. **c–d** 3D printed model of left humerus with proposed osteotomy sites and after alignment post simulation; internal and external rotation radiographs of left humerus post-surgery (Brown et al. 2003) (Copyright © 2003, by The Journal of Bone and Joint Surgery, Incorporated)

of pedicle screw misplacement, and its consequent complications in a number of cases of scoliosis treated with pedicle screw implants post hemivertebra resection. The authors concluded that although 3D CT (virtual reconstruction) allowed accurate identification of congenital spine anomalies, determination of many tangible skeletal parameters such as screw entry point, trajectory, and length, and pedicle diameter and thickness was possible only through 3D printed models.

In another approach, Debarre et al. (2012) assessed the feasibility of 3D printing (FDM) in assisting diagnosis of three cases: epiphyseal malunion, shoulder arthroplasty, and femoral trochleoplasty. In the first case, the FDM model helped in understanding the anatomy and in planning of the osteotomy lines. In the second case, the scapulohumeral 3D printed model enabled excellent estimation of available bone capital by manually comparing available bone volume to the intended implant and also allowed for practice of the glenosphere and screw positioning. For the third case, excessive relief of the proximal part of the trochlea was perfectly visible in the model, allowing its size to be quantified with precision. Moreover, are bony outgrowths on a normal bone such as "Spurs" and "double contours" were very accurately shown, facilitating trochleoplasty planning. In a similar setting, Starosolski et al. (2014) published a paper with 3D printing (FDM) application for a patient with a complex proximal femoral deformity from healing Perthes disease. The 3D printed model provided an understanding of the precise location and size of the femoral head and neck deformity as well as the degree of the hip flexion when it came into contact with the acetabular rim. Studying the anatomy on the 3D model allowed for planning of the implant placement as well as rehearsal of the complete surgery on the model. The functionality was fully restored post-surgery.

4.5 Other

There are a few isolated reports published on 3D printing applications in organs areas not covered in the above sections. This section will discuss such special applications of 3D printing.

4.5.1 Lung

Bustamante et al. (2014) published a paper on using 3D printing (polyjet) technology for simulating bronchoscopic anatomy. Unfamiliarity with endoscopic bronchial anatomy is correlated with failure in achieving lung isolation, and the authors believe that these 3D bronchial anatomical models could bridge the knowledge gap. Also, uncommon and rare variants of the tracheobronchial tree can be replicated for training purposes. The authors fabricated two 3-dimensional models of the tracheobronchial tree, one with normal anatomy and the other with an early take off of the right apical bronchus. These models were then attached to mannequins and examined using flexible fiberoptic bronchoscope (FBB) and shown to be congruent to actual patient anatomy (Fig. 7a–c).

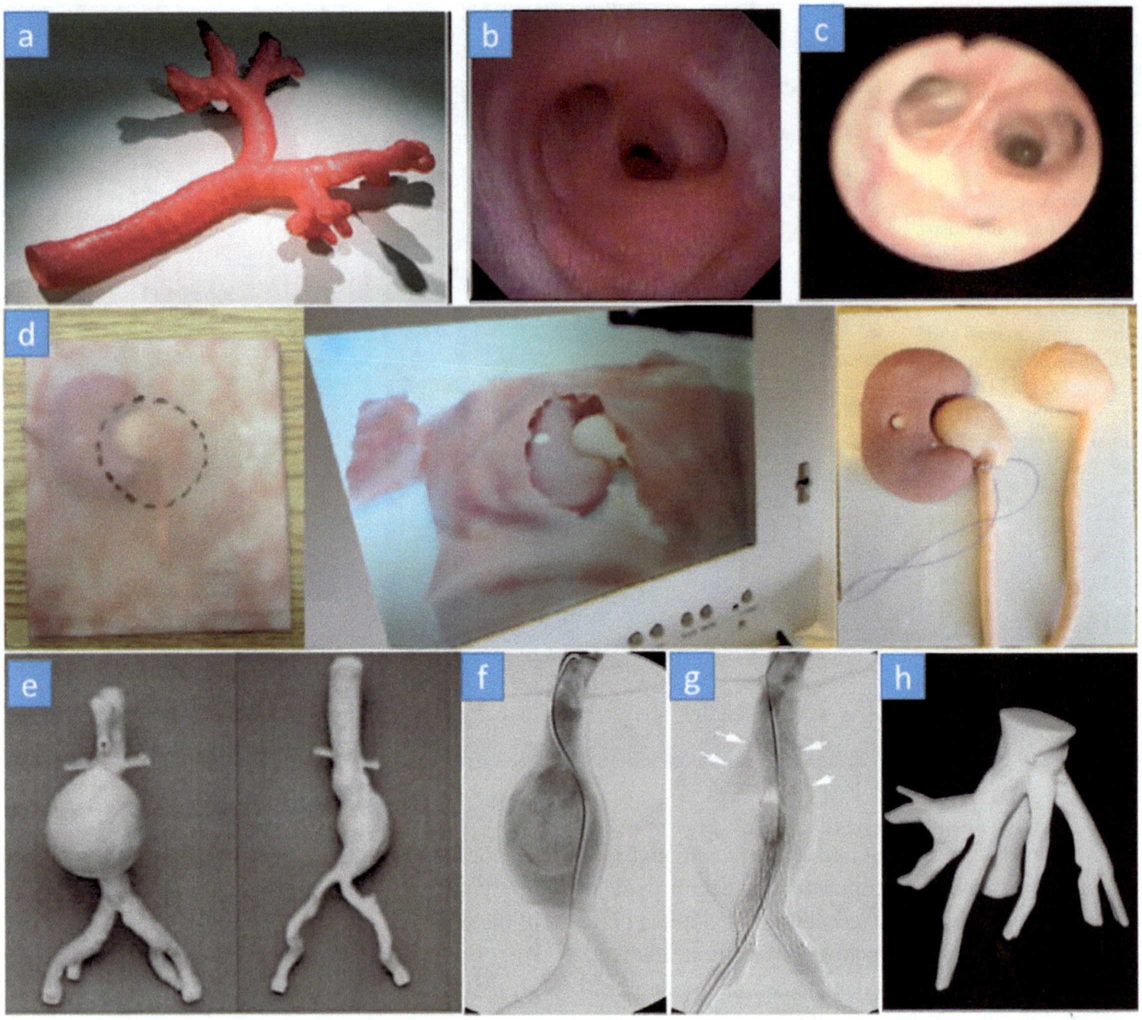

Fig. 7 3D printing applications in other areas. **a–c** 3D printed model of the tracheobronchial tree; fiberoptic view of bronchus intermedius for 3D model; fiberoptic view of bronchus intermedius in real anatomy (Bustamante et al. 2014) (Copyright © 2014 Elsevier Inc.) (**d**) pediatric pyeloplasty simulator with overlying peritoneum; dissection through the peritoneum and anastomosed ureter and renal pelvis, with replacement renal pelvis and ureter (Cheung et al. 2014) (Copyright © 2014 Association of Program Directors in Surgery. Published by Elsevier B.V.) (**e–g**) flexible silicone 3D models of human abdominal aortic aneurysms (AAA); digital subtraction angiogram of a replica perfused with contrast agent with a catheter; angiogram of a replica with a deployed stent (Berry et al. 2002) (Copyright © 2002, © SAGE Publications) h) IVC and hepatic veins of a patient (Watson 2014). (Copyright © 2014 Published by Elsevier B.V.)

4.5.2 Renal

In another report, Cheung et al. (2014) demonstrated 3D printing application in surgical simulation of pediatric laparoscopic pyeloplasty. Due to smaller workspaces, finer sutures, and potentially more delicate tissues that required increased surgical dexterity, pediatric laparoscopy poses unique training challenges. A pediatric pyeloplasty model was created using 3D target organ models whose negative volume molds were 3D printed and cast with silicone rubber. For the simulator, the target organs included kidney, dilated renal pelvis, obstructed ureteropelvic junction, ureter, and overlying peritoneum, and the arrangement allowed for a traditional transmesenteric simulation. The organ shapes were created by combining geometrical shapes, such as spheres and toroids, to approximate the organ. The models were scaled to the size of an approximately 10-month old (Fig. 7d).

4.5.3 Liver

Watson (2014) tested the use of 3D printing for the fabrication of physical hepatic models from CT or MRI data using selective laser sintering (SLS) for surgical resident education. The 3D models of the intrahepatic structures, including portal and hepatic veins, were printed. Hepatobiliary surgery, which is dependent on the understanding of individual, patient-specific variants of general anatomy, can greatly benefit from such models (Fig. 7h).

Table 1 3D printing method mapped to surgical area and their pros and cons

3D printing method	Cardiac surgery application areas	Neurosurgery application areas	Cranio-maxillo-facial application	Orthopaedic surgery application areas	Other application areas	Pros of method	Cons of method
1. SLS	- CHD cases	- Skull base surgery	- Unilateral maxillofacial deformities - Bilateral post-traumatic maxillofacial deformities - Tympanoplasty	- Acetabular fractures	- Hepatic 3D models - Flexible abdominal aortic aneurysms (AAA) with latex material	- High accuracy - Large model size possible and large selection of materials - High strength allows drilling, other simulations - Moderate fabrication time for full sized models - Fabricated models allow device placements simulations (eg stent placement for flexible AAA models)	- Very high cost - Finished models may have a powdery surface - Opaque models
2. SLA	- CHD cases - Pediatric cardiac transplant - Aortic stenosis - Cardiac tumor - Aneurysms	- Brachyther apy - Skull base surgery - Cerebral aneurysms	- Maxillofacial trauma - Congenital malformations - Severe skull defects and craniofacial surgery - Mandibular reconstruction - Dentofacial deformity - Orbital reconstruction	- Shoulder injury	- SLA mold for silicone rubber – abdominal aortic aneurysms	- Very high accuracy - Large model size possible - Fabricated models allow device placement simulations (eg. Occluder device for aneurysms) - Implant/prosthesis design (eg. Distraction device for hemifacial microsomia) - Transparent models possible - Flexible models possible using rubber-like urethane (flexibility not perfect) - Models can be sterilized (USP class VI resin)	- High cost - Moderate strength - Low speed
3. Inkjet (powder based 3D printing)	- CHD cases - Ventricular aneuresectomy and reshaping - Cardiac tumors - Transcatheter AVR and PPVI	- Lesion near motor cortex - Surgical training involving skull drilling etc.	- Cranial reconstruction	- Fractures and deformities of tibia, ankle, foot, spine and acetabulum	- 3D printed mold for silicon rubber – Pediatric laparoscopic pyeloplasty - Inkjet printing + lost wax technique: Arterial pathology models	- Low cost - High speed - Multiple materials in a single model possible - Flexible models with starch as base material	- Relatively lower accuracy - Relatively lower strength - Opaque models
4. FDM	- VSD - Aneurysm - Ulcerations - Prosthetic mitral valve perivalvular leak	- FDM followed by silicone coating for hollow elastic models for cerebral aneurysm clipping simulation	- Cranial reconstruction	- Dysplastic hip - Scoliosis: restoration of straight spine - Epiphyseal malunion - Shoulder arthoplasty - Femoral trochleoplasty	NA	- Low cost - High accuracy - Catheter simulation possible on models - Implant selection based on models - Models can be sterilized for intraoperative use	- Low speed - Removing support structures is time consuming - Opaque models
5. Polyjet	- Left atrial appendage closure - VSD in pediatric heart	NA	NA	- Pelvic reconstruction - Distal tibia comminuted fracture	- Bronchoscopic anatomy	- Using rubber like material allows flexible heart models for simulation - Allows implant customization - High speed - High accuracy	- Opaque models

CHD, Congenital heart disease; *VSD*, Ventricular septal defect; *PPVI*, Percutaneous pulmonary valve implantation; *AVR*, Aortic valve replacement; *SLS*, Selective laser sintering; *SLA*, Stereolithography; *FDM*, Fused deposition modelling

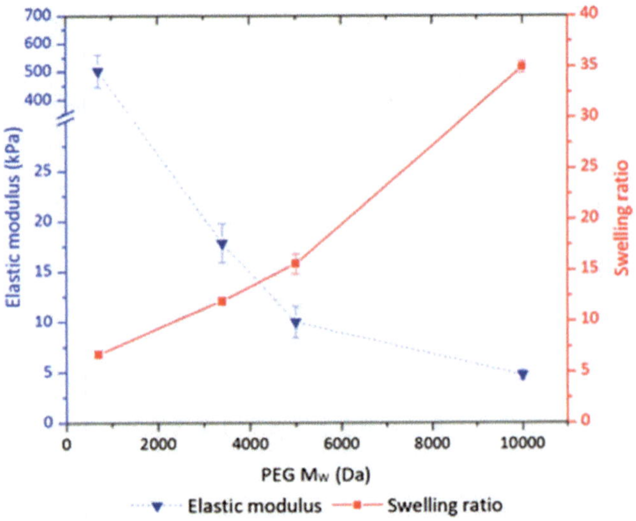

Fig. 8 The variation of elastic modulus and swelling of laser-polymerized PEGDA hydrogels with change in its molecular weight. The experiments were done with 20% PEGDA hydrogels and 0.5% photoinitiator concentration (Chan et al. 2010). (Copyright © 2010, Published by Royal Society of Chemistry)

4.5.4 Vascular (excluding cardiac and brain)

Chong et al. (1997) developed a simulator for endovascular repair of abdominal aortic aneurysms by using 3D printing (followed by silicone rubber molding) to fabricate transparent silicone rubber blocks with enclosed arterial lumens. This system was used to simulate the endovascular stent-graft deployment, a popular technique for repairing diseased vessels. The model provided a good understanding of the aneurysm and its related vasculature, accurate dimensional measurements, and selection of correct device size. It also included renal, iliac, and femoral arteries with a temperature controlled blood–analog fluid perfused under simulated physiological flow conditions. The imaging system also permitted arteriography for determination of accurate stent-graft placement. Similar abdominal aortic models were fabricated by Berry et al. (2002) with SLS and flexible latex, and used for simulating stent placement (Fig. 7e–g). Knox et al. (2005) used 3D printing from CT angiographic data and a lost wax technique to create similar arterial pathology as discussed above. The lost wax technique uses 3D printed lumen anatomy to create a master mold followed by a wax lumen anatomy replica. This wax lumen is coated with silicone and then the wax is thermally and chemically removed, leaving behind a hollow, flexible silicone vessel. These vessels can allow surgeons to teach and practice dangerous interventional techniques. In a recent report, O'Reilly et al. (2015) used 3D printed anatomical models of the lower limb for anatomical teaching and femoral vessel access training in medicine. Using a combination of 3D printing, casting, and lost-wax technique, the authors fabricated an anatomically accurate lower limb model with hard bones, silicone muscles, and perfusable blood vessels, all held together by rare earth magnets. Vascular simulation is possible through application of pumps and artificial blood. These models allow creation of anatomically diverse samples which are very rare and permits their use for surgeon training (Table 1).

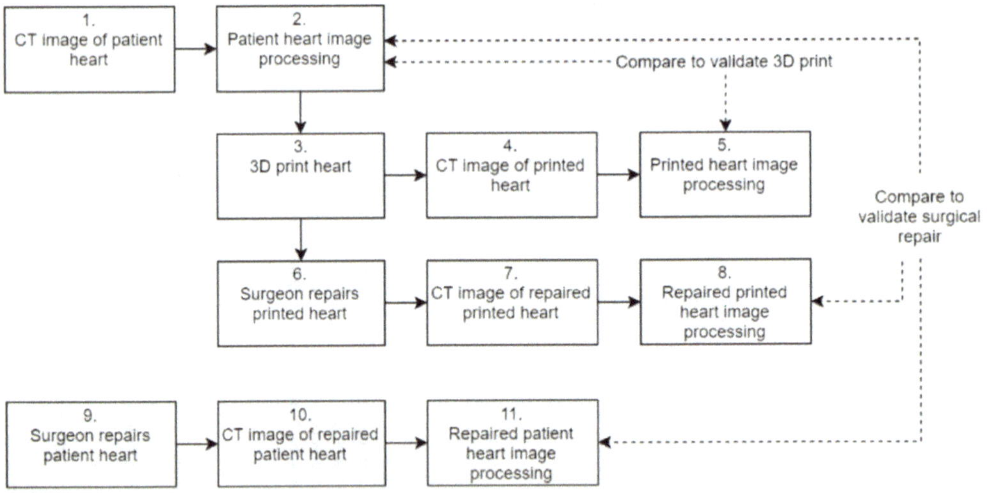

Fig. 9 Proposed process of 3D model fabrication, validation, and use for surgical planning, with an example from cardiology. (1) The patient heart is imaged using CT. (2) The CT image is processed through techniques such as segmentation, then converted into the appropriate STL file format for 3D printing. (3) The patient-specific model is 3D printed. (4) The printed model is imaged using CT. (5) The CT image of the print is processed, and then compared to the processed image of the patient heart in order to validate the 3D print. (6) Once assured that the printed model is identical to the patient heart, the surgeon operates on the printed model. (7) The repaired model heart is CT imaged. (8) The CT image is processed, and then compared to the processed image of the patient heart in order to validate the surgeon's repair of the printed heart. (9) Once the surgeon has successfully operated on the printed heart, he will operate on the patient heart. (10) The repaired patient heart will be CT imaged. (11) The CT image of the repaired patient heart will be processed and then compared to the image of the repaired printed heart as well as the image of the patient heart prior to surgery in order to evaluate the surgical repair

5 Discussion and future perspective

This review highlights the applications of 3D printing in various areas of surgical simulation, including preoperative planning, intraoperative guidance, and surgeon training. 3D printed models have been used to replicate a broad spectrum of tissues, ranging from hard tissue structures such as the skull, spine, limbs, and ear bones, to soft tissue structures including the heart, lungs, liver, and vasculature. 3D printing in surgical assistance domain has seen a tremendous rise in the last two decades. However, despite significant progress, we find that the use of 3D printing to form soft tissue structures has been largely unexplored. Very few attempts have been made to capture the mechanical properties of soft tissue along with high resolution anatomical detail. Previous attempts utilized flexible polyurethane or silicone, but did not achieve the correct mechanical properties such as elastic modulus, and do not provide a true feel of the wet and soft biological tissue. Moreover, in the area of neurosurgery, a combination of hard and soft materials must be utilized to exactly simulate the skull as well as the underlying brain tissue. Such combinational approaches can also be very useful for other surgical areas. We have previously reported the stereolithographic printing of poly(ethylene) glycol-based hydrogels with control over mechanical properties (Fig. 8) (Chan et al. 2010). The wet nature of hydrogels can also add to the true feel of biological tissue in the printed model, and thus serve as an excellent material for soft tissue surgical simulations. The ability to 3D print hydrogel materials which mimic *in vivo* soft tissue properties has been widely utilized for tissue engineering applications, but these techniques have yet to be applied to the fabrication of anatomical models for surgical simulation (Farina et al. 2018; Moroni et al. 2018). An additional need is the ability to use materials that can be sutured, similar to real tissue. Figure 9 shows a proposed process of 3D model fabrication, validation, and use for surgical planning. If heart defects, for example, can be recapitulated via printing, then the surgeons can fix the defects via surgical suturing. The surgical fix can be imaged again and the performance of the surgeons can be quantified precisely and surgeons can improve their skills by repeating the procedure multiple times. We believe that with future developments in these printed materials, the next generation of surgical simulations will possess an added realism which will significantly improve surgical training and preoperative planning, resulting in increased success rates in surgical operations. Furthermore, in March 2018, Materialise became the first company to receive FDA clearance for 3D printed anatomical model software intended for diagnostic use. As stated by Materialise CEO, Wilfried Vancraen, the FDA approval of Materialise Mimics software will "support the adoption of 3D planning and printing in U.S. hospitals and the creation of point-of-care 3D printing facilities". Considering recent advances in 3D printing softwares and model materials together with the FDA's approval of this technology for diagnostic use, the applications of 3D printed anatomical models will continue to expand and further improve healthcare.

References

P. Abdel-Sayed, L. Von Segesser, in *Adv. Appl. Rapid Prototyp. Technol. Mod. Eng.*. Rapid prototyping for training purposes in cardiovascular surgery (2011)

P. Abdel-Sayed, M. Kalejs, L.K. von Segesser, A new training set-up for trans-apical aortic valve replacement. Interact. Cardiovasc. Thorac. Surg. **8**, 599–601 (2009)

N. Adolphs, W. Liu, E. Keeve, B. Hoffmeister, Craniomaxillofacial surgery planning based on 3D models derived from Cone-Beam CT data. Comput. Aided Surg. **18**(5–6), 101–108 (2013)

D.-G. Ahn, J.-Y. Lee, D.-Y. Yang, Rapid prototyping and reverse engineering application for orthopodeic surgery planning. J. Mech. Sci. Technol. **20**(1), 19–28 (2006)

M. Akaike et al., Simulation-based medical education in clinical skills laboratory. J. Med. Invest. **59**(1, 2), 28–35 (2012)

O. Al-Mefty, J.L. Fox, A. Rifai, R.R. Smith, A combined infratemporal and posterior fossa approach for the removal of giant glomus tumors and chondrosarcomas. Surg. Neurol. **28**(6), 423–431 (Dec. 1987)

T. Andrew and H. Piggott, Growth arrest for progressive scoliosis. Combined anterior and posterior fusion of the convexity. J. Bone Jt. Surg. …, (1985)

E. Angeli, D. Pacini, S. Martin-Suarez, A. Dell'Amore, R. Fattori, R. Di Bartolomeo, Stent repair of aortic perianastomotic leak after aortic arch and descending aorta replacement. Ital. Heart J. Off. J. Ital. Fed. Cardiol. **5**(12), 951–953 (Dec. 2004)

D. Bakhos, S. Velut, A. Robier, M. Al Zahrani, E. Lescanne, Three-dimensional modeling of the temporal bone for surgical training. Otol. Neurotol. Off. Publ. Am. Otol. Soc. Am. Neurotol. Soc. Eur. Acad. Otol. Neurotol., 328–334 (2009)

E. Berry et al., Preliminary experience with medical applications of rapid prototyping by selective laser sintering. Med. Eng. Phys. **19**(1), 90–96 (1997)

E. Berry, A. Marsden, K.W. Dalgarno, D. Kessel, D.J. Scott, Flexible tubular replicas of abdominal aortic aneurysms. Proc. Inst. Mech. Eng. H J. Eng. Med. **216**(3), 211–214 (2002)

Bertalanffy, The dorsolateral, suboccipital, transcondylar approach to the lower clivus and anterior portion of the craniocervical junction. Neurosurgery **29**(6), 815–821 (1991)

J.S. Bill et al., Stereolithography in oral and maxillofacial operation planning. Int. J. Oral Maxillofac. Surg. **24**(1), 98–103 (Feb. 1995)

G.A. Brown, K. Firoozbakhsh, T. a DeCoster, J.R. Reyna, M. Moneim, Rapid prototyping: the future of trauma surgery? J. Bone Joint Surg. Am. **85**(A Suppl), 49–55 (2003)

R. Bryan, J. Rand, *Revision total knee arthroplasty* (Clin. Orthop, 1982)

S. Bustamante, S. Bose, P. Bishop, R. Klatte, F. Norris, Novel application of rapid prototyping for simulation of bronchoscopic anatomy. J. Cardiothorac. Vasc. Anesth. **28**(4), 1134–1137 (2014)

R.M. Carr, R.H. Mathog, Early and delayed repair of orbitozygomatic complex fractures. J. Oral Maxillofac. Surg. **55**(3), 253–258 (Mar. 1997)

V. Chan, P. Zorlutuna, J.H. Jeong, H. Kong, R. Bashir, Three-dimensional photopatterning of hydrogels using stereolithography for long-term cell encapsulation. Lab. Chip **10**(16), 2062–2070 (Aug. 2010)

C.L. Cheung, T. Looi, T.S. Lendvay, J.M. Drake, W.a. Farhat, Use of 3-dimensional printing technology and silicone modeling in surgical simulation: Development and face validation in pediatric laparoscopic pyeloplasty. J. Surg. Educ. **71**(5), 762–767 (2014)

C. K. Chong, J. Brennan, T. V How, R. Edwards, G. L. Gilling-Smith, and P. L. Harris, "A prototype simulator for endovascular repair of abdominal aortic aneurysms.," Eur. J. Vasc. Endovasc. Surg. Off. J. Eur. Soc. Vasc. Surg., vol. 13, no. 3, pp. 330–333, 1997.

A. Cohen, A. Laviv, P. Berman, R. Nashef, J. Abu-Tair, Mandibular reconstruction using stereolithographic 3-dimensional printing modeling technology. Oral Surg. Oral Med. Oral Pathol. Oral Radiol. Endodontology **108**(5), 661–666 (2009)

J.J. Collins, S.F. Aranki, Management of mild aortic stenosis during coronary artery bypass graft surgery. J. Card. Surg. **9**(s2), 145–147 (Mar. 1994)

J.P. Costello et al., Utilizing three-dimensional printing technology to assess the feasibility of high-fidelity synthetic ventricular septal defect models for simulation in medical education. World J. Pediatr. Congenit. Heart Surg. **5**(3), 421–426 (2014)

J.P. Costello et al., Incorporating three-dimensional printing into a simulation-based congenital heart disease and critical care training curriculum for resident physicians. Congenit Heart Dis **10**(2), 185–190 (2015)

J. Cui, L. Chen, X. Guan, L. Ye, H. Wang, L. Liu, Surgical planning, three-dimensional model surgery and preshaped implants in treatment of bilateral craniomaxillofacial post-traumatic deformities. J. Oral Maxillofac. Surg. **72**(6), 1138.e1–1138.e14 (2014)

L.L. Cunningham, M.J. Madsen, G. Peterson, Stereolithographic modeling technology applied to tumor resection. J. Oral Maxillofac. Surg. **63**(6), 873–878 (2005)

P.S. D'Urso et al., Stereolithographic (SL) biomodelling in craniofacial surgery. Br. J. Plast. Surg. **51**(7), 522–530 (Oct. 1998)

P.S. D'Urso et al., Custom cranioplasty using stereolithography and acrylic. Br. J. Plast. Surg. **53**(3), 200–204 (Apr. 2000)

E. Debarre, P. Hivart, D. Baranski, P. Déprez, Speedy skeletal prototype production to help diagnosis in orthopaedic and trauma surgery. Methodology and examples of clinical applications. Orthop. Traumatol. Surg. Res. **98**(5), 597–602 (2012)

R. Dhakshyani, Y. Nukman, N.A.A. Osman, A.M. Merican, J. George, Rapid prototyping medical models for dysplastic hip orthopaedic surgery. Proc. Inst. Mech. Eng. Part B J. Eng. Manuf. **224**(5), 769–776 (2010)

R. Dhakshyani, Y. Nukman, N.A. Abu Osman, Rapid prototyping models for dysplastic hip surgeries in Malaysia. Eur. J. Orthop. Surg. Traumatol. **22**(1), 41–46 (2012)

B.O. Erbano et al., Rapid prototyping of three-dimensional biomodels as an adjuvant in the surgical planning for intracranial aneurysms. Acta Cirúrgica Bras. Soc. Bras. Para Desenvolv. Pesqui. Em Cir. **28**(11), 756–761 (2013)

X. Fan, H. Zhou, M. Lin, Y. Fu, J. Li, Late reconstruction of the complex orbital fractures with computer-aided design and computer-aided manufacturing technique. J. Craniofac. Surg. **18**(3), 665–673 (2007)

M. Farina, J.F. Alexander, U. Thekkedath, M. Ferrari, A. Grattoni, Cell encapsulation: Overcoming barriers in cell transplantation in diabetes and beyond. Adv. Drug Deliv. Rev. (2018)

S.F. Fighali et al., Early and late mortality of patients undergoing aortic valve replacement after previous coronary artery bypass graft surgery. Circulation **92**(9), 163–168 (Nov. 1995)

U. Fisch, The infratemporal approach to glomus jugulare tumors. Neurochirurgie. **31**(5), 367–376 (Jan. 1985)

D.H. Frakes, M.J.T. Smith, J. Parks, S. Sharma, S.M. Fogel, A.P. Yoganathan, New techniques for the reconstruction of complex vascular anatomies from MRI images. J. Cardiovasc. Magn. Reson. Off. J. Soc. Cardiovasc. Magn. Reson. **7**(2), 425–432 (2005)

M. Frame, J.S. Huntley, Rapid prototyping in orthopaedic surgery: a user's guide. ScientificWorldJournal **2012**, 838575 (2012)

K. Futami, M. Nakada, M. Iwato, D. Kita, T. Miyamori, J. Yamashita, Simulation of clipping position for cerebral aneurysms using three-dimensional computed tomography angiography. Neurol. Med. Chir. (Tokyo) **44**(1), 6–13 (Mar. 2004)

J. Gateno, M.E. Allen, J.F. Teichgraeber, M.L. Messersmith, An *in vitro* study of the accuracy of a new protocol for planning distraction osteogenesis of the mandible. J. Oral Maxillofac. Surg. Off. J. Am. Assoc. Oral Maxillofac. Surg. **58**(9), 985–990; discussion 990–1 (2000)

J. Geerts et al., Functional magnetic resonance imaging for preoperative localisation of eloquent brain areas relative to brain tumours: clinical implementation in a regional hospital. JBR-BTR Organe Société R. Belge Radiol. SRBR Orgaan Van K. Belg. Ver. Voor Radiol. KBVR **90**(4), 258–263 (Jan. 2007)

K.K. Gnanalingham, V. Apostolopoulos, S. Barazi, K. O'Neill, The impact of the international subarachnoid aneurysm trial (ISAT) on the management of aneurysmal subarachnoid haemorrhage in a neurosurgical unit in the UK. Clin. Neurol. Neurosurg. **108**(2), 117–123 (Feb. 2006)

G.F. Greil et al., Stereolithographic reproduction of complex cardiac morphology based on high spatial resolution imaging. Clin. Res. Cardiol. **96**, 176–185 (2007)

J. Guarino, S. Tennyson, G. McCain, L. Bond, K. Shea, H. King, Rapid prototyping technology for surgeries of the pediatric spine and pelvis: benefits analysis. J. Pediatr. Orthop. **27**(8), 955–960 (2007)

H.K. Hahn, W.S. Millar, O. Klinghammer, M.S. Durkin, P.K. Tulipano, H.-O. Peitgen, A reliable and efficient method for cerebral ventricular volumetry in pediatric neuroimaging. Methods Arch. **43**(4), 376–382 (2004)

H.M. Hidalgo, G.W. Romo, R.T.R. Estolano, Stereolithography: a method for planning the surgical correction of the hypertelorism. J. Craniofac. Surg., vol. **5**, 20 (2009)

J. Hirsch et al., An integrated functional magnetic resonance imaging procedure for preoperative mapping of cortical aread associated with tactile, motor, language, and visual functions. Neurosurgery **47**(3), 711–722 (2000)

J.I.. Hoffman, S. Kaplan, The incidence of congenital heart disease. J. Am. Coll. Cardiol. **39**(12), 1890–1900 (Jun. 2002)

D.E. Holck, E.M. Boyd, J. Ng, R.O. Mauffray, Benefits of stereolithography in orbital reconstruction. Ophthalmology **106**(6), 1214–1218 (1999)

C. Hurson, B.O.'.D. a Tansey, P. Nicholson, J. Rice, J. McElwain, Rapid prototyping in the assessment, classification and preoperative planning of acetabular fractures. Injury **38**(10), 1158–1162 (2007)

S. Jacobs, R. Grunert, F.W. Mohr, V. Falk, 3D-Imaging of cardiac structures using 3D heart models for planning in heart surgery: a preliminary study. Interact. Cardiovasc. Thorac. Surg. **7**(1), 6–9 (2008)

A.D. Jatene, Left ventricular aneurysmectomy. Resection or reconstruction. J. Thorac. Cardiovasc. Surg. **89**(3), 321–331 (Mar. 1985)

J.F. John, R.E. Talbert, J.K. Taylor, W.L. Bargar, Use of acetabular models in planning complex acetabular reconstructions. J. Arthroplasty **10**(5), 661–666 (Oct. 1995)

M. Kalejs, L.K. von Segesser, Rapid prototyping of compliant human aortic roots for assessment of valved stents. Interact. Cardiovasc. Thorac. Surg. **8**, 182–186 (2009)

T. Kaminaga, T. Takeshita, I. Kimura, Role of magnetic resonance imaging for evaluation of tumors in the cardiac region. Eur. Radiol., vol. 13 Suppl **6**, L1–L10 (Dec. 2003)

B. Kavanagh, Cemented revision hip arthroplasty: results, Jt. Replace. Arthroplasty N. Y. Etc Churchill … 1991 Paperpile.

C.J. Kellenberger, S.-J. Yoo, E.R.V. Büchel, Cardiovascular MR imaging in neonates and infants with congenital heart disease. Radiogr. Rev. Publ. Radiol. Soc. N. Am. Inc **27**(1), 5–18 (2007)

C. Kermer, a. Lindner, I. Friede, a. Wagner, W. Millesi, Preoperative stereolithographic model planning for primary reconstruction in craniomaxillofacial trauma surgery. J. Craniomaxillofac. Surg. **26**(3), 136–139 (1998)

J. Kettenbach et al., Computer-based imaging and interventional MRI: applications for neurosurgery. Comput. Med. Imaging Graph. **23**(5), 245–258 (1999)

M.S. Kim, A.R. Hansgen, O. Wink, R.A. Quaife, J.D. Carroll, Rapid prototyping: a new tool in understanding and treating structural heart disease. Circulation **117**(18), 2388–2394 (2008)

T. Kimura et al., Simulation of and training for cerebral aneurysm clipping with 3-dimensional models. Neurosurgery **65**(4), 719–726 (2009)

K. Knox, C.W. Kerber, S. a Singel, M.J. Bailey, S.G. Imbesi, Rapid prototyping to create vascular replicas from CT scan data: making tools to teach, rehearse, and choose treatment strategies. Catheter. Cardiovasc. Interv. Off. J. Soc. Card. Angiogr. Interv. **65**(1), 47–53 (2005)

M. Kozakiewicz et al., Clinical application of 3D pre-bent titanium implants for orbital floor fractures. J. Cranio-Maxillofac. Surg. **37**(4), 229–234 (2009)

J.P. Kruth, Material incress manufacturing by rapid prototyping techniques. CIRP Ann. - Manuf. Technol. **40**(2), 603–614 (1991)

P. Lachiewicz and O. Hussamy, Revision of the acetabulum without cement with use of the Harris-Galante porous-coated implant. Two to eight-year results. J. Bone Jt. Surg., (1994)

R. Lazar and J. Hall, Simultaneous anterior and posterior hemivertebra excision. Clin. Orthop., (1999)

G.M. Lemole, P.P. Banerjee, C. Luciano, S. Neckrysh, F.T. Charbel, Virtual reality in neurosurgical education. Neurosurgery **61**(1), 142–149 (2007)

E.A. Longfield, T.M. Brickman, A. Jeyakumar, 3D printed pediatric temporal bone: a novel training model. Otol Neurotol, 793–795 (2015)

E. Maravelakis, K. David, A. Antoniadis, A. Manios, N. Bilalis, Y. Papaharilaou, Reverse engineering techniques for cranioplasty: a case study. J. Med. Eng. Technol. **32**(2), 115–121 (2008)

T. Mashiko et al., Development of three-dimensional hollow elastic model for cerebral aneurysm clipping simulation enabling rapid and low cost prototyping. World Neurosurg **83**(3), 351–361 (2015)

B. Mavčič, B. Pompe, and V. Antolič, Mathematical estimation of stress distribution in normal and dysplastic human hips, J. …, (2002)

M. McGurk, A.A. Amis, P. Potamianos, N.M. Goodger, Rapid prototyping techniques for anatomical modelling in medicine. Ann. R. Coll. Surg. Engl. **79**(3), 169–174 (May 1997)

S. Mohammadi et al., Reoperation for false aneurysm of the ascending aorta after its prosthetic replacement: surgical strategy. Ann. Thorac. Surg. **79**(1), 147–152; discussion 152 (Jan. 2005)

K. Mori, T. Yamamoto, K. Oyama, H. Ueno, Y. Nakao, K. Honma, Modified three-dimensional skull base model with artificial dura mater, cranial nerves, and venous sinuses for training in skull base surgery: technical note. Neurol. Med. Chir. (Tokyo) **48**(12), 582–587; discussion 587–588 (2008)

L. Moroni et al., Biofabrication strategies for 3D in vitro models and regenerative medicine, Nature Reviews Materials. **3**(5), 21–37, (2018)

S. Mottl-Link et al., Physical models aiding in complex congenital heart surgery. Ann. Thorac. Surg. **86**(1), 273–277 (2008)

B. Mueller, D. Kochan, Laminated object manufacturing for rapid tooling and patternmaking in foundry industry. Comput. Ind. **39**(1), 47–53 (1999)

A. Müller, K.G. Krishnan, E. Uhl, G. Mast, The application of rapid prototyping techniques in cranial reconstruction and preoperative planning in neurosurgery. J. Craniofac. Surg. **14**, 899–914 (2003)

M.C. Murphy et al., Surgical treatment of cardiac tumors: A 25-year experience☆. Ann. Thorac. Surg. **49**(4), 612–618 (Apr. 1990)

S.F. Mustafa, P.L. Evans, A. Bocca, D.W. Patton, A.W. Sugar, P.W. Baxter, Customized titanium reconstruction of post-traumatic orbital wall defects: a review of 22 cases. Int. J. Oral Maxillofac. Surg. **40**(12), 1357–1362 (2011)

E.M. Ngan et al., The rapid prototyping of anatomic models in pulmonary atresia. J. Thorac. Cardiovasc. Surg. **132**(2), 264–269 (2006)

A.M. Noecker et al., Development of patient-specific three-dimensional pediatric cardiac models. *ASAIO J.* **52**(3), 349–353 (2006)

M.K. O'Reilly et al., Fabrication and assessment of 3D printed anatomical models of the lower limb for anatomical teaching and femoral vessel access training in medicine. Anat. Sci. Educ. **00**(2010), 1–9 (2015)

J.A. Odell, C.J. Mullany, H.V. Schaff, T.A. Orszulak, R.C. Daly, J.J. Morris, Aortic valve replacement after previous coronary artery bypass grafting. Ann. Thorac. Surg. **62**(5), 1424–1430 (Nov. 1996)

M. Oishi, M. Fukuda, N. Yajima, and K. Yoshida, Interactive presurgical simulation applying advanced 3D imaging and modeling techniques for skull base and deep tumors: Clinical article, J. Of, (2013).

M. Oliveira et al., 3-D biomodelling technology for maxillofacial reconstruction. Mater. Sci. Eng. C **28**(8), 1347–1351 (2008)

L.J. Olivieri, A. Krieger, Y.-H. Loke, D.S. Nath, P.C.W. Kim, C.A. Sable, Three-dimensional printing of intracardiac defects from three-dimensional echocardiographic images: feasibility and relative accuracy. J. Am. Soc. Echocardiogr. **28**(4), 392–397 (2015)

J.M. Otton et al., Left atrial appendage closure guided by personalized 3d-printed cardiac reconstruction. JACC Cardiovasc. Interv. **8**(7), 1004–1006 (2015)

P. Ou, D.S. Celermajer, G. Calcagni, F. Brunelle, D. Bonnet, D. Sidi, Three-dimensional CT scanning: a new diagnostic modality in congenital heart disease. Heart **93**(8), 908–913 (2007)

S.W. Park, J.W. Choi, K.S. Koh, T.S. Oh, Mirror-imaged rapid prototype skull model and pre-molded synthetic scaffold to achieve optimal orbital cavity reconstruction. J. Oral Maxillofac. Surg. Off. J. Am. Assoc. Oral Maxillofac. Surg., 1540–1553 (2015)

J.Y. Park, G. Gao, J. Jang, D.-W. Cho, 3D printed structures for delivery of biomolecules and cells: tissue repair and regeneration. J. Mater. Chem. B **4**(47), 7521–7539 (2016)

B.C. Patel, J. Hoffmann, Management of complex orbital fractures. Facial Plast. Surg. FPS **14**(1), 83–104 (Jan. 1998)

M. Perry, P. Banks, R. Richards, E.P. Friedman, P. Shaw, The use of computer-generated three-dimensional models in orbital reconstruction. Br. J. Oral Maxillofac. Surg. **36**(4), 275–284 (1998)

P. Potamianos, A.A. Amis, A.J. Forester, M. Mcgurk, M. Bircher, Rapid prototyping for orthopaedic surgery. Proc Inst Mech Eng Part H **212**, 383–393 (2015)

M. Poulsen, C. Lindsay, T. Sullivan, P. D'Urso, Stereolithographic modelling as an aid to orbital brachytherapy. Int. J. Radiat. Oncol. Biol. Phys. **44**(3), 731–735 (1999)

M.D. Reller, M.J. Strickland, T. Riehle-Colarusso, W.T. Mahle, A. Correa, Prevalence of congenital heart defects in metropolitan Atlanta, 1998-2005. J. Pediatr. **153**(6), 807–813 (Dec. 2008)

F. Rengier et al., 3D printing based on imaging data: review of medical applications. Int. J. Comput. Assist. Radiol. Surg. **5**(4), 335–341 (2010)

M. Robiony et al., Accuracy of virtual reality and stereolithographic models in maxillo-facial surgical planning. J. Craniofac. Surg. **19**(2), 482–489 (2008)

D. Rohner, R. Guijarro-Martínez, P. Bucher, B. Hammer, Importance of patient-specific intraoperative guides in complex maxillofacial reconstruction. J. Cranio-Maxillofac. Surg. **41**(5), 382–390 (2013)

M. Ruf and J. Harms, Hemivertebra resection by a posterior approach: innovative operative technique and first results. Spine, (2002)

H.F. Sailer, P.E. Haers, C.P. Zollikofer, T. Warnke, F.R. Carls, P. Stucki, The value of stereolithographic models for preoperative diagnosis of craniofacial deformities and planning of surgical corrections. Int. J. Oral Maxillofac. Surg. **27**(5), 327–333 (1998)

E.K. Sannomiya, J.V.L. Silva, A.A. Brito, D.M. Saez, F. Angelieri, G. da Silva Dalben, Surgical planning for resection of an ameloblastoma and reconstruction of the mandible using a selective laser sintering 3D biomodel. Oral Surg. Oral Med. Oral Pathol. Oral Radiol. Endod **106**, 36–40 (2008)

C. Santler, H. Karcher, C. Ruda, Indications and limitations of three-dimensional models in cranio-maxillofacial surgery. J. Craniomaxillofac. Surg. **26**(1), 11–16 (1998)

R.M. Satava, Accomplishments and challenges of surgical simulation. Surg. Endosc. **15**(3), 232–241 (Mar. 2001)

S. Schievano et al., Percutaneous pulmonary valve implantation based on rapid prototyping of right ventricular outflow tract and pulmonary trunk from MR data. Radiology **242**(2), 490–497 (2007)

D. Schmauss et al., Three-dimensional printing of models for preoperative planning and simulation of transcatheter valve replacement. Ann. Thorac. Surg. **93**(2), e31–e33 (2012)

D. Schmauss, N. Gerber, R. Sodian, Three-dimensional printing of models for surgical planning in patients with primary cardiac tumors. J Thorac Cardiovasc Surg **145**(5), 1407–1408 (2013)

I. Shiraishi, M. Yamagishi, K. Hamaoka, M. Fukuzawa, T. Yagihara, Simulative operation on congenital heart disease using rubber-like urethane stereolithographic biomodels based on 3D datasets of multislice computed tomography. Eur. J. Cardio-Thorac. Surg. Off. J. Eur. Assoc. Cardio-Thorac. Surg. **37**(2), 302–306 (2010)

D.P. Sinn, J.E. Cillo, B. a Miles, Stereolithography for craniofacial surgery. J. Craniofac. Surg. **17**(5), 869–875 (2006)

R. Sodian et al., Stereolithographic models for surgical planning in congenital heart surgery. Ann. Thorac. Surg. **83**, 1854–1857 (2007)

R. Sodian et al., Pediatric cardiac transplantation: Three-dimensional printing of anatomic models for surgical planning of heart transplantation in patients with univentricular heart. J. Thorac. Cardiovasc. Surg. **136**(4), 1098–1099 (2008a)

R. Sodian et al., Three-dimensional printing creates models for surgical planning of aortic valve replacement after previous coronary bypass grafting. Ann. Thorac. Surg. **85**(6), 2105–2108 (2008b)

R. Sodian et al., 3-dimensional printing of models to create custom-made devices for coil embolization of an anastomotic leak after aortic arch replacement. Ann. Thorac. Surg. **88**(3), 974–978 (2009)

P.J. Spevak, P.T. Johnson, E.K. Fishman, Surgically corrected congenital heart disease: utility of 64-MDCT. AJR Am. J. Roentgenol. **191**(3), 854–861 (2008)

B.S. Spottiswoode et al., Preoperative three-dimensional model creation of magnetic resonance brain images as a tool to assist neurosurgical planning. Stereotact. Funct. Neurosurg. **91**(3), 162–169 (2013)

A.T. Stadie et al., Virtual reality system for planning minimally invasive neurosurgery. Technical note. J. Neurosurg. **108**(2), 382–394 (Feb. 2008)

Z.A. Starosolski, J.H. Kan, S.D. Rosenfeld, R. Krishnamurthy, A. Annapragada, Application of 3-D printing (rapid prototyping) for creating physical models of pediatric orthopedic disorders. Pediatr. Radiol. **44**(2), 216–221 (2014)

Y. Su, M. Wang, and W. Chang, Slotted acetabular augmentation in the treatment of painful residual dysplastic hips in adolescents and young adults. J. Formos. Med. …, (2008)

M. Suzuki, Y. Ogawa, A. Hagiwara, H. Yamaguchi, H. Ono, Rapidly prototyped temporal bone model for otological education. ORL J. Oto-Rhino-Laryngol. Its Relat. Spec. **66**(2), 62–64 (2004a)

M. Suzuki, Y. Ogawa, A. Kawano, A. Hagiwara, H. Yamaguchi, H. Ono, Rapid prototyping of temporal bone for surgical training and medical education. Acta Otolaryngol. (Stockh.) **124**(4), 400–402 (2004b)

M. Suzuki, A. Hagiwara, S. Kawaguchi, H. Ono, Application of a rapid-prototyped temporal bone model for surgical planning. Acta Otolaryngol. (Stockh.) **125**(1), 29–32 (Jan. 2005)

A. Thompson and D. Marks, Long-term results of combined anterior and posterior convex epiphysiodesis for congenital scoliosis due to hemivertebrae. Spine, (1995)

J.W.M. Tyl, L.E.C.M. Blank, L. Koornneef, Brachytherapy in orbital tumors. Ophthalmology **104**(9), 1475–1479 (Sep. 1997)

M. Umer, A. Thambyah, W. Tan, and S. De, Acetabular morphometry for determining hip dysplasia in the Singaporean population. J. Orthop. ..., (2006)

P. S. D. Urso et al., A technical note, (1999).

I. Valverde et al., Three-dimensional printed models for surgical planning of complex congenital heart defects: an international multicentre study. Eur. J. Cardio-Thorac. Surg. Off. J. Eur. Assoc. Cardio-Thorac. Surg. **52**(6), 1139–1148 (Dec. 2017)

M. Vranicar, W. Gregory, W.I. Douglas, P. Di Sessa, T.G. Di Sessa, The use of stereolithographic hand held models for evaluation of congenital anomalies of the great arteries. Stud. Health Technol. Inform. **132**, 538–543 (Jan. 2008)

V. Waran et al., Injecting realism in surgical training—initial simulation experience with custom 3d models. J. Surg. Educ. **71**(2), 193–197 (2014a)

V. Waran, V. Narayanan, R. Karuppiah, S.L.F. Owen, T. Aziz, Utility of multimaterial 3D printers in creating models with pathological entities to enhance the training experience of neurosurgeons. J. Neurosurg. **120**(2), 489–492 (2014b)

V. Waran et al., Neurosurgical endoscopic training via a realistic 3-dimensional model with pathology. Simul. Healthc. J. Soc. Simul. Healthc. **10**(1), 43–48 (2015)

R.A. Watson, A Low-Cost Surgical Application of Additive Fabrication. J. Surg. Educ. **71**(1), 14–17 (2014)

J. Winder, R. Bibb, Medical rapid prototyping technologies: state of the art and current limitations for application in oral and maxillofacial surgery. J. Oral Maxillofac. Surg. **63**(7), 1006–1015 (2005)

R. Winter, J. Moe, and J. Lonstein, Posterior spinal arthrodesis for congenital scoliosis. An analysis of the cases of two hundred and ninety patients, five to nineteen years old. J. Bone Jt. Surg., (1984)

Z.-X. Wu et al., Accuracy and safety assessment of pedicle screw placement using the rapid prototyping technique in severe congenital scoliosis. J. Spinal Disord. Tech. **24**(7), 444–450 (2011)

G. Wurm, B. Tomancok, P. Pogady, K. Holl, J. Trenkler, Cerebrovascular stereolithographic biomodeling for aneurysm surgery. Technical note. J Neurosurg **100**(1), 139–145 (2004)

W.-H. Xu, J. Liu, M.-L. Li, Z.-Y. Sun, J. Chen, J.-H. Wu, 3D printing of intracranial artery stenosis based on the source images of magnetic resonance angiograph. Ann. Transl. Med. **2**(8), 74 (2014)

P. Zardini, P. Marino, G. Golia, M. Anselmi, M. Castelli, Ventricular remodeling and infarct expansion. Am. J. Cardiol. **72**(19), G98–G106 (Dec. 1993)

L. Zhou, L. He, H. Shang, G. Liu, J. Zhao, Y. Liu, Correction of hemifacial microsomia with the help of mirror imaging and a rapid prototyping technique: case report. Br. J. Oral Maxillofac. Surg. **47**(6), 486–488 (2009)

Part IV
Microfluidics

Biomedical Microdevices (2019) 21: 35
https://doi.org/10.1007/s10544-019-0381-1

Microfluidic device for the analysis of MDR cancerous cell-derived exosomes' response to nanotherapy

Ruogu Qi[1] · Guixian Zhu[1] · Yu Wang[1] · Suhong Wu[1] · Shengliang Li[1] · Dechen Zhang[1] · Yang Bu[1] · Gauri Bhave[1] · Ruixuan Han[1] · Xuewu Liu[1]

Published online: 25 March 2019

Author's Personal Note In May 2003, an inconspicuous Ohio State University job listing on a local website drew my attention while I was seeking a career transition from the optics industry to biomedical research. Though at the time I knew nothing about the listed PI, Dr. Mauro Ferrari, the position seemed like a valuable opportunity to interface my industry experience with nanobiotechnology research. Despite the lack of any biomedical background, I sent an email expressing my interest. To my surprise, I received an interview arrangement just a couple days later. Little did I know at the time that this one interview would serve as the impetus for the next 16 years of my career. I met the charismatic—yet personable—Dr. Ferrari for the first time. We chatted for less than 10 minutes, during which he briefed the concept for NCI Alliance for Nanotechnology in Cancer. It's an instant decision for my life, and I started my new job a few days later. I have worked with Mauro ever since, from Columbus to Houston, from OSU to the UT Health Science Center to Houston Methodist.

I still remember Mauro asking about my career target during our initial 10-minute meeting. At the time, I answered that I wanted to be known as an inventor. And thanks to Dr. Ferrari's guidance, support, trust for the past decades, I have been able to achieve that goal. Our inventions became primary nanomedicine platforms, supporting hundreds of research papers. The following paper describes an example of such platform that integrates microfabrication, microfluidics and hierarchical porous silicon nanostructures, as applied for cancer research.

Xuewu Liu

Abstract

Exosomes are membrane-enclosed extracellular vesicles which have been indicated as important biomarkers of cancerous cell functionality, such as multiple drug resistance (MDR). Nanoparticles based chemotherapy is a promising strategy to overcome MDR by interfering the production and composition of exosomes. Therefore, tumor-derived exosomes post-treatment by nanotherapy are implied to play critical roles of biomarkers on cancer MDR analysis. However, the efficient isolation of such exosomes from extracellular environment for their therapeutic response analysis remains challenging. In this study, we presented a microfluidic device featured exosome specific anti-CD63 immobilized ciliated micropillars, which were capable to isolate cancer-derived exosomes from cell culture medium. The captured exosomes can be recovered intact by dissolving the cilia on the micropillars using PBS soaking. Owing to the immobilized antibody in the microfluidic device, nearly 70% of exosome from the biofluid could be isolated. So the secreted exosomes of the MDR and ordinary human breast cancer cells pre-treated by free drug or nanotherapy could be isolated with high purity. The drug contents of the isolated exosomes were measured to analysis of the exosomal pathway response of MDR cells to different chemotherapeutic formulations. Such analyses and further definition of the biomarkers of these exosomes could benefit the future investigations of accurately and reliably determine design principle, functional activity, and mechanisms of nanotherapy for MDR overcoming.

Keywords Exosome · Microfluidic device · Isolation · Nanoparticles therapy · Multiple drug resistance

1 Introduction

Nowadays, chemotherapy is playing an important role to against various cancers, such as breast cancer, ovarian cancer and lung cancer etc. (Fisher et al. 1998; Le Chevalier et al. 2004; Parmar et al. 2003). However, the efficiency of chemotherapy is mainly restricted by multiple drug resistance (MDR) acquired by cancerous cells under long-time drug exposure (Foo and Michor 2014; Gottesman 2002). To overcome the MDR, a lot of treatment methods are developed: one of the effective strategies is the introducing of nanotherapy which refer to technologies of delivering chemotherapeutic agents by loading them into nanoformulations such as liposomes, polymeric micelles and other nanomateirals (Hu and Zhang 2012; Yuan et al. 2016; Zhang et al. 2017). The advantages of such strategy are obvious. Briefly, small molecule chemo drugs are transferred into cancerous cells via membrane translocators or passively diffuses, which make them easily to be refluxed back to the extracellular environment by the MDR cell membrane over-expressed P-glycoprotein (P-gp). Meanwhile, drug-loaded nanoparticles are internalized into cells through endocytosis pathways to overcome the MDR P-gp related effluxing mechanism (Bannunah et al. 2014; Oh and Park 2014). Thus, the involving of nanoparticles therapy (nanotherapy) into MDR cancer treatment did result in improved chemotherapy efficacy. However, for the further clinical translation and development of nanotherapy, it is essential to elaborate the specific process of nanoparticles on the MDR overcoming. The recent studies have shown that the cancer-derived exosome secreting pathway plays an important role during this process (Soekmadji and Nelson 2015; Wang et al. 2016).

Exosomes are extra-cellular double-layer lipid vesicles with 40–120 nm in diameter secreted by the majority of viable cells to the bodily fluids, including blood, urine, and saliva.

Electronic supplementary material The online version of this article (https://doi.org/10.1007/s10544-019-0381-1) contains supplementary material, which is available to authorized users.

✉ Xuewu Liu
xliu@houstonmethodist.org

[1] Department of Nanomedicine, Houston Methodist Research Institute, 6670 Bertner Ave, Houston, TX 77030, USA

(Fevrier and Raposo 2004; van der Pol et al. 2012; Yang et al. 2014). Typical exosomes sourced from epithelial cells, lymphoid cells, or tumor cells contents functional proteins, lipids and RNAs and act significant roles in intercellular communication (Kucharzewska and Belting 2013; Rak 2013). Moreover, due to the universal existence and stability of exosomes in most of body fluid and resemblance of their contents to parental owners, exosomes have great potential to perform as traceable biomarkers for various diseases (Melo et al. 2015; Tang and Wong 2015). For example, the exosome derived by cancerous cells impact in many cancer-related pathways, such as tumor growth, metastasis, and angiogenesis (Kalluri 2016; Yang and Robbins 2011). Furthermore, cancer-derived exosomes have been found to participate in the MDR generating and acted as a major role in the process (Giallombardo et al. 2016; Santos et al. 2018). Corcoran et al. discovered exosomes contributed to docetaxel excretion of prostate cancer cells, and caused phenotypic change to non-resistant cell (Corcoran et al. 2012). Roohangiz et al. documented that the exosomes secreted by cisplatin-resistant cancer cells contained 2.5 times more cisplatin than of that from normal cancer cells (Safaei et al. 2005). These studies demonstrated the exosomal pathway effects through small drug efflux during the occurrences of MDR. However, the performance of exosomes in MDR overcoming nanotherapy and the interfering of nanoparticles on exosome generation and secretion still need to be evaluated (Oves et al. 2018) .

To analyze the cancer-derived exosomes, it is important to develop the efficient technique to isolate the exosomes from body fluid. Different methods have been introduced. One of the conventional procedures includes the differential centrifugation which involves a series of centrifugation and filtration steps suffered of inconsistent exosome recovery rates, the breakage of exosome membrane and contamination of co-sedimentation of protein aggregates (Lamparski et al. 2002). Another procedure for exosome isolation is immuno-affinity capturing which applies the materials such as magnetic particles that are coated with a monoclonal antibody to specifically bind the exosome. This method has been implement as less time consuming and high specific exosome purity (Szatanek et al. 2015). Recently, research efforts have increased dramatically on developing immuno-affinity capturing microfluidic devices in order to isolate biological exosomes. Compare to the other procedure used as immuno-affinity capturing, microfluidic technology offers high isolation yield and functional integration for target exosomes (Kanwar et al. 2014). Chen et al. reported an effective microfluidic exosome isolation platform, whose surface is immobilized with anti-CD63 antibody for capturing of exosomes from human serum (Chen et al. 2010). Kanwar et al. developed a similar microfluidic device called ExoChip, which utilized antibody immobilized circular chamber to capture exosomes, following by fluorescent staining for on-chip isolation, quantification and characterization of circulating exosomes (Kanwar et al. 2014). Therefore, antibody immobilized exosome isolation procedures were configured to enhance the capture efficiency. However, it is quite difficult to elute the captured exosomes from the antibody immobilized magnetic particles or the microfluidic device, which can present a major limitation of the possibilities for further experimentation on the extracted exosomes (Tauro et al. 2012).

Our group has presented a microfluidic device consisting of ciliated micropillars with 30–200 nm interspaced porous silicon nanowires through simply electroless etching using electro-deposited silver nanoparticles (Wang et al. 2013). This microfluidic device was able to preferentially physical trap exosome-like lipid vesicles by the silicon nanowire cilia while allowing smaller proteins and larger nanoparticles with the size of cellular debris to pass through unhindered. Moreover, the captured liposome can be released by dissolving the porous silicon nanowires in PBS buffer, thus allowing for the high integrity and purity of isolation contents recovery. However, this platform based on physical vesicle size isolation was restricted by surface lipid saturation limit. In this paper, the alternatively improvements and application of this microfluidic device are developed. The porous silicon nanowires of the device were fabricated by depositing gold nanopattern on the sidewalls of ciliated micropillars which is proved to be more effective and less time consuming. The nanowires of the ciliated micropillars were immobilized by the exosome-specific antibody, which allows efficient and non-invasive rapid isolation of specific exosomes with high purity. We hypothesized such improvements will significantly speed up the exosome isolation and analysis; benefit to acquire more accurate biological information. Moreover, The design of this microfluidic device provide the opportunities to isolate and quality the exosomes secreted from MDR cancerous cells and evaluate the interfering of drug-loaded nanoparticles on exosomes production and composition procedure. The outcome of this study would improve our understanding the mechanisms of exosomal responding to MDR profile during traditional chemotherapy or nanotherapy, thus leading to more efficient MDR cancer treatment.

2 Materials and methods

2.1 Materials

Chloroauric acid ($HAuCl_4$), (3-Aminopropyl)triethoxysilane (APTES), doxorubicin (Dox) and paclitaxel(PTX) were purchased from Sigma-Aldrich (St. Louis, MO). P-type (100) silicon wafers with resistivity of 0.005 Ω-cm were purchased from Silicon Quest (Santa Clara, CA). Polydimethylsiloxane (PDMS) and curing agent were obtained from Dow Corning (Midland, MI). Phosphate buffered saline (PBS, pH 7.2) was obtained from Gibco (Thermo-Fisher, Waltham, MA). Hydrofluoric acid was from Honeywell International Inc. N-maleimidobutyryl-oxysuccinimide ester (GMBS) and thiol functionalized streptavidin were purchased from Nanocs Inc. (Woburn, MA). The biotinylated anti-CD63 antibody and control IgG was purchased from Abcam Co. (Boston, MA). Human breast cancer cell line MDA-MB-231 cells were purchased from the American Type Culture Collection (Rockville, MD, USA). The multidrug resistant MDA-MB-231/MDR cell line was cultured in our lab. All the cells were cultured in DMEM (Thermo-Fisher, Waltham, MA) supplemented with 10% FBS. The exosome Labelling Kit (Exo-Glow™) was purchased from System Bioscience (Palo Alto, CA). Amphiphilic copolymer methoxy-poly (ethylene glycol)-poly (ε-caprolactone) (mPEG-PCL, Mw: 5000:5000) was purchased from Polymer Source Inc. (Canada).

2.2 Device fabrication

The ciliated micropillar arrays on the microfluidic device were fabricated by modification of previously described procedure (Wang et al. 2013). We designed a large capacity microfluidic devices with 627mm^2 micropillar area (as shown in Supporting Information Fig. S1B). A close-pack arrangement of hexagonal micropillars was adapted in the devices to increase the exosome capture efficiency. As seen in Fig. 1a, an example of fabricated ciliated hexagonal micropillar array, the distance between pillars was set as constant ~1.1 μm which will block large objects like cells existing in biological fluids. The devices were fabricated in two steps. First, the fluidic channels with high aspect ratio micropillars were fabricated by the deep silicon etching on an Inductively Coupled Plasma (ICP) etcher. Second, porous silicon nanowires were etched all around the micropillars by using metal assisted nanowire etching. To do so, silicon wafer with micropillars was first dipped into HF solution to remove native oxide. Then the wafer was assembled on the Teflon electroplating cell, and sealed with a Viton O-ring. An aluminum foil was pressed on wafer backside as electrode. A platinum mesh immersed into 0.5 M $HAuCl_4$ plating solution was applied as anode. The reverse pulse electroplating is performed to deposit uniform gold nanopattern on the sidewalls of the micropillars. After that, the wafer was chemically etched in a mixture of 0.1 M H_2O_2 and 2.9 M HF solution for 2 min to generate ~0.5 μm long nanowire cilia on the micropillars. The gold nanopattern was then stripped by immersing into gold etchant (Transene Inc., Denvers, MA) for 2 min. The substrate was then rinsed in excess water, ethanol and dried by blow air.

2.3 Device construction

The ciliated micropillars on the wafer were first functionalized with the APTES by immersing into 30% H_2O_2 at 90 °C overnight for oxidation and subsequently in 1% APTES isopropanol solution at 65 °C for 24 h as reported (Gunda et al. 2014). And the polydimethylsiloxane (PDMS) caps were

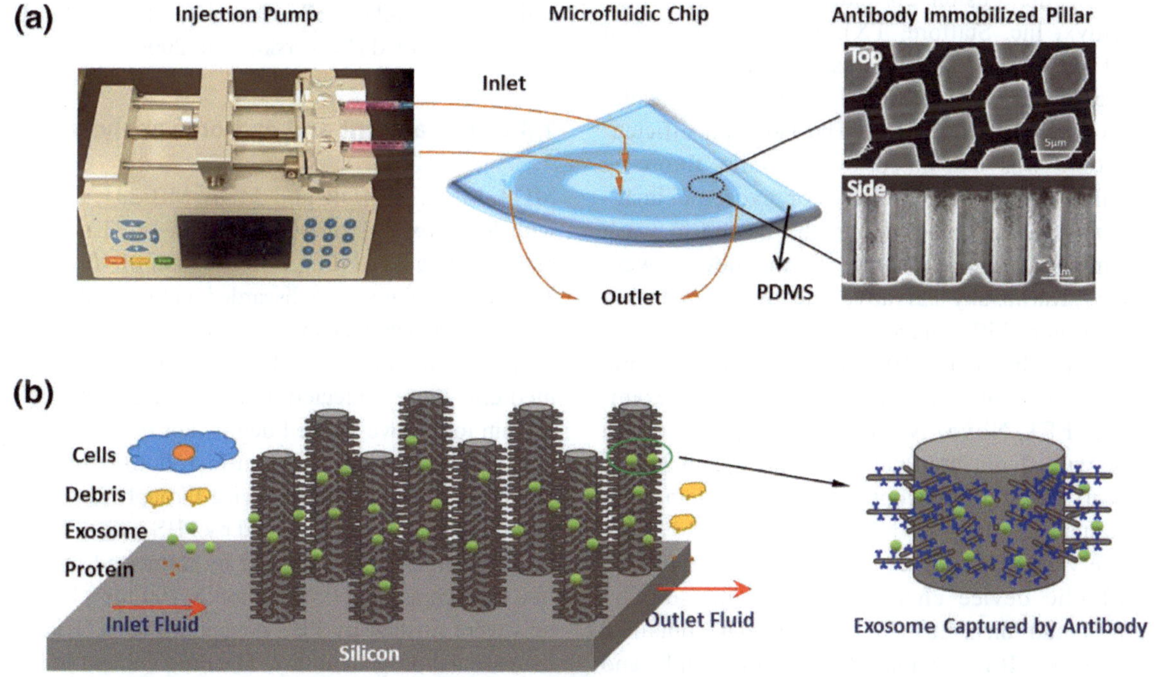

Fig. 1 Schematic of the antibody immobilized ciliated micropillar array for exosome isolation. **a** Experimental set-up of the ciliated microfluidic device system for exosomes isolation. **b** The exosomes are captured by antibody immobilized micropillars when passing the microfluidic system

made in a 10:1 elastomer base-curing agent ratio and incubated at 70 °C for 4 h under vacuum followed by cleaning with isopropanol and deionized water. The APTES functionalized wafer and PDMS cap were treated with oxygen plasma for 30 s, and then manually bonded and subsequently baked at 37 °C for 10 min to get the microfluidic device with the ciliated micropillars.

2.4 Surface modification

For antibody immobilizing, the device was rinsed in the 0.1 mg/ml GMBS solution for 30 min followed by washing with purified water. Then the microfluidic device was sequential dipped into thiol functionalized streptavidin and biotinylated anti-CD63 antibody PBS solution at 4 °C. The antibody immobilized microfluidic device was obtained after washing by 1% BSA solution and PBS. The control IgG immobilized microfluidic device was achieved as the similar method.

2.5 Exosome isolation and recovery

The cells were cultured DMEM supplemented with 10% exosome depleted FBS (System Bioscience Palo Alto, CA) for 12 h. Then the culture medium were collected and centrifuged at 2000 rpm for 10 min to remove cell debris. The exosome in the medium was labelled by Exo-Green Exosome Labelling Kit according to protocol. The labelling exosome solution was continuously injected into the anti-CD63 or control IGG immobilized microfluidic device through the inlet at a flow rate of 10 μL/min by the Fusion 200 high precision syringe pump (Chemyx, Inc. Stafford, TX). The waste was then collected at the outlet. The fluoresce intensity of inlet and outlet exosome was measured by BD plate reader (BD Biosciences, Billerica, MA). The isolation efficiency of exosome was calculated by the ratio fluoresce intensity value of inlet and outlet. For exosome imaging, PBS solution was injected into the device to wash untapped impurity. The exosomes in the device were fixed by 0.5 × Karnovsky's fixative and then sequencing dipped with water, 25% ethanol, 50% ethanol, 70% ethanol and 100% ethanol for 10 min each. The exosome on the micropillars of microfluidic device were examined using FEI Nova Nano Scanning Electron Microscope with 2 nm gold-palladium layer and Nikon A1 Confocal Laser Microscopy Imaging System with 488 nm excitation (Melville, NY, USA).

To recovery the isolated exosome from micropillars, the microfluidic device channels were immersing by PBS for 8 h and then flushed by PBS. The obtained the exosomes in PBS was placed on a mica slide and observed by Multimode Atomic Force Microscope with tapping mode (Bruker, Billerica, MA).

2.6 The preparation of drug loading nanoparticles

The doxorubicin encapsulated liposome (Doxoves®, 83 nm) was purchased from FormuMax Scientific, Inc. (Sunnyvale, CA). The paclitaxel containing micellar nanoparticles was prepared by co-solvent evaporation method. Briefly, mPEG-PCL (10 mg) and 1 mg of PTX were dissolved in 0.2 mL acetone. This solution was added to 1 mL of water. The mixture was then sonicated in ice bath for 2 min. The excess organic solvent was removed under vacuum. The micellar nanoparticles solution was filtered through a 0.22 μm Minisart syringe filter (Sartorius, Germany) to remove free PTX and subsequently lyophilized. The size and Zeta potential of PTX loaded nanoparticles were characterized by Zetasizer Nano (Malvern, UK) and the morphology of the nanoparticles was observed by Atom Force Microscopy in tapping mode. The PTX content in the nanoparticles was measured by Agilent 1200 High Performance Liquid Chromatography System (HPLC) (Agilent, Santa Clara, CA).

2.7 Multiple drug resistance characterization

MDA-MB-231/MDR and MDA-MB-231 were seeded in a 96 well plate at 5×10^3 cells/well for 24 h. Then the cell culture medium was replaced with new medium containing free Dox, Doxoves®, free PTX (dissolved in DMSO) or PTX-micellar nanoparticles with different concentration for culturing another 72 h. The chemo-sensitivity of the MDA-MB-231/MDR and MDA-MB-231 cells were measured by MTT assay as literature reported (Napierska et al. 2009)

2.8 Cell treatment and exosome analysis

MDA-MB-231/MDR and MDA-MB-231 were pretreated with free Dox, Doxoves®, free PTX or PTX-micellar nanoparticles by a fixed drug concentration of 40 μg/mL for 4 h. Then the medium were discarded and the cells washed twice with cold PBS and incubated with DMEM supplemented with 10% exosome depleted FBS for another 12 h. The culture mediums were collected and centrifuged at 2000 rpm for 10 min to remove the cell debris. For exosome isolation, the collected mediums were injected into the anti-CD63 immobilized microfluidic device through the inlet at a flow rate of 10 μL/min and washed by PBS twice. Then the isolation exosomes were collected by immersed the micropillars into PBS containing 0.1% Triton X-100 for 8 h and the channels were flushed by PBS containing 0.1% Triton X-100 to increase the drug solubility. The drug content in the obtained exosomes was tested by Agilent 1200 High Performance Liquid Chromatography System.

3 Results and discussion

3.1 Design of the anti-CD63 antibody immobilized microfluidic device

A microfluidic device capable of simultaneously multi-dimensional filtration of the biofluid is highly desirable to fulfill the mission to isolate the exosomes. We designed and fabricated the microfluidic devices with hierarchical ciliated micropillar structure (Fig. S1) as we reported previously (Wang et al. 2013). In order to increase the capture efficacy, we laid out a large capacity microfluidic devices with 627mm^2 micropillar area (Fig. S1B, C). As illustrated in Fig. 1a, the assembled microfluidic device consists of a PDMS caps and a concentric ellipses-flow fluidic channel containing antibody immobilized ciliated hexagonal micropillars with 5 μm across and 15 μm in height. The gaps among the micropillars are set as constant ~1.1 μm for a steady fluid-micropillar interaction. Circular inlet and outlet ports were connected to the micropillar area with a syringe pump for sample introduction and extraction. We hypothesized that the interstitial sites of the micropillars capable of physically (1) bypassing the unwanted components of sample, such as small molecular protein and submicron-size cellular debris; (2) filtering the larger sample components, such as cells. Moreover, the antibody immobilized nanowires on the sidewalls of ciliated micropillars are dedicated to selectively capture the exosomes flowing through the device (Fig. 1b), and the porous silicon nanowires could be dissolved in buffers to integrity isolate the captured exosomes for their further analysis. Thus, such microfluidic device is capable to apply for the MDR cancerous cell-derived exosomes isolation and the analysis of their response to nanotherapy.

The nanowire cilia were fabricated on micropillars to increase the contacting area between micropillars and bypassing fluid, in order to enhance exosome capturing efficiency. Briefly, the nanowire was fabricated by the combination of electroplating, electroless metal-assisted etching techniques. First, by the governing of the electrochemical reaction and reactant diffusion, as well as the properties of the reactants and silicon substrates, a uniform gold nanopattern is deposited onto the sidewalls of hexagonal micropillars (Fig. 2a). Second, the nanowire cilia were obtained by subsequently metal-assisted chemical etching of the micropillar surface and stripping the excess gold nanopartterns (Fig. 2b). Finally, the nanostructured silicon surface was functionalized with amine groups followed by decorating of streptavidin, thus the biotinylation anti-CD63 antibody could be stable immobilized on the nanowire cilia of micropillars for exosome capturing (Fig. 2c).

3.2 Isolation of exosomes by the microfluidic device

Fluorescein-labelled cancer cell-derived exosomes were labelled and used as the traceable samples to verify the functionality and isolation efficacy of our antibody immobilized microfluidic device. The samples was injected to the inlet through the PDMS cap and extracted at the outlet. After that, the captured exosomes by the ciliated micropillars on the microfluidic device were observed by SEM. As shown in Fig. 3a, b, the captured exosomes are visualized on the sidewalls of micropillars and trapped in the porous of the nanowire, which demonstrated the capacity of microfluidic device with ciliated micropillars to capture the exosomes from body fluid. To quantity of the efficacy on exosomes isolation, the eluting outlet samples were collected and the fluorescence intensity was divided versus that of the totals injected samples. The unspecific IgG immobilized microfluidic device was used as control. As shown in Fig. 3c, 75% of the total exosome retention was observed by the anti-CD63 immobilized microfluidic device, while only less than 20% of exosome retention rate was shown for no-selective IgG immobilized one. The functionalization of antibodies against CD63, an antigen commonly overexpressed in exosomes greatly contributes to this significant efficacy of exosome capturing (Szatanek et al. 2015). The captured exosome by the antibody immobilized microfluidic device was further visualized by confocal laser microscopy. As shown in Fig. 4a, b, the continuously green fluoresce dots which represent the captured exosomes clusters were observed to absorb on the sidewalls of micropillars. Notably, the three-dimension imaging showed the majority of the exosomes was trapped around the micropillars rather than the bottom of the microfluid device, which demonstrated the forming of turbulence among the micropillars enhanced the absorbing chances of exosome and anti-CD63 leading the high isolation efficacy. These revealed results showed that our microfluidic device with anti-CD63 immobilized ciliated micropillars offers obvious advantages on exosome isolating.

3.3 The recovery of captured exosomes

It has been reported that the porous silicon nanowires degraded in the PBS buffer (Chiappini et al. 2010). Thus, the microfluidic device prototype we developed is not only applied for exosome capturing but also recovering the captured exosomes through the degradation of the porous nanowire of the ciliated micropillars. The degradation of nanowire was accomplished by soaking the channels of microfluidic device with PBS for 8 h after capturing exosomes. As shown in Fig. 5a, the nanowires of ciliated micropillars in the

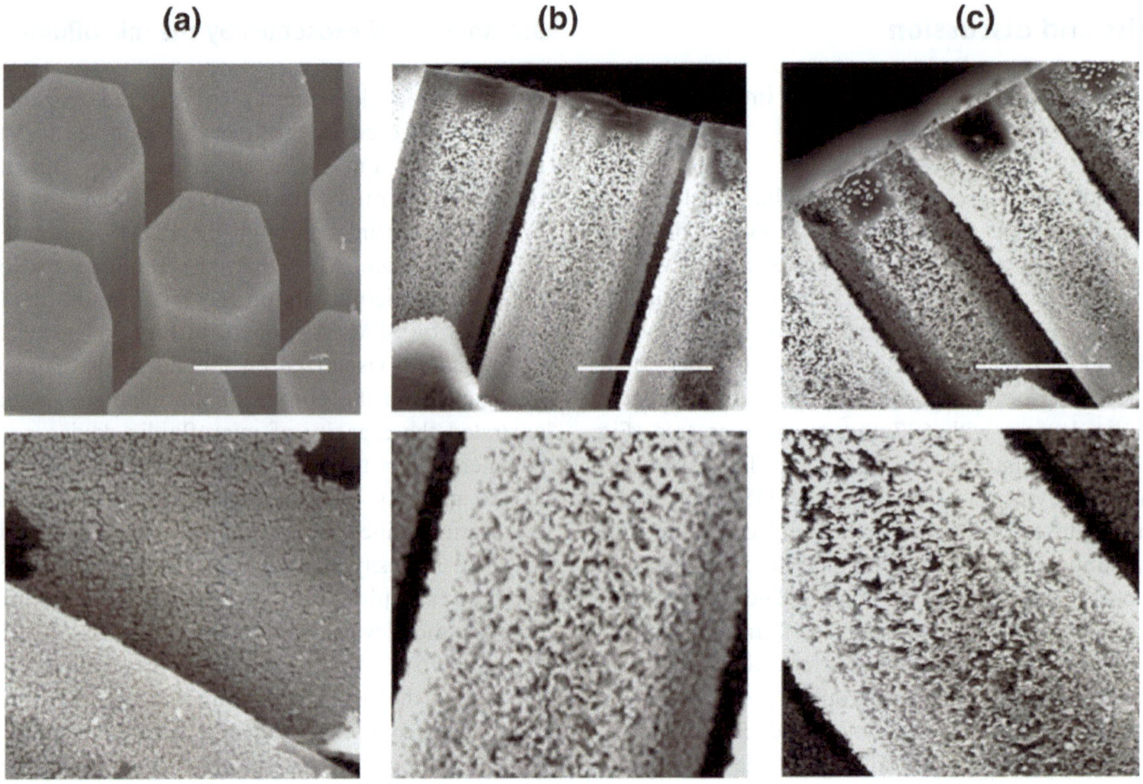

Fig. 2 SEM images of the antibody immobilized ciliated micropillars preparation stages. **a** Hexagonal silicon micropilliars before(top) and after (bottom) gold deposition(bottom). **b** Porous silicon nanowire cilia are formed on the sidewalls of the micropillars after etching (top) and its close-up view (bottom). **c** Anti-CD63 antibody were immobilized on the ciliated micropillars(top) and its close-up view (bottom). All scale bars are 5 μm

microfluidic device fully disappeared, which lead the stripping of the anti-CD63 immobilizing layer and trapping sites to release the capturing exosomes. Commonly, the traditional methods of isolating exosomes always interfere with the structure of the exosomes and hardly to show the morphology of native exosomes (Szatanek et al. 2015). However, the recovered exosome from our microfluidic device maintained its 100 nm sphere morphology and the integrity of surface structure as characterized by Atom Force Microscopy (Fig. 5b). Overall, the microfluidic device we designed presents the significantly capacity on high purity exosomes selectively isolating and recovering from biofluid.

3.4 Measuring of drug content of isolated exosomes

Intrinsic or acquired MDR which considerably limits the efficacy of chemotherapeutics, is mediated by different mechanisms, such as P-gP effluxing (Hayeshi et al. 2006; Kim et al. 2016). Recently, nanotechnology holds great promising development for drug delivery systems with controlled-release and drug-resistance overcoming profiles for MDR cancer treatment (Hayeshi et al. 2006). It is reported that the altering biophysical properties of anticancer drugs by nanoparticles conjugation or encapsulation was able to bypass the recognition of

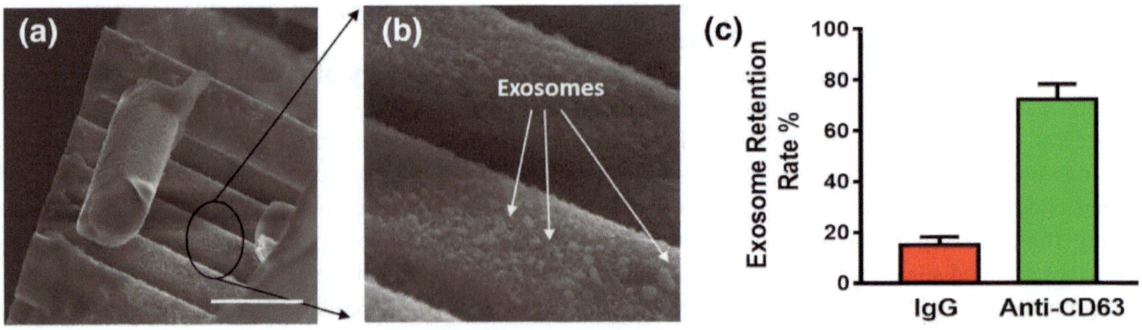

Fig. 3 Exosome retention by the antibody immobilized micropillar array. **a** SEM images of exosome captured by ciliated micropillars (Scale bar 5 μm). **b** Close-up view of captured exosome on the micropillar (Scale bar 500 nm). **c** Comparison of exosome retention rate of anti-CD63 antibody immobilized micropillars and IgG immobilized micropillars

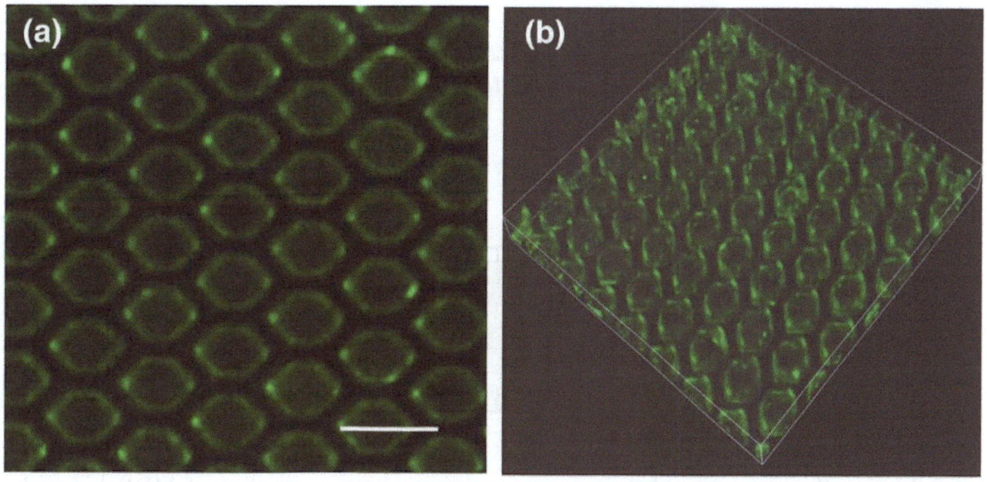

Fig. 4 Confocal microscopy images of fluorescein-labelled exosomes captured by **a** ciliated micropillar and **b** its 3D imaging. Scale bars represent 10 μm

P-gP (Brigger et al. 2012). Moreover, the paramount overlap of exosome expelling and the uptake of nanoparticles are still unclear. Especially considering that the secreting of cancerous cell exosomes plays critical roles in drug resistance generating and transferring (Soekmadji and Nelson 2015; Wang et al. 2016), the effects of exosomal pathway related to the drug-loaded nanoparticles MDR cellular internalization, further to the MDR overcoming need to be well evaluated.

As the results above, our microfluidic device with antibody immobilized ciliated micropillars is a well-suited platform for effective and non-destructive exosome isolation. By isolating exosomes from MDR cells treated with nanotherapy with this device, the drug efflux profile in the exosomal pathway can be determined, which lead a better understanding of drug expulsion mechanism through the exosomal pathway and circumventing MDR effect. For this purpose, two different of drug-loaded nanoformulations were prepared: the commercialized doxorubicin-loaded liposome (Doxoves, 83 nm) and synthetized paclitaxel-loaded mPEG-PCL micellar nanoparticle with sphere morphology of 74 nm diameter and −4.12 mV surface potential (Fig. S2). The normal human breast cancer cell MDA-MB-231 and its MDR type MDA-MB-231/MDR were both treated by free small drugs and nanotherapeutic formulations. After that, the expelled exosomes in the cultured mediums were isolated through anti-CD63 immobilized microfluidic device and the total drug contents in the exosomes was measured to study exosomal pathway corresponding to different forms of drug treatment. As shown in Fig. 6a, b, both the exosomes from nanoparticles treated non-resistant and MDR cell showed much lower drug content than free drug treated counterpart. Moreover, it is notably that free drug treated MDR cells showed more drug efflux than its counterpart sensitive cell due to the MDR profiles (approximately 2 times and 1.5 times higher of DOX and PTX, respectively), while nanoformulations treated MDR cells showed comparable drug efflux to its sensitive counterpart.

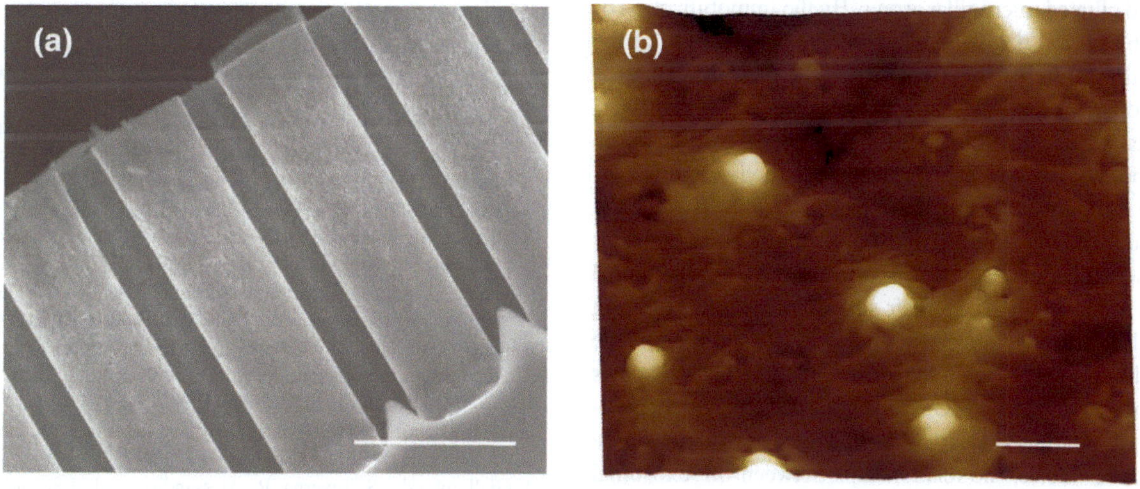

Fig. 5 **a** SEM image of the micropillars after soaking in PBS buffer to recover the captured exosome. (Scale bar represents 5 μm) **b** Atom forced microscopy image of the recovered exosomes from micropillars. (Scale bar represents 200 nm)

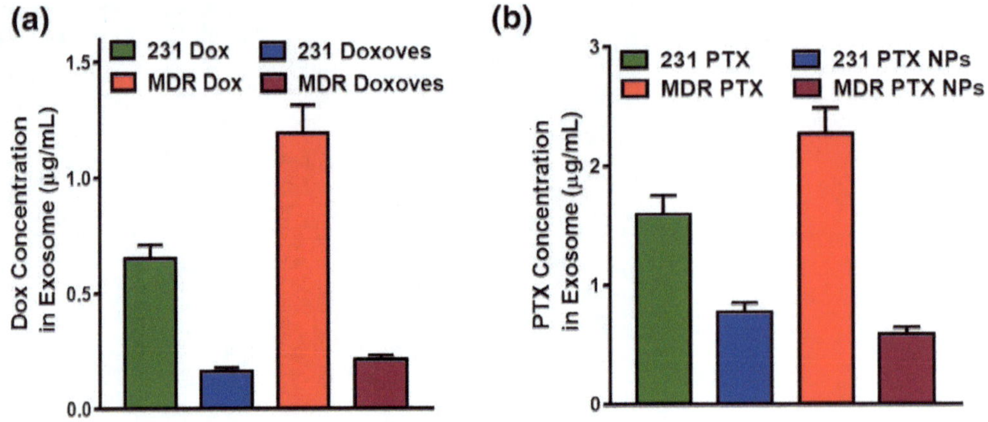

Fig. 6 **a** Doxorubicin content in the exosomes isolated from MDA-MB-231 human breast cancer cells (231) and MDR MDA-MB-231 human breast cancer cells after treating by different drug formulations. **b** Paclitaxel content in the exosomes isolated from MDA-MB-231 human breast cancer cells (231) and MDR MDA-MB-231 human breast cancer cells after treating by different drug formulations

Considering the same treated drug concentration on both sensitive/MDR cancer cells, these results demonstrated the existing overlap of the cellular exosomal expelling pathway and bypassing P-gP effluxing effect of nanoparticles. However, the enhanced cellular internalization of different nanoformulations were not inducing the actual MDR overcoming (Fig. S3, Table S1). The slow drug releasing from the nanoparticles or other mechanisms related MDR such as anti-apoptosis may be responsible to the nanotherapy insensitivity of cancerous cells (Hu and Zhang 2009; Notarbartolo et al. 2002; Song et al. 2012, 2013, 2014). Thus, developing a novel drug-loading nanoplatform not only bypassing exosomal effluxing pathways when internalized by the cancerous cells but also showing intracellular burst release profiles and personalizing capacity against the multi-mechanisms of MDR is essential for the future clinical applications of nanotherapy.

4 Conclusion

We developed a microfluidic device consisting of nanowire enhanced ciliated micropillar arrays. By the immobilization of anti-CD63 on the nanowires, this device was capable of the specifically capturing the exosomes in the biofluid, while simultaneously bypassing smaller proteins/cell debris and filtering larger particles such as cells. The captured exosomes were able to be recovered without structure destructive by dissolving the porous nanowires in buffers. Therefore, this microfluidic device could be used as a rapidly and efficiently tool to selectively isolate exosomes from complex biological samples while maintaining their structural integrity, offering potential benefit to a range of both diagnostic and therapeutic applications. This device was further applied for the analysis of exosomes derived by MDR cancerous cells treated by drug-loading nanoparticles. Due to the superior performance of the microfluidic device, the relationships between the exosomal pathway and the MDR-related drug expulsion under the nanotherapy were identified. These results insure the future studies of MDR biomarkers on secreted exosomes under varies treatment procedure, as which will contribute to our further mechanism understandings on MDR acquiring and overcoming.

Acknowledgements The authors would like to thank Dr. Jianhua Gu of the HMRI Microscopy-SEM/AFM core on AFM imaging. The authors acknowledge financial support from the following sources: NIH 1R21CA190024-01, 1R01GM127558-01, DOD W81XWH-17-1-0389 and Houston Methodist Research Institute.

References

A.M. Bannunah, D. Vllasaliu, J. Lord, et al., Mechanisms of nanoparticle internalization and transport across an intestinal epithelial cell model: Effect of size and surface charge. Mol. Pharm. **11**(12), 4363–4373 (2014)

I. Brigger, C. Dubernet, P. Couvreur, Nanoparticles in cancer therapy and diagnosis. Adv Drug Deliver Rev **64**, 24–36 (2012)

C. Chen, J. Skog, C.H. Hsu, et al., Microfluidic isolation and transcriptome analysis of serum microvesicles. Lab Chip **10**(4), 505–511 (2010)

C. Chiappini, X.W. Liu, J.R. Fakhoury, et al., Biodegradable porous silicon barcode nanowires with defined geometry. Adv. Funct. Mater. **20**(14), 2231–2239 (2010)

C. Corcoran, S. Rani, K. O'Brien, et al., Docetaxel-resistance in prostate Cancer: Evaluating associated phenotypic changes and potential for resistance transfer via exosomes. PLoS One **7**(12), e50999 (2012)

B. Fevrier, G. Raposo, Exosomes: Endosomal-derived vesicles shipping extracellular messages. Curr. Opin. Cell Biol. **16**(4), 415–421 (2004)

B. Fisher, J. Bryant, N. Wolmark, et al., Effect of preoperative chemotherapy on the outcome of women with operable breast cancer. J. Clin. Oncol. **16**(8), 2672–2685 (1998)

J. Foo, F. Michor, Evolution of acquired resistance to anti-cancer therapy. J. Theor. Biol. **355**, 10–20 (2014)

M. Giallombardo, S. Taverna, R. Alessandro, et al., Exosome-mediated drug resistance in cancer: The near future is here. Ther Adv Med Oncol **8**(5), 320–322 (2016)

M.M. Gottesman, Mechanisms of cancer drug resistance. Annu. Rev. Med. **53**, 615–627 (2002)

N.S.K. Gunda, M. Singh, L. Norman, et al., Optimization and characterization of biomolecule immobilization on silicon substrates using (3-aminopropyl)triethoxysilane (APTES) and glutaraldehyde linker. Appl. Surf. Sci. **305**, 522–530 (2014)

R. Hayeshi, C. Masimirembwa, S. Mukanganyama, et al., The potential inhibitory effect of antiparasitic drugs and natural products on P-glycoprotein mediated efflux. Eur. J. Pharm. Sci. **29**(1), 70–81 (2006)

C.M. Hu, J.L.F. Zhang, Therapeutic nanoparticles to combat Cancer drug resistance. Curr. Drug Metab. **10**(8), 836–841 (2009)

C.M. Hu, J.L.F. Zhang, Nanoparticle-based combination therapy toward overcoming drug resistance in cancer. Biochem. Pharmacol. **83**(8), 1104–1111 (2012)

R. Kalluri, The biology and function of exosomes in cancer. J. Clin. Invest. **126**(4), 1208–1215 (2016)

S.S. Kanwar, C.J. Dunlay, D.M. Simeone, et al., Microfluidic device (ExoChip) for on-chip isolation, quantification and characterization of circulating exosomes. Lab Chip **14**(11), 1891–1900 (2014)

M.S. Kim, M.J. Haney, Y. Zhao, et al., Development of exosome-encapsulated paclitaxel to overcome MDR in cancer cells. Nanomedicine **12**(3), 655–664 (2016)

P. Kucharzewska, M. Belting, Emerging roles of extracellular vesicles in the adaptive response of tumour cells to microenvironmental stress. J Extracell Vesicles **2**, 10 (2013)

H.G. Lamparski, A. Metha-Damani, J.Y. Yao, et al., Production and characterization of clinical grade exosomes derived from dendritic cells. J. Immunol. Methods **270**(2), 211–226 (2002)

T. Le Chevalier, R. Arriagada, C. Le Pechoux, et al., Cisplatin-based adjuvant chemotherapy in patients with completely resected non-small-cell lung cancer. New Engl J Med **350**(4), 351–360 (2004)

S.A. Melo, L.B. Luecke, C. Kahlert, et al., Glypican-1 identifies cancer exosomes and detects early pancreatic cancer. Nature **523**(7559), 177–182 (2015)

D. Napierska, L.C.J. Thomassen, V. Rabolli, et al., Size-dependent cytotoxicity of monodisperse silica nanoparticles in human endothelial cells. Small **5**(7), 846–853 (2009)

M. Notarbartolo, M. Cervello, L. Dusonchet, et al., Resistance to diverse apoptotic triggers in multidrug resistant HL60 cells and its possible relationship to the expression of P-glycoprotein, Fas and of the novel anti-apoptosis factors IAP (inhibitory of apoptosis proteins). Cancer Lett. **180**(1), 91–101 (2002)

N.J. Oh, H. Park, Endocytosis and exocytosis of nanoparticles in mammalian cells. Int. J. Nanomedicine **9**, 51–63 (2014)

M. Oves, H.A. Qari, N.M. Felemban, et al., in *J Nanomater.* Exosomes: A paradigm in drug development against Cancer and infectious diseases, vol 2018 (2018), p. 1

M.K.B. Parmar, J.A. Ledermann, N. Colombo, et al., Paclitaxel plus platinum-based chemotherapy versus conventional platinum-based chemotherapy in women with relapsed ovarian cancer: The ICON4/AGO-OVAR-2.2 trial. Lancet **361**(9375), 2099–2106 (2003)

J. Rak, Extracellular vesicles - biomarkers and effectors of the cellular interactome in cancer. Front. Pharmacol. **4**, 21 (2013)

R. Safaei, B.J. Larson, T.C. Cheng, et al., Abnormal lysosomal trafficking and enhanced exosomal export of cisplatin in drug-resistant human ovarian carcinoma cells. Mol. Cancer Ther. **4**(10), 1595–1604 (2005)

J.C. Santos, N.D. Lima, L.O. Sarian, et al., Exosome-mediated breast cancer chemoresistance via miR-155 transfer. Sci. Rep. **8**(1), 829 (2018)

C. Soekmadji, C.C. Nelson, The emerging role of extracellular vesicle-mediated drug resistance in cancers: Implications in advanced prostate Cancer. Biomed Res Int **2015**, 1 (2015)

Y. Song, Y. Zhang, P.E. Bernard, et al., Multiplexed volumetric bar-chart chip for point-of-care diagnostics. Nat. Commun. **3**, 1283 (2012)

Y. Song, Y. Wang, L. Qin, A multistage volumetric bar chart chip for visualized quantification of DNA. J. Am. Chem. Soc. **135**(45), 16785–16788 (2013)

Y. Song, X. Xia, X. Wu, et al., Integration of platinum nanoparticles with a volumetric bar-chart chip for biomarker assays. Angew Chem Int Ed Engl **53**(46), 12451–12455 (2014)

R. Szatanek, J. Baran, M. Siedlar, et al., Isolation of extracellular vesicles: Determining the correct approach (review). Int. J. Mol. Med. **36**(1), 11–17 (2015)

M.K.A. Tang, S. Wong, Exosomes: Emerging biomarkers and targets for ovarian cancer. Cancer Lett. **367**(1), 26–33 (2015)

B.J. Tauro, D.W. Greening, R.A. Mathias, et al., Comparison of ultracentrifugation, density gradient separation, and immunoaffinity capture methods for isolating human colon cancer cell line LIM1863-derived exosomes. Methods **56**(2), 293–304 (2012)

E. van der Pol, A.N. Boing, P. Harrison, et al., Classification, functions, and clinical relevance of extracellular vesicles. Pharmacol. Rev. **64**(3), 676–705 (2012)

Z.X. Wang, H.J. Wu, D. Fine, et al., Ciliated micropillars for the microfluidic-based isolation of nanoscale lipid vesicles. Lab Chip **13**(15), 2879–2882 (2013)

X.H. Wang, C.F. Xu, Y.T. Hua, et al., Exosomes play an important role in the process of psoralen reverse multidrug resistance of breast cancer. J Exp Clin Canc Res **35**(1), 186 (2016)

C.P. Yang, D. Robbins, The roles of tumor-derived exosomes in cancer pathogenesis. Clin Dev Immunol **2011**, 842849 (2011)

X. Yang, Z.Q. Weng, D.L. Mendrick, et al., Circulating extracellular vesicles as a potential source of new biomarkers of drug-induced liver injury. Toxicol. Lett. **225**(3), 401–406 (2014)

Y.L. Yuan, T.G. Cai, X. Xia, et al., Nanoparticle delivery of anticancer drugs overcomes multidrug resistance in breast cancer. Drug Deliv **23**(9), 3350–3357 (2016)

M. Zhang, E.G. Liu, Y.N. Cui, et al., Nanotechnology-based combination therapy for overcoming multidrug-resistant cancer. Cancer Biol Med **14**(3), 212–227 (2017)

Biomed Microdevices (2019) 21: 30
https://doi.org/10.1007/s10544-019-0382-0

Glioblastoma adhesion in a quick-fit hybrid microdevice

Hsieh-Fu Tsai[1] · Kazumi Toda-Peters[1] · Amy Q. Shen[1]

Published online: 21 March 2019

Author's Personal Note

Dear Mauro,

Your passion and imagination for science and life inspire us all.

Best,
Amy Q. Shen

Abstract

Translational research requires reliable biomedical microdevices (BMMD) to mimic physiological conditions and answer biological questions. In this work, we introduce a reversibly sealed quick-fit hybrid BMMD that is operator-friendly and bubble-free, requires low reagent and cell consumption, enables robust and high throughput performance for biomedical experiments. Specifically, we fabricate a quick-fit poly(methyl methacrylate) and poly(dimethyl siloxane) (PMMA/PDMS) prototype to illustrate its utilities by probing the adhesion of glioblastoma cells (T98G and U251MG) to primary endothelial cells. In static condition, we confirm that angiopoietin-Tie2 signaling increases the adhesion of glioblastoma cells to endothelial cells. Next, to mimic the physiological hemodynamic flow and investigate the effect of physiological electric field, the endothelial cells are pre-conditioned with concurrent shear flow (with fixed 1 Pa shear stress) and direct current electric field (dcEF) in the quick-fit PMMA/PDMS BMMD. With shear flow alone, endothelial cells exhibit classical parallel alignment; while under a concurrent dcEF, the cells align perpendicularly to the electric current when the dcEF is greater than 154 V m^{-1}. Moreover, with fixed shear stress of 1 Pa, T98G glioblastoma cells demonstrate increased adhesion to endothelial cells conditioned in dcEF of 154 V m^{-1}, while U251MG glioblastoma cells show no difference. The quick-fit hybrid BMMD provides a simple and flexible platform to create multiplex systems, making it possible to investigate complicated biological conditions for translational research.

Keywords Glioblastoma adhesion · Endothelium · Bubble-free · Electric field · Shear flow · Multiplexing

1 Introduction

In biomedical microdevices (BMMDs), biochemical and biophysical microenvironments can be manipulated by combining device design and operation conditions to probe biochemical and biophysical properties or to investigate biological phenomena. Practical BMMD designs must consider system integration and fabrication process to reduce the production volume and cost, ensure the sterilization and user-friendly operation (McRae et al. 2016). Common materials used in fabrication of BMMDs include different types of thermoplastics, transparent and gas permeable silicone rubbers, glasses, and perfluoropolymers (Ren et al. 2013). Transparent rigid thermoplastics, such as poly(methyl methacrylate) (PMMA) and cyclic olefin copolymers (COC), and transparent elastomeric silicone rubber poly(dimethyl siloxane) (PDMS) are particularly favored in BMMD production in laboratory settings (Cheng et al. 2004; Prokop et al. 2004; Steigert et al. 2007). Each material has its distinct advantages and disadvantages (see summary in Table 1). By using thermoplastics, fabrication cost is low while bonding strategies and interconnect choices are more readily available for constructing the world-to-chip interface. Fabrication of complex three dimensional (3D) internal structures is also possible, but the spatial resolution of direct-writing methods is inferior in comparison to lithographic methods (Cheng et al. 2004). Alternatively, micrometer-precision with PDMS substrates

This work is supported by JSPS KAKENHI JP1700362 and OIST Graduate University with subsidy funding from the Cabinet Office, Government of Japan.

Hsieh-Fu Tsai is a Research Fellow of Japan Society of Promotion for Science

This article is part of the Topical Collection on *Biomedical Micro-Nanotechnologies toward Translation*

Electronic supplementary material The online version of this article (https://doi.org/10.1007/s10544-019-0382-0) contains supplementary material, which is available to authorized users.

✉ Amy Q. Shen
amy.shen@oist.jp

✉ Hsieh-Fu Tsai
hsieh-fu.tsai@oist.jp

[1] Micro/Bio/Nanofluidics Unit, Okinawa Institute of Science and Technology Graduate University, 1919-1 Tancha, Onna-son, Okinawa 9040495, Japan

Table 1 Common materials to fabricate BMMDs

	Thermoplastics (PMMA, COC)	PDMS	Glass (Industrial)
Advantages	1. 3D fluidics by stacking 2. Low gas permeability 3. Versatile world-to-chip connection	1. High spatial resolution by lithography (1 μm) 2. High gas permeability 3. Active components possible 4. Easy priming to remove bubbles	1. High spatial resolution by etching (10 nm) 2. Excellent chemical stability 3. High transparency 4. High rigidity to sustain high flow rates
Disadvantages	1. Low spatial resolution by laser cutting (100 μm) 2. High μm-scale roughness by micromilling 3. Birefringent 4. Rigid structures only 5. Prone to trap air bubbles 6. Medium solvent compatibility	1. Limited to 2.5D structures 2. Adsorption/absorption of hydrophobic chemical species 3. Limited bonding strategies and substrate choice due to low surface energy 4. Limited world-to-chip interface choice 5. Poor solvent compatibility	1. Limited to 2.5D structures 2. Facility requirement for wet or dry etching 3. Special bonding requirement 4. Brittle

is possible through the soft lithography process, although the process is usually limited to quasi-planar (2.5D) microstructures (Xia and Whitesides 1998). PDMS is also flexible and highly gas permeable, making it suitable to fabricate BMMDs with active components such as microvalves, and to create gas concentration gradients for automated lab-on-chip applications. Although fabrication of complex 3D microstructures in PDMS BMMDs is possible, it requires complicated multi-step fabrication, alignment, and off-stoichiometry bonding (Thorsen et al. 2002). Moreover, PDMS has very low surface energy, which limits its bonding and world-to-chip interconnect choices.

For successful and reliable cell culture in BMMDs, it is essential to ensure sterility, controlled microenvironment, and guided reagent delivery with bubble-free conditions (Kim et al. 2007). Bubble prevention is crucial because microbubbles can impose strong interfacial tensions that shear adherent cells or cause cell damage in the microchannel. In thermoplastic BMMDs, to minimize air bubbles, the device is often pre-assembled with reagent delivery components and a world-to-chip interface prior to seeding cells into the device. As a result, the dead volume tends to be large in pre-assembled thermoplastic BMMDs, which leads to longer reagent delivery time to cells (Tsai et al. 2012). In PDMS BMMDs, although a delicate buffer exchange procedure can be adapted to prevent air bubbles in microchannels (Wang et al. 2012), the hydrophobic nature of PDMS often causes air bubble accumulation at world-to-chip interconnects, which can easily propagate into microchannels. Moreover, the elastic nature of PDMS is prone to leakage at interconnects (Christensen et al. 2005).

In this work, we present a hybrid quick-fit PMMA/PDMS BMMD, combining the advantages and mitigating the disadvantages of PMMA and PDMS to create an air-tight but reversibly sealed cell culture platform with low reagent dead volumes. Reversibly sealed BMMDs are empowered by mechanical or vacuum sealing of an elastomeric PDMS chip to a rigid substrate, thus allowing the BMMD to be sealed and dismantled quickly before and after each experiment (Khademhosseini et al. 2005; Bang et al. 2006; Skafte-Pedersen et al. 2012; Uzel et al. 2016; Abhyankar et al. 2016). This hybrid device enables further cellular treatment (i.e., fixation and immuno-staining) in the BMMD. To validate this setup, we choose glioblastoma and endothelial cells as a cell-cell interaction model system. Glioblastoma, the most common primary high-grade brain tumor type in adults, can diffuse and metastasize intracranially through white matter tracts or defined perivascular structures, such as blood vessels and the subarachnoid space (Holland 2000; Segarra et al. 2015). Although intravasation and extracranial metastasis in glioblastoma seldom occur (Bernstein and Woodard 1995; Beauchesne 2011), endothelial cells and the associated blood-brain barrier contribute greatly to establish and maintain the tumor microenvironment of glioblastoma (Cuddapah et al. 2014; Watkins et al. 2014). An *in vitro* cell-cell interaction model of glioblastoma and endothelial cells could further our understanding of the perivascular tumor microenvironment of glioblastoma (Tsai et al. 2017). Endothelial cells cultured *in vitro* can be conditioned chemically or physically through mechanical or electrical stimulation to up-regulate expression of cell adhesion molecules or to promote cell morphology with more natural physiological conditions (Sheikh et al. 2003; Zhao et al. 2004; Bai et al. 2011; Khan and Sefton 2011; Uzarski et al. 2013; Jaczewska et al.

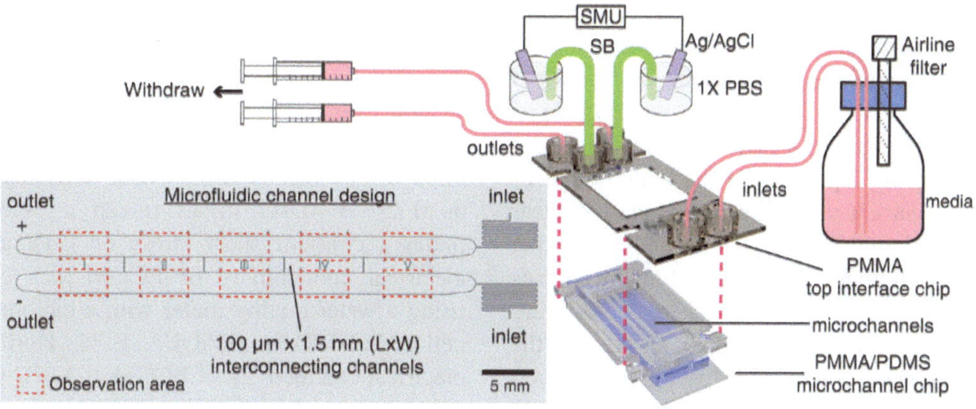

Fig. 1 Experimental setup of concurrent shear flow and electric field conditioning of endothelial cells in a *shear flow and electric field co-stimulation microfluidic chip* (SFEFC). Endothelial cells are cultured in the bottom PMMA/PDMS microchannel device in a user-friendly manner. To pre-condition the cells, the PMMA/PDMS chip is reversibly sealed with the top PMMA interface chip before applying the shear flow and electric field. After conditioning, the chip can be easily recovered. SMU: source measure unit; SB: salt bridge. Detailed configuration of PMMA top interface chip and PMMA/PDMS microchannel chip can be found in Fig. 3

2014; Davis et al. 2015). We employ the quick-fit BMMD to demonstrate its robustness by applying concurrent electrical and mechanical conditioning on the endothelial cells and further investigate the adherence of glioblastoma cells on the conditioned endothelium. The design and fabrication of a shear flow and electric field co-stimulation microfluidic chip with the quick-fit hybrid BMMD is discussed in Sections 2.1–2.2. The adhesion of glioblastoma to endothelial cells in a static condition and in a coexisting shear flow and electric field microenvironment are discussed in Sections 2.3–2.7 and Section 3. Conclusion is provided in Section 4.

2 Materials and methods

2.1 Concurrent shear flow and electric field chip design

The shear flow and electric field co-stimulation microfluidic chip (SFEFC) was designed to quick-fit a top PMMA interface chip with a bottom PMMA/PDMS microchannel device where cells were cultured (see schematic in Fig. 1). The SFEFC was constructed to create multiple electric fields in an R-2R resistor ladder configuration (Tsai et al. 2012; Zhao et al. 2014). Two 2 mm-wide main

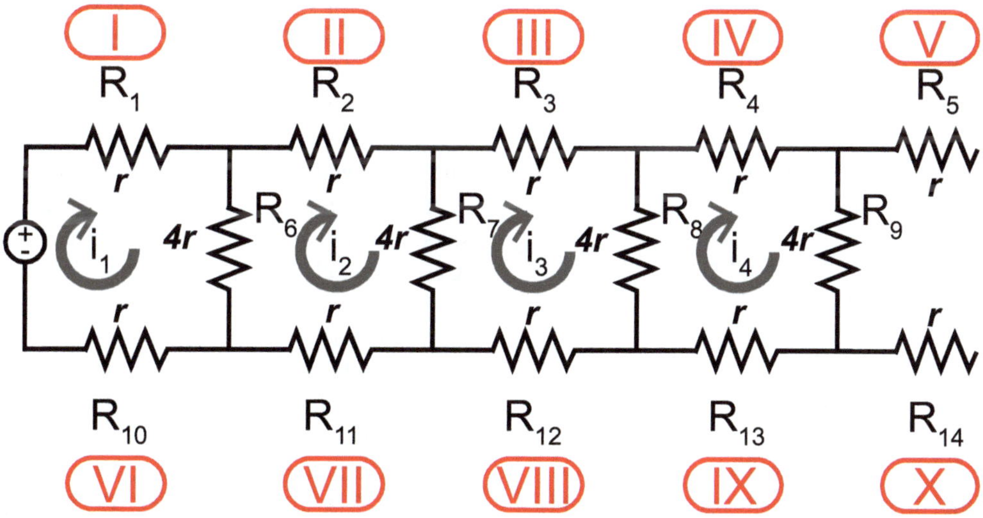

Fig. 2 The equivalent circuit of electric field in the *shear flow and electric field co-stimulation microfluidic chip* (SFEFC). Each microfluidic channel segment can be regarded as a flow resistor and an electrical resistor with relative electrical resistances can be calculated according to the Ohm's law

microchannels with interconnected 100 μm by 1.5 mm (L×W) channels at a spatial interval of 7.5 mm in SFEFC created 10 channel segments with various electric field strengths (EFSs) (Fig. 1). The segments on the top side of SFEFC had electric current vector flowing against the shear flow direction, while the segments on the bottom side of SFEFC had electric current vector flowing along the shear flow direction. This multiplex configuration provided a platform for high-throughput screening of cellular responses to coexisting shear flow and electric field. Cells were observed only in the 4.5 mm-long observation area, in the middle of each segment, 1.5 mm from the interconnection channels (where the electric fields were stable).

The electrical equivalent circuit of SFEFC is shown in Fig. 2. Each segment of the microfluidic channel network was regarded as an electrical resistor in which relative electrical resistances were calculated and modeled by Ohm's law and Kirchhoff's circuit laws. In the equivalent circuit, the endpoint of R_5 and R_{14}, the adjacent segments from both inlets, were open in the electric circuit. No electric current was flowing through them. Cells in the two segments were only subjected to shear flow.

According to Ohm's law, the electrical resistance of a resistor, R, is proportional to the length and inversely proportional to the cross-sectional area:

$$R = \rho \frac{L}{A} = \frac{\rho L}{WH}, \tag{1}$$

where ρ, L, A, W, and H are the electrical resistivity of the medium, the length, the cross-sectional area, the width, and the height of the microchannel, respectively. Assuming the electrical resistance of R_1 being r, the relative electrical resistances of other segments (R_2 : R_{14}) can be calculated accordingly.

The electric current flowing through each resistor was calculated by Kirchhoff's circuit law and simulated in the electronic design automation software (OrCAD Lite, Cadence Design Systems, USA) by:

$$10r \times i_2 - 4r \times i_1 - 4r \times i_3 = 0, \tag{2a}$$

$$10r \times i_3 - 4r \times i_2 - 4r \times i_4 = 0, \tag{2b}$$

$$10r \times i_4 - 4r \times i_3 = 0. \tag{2c}$$

By solving the system of equations in Eq. 2, the ratio of electric currents between each segment, hence the ratio of electric field strengths, was derived in Eq. 3:

$$i_1 : i_2 : i_3 : i_4 = E_I : E_{II} : E_{III} : E_{IV} : E_V \tag{3a}$$

$$= E_{VI} : E_{VII} : E_{VIII} : E_{IX} : E_X \tag{3b}$$

$$= 10.5 : 5.2 : 2.5 : 1 : 0 \tag{3c}$$

To calculate the flow rate and electric current applied to condition endothelial cells, the dynamic shear viscosity and the electrical conductivity of the endothelial cell growth medium (ECGM, C22110, PromoCell GmbH, Germany) were characterized experimentally. The dynamic shear viscosity was measured by using an A05 sensor in a microfluidic viscometer (m-VROC, Rheosense, USA). The ECGM was verified as a Newtonian fluid within the flow rate regimes of interest (shear rate ranging 293–1334 s^{-1}). The dynamic shear viscosity was measured as 0.757 ± 0.012 mPa.s (mean ± standard deviation) under a constant shear stress of 1 Pa and 37 °C. The electrical conductivity of ECGM was characterized by using a water quality meter with a calibrated conductivity cell (3552-10D on LAQUA F-74, Horiba, Japan). The electrical conductivity of ECGM at 37 °C was measured as 1.444 ± 0.001 S m^{-1} (mean ± standard deviation).

The Stokes flow and electric field in SFEFC were simulated by solving steady-state Navier-Stokes equations and Maxwell's equations in COMSOL Multiphysics software (v5.2, COMSOL, USA). An aqueous medium with dynamic shear viscosity of 0.75 mPa.s, electrical conductivity of 1.44 S m^{-1}, density of 1000 Kg m^{-3}, and a dielectric constant of 80 were used for the ECGM (Tsai et al. 2016).

The simulated EFS ratio, when an electric current of 86.4 μA was carried from one outlet of the SFEFC to the other outlet, is shown in Eq. 4:

$$E_I : E_{II} : E_{III} : E_{IV} : E_V = E_{VI} : E_{VII} : E_{VIII} : E_{IX} : E_X \tag{4a}$$

$$= 9.72 : 4.97 : 2.44 : 1 : 0 \tag{4b}$$

2.2 Device fabrication

The PMMA interface chip (Figs. 1 and 3) was fabricated by patterning fluidic connections, inlets/outlets, salt bridge (SB) connections, and M4 screw clamp slots on three 0.5 mm thick PMMA sheets (CM-205, Chi-Mei Corp, Taiwan), by using a CO_2 laser scriber (VLS3.60, Universal Laser Systems, USA). The PMMA sheets were joined by thermal bonding, as previously described by Tsai et al. (2012, 2016). Adapters with M6 threads (SPC-M6-C, Nabeya, Japan) were glued onto the PMMA chip using a UV adhesive (3301, Loctite, USA). PDMS slabs, 2 mm thick, punched with inlet and outlet holes were affixed to the interface chip using a dual-energy silicone/acrylic double-sided tape (85 μm thick, No. 5302A, Nitto Denko, Japan) (Carlborg et al. 2010).

Similar to the interface chip, the PMMA component of the PMMA/PDMS microchannel chip (Figs. 1 and 3) was fabricated by patterning liquid reservoirs, inlets, outlets, and screw clamp slots on two 2 mm-thick PMMA sheets. The PMMA pieces were again thermally bonded. The PDMS channel chip was fabricated by the soft lithography method (Xia and Whitesides 1998). Briefly, a chrome photomask was patterned with microchannels through a maskless writer (DL-1000, Nanosystem Solutions, Japan), developed, and etched. A 100 μm-thick SU-8 3025 template

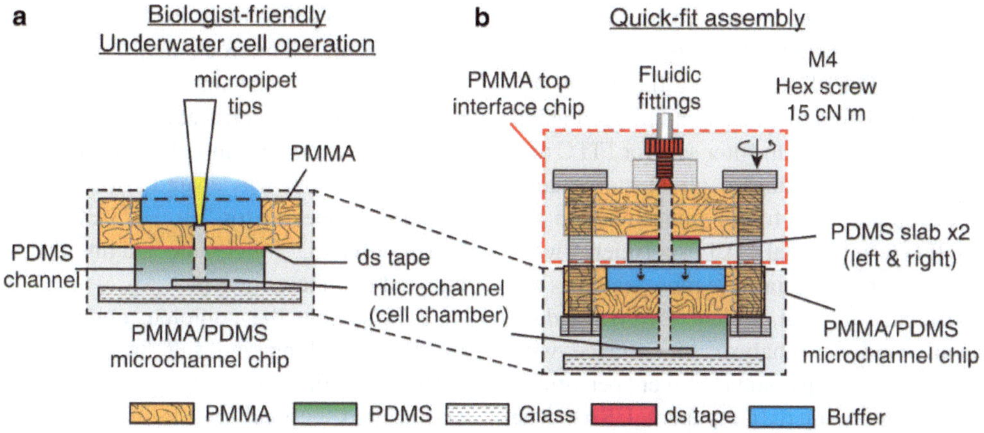

Fig. 3 Liquid manipulation on the SFEFC. **a** Operator-friendly underwater fluid manipulation at early stage of chip preparation using the PMMA/PDMS microchannel chip. Cells are cultured in the microchannels in the PMMA/PDMS chip; **b** The complete chip can be assembled just prior to on-chip experiments. The PMMA top interface chip is shown in red dashed box. PDMS slabs on the PMMA top interface chip compress and seal the interface to the PMMA/PDMS microchannel chip (inside the black dashed box). After each experiment, the cells in the PMMA/PDMS microchannel chip can be easily recovered after disassembly

of microchannels was patterned on a silicon wafer using a mask aligner (MA/BA6, SUSS MicroTec, Germany) with the chrome photomask. After the template was treated with perfluorosilane, PDMS resin (Sylgard 184, Dow Corning, USA) was sandwiched between the SU-8 3025 template and a 15 mm-thick PMMA sheet with a 4 mm-thick PMMA spacer in between to ensure parallelism and flatness of both surfaces. The PDMS resin was cured at 60 °C for 4 h to fabricate the PDMS component. The PMMA/PDMS chip was assembled by joining the bonded PMMA component and the PDMS microchannel component with the silicone/acrylic double-sided tape. The PMMA/PDMS microchannel chip was completed by bonding the joined PMMA/PDMS piece to a glass substrate (60 × 24 mm, No. 1.5H, Paul Marienfeld GmbH, Germany) after O_2 plasma treatment (AP-300, Nordson MARCH, USA). The glass substrate was washed thoroughly in an industrial cleaning solution (1% TFD4, Franklab, France), deionized water, and dried under nitrogen gas prior to bonding.

2.3 Cell culture and maintenance

Primary human umbilical vein endothelial cells (HUVECs) were cultured in the endothelial cell growth medium (C12203, PromoCell GmbH, Germany). T98G (CRL-1690, ATCC, USA) and U251MG (IFO50288, JCRB, Japan) glioblastoma cells were cultured in minimum essential medium α (MEMα, 12000022, Gibco, USA) supplemented with 2.2 gL^{-1} $NaHCO_3$ and 10% fetal bovine serum (FBS, Invitrogen, USA). All cells were cultured in a humidified 5% CO_2 atmosphere (Forma Steri-cycle i160, Thermofisher, USA) and passaged whenever confluency reached 80%. Cells from passages 2 to 8 were used in the adhesion study. Red fluorescent protein-expressing clones of T98G (T98G-dsRed) and U251MG (U251MG-dsRed) cells were established by transfecting a plasmid which expressed translation elongation factor-1 alpha conjugated with a red fluorescent protein into the two cell lines (EF1alpha-DsRed-Express2, 631979, Takara Bio, Japan). Stable fluorescent clones were selected by subculture in media containing 500 $\mu g\ mL^{-1}$ G418 and aseptically sorted by flow cytometry (Aria II, BD, USA).

2.4 Glioblastoma adhesion on statically cultured endothelium

In a static adhesion experiment, glass bottoms of 8-well chambered slides (Lab-tek II, Nunc, USA) were coated with 10 $\mu g\ mL^{-1}$ of human plasma fibronectin (Invitrogen, USA). HUVEC cells of 1×10^5 were seeded in each well and allowed to adhere for 24 h. Angiopoietin 1 (Ang1), an endothelial cell-specific tyrosine kinase receptor ligand that upregulates adhesion molecules on glioblastoma cells was purchased from Peprotech USA. Both glioblastoma cells were incubated with 100 ng mL^{-1} of Ang1 for 24 h prior to adhesion experiments. Activated endothelial cells were prepared by incubating HUVECs with 10 ng mL^{-1} tumor necrosis factor α (TNFα) for 24 h (Invitrogen, USA). Prior to adhesion experiments, confluent HUVECs were labeled by incubation with 1 μM CellTracker Green CMFDA dye (C2925, Invitrogen, USA) for 15 min and washed once with Dulbecco's phosphate buffered saline (D-PBS).

To commence static adhesion experiments, T98G-dsRed and U251MG-dsRed cells were trypsinized and resuspended in ECGM. The two glioblastoma cell lines were seeded at glioblastoma (2) : HUVEC (1) ratio.

Glioblastoma cells were allowed to adhere to endothelial cells for 1 h at 37 °C. Unadhered cells were washed away by gentle pipetting with D-PBS. Cells were fixed with 4% paraformaldehyde in D-PBS immediately and imaged using an epifluorescence microscope with filter sets for FITC and TRITC channels. Densities of adhered glioblastoma cells were counted and averaged over four 1 mm^2 fields in the middle of each well (16 fields total). All experiments were performed in triplicates and the data were represented as the mean ± 95% confidence interval, which is 1.96 standard errors of the mean. One-way analysis of variance with Tukey's post-hoc multiple-comparison tests were performed on collected data using Prism 6 (Graphpad, USA). The confidence level to reject a null hypothesis between two data sets was set at 95%. A p-value (P), the probability for a true null hypothesis less than 0.05 represents a statistical significance at 95% confidence.

2.5 SFEFC assembly and endothelium conditioning by external fields

The usage of quick-fit hybrid devices consists of two steps: user-friendly on-chip cell culture and quick-fit assembly. Briefly, in the first step, users can prepare cells and culture them in the microchannels on the PMMA/PDMS microchannel chip (Fig. 3a). Bubble-free cell culture is achieved by an underwater cell operation workflow. In the second step, prior to initiation of shear flow conditioning and electric field stimulation, the two PDMS slabs on the PMMA top interface chip are aligned to the inlets and outlets of the PMMA/PDMS microchannel chip. Through the compressive force applied by tightening hex screws, the PDMS slabs deform slightly to seal the interface to the PMMA piece of the PMMA/PDMS microchannel chip (Fig. 3b). Moreover, the reagents are primed in the channels of the PMMA interface chip, thus, after quick-fit assembly, the dead volume is reduced and the reagents can be delivered to cells with reduced delivery time.

To prepare confluent HUVEC culture, the PMMA/PDMS chip was filled with 99.5% ethanol (Wako, Japan) to remove bubbles, as described by Wang et al. (2012). The solution in the microchannels was next replaced by deionized water and D-PBS. The buffer held in fluid reservoirs on the PMMA part of the PMMA/PDMS chip (Fig. 3a) ensured that no bubbles were trapped at the interface of the inlets and outlets and prevented accidental bubble injection into the microchannels that would cause disruption of microfluidic flow and cell death.

At an early stage of preparing cells, no complex tubing connections were required. Manipulation of the PMMA/PDMS chip was user-friendly because micropipet tips could be used to deliver fluids and cells by either active pressure delivery or by gravity driven flow (see Video S1 in supplementary information). Furthermore, the fluid volume needed to fill the microchannels was very low (on the order of microliters), so the amount of extracellular matrix protein and the number of cells required were limited.

To start endothelium culture in the PMMA/PDMS chip, the glass bottom was coated with 10 $\mu g\ mL^{-1}$ of human plasma fibronectin for 2 h. HUVEC cells at concentration of 10^7 cells mL^{-1} were injected into the microchannels and allowed to adhere for 3 h.

To assemble the complete quick-fit microfluidic chip, the fluidic tubing and salt bridges (1.5% agarose in D-PBS) were first connected to the PMMA interface chip and primed with cell culture media. The PMMA interface chip was then brought into contact with PMMA/PDMS microchannel chip and sealed tight with four M4 stainless hex screws and nuts. The four screw clamp slots are located on both the PMMA interface chip and PMMA/PDMS microchannel chip so that even pressure can be applied to the PDMS slabs on the PMMA interface chip. The SFEFC was assembled after a tight seal was achieved by applying 15 cN m of torque on the four hex screws with a torque driver (RTD60CN, Tonichi, Japan) (Fig. 3b). The pressure applied on the PDMS slabs deforms the slabs and seals the interface between the top interface chip and the microchannel chip. After each experiment, the cells in the PMMA/PDMS chip could be quickly recovered by removing the hex screws. In addition, the PMMA interface chip could be reused to assemble with another PMMA/PDMS chip to increase experimental throughput with low liquid dead volume.

2.5.1 Endothelium conditioned by shear flow

The assembled SFEFC with a confluent HUVEC cell layer in the microchannel was set up in the incubator as shown in Fig. 1 and supplementary Fig. S.1. We conditioned the endothelium by shear flow mimicking the physiological conditions in hemodynamic flows (Malek et al. 1999). The shear stress imposed on the endothelial cells in a rectangular channel can be calculated according to the modified Hagen-Poiseuille equation Eq. 5 (Gaver and Kute 1998):

$$(\tau_s^*)_{max} = 2.95 \times \frac{6\mu Q}{W \times H^2}, \tag{5}$$

where μ, Q, W, and H are the dynamic shear viscosity, volumetric flow rate, width, and height of the rectangular microchannel, respectively. Primary endothelial cells like HUVECs are reported to respond to shear flow and align parallel to the flow direction when shear stress exceeds 10 dynes cm^{-2} (equivalent to 1 Pa) (Buchanan et al. 2014; Abaci et al. 2014). Based on Eq. 5, to impose 1 Pa shear stress on the endothelial cells in a 2 mm-wide and 100 μm-high microchannel filled with a cell culture medium with dynamic viscosity of 7.57×10^{-4} Pa s, the flow rate should be 1.49 μL s^{-1} (equivalent to 5.4 mL h^{-1}).

Specifically, the flow was delivered by pulling ECGM supplemented with antibiotics cocktail (1X PSN, Invitrogen, USA) from a 250 mL serum bottle by two 50 mL syringes (Terumo, Japan) using a dual channel syringe pump (YSP-202, YMC, Japan). The syringe pump was controlled remotely with the control software (Syringepump Pro, USA). The flow rate was set starting at 0.1 mL h^{-1} and doubled every 3 h until the flow rate reached 5.4 mL h^{-1}.

2.5.2 Endothelium conditioned by both shear flow and electric field

Once the flow rate reached 5.4 mL h^{-1} (corresponding to the shear stress of 1 Pa), a direct current electric field (dcEF) was applied to HUVECs in sections I to V and VI to X at 300:153.6:75.5:30.9:0 respectively based on the simulation results (described in Section 2.1). Specifically, the dcEF was applied by a source measure unit (2410, Keithley, USA) through two silver/silver chloride electrodes in D-PBS where the salt bridges were immersed (Fig. 1). The simultaneous conditioning of the HUVECs with shear flow and electric field were carried out for 24 h prior to glioblastoma cell adhesion experiments. Upon filling of the syringes, the content was discarded prior to restarting the syringe pump.

2.6 Glioblastoma adhesion experiments on conditioned endothelium

After shear flow and dcEF conditioning, the PMMA interface chip was immediately removed by unscrewing M4 Hex screws. The confluent HUVECs were labeled by incubation with 1 μM CellTracker Green for 15 min and washed once with 1X D-PBS before the glioblastoma cell adhesion experiment. T98G-dsRed and U251MG-dsRed cells were trypsinized and suspended in ECGM. The two types of glioblastoma cells were seeded at 2×10^6 cells mL^{-1} in 200 μL by gravity-driven flow. Glioblastoma cells were allowed to adhere for 1 h at 37 °C. Unadhered cells were washed away with 200 μL of D-PBS by gravity flow. Cells were fixed with 4% paraformaldehyde in D-PBS and imaged with an epifluorescence microscope. Cell counting and statistical inferences in these adhesion experiments folllowed the same statistical measurements as those on statically cultured endothelium (more details in Section 2.4).

2.7 Immunofluorescence staining

The presence of glial fibrillary acidic protein (GFAP) and Tie2 receptor in T98G-dsRed and U251MG-dsRed cells were examined by immunofluorescence staining. Both cells were fixed with 4% paraformaldehyde and permeabilized with 0.1% Triton X-100 in D-PBS. Fluorophore-conjugated species-specific secondary antibodies against the primary antibodies were used to detect the two proteins.

The expression of CD31 membrane adhesion molecule (PECAM-1) and cytoskeleton F-actin of conditioned endothelial cells was also characterized by immunofluorescence staining on chip following the same fixation and permeabilization protocol described above. The detection of CD31 and F-actin was conducted with an anti-CD31 primary antibody and a fluorophore-conjugated secondary antibody (NBP1-71663SS, Novus Biological, USA & A21467, Invitrogen, USA) as well as the fluorophore-conjugated phalloidin (A12380, Invitrogen, USA). The stained cells were scanned under a confocal laser scanning microscope with a 10X objective (A1R+, Nikon, Japan).

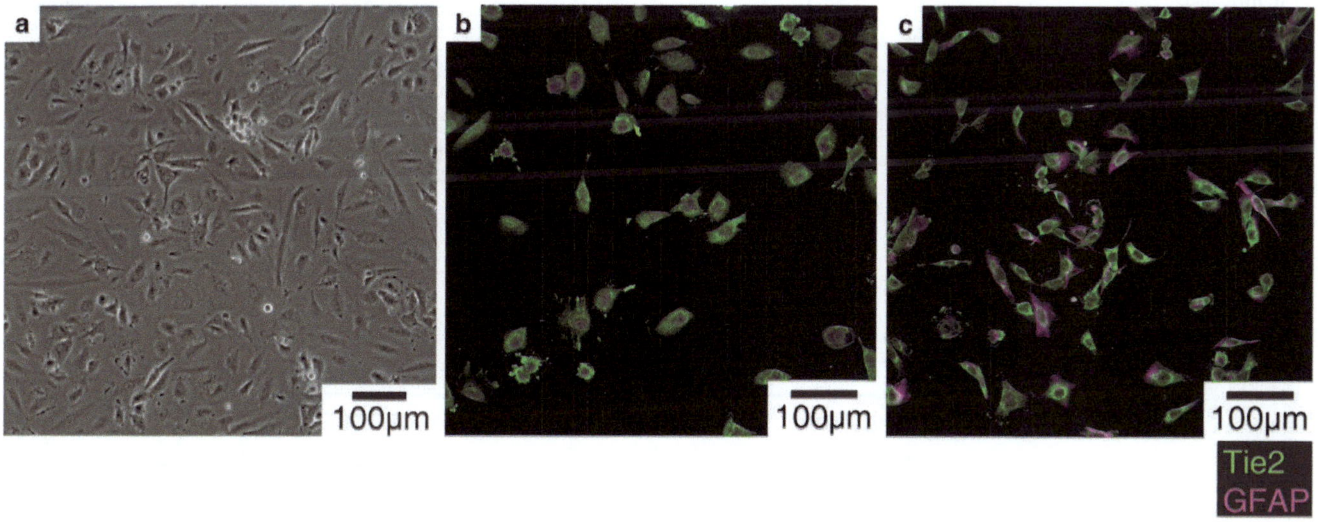

Fig. 4 Microimages of endothelial cells and glioblastoma cells: **a** Endothelial cells cultured in static condition show random orientation. Both **b** T98G and **c** U251MG glioblastoma cells exhibit both GFAP and Tie2 expressions

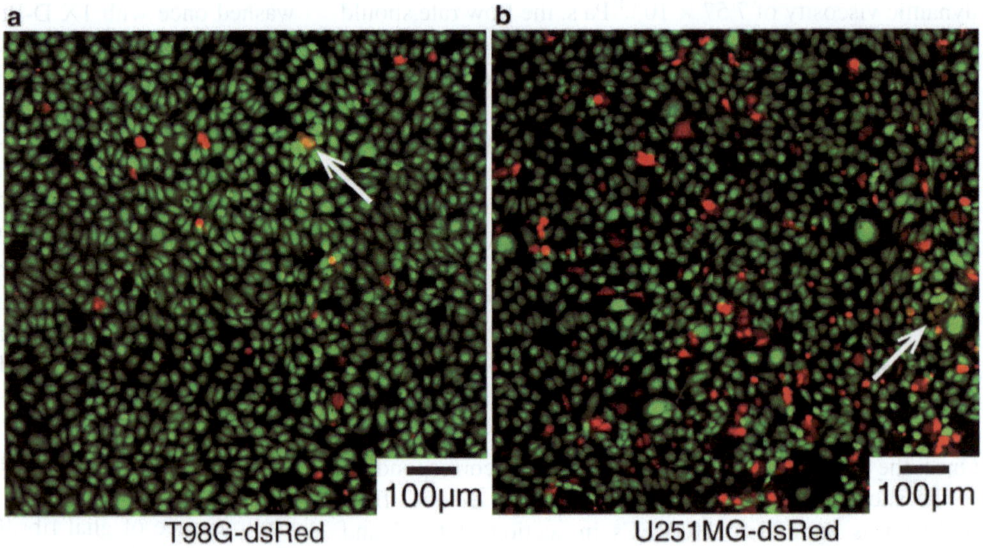

Fig. 5 The adhesion of **a** T98G-dsRed, and **b** U251MG-dsRed, to confluent endothelium (fluorescently labeled green). The cells with yellow color (next to the white arrows) indicate the colocalization of CellTracker Green dye in endothelial cells and dsRed fluorescent protein in glioblastoma cells. The colocalization implies cell fusion or intercellular exchange events that may be important in the perivascular microenvironment of glioblastoma

3 Results and discussion

3.1 Glioblastoma adhesion to endothelium in static condition

Phenotyping endothelial cells and glioblastoma cells cultured in static condition: Prior to adhesion experiment, the phenotypes of endothelial cells and glioblastoma cells in static condition were verified. Endothelial cells cultured in static condition expressed random orientation (Fig. 4a). The immunofluorescence staining confirmed that both T98G and U251MG glioblastoma cells exhibited both GFAP and Tie2 expressions, indicating a glial phenotype with potential angiopoietin/Tie2 signaling (Fig. 4b and c).

Glioblastoma cell adhesion to endothelial cells in static condition: Liu et al. (2010) reported that U87MG and U251MG cells adhered to endothelial cells through Ang1/Tie2 signaling. We first tested the adhesion of T98G and U251MG glioblastoma cells under similar conditions. The adhesion of T98G-dsRed and U251MG-dsRed glioblastoma cells to

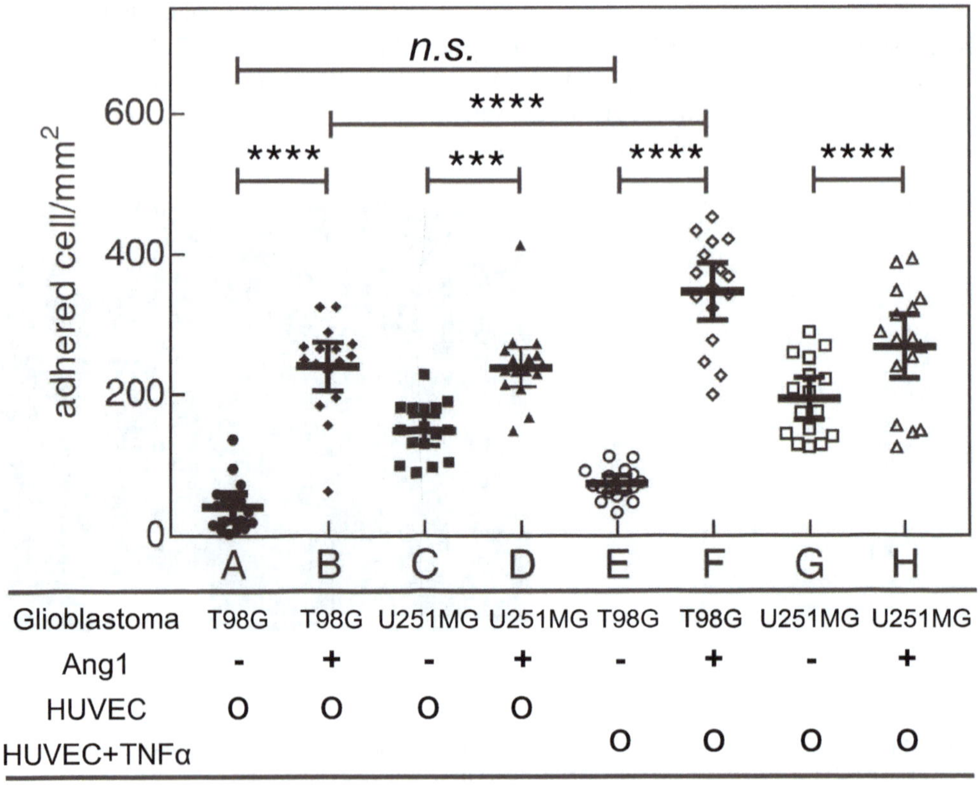

Fig. 6 The adhesion of T98G-dsRed and U251MG-dsRed to endothelium under the stimulation with Ang1 and TNFα in static condition. *** denotes $P < 0.001$; **** denotes $P < 0.0001$; n.s. denotes no significance

confluent endothelial cells was examined (Fig. 5). After 1 h of adhesion, glioblastoma cells adhered to the endothelial and displaced them. The adhesion of glioblastoma cells with and without Ang1 to confluent HUVECs with and without activation with TNFα were quantified and statistically examined in Fig. 6 (microscopy images are shown in supplementary Figs. S.2 & S.3).

Without Ang1 treatment, adhesion of U251MG-dsRed to endothelium was greater than that of T98G-dsRed ($P < 0.0001$). Ang1 treatment increased the adhesion of both T98G-dsRed and U251MG-dsRed to endothelium ($P < 0.001$). This is consistent with existing reports that glioblastoma cells demonstrate enhanced interaction with endothelial cells via Ang1/Tie2 signaling (Liu et al. 2010). TNFα is a cytokine known to upregulate adhesion molecules on endothelial cells (Mackay et al. 1993; Jaczewska et al. 2014). However, only T98G cells activated with Ang1 showed increased adhesion to TNFα-activated HUVECs but not U251MG cells. This result suggests that the Ang1/Tie2 signaling of glioblastoma may not directly cross-talk with TNFα signaling-related adhesion molecules expressed on HUVECs. Further identification of the adhesion molecules is needed.

In addition to glioblastoma adhesion, colocalization events were observed between the dsRed fluorescence of glioblastoma and CellTracker Green fluorescence of endothelial cells *in vitro* (Fig. 5). Colocalization events were quantified and calibrated by adhesion events as shown in Fig. 7 and supplementary Fig. S4. The colocalization events did not increase or decrease ($P > 0.05$) due to Ang1 stimulation or TNFα activation, although Ang1-stimulation increased the adhesion in both glioblastoma cell lines. Interestingly, more colocalization events were detected in Ang1-stimulated T98G-dsRed but not in U251MG-dsRed (see Fig. S.4, $P < 0.0001$).

Colocalization events suggest that glioblastoma cells and endothelial cells underwent a fusion event or an intercellular transport event (Lu and Kang 2009; Pasquier et al. 2013). Cell fusion may contribute to the origin of cancer stem cells and acquisition of drug resistance. Specifically, glioblastoma-endothelial cell hybrids have been proposed to play a pivotal role in the perivascular microenvironment of glioblastoma and the maintenance of glioblastoma cancer stem cells (El Hallani et al. 2014). Further identification of the molecular signaling underlying the intercellular transport and elucidation of glioblastoma-endothelial hybrid function in glioblastoma progression can further our understanding of the angiogenesis of glioblastoma. However, it has been reported that brain microvascular endothelial cells show a distinct phenotype

Fig. 7 The calibrated intercellular transport event ratio (CITER) observed in static condition calibrated by adhesion count

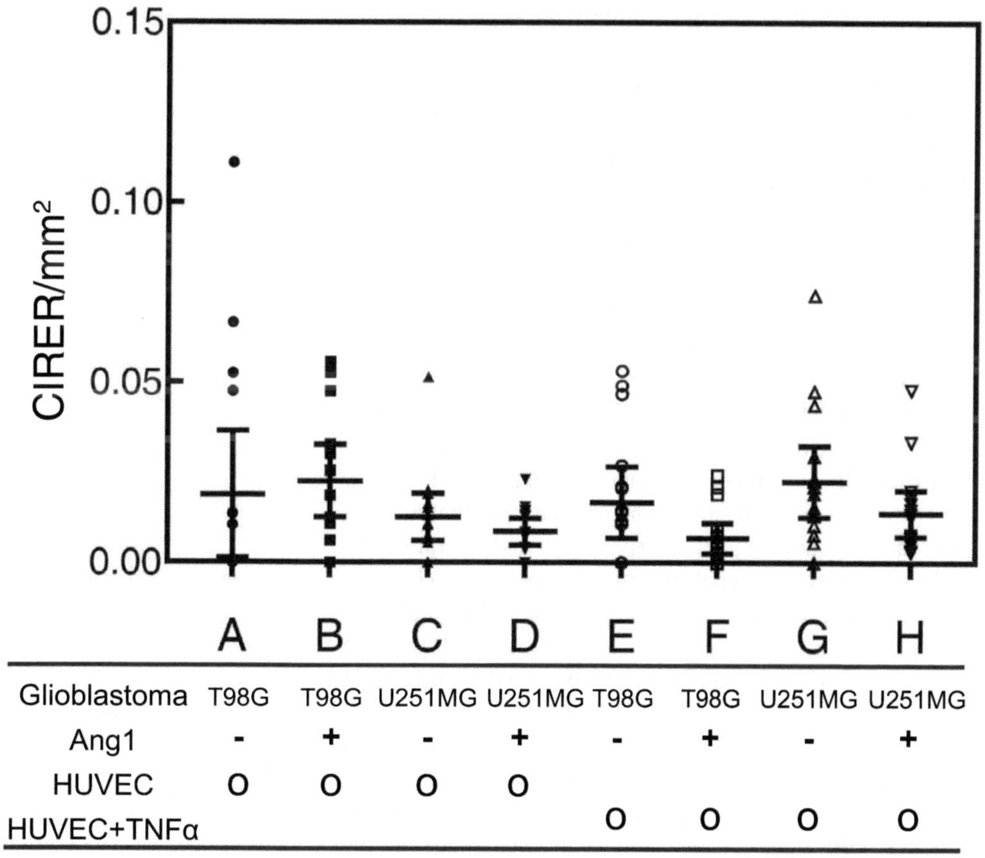

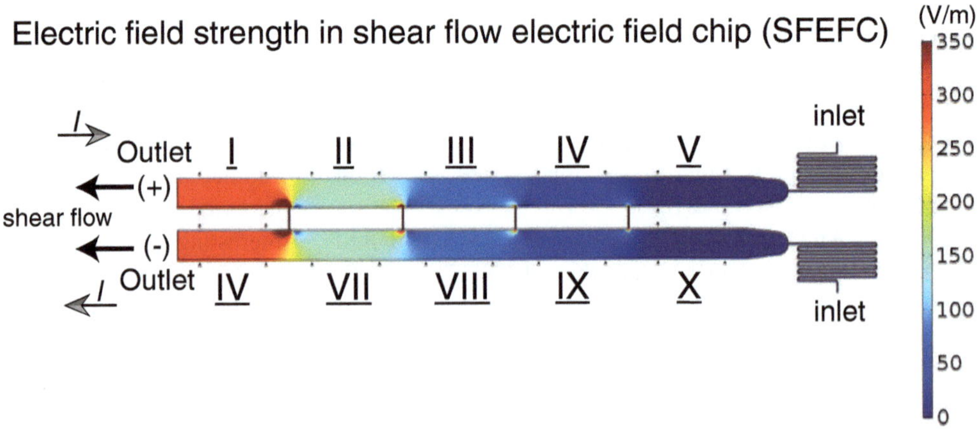

Fig. 8 Numerical results of electric field strength in the shear flow and electric field co-stimulation microfluidic chip (SFEFC). SFEFC features 10 segments with coexistent electric field and shear flow. *I* denotes electric current. From sections I to IV, the shear flow flows against the electric current direction, while in sections VI to IX, the shear flow flows along the electric current vector

compared to endothelial cells isolated from umbilical cord, suggesting that closer physiological models may require the use of tissue-specific endothelial cells (Ye et al. 2014; Reinitz et al. 2015).

3.2 Quick-fit shear flow and electric field co-stimulation chip for high throughput experiments

In the shear flow and electric field co-stimulation chip (SFEFC), 10 segments with combinations of shear flow (fixed shear stress of 1 Pa) and varying electric field strengths have provided possibilities for high-throughput experiments (see Fig. 8). Some key advantages of quick-fit SFEFC include: operator-friendly workflow, low dead volume, low reagent usage, air bubble free operation, air-tight sealing to sustain high flow rates, and reusable top PMMA interface chip that can be assembled with different PMMA/PDMS microchannel chips.

Although the sealing of SFEFC is reversible, it can withstand high flow rates necessary in microfluidic flow cells to create shear stress conditions similar to that in physiological hemodynamics (0.1–0.6 Pa in normal vein and 1.0–7.0 Pa in normal artery as reported by Malek et al. 1999). HUVECs cultured in SFEFC under simultaneous shear flow and electric field conditions demonstrated characteristic morphological changes (Fig. 9). The shear flow imposes a shear stress on adhered endothelial cells and induces an aligned morphology and a quiescent anti-inflammatory and anti-thrombotic phenotype (Uzarski et al. 2013). Electric field stimulation on endothelial cells instead induces a perpendicular aligned morphology with upregulation of a pro-angiogenic response and the release of VEGF (Zhao et al. 2004; Bai et al. 2011).

Under concurrent 5.4 mL h^{-1} shear flow with 1 Pa shear stress and 300 V m^{-1} dcEF stimulation, endothelial cells aligned perpendicular to the electrical current vector and the dcEF caused some cell death that rendered the endothelial cell layer patchy as shown in Fig. 9a. Strong perpendicular cell alignment was also evident in cells stimulated with concurrent 5.4 mL h^{-1} shear flow and 153.6 V m^{-1} dcEF (Fig. 9b). Under the fixed shear flow rate (1 Pa), perpendicular cell alignment by electrical field stimulation was not observed in cells conditioned with 75.5 V m^{-1} electric field (Fig. 9c), suggesting that a minimum electric field stimulation on cells is required to induce the perpendicular cell alignment. Under the 5.4 mL h^{-1} shear flow stimulation alone, the 1 Pa shear stress

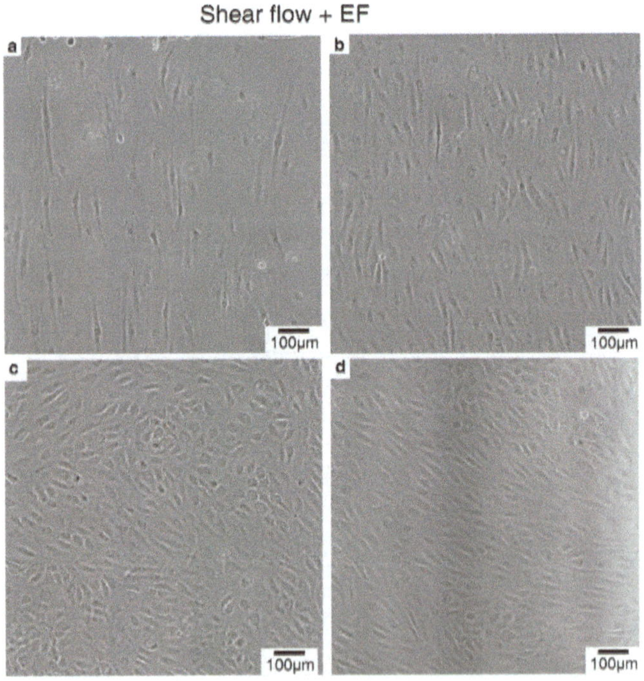

Fig. 9 Microscopy images of endothelial cells cultured in sections I, II, III, and V in SFEFC. **a** Endothelial cells conditioned with shear flow (1 Pa) and 300 V m^{-1}. **b** Endothelial cells conditioned with shear flow (1 Pa) and 153.6 V m^{-1} showing perpendicular alignment as a result of electrical stimulation. **c** Endothelial cells conditioned with shear flow (1 Pa) and 75.5 V m^{-1}. **d** Endothelial cells conditioned with shear flow (1 Pa) demonstrated more parallel alignment phenotypes

induced classical parallel alignment of HUVECs (Fig. 9d), consistent with prior observations (Buchanan et al. 2014; Abaci et al. 2014). Without any shear flow, the HUVECs exhibited no alignment preference (Fig. 4a). These results imply that the SFEFC in our experimental setup can support long-term shear flow and electrical stimulation conditioning of endothelial cells on chip (Fig. 1).

The alignment of HUVECs in SFEFC under simultaneous shear flow and electric field was further characterized by immunofluorescence staining against F-actin and CD31 (supplementary figure S.5). CD31 was expressed on conditioned endothelial cells. Under shear flow conditioning, endothelial cells displayed parallel alignment that was also demonstrated by the parallel orientation of actin stress fibers. When a co-existing electric field increased in strength, electrical stimulation induced cells to undertake a more perpendicular phenotype as the stress fibers became more perpendicularly oriented.

3.3 Glioblastoma adhesion to endothelial cell layer conditioned by shear flow and electric field

We next examined how simultaneous shear flow and electric field stimulation affected the adhesion of the two glioblastoma cell lines on the endothelial cells (Fig. 10). Only T98G-dsRed cells (not U251MG-dsRed cells, $P > 0.05$) showed increased adhesion to the endothelial cells preconditioned under 153.6 V m^{-1} dcEF and 1 Pa shear stress when compared to two pre-conditioned control studies: shear-flow conditioned control ($P < 0.01$, section V of Fig. 10) and the static control ($P < 0.01$). As described in Section 3.2, the endothelial cells preconditioned under 300 V m^{-1} dcEF and shear flow with 1 Pa shear stress were patchy due to some cell death at this electric field strength (Fig. 9a), which may contribute to the lower adhesion of glioblastoma cells ($P > 0.05$). Even though the endothelial cells conditioned in shear flow of 1 Pa and electric fields of 75.5 V m^{-1} and 30.9 V m^{-1} were confluent, the adhesion of both glioblastoma cells lines on pre-conditioned endothelial cells were similar when compared to the control groups. This suggests a dose dependent response of electric field conditioning which was observed in the adhesion of osteosarcoma to polyethylene substrate by Naegele et al. (1991) and migration of fibroblasts by Song et al. (2013). Future identification of the molecular targets that contribute to the difference of adhesion could facilitate molecular typing of glioblastoma cells and further our understanding on the cell-cell interaction between glioblastoma and the endothelial cells that may contribute to the metastasis (Wang et al. 2010; Lombard et al. 2015).

In addition, under the static condition, the adhesion of U251MG-dsRed was higher than that of the adhesion of T98G-dsRed cells (Fig. 6), but such difference was not observed in the on-chip static control (Fig. 10). The shear force from the gravity flow in on-chip glioblastoma adhesion experiments could contribute to this discrepancy. It

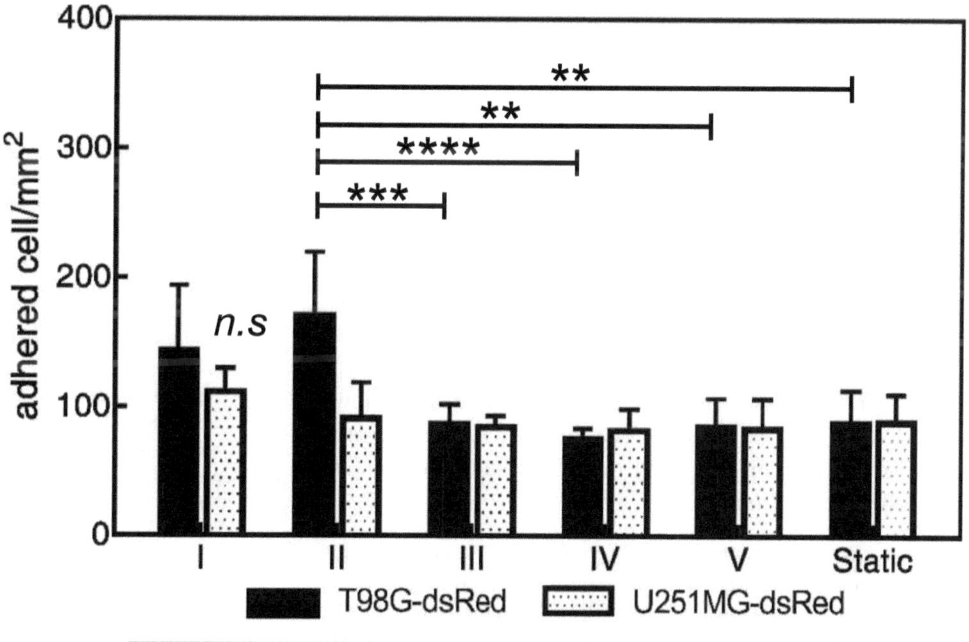

	I	II	III	IV	V	Static
EFS (V/m)	300	153.6	75.5	30.9	-	-
Shear	+	+	+	+	+	-

Fig. 10 Adherence of glioblastoma cells to endothelium conditioned under shear flow and electric field or endothelium cultured in static flow on the SFEFC. ** denotes $P < 0.01$; *** denotes $P < 0.001$; **** denotes $P < 0.0001$

is known that hydrodynamic flow could alter the adhesion dynamics and cell-cell contact. Quantitative study of glioblastoma-endothelial adhesion dynamics under a hydrodynamic flow can elucidate the underlying biophysical mechanism in future studies (Korn and Schwarz 2006).

In this work, applied electrical current induces perpendicular cell alignment and shear flow instigates a parallel alignment on endothelial cells, thus cells are subjected to competing physical cues. In the future, a microfluidic design with orthogonal electric field and shear flow configuration can be useful for investigating the synergistic cell alignment effect in a complex microenvironment.

4 Conclusion

We demonstrated a quick-fit hybrid BMMD made of PMMA and PDMS that provided advantages over single-material counterparts, such as bubble prevention, user-friendliness, low dead volume, and air-tight sealing. The quick-fit design allows high experimental-throughput setup so that by reusing the top interface chip, multiple experiments can be performed sequentially by assembling and disassembling the quick-fit chip. The reagent waste was significantly reduced.

We verified the operation of quick-fit BMMD by conditioning endothelial cells on-chip to concurrent shear flow and electric field. No leakage or bubbles at high volumetric flow rate were observed. T98G-dsRed and U251MG-dsRed glioblastoma cell adhesion under static culture and shear flow with electric field-conditioned endothelium was examined. Angiopoietin 1 activation increased the adhesion of both glioblastoma cell lines on a statically cultured endothelial cell layer. T98G-dsRed glioblastoma cells also showed increased adhesion to an endothelial cell layer conditioned with intermediate electric field and shear flow. On the other hand, U251MG-dsRed showed no adhesion difference. Further identification and typing of adhesion molecules is expected to further our understanding of how glioblastoma interacts with endothelial cells. We envision that the quick-fit hybrid microdevice can be applied to study other cell-tissue interactions in controlled shear flow and electric field conditions. The quick-fit hybrid device can also aid in drug screening on cells conditioned under biomimetic conditions.

Acknowledgements H.F.T is a JSPS DC1 fellow (Japan Society of Promotion for Science) and this work is supported by JSPS KAKENHI [Grant Number JP1700362]. The authors also thank Okinawa Institute of Science and Technology Graduate University (OIST) for its financial support with subsidy funding from the Cabinet Office, Government of Japan. Funders had no role in study design, data collection, the decision to publish, or preparation of the manuscript. The authors thank Ms. Fuka Koja from Image Analysis Section of OIST for technical assistance on flow cytometry. The authors thank Ms. Yi-Ching Tsai for her assistance on illustration preparation. The authors thank Dr. Steven D. Aird from OIST for the proofread.

References

H.E. Abaci, Y.I. Shen, S. Tan, S. Gerecht, Recapitulating physiological and pathological shear stress and oxygen to model vasculature in health and disease. Sci. Rep. **4**, 4951 (2014). https://doi.org/10.1038/srep04951

V.V. Abhyankar, M. Wu, C.Y. Koh, A.V. Hatch, A reversibly sealed, easy access, modular (seam) microfluidic architecture to establish *in vitro* tissue interfaces. PloS One **11**, e0156341 (2016). https://doi.org/10.1371/journal.pone.0156341

H. Bai, J.V. Forrester, M. Zhao, Dc electric stimulation upregulates angiogenic factors in endothelial cells through activation of vegf receptors. Cytokine **55**(1), 110–115 (2011)

H. Bang, W.G. Lee, J. Park, H. Yun, J. Lee, S. Chung, K. Cho, C. Chung, D.C. Han, J.K. Chang, Active sealing for soft polymer microchips: method and practical applications. J. Micromech. Microeng. **16**(4), 708 (2006)

P. Beauchesne, Extra-neural metastases of malignant gliomas: myth or reality? Cancers **3**(1), 461–477 (2011)

J.J. Bernstein, C.A. Woodard, Glioblastoma cells do not intravasate into blood vessels. Neurosurgery **36**(1), 124–132 (1995)

C.F. Buchanan, S.S. Verbridge, P.P. Vlachos, M.N. Rylander, Flow shear stress regulates endothelial barrier function and expression of angiogenic factors in a 3d microfluidic tumor vascular model. Cell Adhes. Migr. **8**, 517–524 (2014). https://doi.org/10.4161/19336918.2014.970001

C.F. Carlborg, K.B. Gylfason, A. Kaźmierczak, F. Dortu, M.J.B. Polo, A.M. Catala, G.M. Kresbach, H. Sohlström, T. Moh, L. Vivien, J. Popplewell, G. Ronan, C.A. Barrios, G. Stemme, W. van der Wijngaart, A packaged optical slot-waveguide ring resonator sensor array for multiplex label-free assays in labs-on-chips. Lab on a Chip **10**(3), 281–290 (2010). https://doi.org/10.1039/B914183A

J.Y. Cheng, C.W. Wei, K.H. Hsu, T.H. Young, Direct-write laser micromachining and universal surface modification of pmma for device development. Sensors Actuators B Chem. **99**(1), 186–196 (2004)

A.M. Christensen, D.A. Chang-Yen, B.K. Gale, Characterization of interconnects used in pdms microfluidic systems. J. Micromech. Microeng. **15**(5), 928 (2005)

V.A. Cuddapah, S. Robel, S. Watkins, H. Sontheimer, A neurocentric perspective on glioma invasion. Nat. Rev. Neurosci. **15**(7), 455–465 (2014)

C.A. Davis, S. Zambrano, P. Anumolu, A.C. Allen, L. Sonoqui, M.R. Moreno, Device-based *in vitro* techniques for mechanical stimulation of vascular cells: a review. J. Biomech. Eng. **137**(4), 040801 (2015)

S. El Hallani, C. Colin, Y. El Houfi, A. Idbaih, B. Boisselier, Y. Marie, P. Ravassard, M. Labussière, K. Mokhtari, J.-L. Thomas, J.-Y. Delattre, A. Eichmann, M. Sanson, Tumor and endothelial cell hybrids participate in glioblastoma vasculature. Biomed. Res. Int. **2014**, 827327 (2014). http://dx.doi.org/10.1155/2014/827327

D.P. Gaver, S.M. Kute, A theoretical model study of the influence of fluid stresses on a cell adhering to a microchannel wall. Biochem. J. **75**, 721–733 (1998). https://doi.org/10.1016/S0006-3495(98)77562-9

E.C. Holland, Glioblastoma multiforme: the terminator. Proc. Natl. Acad. Sci. **97**(12), 6242–6244 (2000)

J. Jaczewska, M.H. Abdulreda, C.Y. Yau, M.M. Schmitt, I. Schubert, P.O. Berggren, C. Weber, R.R. Koenen, V.T. Moy, E.P. Wojcikiewicz, Tnf-α and ifn-γ promote lymphocyte adhesion to endothelial junctional regions facilitating transendothelial migration. J. Leukoc. Biol. **95**(2), 265–274 (2014)

A. Khademhosseini, J. Yeh, G. Eng, J. Karp, H. Kaji, J. Borenstein, O.C. Farokhzad, R. Langer, Cell docking inside microwells within reversibly sealed microfluidic channels for fabricating multiphenotype cell arrays. Lab Chip **5**, 1380–1386 (2005). https://doi.org/10.1039/b508096g

O.F. Khan, M.V. Sefton, Endothelial cell behaviour within a microfluidic mimic of the flow channels of a modular tissue engineered construct. Biomed. Microdevices **13**(1), 69–87 (2011)

L. Kim, Y.C. Toh, J. Voldman, H. Yu, A practical guide to microfluidic perfusion culture of adherent mammalian cells. Lab on a Chip **7**(6), 681–694 (2007)

C. Korn, U. Schwarz, Efficiency of initiating cell adhesion in hydrodynamic flow. Phys. Rev. Lett. **97**(13), 138103 (2006)

D. Liu, V. Martin, J. Fueyo, O.H. Lee, J. Xu, N. Cortes-Santiago, M.M. Alonso, K. Aldape, H. Colman, C. Gomez-Manzano, Tie2/tek modulates the interaction of glioma and brain tumor stem cells with endothelial cells and promotes an invasive phenotype. Oncotarget **1**, 700–709 (2010). https://doi.org/10.18632/oncotarget.101204

A. Lombard, N. Goffart, B. Rogister, Glioblastoma circulating cells: reality, trap or illusion? Stem Cells Int **2015**, 182985 (2015). http://dx.doi.org/10.1155/2015/182985

X. Lu, Y. Kang, Cell fusion as a hidden force in tumor progression. Cancer Res. **69**, 8536–8539 (2009). https://doi.org/10.1158/0008-5472.CAN-09-2159

F. Mackay, H. Loetscher, D. Stueber, G. Gehr, W. Lesslauer, Tumor necrosis factor alpha (tnf-alpha)-induced cell adhesion to human endothelial cells is under dominant control of one tnf receptor type, tnf-r55. J. Exp. Med. **177**, 1277–1286 (1993)

A.M. Malek, S.L. Alper, S. Izumo, Hemodynamic shear stress and its role in atherosclerosis. JAMA **282**(21), 2035–2042 (1999)

M.P. McRae, G. Simmons, J.T. McDevitt, Challenges and opportunities for translating medical microdevices: insights from the programmable bio-nano-chip. Bioanalysis **8**, 905–919 (2016). https://doi.org/10.4155/bio-2015-0023

R. Naegele, J. Lipari, D. Chakkalakal, B. Strates, M. McGuire, Electric field stimulation of human osteosarcoma-derived cells: a dose-response study. Cancer Biochem. Biophys. **12**(2), 95–101 (1991)

J. Pasquier, B.S. Guerrouahen, H. Al Thawadi, P. Ghiabi, M. Maleki, N. Abu-Kaoud, A. Jacob, M. Mirshahi, L. Galas, S. Rafii, F. Le Foll, A. Rafii, Preferential transfer of mitochondria from endothelial to cancer cells through tunneling nanotubes modulates chemoresistance. J. Transl. Med. **11**, 94 (2013). https://doi.org/10.1186/1479-5876-11-94

A. Prokop, Z. Prokop, D. Schaffer, E. Kozlov, J. Wikswo, D. Cliffel, F. Baudenbacher, Nanoliterbioreactor: long-term mammalian cell culture at nanofabricated scale. Biomed. Microdevices **6**, 325–339 (2004). https://doi.org/10.1023/B:BMMD.0000048564.37800.d6

A. Reinitz, J. DeStefano, M. Ye, A.D. Wong, P.C. Searson, Human brain microvascular endothelial cells resist elongation due to shear stress. Microvasc. Res. **99**, 8–18 (2015). https://doi.org/10.1016/j.mvr.2015.02.008

K. Ren, J. Zhou, H. Wu, Materials for microfluidic chip fabrication. Acc. Chem. Res. **46**, 2396–2406 (2013). https://doi.org/10.1021/ar300314s

M. Segarra, B.C. Kirchmaier, A. Acker-Palmer, A vascular perspective on neuronal migration. Mech. Dev. **138**, 17–25 (2015)

S. Sheikh, G.E. Rainger, Z. Gale, M. Rahman, G.B. Nash, Exposure to fluid shear stress modulates the ability of endothelial cells to recruit neutrophils in response to tumor necrosis factor-α: a basis for local variations in vascular sensitivity to inflammation. Blood **102**(8), 2828–2834 (2003)

P. Skafte-Pedersen, M. Hemmingsen, D. Sabourin, F.S. Blaga, H. Bruus, M. Dufva, A self-contained, programmable microfluidic cell culture system with real-time microscopy access. Biomed. Microdevices **14**(2), 385 (2012). https://doi.org/10.1007/s10544-011-9615-6

S. Song, H. Han, U.H. Ko, J. Kim, J.H. Shin, Collaborative effects of electric field and fluid shear stress on fibroblast migration. Lab Chip **13**(8), 1602–1611 (2013)

J. Steigert, S. Haeberle, T. Brenner, C. Müller, C. Steinert, P. Koltay, N. Gottschlich, H. Reinecke, J. Rühe, R. Zengerle et al, Rapid prototyping of microfluidic chips in coc. J. Micromech. Microeng. **17**(2), 333 (2007)

T. Thorsen, S.J. Maerkl, S.R. Quake, Microfluidic large-scale integration. Science **298**(5593), 580–584 (2002)

H.F. Tsai, S.W. Peng, C.Y. Wu, H.F. Chang, J.Y. Cheng, Electrotaxis of oral squamous cell carcinoma cells in a multiple-electric-field chip with uniform flow field. Biomicrofluidics **6**(3), 034116 (2012)

H.F. Tsai, J.Y. Cheng, H.F. Chang, T. Yamamoto, A.Q. Shen, Uniform electric field generation in circular multi-well culture plates using polymeric inserts. Sci. Rep. **6**, 26222 (2016). https://doi.org/10.1038/srep26222

H.F. Tsai, A. Trubelja, A.Q. Shen, G. Bao, Tumour-on-a-chip: microfluidic models of tumour morphology, growth and microenvironment. J. R. Soc. Interface **14**(131), 20170137 (2017)

J.S. Uzarski, E.W. Scott, P.S. McFetridge, Adaptation of endothelial cells to physiologically-modeled, variable shear stress. PloS One **8**(2), e57004 (2013)

S.G.M. Uzel, R.J. Platt, V. Subramanian, T.M. Pearl, C.J. Rowlands, V. Chan, L.A. Boyer, P.T.C. So, R.D. Kamm, Microfluidic device for the formation of optically excitable, three-dimensional, compartmentalized motor units. Sci. Adv. **2**, e1501429 (2016). https://doi.org/10.1126/sciadv.1501429

R. Wang, K. Chadalavada, J. Wilshire, U. Kowalik, K.E. Hovinga, A. Geber, B. Fligelman, M. Leversha, C. Brennan, V. Tabar, Glioblastoma stem-like cells give rise to tumour endothelium. Nature **468**(7325), 829 (2010)

Y. Wang, D. Lee, L. Zhang, H. Jeon, J.E. Mendoza-Elias, T.A. Harvat, S.Z. Hassan, A. Zhou, D.T. Eddington, J. Oberholzer, Systematic prevention of bubble formation and accumulation for long-term culture of pancreatic islet cells in microfluidic device. Biomed. Microdevices **14**(2), 419–426 (2012). https://doi.org/10.1007/s10544-011-9618-3

S. Watkins, S. Robel, I.F. Kimbrough, S.M. Robert, G. Ellis-Davies, H. Sontheimer, Disruption of astrocyte–vascular coupling and the blood–brain barrier by invading glioma cells. Nat. Commun. **5**, 4196 (2014)

Y. Xia, G.M. Whitesides, Soft lithography. Angew. Chem. Int. Ed. **37**(5), 550–575 (1998). https://doi.org/10.1002/(SICI)1521-3773 (19980316)37:5<550::AID-ANIE550>3.0.CO;2-G

M. Ye, H.M. Sanchez, M. Hultz, Z. Yang, M. Bogorad, A.D. Wong, P.C. Searson, Brain microvascular endothelial cells resist elongation due to curvature and shear stress. Sci. Rep. **4**, 4681 (2014). https://doi.org/10.1038/srep04681

M. Zhao, H. Bai, E. Wang, J.V. Forrester, C.D. McCaig, Electrical stimulation directly induces pre-angiogenic responses in vascular endothelial cells by signaling through vegf receptors. J. Cell Sci. **117**, 397–405 (2004). https://doi.org/10.1242/jcs.00868

S. Zhao, K. Zhu, Y. Zhang, Z. Zhu, Z. Xu, M. Zhao, T. Pan, Electrotaxis-on-a-chip (etc): an integrated quantitative high-throughput screening platform for electrical field-directed cell migration. Lab Chip **14**(22), 4398–4405 (2014)

Publisher's note Springer Nature remains neutral with regard to jurisdictional claims in published maps and institutional affiliations.

Biomedical Microdevices (2019) 21: 41
https://doi.org/10.1007/s10544-019-0398-5

A tissue chamber chip for assessing nanoparticle mobility in the extravascular space

Valeria Lusi[1,2] · Thomas L. Moore[1] · Federica Laurino[1,3] · Alessandro Coclite[1] · Rui Perreira[1] · Miguel Ferreira[1] · Ilaria Rizzuti[1] · Roberto Palomba[1] · Paolo Zunino[3] · Marti Duocastella[4] · Shoshy Mizrahy[1,5] · Dan Peer[5] · Paolo Decuzzi[1]

Published online: 6 April 2019

Author's Personal Note The enthusiasm of a child receiving a new toy, the dedication of a leader appreciating the fine balance between power and responsibility, and the accuracy of a scientist seeking the ultimate cure for deadly diseases. I have witnessed all these attributes in Mauro multiple times, regardless of the occasion. It might have been the launch of the Cancer Nanomedicine initiative, webcasted worldwide from a NCI podium in Washington DC, or an informal scientific gathering by the beach on the Ionic sea while sipping some wine in good company. Your enthusiasm, dedication and accuracy, Mauro, will stay the same across the years and continue to inspire many more people. This will fuel your next giant steps! Paolo

Mauro, endless dedication and support to the field of Nanoscience & Nanotechnology. It has been a great honor to meet you for the first time in 2010. With great admiration for your enthusiasm and ability to advance push forward the field of nanomedicine, taking it from the lab to clinical practice has been an inspiration for many scientists (me including). You have educated numerous young graduate students and postdocs who became internationally known in this field. This is by itself an important milestone in your endless achievements. I am sure you will keep inspiring the world with new opportunities and go beyond the imagination to create, innovate and execute outstanding translational science.

Dan Peer and Paolo Decuzzi

Abstract

Although a plethora of nanoparticle configurations have been proposed over the past 10 years, the uniform and deep penetration of systemically injected nanomedicines into the diseased tissue stays as a major biological barrier. Here, a 'Tissue Chamber' chip is designed and fabricated to study the extravascular transport of small molecules and nanoparticles. The chamber comprises a collagen slab, deposited within a PDMS mold, and an 800 μm channel for the injection of the working solution. Through fluorescent microscopy, the dynamics of molecules and nanoparticles was estimated within the gel, under different operating conditions. Diffusion coefficients were derived from the analysis of the particle mean square displacements (MSD). For validating the experimental apparatus and the protocol for data analysis, the diffusion *D* of FITC-Dextran molecules of 4, 40 and 250 kDa was first quantified. As expected, *D* reduces with the molecular weight of the dextran molecules. The MSD-derived diffusion coefficients were in good agreement with values derived via fluorescence recovery after photobleaching (FRAP), an alternative technique that solely applies to small molecules. Then, the transport of six nanoparticles with similar hydrodynamic diameters (~ 200 nm) and different surface chemistries was quantified. Surface PEGylation was confirmed to favor the diffusion of nanoparticles within the collagen slab, whereas the surface decoration with hyaluronic acid (HA) chains reduced nanoparticle mobility in a way proportional to the HA molecular weight. To assess further the generality of the proposed approach, the diffusion of the six nanoparticles was also tested in freshly excised brain tissue slices. In these ex vivo experiments, the diffusion coefficients were 5-orders of magnitude smaller than for the Tissue Chamber chip. This was mostly ascribed to the lack of a cellular component in the chip. However, the trends documented for PEGylated and HA-coated nanoparticles *in vitro* were also confirmed ex vivo. This work demonstrates that the Tissue Chamber chip can be employed to effectively and efficiently test the extravascular transport of nanomedicines while minimizing the use of animals.

Keywords Nanoparticle transport · Nanomedicine · Tissue chamber

Electronic supplementary material The online version of this article (https://doi.org/10.1007/s10544-019-0398-5) contains supplementary material, which is available to authorized users.

✉ Paolo Decuzzi
paolo.decuzzi@iit.it

1 Laboratory of Nanotechnology for Precision Medicine, Italian Institute of Technology, Via Morego 30, 16163 Genoa, Italy

2 DIBRIS, University of Genova, Via Opera Pia 13, 16145 Genoa, Italy

3 MOX, Department of Mathematics, Politecnico di Milano, Piazza Leonardo da Vinci 32, 20133 Milan, Italy

4 Nanophysics, Italian Institute of Technology, Via Morego 30, 16163 Genoa, Italy

5 Laboratory of NanoMedicine, Department of Cell Research and Immunology, Tel-Aviv University, Ramat Aviv, 69978 Tel-Aviv, Israel

1 Introduction

Since approximately the turn of the millennium, nanoparticles have been promoted as a paradigm-shifting approach to early diagnosis and improved treatment of multiple diseases (Ferrari 2005; Rosenblum et al. 2018). The potential of particle-based drug delivery systems, or in a simple word nanomedicines, to protect drugs (small molecules, biologicals, peptides, etc.) from premature degradation; prolong the circulation time in the blood; reduce systemic toxicity; and control release has been documented extensively by the scientific community (Shi et al. 2017). Moreover, nanomedicines enabled the realization of multifunctional delivery systems with combined therapy and diagnostics (e.g. theranostics) (Nabil et al. 2019; Arranja et al. 2017; Moore et al. 2014), co-delivery of drugs (Blanco et al.

2014; Pushpalatha et al. 2017), and targeted delivery (Tietjen et al. 2018; Lazarovits et al. 2015).

Although several nanomedicines, mostly lipid-based formulations, are under clinical investigation in early and advanced trials, there are still significant challenges in fully integrating nanomedicines into clinical practice. Some of these challenges are merely technical, such as the difficulty in the reproducible synthesis and large-scale manufacturing (Shi et al. 2017), while others are due to challenges in fundamental understanding of particle behavior in complex biological systems (Wilhelm et al. 2016; Hua et al. 2018). To this last point, a major interest in nanomedicine is understanding the transport of particles across biological barriers to their final pathological targets and critical to this goal is understanding the dynamics of particles through the extravascular tissue (Blanco et al. 2015). Researchers have attempted to overcome this grand challenge by tailoring particle physico-chemical properties (i.e. size, surface charge, surface functionalization, material density, shape, etc.), utilizing stimuli-responsive materials, masking particles in cellular coatings, using cells to shuttle particles across barriers, or employing multi-stage delivery systems. For example, Cabral et al. (2011) showed *in vivo* that particle size (30 to 100 nm) played a significant role in the accumulation of nanoparticles in tumors of varying permeability. That is, size had little effect on particle-tumor accumulation in highly permeable tumors, however when tumors have low permeability the smaller particles (30 nm) accumulate more compared to medium-sized (50 nm) or larger particles (70 and 100 nm). One approach to capitalize on these phenomena is to develop particles that modulate their size throughout their voyage to the target site. Wong et al. (2011) reported multistage nanoparticles that shrink from approximately 100 nm in diameter to 10 nm as they extravasate from the tumor vasculature into the tumor tissue. This was achieved by having a primary particle which can be actively degraded by proteases in the tumor microenvironment, thereby releasing smaller non-degraded particles (i.e. 10 nm quantum dots). In this case, the larger primary particles enable long circulation half-life and passive tumor targeting, and the release of small particles enables deeper tumor penetration. Another such multistage particle system was reported by Tasciotti et al. (2008), where larger, mesoporous silicon primary microparticles were loaded with quantum dots or single-walled carbon nanotubes (SWCNTs). This system takes advantage of the micron-sized and specifically-shaped primary particles' ability to marginate in the vasculature to better deliver the nanosized quantum dots or SWCNTs. Finally, biomimetic approaches are attempting to utilize naturally occurring cellular processes to transport particles across biological barriers. Particles have been coated in "stealth" cellular membranes to avoid immunological detection and clearance, and improve tumor targeting (Hu et al. 2011; Pitchaimani et al. 2019). Furthermore, particles have been attached to cells as "hitchhikers" (Anselmo et al. 2015; Brenner et al. 2018) or transported across biological barriers as intracellular cargo (Choi et al. 2012; Moore et al. 2017; Zhang et al. 2018). Despite all of these innovative approaches to optimize particle transport across biological barriers, there exists a need to understand particle transport through tissues on a fundamental level.

In vivo disease models are the benchmark for studying particle transport processes as they are dynamic, complex, and mimic the biological fate of particles in human biology. However, organs-on-chips technologies are increasingly being looked at as alternatives to animal models (Sontheimer-Phelps et al. 2019). These technologies are developed with increasing complexity (i.e. able to mimic organ and organ systems), while providing precise control over the system engineering and parameters. In fact, previous studies on a simplistic collagen matrix have shown that the ionic strength of media, particle surface charge, and biophysical properties of the matrix heavily influence the diffusivity of quantum dots into a collagen gel (Stylianopoulos et al. 2010). Here, a Tissue Chamber capable of studying the diffusivity of particles through an extravascular tissue-mimic in real-time is reported. Importantly, this research aimed to show as a proof-of-principle that the Tissue Chamber chip platform could quickly evaluate particle diffusivity and observe how particle properties could be optimized to facilitate diffusion through the tissue. The Tissue Chamber was realized by fabricating a collagen gel containing a cylindrical tube throughout the center. The transport properties of small-molecule fluorescent dyes (FITC-labelled dextran with varying molecular weights), 200 nm polystyrene beads or spherical PLGA nanoparticles coated with different amounts of poly(ethylene glycol) (PEG), and liposomes coated with different molecular weight hyaluronic acids (HA) were determined through fluorescent microscopy. Image analysis was then performed to evaluate the Mean Square Displacement (MSD) and subsequently the diffusivity of these different agents. Finally, ex vivo diffusion studies were performed in freshly excised brain tissue slices.

2 Materials and methods

Materials Collagen type I (4 mg/ml) from bovine origin and dextran with different molecular weights 4, 40, and 250 kDa were purchased from Sigma Aldrich (Sigma Aldrich, USA). Dulbecco's Phosphate Buffered Saline (DPBS) was acquired from Gibco Life technologies UK and Fluoresbrite™ carboxy nyo 0.20 μm microspheres were purchased from Polysciences, Inc. (Warrington PA). Poly(lactic-co-glycolic acid) PLGA (50:50, carboxy-terminated, MW 38,000–54,000 Da) was purchased from Sigma Aldrich (St. Louis, MO, USA); 1,2-dipalmitoyl-sn-glycero-3-phosphocholine (DPPC), DSPE-PEG(2000) Carboxylic Acid, DSPE-RhB (Liss Rhod PE), DSPE-EGG, 1,2-Dipalmitoyl-*sn-glycero*-3-

phoshoethanolamine (DPPE) and cholesterol (Chol) were obtained from Avanti Polar Lipids (Alabaster, Alabama). Pure soybean phosphatidylcholine (Phospholipon 90G) (PC) was a kind gift from Phospholipid GmbH (Germany). Sodium hyaluronate was purchased from Lifecore Biomedical, LLC (MN,USA). O.C.T. mounting medium compound for cryotomy was purchased from VWR Chemicals. Permount Mounting Media was purchased from Fisher Scientific.

Synthesis of spherical polymeric nanoconstructs (SPNs) Spherical Polymeric Nanoparticles (SPNs) tagged with different amounts of PEG were prepared by a slightly modified sonication-emulsion technique according to previously described procedures (Lee et al. 2016). Two different configurations were prepared to obtain SPNs tagged with 80% or 20% of PEG on the outside surface. On both preparations, 20% of DSPE-RhB was included. Briefly, carboxyl-terminated PLGA and DPPC, in a 10:1 ratio, were dissolved in chloroform to obtain a homogeneous solution (oil phase). For the surface lipid monolayer with 80% PEG, two lipids were used (DSPE-PEG-COOH and DSPE-RhB) with a molar ratio of 5:1, dissolved in aqueous phase (4% ethanol). For the surface lipid monolayer with 20% PEG, three lipids were used (DSPE-PEG-COOH, DSPE-RhB and DSPE-EGG) with a molar ratio of 1:1:3, dissolved in aqueous phase (4% ethanol). The ratio between the oil phase and the aqueous phase was 1 to 5. Afterwards, the oil phase was added drop wisely to the aqueous phase under ultrasonication. The obtained emulsion was then placed under magnetic stirring to facilitate solvent evaporation. SPNs were centrifuged first for 5 min at 254 g to settle down any possible debris and then the supernatant was centrifuged 3 more times for 20 min at $1{,}8 \cdot 10^4$ g.

Preparation of liposomes Liposomes were synthesized according to previously published procedures (Mizrahy et al. 2014). With minor modifications, Multilamellar Vesicles (MLVs) were made by a mixture of PC, Chol and DPPE in a molar ratio of 60:20:20 (Peer and Margalit 2000; Peer and Margalit 2004; Peer et al. 2008; Kedmi et al. 2010). In brief, lipids were dissolved in ethanol, dried under reduced pressure in a rotary evaporator (Buchi Rotary Evaporator Vacuum System Flavil, Switzerland) and afterwards were hydrated using a PBS solution at pH 7.4. MLVs also contained 0.5% Cy5 labeled DPPE (Mizrahy et al. 2014). At the end, MLVs were vortexed followed by 2 h of incubation in a shaker bath at 37 °C. Then, MLVs were extruded by the use of Lipex extrusion device (Northern lipids, Vancouver, Canada) at 65 °C and under 200–500 psi nitrogen pressure. Extrusion was accomplished in several steps using decreasing pore-size polycarbonate membranes (Whatman Inc., UK), performing numerous cycles per pore-size, in order to obtain unilamellar vescicles with a size ranging between 100 and 200 nm in diameter.

Surface modifications of liposomes The liposome surface modification was made according to previous reported procedures (Landesman-Milo et al. 2013). In brief, high and low molecular weight hyaluronic acid (700 and 5 kDa, respectively) were dissolved in 0.2 M MES buffer (pH 5.5) to reach a final concentration of 5 mg/ml or 40 mg/ml. Hyaluronic Acid (HA) was activated for 30 min with ethyl-dimethyl-aminopropyl-carbodiimide (EDC) (Sigma-Aldrich, USA) and sulfo-NHS (Proteochem) using a molar ratio of 1:1:6 between HA, EDC and sulfo-NHS. Afterwards, liposomes was added and pH was adjusted to 7.4. The solution was then incubated at room temperature for 2 h and the free HA was removed by washing via centrifugation at 4 °C for 60 min at $1.3 \cdot 10^5$ g.

Size and stability characterization of nanoparticles Dynamic light scattering (DLS, Malvern Zetasizer Nano S) was employed to characterize the size and zeta (ζ) potential of nanoparticles under hydrated conditions at pH 7.0. Nanoparticle stability was performed in water at 37 °C, following the size variation through DLS measurement. The same instrument was used to determine liposomes size and ζ potential for 5 days at 37 °C using HEPES 1 M for the size and DI water for the ζ potential.

Tissue chamber fabrication The Tissue Chamber device was fabricated with polydimethylsiloxane (PDMS) using a pre-polymer solution of Sylgard 182 mixed with the curing agent in a 10:1 ratio *w*/w. This solution was poured in a petri dish to create a 3.5 mm thick PDMS sheet. The petri dish with PDMS was degassed in a vacuum chamber and cured in the oven at 60 °C overnight. After 15 min at room temperature the petri dish was cooled to −20 °C for 1 h before removing the sheet from the base. The PDMS sheets were cut into $15 \times 15 \times 3.5$ mm^3 parallelepipeds, and a $5 \times 5 \times 3.5$ mm^3 empty space for the Tissue Chamber was cut into the center. The PDMS Tissue Chamber was irreversibly bonded to a glass microscope slide (VWR) using oxygen plasma. At the end of the process, a 21 G (0.8 mm diameter) needle (Sterican B. Braun) was used to generate a channel in the Tissue Chamber from side to side. At the very end, chips were sterilized by autoclave and dried in an incubator overnight. Figure 1 illustrates the Tissue Chamber fabrication and final geometry.

Collagen type I gel preparation Collagen type I pH and ionic strength were adjusted by addition of the buffer solution (pH 7.4) to achieve a final pH between 7 and 7.4, with a final concentration of 1.5 mg/ml. 150 μl of collagen type I solution was injected inside the Tissue Chamber and polymerization occurred at 37 °C for 1 h. Next, gels were covered with 50 μl of DPBS to keep them in a hydrated state. After 1 h, the needle was removed and the channel was confirmed by optical microscope images (Fig. 1).

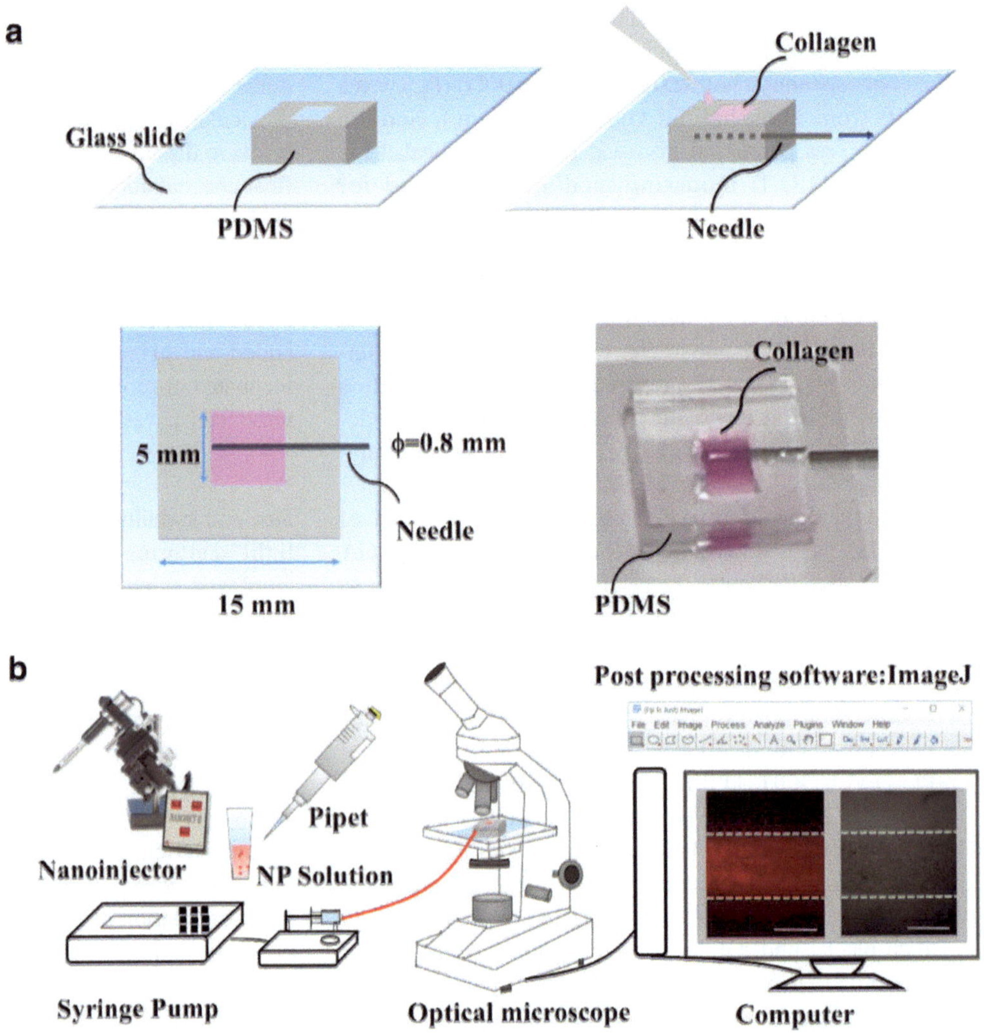

Fig. 1 **Tissue Chamber microfluidic chip. a** 3D schematic representation of the Tissue Chamber PDMS chip with a description of channel fabrication process, an overhead view of the Tissue Chamber showing the dimensions, and a photographic image of the Tissue Chamber bonded on a glass slide and filled with collagen type I. The photograph shows the needle inserted across the chamber, which is used to create the channel for injecting tracers. **b** Schematic of the entire experimental setup with representative optical microscopy images of 200 nm beads injected inside the channel: brightfiled (left) and fluorescent (right) images of a channel portion. Scale bars are 500 μm

Diffusion assays Samples were placed under an inverted optical microscope (Leica DMI 6000 B) equipped with a DFC360 FX digital fluorescence camera. Different molecules and particles were injected to study the diffusion through the collagen hydrogel: Dextran 4, 40 and 250 kDa, 200 nm Fluoresbrite® carboxylate microspheres, SPNs and liposomes with different PEG and HA amounts. Fluoresbrite® carboxylate microspheres (from now on called NP 200) are commercially available monodispersed fluorescent polystyrene microspheres. These microspheres were provided as a 2.5% aqueous suspensions and injected in the Tissue Chamber at 0.05% (*w*/*v*), SPNs 20% PEG and SPNs 80% PEG were injected at the same ratio (0.05% w/v). Solutions of Dextran 4, 40 and 250 kDa were injected at 0.5 mg/ml and the 3 different kinds of liposomes were injected at stock concentrations (lipid 40 mg/ml). Images were collected every 30 s for 4.5 min for the Dextran 4, 40 and 250 kDa (5 μl); every minute for 10 min in the case of NP 200; every 4 s for 44 s for all SPNs and liposomes (5 μl).

Mean square displacement (MSD) By tracking the colored wave-front inside the Tissue Chamber collagen over time, the overall diffusion coefficient D in one dimension (1D) was determined as

$$D = \frac{MSD}{2T} \tag{1}$$

where T is the time interval and the MSD were calculated from:

$$MSD_{1D} = < \Delta x(j\Delta t)^2 > \tag{2}$$

While the more general formula, for the 3D diffusion is:

$$MSD_{3D} = < \Delta x(j\Delta t)^2 + \Delta y(j\Delta t)^2 + \Delta z(j\Delta t)^2 > \tag{2a}$$

With the j index running on the total number of Δt in T and x represents the height of colored channel at time t (Valentine et al. 2004). The area of the channel perfused by the working solution is rectangular with the long edge aligned with the

800 μm injection channel and the short edge corresponding to the distance traveled by the working solution into the collagen matrix. This distance is calculated by post processing the fluorescence pics via Fiji. Notice that this distance is also the distance traveled by the molecules, or nanoparticles, over time and corresponds to the displacement in the MSD analyses. Following Xavier (2010), the MSD is calculated on all the experimental points while the diffusion coefficient is obtained by linearly interpolating the points falling in the first quartile of the MSD curves. Using the conventional formula in Eq. (1), the diffusion coefficients were determined as half of the slope of the curve fitting the MSD points in the first quartile.

Fluorescence recovery after photobleaching (FRAP) In this analysis, experiments were conducted using a Nikon A1R confocal laser scanning microscope with a Plan Apo DIC N2 20×/ 0.75 objective. A 488 nm Argon ions laser line was used to excite the samples and a bandpass 500–550 filter was placed before the photomultiplier tube. The collagen solution was placed in μ-Dish Micro-Insert 4 well (Ibidi, Germany) and 10 μl of neutralized collagen type I was polymerized at 37 °C for 1 h. Each tissue construct was injected with 23 nl using a nanoinjector (Drummond "Nanoject II" automatic nanoliter injector), following manufacturer's instructions. Injection velocity was 23 nl/s. The Nanoject II required pulled micropipets with a capillary outer diameter of 1140 μm and inner diameter of 530 μm. Experiments were performed at equilibrium, generally 12–48 h post injection depending on the MW of the molecule. Using a circular region of interest (ROI) with a radius of 92.5 μm, gels were first photobleached with a laser for 20 min and then the recovery of fluorescence was observed in the following 20 min. The recovery of fluorescence in the bleached ROIs gave the diffusion coefficient. The software provides the curve associated with bleaching and recovery as well as the parameter τ_D, which is related to the diffusion coefficient through the following formula:

$$\tau_D = \frac{\omega^2}{4D} \tag{3}$$

where ω is the radius of the ROI and D is the diffusion coefficient (Erikson et al. 2008).

Quantification of the diffusion coefficient via error minimization algorithm (EMA) This technique is useful when FRAP and MSD cannot be applied. It is rather general and robust as it takes as an input the raw data from the diffusion assays (Fig. 6). An efficient algorithm for fitting a vector of parameters on a given dataset is developed (Tröltzsch 2010). The parameters are interpreted as control variables in an optimization problem that minimizes a functional representing the difference between the experimental observations and the model predictions. The collagen gel in the Tissue Chamber chip is described as a rectangular domain Ω in which, at the initial time, a concentration $c(x,y,0)=c_{inj}(x, y)$ of molecules is prescribed. The temporal evolution of the concentration $c = c(x,y,t)$ in the time interval $[0, T]$ is described by:

$$\begin{cases} \frac{\partial c}{\partial t} - D\Delta c = 0 & \text{in } \boldsymbol{\Omega}, t > 0, \\ D\nabla c \cdot n = 0 & \text{on } \partial\boldsymbol{\Omega} \times (0, T), \\ c = c_{inj} & \text{in } \boldsymbol{\Omega}, t = 0, \end{cases} \tag{4}$$

where $\Delta u = \partial_{xx}u + \partial_{yy}u$ is the Laplace operator and $c_{inj} = c_{inj}(x, y)$ is an assigned function obtained processing the fluorescence images at time $t = 0$ using the algorithm summarized in steps 1, 2, 3, 4 described in what follows. The system can be readily solved numerically if D is known (direct problem). For the inverse problem, D is not known a priori. In this case, the solution is obtained by minimizing the following functional J:

$$\min J(D) = \frac{1}{2} \left\| c(T) - \overline{c}(T) \right\|^2_{L^2(\Omega)} + \frac{\lambda}{2} |D|^2, \tag{5}$$

subject to the Eq. (4). The first term in J gives the difference between the experimentally observed concentration field $\overline{c}$ at T and the concentration field c , which is computed at time T using Eq. (4) under a specific assumption for D. The second term $\frac{\lambda}{2}|D|^2$ is a regularization operator. For the Lagrangian principle, the optimal diffusion parameter is the solution of the unconstrained minimization problem $\min\mathcal{L}(c, D, p)$ where $\mathcal{L}$ is the Lagrangian functional, defined as follows

$$\mathcal{L}(c, D, p) = J(c, D) - \int_0^T \int_\Omega \frac{\partial c}{\partial t} p - \int_0^T \int_\Omega D\nabla c \cdot \nabla p \tag{6}$$

where p is the so-called Lagrange multiplier. For the space discretization, an admissible triangulation of the domain Ω is introduced and the linear finite element method is applied (Fig. 6). For time discretization, the backward Euler scheme is chosen. To evaluate the functional J, the finite element approximation of the state equations must be compared with the observed concentration field $\overline{c}$. To this end, the pixel map of $\overline{c}$ is converted into a finite element function. This is achieved by means of image analysis tools as per the following protocol: 1. Conversion of the RGB picture for $\overline{c}$ into a gray scale image; 2. generation of a mesh with triangular elements having the same dimensions as pixel number of the image; 3. definition of a finite element variable that accounts for the grey level for each pixel; 4. projection of the previous gray scale map on a coarse mesh of triangular elements used for the solution of the finite element method applied to Eq. (4).The minimization of the Lagrangian functional is achieved by means of a Non Linear Conjugate Gradient iterative method, which, given a starting value for D, solves Eq. (4), then computes the functional J, solves for the adjoint equation to estimate p, and evaluates the derivative of $\mathcal{L}$ with respect to D. If this is sufficiently close to zero then the correct value for D is returned. If this is not close to zero, then it updates the diffusion coefficient estimation and it runs again the full analysis starting a new iteration.

Ex vivo experiments Wild-type C57BL/6 mice aged 2–6 months were sacrificed to obtain 2.0 mm brain tissue slices via a Zivic Mouse brain slicer. The striatum of these tissues was injected with 69 nl of a solution containing approximately 1000 nanoparticles for each experiment using the nanoinjector with a velocity of 23 nl/s. Injected slices were placed in a custom-made, three-dimensional (3D) particle tracking system consisting of a wide-field inverted microscope (Nikon Ti) with an oil-immersion objective (Nikon Plan Apo VC 100×/1.4 oil DIC N2), a piezoelectric stage (Mad City Lab) and a single EMCCD camera (DU897DCS-BV Andor Technology) (Sancataldo et al. 2017; Zhang et al. 2017). Sequential images of multiple nanoparticles were recorded at 10 frames per second (fps) in time slots of 40 s (Zhang et al. 2017). At the end, the 3D particle trajectories and the corresponding MSD were calculated by post processing the movies with Fiji (Schindelin et al. 2012) and a custom MATLAB script (Supporting Information).

3 Results

The tissue chamber apparatus for the diffusion experiments The Tissue Chamber chip forms a $5 \times 5 \times 3.5$ mm^3 parallelepiped. Figure 1a shows a schematic representation of the fabrication process to realize the Tissue Chamber chip. The pink fluid is related to a collagen solution that is cast around a 21G needle (black line), which is used to realize the channel for dispersing small molecules and nanoparticles within the chip itself. The needle outer diameter is equal to 800 μm (ϕ = 0.8 mm). This size was chosen in order to ensure the mechanical stability of the channel, and thus prevent its collapse and closure. Furthermore, it should be highlighted that the overall width of the Tissue Chamber (5 mm) is significantly larger than the channel diameter (0.8 mm). This size ratio ensures that any boundary effects in the transport process can be neglected, while still realizing a working region that could be observed with a 4× objective. Fig. 1b shows the different steps in the analysis, which includes the chip fabrication; the injection of the test solution (e.g. dextran-dye or particles) via a pipette NanoInjector or syringe pump; the microscopy acquisition, and image post-processing and mathematical analysis. Specifically, in order to perform the diffusion measurements, the Tissue Chamber is placed on the stage of an inverted optical microscope, and images were acquired following the administration of fluorescent dextran or particle. Images are post-processed using Fiji (https://imagej.nih.gov/).

Analysis of the diffusion of small molecules within the tissue chamber chip In order to test the chip and the whole measurement apparatus, dextran molecules of different molecular weights were used. Specifically, 5 μl of an aqueous solution containing 4, 40 or 250 kDa FITC-dextran molecules were injected into the Tissue Chamber. The progressive spreading of the green fluorescent solution within the gel is recorded over time. The three top rows of Fig. 2a–c present the diffusing front of the dextran solutions at different time points, namely 0, 90 and 240 s. At each time point, the size of the colored area is estimated by post-processing the fluorescent images via the Fiji software. Note that the size of the channel is about 800 μm, which corresponds to the outer diameter of the needle used for realizing the channel itself within the collagen matrix. The lowest row in Fig. 2a–c shows the variation over time of the averaged MSD, calculated from the experimental data using Eq. (2) in the Methods section. Each point on these plots corresponds to the MSD at that time t, averaged over multiple experiments ($n \geq 5$). The slope of the MSD (t) curves provides, through Eq. (1), the actual diffusion coefficient. Figure 2d summarizes the experimental results providing the diffusion coefficient of the three tested molecules (4, 40 and 250 FITC-dextran) as derived from the MSD measurements in the Tissue Chamber chip (blue bars). As expected, the diffusion coefficient reduces as the molecular weight of dextran increases. For the 4 kDa dextran, a coefficient of diffusion $D = 44.20 \pm 6.65$ μm^2/s is derived. This number reduces by about 49% ($D = 22.4 \pm 16.8$ μm^2/s) for 40 kDa dextran, and by about 78% ($D = 9.9 \pm 3.71$ μm^2/s) for 250 kDa dextran.

The Einstein-Stoke relation was used to estimate the diffusion coefficients of the dextran molecules in pure water:

$$D_w = \frac{k_B T}{(6\pi\mu R_H)} \tag{7}$$

from which it results that $D_w < 270$ μm^2/s for 4 kDa dextran, < 40 μm^2/s for 40 kDa dextran, and < 20 μm^2/s for 250 kDa dextran. Indeed, as expected, the diffusion coefficient in water is significantly higher than in the collagen matrix for all dextran molecular weights. However, interestingly, it is the dextran with the smaller molecular weight (4 kDa) that is subjected to the largest reduction (6-fold) in diffusion from <270 to <45 μm^2/s. In Eq. (7), $k_B T$ is the Boltzmann energy at room temperature ($4.11 \cdot 10^{-21}$ J), μ is the dynamic viscosity of water (10^{-3} Pa·s), and R_H is the hydrodynamic radius of the molecule ($R_H < 0.8$ nm for 4 kDa dextran, < 5 nm for 40 kDa dextran, and < 11.5 nm for 250 kDa dextran) (Armstrong et al. 2004). To validate the experimental set-up, the diffusion of dextran molecules in collagen was also assessed via Fluorescence Recovery After Photobleaching (FRAP) returning values in good agreement with those quantified within the Tissue Chamber chip. Specifically, the diffusion

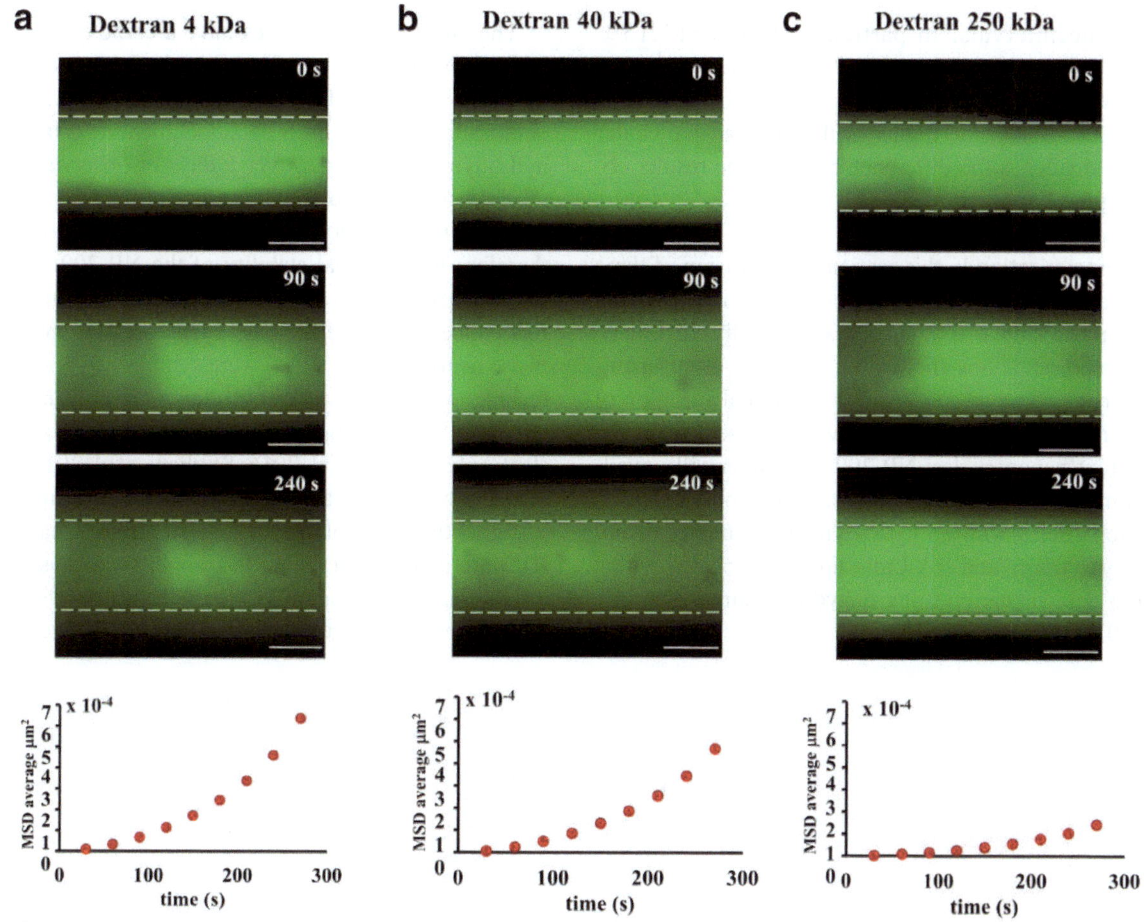

d

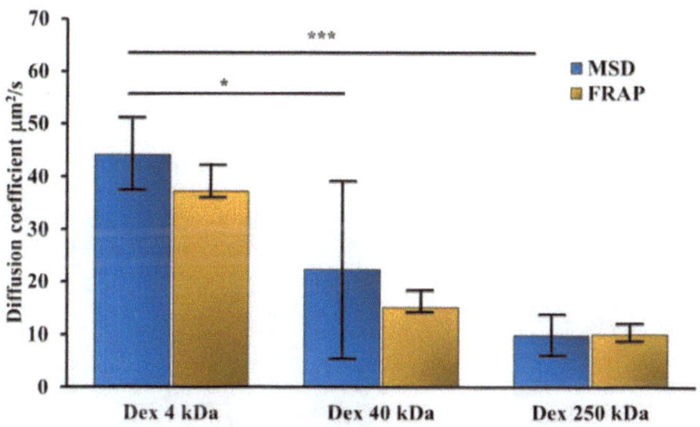

e

Dextran	D_{exp} (MSD) [μm²/s]	D_{exp}(FRAP) [μm²/s]	$(D_{MSD}/D_{SE})_{av}$
4 kDa	44.2±6.65	37.3±5.06	0.164
40 kDa	22.4±16.8	15.2±3.14	0.80
250 kDa	9.9±3.71	10.1±2.06	0.495

Fig. 2 Optical fluorescent microscopy images and quantification of different molecular weight Dextran diffusion. a Fluorescent images of the channel in the Tissue Chamber filled up with Dextran 4 kDa acquired at different time point (0 s, 90 s, 240 s) and averaged values of the Mean Square Displacement (MSD). **b** Fluorescent images of the channel in the Tissue Chamber filled up with Dextran 40 kDa acquired at different time point (0 s, 90 s, 240 s) and averaged values of the MSD. **c** Fluorescent images of the channel in the Tissue Chamber filled up with Dextran 250 kDa acquired at different time point (0 s, 90 s, 240 s) and averaged values of the MSD. Scale bars are 500 μm. **d** Bar chart of molecular diffusion coefficients obtained for molecules using two different analysis (MSD and Fluorescent Recovery After Photobleaching, FRAP). **e** Summarized table of MSD and FRAP derived molecular diffusion coefficients for same dye as well as their experimental and theoretical diffusion ratio. Number of repetitions $n > 5$ for MSD and $n = 4$ for FRAP. * Symbol denotes statistically significant difference $p < 0.05$, *** $p < 0.001$

coefficients estimated via FRAP were equal to $D_{FRAP} = 37.30 \pm 5.06$ μm²/s for 4 kDa dextran, 15.20 ± 3.14 μm²/s for 40 kDa dextran, and 10.10 ± 2.06 μm²/s for 250 kDa dextran. Finally, the diffusion coefficient was also estimated via the EMA approach and compared with the MSD and FRAP results. The EMA was applied using a regularization parameter $\lambda = 10^{-2}$. The diffusion coefficient was computed as the average of the diffusion coefficients resulting from the analysis of different time intervals. In particular, $D_{EMA} = 81.15 \pm 63.09\ \mu m^2/s$ for 4 kDa Dextran, $49.15 \pm 19.73\ \mu m^2/s$ for 40 kDa Dextran and $13.54 \pm 5.38\ \mu m^2/s$ for 250 kDa Dextran. Interestingly, the higher is the Dextran molecular weight, the smaller is the difference between the results obtained using the EMA and MSD.

Particle characterization In this section, the diffusion of three different nanoparticles within the collagen matrix of the Tissue Chamber chip was studied systematically. In Fig. 3, schematic representations (first row); scanning electron microscopy images (second row); and the temporal variation of size and surface ζ potential (third row) are presented for the three different nanoparticles: NP 200 – commercially available 200 nm polystyrene nanoparticles; SPN – spherical polymeric nanoparticles with all hydrodynamic diameter of approximately 200 nm; and HA-Lip – liposomes coated with a hyaluronic acid layer returning a hydrodynamic diameter around 200 nm.

The NP 200 nanoparticles have a carboxylated surface, a hydrodynamic diameter of 187.96 ± 2.42 nm and a ζ potential of −42.5 ± 2.12 mV. The stability of these particles was clearly demonstrated by the DLS data documenting a fairly constant hydrodynamic diameter over 5 days of observation (Fig. 3a, third row). The ζ potential consistently stayed between −40 and − 50 mV contributing to the electrostatic repulsion and therefore the colloidal stability of the NP 200. The electron microscopy images of Fig. 3a (second row) confirm the uniform spherical shape of these particles. The spherical polymeric nanoparticles (SPNs) were synthesized via an emulsion technique and possess a poly(lactic-co-glycolic acid) (PLGA) core stabilized by a lipid monolayer, with polyethylene glycol (PEG) chains (Lee et al. 2016; Stigliano et al. 2015). Figure 3b (first row) shows a schematic representation of SPNs documenting the polymeric/lipid structure. Chains of lipid-RhB are included in the surface monolayer thus introducing a fluorescent reporting molecules in the nanoparticle structure. Two different configurations of SPNs are realized depending

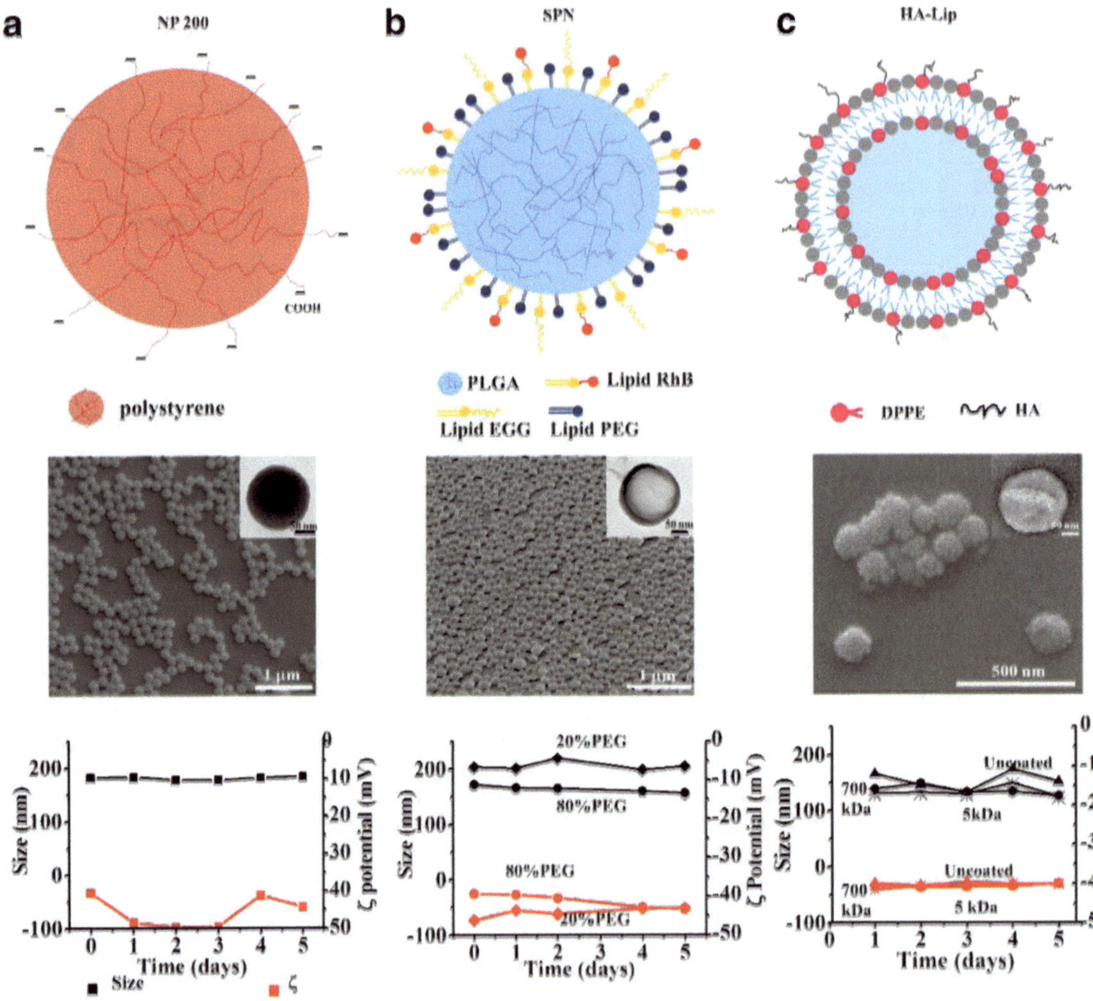

Fig. 3 Schematic representations, electron microscopy images, dynamic light scattering (DLS) hydrodynamic size and ζ-potential measurements. a Commercially available polystyrene carboxylate beads, **b** spherical PEGylated PLGA nanoparticles, and **c** liposomes coated with varying molecular weight hyaluronic acid. DLS and ζ-potential measurements were made over 5 days in water

on the initial PEG content. For 20% SPN, the initial concentration of lipid-PEG is 20% of the total lipid mass while, for 80% SPN, the initial concentration of lipid-PEG is 80% of the total lipid mass. In both cases, the electron microscopy analysis confirms the spherical shape and monodisperse size distribution of the nanoparticles (Fig. 3b, second row). At time zero, the DLS returns a hydrodynamic size of 186 ± 13 nm and a ζ potential of -45.9 ± 0.79 mV for the 20% SPNs. The 80% SPNs are slightly smaller with a hydrodynamic size of 170 ± 3 nm and a ζ potential of -39.1 ± 0.87 mV. Indeed, the higher percent of lipid-PEG on the SPN surface fosters stabilization, thus reducing the overall hydrodynamic diameter (Fig. 3c, third row). A similar behavior is observed of the HA-Lip particles. Notice that particle stabilization and size reduction have been well documented in the literature also by other authors (see for instance Essa et al. 2011; Wolfram et al. 2014). The spherical polymeric nanoconstructs are stable over the period of 5 days with an overall variation in hydrodynamic size limited to 10% for both 20% and 80% SPNs (Fig. 3c, third row).

Finally, the third type of nanoparticles schematically presented in the first row of Fig. 3c are the liposomes with an aqueous core and a lipid bilayer, which is externally coated at different degrees with hyaluronic acid (HA) (Mizrahy et al. 2014). Three different configurations of this HA-Lip are considered: liposomes not coated with any HA layer (uHA-Lip); liposomes coated with a 5 kDa HA layer (5HA-Lip); and liposomes coated with a 700 kDa HA layer (700HA-Lip). For the uHA-Lip, the initial hydrodynamic diameter was 146.1 ± 4.271 nm. This decreases to 138.5 ± 4.845 and 132.4 ± 6.493 nm for the 5HA-Lip and 700Ha-Lip, respectively. The ζ potential is stable at -37 ± 4 mV for all three configurations. SEM and TEM images confirm the DLS data. Specifically, the electron microscopy images in Fig. 3c (second row) are for the 5HA-Lip (SEM) and for uHA-Lip (TEM). Furthermore, these nanoparticles present good colloidal stability over a 5-days period Fig. 3c (third row).

Analysis of the diffusion of nanoparticles within the tissue chamber chip After confirming the size and stability of the nanoparticles, diffusion experiments were conducted in the Tissue Chamber chip. Specifically, a 5 μl solution with SPNs or HA-Lip was introduced into the chip and the progressive diffusion within the collagen matrix was recorded over time. The bar charts in Fig. 4a provides the diffusion coefficients in the collagen matrix for NP 200, the two SPN configurations, and the three HA-Lip configurations. Representative images of NP 200 in the chip, in both brightfield and fluorescence, are given in Fig. 4b. It is shown that the diffusion coefficient increases from 1.75 ± 0.65 $\mu m^2/s$ for the NP 200, un-PEGylated nanoparticles to 3.38 ± 1.89 $\mu m^2/s$ for the 20% SPNs and 4.89 ± 0.25 $\mu m^2/s$ for the 80% SPNs, which are characterized by the highest surface density of PEG. For the liposomes, the presence of HA over the surface reduces the diffusion coefficient from 5.01 ± 1.96 $\mu m^2/s$ for the uHA-Lip; to 3.83 ± 1.80 $\mu m^2/s$ for the 5HA-Lip and 2.21 ± 1.30 $\mu m^2/s$ for the 700HA-Lip. The theoretical diffusion coefficient through the equation of Einstein-Stokes was also calculated. The diffusion coefficients for the NP 200, 20% SPNs 20% PEG and SPNs 80% PEG using the EMA approach were also computed. In particular, $D_{EMA} = 1.75 \pm 0.65$ $\mu m^2/s$ for the NP 200, 21.40 ± 7.41 $\mu m^2/s$ for the SPNs 20% PEG and 6.68 ± 2.56 $\mu m^2/s$ for the SPNs 80% PEG. While the D_{EMA} values for NP 200 and 80% SPNs are in good agreement with the MSD-derived diffusion coefficients, EMA fails in predicting accurately the diffusion for the SPNs 20% PEG.

These results emphasize that an increase in PEGylation can more efficiently lubricate the particle-tissue interface and favor the diffusion and tissue penetration (Dancy et al. 2016). For the second group of nanoparticles studied, i.e. liposomes, the diffusion coefficient within the type I collagen gel calculated with the MSD shows that the uHA-Lip diffuse more than the HA coated liposomes, ostensibly due to the interaction of collagen with hyaluronic acid (Annabi et al. 2004). Moreover, the 5HA-Lip may diffuse more than 700HA-Lip due to for their greater lubricating effect (Mizrahy et al. 2011).

Analysis of the diffusion of nanoparticles within brain tissue slices The diffusion of nanoparticles was also assessed in 2 mm thick, freshly excised brain slices from C57BL/6 wild-type mice. To avoid tissue death, brain slices were kept on ice and hydrated with cold PBS. A 69 nl of a solution containing the different types of nanoparticles was introduced in the striatum of the brain slices (Fig. 5a–c). The Nanoinjector was placed at 1.5 mm from the top of the slice and the desired volume was introduced at 0.5 mm from the bottom of the slice (Zhang et al. 2017). The dynamics of the nanoparticles was monitored over time using an inverted microscope specifically modified for single particle tracking, using an oil immersion 100× objective. Representative images are provided in Fig. 5c. The resulting movies were post-processed through the Fiji trackmate tool (Tinevez et al. 2017) and analyzed with a custom Matlab script, which was specially developed to extrapolate the MSD (Sancataldo et al. 2017; Xavier 2010). This custom MATLAB (version R2015a) script, found in the Supporting Information, was used to estimate the MSD and the diffusion coefficient of the different molecules and nanoparticles in ex vivo brain tissue.

From this analysis, 3D (x,y,z) trajectories of the individual particles over time can be extracted, as shown in Fig. 5d. Finally, the MSD of the nanoparticles can be computed and, consequently, the diffusion coefficient is derived as described in the previous paragraphs. The diffusion coefficient for the six different particles are provided by the bar chart in Fig. 5e. Notably, the trends are similar to the one derived for the same

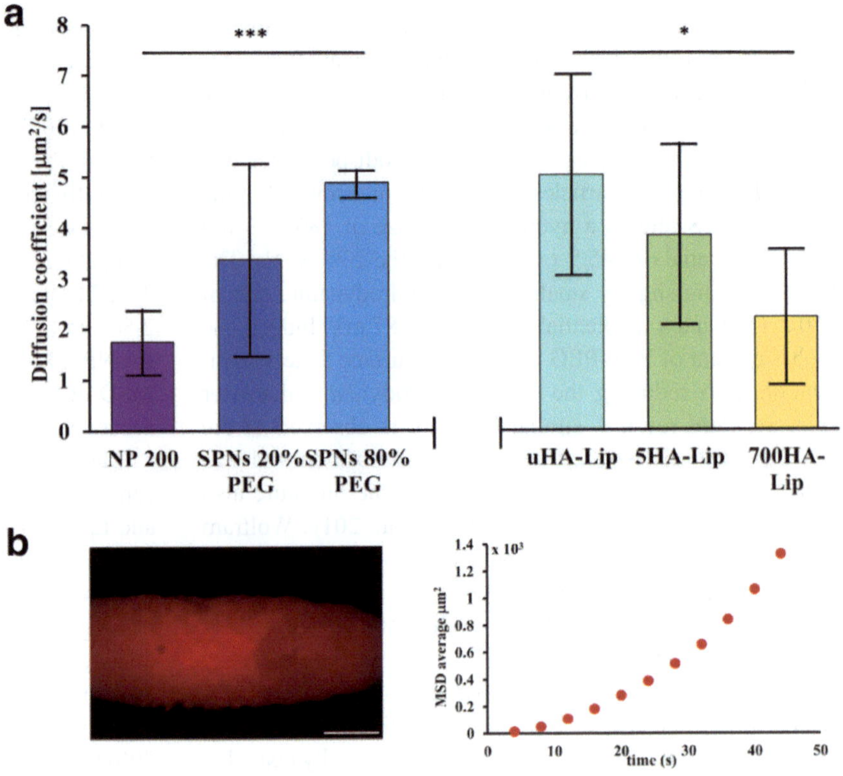

c

	Diameter (nm)	D_{MSD} [μm²/s]	D_{SE} [μm²/s]	$(D_{MSD}/D_{SE})_{av}$
NP 200	187.96±2.42	1.75±0.645	2.325	0.753
SPNs 20%PEG	186.2±13	3.38±1.89	2.347	1.440
SPNs 80%PEG	170.4±2.5	4.89±0.25	2.564	1.907
uHA-Lip	146.1 ±4.271	5.01±1.96	2.986	1.678
5HA-Lip	138.5 ±4.845	3.83±1.8	3.127	1.225
700HA-Lip	132.4 ±6.493	2.21±1.30	3.199	0.691

Fig. 4 Nanoparticles molecular diffusion coefficient in collagen gel. a Quantification of nanoparticles molecular diffusion coefficient in collagen gel was determined for nanoparticles using the MSD. **b** Representative fluorescent images of 5HA-Lip injected inside the Tissue Chamber and the corresponding the MSD plot. Scale bar represents 500 μm. **c** A table summarizing the nanoparticles diameters, and the MSD and Stokes-Einstein-derived molecular diffusion coefficients for the particles. $N \geq 4$. * Denotes statistically significant difference $p < 0.05$, *** $p < 0.001$

particle *in vitro* within the collagen matrix of the Tissue Chamber chip. In other words, the diffusion coefficient increase moving from un-PEGylated nanoparticles (NP 200) to PEGylated SPNs, and reduces with the presence of HA on liposomes. Specifically, the diffusion was $2.68 \pm 1.33 \cdot 10^{-5}$ μm²/s for the NP 200; $4.84 \pm 2.41 \cdot 10^{-5}$ μm²/s for the 20% SPNs; $5.36 \pm 2.49 \cdot 10^{-5}$ μm²/s for the 80% SPNs; $6.32 \pm 2.81 \cdot 10^{-5}$ μm²/s for the uHA-Lip; $3.50 \pm 1.91 \cdot 10^{-5}$ μm²/s for the 5HA-Lip; and $2.48 \pm 1.35 \cdot 10^{-5}$ μm²/s for the 700HA-Lip.

Comparing these results with the one obtained in the Tissue Chamber chip, a decrease in the diffusion coefficient of five orders of magnitude was noticed. This is indeed expected given that the Tissue Chamber chip does not include cells in its current configuration.

4 Discussion

Here, a Tissue Chamber chip has been demonstrated for estimating the diffusion ability of nanoparticles under controlled biophysical conditions. First, the diffusion of FITC-Dextran molecules with three different molecular weights (4, 40 and 250 kDa) was assessed in a collagen gel using two different and independent techniques. The first approach is based on extracting the diffusion coefficient from the Mean Square Displacement (MSD). This is a very general technique that can be applied to molecules and nanoparticles. The second approach is based on the use of Fluorescence recovery After Photo-bleaching (FRAP), but this technique cannot be applied to nanoparticles. Figure 2d shows that the diffusion coefficients estimated with both techniques are in

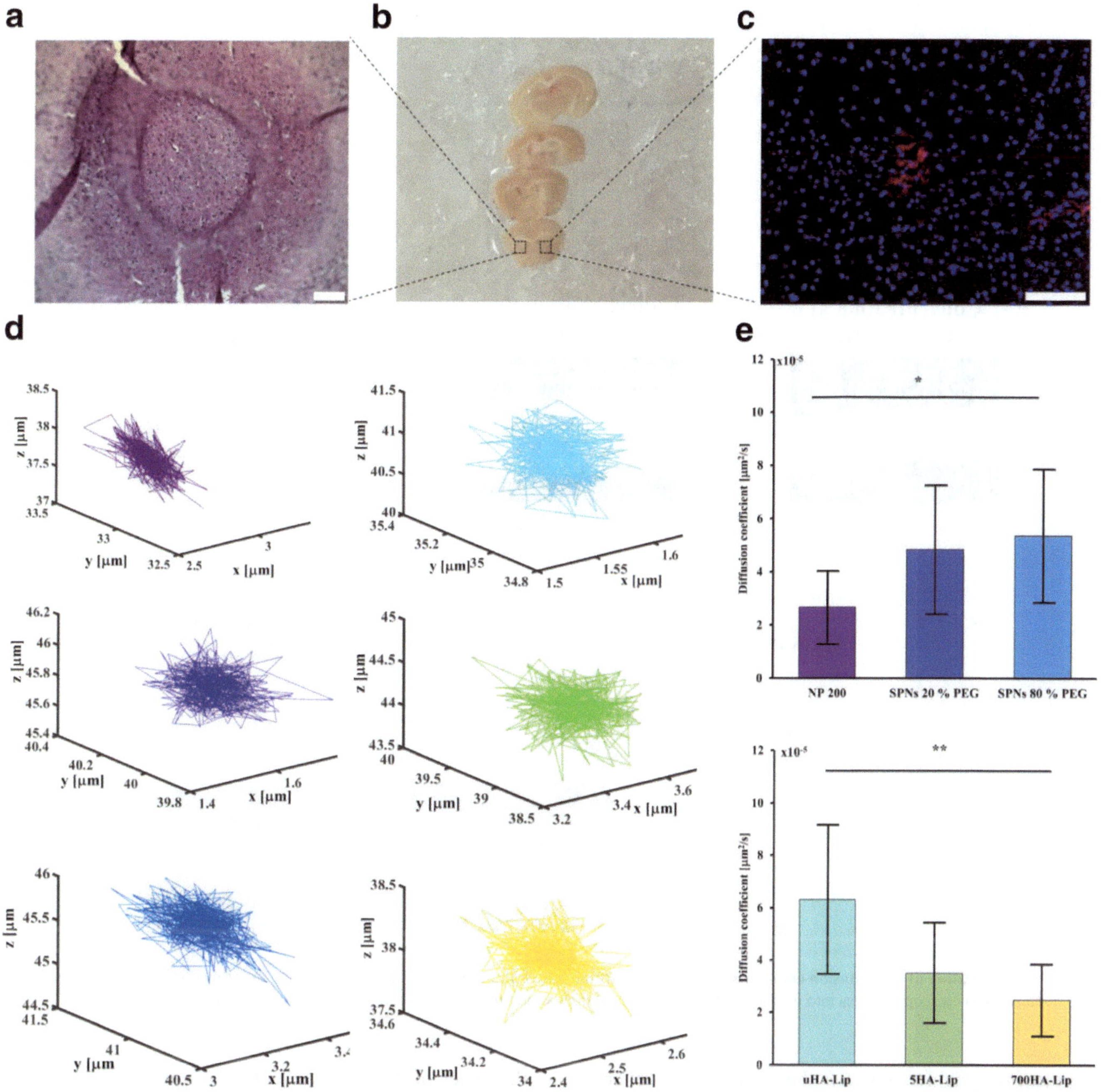

Fig. 5 Ex-vivo single tracking and quantification of nanoparticles molecular diffusion coefficient. **a** Representative image of a 12 μm mouse brain slice stained with H&E with recognizable sign of the injection needle. Scale bar represents100 μm. **b** Representative image of different brain sections. Squares highlight the striatum. **c** Confocal fluorescent microscopy image of a brain slice shows the NPs as red spots and the nuclei stained in DAPI. Scale bar represents 100 μm. **d** 3D trajectories of the particles diffusing through ex vivo brain slices. **e** Diffusion coefficients obtained for nanoparticles in the ex vivo brain tissue derived using MSD analysis. * Denotes a statistically significant difference $p < 0.05$, ** $p < 0.01$

good agreement and no statistically significant difference is documented for all three tested Dextran molecules ($p = 0.118$ for 4 kDa dextran; $p = 0.426$ for 40 kDa dextran; and $p = 0.926$ for 250 kDa dextran). The FRAP technique is easier and more reliable then the MSD approach, as suggested by the smaller variations among the different measurements. The MSD technique is more cumbersome and reproducibility is more affected by the intrinsic higher difficulty associated with performing the experiments. Specifically, the channel diameter may vary slightly between experiments due to the fabrication process, the connection with external tubing or pipettes to the channel is not always optimal, and so on. Yet, FRAP and MSD return values for the diffusion coefficient that are in close

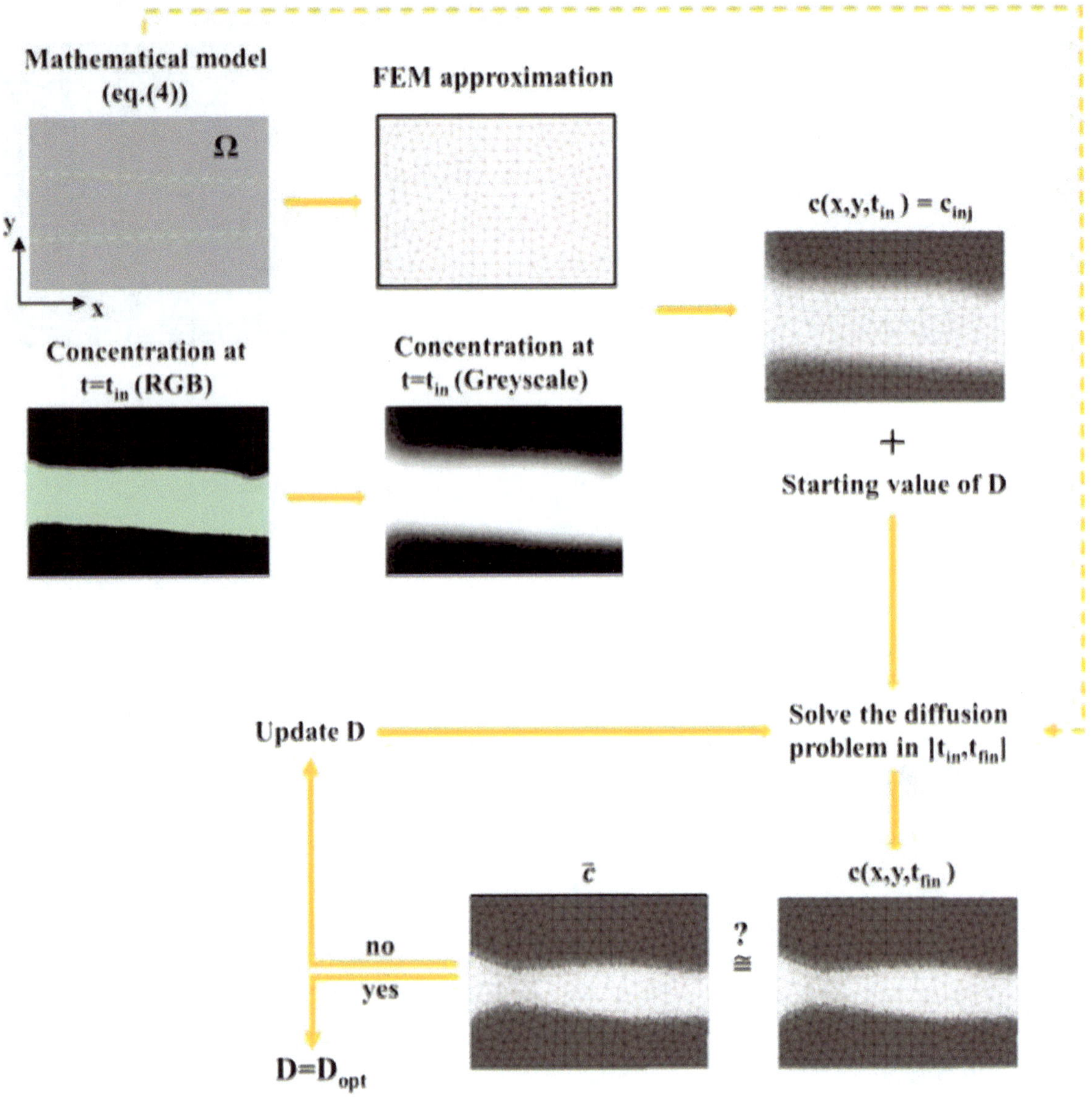

Fig. 6 Quantification of the diffusion coefficient via an error minimization algorithm. Post-processing of the experimental data with conversion of the fluorescent microscopy images into greyscale maps and error minimization procedure implemented in the computational algorithm

agreement. This demonstrates the accuracy and reliability of the whole measurement protocol and the Tissue Chamber chip.

As FRAP cannot be applied to nanoparticles, the following measurements on SPNs and HA-Lip particles were performed using the MSD (Fig. 4) and EMA (Fig. 6) approaches. Figure 4 shows that the decoration of the nanoparticle surface with short PEG chains (2 kDa) increases their diffusivity in the collagen gel. This trend is indeed in agreement with the experimental results presented by Zhang et al. (2017). In their work, the diffusion of two nanoparticles (65 ± 3 nm bare cisplatin-loaded poly(aspartic acid) (PAA) particles and 74 ± 2 nm PEGylated cisplatin-loaded PAA particles) was assessed within freshly excised healthy rat brain tissue slices. Under these conditions, Zhang and colleagues derived diffusion coefficients of the order of 10^{-3} μm²/s and 10^{-1} μm²/s for naked and PEGylated particles, respectively. Other authors have estimated the diffusion of PEGylate particles into mucus (Cu and Saltzman 2009; Xu et al. 2015). Even in these works, the addition of PEG chains over a PLGA particle core improved diffusion. This was mainly ascribed to steric interactions arising at the interface between PEG chains and the surrounding mucus structure. More recently, Labouta et al. (2018) showed that PEG density was the contributing factor in determining the penetration depth of liposome into collagen gels. Similarly, in this work the coating of spherical, solid polymeric nanoparticles with PEG (2 kDa) increases the mobility within the collagen gel, as compared to bare nanoparticles.

Following the same procedure of Zhang and colleagues, the diffusion of bare and PEGylated nanoparticles was also assessed in freshly excised brain tissue slides. The same behavior observed for the *in vitro* experiments were also documented ex vivo. Specifically, the nanoparticle diffusion increases with the surface density of the PEG chains. Indeed, the absolute values of the ex vivo diffusion coefficient are significantly smaller than those documented *in vitro*. This difference should be mostly attributed to three differences between the excised brain tissue and the Tissue Chamber chip, namely the presence of cells, the extracellular matrix composition, and the limited extracellular space that characterizes brain tissues (Syková and Nicholson 2008). It should be emphasized that it is difficult to perform a direct and objective comparison with previous results available in the literature because of differences in particle size and surface properties. Thorne and Nicholson (2006) measured, in living animals, the diffusivity of 35 nm quantum dots to be in the order of 10^{-1} $\mu m^2/s$. Zhang and colleagues (Zhang et al. 2017) measured the diffusivity of 60 nm PAA particles to be in the order of 10^{-3} to 10^{-1} $\mu m^2/s$. In the current manuscript, the authors estimated the diffusivity of 200 nm particles to be in the order of 10^{-5} to 10^{-4} $\mu m^2/s$. These significant differences in diffusivity could be also ascribed to the significant difference in particle size.

In the case of hyaluronic acid-functionalized liposomes, the opposite trend was observed. Specifically, HA decoration over the liposome surface was responsible for a significant reduction in mobility. This decrease was directly related to HA molecular weight and possibly due to the entanglement of the long HA chains with the surrounding extracellular matrix. This macromolecular phenomenon may be driven by electrostatic forces (Gelman and Blackwell 1974) or by specific HA-collagen interactions (Annabi et al. 2004; McDevitt et al. 1991). Importantly, even in the case of the HA-Lip, the *in vitro* trends are in full agreement with the ex vivo observations.

5 Conclusions

Taken all together, the data presented in this manuscript show that the proposed Tissue Chamber chip provides a versatile platform that can realize fundamental studies on particle diffusion in a tissue-like environment. Importantly, multiple biophysical parameters can be accurately controlled. For example, the tissue composition can be tailored to include collagen type I, type IV as well as hyaluronic acid, matrigel, other biomacromolecules, and combinations thereof. Furthermore, the ECM can be modified to express specific adhesion molecules (e.g. integrins) or the density can be modified to affect the porosity of the tissue. Finally, cells could be included which could dynamically remodel the ECM, interact with the particles (uptake and trafficking) or act as physical barriers to particle transport. This platform can therefore be employed to conduct systematic, comparative studies to evaluate nanoparticle transport processes in different recapitulated tissues. With such a tool it would be possible to optimize the geometrical and surface properties of nanoparticles to achieve high and uniform tissue penetration.

Acknowledgments All the authors have read and approved the manuscript. This project was partially supported by the European Research Council, under the European Union's Seventh Framework Programme (FP7/2007-2013)/ERC grant agreement no. 616695, by the Italian Association for Cancer Research (AIRC) under the individual investigator grant no. 17664, and by the European Union's Horizon 2020 research and innovation programme under the Marie Skłodowska-Curie grant agreement no. 754490.

References

B. Annabi, S. Thibeault, R. Moumdjian, R. Béliveau, J. Biol. Chem. **279**, 21888 (2004)

A.C. Anselmo, J.B. Gilbert, S. Kumar, V. Gupta, R.E. Cohen, M.F. Rubner, S. Mitragotri, J. Control. Release **199**, 29 (2015)

J.K. Armstrong, R.B. Wenby, H.J. Meiselman, T.C. Fisher, Biophys. J. **87**, 4259 (2004)

A.G. Arranja, V. Pathak, T. Lammers, Y. Shi, Pharmacol. Res. **115**, 87 (2017)

E. Blanco, T. Sangai, S. Wu, A. Hsiao, G.U. Ruiz-Esparza, C.A. Gonzalez-Delgado, F.E. Cara, S. Granados-Principal, K.W. Evans, A. Akcakanat, Y. Wang, K.A. Do, F. Meric-Bernstam, M. Ferrari, Mol. Ther. **22**, 1310 (2014)

E. Blanco, H. Shen, M. Ferrari, Nat. Biotechnol. **33**, 941 (2015)

J.S. Brenner, D.C. Pan, J.W. Myerson, O.A. Marcos-Contreras, C.H. Villa, P. Patel, H. Hekierski, S. Chatterjee, J.-Q. Tao, H. Parhiz, K. Bhamidipati, T.G. Uhler, E.D. Hood, R.Y. Kiseleva, V.S. Shuvaev, T. Shuvaeva, M. Khoshnejad, I. Johnston, J.V. Gregory, J. Lahann, T. Wang, E. Cantu, W.M. Armstead, S. Mitragotri, V. Muzykantov, Nat. Commun. **9**, 2684 (2018)

H. Cabral, Y. Matsumoto, K. Mizuno, Q. Chen, M. Murakami, M. Kimura, Y. Tereda, M.R. Kano, K. Miyazono, M. Uesaka, N. Nishiyama, K. Kataoka, Nat. Nanotechnol. **6**, 815 (2011)

M.-R. Choi, R. Bardhan, K.J. Stanton-Maxey, S. Badve, H. Nakshatri, K.M. Stantz, N. Cao, N.J. Halas, S.E. Clare, Cancer Nanotechnol. **3**, 47 (2012)

Y. Cu, W.M. Saltzman, Mol. Pharm. **6**, 173 (2009)

J.G. Dancy, A.S. Wadajkar, C.S. Schneider, J.R.H. Mauban, O.G. Goloubeva, G.F. Woodworth, J.A. Winkles, A.J. Kim, J. Control. Release **238**, 139 (2016)

A. Erikson, H.N. Andersen, S.N. Naess, P. Sikorski, C.d.L. Davies, Biopolymers **89**, 135 (2008)

S. Essa, J.M. Rabanel, P. Hildgen, Int. J. Pharm. **411**, 178 (2011)

M. Ferrari, Nat. Rev. Cancer **5**, 161 (2005)

R.A. Gelman, J. Blackwell, Biochim. Biophys. Acta **342**, 254 (1974)

C.-M.J. Hu, L. Zhang, S. Aryal, C. Cheung, R.H. Fang, L. Zhang, Proc. Natl. Acad. Sci. **108**, 10980 (2011)

S. Hua, M.B.C. de Matos, J.M. Metselaar, G. Storm, Front. Pharmacol. **9**, 790 (2018)

R. Kedmi, N. Ben-Arie, D. Peer, Biomaterials **31**, 6867 (2010)

H.I. Labouta, M.J. Gomez-Garcia, C.D. Sarsons, T. Nguyen, J. Kennard, W. Ngo, K. Terefe, N. Iragorri, P. Lai, K.D. Rinker, D. Cramb, RSC Adv. **8**, 7697 (2018)

D. Landesman-Milo, M. Goldsmith, S. Leviatan-Ben-Arye, B. Witenberg, E. Brown, S. Leibovitch, S. Azriel, S. Tabak, V. Morad, D. Peer, Cancer Lett. **334**, 221 (2013)

J. Lazarovits, Y.Y. Chen, E.A. Sykes, W.C.W. Chan, Chem. Commun. **51**, 2756 (2015)

A. Lee, D. di Mascolo, M. Francardi, F. Piccardi, T. Bandiera, P. Decuzzi, Nanomed.: Nanotechnol. Biol. Med. **12**, 2139 (2016)

C.A. McDevitt, J. Marcelino, L. Tucker, FEBS Lett. **294**, 167 (1991)

S. Mizrahy, S.R. Raz, M. Hasgaard, H. Liu, N. Soffer-Tsur, K. Cohen, R. Dvash, D. Landsman-Milo, M.G.E.G. Bremer, S.M. Moghimi, D. Peer, J. Control. Release **156**, 231 (2011)

S. Mizrahy, M. Goldsmith, S. Leviatan-Ben-Arye, E. Kisin-Finfer, O. Redy, S. Srinivasan, D. Shabat, B. Godin, D. Peer, Nanoscale **6**, 3742 (2014)

T.L. Moore, H. Chen, R. Morrison, F. Wang, J.N. Anker, F. Alexis, Mol. Pharm. **11**, 24 (2014)

T.L. Moore, D. Hauser, T. Gruber, B. Rothen-Rutishauser, M. Lattuada, A. Petri-Fink, R. Lyck, ACS Appl. Mater. Interfaces **9**, 18501 (2017)

G. Nabil, K. Bhise, S. Sau, M. Atef, H.A. El-Banna, A. Iyer, Drug Discov. Today **24**, 462 (2019)

D. Peer, R. Margalit, Arch. Biochem. Biophys. **383**, 185 (2000)

D. Peer, R. Margalit, Int. J. Cancer **108**, 780 (2004)

D. Peer, E.J. Park, Y. Morishita, C.V. Carman, M. Shimaoka, Science **319**, 627 (2008)

A. Pitchaimani, T.D.T. Nguyen, R. Marasini, A. Eliyapura, T. Azizi, M. Jaberi-Douraki, S. Aryal, Adv. Funct. Mater. **29**, 1806817 (2019)

R. Pushpalatha, S. Selvamuthukumar, D. Kilimozhi, J. Drug Delivery Sci. Technol. **39**, 362 (2017)

D. Rosenblum, N. Joshi, W. Tao, J.M. Karp, D. Peer, Nat. Commun. **9**, 1410 (2018)

G. Sancataldo, L. Scipioni, T. Ravasenga, L. Lanzanò, A. Diaspro, A. Barberis, M. Duocastella, Optica **4**, 367 (2017)

J. Schindelin, I. Arganda-Carreras, E. Frise, V. Kaynig, M. Longair, T. Pietzsch, S. Preibisch, C. Rueden, S. Saalfeld, B. Schmid, J.Y. Tinevez, D.J. White, V. Hartenstein, K. Eliceiri, P. Tomancak, A. Cardona, Nat. Methods **9**, 676 (2012)

J. Shi, P.W. Kantoff, R. Wooster, O.C. Farokhzad, Nat. Rev. Cancer **17**, 20 (2017)

A. Sontheimer-Phelps, B.A. Hassell, D.E. Ingber, Nat. Rev. Cancer **19**, 65 (2019)

C. Stigliano, J. Key, M. Ramirez, S. Aryal, P. Decuzzi, Adv. Funct. Mater. **25**, 3371 (2015)

T. Stylianopoulos, M.-Z. Poh, N. Insin, M.G. Bawendi, D. Fukumara, L.L. Munn, R.K. Jain, Biophys. J. **99**, 1342 (2010)

E. Syková, C. Nicholson, Physiol. Rev. **88**, 1277 (2008)

E. Tasciotti, X. Liu, R. Bhavane, K. Plant, A.D. Leonard, B.K. Price, M.M.-C. Cheng, P. Decuzzi, J.M. Tour, F. Robertson, M. Ferrari, Nat. Nanotechnol. **3**, 151 (2008)

R.G. Thorne, C. Nicholson, Proc. Natl. Acad. Sci. **103**, 5567 (2006)

G.T. Tietjen, L.G. Bracaglia, W.M. Saltzman, J.S. Pober, Trends Mol. Med. **24**, 598 (2018)

J.-Y. Tinevez, N. Perry, J. Schindelin, G.M. Hoopes, G.D. Reynolds, E. Laplantine, S.Y. Bednarek, S.L. Shorte, K.W. Eliceiri, Methods **115**, 80 (2017)

F. Tröltzsch, Optimal control of partial differential equations, vol. 112 (Graduate Studies in Mathematics, 2010)

M.T. Valentine, Z. E: Perlman, M.L. Gardel, J.H. Shin, P. Matsudaira, T.J. Mitchison, D.A. Weitz, Biophys. J. **86**, 4004 (2004)

S. Wilhelm, A.J. Tavares, Q. Dai, S. Ohta, J. Audet, H.F. Dvorak, W.C.W. Chan, Nat. Rev. Mater. **1**, 16014 (2016)

J. Wolfram, K. Suri, Y. Yang, J. Shen, C. Celia, M. Fresta, Y. Zhao, H. Shen, M. Ferrari, Colloid Surf. B Biointerfaces **114**, 294 (2014)

C. Wong, T. Stylianopoulos, J. Cui, J. Martin, V.P. Chauhan, W. Jiang, Z. Popović, R.K. Jain, M.G. Bawendi, D. Fukumura, Proc. Natl. Acad. Sci. **108**, 2426 (2011)

M. Xavier, Phys. Rev. **E82**, 041914 (2010)

Q. Xu, L.M. Ensign, N.J. Boylan, A. Schön, X. Gong, J.-C. Yang, N.W. Lamb, S. Cai, T. Yu, E. Freire, J. Hanes, ACS Nano **9**, 9217 (2015)

C. Zhang, E.A. Nance, P. Mastorakos, J. Chisholm, S. Berry, C. Eberhart, B. Tyler, H. Brem, J.S. Suk, J. Hanes, J. Control. Release **263**, 112 (2017)

W. Zhang, M. Wang, W. Tang, R. Wen, S. Zhou, C. Lee, H. Wang, W. Jiang, I.M. Delahunty, Z. Zhen, H. Chen, M. Chapman, Z. Wu, E.W. Howerth, H. Cai, Z. Li, J. Xie, Adv. Mater. **30**, 1805557 (2018)

Part V
Tissue Regeneration

Biomedical Microdevices (2019) 21: 42
https://doi.org/10.1007/s10544-019-0372-2

In situ three-dimensional printing for reparative and regenerative therapy

Nureddin Ashammakhi[1,2,3,4,5,6] · **Samad Ahadian**[1,2,4] · **Ippokratis Pountos**[7,8] · **Shu-Kai Hu**[1,2,3] · **Nazzar Tellisi**[7] · **Praveen Bandaru**[1,2,3] · **Serge Ostrovidov**[1,2,3] · **Mehmet Remzi Dokmeci**[1,2,3] · **Ali Khademhosseini**[1,2,3,4,9,10,11]

Published online: 6 April 2019

Author's Personal Note It's an honor to write this note about one of the pioneering biomedical engineers today, Mauro Ferrari. Mauro has always been such a wonderful and supportive colleague and friend. I remember when I was just starting my lab, Mauro supported and mentored me even though I was not one of his students. He learned about our work and championed our science when very few people had even heard about us. He invited me to join the Editorial Board of Biomedical Microdevices and also opened his door to my wife and I as one of his guests during one of Nanogagliato conferences in a small village in Italy. Being there and seeing his generosity and the way that he communicates with the locals really made me appreciate his humility and humanity. Mauro is also an amazing leader as shown by his many years of leadership at various institutions. He is a champion of translational research and has done some really incredible work in trying to make a lasting impact on patients' lives. His contributions in starting the Cancer Nanotechnology program cannot be overstated. Amazingly despite being a dedicated 'family man' Mauro finds time to mentor and inspire many young students. I'm sure that he will be making lasting contributions for many more years to come. Happy birthday Mauro!

Ali Khademhosseini

Abstract

Three-dimensional (3D) bioprinting is an emerging biofabrication technology, driving many innovations and opening new avenues in regenerative therapeutics. The aim of 3D bioprinting is to fabricate grafts *in vitro,* which can then be implanted *in vivo*. However, the tissue culture ex vivo carries safety risks and thereby complicated manufacturing equipment and practice are required for tissues to be implanted in the humans. The implantation of printed tissues also adds complexities due to the difficulty in maintaining the structural integrity of fabricated constructs. To tackle this challenge, the concept of *in situ* 3D bioprinting has been suggested in which tissues are directly printed at the site of injury or defect. Such approach could be combined with cells freshly isolated from patients to produce custom-made grafts that resemble target tissue and fit precisely to target defects. Moreover, the natural cellular microenvironment in the body can be harnessed for tissue maturation resulting in the tissue regeneration and repair. Here, we discuss literature reports on *in situ* 3D printing and we describe future directions and challenges for *in situ* 3D bioprinting. We expect that this novel technology would find great attention in different biomedical fields in near future.

Keywords 3D bioprinting · Biofabrication · Bioinks · *In situ* 3D printing · Regeneration

1 Introduction

Tissue defects resulting from disease, trauma, or surgical resection require treatment to restore tissue structure and function. There are certain challenges related to reconstructing complex tissue defects, such as those affecting the skeleton or craniomaxillofacial area and those resulting after surgical debridement (Ashammakhi and Kaarela 2018; Datta et al. 2017). Such defects are difficult to treat using conventional and standard size implants. In particular, they need to be treated by using modified or trimmed implants or constructs. Three-dimensional (3D) printing offers the possibility to provide constructs that can be predesigned to precisely fit defect size and shape. These constructs are usually designed and engineered *ex vivo* and subsequently implanted into the body (Tellisi et al. 2018). However, this process has several

✉ Nureddin Ashammakhi
n.ashammakhi@ucla.edu

✉ Ali Khademhosseini
khademh@ucla.edu

1 Center for Minimally Invasive Therapeutics (C-MIT), University of California - Los Angeles, California, Los Angeles, USA

2 California NanoSystems Institute (CNSI), University of California - Los Angeles, 570 Westwood Plaza, Building 114, Room 4528, Los Angeles, CA 90095, USA

3 Department of Radiological Sciences, University of California - Los Angeles, California, Los Angeles, USA

4 Department of Bioengineering, University of California - Los Angeles, California, Los Angeles, USA

5 Division of Plastic Surgery, Department of Surgery, Oulu University, Oulu, Finland

6 School of Technology and Innovations, University of Vaasa, Vaasa, Finland

7 Department of Trauma and Orthopaedics, University of Leeds, Leeds, UK

8 Chapel Allerton Hospital, Leeds Teaching Hospitals, Leeds, UK

9 Department of Chemical and Biomolecular Engineering, University of California - Los Angeles, California, Los Angeles, USA

10 Center of Nanotechnology, Department of Physics, King Abdulaziz University, Jeddah, Saudi Arabia

11 Department of Bioindustrial Technologies, College of Animal Bioscience and Technology, Konkuk University, Seoul, Republic of Korea

challenges related to logistics (3D printing in a location far from the operating room), sterility, and necessity to modify, trim, adjust, or combine different pieces to fit the shape and size of treated defect. In addition, the size and shape changes and thus ready-printed construct may not always fit following the debridement of defect. Alternatively, 3D printed implant may be installed on a subsequent second operation and this 2-stage procedure carries higher risks and problems.

Ideally, a customized implant should be produced and implanted during the same setting. However, this approach has been limited by various technical, procedural, and regulatory issues, thus far. This ideal solution would employ a commonly used 3D printing technique in the operating room and 3D print constructs *in situ*. This concept may involve the use of either handheld printers (Duchi et al. 2017; Di Bella et al. 2018; Hakimi et al. 2018) (Fig. 1) or robotic arms carrying printer nozzles, that are controlled by computers and scanners, which constantly measure the exact size of the defect (Wang et al. 2015a) (Fig. 2). The integration of developments made in robotics and computer-assisted interventions will also enable achieving more precision in printing procedure. These advances will enable developing more innovative solutions for *in situ* 3D printing to be employed in operating rooms. When fully deployed, the technology of *in situ* 3D printing will result in more accurate reconstruction of tissue defects and result in faster and more efficient healing of tissue defects.

In this work, we discuss technologies and materials used for *in situ* 3D (bio)printing. Applications of this novel technology to fabricate different tissues are described. We then highlight future directions and challenges toward wide applications of *in situ* 3D bioprinting in tissue regeneration.

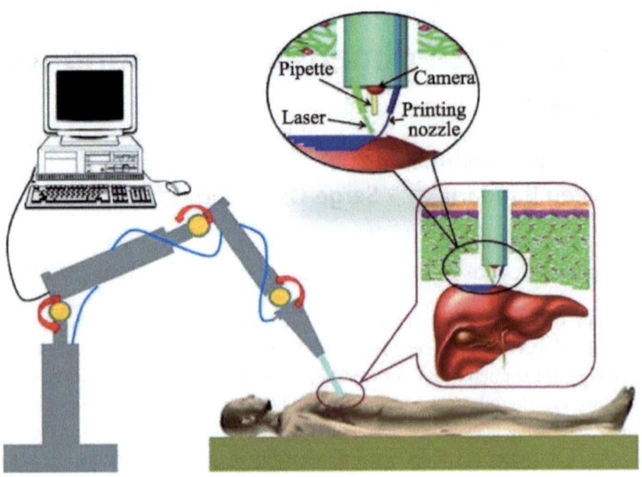

Fig. 2 Robot-assisted *in situ* 3D bioprinting. Schematic illustration showing the concept of robot-assisted *in vivo* 3D bioprinting in the frame of minimally invasive surgery. Reproduced with permission from Wang et al. (Wang et al. 2015b)

2 Technologies for *in situ* 3D (bio)printing

Although *in situ* 3D bioprinting is a rather novel approach, some facile and portable printing devices adapted for *in situ* applications have recently been reported (Di Bella et al. 2018; Keriquel et al. 2010; Sofokleous et al. 2013; Keriquel et al. 2017; Binder et al. 2010) (Di Bella et al. 2018; Keriquel et al. 2010; Sofokleous et al. 2013; Keriquel et al. 2017; Binder et al. 2010). For example, Sofokleous et al. (2013) developed a motorized and handheld spray gun equipped with electrohydrodynamic multi-needles capable of producing polymers as solid particles and fibers. The polymeric particles

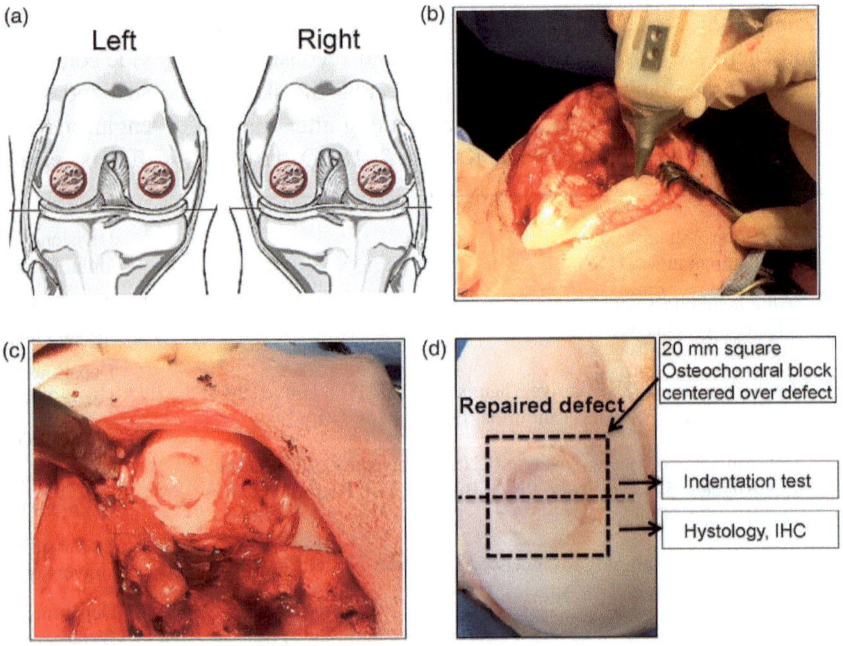

Fig. 1 Use of hand-held 3D bioprinting device for cartilage tissue regeneration. **a** A full-thickness cartilage defect was made in the medial and lateral femoral condyles of both stifle joints in sheep. **b** Images of intra-operative in situ 3D printing using a hand-held device. **c** Defect filled with HA-GelMA and MSCs and coated with fibrin glue spray. The defect was fully filled exhibiting curvature of the femoral condyle. **d** Macroscopic image of the retrieved treated cartilage defect. Reproduced with permission from Di Bella et al. (Di Bella et al. 2018)

and fibers were produced in sub-micrometer to micrometer range, which is physiologically relevant to mimic the extracellular matrix (ECM) structure. The use of multi-needles would enable the device to extrude multiple bioinks in a continuous manner for rapid fabrication of complex and heterogeneous tissue constructs (Liu et al. 2017; Miri et al. 2018). In another work, Cohen et al. (2010) used a robocasting-based additive manufacturing system to print alginate hydrogels onto cadaver calf femoral chondral and osteochondral defects. The ionic crosslinking of the gels is compatible with *in situ* repair. Cells can also be cultured in the hydrogels for potential applications in trauma surgery and facial reconstruction. However, other naturally-derived polymers (e.g., gelatin (Taira et al. 2018; Ahadian et al. 2015), gelatin methacryloyl (GelMA) (Colosi et al. 2016), and autologous platelet-rich plasma (Faramarzi et al. 2018)) should be added to the alginate to enhance its cell-responsiveness. Keriquel et al. proposed a laser-based fabrication technology adapted for *in situ* bone tissue regeneration with high precision and resolution using mesenchymal stromal cells (MSCs) (Fig. 3) (Keriquel et al. 2017). However, the technology was suitable to print on flat surfaces only and thereby new technological developments are required to obtain complex structures using such laser-based bioprinter.

3 *In situ* 3D (bio)printing in tissue regeneration

There are several reports on using *in situ* 3D (bio)printing to regenerate different tissues in the body (Table 1). An early report showed that nano-hydroxyapatite (n-HAp) could directly fill calvarial defects in mice in a minimally invasive manner (Keriquel et al. 2010). Later on, the same group

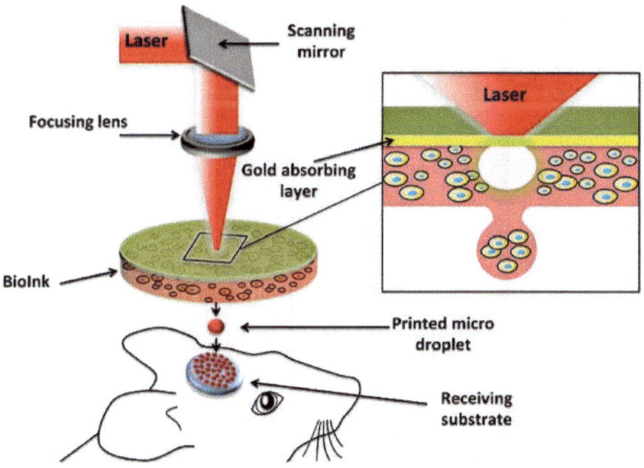

Fig. 3 *In situ* laser-assisted 3D bioprinting. Proof of concept for *in situ* printing using laser-assisted bioprinting for depositing of HAp into a critical-size mouse calvarial defect. Different geometric cell patterns were 3D bioprinted for guiding the regeneration of bone tissue. Reproduced with permission from Keriquel et al. (Keriquel et al. 2017)

printed MSCs either on the peripheral or central regions of discs made of n-HAp and collagen type I and applied to calvarial bone defects in mice (Keriquel et al. 2017). The deposition site of cell-laden hydrogel significantly affected bone tissue regeneration. The authors highlighted that the cells are a vital element of the 3D bioprinting process because poor bone regeneration was observed in the control groups that osteoprogenitor cells were not included in the bioink. In another work, Li et al. obtained the precise size of defect areas of bone and cartilage by the aid of high-resolution 3D scanning and then used *in situ* 3D bioprinting for defect regeneration *ex vivo* (Li et al. 2017). They used alginate and alginate-poly(-ethylene glycol) (PEG) hydrogels. The alginate-PEG hydrogels were polymerized using UV light, which may limit their clinical applications. In another study, Cohen et al. used a demineralized bone matrix and an alginate hydrogel for *in situ* repair of bone and cartilage defects, respectively (Cohen et al. 2010). Di Bella et al. treated critical-sized osteochondral defects in sheep (Di Bella et al. 2018). The bioink was composed of MSCs in a scaffold of GelMA and hyaluronic acid (HA) methacrylate hydrogels. Interestingly, *in situ* 3D bioprinting did not cause any intra- or postoperative complications. The abovementioned proof-of-concept works on *in situ* bioprinting of bone and osteochondral tissues provide evidence that geographically complex pathologies could be treated with further advancement of this technology instead of salvage procedures such as joint replacement or joint fusion procedures.

Binder et al. used a model of full-thickness skin defect and bioprinted constructs containing keratinocytes and fibroblasts (Binder et al. 2010). They found that the produced skin was similar to the normal skin, and complete wound closure was noted. Next, Skardal et al. bioprinted amniotic fluid-derived stem cells and MSCs suspended in fibrin/collagen gel onto a full-thickness skin wound (2 × 2 cm). They reported a higher level of re-epithelialization and increased microvessel density and capillary diameters in wounds treated with the amniotic fluid-derived stem cells as compared to those treated with MSCs. The bioprinted cells did not permanently integrate with the tissue. Therefore, secreted trophic factors may be responsible for a favorable effect rather than direct cell-cell interactions. In a recent study, the same group bioprinted amniotic fluid-derived stem cells embedded in photocrosslinkable heparin conjugated HA onto full-thickness skin wounds in a murine model. They reported re-epithelialization, vascularization, ECM production, and wound closure in treated wounds (Skardal et al. 2017). In another study, Albanna et al. (2012) explored the healing potential of fibroblasts and keratinocytes suspended in fibrogen/collagen and compared the overall potential of using autologous versus allogeneic cells. They used 10 × 10 cm skin defects in a porcine model. In this preclinical study, the authors demonstrated that 3D bioprinters can provide precise and immediate cover to a large skin defect. In addition, supplementing the skin graft with cells involved in

Table 1 Summary of *in situ* 3D (bio)printing works for tissue regeneration applications

Animal model	Printing technology	(Bio)ink	Target tissue	Outcomes/comments	Ref.
Mouse ($n = 36$). Full-thickness excisional skin defect	Inkjet bioprinter	Human keratinocytes and fibroblasts	Skin	(1) Acceptable survival rate of cells after printing (2) Fast healing rate of the treated defects	(Binder et al. 2010)
Mouse ($n = 30$). Critical size calvarial defect	Laser-adapted printer	n-HAp	Bone	(1) *In vivo* bioprinting is possible (2) No effect on the animals' brain (3) Bone formation only occurred in some defects	(Keriquel et al. 2010)
Pig ($n = 6$). Skin defect	Inkjet printer	Autologous vs allogeneic fibroblasts and keratinocytes suspended in fibrinogen/collagen	Skin	(1) *In situ* skin bioprinting is a viable option for treatment of large skin defects (2) The utilization of autologous cells outperformed in healing potential compared to allogeneic cells	(Albanna et al. 2012)
Mouse ($n = 5$). Full-thickness skin wound	Inkjet bioprinter	Amniotic fluid cells and bone marrow MSCs suspended in fibrin/collagen gel	Skin	(1) The graft improved re-epithelialization with increased micro-vessel density and capillary diameters (2) Secreted trophic factors were responsible for the favorable effect, rather than direct cell-cell interactions	(Skardal et al. 2012)
Sheep ($n = 6$). Full-thickness osteochondral defect	Hand-held bioprinter (Biopen)	MSCs in GelMA and HA-GelMA	Cartilage	Bioprinting with the Biopen is feasible and resulted in early cartilage regeneration	(Di Bella et al. 2018)
Mouse ($n = 64$). Critical size calvaria defect	Laser-assisted bioprinter	MSCs on n-HAp and collagen	Bone	(1) The technology could print complex structures, which favored bone regeneration (2) Cell arrangement had a significant impact on bone regeneration	(Keriquel et al. 2017)
Rats. Circular skin defect	Droplet-based bioprinter	Rat primary dermal fibroblasts in collagen/fibrinogen	Skin	(1) Healing of defects was accelerated (2) Regeneration of functional skin with anatomically correct features was not achieved	(Gudapati et al. 2017)

the skin healing process could enhance the wound healing. These favorable results were demonstrated by 90% decrease in wound size, wound re-epithelization, and 20% reduction of the wound contractures compared to the controls. The authors reported that the use of autologous cells outperforms the use of allogeneic cells in terms of their healing potential. Despite these encouraging results from the aforementioned studies, the wounds were covered by scar tissue rather than normal skin. Nevertheless, they represent a promising option for developing treatment of large skin defects. Compared to other cell-based skin regenerative therapies, such as spraying keratinocytes (Sood et al. 2015) or keratinocyte sheet (Kirby et al. 2018), *in situ* printing shows the higher potential in addressing deeper wounds (i.e., deep into basal layer) by combining multi-material printing system. In addition, *in situ* printing can perform precise cell deposition for more efficient use of kerotinocytes since it is difficult to culture for large-scale production (Anderson et al. 2018).

4 Challenges and future directions

Three-dimensional bioprinting technologies have emerged as a method to create complex constructs for tissue regeneration and repair (Ahadian et al. 2018). It is now feasible to print large-scale and clinically relevant tissue constructs in a relatively short time (Gungor-Ozkerim et al. 2018). Other micro-scale technologies (e.g., microfluidic (Zhang et al. 2017) and fiber (Zhang et al. 2016) technologies) can also be combined with 3D bioprinting to achieve automated and controllable multimaterial bioprinting to recapitulate the heterogeneous structure and composition of native tissues. However, printed tissues still need to be matured using bioreactors *in vitro* and may require modifications in size and shape prior to their implantation *in vivo*. *In situ* 3D bioprinting has been developed to tackle these problems by directly printing tissues *in vivo*, where the natural environment in the body is used for tissue survival and functionality.

Current evidence on using *in situ* 3D (bio)printing for tissue regeneration is still limited. However, there is enough evidence to show that this procedure is possible, and has huge potential for use in tissue regeneration. It is expected that maturation, functionality, and integration of bioprinted constructs *in situ* are improved compared to engineered tissues *in vitro* because the microenvironment in the body is richer in signaling molecules than *in vitro* microenvironment. Furthermore, *in situ* 3D bioprinting procedures, within the framework of minimally invasive techniques, should lead to better patient outcomes, less pain, shorter time-to-recovery, reduced nosocomial infection, and reduced hospitalization (Ashammakhi et al. 2019). However, several issues still remain and should be tackled. For example, *in situ* 3D bioprinting requires fast gelation process, sufficient stability of printed construct *in vivo*, and friendly bioprinting environment (e.g., drying step by airflow). In particular, large size constructs need early integration with the host tissue in order to survive and function *in vivo* (Khademhosseini and Langer 2016). Hence, continuous development and optimization of bioinks and bioprinters, adapted for *in situ* printing are needed.

Development of portable bioprinters and mobile units may be a viable option for use in emergency situations and remote locations. Moreover, *in situ* bioprinting can be used primarily by surgeons for regenerative and reconstructive applications. Therefore, integrating *in situ* bioprinting with advanced robotic systems that utilize user-friendly interfaces is necessary for translation into common surgical procedures. In addition, advanced real time imaging with computational modeling will be useful for fast morphological assessment of defects and fabrication of tissues *in vivo*. In the future, various imaging modalities would be combined with advanced robotics to provide input for *in situ* printing interfaces. The use of stimuli responsive bioinks in *in situ* 3D bioprinting would also enable us to produce the so-called four-dimensional constructs (Li et al. 2016; Ashammakhi et al. 2018), which would be capable of recapitulating the natural morphological and structural changes in the tissues. All these developments will have impact on the future of *in situ* 3D bioprinting and its translation to clinic.

5 Conclusions

Thus far, research into the field of *in situ* 3D bioprinting has shown feasibility and usefulness. This novel research field provides great potential for tissue engineering and regenerative medicine applications where preoperative planning of construct size and shape is difficult or impossible to predict. Advances in robotic surgery, fused imaging, and computer-assisted medical interventions should also be integrated to develop future clinical *in situ* 3D bioprinting processes, which can be translated to products for a variety of surgical applications.

Acknowledgements The authors acknowledge funding from the National Institutes of Health (EB021857, AR066193, AR057837, CA214411, HL137193, EB024403, EB023052, EB022403, and EB021857), and Air Force Office of Sponsored Research under award #FA9550-15-1-0273.

Compliance with ethical standards

Conflict of interest The authors have no competing interests.

References

S. Ahadian, R.B. Sadeghian, S. Salehi, S. Ostrovidov, H. Bae, M. Ramalingam, A. Khademhosseini, Bioconjugated hydrogels for tissue engineering and regenerative medicine. Bioconjugate Chem. **26** 1984–2001 (2015)

S. Ahadian, R. Civitarese, D. Bannerman, M.H. Mohammadi, R. Lu, E. Wang, L. Davenport-Huyer, B. Lai, B. Zhang, Y. Zhao, S. Mandla, A. Korolj, M. Radisic, Organ-on-a-chip platforms: a convergence of advanced materials, cells, and microscale technologies. Adv. Healthc. Mater. **7**, 1700506 (2018)

M. Albanna, S. Murphy, W. Zhao, I. El-Amin, J. Tan, D. Dice, H.-W. Kang, J. Jackson, A. Atala, J. Yoo, In situ bioprinting of skin for reconstruction. J. Am. Coll. Surg. **187**, e8 (2012)

E.D. Anderson, I. Sastalla, N.J. Earland, M. Mahnaz, I.N. Moore, F. Otaizo-Carrasquero, T.G. Myers, C.A. Myles, S.K. Datta, I.A. Myles, Prolonging culture of primary human keratinocytes isolated from suction blisters with the Rho kinase inhibitor Y-27632. Plos One **13**, e0198862 (2018)

N. Ashammakhi, O. Kaarela, Three-dimensional bioprinting can help bone. J. Craniofac. Surg. **29**, 9–11 (2018)

N. Ashammakhi, S. Ahadian, F. Zengjie, K. Suthiwanich, F. Lorestani, G. Orive, S. Ostrovidov, A. Khademhosseini, Advances and future perspectives in 4D bioprinting. Biotechnol. J. **13**, 1800148 (2018)

N. Ashammakhi, S. Ahadian, M.A. Darabi, M. El Tahchi, J. Lee, K. Suthiwanich, A. Sheikhi, M.R. Dokmeci, R. Oklu, A. Khademhosseini, Minimally invasive and regenerative therapeutics. Adv. Mater. **31**, 1804041 (2019)

K.W. Binder, W. Zhao, T. Aboushwareb, D. Dice, A. Atala, J. Yoo, In situ bioprinting of the skin for burns. J. Am. Coll. Surg. **211**, S76 (2010)

D.L. Cohen, J.I. Lipton, L.J. Bonassar, H.J.B. Lipson, Additive manufacturing for in situ repair of osteochondral defects. Biofabrication **2**, 035004 (2010)

C. Colosi, S.R. Shin, V. Manoharan, S. Massa, M. Costantini, A. Barbetta, M.R. Dokmeci, M. Dentini, A. Khademhosseini, Microfluidic bioprinting of heterogeneous 3D tissue constructs using low-viscosity bioink. Adv. Mater. **28** 677–684 (2016)

P. Datta, V. Ozbolat, B. Ayan, A. Dhawan, I.T. Ozbolat, Bone tissue bioprinting for craniofacial reconstruction. Biotechnol. Bioeng. **114**, 2424–2431 (2017)

C. Di Bella, S. Duchi, C.D. O'Connell, R. Blanchard, C. Augustine, Z. Yue, F. Thompson, C. Richards, S. Beirne, C. Onofrillo, S.H. Bauquier, S.D. Ryan, P. Pivonka, G.G. Wallace, P.F. Choong, In situ handheld three-dimensional bioprinting for cartilage regeneration. J. Tissue. Eng. Regen. Med. **12**, 611–621 (2018)

S. Duchi, C. Onofrillo, C. O'Connell, R. Blanchard, C. Augustine, A. Quigley, R. Kapsa, P. Pivonka, G. Wallace, C. Di Bella, Handheld co-axial bioprinting: Application to in situ surgical cartilage repair. Sci. Rep. **7**, 5837 (2017)

N. Faramarzi, I.K. Yazdi, M. Nabavinia, A. Gemma, A. Fanelli, A. Caizzone, L.M. Ptaszek, I. Sinha, A. Khademhosseini, J.N. Ruskin, A. Tamayol, Patient-specific bioinks for 3D bioprinting of tissue engineering scaffolds. Adv. Healthc. Mater. **7**, 1701347 (2018)

H. Gudapati, M. Hospodiuk, V. Ozbolat, D. Ravnic, I. Ozbolat, In-situ droplet-based bioprinting for functional skin regeneration. In: New Jersy Symposium on Biomaterials Science, (2017)

P.S. Gungor-Ozkerim, I. Inci, Y.S. Zhang, A. Khademhosseini, M.R. Dokmeci, Bioinks for 3D bioprinting: an overview. Biomater. Sci. **6**, 915–946 (2018)

N. Hakimi, R. Cheng, L. Leng, M. Sotoudehfar, P.Q. Ba, N. Bakhtyar, S. Amini-Nik, M.G. Jeschke, A. Günther, Handheld skin printer: in situ formation of planar biomaterials and tissues. Lab. Chip. **18**, 1440–1451 (2018)

V. Keriquel, F. Guillemot, I. Arnault, B. Guillotin, S. Miraux, J. Amédée, J.-C. Fricain, S.J.B. Catros, *In vivo* bioprinting for computer-and robotic-assisted medical intervention: preliminary study in mice. Biofabrication **2**, 014101 (2010)

V. Keriquel, H. Oliveira, M. Rémy, S. Ziane, S. Delmond, B. Rousseau, S. Rey, S. Catros, J Amédée, F. Guillemot, J.C. Fricain, In situ printing of mesenchymal stromal cells, by laser-assisted bioprinting, for *in vivo* bone regeneration applications. Sci. Rep. **7**, 1778 (2017)

A. Khademhosseini, R. Langer, A decade of progress in tissue engineering. Nat. Protoc. **11**, 1775 (2016)

G.T. Kirby, A. Michelmore, L.E. Smith, J.D. Whittle, R.D. Short, Cell sheets in cell therapies. Cytotherapy **20**, 169–180 (2018)

Y.-C. Li, Y.S. Zhang, A. Akpek, S.R. Shin, A. Khademhosseini, 4D bioprinting: the next-generation technology for biofabrication enabled by stimuli-responsive materials. Biofabrication **9**, 012001 (2016)

Z. Li, C.C. Tasan, H. Springer, B. Gault, D. Raabe, Interstitial atoms enable joint twinning and transformation induced plasticity in strong and ductile high-entropy alloys. Sci. Rep. **7**, 40704 (2017)

W. Liu, Y.S. Zhang, M.A. Heinrich, F.D. Ferrari, H.L. Jang, S.M. Bakht, M.M. Alvarez, J. Yang, Y.C. Li, G.T. Santiago, A.K. Miri, K. Zhu, P. Khoshakhlagh, G. Prakash, H. Cheng, X. Guan, Z. Zhong, J. Ju, G.H. Zhu, S.R. Shin, M.R. Dokmeci, A. Khademhosseini, Rapid continuous multimaterial extrusion bioprinting Adv. Mater. **29**, 1604630 (2017)

A.K. Miri, D. Nieto, L. Iglesias, H. Goodarzi Hosseinabadi, S. Maharjan, G.U. Ruiz-Esparza, P. Khoshakhlagh, A. Manbachi, M.R. Dokmeci, S. Chen, S.R. Shin, Y.S. Zhang, A. Khademhosseini, Microfluidics-enabled multimaterial maskless stereolithographic bioprinting. Adv. Mater. 30, 1800242 (2018)

A. Skardal, D. Mack, E. Kapetanovic, A. Atala, J.D. Jackson, J. Yoo, S. Soker, Bioprinted amniotic fluid-derived stem cells accelerate healing of large skin wounds. Stem. Cells Transl. Med. **1**, 792–802 (2012)

A. Skardal, S.V. Murphy, K. Crowell, D. Mack, A. Atala, S. Soker, A tunable hydrogel system for long-term release of cell-secreted cytokines and bioprinted in situ wound cell delivery. J. Biomed. Mater. Res. B Appl. Biomater. **105**, 1986–2000 (2017)

P. Sofokleous, E. Stride, W. Bonfield, M.J.M.S. Edirisinghe, Design, construction and performance of a portable handheld electrohydrodynamic multi-needle spray gun for biomedical applications. Mater. Sci. Eng. C Mater. Biol. Appl. **33**, 213–223 (2013)

R. Sood, D.E. Roggy, M.J. Zieger, M. Nazim, B.C. Hartman, J.T. Gibbs A comparative study of spray keratinocytes and autologous meshed split-thickness skin graft in the treatment of acute burn injuries, Wounds **27** 31–40 (2015)

N. Taira, K. Ino, J. Robert, H. Shiku, Electrochemical printing of calcium alginate/gelatin hydrogel. Electrochem. Acta **281**, 429–436 (2018)

N. Tellisi, N.A. Ashammakhi, F. Billi, O. Kaarela, Three dimensional printed bone implants in the clinic. J. Craniofac. Surg. **29**, 2363–2367 (2018)

M. Wang, J. He, Y. Liu, M. Li, D. Li, Z. Jin, The trend towards *in vivo* bioprinting. Int. J. Bioprinting **1**, 15–26 (2015a)

S. Wang, J.M. Lee, W.Y. Yeong, Smart hydrogels for 3D bioprinting. Int. J. Bioprinting **1**, 3–14 (2015b)

Y.S. Zhang, A. Arneri, S. Bersini, S.-R. Shin, K. Zhu, Z. Goli-Malekabadi, J. Aleman, C. Colosi, F. Busignani, V. Dell'Erba, C. Bishop, T. Shupe, D. Demarchi, M. Moretti, M. Rasponi, M.R. Dokmeci, A. Atala, A. Khademhosseini, Bioprinting 3D microfibrous scaffolds for engineering endothelialized myocardium and heart-on-a-chip. Biomaterials **110**, 45–59 (2016)

Y.S. Zhang, K. Yue, J. Aleman, K. Mollazadeh-Moghaddam, S.M. Bakht, J. Yang, W. Jia, V. Dell'Erba, P. Assawes, S.R. Shin, M.R. Dokmeci, R. Oklu, A. Khademhosseini, 3D bioprinting for tissue and organ fabrication. Ann. Biomed. Eng. **45**, 148–163 (2017)

Biomedical Microdevices (2019) 21: 34
https://doi.org/10.1007/s10544-019-0373-1

Chitosan-based composite bilayer scaffold as an *in vitro* osteochondral defect regeneration model

Ariane E. Erickson[1] · Jialu Sun[1] · Sheeny K. Lan Levengood[1] · Shawn Swanson[1] · Fei-Chien Chang[1] · Ching T. Tsao[1] · Miqin Zhang[1]

Published online: 25 March 2019

Abstract

Prolonged osteochondral tissue damage can result in osteoarthritis and decreased quality of life. Multiphasic scaffolds, where different layers model different microenvironments, are a promising treatment approach, yet stable joining between layers during fabrication remains challenging. Here, a bilayer scaffold for osteochondral tissue regeneration was fabricated using thermally-induced phase separation (TIPS). Two distinct polymer solutions were layered before TIPS, and the resulting porous, bilayer scaffold was characterized by seamless interfacial integration and a mechanical stiffness gradient reflecting the native osteochondral microenvironment. Chitosan is a critical component of both scaffold layers to facilitate cell attachment and the formation of polyelectrolyte complexes with other biologically relevant natural polymers. The articular cartilage region was optimized for hyaluronic acid content and stiffness, while the subchondral bone region was defined by higher stiffness and osteoconductive hydroxyapatite content. Following co-culture with chondrocyte-like (SW-1353 or mesenchymal stem cells) and osteoblast-like cells (MG63), cell proliferation and migration to the interface along with increased gene expression associated with relevant markers of osteogenesis and chondrogenesis indicates the potential of this bilayer scaffold for osteochondral tissue regeneration.

Keywords Chitosan · Hyaluronic acid · Alginate · Hydroxyapatite · Osteochondral defect · Bilayer scaffold

1 Introduction

Osteochondral tissue is composed of an articular cartilage region and a subchondral bone region with a calcified cartilage interface linking the two distinct microenvironments. Osteochondral defects arise when damage or disease affect both regions (Hunziker 2002) and surgical treatment options vary from bone marrow stimulation to chondrocyte and/or graft transplantation (Bekkers et al. 2012; Cole et al. 2009). However, these treatments are associated with inherent complications (Bexkens et al. 2017; Hunziker 2002; Lopez-Ruiz et al. 2016; Smith et al. 2005). For example, therapies only targeting the cartilage zone without consideration of the interface or subchondral bone result in increased defect size, a weak tissue interface, and structural degradation of the underlying bone (Madry et al. 2010; Schinhan et al. 2012). Regeneration of osteochondral defects is nontrivial and requires synchronized vascularization and calcification in the bone region concurrently with avascular collagen deposition in the cartilage region (Hunziker 2002).

Tissue engineering (TE) scaffolds provide a three-dimensional (3D), biodegradable template for cell attachment/differentiation. TE scaffolds can be utilized for treatment of osteochondral defects by providing synchronized, yet segregated support of two or more distinct cell populations with differing biochemical and biomechanical requirements. In monophasic scaffolds, extent of osteochondral tissue regeneration is low relative to multiphasic scaffolds indicating that monophasic materials may be inadequate mimics of the native microenvironment and therefore unable to simultaneously support chondrogenesis and osteogenesis (Seo et al. 2014; Singh et al. 2018). Thus, various bilayer and multilayer (3+ layers) 3D, porous scaffolds have been fabricated to address the unique characteristics of the osteochondral microenvironment. Porous scaffolds that mimic osteochondral extracellular matrix (ECM)

Electronic supplementary material The online version of this article (https://doi.org/10.1007/s10544-019-0373-1) contains supplementary material, which is available to authorized users.

✉ Miqin Zhang
mzhang@u.washington.edu

[1] Department of Materials Science & Engineering, University of Washington, Seattle, WA 98195, USA

are fabricated with a wide range of physical properties including tailored pore size (Di Luca et al. 2016; Nie et al. 2016; Oh et al. 2007), pore shape, (Aydin 2011) mineral content (Kon et al. 2010; Yusong et al. 2013), or biochemical gradients (Levingstone et al. 2014; Oh et al. 2011). These studies demonstrate the efficacy of multilayer or multiphasic scaffolds *in vitro* (Galperin et al. 2013; Levingstone et al. 2014) and *in vivo* (He et al. 2017; Levingstone et al. 2016). Although promising, limitations of multiphasic scaffolds include time-consuming, iterative fabrication or joining procedures, as well as inadequate or unstable interfacial regions between layers (Hunziker and Driesang 2003; Levingstone et al. 2014). These significant limitations indicate a need for further innovation in the design and optimization of such scaffolds for osteochondral tissue engineering.

Synthetic polymers are often utilized in bone tissue engineering due to their superior mechanical integrity relative to natural polymers. However, synthetic polymers have slow degradation rates, harmful degradation products, and lack cell attachment and differentiation cues (Schaefer et al. 2002). Natural polymers are more biocompatible, but their mechanical properties are suboptimal for many tissue engineering applications. Polyelectrolyte complexes (PECs), such as chitosan-alginate (Florczyk et al. 2011; Li and Zhang 2005) and chitosan-hyaluronic acid, (Florczyk et al. 2013), formed via ionic interactions between natural cationic and anionic polymers, possess enhanced integrity and stiffness. Further, the integration of bioactive cues such as hyaluronic acid (HA), a glycosaminoglycan (GAG) found in articular cartilage, improves chondrocyte attachment and chondrogenic differentiation (Galperin et al. 2013; Kim et al. 2003; Yoo et al. 2005). Bone scaffolds frequently contain osteoconductive materials, such as tricalcium phosphate or hydroxyapatite (HAp), to stimulate osteoblastic activity, increase scaffold stiffness, and induce influence phenotypic expression in osteoblast-like cells (Galperin et al. 2013). The incorporation of bioactive cues specific to the distinct microenvironments present in native osteochondral tissue can promote tissue regeneration (Galperin et al. 2013; Kim et al. 2003; Shu et al. 2003; Yoo et al. 2005).

In this study, we present a bilayer scaffold with a stable, calcified transition zone fabricated via a thermally-induced phase separation (TIPS) fabrication process. The scaffold has two regions. The first region, a chitosan-hyaluronic acid cartilage layer, is defined by relatively lower stiffness and the presence of hyaluronic acid (HA). The second region, a chitosan-alginate bone layer, was optimized for higher stiffness and osteoconductive hydroxyapatite nanorod (6% CA + HAp) content. Chitosan was selected as a major component of the bilayer scaffold as it supports attachment and proliferation of both osteoblasts (Levengood and Zhang 2014; Li et al. 2005) and chondrocytes (Li and Zhang 2005). This report describes how each layer of the scaffold, or each base scaffold, was optimized to reflect structural, physical, and biological properties unique to the osteochondral microenvironment. The scaffold design process required a multi-step optimization involving material characterization and *in vitro* cell-based evaluation. By combining two distinct scaffolds optimized for cartilage and bone, respectively, using TIPS fabrication processes, a region between the two layers was established, resulting in a gradient transition zone with increasing stiffness and calcium content resembling that of native osteochondral tissue. The biological performance of the optimized scaffold was characterized to investigate scaffold potential for osteochondral tissue engineering applications.

2 Experimental

2.1 Materials

Unless otherwise stated, all chemicals were purchased from Sigma Aldrich and used as received. Scaffolds were synthesized using chitosan (from shrimp shells, practical grade, MW = 190–375 kDa, $\geq$ 75% deacetylated), hyaluronic acid sodium salt (from *Streptococcus equi*, MW = 1.5–1.8 × 10^6 Da) and alginic acid sodium salt (from brown algae, medium viscosity). All tissue culture reagents (antibiotic-antimycotic (AA), Dulbecco's phosphate buffered saline (D-PBS), Dulbecco's modified Eagle medium (DMEM), TripLE, trypsin, and fetal bovine serum (FBS)) were purchased from Life Technologies. Human chondrocyte-like (SW-1353) and human osteoblast-like (MG63) cells were used as received (American Type Culture Collection, Manassas, VA). Rat bone marrow derived mesenchymal stem cells (MSC) were isolated from Sprague-Dawley rat bone marrow aspirate as previously described (Florczyk et al. 2012) and cultured in fully supplemented DMEM. Cells were maintained according to manufacturer's instructions in fully supplemented DMEM with 10% FBS and 1% AA at 37 °C and 5% CO_2 in a fully humidified incubator.

2.2 Scaffold fabrication

2.2.1 Base scaffold

Six different scaffolds (Table 1) were fabricated for optimization of the base scaffolds for the cartilage or bone regions. Pure chitosan (C) (Jana et al. 2012), chitosan-alginate (CA) (Florczyk et al. 2016; Kievit et al. 2010), and chitosan-hyaluronic acid (CHA) (Florczyk et al. 2013; Wang et al. 2016) were prepared as previously reported. During scaffold fabrication, all homogenization steps were performed three times at 2000 RPM for 3 min using a Thinky mixer (ARM-300, Thinky USA, Laguna Hills, CA).

4 wt% pure chitosan scaffolds (4% C) were fabricated by first adding 4 g of chitosan to a 1 wt% acetic acid solution. For 6 wt% pure chitosan scaffolds (6% C), 6 g of chitosan were added to a 2 wt% acetic acid solution. The higher acetic acid

Table 1 Scaffold naming nomenclature and composition

Scaffold Type	Solution Fabrication (wt%)		Total Polymer (wt%)
	Step 1	Step 2	
4% C	4% C in 1% AA	–	4% C
6% C	6% C in 2% AA	–	6% C
4% CA	2% A in DI water	2% C in 0.5% AA	2% A + 2% C = 4% CA
6% CA	3% A in DI water	3% C in 1% AA	3% A + 3% C = 6% CA
4% CHA	4% C in 1% AA	1% HA in 1% AA	2% C + 0.5% HA = 2.5% CHA
6% CHA	6% C in 2% AA	1% HA in 1% AA	3% C + 0.5% HA = 3.5% CHA

C, chitosan; A, alginate; HA, hyaluronic acid; AA, acetic acid; CA, chitosan+alginate; CHA, chitosan+hyaluronic acid

concentration was used to ensure dissolution of chitosan at the higher polymer concentration. Both solutions were homogenized, aged overnight at room temperature to ensure complete dissolution, and cast into a 24-well plate at 3 mL per well.

To fabricate 4 wt% chitosan-alginate scaffolds (4% CA), 4 g of alginate were dissolved in 199 mL of DI water and stirred using a Thinky mixer. Upon alginate dissolution, chitosan (4 g) and acetic acid (1 g) were added and the solution was homogenized until dissolved. For 6% CA scaffolds, 6 g of alginate was dissolved in 198 mL of DI water followed by the addition of 6 g of chitosan and 2 g of acetic acid. Upon chitosan dissolution, the polymer solutions were mixed in a blender for 5 min, cooled in a cold-water bath to prevent polymer solution overheating, and mixed in a blender again for 5 min. The scaffold solution was cast into a 24 well-plate at 3 mL per well.

4 wt% chitosan – hyaluronic acid (4% CHA) scaffolds were fabricated by first preparing the chitosan and HA polymer solutions separately. Chitosan (4 wt%) and HA (1 wt%) were separately dissolved in 1 wt% acetic acid. For 6% CHA scaffolds, chitosan (6 wt%) was dissolved in 2 wt% acetic acid, to ensure dissolution of chitosan at the higher polymer concentration. The HA solution was created by dissolving 1 wt% HA in a 1 wt% acetic acid solution. Chitosan and HA polymer solutions were individually homogenized and aged overnight at room temperature. After dissolution, chitosan and HA polymer solutions were combined, homogenized, and mixed in a blender for one min. The resultant polymer solution was cast into a 24-well plate at 3 mL per well.

All 24-well plates containing the polymer solutions were centrifuged for 1–5 min at 2000 RPM to remove air bubbles, refrigerated for >12 h, frozen at −20 °C for >24 h, and lyophilized using a Labconco Freezone 6 freeze dryer.

Table 1 summarizes the scaffold naming conventions. Naming conventions are based on a ratio of polymer mass to total solution volume. Specifically, for pure chitosan (C) scaffolds, the total polymer content (wt%) was equivalent to the amount of solubilized chitosan. A 6% pure chitosan scaffold contains 6 wt% chitosan. For CA scaffolds, the total polymer content was equivalent to the amount of solubilized chitosan plus the amount of solubilized alginate. A 6% CA scaffold contained 3 wt% chitosan and 3 wt% alginate for a 6 wt% total polymer content.

HA has poor solubility and high viscosity at moderate concentrations. Therefore, 1 wt% HA was the highest concentration used. The association between the naming convention of CHA scaffolds and the total polymer content is different from the C and CA scaffolds. Importantly, the naming convention, which is based on the chitosan concentration prior to blending, remains consistent with scaffolds presented in literature (Florczyk et al. 2013). A 6% CHA scaffold contains a 6 wt% chitosan solution blended with a 1 wt% HA solution for a total of 3.5 wt% polymer.

2.2.2 Composite chitosan-alginate + hydroxyapatite scaffolds

Hydroxyapatite nanorods (HAp) were synthesized based on a previously reported procedure (Jin et al. 2015) with slight modifications to obtain processing conditions encompassing ambient pressure and body temperature. Briefly, 0.75 M sodium citrate was added dropwise to 0.1 M calcium nitrate. Sodium hydroxide was used to adjust pH to 9 and the solution was aged for 30 min at room temperature. A 0.06 M sodium phosphate monobasic solution was prepared, adjusted to pH 9, and then aged for 30 min. Under constant stirring, the solutions were mixed to achieve a precipitate with a Ca/P ratio of 1.67. The mixture was heated to 37 °C under constant stirring for 2 days while maintaining a pH of 9. The nanorod precipitate (HAp) was filtered, washed with DI water, and dried at 70 °C overnight. The length of the nanorods was 80.0 ± 25.4 nm and width was 6.7 ± 4.0 nm (n = 20 nanorods). Nanorod morphology was observed with transmission electron microscopy (TEM). HAp was added to a 6% CA polymer solution (synthesis outlined above) to obtain concentrations of 0.1 wt%, 0.25 wt%, 0.5 wt%, 1 wt%, 2 wt%, and 3 wt% HAp. The composite polymer solution was blended, homogenized, and cast in a 24-well plate at 3 mL per well. The plate was centrifuged at 1000 RPM for 5 min, refrigerated for >12 h, placed in a − 20 °C freezer for >24 h, and lyophilized using a Labconco freeze dryer.

2.2.3 Bilayer scaffolds

The complete bilayer scaffold fabrication process is summarized in Fig. 1a. During scaffold fabrication, all homogenization steps were conducted using a Thinky mixer three times at 2000 RPM for 3 min. Solutions for each layer of the bilayer scaffold were prepared separately. For the cartilage layer, a 4% CHA solution was produced by dissolving chitosan (4 wt%) and HA (1 wt%) separately in 1 wt% acetic acid. Chitosan and HA solutions were individually homogenized and aged overnight at room temperature. The polymer solutions were then homogenized and mixed in a blender for 1 min. Wells of a 24-

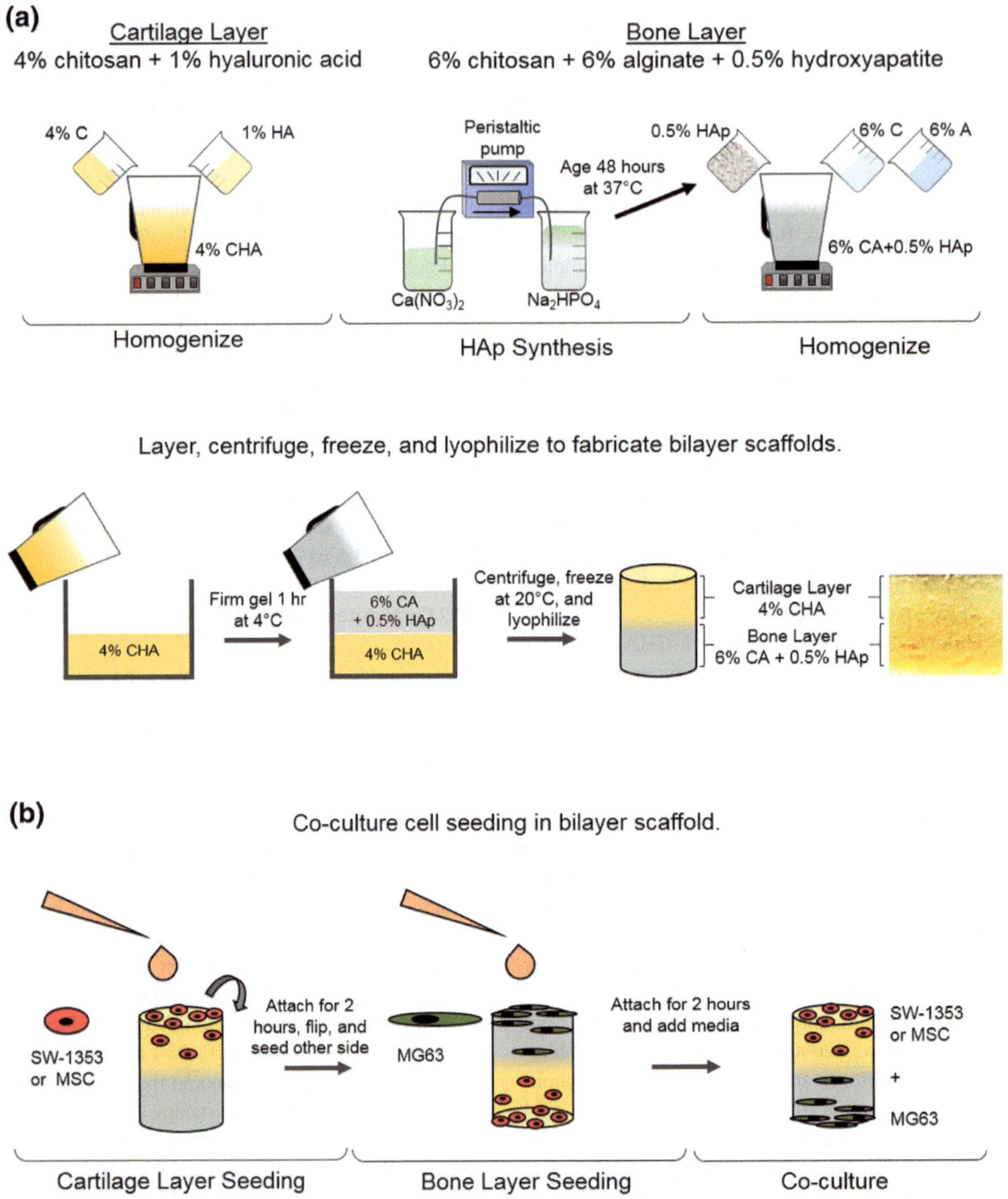

Fig. 1 Bilayer scaffold synthesis and co-culture cell seeding. **a** The three-step fabrication process includes: (1) synthesis of HAp nanorods using a chemical precipitation method, (2) individual preparation of the 4% CHA and 6% CA + HAp solutions, and (3) layering the two polymer solutions, centrifuging, freezing, and lyophilizing to achieve a gradient scaffold with an integrated interface region. **b** Co-culture seeding scheme in the bilayer scaffold where SW-1353 or MSC cells are first seeded in the cartilage layer (4% CHA), the scaffold is then flipped, and MG63 cells are seeded in the bone layer (6% CA + HAp)

well plate were filled with 1.5 mL of the 4% CHA polymer solution per well. The plate was then centrifuged and refrigerated for 1 h to yield a firmer and more viscous solution. For the bone layer, a 6% CA + 0.5% HAp material solution was prepared by dissolving 6 g of alginate in 198 mL of DI water and homogenizing. Upon dissolution, chitosan (6 g) and acetic acid (2 g) were added and the solution was homogenized, mixed in a blender for 5 min, and cooled in a cold-water bath. Hydroxyapatite nanorods (0.5 wt%) were then dispersed in the CA polymer solution. The 6% CA + HAp polymer solution was blended for 5 min and then mixed at 2000 RPM for 3 min with a Thinky mixer to eliminate bubbles. 1.5 mL of the CA polymer solution was gently cast in each well on top of the refrigerated 4% CHA solution in a 24-well plate. The layered polymer solutions were then centrifuged at 2000 RPM for 5 min. This centrifugation step eliminated air bubbles at the interface thereby integrating the two layers and preventing delamination. The scaffolds were then refrigerated for >12 h, frozen at −20 °C for >24 h and lyophilized using a Labconco freeze dryer.

2.2.4 Scaffold neutralization and sterilization

All base and HAp-integrated single-phase scaffolds had a diameter of ~15 mm and a thickness of 10 mm and were cut into 3 mm-thick disks, and neutralized and/or crosslinked as summarized in Table S1. The cylindrical bilayer scaffolds were 6 mm tall with an approximate diameter of 15 mm. Scaffolds were cut into four quarters, and simultaneously neutralized and crosslinked. Neutralization, crosslinking, and sterilization steps were conducted under vacuum. All crosslinked/neutralized scaffolds were rinsed 3 times with DI water, and soaked in D-PBS or DI water on a shaker at 60 RPM overnight. After ethanol sterilization, all scaffolds were rinsed three times and soaked overnight in sterile D-PBS to remove residual ethanol. Before cell seeding, the scaffolds were transferred to fully supplemented cell culture media for at least 2 h.

2.3 Microscopy

Scaffold pore morphology was visualized with a scanning electron microscope (SEM, JEOL JSM-7000F). Specifically, dry scaffolds were mounted on stubs using carbon tape and coated with Au/Pd for 60 s before imaging. HAp nanorod samples were prepared using a copper grid and imaged with a Transmission Electron Microscope (TEM, Philips EM430 TEM).

2.4 Pore size measurement

Individual pore areas were measured from SEM micrographs using ImageJ software. At least 14 pores were measured per base scaffold condition to determine mean pore area. For the bilayer scaffold, > 35 pores were measured in each layer. Pores were outlined using the freehand selection function in ImageJ and the area measured. Pore diameter was estimated using the mean pore area and assuming pores were a perfect circle.

2.5 Mechanical testing

Compression testing was conducted with a Shimadzu Universal tester (AGS-X) using a 100 N load cell at a strain rate of 0.4 mm/min. Scaffolds were cut into 2-mm thick disks and tested in dry and hydrated states. Hydrated samples were crosslinked and/or neutralized and soaked in D-PBS for >1 h prior to testing. Testing of bilayer scaffolds was conducted using full thickness (6-mm height) hydrated scaffolds. At least four samples were tested per condition.

2.6 Porosity and density measurements

Bulk density of each scaffold was determined by measuring the scaffold mass to the ten-thousandth of a gram and the scaffold diameter using micrometers ($n = 12$). Porosity of bilayer scaffolds was measured using a modified liquid displacement method.(Hsu et al. 1997; Zhang and Ma 1999) Specifically, the weight (W_i) and volume (V_i) of a dry scaffold was recorded. The scaffold was fully submerged in isopropanol for 20 min under vacuum. The isopropanol-impregnated scaffold was weighed (W_f). Scaffold porosity ($n = 5$) was calculated as a ratio of the volume of the solvent in the scaffold pores to the volume of the dry scaffold as shown in Eq. 1, where ρ is the density of isopropanol. Weight was measured in mg, and volume was measured in mm^3.

$$\text{Porosity} = \frac{(W_f - W_i)/\rho}{V_i} \times 100 \tag{1}$$

2.7 Cell seeding and cell proliferation in base scaffolds

Base scaffolds were seeded with 20,000 SW-1353, MSC, or MG63 cells suspended in 50 μL of fully supplemented DMEM. Scaffolds were incubated for 2 h to allow for cell infiltration and attachment before adding fully supplemented media. Media were changed every 2–3 days for the 14-day culture period. Cell proliferation was evaluated at 2, 7, and 10 days using 10% alamarBlue® reagent in fully supplemented media ($n = 4$ per condition). Scaffolds were incubated for 4 h in the alamarBlue® solution before an aliquot of the solution was transferred to an opaque black 96-well plate. Fluorescence intensity was measured using a SpectraMax M5 microplate reader (Molecular Devices, Union City, CA) at an excitation wavelength of 560 nm and emission wavelength of 590 nm. Fluorescence intensity was converted to cell number based on standard curves created for each scaffold type/cell line. Excess alamarBlue® was aspirated and scaffolds were washed with D-PBS before fresh media was added after each timepoint.

2.8 Cell seeding, cell proliferation, and live cell imaging in bilayer scaffolds

One cell type was seeded into each layer of the bilayer scaffolds to evaluate metabolic activity of co-cultures (SW-1353 + MG63 or MSC + MG63) (Fig. 1b). This culture condition is referenced throughout the text as a "co-culture." In the cartilage region, 200,000 cells (SW-1353 or MSC) suspended in 100 μL of fully supplemented DMEM were seeded into the 4% CHA layer and incubated for 2 h to allow cell infiltration and attachment. Scaffolds were then flipped and 200,000 MG63 cells in 100 μL of fully supplemented DMEM were seeded in the 6% CA + HAp layer. The scaffold was again incubated for 2 h before the addition of fully supplemented media. Media were changed every 2–3 days over the 14-day culture period.

Proliferative capacity (n = 4) was determined at 2, 4, 7, 10, and 14 days using the alamarBlue® assay. Scaffolds containing cells and blank scaffolds were incubated for 4 h in alamarBlue® solution. Then an aliquot of the solution was transferred to an opaque black 96-well plate. Fluorescence intensity was measured using a SpectraMax M5 microplate reader (Molecular Devices, Union City, CA) at an excitation wavelength of 560 nm and emission wavelength of 590 nm. The measured fluorescence intensity from scaffolds cultured with cells was normalized to the measured fluorescence intensity associated with blank scaffolds. The fluorescence intensity correlates with the metabolic activity of each culture condition. Excess alamarBlue® was aspirated and scaffolds were washed with D-PBS before fresh media was added after each timepoint.

To view cell distribution and infiltration into the bilayer scaffold, live cell images were acquired using a Nikon inverted fluorescent microscope with appropriate filters on a Nikon Ri1 Color Cooled Camera system (Nikon instruments, Melville, NY). To differentiate cells seeded in each layer of the bilayer scaffolds, the SW-1353 or MSC cells were stained with red Vybrant® DiI cell-labeling (Molecular Probes, Thermo Fisher) rendering the cells red-orange when excited at 549 nm (565 nm emission). Cells (SW-1353 or MSC) were incubated for 30 min in red Vybrant® DiI at 1 μg/mL in D-PBS prior to seeding in the cartilage layer. MG63 cells were seeded in the bone layer with no prior staining. Before imaging after 1 and 14 days of culture, the entire bilayer scaffold was incubated for 30 min in 5 μM calcein AM (Thermo Fisher) in D-PBS. Calcein AM is a cell-permeant dye that is converted from colorless to a green-fluorescent calcein after hydrolysis when excited at 495 nm (emission 515 nm), and therefore can be used for determination of live cells. Cells in the cartilage layer, SW-1353 or MSC, were double stained and emitted both red and green fluorescence, depending on the excitation wavelength, and in the bone layer, MG63 were identified as green with no red signal. After merging images using ImageJ and Adobe Photoshop, differences in excitation wavelength between DiI (red) and calcein AM (green) stained cells allowed cells in the cartilage layer (SW-1353 or MSC) to be visualized as red-orange and bone layer cells (MG63) to be visualized as green.

2.9 PCR

After two weeks of culture, cells were detached from the bilayer scaffolds using TripLE Express and ribonucleic acid (RNA) was extracted following the manufacturer's protocol using the Qiagen RNeasy kit (Qiagen, Valencia, CA). Following the manufacturer's instructions, iScript cDNA synthesis kit (Bio-Rad, Hercules, CA) was used for reverse transcription (RT) to prepare cDNA. DNA transcripts were probed using SoAdvanced Universal SYBR Green Supermix (Bio-Rad) with the primers listed in Table S2. Thermocycling was performed on 10 μL of solution containing 5 μL SYBR Supermix, 300 nM primers (Integrated DNA Technologies, Coralville, IA), and cDNA concentrated at 4 ng/μL. The thermocycling was performed using the BioRad CFX96 System at 95 °C for 3 min, 40 cycles at 95 °C for 15 s, and 60 °C for 1 min. All qRT-PCR data was analyzed with the CFX Manager software (Bio-Rad) where expression levels were normalized to a reference gene, GAPDH. Gene expression between conditions was normalized to MG63 expression on 2D tissue culture polystyrene (TCPS) (set to 1-fold) for comparison.

2.10 Statistical analysis

All data was analyzed to express mean ± standard deviation unless stated otherwise. The data were analyzed using one-sided or two-sided analysis of variance (ANOVA) with post-hoc Tukey multiple comparisons testing for statistical significance ($p < 0.05$) in GraphPad Prism 7 (Prism version 7.04, Graph Pad Software, San Diego, CA).

3 Results and discussion

3.1 Base scaffold microstructure and other material properties

Repair of osteochondral defects requires simultaneous support of two cell populations, chondrogenic and osteogenic cells, that reside within distinct microenvironments – articular cartilage and subchondral bone. Differing biochemical and biomechanical cues in each region makes scaffold design challenging, necessitating a multiphasic scaffold approach. To accurately mimic the layered osteochondral ECM, biochemical and biomechanical optimization of the cartilage and bone regions of a bilayer scaffold was necessary. Pure chitosan (C), chitosan-alginate (CA), and chitosan- hyaluronic acid (CHA) base scaffolds were each fabricated with two polymer concentrations (4 wt% and 6 wt%) per scaffold composition and screened as candidate scaffolds for the cartilage or bone layers of the target osteochondral bilayer scaffold. The polymer concentrations were selected based on previous studies showing 4 wt% CA scaffolds support osteogenic and chondrogenic cell culture (Li et al. 2005; Li and Zhang 2005). Scaffold pore structure and morphology, observed with SEM (Fig. 2a), displayed a defined, yet interconnected, open pore network for all scaffolds. Despite a similar overall polymer content, the 6% CA scaffolds had a significantly smaller mean pore cross-sectional area ($1.10 \pm 0.64 \times 10^4$ μm^2) than the 6% C scaffolds ($2.93 \pm 1.63 \times 10^4$ μm^2) likely due to the increased viscosity of the CA PEC ($p \leq 0.05$) (Fig. S1a). Scaffold density (Fig. 2b), which increased with polymer content, was different among all scaffold compositions, ranging from 0.04 g/cm^3 to 0.08 g/cm^3, and similar to values cited for other highly porous polymeric foams (Dorati et al. 2014).

Polymer content and density correlated with the compressive modulus or stiffness of the bulk, porous scaffold. The compressive modulus of dry and hydrated scaffolds increased as scaffold density increased (Fig. 2c, d). Overall, the CA scaffolds were the stiffest followed by the C scaffolds, and CHA scaffolds. The low compressive modulus of both dry and hydrated CHA scaffolds was due to low total polymer content relative to the other scaffold types.

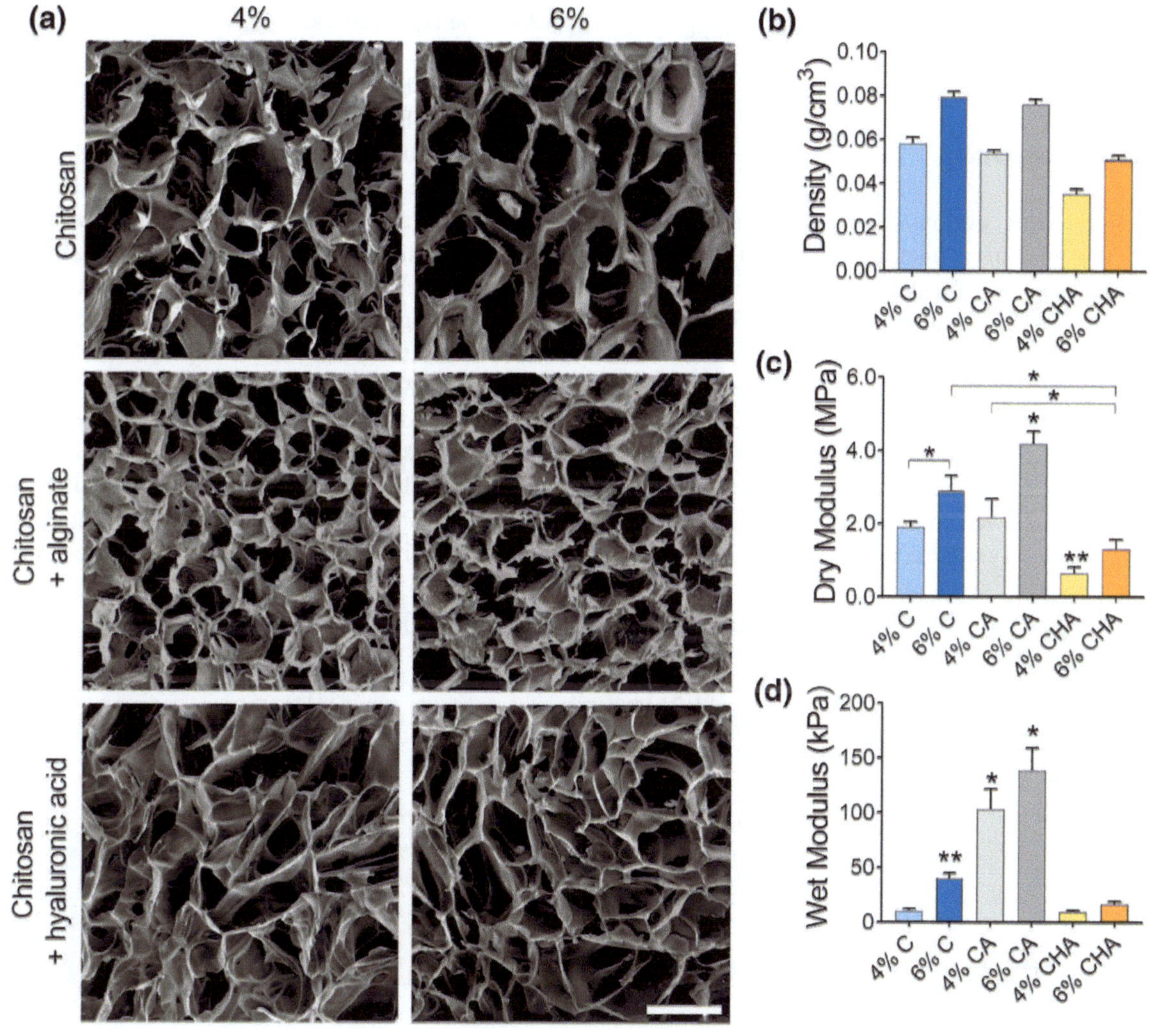

Fig. 2 Base scaffold microstructure and physical property characterization. **a** SEM images showing pore morphology. Scale bar is 200 μm. **b** Dry bulk density of all scaffolds significantly different ($p \leq 0.05$, $n = 12$). Compressive modulus of (**c**) dry scaffolds in MPa ($n = 4$) and (**d**) hydrated scaffolds in kPa ($n = 4$). *Statistically significant from all or specified conditions ($p \leq 0.05$). **Statistically significant from all conditions, except 6% CHA. ($p \leq 0.05$)

3.2 Cell attachment and proliferation on base scaffolds

In addition to characterizing and optimizing the physical properties of the scaffold to reflect the *in vivo* microenvironment, the response of relevant cells to the scaffold in terms of attachment and proliferation is also critical. Growth profiles for chondrocyte-like cells (SW-1353), rat bone marrow mesenchymal stem cells (MSC), and osteoblast-like cells (MG63) were established for each base scaffold by using the metabolic assay alamarBlue® (Fig. 3). The 4% CHA and 6% CHA scaffolds were seeded with chondrocyte-like SW-1353 or MSCs to evaluate their support of chondrogenic cells. Human SW-1353 cells have been utilized previously to evaluate cartilage regeneration (Cecen et al. 2016; Chien et al. 2012; Lu et al. 2015) and MSCs possess chondrogenic potential. The 4% and 6% C and CA scaffolds were seeded with osteoblast-like MG63, a cell line frequently used to evaluate osteogenic cell behavior on scaffolds (Gupta et al. 2016; Trombetta et al. 2017).

3.2.1 Cartilage layer scaffold

CHA scaffolds were evaluated as a potential chondral layer with the goal of mimicking native cartilage ECM compromised heavily of Glycosaminoglycans (GAGs). The structure of chitosan is similar to GAGs (Kim et al. 2008) and HA, a component of native articular cartilage, is important in tissue homeostasis and chondrogenesis (Kim et al. 2003; Knudson and Knudson 2004; Yoo et al. 2005). Combining chitosan and HA results in a stable PEC, blending the beneficial properties of both natural polymers. Lower polymer concentration scaffolds relative to bone layer scaffolds, as summarized in Table 1, were investigated for the cartilage layer because chondrocytes prefer lower stiffness substrates relative to osteogenic cells. Additionally, lower stiffness substrates may influence mesenchymal stem cell fate toward a chondrogenic phenotype (Olivares-Navarrete et al. 2017).

In native articular cartilage, chondrocytes populate and deposit ECM to maintain healthy tissue. Cartilage layer bioactivity was evaluated using chondrocyte-like SW-1353 and MSCs with chondrogenic potential. An increase in metabolic activity from day 2 to day 10 was observed for SW-1353 and MSC cells in both 4% and 6% CHA scaffolds (Fig. 3a, b). The number of MSC cells on 4% CHA scaffolds was significantly higher than the number of MSCs on 6% CHA scaffolds at day 10. Based on proliferation and metabolic activity of the SW-1353 and MSC cells, 4% CHA scaffolds were selected for the cartilage layer of the osteochondral scaffold.

3.2.2 Bone layer scaffold

Subchondral bone, composed of type I collagen and calcium phosphate, is a stiffer tissue than cartilage. Presentation of appropriate stiffness cues is crucial to regeneration of healthy tissue because mechanosensing by stem cells is a key factor in lineage mediation. Determining the optimal bone layer base scaffold required a two-phase evaluation: (1) characterization of cell proliferation on higher stiffness scaffolds to identify the base scaffold material, and (2) addition of hydroxyapatite nanorods (HAp) and optimization of HAp content in the base scaffold to enhance scaffold osteoconductivity. Osteoblast-like MG63 cells were cultured on the four stiffest scaffolds (4% C, 6% C, 4% CA, and 6% CA summarized in Table 1) (Fig. 2c, d) resulting in increased metabolic activity for all scaffolds over the 10-day culture

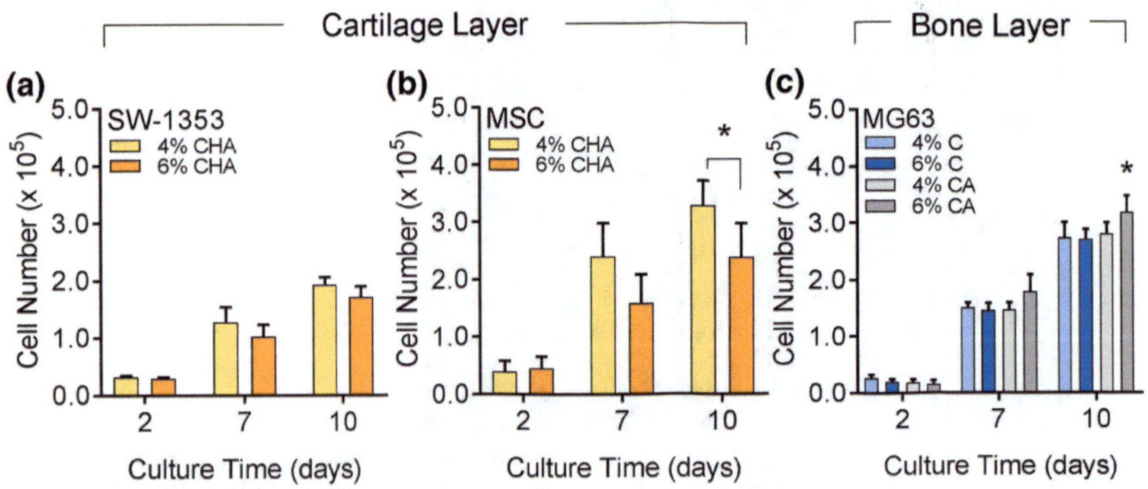

Fig. 3 Cell proliferation on base scaffolds. **a** SW-1353 and (**b**) MSC cultured on 4% and 6% CHA, and (**c**) MG63 cultured on 4% C, 6% C, 4% CA, and 6% CA for 10 days (n = 4). *Statistical significance within specified time point. ($p \leq 0.05$)

period (Fig. 3c). At day 10, the number of MG63 cells on 6% CA scaffolds was significantly higher than on the C scaffolds or 4% CA scaffolds. The 6% CA scaffold was selected as the base material for the bone layer because of increased scaffold stiffness and superior support of osteogenic cell proliferation.

The addition of HAp, similar in composition to the inorganic component of bone (Batchelar et al. 2006; Malmberg and Nygren 2008), promotes osteogenesis when integrated into chitosan scaffolds (Chesnutt et al. 2009), and has shown promising results in bone TE *in vitro* and *in vivo*. Additionally, the scaffold with embedded nanorods mimics the native composite bone structure where mineralized calcium phosphate platelets and rods are embedded in a collagen matrix (Weiner and Wagner 1998). Therefore, to further enhance the bioactivity of the bone layer, hydroxyapatite nanorods were added to the 6% CA base scaffold (Fig. 4a, b). HAp nanorods were added in concentrations ranging from 0.1 wt% to 3 wt%. Pore morphologies in the HAp-integrated scaffolds was visualized with SEM (Fig. 4c). All HAp-incorporated scaffolds possessed an open, porous structure indicating that HAp addition does not hinder pore formation during TIPS. As the HAp concentration within the CA scaffold increased up to 3%, the dry bulk density of the scaffolds also increased (Fig. 4d). Homogeneous HAp dispersal throughout the CA matrix was observed for all conditions, despite dispersal challenges cited in previous works (Venkatesan and Kim 2010). In the CA scaffold, the strong ionic interaction between alginate and Ca^{+2} in HAp acted as a dispersant (Kim et al. 2015).

The proliferative capacity of MG63 cells in HAp-integrated scaffolds was determined by evaluating metabolic activity (Fig. 4e). At day 7, no significant difference in metabolic activity was observed for cells cultured on CA scaffolds with or without HAp. By day 14, scaffolds with ≥1 wt% HAp began to disintegrate and release cells into suspension, resulting in decreased cell number. After two-weeks of culture, the 6% CA + 0.5% HAp scaffolds demonstrated aqueous stability while supporting proliferation of osteogenic cells. Therefore, this scaffold was selected for the osteochondral scaffold bone layer.

3.3 Bilayer scaffold architecture and mechanical properties

TE scaffolds must be biochemically suited for attachment and proliferation of the desired cell type. Nevertheless, appropriate scaffold structural cues are also essential to the generation of functional tissue. After evaluating six different scaffolds for supporting attachment and proliferation of the appropriate cell type, we found that SW-1353 and MSC cells prefer a higher HA content scaffold with lower elastic modulus. The MG63 cells favor scaffolds with a higher elastic modulus and the highest HAp content while maintaining scaffold aqueous stability. By layering two polymer solutions with differing compositions and polymer contents (4% CHA and 6% CA + 0.5% HAp) before a single TIPS-lyophilization process, a gradient interface was achieved.

The resultant bilayer scaffold pore structure, including the interface region, was visualized via SEM (Fig. 5a–d). The interface between the two distinct layers of the scaffold was amalgamated and seamless (Fig. 5c) and the cartilage region (4% CHA) was less dense than the bone region (6% CA + HAp) (Fig. 5e). Scaffold porosity followed a similar trend, where the 4% CHA layer porosity was higher than the 6% CA + HAp layer porosity. Importantly, both regions were highly porous with >87% porosity. High porosity encourages recruitment and infiltration of native cells where the bioactive cues in each scaffold layer (i.e., HA and HAp) can help direct layer-specific ECM and tissue development. Pore size is a critical factor in cell attachment and infiltration, and can play a deterministic role in cell behavior (Di Luca et al. 2016). Due to the higher polymer content of the bone layer, 6% CA + HAp scaffold, the mean pore area was significantly smaller ($1.16 \pm 0.87 \times 10^4$ μm^2) than the cartilage layer 4% CHA scaffold ($2.36 \pm 1.99 \times 10^4$ μm^2) (Fig. S1b). Based on mean pore area and the assumption of circular pores to estimate mean pore diameter, the pore diameters would be approximately 145 μm (cartilage) and 110 μm (bone), respectively. These diameters fall in the range of relevant pore sizes for cell growth and tissue regeneration (Yang et al. 2001).

Pore size, shape, and interconnectivity in TIPS-fabricated scaffolds is dependent on the solution properties and processing temperatures. Development of a homogenous, isotropic pore structure requires uniform thermal energy transfer during freezing and lyophilizing. Layering different polymer solutions can result in non-uniform pore structures as the overlying solution has no direct contact with the freezing surface. Uniform conduction between the two polymer solutions is particularly important to ensure a consistent and integrated interface to mimic the calcified cartilage region, which should help prevent bone ingrowth in the cartilage region (Hunziker and Driesang 2003). Despite differing polymer solution densities and compositions, the freezing procedure reported here enabled ice crystal formation throughout the scaffold, as demonstrated by the consistent pore formation throughout the entirety of the bilayer scaffold. Further, pores bridge both scaffold layers, thereby creating a stable interface. A stable and interconnected interface is critical to prevent delamination, anchor the cartilage to the bone, and prevent zonal segregation from inhibiting integrated regeneration at the calcified cartilage interface (Kreklau et al. 1999; Niederauer et al. 2000).

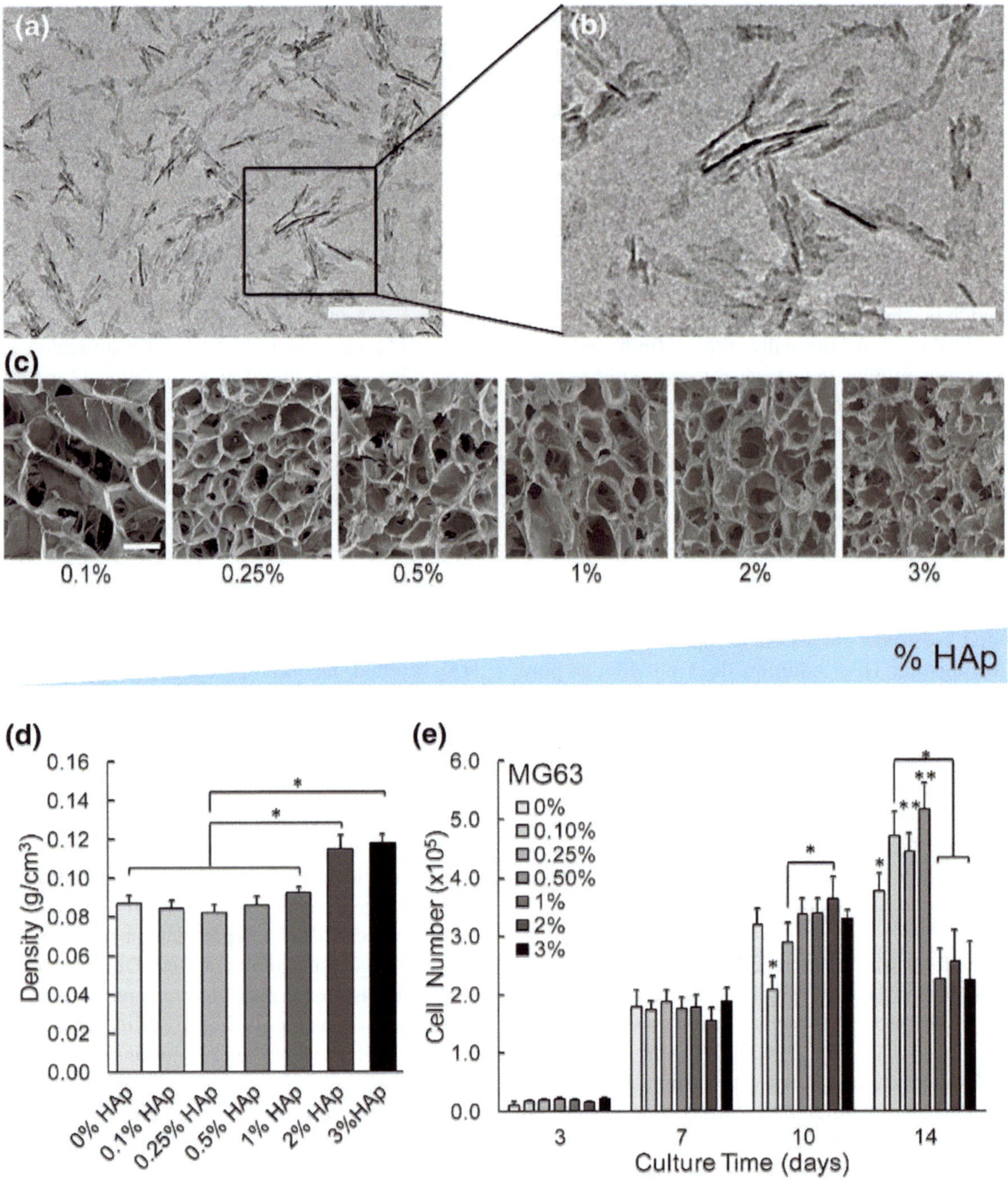

Fig. 4 Effect of HAp concentration on 6% CA scaffold properties. **a**, **b** Visualization of HAp nanorods with TEM. Scale bars represent 200 nm, and 100 nm, respectively, in (**a**) and (**b**). **c** SEM images of scaffold pore morphology with varying HAp (HAp concentration increases from left to right). Scale bar represents 200 μm. **d** Density of CA + HAp composite scaffolds ($n = 8$) ($p \leq 0.05$). (**e**) MG63 proliferation on CA + HAp composite scaffolds over a 2-week culture period ($n = 4$). *Statistically significant from all or specified conditions ($p \leq 0.05$). **Statistically significant from all conditions except 0.1% HAp ($p \leq 0.05$)

The stiffness gradient between the cartilage and bone layers of a multiphasic, osteochondral TE scaffold is hypothesized to help mediate lineage in undifferentiated cells (Discher et al. 2005; Engler et al. 2006). This bilayer scaffold displayed stress-strain behavior typical for porous polymers (Frydrych et al. 2011) (Fig. 5g). Two distinct linear-elastic regions corresponding to each scaffold layer (Fig. 5h) suggest compression of the low stiffness region (4% CHA) prior to the stiffer region (6% CA + HAp). Modulus 1 is defined by a compressive modulus of 6.90 ± 1.29 kPa, whereby Modulus 2 represents a stiffer region with a compressive modulus of 55.84 ± 16.0 kPa (Fig. 5h). Modulus 1 correlates with the modulus of a standalone 4% CHA scaffold (6.81 ± 0.70 kPa). The compressive modulus of a 6% CA + HAp scaffold was 139.0 ± 10.4 kPa, which is a 150% greater than Modulus 2,

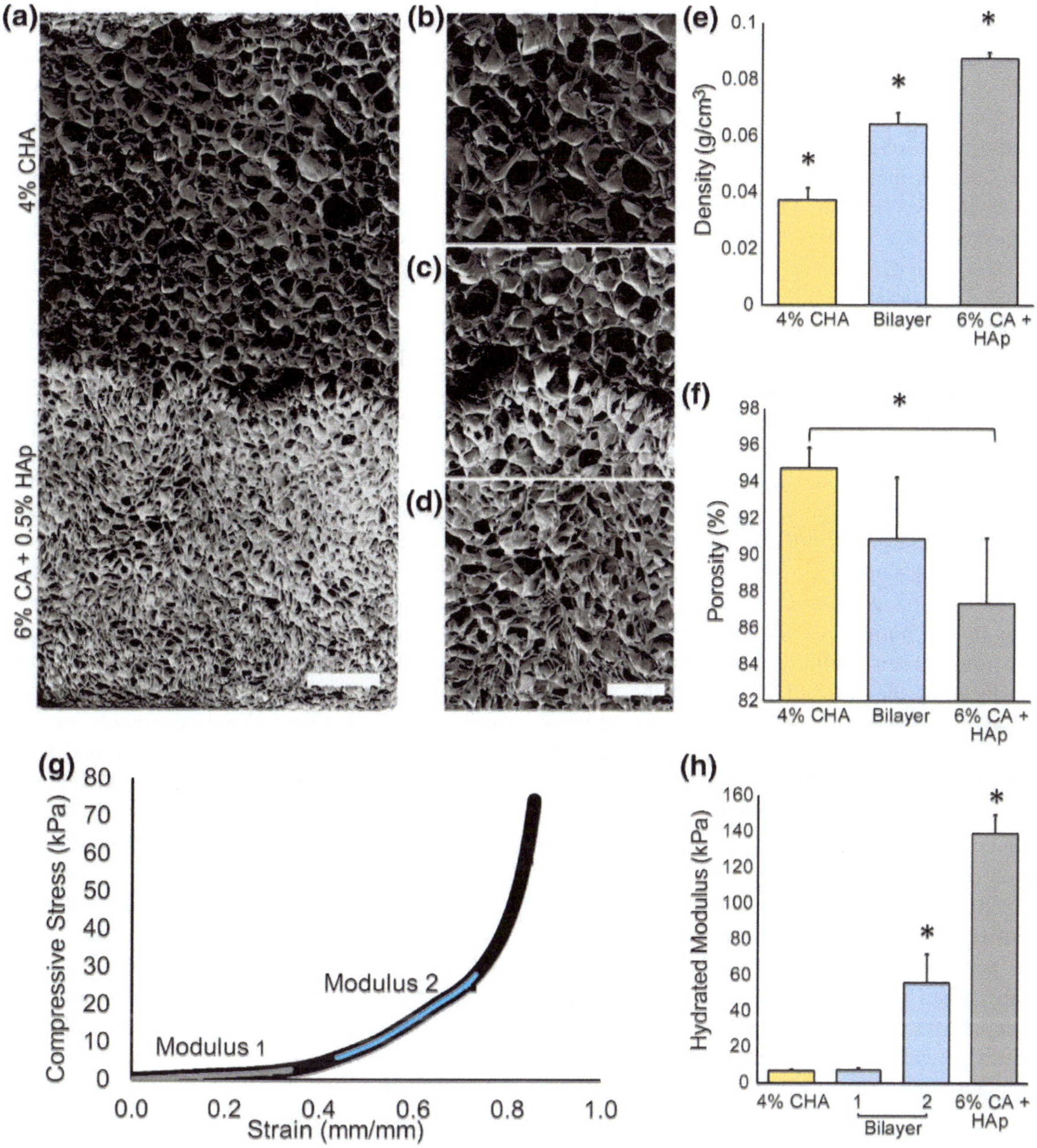

Fig. 5 Bilayer scaffold characterization. **a** SEM of bilayer scaffold with increased magnification of the (**b**) 4% CHA, (**c**) interface, and (**d**) 6% CA + 0.5% HAp layers. Scale bar represents 1 mm in (**a**) and 500 μm in (b-d). **e** Scaffold dry bulk density ($n = 4$). **f** Scaffold porosity ($n = 4$). **g** Representative compressive stress-strain curve for hydrated bilayer scaffold with modulus 1 corresponding to CHA region and modulus 2 corresponding to the CA + HAp region. **h** Compressive modulus of hydrated bilayer scaffold compared to pure components ($n = 4$). *Statistically significant from all or specified conditions ($p \leq 0.05$)

suggesting that modulus 2 in the bilayer scaffold had some contribution from both the cartilage (4% CHA) and bone (6% CA + HAp) layers indicating a stiffness gradient. In the native osteochondral microenvironment, the mechanical properties of the osteochondral region vary with patient age, activity, and location in the body, yet the compressive modulus has been cited at 98–270 MPa for subchondral bone (Schlichting et al. 2008) and 2–20 MPa for cartilage (Shepherd and Seedhom 1999). Despite advancements in materials and synthesis techniques, native osteochondral tissue has not been fully recapitulated using porous scaffolds. Polymeric scaffold stiffness is significantly less than native bone tissue, yet for non-load bearing bone or unique gradient microenvironments like the osteochondral tissue, a polymeric scaffold may be advantageous to promote healing at the interface. The stiffness of this bilayer scaffold is comparable to other natural polymer-based multiphasic scaffolds that have shown promise *in vivo* (Levingstone et al. 2014; Levingstone et al. 2016).

3.4 *In vitro* assessment of bilayer scaffold

Cell attachment, proliferation, and invasion in the bilayer scaffold was determined using two growth factor-free co-culture conditions over a two-week culture period. Scaffold co-cultures included: (1) SW-1353 + MG63 and (2) MSC + MG63. SW-1353 or MSC cells were seeded in the cartilage layer (4% CHA) and MG63 was seeded in the bone layer (6% CA + 0.5% HAp) to mimic native tissue (seeding illustrated in Fig. 1b). A significant increase in metabolic activity over a two-week culture period indicated the bilayer scaffold was amenable to simultaneous proliferation of these two co-cultures conditions (Fig. 6). Direct quantification of the specific contributions of each cell type to the increased metabolic activity or specific cell numbers associated with each cell line could not be determined using this assay model.

Cell migration and proliferation in each scaffold layer was visualized using fluorescence microscopy (Fig. 7). In the SW-1353 + MG63 co-culture condition (Fig. 7a, b), an increase in the number of SW-1353 (top, red) and MG63 (bottom, green) cells was observed near the seeding surfaces of the scaffold relative to the scaffold interior at day 1. By day 14, cell density increased in all scaffold layers, and cells appeared to invade the scaffold interior relative to day 1. SW-1353 cells were mostly present as single cells and small cell clusters whereas the MG63 cells formed larger cell clusters. Although sparse, both green and red cells were observed at the interface at day 1. By day 14, the interface displayed a higher density of both cell types, suggesting that both cell types penetrated the scaffold to the interface. At the interface, small cell clusters of SW-1353 and larger cell clusters of MG63 were observed.

An increase in cell number in both cartilage and bone regions was observed at day 14 in the bilayer scaffold co-cultured with MSC + MG63 (Fig. 7c, d) as compared to the condition in the bilayer scaffold cocultured with SW-1353 + MG63, although the cell morphologies were similar under both conditions. MG63 cells in the bone layer (bottom, green) formed larger cell clusters than MSCs in the cartilage layer (top, red). Both cell types were present at the interface of the bilayer scaffold. SEM was used to examine cell morphology (Fig. 8). In the cartilage layer (CHA) (Fig. 8a, c), both SW-1353 and MSC cells in two different bilayer scaffolds assumed a rounded or spherical morphology. SW-1353 cells formed cell clusters, whereas MSC cells are present both in single cell state and in clusters. In the bone layer (CA + HAp) (Fig. 8b, d), MG63 cells showed their characteristic, spindle-like or elongated morphology, and spread along the pore walls. The MG63 cells formed large cell clusters as shown in the fluorescence images. In addition, it appears that a layer of extracellular fibrous matrix material was deposited on the highly populated cell areas of the scaffold making it difficult to identify the cell morphology.

To try to correlate increased proliferation with increased functionality, early chondrogenic (collagen type II) and osteogenic (osteocalcin) markers were evaluated using qRT-PCR. Type II collagen plays a critical role in stem cell and chondrocyte maturation and is vital to the development and formation of the GAG-rich articular cartilage (Bobick et al. 2009). For bone remodeling, osteocalcin is a known osteogenic marker (Marom et al. 2005) where increased expression signifies osteoblast maturation (Fig. 9). Each cell type was cultured separately on TCPS for comparison of gene expression in co-cultures on the bilayer scaffold relative to 2D TCPS. Gene expression for each condition was normalized to expression of the reference gene GAPDH. For comparison among 2D and co-culture conditions, gene expression was normalized to the 2D TCPS culture of MG63 (set to 1-

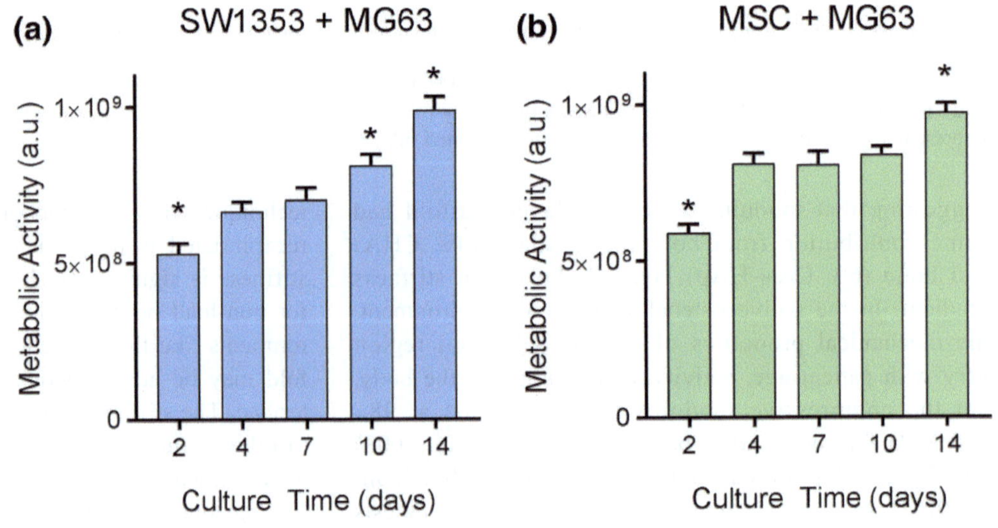

Fig. 6 Proliferation of chondrogenic and osteogenic cells in bilayer scaffold evaluated by measuring metabolic activity. Evaluation of (**a**) SW-1353 + MG63 and (**b**) MSC + MG63 metabolic activity measured with alamarBlue® ($n = 4$) *Statistical significance relative to all other time points. ($p \leq 0.05$)

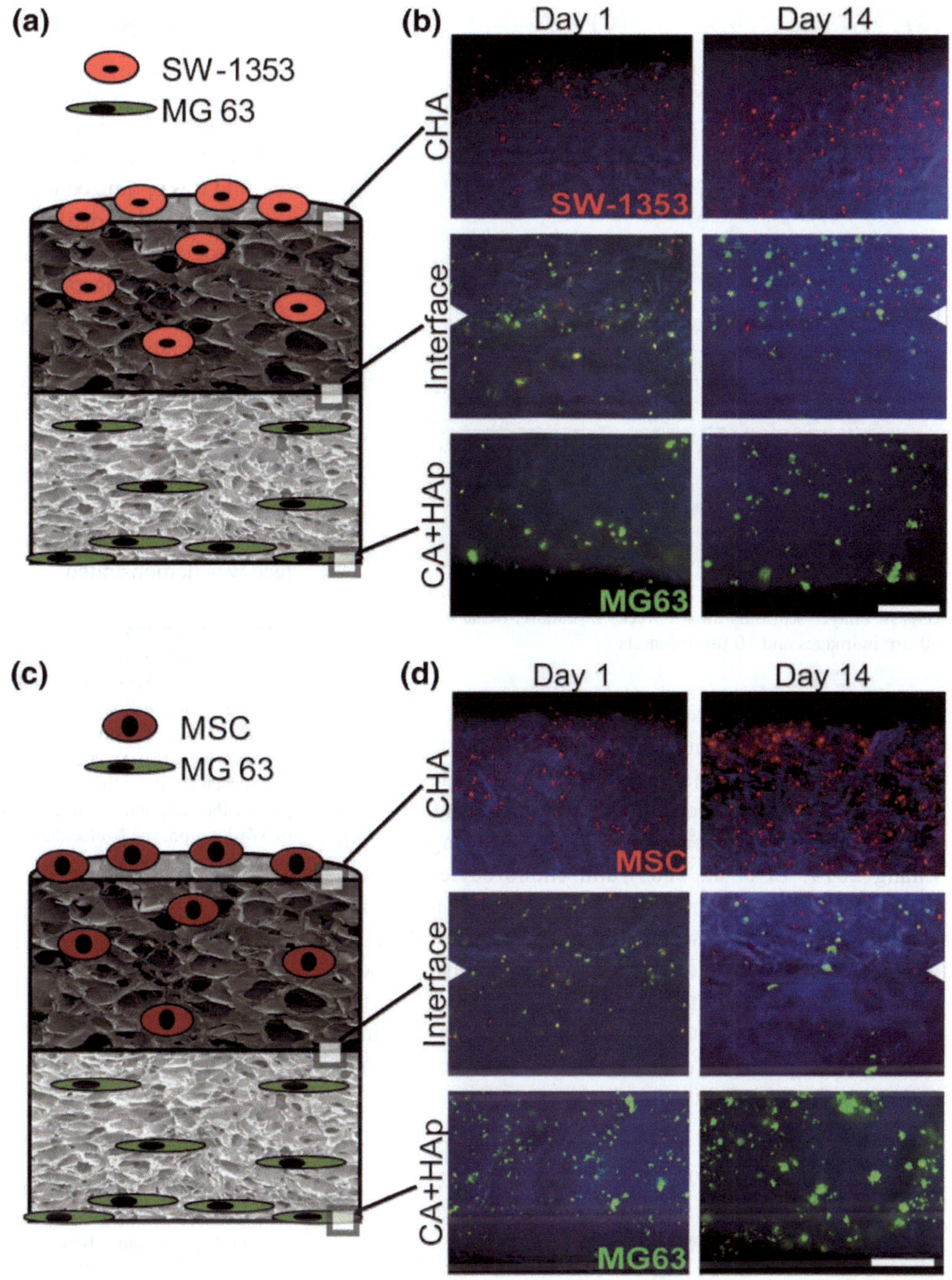

Fig. 7 Proliferation and migration of chondrogenic and osteogenic cells in bilayer scaffold cross-sections. Evaluation of SW-1353 + MG63 with (**a**) illustration of seeding condition and (**b**) fluorescence imaging of SW-1353 (red) and MG63 (green) in cartilage layer (top), interface (middle, marked by white triangles), and bone layer (bottom) of bilayer scaffold (blue auto fluorescence). Evaluation of MSC + MG63 with (**c**) illustration of seeding condition and (**d**) fluorescence imaging of MSC (red) and MG63 (green) in cartilage layer (top), interface (middle, marked by white triangles), and bone layer (bottom) of bilayer scaffold (blue auto fluorescence). Scale bar is 500 μm in (**b**) and (**d**)

fold) because both co-cultures contained MG63 cells. Increased expression of chondrogenic and osteogenic markers was observed for both co-cultures on the bilayer scaffolds relative to all 2D cultures indicating that at early timepoints, chitosan-based bilayer scaffolds effectively support relevant cells.

Multiphasic scaffolds have shown promise in regeneration of layered tissues, yet current multiphasic scaffold design is hindered by time-consuming, iterative fabrication techniques and insufficient amalgamation between layers. Here, we designed a biomimetic, bilayer scaffold using a straightforward, fast, and pH-neutral, single-stage fabrication approach. Our layered TIPS approach overcomes the onerous, iterative fabrication requirements of previous technologies while maintaining tunability of each layer. Tailoring bioactive factors and mechanical gradients in multiphasic scaffold layers is vital to support synchronized regeneration. In osteochondral TE, specialized integration and segregation of bioactive factors specific to the bone region or the cartilage region is crucial to enhance zonal organization of the regenerated osteochondral tissue. By exploiting TIPS and the enhanced aqueous stability of PECs, a highly porous, bilayer scaffold was fabricated entirely using natural polymers.

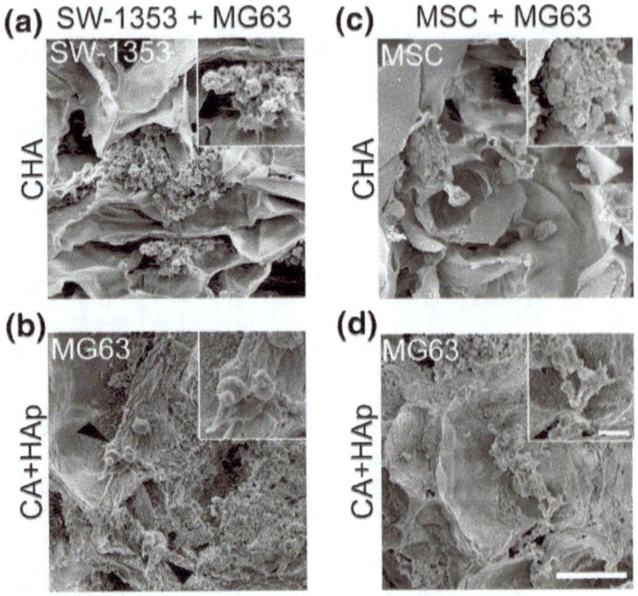

Fig. 8 SEM images of (**a, b**) SW1355 + MG63 and (**c, d**) MSC + MG63 **cells in** bilayer scaffolds after 2 weeks in culture. Scale bars represent 50 μm in images and 10 μm for insets

Natural polymers are advantageous over synthetic polymers because of their biocompatibility and biodegradability and chitosan, a major component of the bilayer scaffold, has previously been shown to support attachment and proliferation of both osteoblasts (Levengood and Zhang 2014; Li et al. 2005) and chondrocytes (Li and Zhang 2005). Importantly, the TIPS approach to layered scaffold fabrication has potential in osteochondral TE as well as for other complex layered tissues. By using a non-toxic, pH-neutral synthesis, future studies involving tailored growth factors integration could allow for sustained delivery and potential chemotactic recruitment of native cells directly to the defect site.

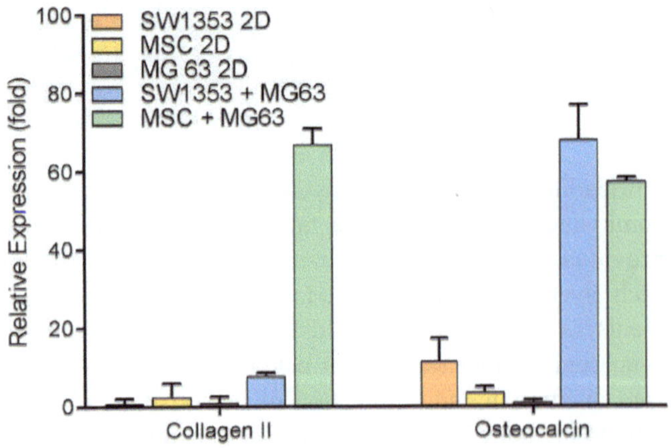

Fig. 9 Relative expression of RNA content in cells cultured on 2D TCPS (SW1353, MSC, and MG63) versus co-cultures in bilayer scaffolds (SW-1353 + MG63 and MSC + MG63). Gene expression was normalized to a reference gene, GAPDH, and then normalized to expression of MG63 on 2D TCPS (set as 1-fold). ($n = 3$)

4 Conclusion

We developed a bilayer, polysaccharide and GAG-based scaffold for osteochondral tissue regeneration using a simple, fast, and pH-neutral synthesis process. Each region of the bilayer scaffold was optimized with biomechanical and bioactive cues to enhance the proliferation of intended cell types (chondrogenic or osteogenic cells). Fabrication of this layered scaffold is advantageous over previously reported techniques due to the tunability of the elastic moduli of each layer as well as the integration of bioactive factors segregated to each specific region (HA in the cartilage region and HAp in the bone region). Additionally, the seamless gradient zone at the interface ensures stability by inhibiting delamination. The potential of this bilayer scaffold for osteochondral tissue engineering applications was characterized during a two-week co-culture where cell proliferation and infiltration to the scaffold interface was demonstrated.

Acknowledgements The authors acknowledge support for this work by the Kyocera Professor Endowment and NIH grant (R01CA172455) to Miqin Zhang. Ariane E. Erickson acknowledges support from the National Science Foundation Graduate Research Fellowship Program (DGE–1256082]. Part of this work was conducted at the Molecular Analysis Facility, a National Nanotechnology Coordinated Infrastructure site at the University of Washington which is supported in part by the National Science Foundation (ECC-1542101), the University of Washington, the Molecular Engineering & Sciences Institute, and the Clean Energy Institute.

References

H.M. Aydin, A three-layered osteochondral plug: Structural, mechanical, and in vitro biocompatibility analysis. Adv. Eng. Mater. **13**(12), B511–B517 (2011)

D.L. Batchelar, M.T.M. Davidson, W. Dabrowski, I.A. Cunningham, Bone-composition imaging using coherent-scatter computed tomography: Assessing bone health beyond bone mineral density. Med. Phys. **33**(4), 904–915 (2006)

J.E. Bekkers, T.S. de Windt, M. Brittberg, D.B. Saris, Cartilage repair in football (soccer) athletes: What evidence leads to which treatment? A critical review of the literature. Cartilage **3**(1 Suppl), 43S–49S (2012)

R. Bexkens, P.T. Ogink, J.N. Doornberg, G. Kerkhoffs, D. Eygendaal, L.S. Oh, M.P.J. van den Bekerom, Donor-site morbidity after osteochondral autologous transplantation for osteochondritis dissecans of the capitellum: A systematic review and meta-analysis. Knee Surg. Sports Traumatol. Arthrosc. **25**(7), 2237–2246 (2017)

B.E. Bobick, F.H. Chen, A.M. Le, R.S. Tuan, Regulation of the Chondrogenic phenotype in culture. Birth Defects Res. C. Embryo Today **87**(4), 351–371 (2009)

B. Cecen, L.D. Kozaci, M. Yuksel, O. Ustun, B.U. Ergur, H. Havitcioglu, Biocompatibility and biomechanical characteristics of loofah based scaffolds combined with hydroxyapatite, cellulose, poly-l-lactic acid with chondrocyte-like cells. Mater. Sci. Eng. C **69**, 437–446 (2016)

B.M. Chesnutt, A.M. Viano, Y. Yuan, Y. Yang, T. Guda, M.R. Appleford, et al., Design and characterization of a novel chitosan/nanocrystalline calcium phosphate composite scaffold for bone regeneration. J. Biomed. Mater. Res. A **88**(2), 491–502 (2009)

C.-S. Chien, H.-O. Ho, Y.-C. Liang, P.-H. Ko, M.-T. Sheu, C.-H. Chen, Incorporation of exudates of human platelet-rich fibrin gel in biodegradable fibrin scaffolds for tissue engineering of cartilage. J. Biomed. Mater. Res. B Appl. Biomater. **100B**(4), 948–955 (2012)

B.J. Cole, C. Pascual-Garrido, R.C. Grumet, Surgical management of articular cartilage defects in the knee. J. Bone Joint Surg. Am. **91**(7), 1778–1790 (2009)

A. Di Luca, B. Ostrowska, I. Lorenzo-Moldero, A. Lepedda, W. Swieszkowski, C. Van Blitterswijk, L. Moroni, Gradients in pore size enhance the osteogenic differentiation of human mesenchymal stromal cells in three-dimensional scaffolds. Sci. Rep. **6**, 22898 (2016)

D.E. Discher, P. Janmey, Y.L. Wang, Tissue cells feel and respond to the stiffness of their substrate. Science **310**(5751), 1139–1143 (2005)

R. Dorati, C. Colonna, C. Tomasi, I. Genta, G. Bruni, B. Conti, Design of 3D scaffolds for tissue engineering testing a tough polylactide-based graft copolymer. Mater. Sci. Eng. C Mater Biol Appl. **34**, 130–139 (2014)

A.J. Engler, S. Sen, H.L. Sweeney, D.E. Discher, Matrix elasticity directs stem cell lineage specification. Cell **126**(4), 677–689 (2006)

S.J. Florczyk, D.J. Kim, D.L. Wood, M. Zhang, Influence of processing parameters on pore structure of 3D porous chitosan-alginate polyelectrolyte complex scaffolds. J. Biomed. Mater. Res. A **98**(4), 614–620 (2011)

S.J. Florczyk, M. Leung, S. Jana, Z.S. Li, N. Bhattarai, J.I. Huang, et al., Enhanced bone tissue formation by alginate gel-assisted cell seeding in porous ceramic scaffolds and sustained release of growth factor. J. Biomed. Mater. Res. A **100A**(12), 3408–3415 (2012)

S.J. Florczyk, K. Wang, S. Jana, D.L. Wood, S.K. Sytsma, J.G. Sham, et al., Porous chitosan-hyaluronic acid scaffolds as a mimic of glioblastoma microenvironment ECM. Biomaterials **34**(38), 10143–10150 (2013)

S.J. Florczyk, F.M. Kievit, K. Wang, A.E. Erickson, R.G. Ellenbogen, M.Q. Zhang, 3D porous chitosan-alginate scaffolds promote proliferation and enrichment of cancer stem-like cells. J. Mater. Chem. B **4**(38), 6326–6334 (2016)

M. Frydrych, C.Y. Wan, R. Stengler, K.U. O'Kelly, B.Q. Chen, Structure and mechanical properties of gelatin/sepiolite nanocomposite foams. J. Mater. Chem. **21**(25), 9103–9111 (2011)

A. Galperin, R.A. Oldinski, S.J. Florczyk, J.D. Bryers, M.Q. Zhang, B.D. Ratner, Integrated bi-layered scaffold for osteochondral tissue engineering. Adv. Healthc. Mater. **2**(6), 872–883 (2013)

P. Gupta, M. Adhikary, J.C. M, M. Kumar, N. Bhardwaj, B.B. Mandal, Biomimetic, Osteoconductive non-mulberry silk Fiber reinforced Tricomposite scaffolds for bone tissue engineering. ACS Appl. Mater. Interfaces **8**(45), 30797–30810 (2016)

A. He, L. Liu, X. Luo, Y. Liu, F. Liu, X. Wang, et al., Repair of osteochondral defects with in vitro engineered cartilage based on autologous bone marrow stromal cells in a swine model. Sci. Rep. **7**, 40489 (2017)

Y.Y. Hsu, J.D. Gresser, D.J. Trantolo, C.M. Lyons, P.R.J. Gangadharam, D.L. Wise, Effect of polymer foam morphology and density on kinetics of in vitro controlled release of isoniazid from compressed foam matrices. J. Biomed. Mater. Res. **35**(1), 107–116 (1997)

E.B. Hunziker, Articular cartilage repair: Basic science and clinical progress. A review of the current status and prospects. Osteoarthr. Cartil. **10**(6), 432–463 (2002)

E.B. Hunziker, I.M. Driesang, Functional barrier principle for growth-factor-based articular cartilage repair. Osteoarthr. Cartil. **11**(5), 320–327 (2003)

S. Jana, S.J. Florczyk, M. Leung, M.Q. Zhang, High-strength pristine porous chitosan scaffolds for tissue engineering. J. Mater. Chem. **22**(13), 6291–6299 (2012)

X. Jin, J. Zhuang, Z. Zhang, H. Guo, J. Tan, Hydrothermal synthesis of hydroxyapatite nanorods in the presence of sodium citrate and its aqueous colloidal stability evaluation in neutral pH. J. Colloid Interface Sci. **443**, 125–130 (2015)

F.M. Kievit, S.J. Florczyk, M.C. Leung, O. Veiseh, J.O. Park, M.L. Disis, M. Zhang, Chitosan-alginate 3D scaffolds as a mimic of the glioma tumor microenvironment. Biomaterials **31**(22), 5903–5910 (2010)

G. Kim, M. Okumura, T. Ishiguro, T. Kadosawa, T. Fujinaga, Preventive effect of hyaluronic acid on the suppression of attachment and migration abilities of bovine chondrocytes by IL-1 alpha in vitro. J. Vet. Med. Sci. **65**(3), 427–430 (2003)

I.Y. Kim, S.J. Seo, H.S. Moon, M.K. Yoo, I.Y. Park, B.C. Kim, C.S. Cho, Chitosan and its derivatives for tissue engineering applications. Biotechnol. Adv. **26**(1), 1–21 (2008)

H.L. Kim, G.Y. Jung, J.H. Yoon, J.S. Han, Y.J. Park, D.G. Kim, et al., Preparation and characterization of nano-sized hydroxyapatite/alginate/chitosan composite scaffolds for bone tissue engineering. Mater. Sci. Eng. C Mater. Biol. Appl. **54**, 20–25 (2015)

C.B. Knudson, W. Knudson, Hyaluronan and CD44: Modulators of chondrocyte metabolism. Clin. Orthop. Relat. Res. (427 Suppl), S152–S162 (2004)

E. Kon, M. Delcogliano, G. Filardo, M. Fini, G. Giavaresi, S. Francioli, et al., Orderly osteochondral regeneration in a sheep model using a novel nano-composite multilayered biomaterial. J. Orthop. Res. **28**(1), 116–124 (2010)

B. Kreklau, M. Sittinger, M.B. Mensing, C. Voigt, G. Berger, G.R. Burmester, et al., Tissue engineering of biphasic joint cartilage transplants. Biomaterials **20**(18), 1743–1749 (1999)

S.L. Levengood, M. Zhang, Chitosan-based scaffolds for bone tissue engineering. J. Mater. Chem. B **2**(21), 3161–3184 (2014)

T.J. Levingstone, A.C. Ramesh, R.T. Brady, P. Brama, J.P. Gleeson, F.J. O'Brien, Collagen-based multilayered scaffold shows potential for osteochondral defect repair. J. Tissue Eng. Regen. Med. **8**, 82–83 (2014)

T.J. Levingstone, E. Thompson, A. Matsiko, A. Schepens, J.P. Gleeson, F.J. O'Brien, Multi-layered collagen-based scaffolds for osteochondral defect repair in rabbits. Acta Biomater. **32**, 149–160 (2016)

Z.S. Li, M.Q. Zhang, Chitosan-alginate as scaffolding material for cartilage tissue engineering. J. Biomed. Mater. Res. A **75A**(2), 485–493 (2005)

Z.S. Li, H.R. Ramay, K.D. Hauch, D.M. Xiao, M.Q. Zhang, Chitosan-alginate hybrid scaffolds for bone tissue engineering. Biomaterials **26**(18), 3919–3928 (2005)

E. Lopez-Ruiz, G. Jimenez, M.A. Garcia, C. Antich, H. Boulaiz, J.A. Marchal, M. Peran, Polymers, scaffolds and bioactive molecules with therapeutic properties in osteochondral pathologies: what's new? Expert Opin. Ther. Pat. **26**(8), 877–890 (2016)

H.-T. Lu, M.-S. Hsieh, C.-W. Cheng, L.-F. Yao, T.-Y. Hsu, J. Lan, et al., Alterative effects of an oral alginate extract on experimental rabbit osteoarthritis. J. Biomed. Sci. **22**(1), 64 (2015)

H. Madry, C.N. van Dijk, M. Mueller-Gerbl, The basic science of the subchondral bone. Knee Surg. Sports Traumatol. Arthrosc. **18**(4), 419–433 (2010)

P. Malmberg, H. Nygren, Methods for the analysis of the composition of bone tissue, with a focus on imaging mass spectrometry (TOF-SIMS). Proteomics **8**(18), 3755–3762 (2008)

R. Marom, I. Shur, R. Solomon, D. Benayahu, Characterization of adhesion and differentiation markers of osteogenic marrow stromal cells. J. Cell. Physiol. **202**(1), 41–48 (2005)

T. Nie, L. Xue, M. Ge, H. Ma, J. Zhang, Fabrication of poly(L-lactic acid) tissue engineering scaffolds with precisely controlled gradient structure. Mater. Lett. **176**, 25–28 (2016)

G.G. Niederauer, M.A. Slivka, N.C. Leatherbury, D.L. Korvick, H.H. Harroff, W.C. Ehler, et al., Evaluation of multiphase implants for repair of focal osteochondral defects in goats. Biomaterials **21**(24), 2561–2574 (2000)

S.H. Oh, I.K. Park, J.M. Kim, J.H. Lee, In vitro and in vivo characteristics of PCL scaffolds with pore size gradient fabricated by a centrifugation method. Biomaterials **28**(9), 1664–1671 (2007)

S.H. Oh, T.H. Kim, J.H. Lee, Creating growth factor gradients in three dimensional porous matrix by centrifugation and surface immobilization. Biomaterials **32**(32), 8254–8260 (2011)

R. Olivares-Navarrete, E.M. Lee, K. Smith, S.L. Hyzy, M. Doroudi, J.K. Williams, et al., Substrate stiffness controls osteoblastic and Chondrocytic differentiation of mesenchymal stem cells without exogenous stimuli. PLoS One **12**(1), e0170312 (2017)

D. Schaefer, I. Martin, G. Jundt, J. Seidel, M. Heberer, A. Grodzinsky, et al., Tissue-engineered composites for the repair of large osteochondral defects. Arthritis Rheum. **46**(9), 2524–2534 (2002)

M. Schinhan, M. Gruber, P. Vavken, R. Dorotka, L. Samouh, C. Chiari, et al., Critical-size defect induces unicompartmental osteoarthritis in a stable ovine knee. J. Orthop. Res. **30**(2), 214–220 (2012)

K. Schlichting, H. Schell, R.U. Kleemann, A. Schill, A. Weiler, G.N. Duda, D.R. Epari, Influence of scaffold stiffness on subchondral bone and subsequent cartilage regeneration in an ovine model of osteochondral defect healing. Am. J. Sports Med. **36**(12), 2379–2391 (2008)

S.-J. Seo, C. Mahapatra, R.K. Singh, J.C. Knowles, H.-W. Kim, Strategies for osteochondral repair: Focus on scaffolds. J. Tissue Eng. **5**, 2041731414541850–2041731414541850 (2014)

D.E. Shepherd, B.B. Seedhom, The 'instantaneous' compressive modulus of human articular cartilage in joints of the lower limb. Rheumatology (Oxford) **38**(2), 124–132 (1999)

R. Shu, R. McMullen, M.J. Baumann, L.R. McCabe, Hydroxyapatite accelerates differentiation and suppresses growth of MC3T3-E1 osteoblasts. J. Biomed. Mater. Res. A **67A**(4), 1196–1204 (2003)

Y.P. Singh, J.C. Moses, B.K. Bhunia, S.K. Nandi, B.B. Mandal, Hierarchically structured seamless silk scaffolds for osteochondral interface tissue engineering. J. Mater. Chem. B **6**(36), 5671–5688 (2018)

G.D. Smith, G. Knutsen, J.B. Richardson, A clinical review of cartilage repair techniques. J. Bone Joint Surg. Br. Vol. **87B**(4), 445–449 (2005)

R. Trombetta, J.A. Inzana, E.M. Schwarz, S.L. Kates, H.A. Awad, 3D printing of calcium phosphate ceramics for bone tissue engineering and drug delivery. Ann. Biomed. Eng. **45**(1), 23–44 (2017)

J. Venkatesan, S.-K. Kim, Chitosan composites for bone tissue engineering—An overview. Mar. Drugs. **8**(8), 2252–2266 (2010)

K. Wang, F.M. Kievit, A.E. Erickson, J.R. Silber, R.G. Ellenbogen, M.Q. Zhang, Culture on 3D chitosan-hyaluronic acid scaffolds enhances stem cell marker expression and drug resistance in human glioblastoma Cancer stem cells. Adv. Healthc. Mater. **5**(24), 3173–3181 (2016)

S. Weiner, H.D. Wagner, THE MATERIAL BONE: Structure-mechanical function relations. Annu. Rev. Mater. Sci. **28**(1), 271–298 (1998)

S. Yang, K.F. Leong, Z. Du, C.K. Chua, The design of scaffolds for use in tissue engineering. Part I. Traditional factors. Tissue Eng. **7**(6), 679–689 (2001)

H.S. Yoo, E.A. Lee, J.J. Yoon, T.G. Park, Hyaluronic acid modified biodegradable scaffolds for cartilage tissue engineering. Biomaterials **26**(14), 1925–1933 (2005)

P. Yusong, S. Qianqian, P. Chengling, W. Jing, Prediction of mechanical properties of multilayer gradient hydroxyapatite reinforced poly(vinyl alcohol) gel biomaterial. J. Biomed. Mater. Res. B Appl. Biomater. **101B**(5), 729–735 (2013)

R.Y. Zhang, P.X. Ma, Poly(alpha-hydroxyl acids) hydroxyapatite porous composites for bone-tissue engineering. I. Preparation and morphology. J. Biomed. Mater. Res. **44**(4), 446–455 (1999)

Biomedical Microdevices (2019) 21: 43
https://doi.org/10.1007/s10544-019-0394-9

Hang on tight: reprogramming the cell with microstructural cues

Long V. Le[1] · Michael A. Mkrtschjan[2] · Brenda Russell[3] · Tejal A. Desai[1]

Published online: 6 April 2019

Author's Personal Note

It is hard to describe how much Mauro has shaped my career trajectory. As one of his first graduate students in bioengineering, he taught me the importance of bringing together multiple fields in order to solve the most pressing needs in health. He is not afraid of thinking outside the box and sharing his enthusiasm and passion with others. He taught me to stay true to myself and never give up. For this, I am forever thankful! Congratulations on this milestone and may you achieve all your dreams!

Tejal A. Desai

Abstract

Cells interact intimately with complex microdomains in their extracellular matrix (ECM) and maintain a delicate balance of mechanical forces through mechanosensitive cellular components. Tissue injury results in acute degradation of the ECM and disruption of cell-ECM contacts, manifesting in loss of cytoskeletal tension, leading to pathological cell transformation and the onset of disease. Recently, microscale hydrogel constructs have been developed to provide cells with microdomains to form focal adhesion binding sites, which enable restoration of cytoskeletal tension. These synthetic anchors can recapitulate the complex 3D architecture of the native ECM to provide microtopographical cues. The mechanical deformation of proteins at the cell surface can activate signaling cascades to modulate downstream gene-level transcription, making this a unique materials-based approach for reprogramming cell behavior. An overview of the mechanisms underlying these mechanosensitive interactions in fibroblasts, stem and other cell types is provided to review their effects on cellular reprogramming. Recent investigations on the fabrication, functionalization and implementation of these materials and microtopographical features for drug testing and therapeutic applications are discussed.

Keywords Microtopography · Microfabrication · Mechanotransduction · Extracellular matrix · Hydrogel · Drug delivery · Focal adhesion · Cytoskeleton

1 Introduction

Cells exist in a sensitive biochemical and mechanical equilibrium presented by their surrounding extracellular matrix (ECM) and other tissue components. Disruption of this equilibrium due to acute injury or progressive pathology manifests in the development of chronic disease (Jaalouk and Lammerding 2009; Lindsey et al. 2003; Midwood et al. 2004; Prabhu and Frangogiannis 2016; Raposo and Schwartz 2014). Classical therapeutic approaches to regenerate function to the affected area involve the administration of pharmacological agents or biologics to modulate cell behavior or to recruit cells to the region. However, it is becoming increasingly recognized that treatment with soluble factors is insufficient for comprehensive tissue repair. The native tissue environment is a complex milieu of signaling factors and biophysical microdomains with specific spatiotemporal distribution. Therefore, failure to recapitulate these myriad factors has significantly limited the efficacy of current tissue regeneration strategies.

An emerging paradigm in tissue engineering focuses on the generation of appropriate mechanical environments for cell attachment, growth and differentiation. Stem cells, for example, can be instructed to differentiate into specific lineages based on the stiffness of their substrate alone (Engler et al. 2006; Huebsch et al. 2010). These mechanical factors are so crucial that without an appropriate substrate stiffness, introduction of osteogenic factors is unable to instruct these cells towards an osteogenic lineage, highlighting the complex interplay between growth factor signaling and mechanical stimuli and its crucial role in dictating cell behavior. Researchers have focused on engineering matrices that recapitulate the topography and bulk mechanical properties of the native

✉ Tejal A. Desai
Tejal.Desai@ucsf.edu

Long V. Le
Long.Le@ucsf.edu

Michael A. Mkrtschjan
mkrts@uw.edu

Brenda Russell
russell@uic.edu

[1] Department of Bioengineering and Therapeutic Sciences, University of California, 1700 4th St Rm 204, San Francisco, CA 94158, USA

[2] Department of Bioengineering, University of Illinois, Chicago, 835 S. Wolcott, Chicago, IL 60612, USA

[3] Department of Physiology and Biophysics, University of Illinois, Chicago, 835 S. Wolcott, Chicago, IL 60612, USA

ECM to promote proper tissue growth. The development of these materials has been supported by advancements in fabrication technologies, such as photolithography, microfluidics and 3D printing, which have been extensively reviewed (Annabi et al. 2010; Khademhosseini et al. 2006b; Khademhosseini and Langer 2007; Leijten et al. 2017; Ovsianikov et al. 2012; Slaughter et al. 2009; Yanagawa et al. 2016; Zorlutuna et al. 2012). Unfortunately, complex matrices with tailored microchannels and other topographies cannot be injected *in vivo* and are limited to *in vitro* studies. Injectable materials, on the other hand, are easily administered but typically comprise of bulk hydrogels that are homogeneous and isotropic in nature, failing to recapitulate the complex architecture of the ECM. Due to these challenges, the use of mechanically instructive materials for therapeutic applications remain quite limited.

Recently, discrete topographical features have been used to reprogram cell phenotype. These materials can be fabricated with a wide variety of stiffnesses, geometries and porosities and are fully formed and characterized before use. Introduction of these materials into the extracellular matrix provides cells with anchor points that enable restoration of intracellular tension and subsequent modulation of signaling pathways via complex mechanisms (Ayala et al. 2010; Curtis et al. 2010, 2013; Le et al. 2018; Norman et al. 2008; Pinney et al. 2014a). Further, these materials are a versatile technology that can be modified and tuned in a variety of ways, such as drug loading, surface modification or physical alignment to meet the demands of varying disease states. Here, we review the mechanisms by which cells interact with discrete topographical cues and highlight novel developments in this technology for tissue engineering applications.

2 Cellular machinery to sense the physical microenvironment

Cells intimately interact with their local microenvironment and are exposed to a variety of externally applied stresses. As such, they are equipped with an abundance of cellular machinery that allow them to detect and respond to these stimuli (Fig. 1). Cells interpret such mechanical forces by mechanotransduction, the process by which extracellular physical stimuli deform molecular structure permitting post translational modification of proteins in the cell, and leading to gene expression changes in the nucleus. Taken together, cellular behavior is regulated. These processes of mechanosensing, mechanotransduction and response occur at multiple locations in the cell, including the plasma membrane, the cytoskeleton and the nucleus.

The cell membrane serves as the most direct link between the cell and the surrounding environment. Mechanosensitive ion channels, for example, are transmembrane proteins that can be physically stretched by external forces, causing an immediate influx of ions that distorts previously established ionic equilibrium (Martinac 2004). These ion currents are capable of initiating signaling cascades and modulating the expression of key transcriptional factors that reprogram cell behavior. In fibroblasts, stretch activation of calcium ion channels has been linked to short-term increases in the production of alpha smooth muscle actin (αSMA), which promotes the differentiation of fibroblasts into a contractile myofibroblast phenotype (Asazuma-Nakamura et al. 2009).

Cells primarily interact with the surrounding ECM through transmembrane receptors on the cell surface. While receptors for a variety of ECM components exist, such as CD44 and syndecans, the most studied receptors are the family of integrins, which are able to bind to fibronectin, laminins and collagens present in the ECM (Geiger et al. 2009; Katsumi et al. 2004; Martino et al. 2018; Orr et al. 2006; Puklin-Faucher and Sheetz 2009). Activation of integrins enables cells to form intimate attachments to the surrounding matrix for cell adhesion and migration. Further maturation of these attachments is characterized by the recruitment of key adapter proteins, such as talin, paxillin and vinculin, that make up a focal adhesion complex (FAC). This assembly of proteins couples the extracellular environment to the intracellular cytoskeleton, enabling cells to experience externally applied forces (Parsons et al. 2010; Ziegler et al. 2008). Further, these complexes act as a hub for activation of downstream biochemical signaling pathways. This is typically initiated by phosphorylation of signaling molecules at the integrins' cytosolic domains by protein tyrosine kinases such as focal adhesion kinase (FAK) and Src which induce signaling cascades involving ERK1/2 and MAPK, culminating in altered cell migration, proliferation and differentiation (Westhoff et al. 2004).

Lipid signaling at the cell membrane has also been shown to rapidly affect focal adhesion assembly, facilitating conversion of mechanical cues to signaling events. Phosphatidylinositol 4,5-biphosphate (PIP2) signaling has been shown to modulate organization of the actin cytoskeleton and lamellar architecture (Mkrtschjan et al. 2018a). In the context of focal adhesions and the actin cytoskeleton, PIP2 plays key roles in the recruitment of adaptor proteins, such as vinculin, talin and paxillin, to the focal adhesions enabling propagation of forces to the cytoskeleton (Zhang et al. 2012). While focal adhesions are an integral component of mechanosensing discussed in this review, a more detailed analysis of the intricacies of focal adhesion formation and maturation are found elsewhere (Geiger et al. 2009; Katsumi et al. 2004; Puklin-Faucher and Sheetz 2009; Wozniak et al. 2004).

Focal adhesions are the anchoring complex to which the cytoskeleton attaches. The actin cytoskeleton regulates cell shape, function, and overall cellular mechanics and is able to orchestrate responses to mechanical perturbations. Cells exist in a state of cytoskeletal pre-stress that balances the

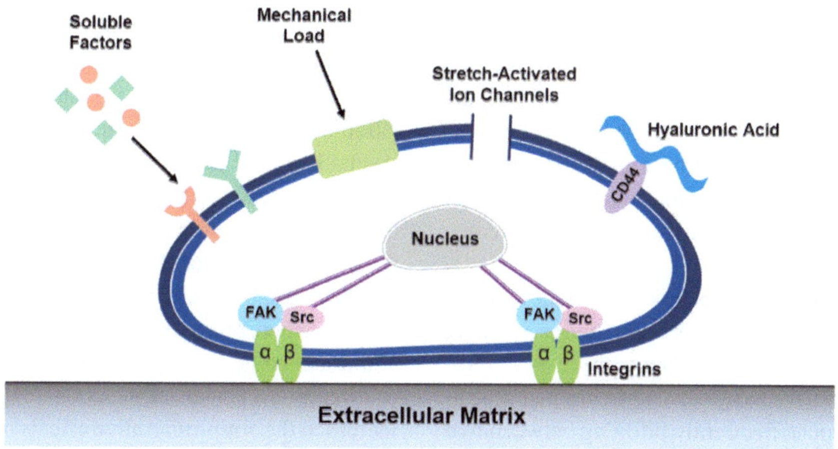

Fig. 1 Cellular mechanisms for detecting external mechanical stimuli. Cells are equipped with a wide range of cellular machinery to interact with the extracellular environment, including focal adhesions, ion channels and cell surface receptors. Once adhered to the extracellular matrix, activation of focal adhesion kinases and other signaling pathways, or direct propagation of force to the nucleus can initiate signaling cascades to modulate gene expression and cellular phenotype

intracellular tension with the physical properties of the ECM. Upon encountering external forces that perturb this pre-stress, activation of contractility dependent mechanisms enacted by myosin-II motors act to crosslink, organize and affect sliding between actin microfilaments to restore cytoskeletal tension (Ingber et al. 2014; Ingber 1997, 2006). These responses are dependent on activation of small Rho GTPases, such as RhoA, and subsequent phosphorylation of rho-associated protein kinase (ROCK). The actin cytoskeleton also facilitates propagation of extracellular stimuli from the outside of the cell to the nuclear membrane with its own skeleton comprised of lamins formed into filaments that provide structural support for the nucleus (Wang et al. 2009). Chromosomes are anchored to the lamina at the periphery of the nucleus and are susceptible to physical perturbations that are propagated from the cytoskeleton. Force propagation can mechanically distort the cell nucleus, changing overall nuclear shape and reorganizing residing chromosomes to ultimately affect gene transcription (McNamara et al. 2012).

Cardiomyocytes (CMs) are extremely responsive to mechanical cues in order to accommodate rapid changes in blood volume and subsequent altered contractile demand, as well as long-term adaptation to static load. Cardiomyocytes have evolved to respond to changes in tensions through a variety of internal sensors, ranging from components of the focal adhesion complex such as vinculin and talin to proteins further downstream within the sarcomere, including titin and Z-disc components (Hoshijima 2006; Pandey et al. 2018). The near immediate response to functional demand is reliant primarily on length-dependent activation, widely considered the predominant effector in the Frank-Starling Law of the heart, via increased myofilament Ca^{2+} sensitivity (de Tombe et al. 2010). Long-term alteration of myocyte contractility is often a product of changes to the cytoskeleton and structural remodeling of the myocyte. In particular, cyclic mechanical stretch and increasing stiffness of the underlying substrata have been used as a mechanical, hypertrophic stimuli, resulting in increased cell size and sarcomere assembly dynamics (Li et al. 2014; Lin et al. 2013; Mkrtschjan et al. 2018b). Upon presentation of a stimulus, this signal is transduced to the sarcomere where the actin capping protein, CapZ, is phosphorylated (e.g. PKCε), and its association with the barbed end of actin is decreased. The resultant effect is increased assembly rate of sarcomeric actin and overall cellular and tissue hypertrophy. Moreover, due to the individual cardiomyocyte's limited capacity to undergo cell division, myofibril addition is the primary way the heart muscle mass can increase.

In sum, cells are equipped with a wide array of mechanisms to sense, interact and respond to their local physical environments. These biophysical events are capable of reprogramming cell behavior and as noted above, mechanical cues have gained much attention as an adjustable tool for tissue engineering applications. In this review, we will focus on the influence of topography in modulating cellular phenotype.

3 Microtopographical regulation of cellular phenotype

Cells are capable of detecting environmental geometries through a complex mechanism known as contact guidance (Al-Haque et al. 2012; Rørth 2011). These intimate interactions with local topographical features provide instructive cues for directional cell spreading, migration and in some cases, differentiation (Cortese et al. 2013; Jiang et al. 2011; Kilian et al. 2010; Kim et al. 2009; Teixeira 2003). Researchers have shown that simply culturing cells on scaffolds of confined micron-scale geometries can promote the reprogramming of epigenetic state for generation of induced pluripotent stem cells (iPSCs), direct stem cells towards specific lineages, or even guide multi-cell clusters to polarize and form lumenized microtissues (Fig. 2) (Cerchiari et al. 2015; Downing et al. 2013; Kilian et al. 2010). Despite significant interest in utilizing topography for tissue engineering applications, the underlying mechanisms behind cell-microtopography interactions remain incompletely understood.

Seminal work from the Desai group described the interaction of cardiomyocytes with topographical features utilizing microtextured silicone membranes (Deutsch et al. 2000). Myocytes were found to have terminated with blunted protrusions ending on a micropost, frequently bridging between two microposts. This was in contrast to untextured membranes that have randomly oriented myocytes with tapered protrusions. The textured membranes had 4-fold increases in cellular attachment with cells adopting more *in vivo*-like morphologies despite equivalent adhesion motifs present on the surface, suggesting increased interaction via alternative mechanisms. The incorporation of microposts likely presents the cell with 3D cues not present on flat surfaces used in culture dishes and, by mechanisms to be discussed, is able to recapitulate a complex 3D mechanical environment to generate more *in vivo*-like phenotypes.

Fibroblasts were similarly grown in the presence of microposts and exhibited high preference for micropost attachments as well (Boateng et al. 2003). After contacting a micropost, fibroblast migration velocity was significantly decreased and the attachments matured with extensive accumulation of FAK around the microposts. These interactions manifested in global reduction of cell proliferation, although these effects are believed to occur on the single-cell level. If this is the case, the observed effects on migration and proliferation would not be affected by micropost size, but rather by the ability to establish contact and exert force on the microposts. Indeed, when presented with large microposts that provided additional contact area, there were no differences in fibroblast proliferation (Thakar et al. 2008). Higher micropost density increased the antiproliferative effects in culture, supporting the hypothesis that single-cell level transcriptional changes were occurring. This was confirmed by BrdU staining, where cells in direct contact with the microposts had reduced BrdU incorporation, whereas non-contacting cells were similar to cells on non-patterned substrates. These results were consistent with C2C12 mouse skeletal myoblasts and mouse embryonic stem cells as well (Fig. 3a) (Biehl et al. 2009; Thakar et al. 2008).

Such mechanisms also carried over to fully three-dimensional environments in which cells experience forces from all directions, leading to significantly different experience of extracellular forces and the adoption of more physiological cell morphologies (Cukierman et al. 2001; Geiger 2001). This is particularly relevant in damaged tissues where

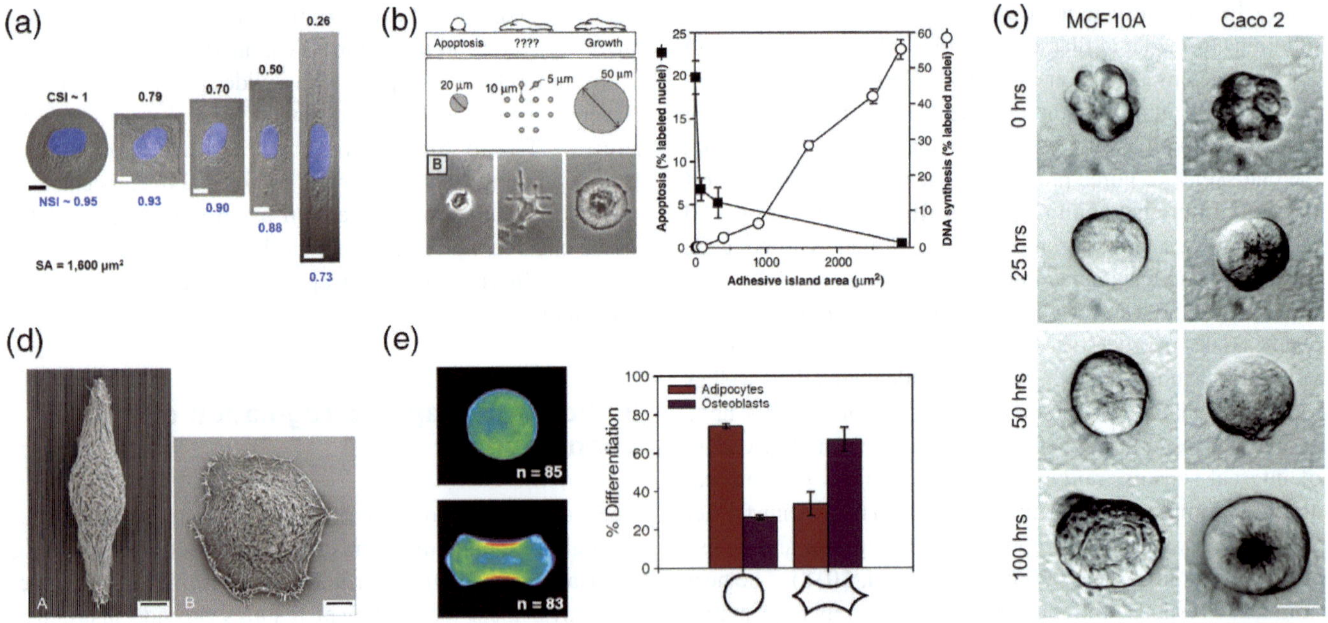

Fig. 2 Topographic cues are potent modulators of cell proliferation, migration and differentiation. a Endothelial cell and nuclear shape can be controlled by culturing cells on islands of varying shape and size (scale bars = 10 μm) (Versaevel et al. 2012). **b** Control of endothelial cell size can induce apoptosis or cell growth (Chen et al. 1997). **c** Geometric confinement of multi-cell clusters within microwells can induce cell polarization into fully lumenized spheroids (scale bar = 20 μm) (Cerchiari et al. 2015). **d** Corneal epithelial cells stretch and align with underlying microchannels compared to cells cultured on flat, which remain rounded (Teixeira 2003). **e** Changes in stem cell shape modulates myosin II activity and cell contractility to guide differentiation of stem cells into adipocytes or osteoblasts (Kilian et al. 2010). Figures reproduced with permission

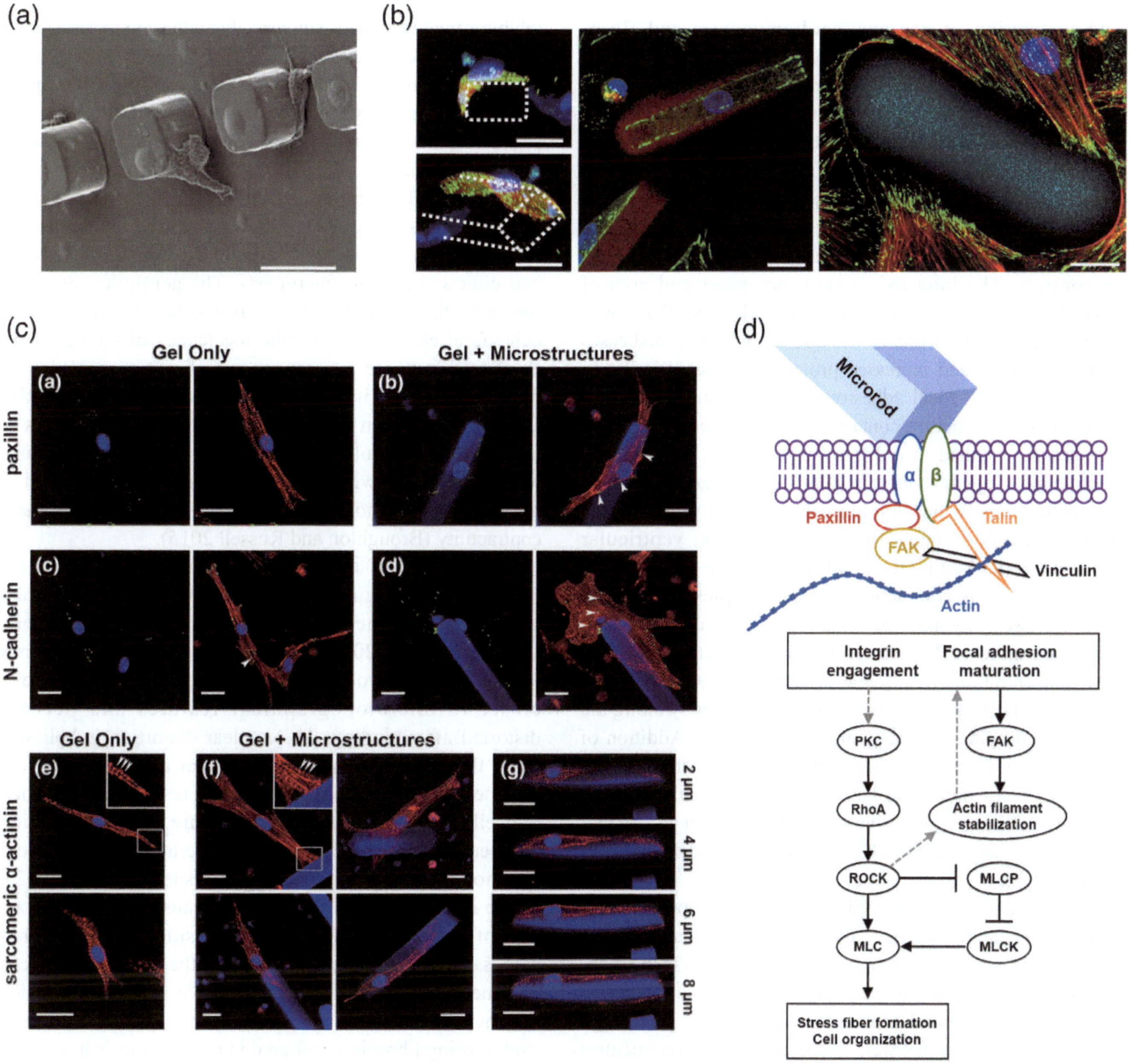

Fig. 3 Cellular interactions with discrete topographical cues. a Fibroblasts form intimate interactions with microscale posts by stretching to the top of microscale posts or pushing up against the base (scale bars = 25 μm) (Thakar et al. 2008). **b** Primary cardiac fibroblasts wrap around microrods and spread across the entire structure, accumulating focal adhesions on the edges of the rods (Green = paxillin, red = actin, blue = nuclei, cyan = microrod, scale bars = 20 μm) (Le et al. 2018). **c** Primary ventricular myocytes form strong associations with microrods in 3D that result in accumulation of focal adhesions at the cell-microrod interface. Clusters of cells are also able to interact with the microrods as well (red = α-actinin, green = paxillin, N-cadherin, blue = nuclei, scale bars = 20 μm) (Curtis et al. 2010). **d** Possible mechanism of microrod-affected mechanics. The schematic depicts the signaling pathways involved upon cell binding to microrods. Addition of these microrods introduces local heterogeneity, which guides the cells towards integrin engagement and focal adhesion formation. The maintenance of PKC and RhoA/ROCK signaling mediate the microrod-related changes in contractile behavior. Detailed pathways and actin stress fiber assembly vary with cell type. Figures reproduced with permission

the softened microenvironment and disruption of cell-cell and cell-ECM connections removes native instructional cues for development (Lindsey et al. 2003; Midwood et al. 2004; Prabhu and Frangogiannis 2016). Discrete micron-scale hydrogels, termed microrods, were incorporated into these soft 3D matrices to act as anchors for cell attachment and recapitulate native ECM mechanics (Fig. 3b-c) (Ayala et al. 2010; Collins et al. 2010a; Curtis et al. 2010, 2013; Pinney et al. 2014a). The structural guidance provided by these materials can instruct specific alignment and maturation of neurons, cardiomyocytes and fibroblasts in three dimensions (Ayala et al. 2010; Curtis et al. 2010; Le et al. 2018; Norman et al. 2008; Pinney et al. 2014a; Rose et al. 2017, 2018).

Myocytes cultured with these microrods formed intimate attachment to the microstructures and were able to pull and displace the microrods within the soft 3D matrix (Curtis et al. 2010, 2013). As a whole, myocytes generated greater magnitude of displacement in the presence of microrods compared to controls. The long axis length, perimeter and area of myocytes grown with microrods was much larger than those grown without. This interaction also significantly decreases shortening time and increased protein synthesis. Similar to previous studies, focal adhesions were accumulated at the myocyte-microstructure contact regions. Interestingly, these interactions resulted in a 2-fold higher rate of spontaneous contraction. Together, these results suggest that addition of microstructures, which introduce heterogeneity in stiffness and topography when culturing neonatal rat ventricular myocytes (NRVMs) in 3D, is sufficient to alter the hypertrophy, spontaneous contraction and gene expression. Neural cells are also responsive to these topographical cues in 3D, growing parallel to the discrete features (Rose et al. 2017). At low concentrations of discrete microstructures, dorsal root ganglions (DRGs) orient and grow unidirectionally along the structures despite minimal geometric constraint. Addition of adhesive motifs to the surface of these microstructures facilitated the interactions with DRGs, which were able to better align with those instructive features (Rose et al. 2018). These results suggest a mechanism that is dependent on adhesion strength.

A contractility-based mechanism was thus proposed to explain these changes. While challenge with ROCK and myosin light-chain kinase (MLCK) inhibitors did not reverse the global effect of microposts on cell proliferation, closer examination revealed that inhibitor treatment increased the percentage of proliferating cells that were in contact with microstructures (Thakar et al. 2008). Additionally, inhibitor treatment reduced the number of adherent cells, likely due to reduced ability to pull on the substrates and form fully matured focal adhesions (Boateng et al. 2003). Patel et al utilized time lapse imaging to examine the interaction of single cells with microtopographical cues (Patel et al. 2010). It was observed that cells cultured on micropost-textured scaffolds had greater mean tether lengths compared to cells cultured on flat substrates. After coming into contact with these microstructural anomalies, cell de-adhesion from the microposts was also slowed. Both results suggest that micropost interactions alter the balance between cell contractility and adhesion strength. While ROCK inhibition abolished the differences between micropost versus flat culture, the differences persisted during treatment with MLCK inhibitor, with micropost-textured scaffolds having similar tether lengths to flat scaffolds without inhibitor treatment. This suggests that adhesion to microposts make up for the reduced contractility induced by MLCK treatment, but is not sufficient to affect upstream ROCK activity (Fig. 3d). The dependence on contractility and exertion of active forces was confirmed by comparing the interaction of fibroblasts with immobilized and free microspheres on a flat 2D substrate (Boateng et al. 2003). Free microspheres were easily moved around by the fibroblasts and had no impact on fibroblast proliferation, while immobilized microspheres elicited effects similar to microposts. The generation of active forces by the cell must be responsible for changes in these behavioral changes. While the mechanics of the cell were unaffected by these interactions, significant upregulation of myosin contractility pathway elements, including Rho GTPase and myosin II heavy chain suggest local changes in cellular mechanobiology at the adhesion sites. These results were corroborated with later studies investigating the effect of microenvironment topography on cardiomyocyte subdomain contractility (Broughton and Russell 2015).

While many of these responses are thought to be focused at the cell surface, several reports have suggested modulation of gene expression by altering nuclear shape (Downing et al. 2013; McNamara et al. 2012; Thomas et al. 2002). Fibroblast nuclei were observed to move closer to microtopographical features and become distorted after binding. The nuclear distortion is believed to be the result of reactive forces that are exerted on the cell membrane and transmitted to the rest of the cell when the cell pushes against a microstructure. This deformation has been shown to affect cell proliferation and gene expression (Thomas et al. 2002; Versaevel et al. 2012). These responses are believed to be caused by reorganization of the nuclear lamina and repositioning of chromosomes. Depending on new position, the chromosomes can become more heterochromatic and have reduced expression of genes (Finlan et al. 2008; McNamara et al. 2012). Previous studies using fibroblasts cultured in microchannels have demonstrated that nuclear shape, chromosome localization, and histone modifications are affected by topography and can be used to facilitate the restoration of epigenetic state for cell reprogramming (Downing et al. 2013). These mechanical effects are strong enough to replace the use of histone deacetylase and histone demethylase inhibitors for generation of iPSCs.

These observations suggest that interaction with topography can modulate cell gene expression via both direct chromosomal modifications and by transcriptional regulation of genes initiated at focal adhesions. Despite the complex mechanisms at play, it is clear that topography is an effective tool for modulating cell behavior. The ability to reprogram cells to a more physiological phenotype would enable us to harness intrinsic healing mechanisms and promote healthy repair of tissues.

4 Using microstructural cues for therapeutic applications

Modifying the physical and biochemical properties of extracellular matrix of diseased tissues at the micron scale is an emerging strategy for enhancing our currently limited capabilities in regenerating complex tissues. It is widely recognized that the success of cellular replacement therapies has largely been hampered by poor engraftment with the local microenvironment. For example, after acute myocardial infarction (MI), widespread cell death and the release of intracellular components initiate a complex cascade of signaling events that lead to pathological matrix remodeling and disruption of the physicochemical and biochemical homeostasis of the ECM (Frangogiannis 2015; Li et al. 2018; Segura et al. 2014; Talman and Ruskoaho 2016). While these perturbations are often resolved through intrinsic wound healing mechanisms in some tissues, myocardial tissue damage is characterized by continued aberrant tissue remodeling and the formation of a stiff myocardial scar that permanently disrupts cell-cell and cell-ECM interactions. Further, the lack of adequate nutrient supply in this avascular scar challenges the survival of transplanted stem or differentiated cell types seeking to replace the lost tissue.

Injectable hydrogel materials have been the focus of bioengineering approaches to create suitable extracellular environments that promote cell engraftment (Dimatteo et al. 2018; Tibbitt and Anseth 2009; Zhu et al. 2017). While it is possible to engineer the biophysical and biochemical makeup of these materials to support cell attachment, growth and differentiation, these materials tend to be physically and chemically homogeneous and anisotropic in nature. This is unrepresentative of the native ECM which consists of chemical gradients and biophysical microdomains that facilitate directional cell migration and physiological behavior. There are, indeed, engineering techniques available to introduce spatial heterogeneity into these hydrogel materials, but these approaches tend to involve foaming agents, porogens, or other cytotoxic materials and are not suitable for coadministration with live cells (Annabi et al. 2010).

To address this, discrete microengineered materials have been used to introduce mechanical anomalies to target tissues, promoting cellular attachment and modulating the behavior of both native and transplanted cell types. Xin et al reported the assembly of poly(ethylene glycol) microgels that can be co-injected with cells and later be annealed into a porous 3D scaffold, termed granular hydrogel, *in vivo* (Xin et al. 2018). These hydrogels were designed to have highly interconnected microporous structure and have Young's moduli that span the range of physiological stiffnesses. Human mesenchymal stem cells (hMSCs) cultured with these microgels spread and surrounded the structures, growing within and throughout the pores to form a 3D cellular network. This was in contrast to bulk gels which had limited cell spreading. Further, it was observed that YAP nuclear intensity was two-fold higher for hMSCs cultured in the stiff (35 kPa) hydrogels compared to the softer hydrogels. Normally, hMSCs are observed to have decrease YAP nuclear staining on stiff hydrogels but in this 3D environment, the effects appear to be reversed. This can be attributed to differences in cytoskeletal tension in 3D matrices compared to 2D scaffolds. In two dimensional environments, YAP and TAZ expression levels are decreased on softer substrates where there is low cellular tension and reduced stress fiber formation (Dupont et al. 2011). Soft 3D environments, however, lead to reactive stiffening of fibroblasts and mechanically stressed cellular responses, such as increased production of ECM proteins. Rigid microstructures within this soft environment can act as anchors to facilitate appropriate generation of cellular tension and reduces production of YAP/TAZ compared to conditions without domains for cell attachment. This highlights the importance of mechanical microdomains within a 3D scaffold to act as anchors for cell attachment and alleviate aberrant cellular tension to decrease stress-induced mechanotransductive pathways.

While the above study illustrates the benefits of providing microdomains of increased stiffness to promote cell engraftment, others have utilized microstructural cues to directly modulate cellular phenotype and prevent pathological conditions. The application of microstructures to provide instructive signals and modulate cell behavior is highlighted in early studies by Norman et al. (2008). Anisotropic microscale hydrogels, termed microrods, were fabricated via photolithography and used to recapitulate the complex 3D patterning and regional architecture of native tissues (Fig. 4a). These microrods measured 15 $\mu m \times 15$ $\mu m \times 100$ μm to approximate the size of the cardiac muscle cell. This anisotropy has also been demonstrated to inhibit phagocytosis by macrophages, a desired trait for therapeutic applications (Champion et al. 2007; Champion and Mitragotri 2006, 2009). It was demonstrated that incorporation of these microrods into a 3D construct provided microdomains of high stiffness facilitated fibroblast attachment. Upon attaching to these microrods, the fibroblasts lost their dendritic morphology that was common to fibroblasts in 3D matrices and took on a morphology that was similar to 2D substrates, with stress fibers forming between the focal adhesions on the microrods. This intimate interaction stunted proliferation of fibroblasts and suggested potential applications for fibrosis.

Ayala et al extended this work, utilizing free poly(ethylene glycol) dimethacrylate (PEGDMA) based microrods in 3D culture, to attenuate fibroblast proliferation and inhibit expression of key myofibroblast markers implicated in cardiac fibrosis (Fig. 4b-c) (Ayala et al. 2010). Similar to the microrods generated by Norman et al, these PEGDMA microrods are reproducibly fabricated on silicon wafers via a scalable photolithography approach and can be easily removed from the

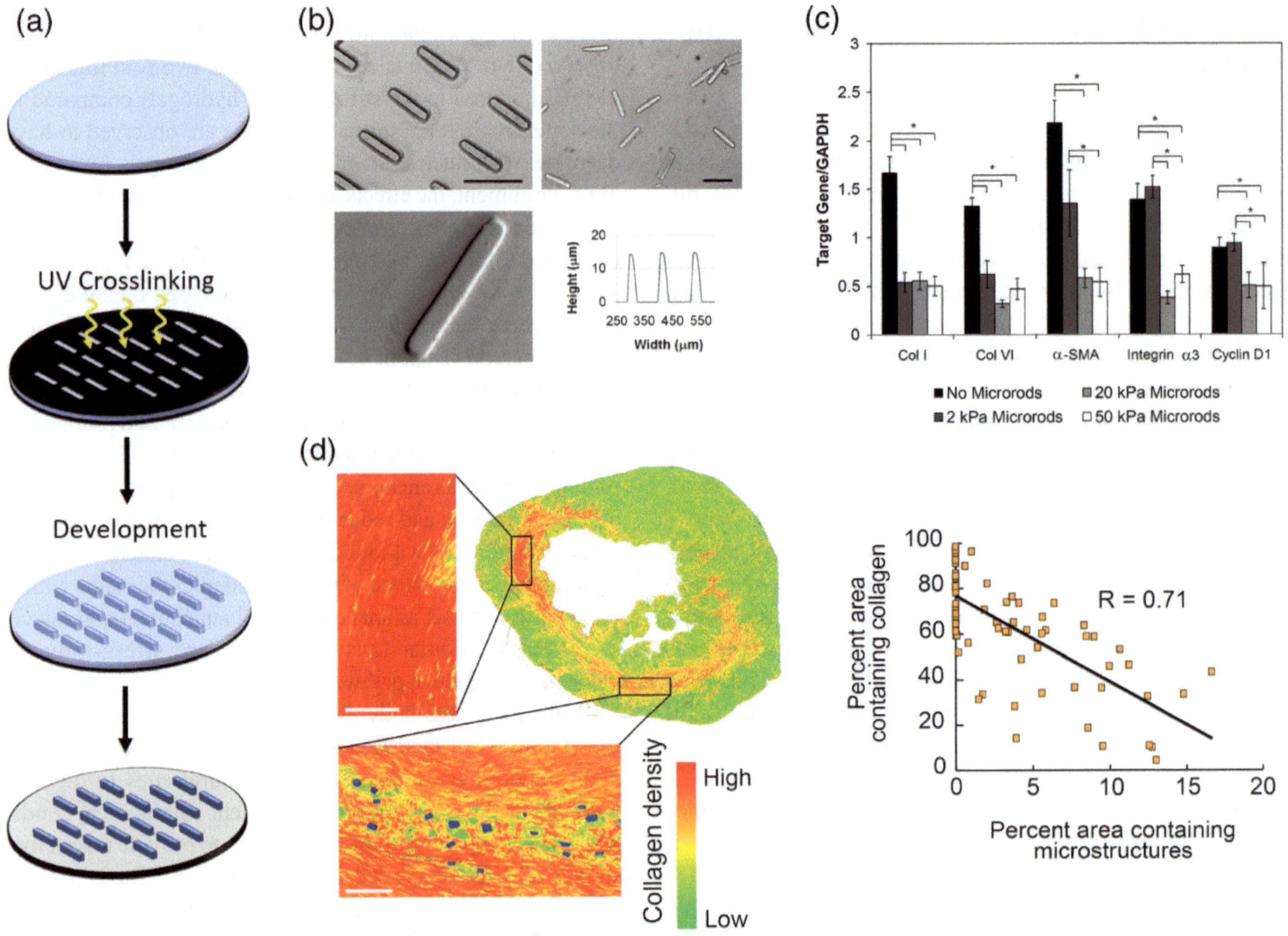

Fig. 4 Discrete microstructural cues attenuate chronic cardiac fibrosis. a Anisotropic micron-scale hydrogels, termed "microrods" can be reproducibly fabricated by UV photolithography (Le et al. 2018). **b** Microrod dimensions can be controlled and are uniform across the wafer (scale bars = 100 μm) (Ayala et al. 2010). **c** PEGDMA-based microrods of different stiffness have varied response on extracellular matrix synthesis and gene expression in 3D gels (Ayala et al. 2010). **d** Collagen production is reduced proximal to microstructures in infarcted rat hearts, with increasing numbers of microstructures reducing total collagen area (scale bar = 100 μm) (Pinney et al. 2014a). Figures reproduced with permission

wafer and stored until use (Norman et al. 2008). When added to fibroblast cell culture in a 3D Matrigel scaffold, fibroblasts interacted directly with the microrods, stretching along individual microrods as well as across multiple structures. Similarly, this interaction led to reduced proliferation of fibroblast cells. Key genes associated with the fibroblast to myofibroblast transformation were examined in the presence of microrods of varying Young's moduli. Cells cultured with microrods that were at least 20 kPa had significantly reduced expression of collagen I, collagen VI and αSMA, genes known to be upregulated in fibrosis (Fig. 4c). Collagen I is the primary form of fibrillar collagen that makes up the bulk of scar tissue and is responsible for pathological stiffening of tissues. Expression of αSMA in fibroblast is a key marker for the transformation into the contractile myofibroblast phenotype and has been shown to be influenced by the mechanical microenvironment (Hinz 2007, 2010; Santiago et al. 2010). These effects were also found to be dependent on contractility, as treatment of the cells with MLCK or ROCK inhibitors (ML7 and Y27632, respectively) abolished the observed effects. This is in agreement with previous studies illustrating that contractile force generation is required for the cells to interact with microtopographical cues (Boateng et al. 2003; Curtis et al. 2010; Patel et al. 2010).

The potential for these PEGDMA microrods to be used as a therapeutic strategy *in vivo* was investigated using a rat model of myocardial infarction (Pinney et al. 2014a). As previously mentioned, the acute tissue damage arising from a MI culminates in the formation of a dense avascular scar that obstructs tissue regeneration, both intrinsic healing mechanisms and exogenous interventions (Frangogiannis 2015; Li et al. 2018; Segura et al. 2014; Talman and Ruskoaho 2016). The ability to attenuate this runaway fibrotic response would promote the survival of native and transplanted cell populations, making way for healthy tissue repair. Pinney et al probed the effect of microstructure administration on matrix remodeling by fibroblasts, namely matrix metalloproteinase 2 (MMP-2) production and transforming growth factor β (TGFβ) signaling (Pinney et al. 2014a). MMP-2 is upregulated within the first few days after MI and is responsible for degradation of the ECM to make way for macrophages to clear out cellular debris (Chen et al. 2005). This ECM degradation releases latent TGFβ that triggers myofibroblast transformation and overproduction of ECM components and pathological tissue repair (Frangogiannis 2012, 2014; Prabhu and Frangogiannis 2016).

When injected into the infarct zone of the myocardium via transthoracic injection, rats treated with PEGDMA microrods had significantly reduced deterioration of cardiac function compared to saline controls (Pinney et al. 2014a). The observed functional improvements appeared to be dose dependent, with increased numbers of delivered microstructures resulting in better cardiac output. These functional improvements were accompanied by reduced collagen deposition in the periphery of the microstructures, increased vascularization of the left ventricle and reduced TGFβ production (Fig. 4d). These outcomes are believed to be the result of introducing artificial anchors for cells to bind to in the softened, post-infarct matrix. This attachment alleviates mechanical stress-induced upregulation of TGFβ release and collagen deposition and promotes deposition of softer ECM components that make up the healthy uninjured microenvironment. Such events would allow for native tissue repair processes to occur and be amenable to cell replacement therapies.

It is important to note that the effects of these anisotropic microstructural cues vary from cell type to cell type. While proliferation appears to be stunted by fibroblast-microstructure interactions, the opposite was observed for hMSCs (Collins et al. 2010a). Similar to studies by Xin et al, hMSCs spread out with fingerlike projections to attach to the PEGDMA microrods (Xin et al. 2018). The hMSCs adopted a more flattened and polarized morphology in the presence of microrods. This change in morphology was accompanied by increased proliferation, leading to larger cluster sizes in 3D culture. Additionally, hMSCs had reduced penchant for osteogenic differentiation when cultured with these microrods, as depicted by reduced bone morphogenic protein 6 (BMP-6) and Col1A1 gene expression. It is unclear whether the topographical interactions or paracrine signaling due to larger clusters are responsible for this change. The many possible mechanisms highlight that topographical cues can affect cells in ways that have yet to be elucidated.

5 Synergistic effects of biochemical signaling and microtopography

The discussion of topographical signaling thus far has focused on the use of bio-inert micromaterials, enabling isolation of biophysical effects alone. While a powerful and instructive mechanism, it is crucial to recognize the complex interplay of biophysical cues and biochemical factors, both soluble and immobilized, that are present in the native ECM contribute to cell function (Adams and Watt 1993; Frantz et al. 2010; Hynes 2009). Among the major components of the ECM is hyaluronic acid (HA), a widely distributed glycosaminoglycan that plays a wide array of roles in inflammation and wound healing (Chen and Abatangelo 1999; Gao et al. 2010; Litwiniuk et al. 2016; Petrey and de la Motte 2014). Many cell types express cell surface receptors, such as CD44 and receptor for hyaluronan-mediated motility (RHAMM), that can directly bind with the HA backbone and undergo changes in cell proliferation and migration (Chen and Abatangelo 1999; Jiang et al. 2007; Litwiniuk et al. 2016; Petrey and de la Motte 2014; Toole 2004; Turley et al. 2002). As such, HA has been used for tissue engineering applications (Burdick and Prestwich 2011; Ifkovits et al. 2010; Tous et al. 2011; Yoon et al. 2009). Sideris et al utilized microfluidics devices to fabricate injectable HA microparticles that can be annealed into a microporous 3D scaffold in the presence of cells with limited cytotoxicity (Sideris et al. 2016). These scaffolds supported the attachment and growth of human dermal fibroblasts, which eventually adopted a spread morphology in only 2 days. HA microparticles with guest-host motifs on the particle surface were developed to generate shear-thinning granular hydrogels that can be injected into tissues and immediately set (Mealy et al. 2018). MMP degradable motifs were incorporated into these hydrogel materials and demonstrated to degrade in 2 weeks when injected into the infarcted myocardium, due to upregulated expression of MMP relative to healthy controls.

Le et al demonstrated the fabrication of discrete HA-based microstructures for direct reprogramming of fibroblasts (Le et al. 2018). Studies have demonstrated the interplay of CD44 signaling with integrin-mediated mechanotransduction, suggesting improved mechanosensitivity on HA substrates compared to bioinert substrates (Chopra et al. 2012, 2014). Indeed, the interactions of fibroblasts with HA microrods appeared to be more intimate than bio-inert PEGDMA microrods, resulting in more dramatic downregulation of fibrosis markers, col1A2 and αSMA, as well as MMP-2 and TGFβ signaling components. These phenotypic alterations *in vitro* were translated into functional outcomes in a rat model of MI. Rats treated with HA microrods were able to attenuate the loss in cardiac function and preserve myocardial wall structure compared to saline controls. Soluble HA was also administered as a material control and was able to demonstrate modest improvements as well. These effects are likely attributed to the anti-inflammatory effects of HA which can alleviate the extreme cellular responses after acute tissue injury. The combination of material and topographical cues is hypothesized to synergistically improve these outcomes. It should be noted that the number/volume of HA microrods administered was much less than PEGDMA microrods, which were previously shown to have a dose dependent effect. Together, these data suggest that the synergistic effects of biochemical signaling of HA with microtopographical cues are more potent than either effect alone.

An additional advantage of using hyaluronic acid is the inherent biodegradability of the material. HA is degraded by endogenous enzymatic mechanisms, releasing small oligosaccharides that have been implicated in promoting cell migration and growth (Cyphert et al. 2015; Gao et al. 2010; Jiang et al. 2007; Litwiniuk et al. 2016; Noble 2002). A key limitation of bio-inert materials is their longevity in tissues after injection. Complete repair of tissue necessitates the clearance of exogenous materials to make way for cell migration and growth to repopulate the damaged tissue. Both HA microrods and granular HA hydrogels were shown to degrade after a few weeks *in vitro* and *in vivo* (Le et al. 2018; Mealy et al. 2018).

Aside from microgel backbone, there is great promise in functionalizing the surfaces of these materials to present binding motifs that are capable of eliciting specific cell responses. Anisotropic PEG based microgels functionalized with RGD motifs encouraged cell attachment via integrin binding, increasing total cell contact area (Rose et al. 2018). Additionally, cells in contact with RGD-PEG microgels had strong stress fiber signals and more elongated cell nuclei, suggesting significant cytoskeletal forces. These forces are likely due to the microdomains of high local stiffness relative to the surrounding matrix. This was supported by studies indicating increased shuttling of YAP into the nucleus in the presence of microgels, with modified microgels having higher YAP localization to the nucleus. When aligned in one uniform direction, PEGDMA microgels that were modified with RGD were able to promote neuron alignment to a greater degree than unmodified PEGDMA microgels. The difference between modified and unmodified microgels was muted when cultured with chicken derived dorsal root ganglions. These cells are able to produce their own fibronectin and generate binding sites on the microstructures. This mechanism is consistent with observations that fibronectin can readily adsorb to hydrogels and facilitate binding of cells with otherwise non-adhesive materials (Chopra et al. 2014; Khademhosseini et al. 2006a).

This complex interplay between surface receptors and topography is not limited to traditionally mechanosensitive cell types. CD8+ T cells have also been shown to have increased activation and proliferation when presented with ellipsoidal, rather than spherical, antigen-presenting microparticles with equivalent antigen dose, co-localization and density (Sunshine et al. 2014). These effects are hypothesized to be a result of contact angle with the microparticles – when a T cell interacts with an isotropic microparticle, the membrane reorganizes and orients to maximize contact with the long axis of the microparticle. This results in increased frequency and size of the T cell-microparticle contact compared to spherical particles. When injected into a B16 melanoma tumor model, the ellipsoidal antigen-presenting microparticles elicited reduced tumor size and improved survival as compared to spherical particles and T cells alone.

6 Discrete microstructures as drug delivery devices

Proteins and small molecule drugs can be incorporated into the microparticle bodies for simultaneous delivery of therapeutics and microtopographical cues. These materials are thus able to modulate both the biomechanical and biochemical microenvironment to encourage healthy tissue repair. Doroudian et al described loading of PEGDMA microrods with mechano-growth factor (MGF) peptide to reduce cell apoptosis after myocardial infarction (Fig. 5) (Doroudian et al. 2014). Native MGF has been shown to block the apoptosis of injured myocytes as well as attract stem cell populations (Ates et al. 2007; Carpenter et al. 2008; Collins et al. 2010b; Mills et al. 2007; Musaro et al. 2004). MGF-peptide was eluted from PEGDMA microrods over the course of 2 weeks and maintained their biological activity, as indicated by reduced TUNEL staining and increased Bcl-2 expression (Fig. 5c-e) (Doroudian et al. 2014). These biochemical effects were maintained and independent from the mechanical effects exerted by topography, which remodeled hMSC morphology and stunted hMSC growth (Fig. 5a, b). Interestingly, this was in contrast to previous studies with hMSCs and PEGDMA microrods, likely due to differences in cytoskeletal tension that is generated in 2D compared to 3D culture (Collins et al. 2010a). Direct intracardiac injection of these MGF-peptide loaded microrods into a rat model of MI resulted in improved cardiac function and long-term survival as compared to saline injections and empty microrod injections (Peña et al. 2015).

Loading of a drug into microrods may also be useful for wound healing applications. PEGDMA microrods were loaded with high concentrations of neomycin, a small molecule drug capable promoting cell migration *in vitro* through modulation of lipid signaling, and delivery was maintained for over 12 h. When cultured with primary rat primary fibroblasts, the drug released to the media significantly increased the migration velocity of cells into the gap in a wound closure assay (Mkrtschjan et al. 2018a). Hence, development of these neomycin microrod devices for a therapeutic application might improve wound healing.

As discrete hydrogel materials, microstructures can also be independently fabricated and loaded with drug. It is possible to combine multiple types of drug-loaded microstructures that exhibit different release profiles to specifically address a variety of disease states (Fig. 6a-d) (Mealy et al. 2018). Further, the spatial distribution of multiple microstructure formulations in tissues can be controlled in order to recapitulate the biochemical gradients observed in native tissues (Fig. 6e)

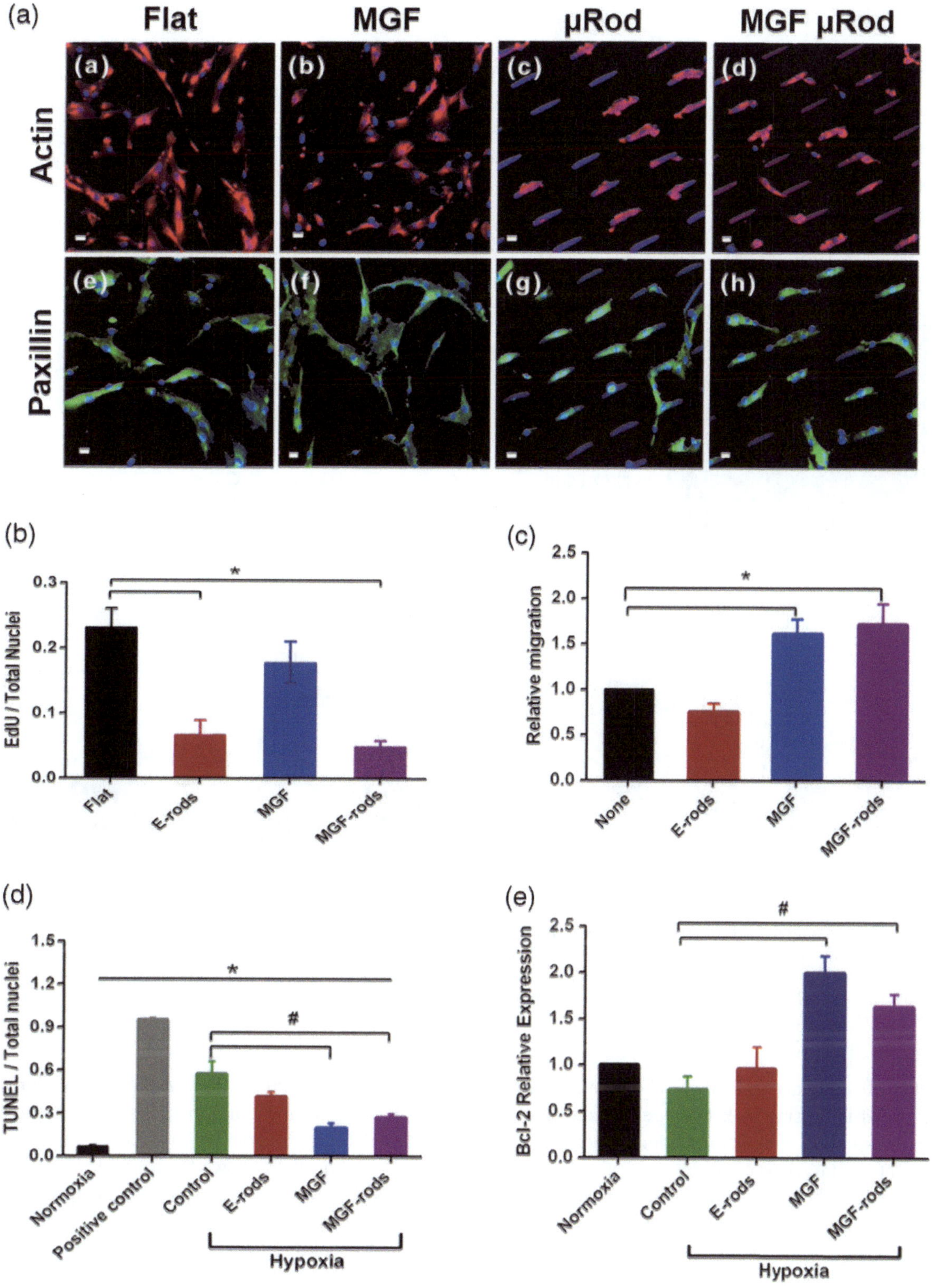

Fig. 5 **MGF-peptide loaded microrods exert biophysical effects of topography and biochemical effects of peptide administration on hMSCs and myocytes** (Doroudian et al. 2014). **a** Microrods remodel hMSC morphology, the actin cytoskeleton and focal adhesions while MGF administration has no effect (red = actin, green = paxillin, blue = nuclei and microrods, scale bar = 20 μm). **b** Microrods blunt the proliferation of hMSCs through mechanical signals and **c** increase migration via eluted MGF, which remains bioactive throughout microrod fabrication. Eluted MGF protected NRVM from apoptosis induced by hypoxia, as assessed by **d** TUNEL and **e** Bcl-2 expression. Figures reproduced with permission

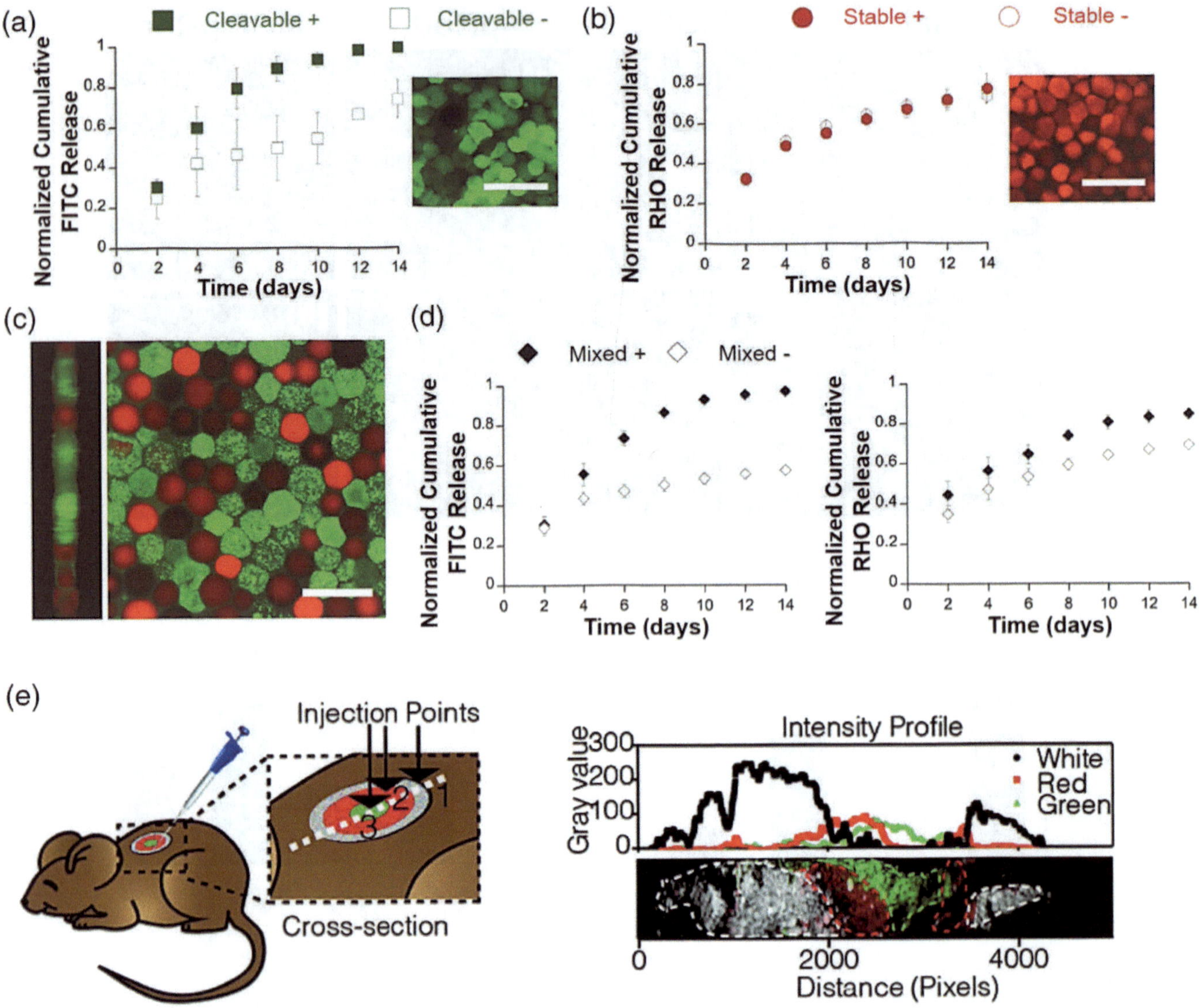

Fig. 6 Drug elution and spatial distribution of injected microparticles can be tuned for therapeutic applications *in vivo*. **a, b** FITC-Albumin and rhodamine-dextran release from hydrogel microparticles can be tuned by inclusion of cleavable linkers (Mealy et al. 2018). **c, d** Microparticles can be combined and simultaneously elute two drugs with differential release kinetics. **e** Spatial distribution of three types of microparticles can be controlled during injection (Darling et al. 2018). Figures reproduced with permission

(Darling et al. 2018; Mealy et al. 2018). These capabilities are otherwise impossible using bulk hydrogel injections which are homogenous and are difficult to tune for multidrug release. In sum, discrete hydrogel materials are a versatile technology that can be used for therapeutic applications where tight spatiotemporal control of biochemical gradients is necessary for healthy tissue repair.

7 Remote control of injectable topographical cues

The random orientation of discrete microstructural cues after injection makes these systems unpredictable directors of cell behavior. Several groups have attempted to address this by incorporating magnetic nanoparticles into the microstructure scaffold. Utilizing anisotropic materials, application of an external magnetic field allows for controlled alignment of the microstructures. This is particularly beneficial for tissues in which proper alignment is crucial for organ function, such as in neural or cardiac spaces. Pinney et al demonstrated the successful alignment of PEGDMA based microstructures that were loaded with PEGDMA-coated superparamagnetic iron oxide nanoparticles (SPIONs) (Pinney et al. 2014b). Covalent attachment of SPIONs to the microstructure matrix allowed for stronger responses to an external magnetic field. Through soft lithography, Rose et al were able to fabricate anisotropic PEG-based microstructures that could also be aligned via an external magnetic field (Rose et al. 2018; Rose et al. 2017). These materials oriented the growth of fibroblasts and dorsal root ganglions via contact guidance. Additionally, these interactions induced the shuttling of YAP into the nucleus. Further conjugation with adhesive ligands did not result in improved neurite growth and alignment, confirming mechanosensory mechanisms as the dominant effector.

The ability to remotely control these topographical cues also have important implications for tissue engineering. Work in our lab has utilized SPION-loaded microrods as force actuators to study the contractile behavior of living cardiomyocytes following acute loading. This application of load was compatible with live cell imaging techniques. Briefly, SPIONs were synthesized via precipitation of iron salts and surface modified to add an additional silica-thiol coating for improved particle stability in solution. The addition of thiol functional groups on the silica shell surface provides reactive handles that enabled covalently crosslinking of SPIONs to a PEGDMA polymer matrix via thiolene-click chemistry (Fig. 7a). Magnetic microstructures, named micromagnets, were then fabricated using previously described photolithographic techniques.

Once the cells were attached, line scans were taken to measure beating before and after loading cardiomyocytes (CM) through the micromagnets (Fig. 7b). Kymographs showed an increase in contractile displacement, as well as permitting measurements for the contractile rising and relaxation falling phases. Changes in contractile metrics due to micromagnet

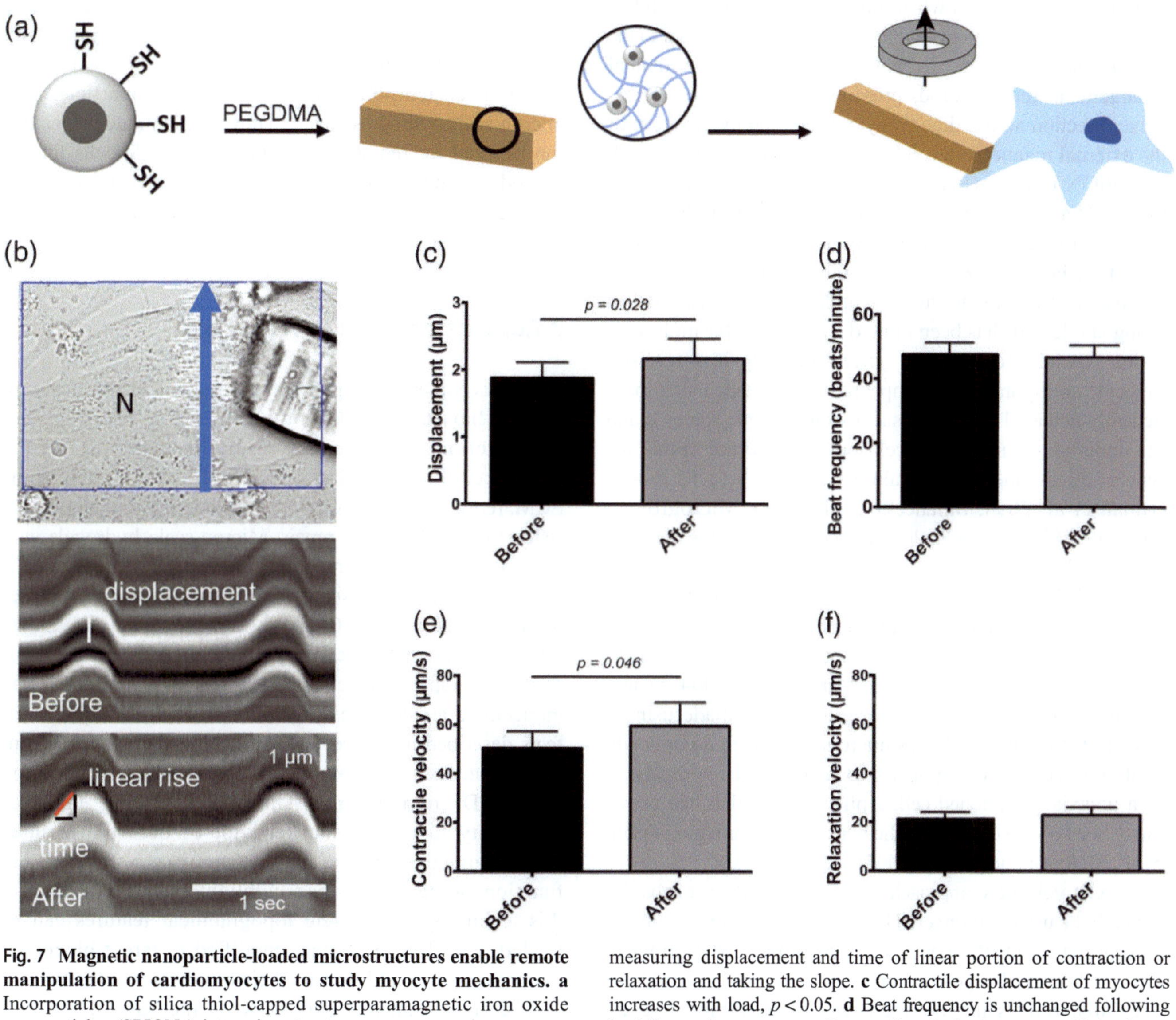

Fig. 7 Magnetic nanoparticle-loaded microstructures enable remote manipulation of cardiomyocytes to study myocyte mechanics. a Incorporation of silica thiol-capped superparamagnetic iron oxide nanoparticles (SPIONs) into microstructures generates micromagnets that enable remote manipulation of cells upon focal-adhesion attachment. **b** Line scan kymographs before and after addition of magnet to provide load. Contractile velocities are obtained by measuring displacement and time of linear portion of contraction or relaxation and taking the slope. **c** Contractile displacement of myocytes increases with load, $p < 0.05$. **d** Beat frequency is unchanged following load from micromagnets. **e** Contractile velocity increases with load, $p < 0.05$. **f** Relaxation velocity is unchanged following load from SPION micromagnets

loading was studied in 30 cells. As determined by an increase in contractile displacement, success rate for eliciting a response through micromagnet loading was approximately 66.6% (20 out of 30 cells), though all experiments were included in the analysis. Contractile displacement was significantly different in cells following loading, increasing from 1.88 μm to 2.17 μm ($p < 0.05$) (Fig. 7c). Additionally, a change in contractile velocity was observed between the two conditions, increasing from 50.47 μm/s to 59.48 μm/s following loading ($p < 0.05$) (Fig. 7e).

CMs interacted directly with the micromagnets such that simple placement of an external neodymium magnet allowed accurate measurement of transverse loading on CM contractility using phase microscopy. The focal adhesions formed between the microrods and the CMs were strong enough to load and displace CMs when the external magnet was introduced. The displacement resulted in significant changes to the rate of rise in tension of the CMs. With this experimental arrangement, the microrods provide load mainly in a transverse direction across the myocyte. Changing the position of the external magnet to the side of the culture dish would pull the SPION microrods along the longitudinal axis of the CM that could allow for sarcomere length dependent studies at the single cell level, potentially recapitulating the Frank-Starling law of the heart (Katz 2002; de Tombe and ter Keurs 2016).

To date, the ability to study the effect of mechanical load on living single cells has been limited by the need for high resolution spatial and temporal optical imaging in a realistic myocyte configuration. Attempts have been made using magnetically actuated micropost surfaces to provide forces along the underlying surface of a cell, but these do not mimic the type of three-dimensional strain experienced by cells *in vivo* (Bidan et al. 2018; Sniadecki et al. 2008). Microgroove-aligned CMs were cyclically strained using the Flexcell device, but these cells could not be imaged while mechanically deformed (Motlagh et al. 2003; Senyo et al. 2007). Signals and remodeling were also studied after static strain or after bouts of exercise, but these were not live cell recordings (Li and Russell 2013; Lin et al. 2013; Mansour et al. 2004; Yu and Russell 2005). More recently, advances were made using a system that combines PDMS microgrooves with an optically ready uniaxial stretching device to capture sarcomerogenesis of a uniaxially stretched cell, a phenomenon that had previously been observed without the benefit of live imaging (Yang et al. 2016; Yu and Russell 2005).

A relatively new approach is to study cardiac mechanics through the use of engineered heart tissues. This approach can seed primary rodent myocytes or human induced pluripotent stem cell-derived myocytes onto protein-based scaffolds or decellularized tissues between two pillars (Schwan et al. 2016). This method is advantageous, as cells derived from patient with specific mutations can be studied in a tissue-like preparation. A similar method has been used recently while adjusting the afterload experienced by the spontaneously contracting construct in order to mature the tissue, highlighting the importance of the mechanical conditions used in cell and tissue culture (Leonard et al. 2018). The application of tissue engineering approaches is advancing rapidly both for replacement organs and for drug testing (Ronaldson-Bouchard and Vunjak-Novakovic 2018; Zhang et al. 2018).

Strain deforms the proteins at the cell membrane, which is known to trigger a variety of signaling pathways such as PKA and PLC and activates electrophysiological channels leading to the contractile changes observed (Pfeiffer et al. 2014). Stretch-activated channels include the L-type calcium channel, which results in an increased intracellular calcium. Troponin C (TnC) binds the calcium allowing the myosin ATPase to produce the force with the actin in the thin filament (Puglisi et al. 2014). The angiotensin II receptor type 1 is also activated by mechanical stimulation and could potentially play a role in altered contractile behavior (Yasuda et al. 2008). The incorporation of magnetic nanoparticles into micron-scale hydrogels creates a novel and unique device that can be used for remote manipulation of single cells for advanced cell and molecular physiological studies. Its use could be extended for studies of mechanotransduction and sarcomere assembly.

8 Conclusions and perspectives

The extracellular matrix provides tissues with mechanical support and bioactive signals for the maintenance of cell survival and function. Tissue injury results in disruption of the established biochemical and mechanical equilibrium in the ECM, requiring restoration of these properties to their original states for healthy tissue repair. Micron-scale hydrogels with tunable geometry and mechanical properties provide cells with anchor points that serve as replacements for binding sites present in the native ECM. While there are still unknown intricacies to the underlying mechanisms of these cell-microtopography interactions, there is a clear dependence on intracellular architecture and tension to elicit the reported effects described in this review. These mechanisms are consistent between multiple cell types and carry over from 2D to 3D environments. This understanding enhances our ability to control the behavior of both transplanted and endogenous cell types for improved survival and function. Recent work has highlighted the versatility of this technology – discrete topographical features can be loaded with drug or surface modified to target biochemical pathways in addition to their biophysical effects.

The significant interplay between mechanical and biochemical cues has been demonstrated but the details of these interactions are still largely elusive. More in-depth study of these synergies would be made possible by simultaneous physical manipulation of cells *in vitro* via micromagnets and treatment with growth factors or signaling ligands. Further, the optimal spatial and temporal distribution of these mechanical and biochemical effects has yet to be determined. When are topographical cues necessary for treating pathological conditions and what time-frame of growth factor release is necessary? The use of degradable materials for these approaches provides additional knobs to tune for fabrication of patient- and disease-specific microstructures. Research in the past two decades using bioengineered microtopography clearly shows that how cells "hang on tight" in three dimensions is a major driver of cell structure and function. However, more work needs to be done to bring this ever-growing knowledge of basic discovery into the practical realm for translation into useful clinical application.

Acknowledgements This work was supported by the National Institutes of Health (P01-HL062426 and R01-HL137209).

Compliance with ethical standards

Conflict of interest The authors declare that they have no conflict of interest.

References

J.C. Adams, F.M. Watt, Development **117**, 1183 (1993)
S. Al-Haque, J.W. Miklas, N. Feric, L.L.Y. Chiu, W.L.K. Chen, C.A. Simmons, M. Radisic, Macromol. Biosci. **12**, 1342 (2012)
N. Annabi, J.W. Nichol, X. Zhong, C. Ji, S. Koshy, A. Khademhosseini, F. Dehghani, Tissue Eng. Part B Rev. **16**, 371 (2010)
Y. Asazuma-Nakamura, P. Dai, Y. Harada, Y. Jiang, K. Hamaoka, T. Takamatsu, Exp. Cell Res. **315**, 1190 (2009)
K. Ates, S.Y. Yang, R.W. Orrell, A.C.M. Sinanan, P. Simons, A. Solomon, S. Beech, G. Goldspink, M.P. Lewis, FEBS Lett. **581**, 2727 (2007)
P. Ayala, J.I. Lopez, T.A. Desai, Tissue Eng. Part A **16**, 2519 (2010)
C.M. Bidan, M. Fratzl, A. Coullomb, P. Moreau, A.H. Lombard, I. Wang, M. Balland, T. Boudou, N.M. Dempsey, T. Devillers, A. Dupont, Sci. Rep. **8**, 1464 (2018)
J.K. Biehl, S. Yamanaka, T.A. Desai, K.R. Boheler, B. Russell, Dev. Dyn. **238**, 1964 (2009)
S.Y. Boateng, T.J. Hartman, N. Ahluwalia, H. Vidula, T.A. Desai, B. Russell, Am. J. Physiol. Physiol. **285**, C171 (2003)
K.M. Broughton, B. Russell, Biomech. Model. Mechanobiol. **14**, 589 (2015)
J.A. Burdick, G.D. Prestwich, Adv. Mater. **23**, H41 (2011)
V. Carpenter, K. Matthews, G. Devlin, S. Stuart, J. Jensen, J. Conaglen, F. Jeanplong, P. Goldspink, S.-Y. Yang, G. Goldspink, J. Bass, C. McMahon, Hear. Lung Circ. **17**, 33 (2008)
A. Cerchiari, J.C. Garbe, M.E. Todhunter, N.Y. Jee, J.R. Pinney, M.A. LaBarge, T.A. Desai, Z.J. Gartner, Tissue Eng. Part C Methods **21**, 541 (2015)
J.A. Champion, S. Mitragotri, Proc. Natl. Acad. Sci. **103**, 4930 (2006)
J.A. Champion, S. Mitragotri, Pharm. Res. **26**, 244 (2009)
J.A. Champion, Y.K. Katare, S. Mitragotri, J. Control. Release **121**, 3 (2007)
W.Y.J. Chen, G. Abatangelo, Wound Repair Regen. **7**, 79 (1999)
C.S. Chen, M. Mrksich, S. Huang, G.M. Whitesides, D.E. Ingber, Science **276**, 1425 (1997)
J. Chen, C.-H. Tung, J.R. Allport, S. Chen, R. Weissleder, P.L. Huang, Circulation **111**, 1800 (2005)
A. Chopra, V. Lin, A. McCollough, S. Atzet, G.D. Prestwich, A.S. Wechsler, M.E. Murray, S.A. Oake, J. Yasha Kresh, P.A. Janmey, J. Biomech. **45**, 824 (2012)
A. Chopra, M.E. Murray, F.J. Byfield, M.G. Mendez, R. Halleluyan, D.J. Restle, D. Raz-Ben Aroush, P.A. Galie, K. Pogoda, R. Bucki, C. Marcinkiewicz, G.D. Prestwich, T.I. Zarembinski, C.S. Chen, E. Puré, J.Y. Kresh, P.A. Janmey, Biomaterials **35**, 71 (2014)
J.M. Collins, P. Ayala, T.A. Desai, B. Russell, Small **6**, 355 (2010a)
J.M. Collins, P.H. Goldspink, B. Russell, J. Mol. Cell. Cardiol. **49**, 1042 (2010b)
B. Cortese, M.O. Riehle, S. D'Amone, G. Gigli, J. Colloid Interface Sci. **394**, 582 (2013)
E. Cukierman, R. Pankov, D.R. Stevens, K.M. Yamada, Science **294**(80), 1708 (2001)
M.W. Curtis, S. Sharma, T.A. Desai, B. Russell, Biomed. Microdevices **12**, 1073 (2010)
M.W. Curtis, E. Budyn, T.A. Desai, A.M. Samarel, B. Russell, Biomech. Model. Mechanobiol. **12**, 95 (2013)
J.M. Cyphert, C.S. Trempus, S. Garantziotis, Int. J. Cell Biol. **2015**, 1 (2015)
N.J. Darling, E. Sideris, N. Hamada, S.T. Carmichael, T. Segura, Adv. Sci. **5**, 1801046 (2018)
P.P. de Tombe, H.E.D.J. ter Keurs, J. Mol. Cell. Cardiol. **91**, 148 (2016)
P.P. de Tombe, R.D. Mateja, K. Tachampa, Y.A. Mou, G.P. Farman, T.C. Irving, J. Mol. Cell. Cardiol. **48**, 851 (2010)
J. Deutsch, D. Motlagh, B. Russell, T.A. Desai, J. Biomed. Mater. Res. **53**, 267 (2000)
R. Dimatteo, N.J. Darling, T. Segura, Adv. Drug Deliv. Rev. **127**, 167 (2018)
G. Doroudian, J. Pinney, P. Ayala, T. Los, T.A. Desai, B. Russell, Biomed. Microdevices **16**, 705 (2014)
T.L. Downing, J. Soto, C. Morez, T. Houssin, A. Fritz, F. Yuan, J. Chu, S. Patel, D.V. Schaffer, S. Li, Nat. Mater. **12**, 1154 (2013)
S. Dupont, L. Morsut, M. Aragona, E. Enzo, S. Giulitti, M. Cordenonsi, F. Zanconato, J. Le Digabel, M. Forcato, S. Bicciato, N. Elvassore, S. Piccolo, Nature **474**, 179 (2011)
A.J. Engler, S. Sen, H.L. Sweeney, D.E. Discher, Cell **126**, 677 (2006)
L.E. Finlan, D. Sproul, I. Thomson, S. Boyle, E. Kerr, P. Perry, B. Ylstra, J.R. Chubb, W.A. Bickmore, PLoS Genet. **4**, e1000039 (2008)
N.G. Frangogiannis, Circ. Res. **110**, 159 (2012)
N.G. Frangogiannis, Nat. Rev. Cardiol. **11**, 255 (2014)
N. G. Frangogiannis, In *Compr. Physiol.* (John Wiley & Sons, Inc., Hoboken, 2015), pp. 1841–1875
C. Frantz, K.M. Stewart, V.M. Weaver, J. Cell Sci. **123**, 4195 (2010)
F. Gao, Y. Liu, Y. He, C. Yang, Y. Wang, X. Shi, G. Wei, Matrix Biol. **29**, 107 (2010)
B. Geiger, Science **294**(80), 1661 (2001)
B. Geiger, J.P. Spatz, A.D. Bershadsky, Nat. Rev. Mol. Cell Biol. **10**, 21 (2009)
B. Hinz, J. Invest. Dermatol. **127**, 526 (2007)
B. Hinz, J. Biomech. **43**, 146 (2010)
M. Hoshijima, Am. J. Physiol. Circ. Physiol. **290**, H1313 (2006)
N. Huebsch, P.R. Arany, A.S. Mao, D. Shvartsman, O.A. Ali, S.A. Bencherif, J. Rivera-Feliciano, D.J. Mooney, Nat. Mater. **9**, 518 (2010)
R.O. Hynes, Science **326**(80), 1216 (2009)
J.L. Ifkovits, E. Tous, M. Minakawa, M. Morita, J.D. Robb, K.J. Koomalsingh, J.H. Gorman, R.C. Gorman, J. a Burdick, Proc. Natl. Acad. Sci. **107**, 11507 (2010)
D.E. Ingber, Annu. Rev. Physiol. **59**, 575 (1997)
D.E. Ingber, FASEB J. **20**, 811 (2006)

D.E. Ingber, N. Wang, D. Stamenović, Rep. Prog. Phys. **77**, 046603 (2014)
D.E. Jaalouk, J. Lammerding, Nat. Rev. Mol. Cell Biol. **10**, 63 (2009)
D. Jiang, J. Liang, P.W. Noble, Annu. Rev. Cell Dev. Biol. **23**, 435 (2007)
Y. Jiang, S. Lu, Y. Zeng, J. Tissue Eng. Regen. Med. **5**, 402 (2011)
A. Katsumi, A.W. Orr, E. Tzima, M.A. Schwartz, J. Biol. Chem. **279**, 12001 (2004)
A.M. Katz, Circulation **106**, 2986 (2002)
A. Khademhosseini, R. Langer, Biomaterials **28**, 5087 (2007)
A. Khademhosseini, G. Eng, J. Yeh, J. Fukuda, J. Blumling, R. Langer, J.A. Burdick, J. Biomed. Mater. Res. Part A **79A**, 522 (2006a)
A. Khademhosseini, R. Langer, J. Borenstein, J.P. Vacanti, Proc. Natl. Acad. Sci. **103**, 2480 (2006b)
K.A. Kilian, B. Bugarija, B.T. Lahn, M. Mrksich, Proc. Natl. Acad. Sci. **107**, 4872 (2010)
D.-H. Kim, K. Han, K. Gupta, K.W. Kwon, K.-Y. Suh, A. Levchenko, Biomaterials **30**, 5433 (2009)
L.V. Le, P. Mohindra, Q. Fang, R.E. Sievers, M.A. Mkrtschjan, C. Solis, C.W. Safranek, B. Russell, R.J. Lee, T.A. Desai, Biomaterials **169**, 11 (2018)
J. Leijten, J. Seo, K. Yue, G. Trujillo-de Santiago, A. Tamayol, G.U. Ruiz-Esparza, S.R. Shin, R. Sharifi, I. Noshadi, M.M. Álvarez, Y.S. Zhang, A. Khademhosseini, Mater. Sci. Eng. R. Rep. **119**, 1 (2017)
A. Leonard, A. Bertero, J.D. Powers, K.M. Beussman, S. Bhandari, M. Regnier, C.E. Murry, N.J. Sniadecki, J. Mol. Cell. Cardiol. **118**, 147 (2018)
J. Li, B. Russell, Am. J. Physiol. Heart Circ. Physiol. **305**, H1614 (2013)
J. Li, E.J. Tanhehco, B. Russell, Am. J. Physiol. Circ. Physiol. **307**, H1618 (2014)
L. Li, Q. Zhao, W. Kong, Matrix Biol. **490**, 68–69 (2018)
Y.-H. Lin, J. Li, E.R. Swanson, B. Russell, J. Appl. Physiol. **114**, 1603 (2013)
M.L. Lindsey, D.L. Mann, M.L. Entman, F.G. Spinale, Ann. Med. **35**, 316 (2003)
M. Litwiniuk, A. Krejner, M.S. Speyrer, A.R. Gauto, T. Grzela, Wounds **28**, 78 (2016)
H. Mansour, P.P. de Tombe, A.M. Samarel, B. Russell, Circ. Res. **94**, 642 (2004)
B. Martinac, J. Cell Sci. **117**, 2449 (2004)
F. Martino, A.R. Perestrelo, V. Vinarský, S. Pagliari, G. Forte, Front. Physiol. **9**, 1 (2018)
L.E. McNamara, R. Burchmore, M.O. Riehle, P. Herzyk, M.J.P. Biggs, C.D.W. Wilkinson, A.S.G. Curtis, M.J. Dalby, Biomaterials **33**, 2835 (2012)
J.E. Mealy, J.J. Chung, H.-H. Jeong, D. Issadore, D. Lee, P. Atluri, J.A. Burdick, Adv. Mater. **30**, 1705912 (2018)
K.S. Midwood, L.V. Williams, J.E. Schwarzbauer, Int. J. Biochem. Cell Biol. **36**, 1031 (2004)
P. Mills, J.-F. Lafrenière, B.F. Benabdallah, E.M. El Fahime, J.-P. Tremblay, Exp. Cell Res. **313**, 527 (2007)
M.A. Mkrtschjan, S.B. Gaikwad, K.J. Kappenman, C. Solís, S. Dommaraju, L.V. Le, T.A. Desai, B. Russell, J. Cell. Physiol. **233**, 3672 (2018a)
M.A. Mkrtschjan, C. Solís, A.Y. Wondmagegn, J. Majithia, B. Russell, Cytoskeleton **75**, 363 (2018b)
D. Motlagh, T.J. Hartman, T.A. Desai, B. Russell, J. Biomed. Mater. Res. **67A**, 148 (2003)
A. Musaro, C. Giacinti, G. Borsellino, G. Dobrowolny, L. Pelosi, L. Cairns, S. Ottolenghi, G. Cossu, G. Bernardi, L. Battistini, M. Molinaro, N. Rosenthal, Proc. Natl. Acad. Sci. **101**, 1206 (2004)
P.W. Noble, Matrix Biol. **21**, 25 (2002)
J.J. Norman, J.M. Collins, S. Sharma, B. Russell, T.A. Desai, Tissue Eng. Part A **14**, 379 (2008)
A.W. Orr, B.P. Helmke, B.R. Blackman, M.A. Schwartz, Dev. Cell **10**, 11 (2006)
A. Ovsianikov, V. Mironov, J. Stampfl, R. Liska, Expert Rev. Med. Devices **9**, 613 (2012)
P. Pandey, W. Hawkes, J. Hu, W.V. Megone, J. Gautrot, N. Anilkumar, M. Zhang, L. Hirvonen, S. Cox, E. Ehler, J. Hone, M. Sheetz, T. Iskratsch, Dev. Cell **44**, 326 (2018)
J.T. Parsons, A.R. Horwitz, M.A. Schwartz, Nat. Rev. Mol. Cell Biol. **11**, 633 (2010)
A.A. Patel, R.G. Thakar, M. Chown, P. Ayala, T.A. Desai, S. Kumar, Biomed. Microdevices **12**, 287 (2010)
J.R. Peña, J.R. Pinney, P. Ayala, T.A. Desai, P.H. Goldspink, Biomaterials **46**, 26 (2015)
A.C. Petrey, C.A. de la Motte, Front. Immunol. **5**, 1 (2014)
E.R. Pfeiffer, J.R. Tangney, J.H. Omens, A.D. McCulloch, J. Biomech. Eng. **136**, 021007 (2014)
J.R. Pinney, K.T. Du, P. Ayala, Q. Fang, R.E. Sievers, P. Chew, L. Delrosario, R.J. Lee, T.A. Desai, Biomaterials **35**, 8820 (2014a)
J.R. Pinney, G. Melkus, A. Cerchiari, J. Hawkins, T.A. Desai, ACS Appl. Mater. Interfaces **6**, 14477 (2014b)
S.D. Prabhu, N.G. Frangogiannis, Circ. Res. **119**, 91 (2016)
J.L. Puglisi, P.H. Goldspink, A.V. Gomes, M.S. Utter, D.M. Bers, R.J. Solaro, Arch. Biochem. Biophys. **50**, 552–553 (2014)
E. Puklin-Faucher, M.P. Sheetz, J. Cell Sci. **122**, 179 (2009)
C. Raposo, M. Schwartz, Glia **62**, 1895 (2014)
K. Ronaldson-Bouchard, G. Vunjak-Novakovic, Cell Stem Cell (2018)
P. Rørth, Dev. Cell **20**, 9 (2011)
J.C. Rose, M. Cámara-Torres, K. Rahimi, J. Köhler, M. Möller, L. De Laporte, Nano Lett. **17**, 3782 (2017)
J.C. Rose, D.B. Gehlen, T. Haraszti, J. Köhler, C.J. Licht, L. De Laporte, Biomaterials **163**, 128 (2018)
J.-J. Santiago, A.L. Dangerfield, S.G. Rattan, K.L. Bathe, R.H. Cunnington, J.E. Raizman, K.M. Bedosky, D.H. Freed, E. Kardami, I.M.C. Dixon, Dev. Dyn. **239**, 1573 (2010)
J. Schwan, A.T. Kwaczala, T.J. Ryan, O. Bartulos, Y. Ren, L.R. Sewanan, A.H. Morris, D.L. Jacoby, Y. Qyang, S.G. Campbell, Sci. Rep. **6**, 32068 (2016)
A.M. Segura, O.H. Frazier, L.M. Buja, Heart Fail. Rev. **19**, 173 (2014)
S.E. Senyo, Y.E. Koshman, B. Russell, FEBS Lett. **581**, 4241 (2007)
E. Sideris, D.R. Griffin, Y. Ding, S. Li, W.M. Weaver, D. Di Carlo, T. Hsiai, T. Segura, ACS Biomater. Sci. Eng. **2**, 2034 (2016)
B.V. Slaughter, S.S. Khurshid, O.Z. Fisher, A. Khademhosseini, N.A. Peppas, Adv. Mater. **21**, 3307 (2009)
N.J. Sniadecki, C.M. Lamb, Y. Liu, C.S. Chen, D.H. Reich, Rev. Sci. Instrum. **79**, 044302 (2008)
J.C. Sunshine, K. Perica, J.P. Schneck, J.J. Green, Biomaterials **35**, 269 (2014)
V. Talman, H. Ruskoaho, Cell Tissue Res. **365**, 563 (2016)
A.I. Teixeira, J. Cell Sci. **116**, 1881 (2003)
R.G. Thakar, M.G. Chown, A. Patel, L. Peng, S. Kumar, T.A. Desai, Small **4**, 1416 (2008)
C.H. Thomas, J.H. Collier, C.S. Sfeir, K.E. Healy, Proc. Natl. Acad. Sci. **99**, 1972 (2002)
M.W. Tibbitt, K.S. Anseth, Biotechnol. Bioeng. **103**, 655 (2009)
B.P. Toole, Nat. Rev. Cancer **4**, 528 (2004)
E. Tous, J.L. Ifkovits, K.J. Koomalsingh, T. Shuto, T. Soeda, N. Kondo, J.H. Gorman, R.C. Gorman, J. a Burdick, Biomacromolecules **12**, 4127 (2011)
E.A. Turley, P.W. Noble, L.Y.W. Bourguignon, J. Biol. Chem. **277**, 4589 (2002)
M. Versaevel, T. Grevesse, S. Gabriele, Nat. Commun. **3**, 671 (2012)
N. Wang, J.D. Tytell, D.E. Ingber, Nat. Rev. Mol. Cell Biol. **10**, 75 (2009)
M.A. Westhoff, B. Serrels, V.J. Fincham, M.C. Frame, N.O. Carragher, Mol. Cell. Biol. **24**, 8113 (2004)
M.A. Wozniak, K. Modzelewska, L. Kwong, P.J. Keely, Biochim. Biophys. Acta, Mol. Cell Res. **1692**, 103 (2004)
S. Xin, O.M. Wyman, D.L. Alge, Adv. Healthc. Mater. **7**, e1800160 (2018)

F. Yanagawa, S. Sugiura, T. Kanamori, Regen. Ther. **3**, 45 (2016)
H. Yang, L.P. Schmidt, Z. Wang, X. Yang, Y. Shao, T.K. Borg, R. Markwald, R. Runyan, B.Z. Gao, Sci. Rep. **6**, 20674 (2016)
N. Yasuda, H. Akazawa, Y. Qin, Y. Zou, I. Komuro, Naunyn Schmiedeberg's Arch. Pharmacol. **377**, 393 (2008)
S.J. Yoon, Y.H. Fang, C.H. Lim, B.S. Kim, H.S. Son, Y. Park, K. Sun, J. Biomed. Mater. Res. Part B Appl. Biomater. **91B**, 163 (2009)
J.-G. Yu, B. Russell, J. Histochem. Cytochem. **53**, 839 (2005)
L. Zhang, Y. S. Mao, P. A. Janmey, and H. L. Yin, In *Subcell. Biochem.* pp. 177–215 (2012)
J. Zhang, W. Zhu, M. Radisic, G. Vunjak-Novakovic, Circ. Res. (2018)
Y. Zhu, Y. Matsumura, W.R. Wagner, Biomaterials **129**, 37 (2017)
W.H. Ziegler, A.R. Gingras, D.R. Critchley, J. Emsley, Biochem. Soc. Trans. **36**, 235 (2008)
P. Zorlutuna, N. Annabi, G. Camci-Unal, M. Nikkhah, J.M. Cha, J.W. Nichol, A. Manbachi, H. Bae, S. Chen, A. Khademhosseini, Adv. Mater. **24**, 1782 (2012)

Publisher's note Springer Nature remains neutral with regard to jurisdictional claims in published maps and institutional affiliations.

Part VI
Modelling

Biomedical Microdevices (2019) 21: 40
https://doi.org/10.1007/s10544-019-0380-2

Mathematical modeling in cancer nanomedicine: a review

Prashant Dogra[1] · Joseph D. Butner[1] · Yao-li Chuang[2] · Sergio Caserta[3] · Shreya Goel[4] · C. Jeffrey Brinker[5,6,7,8]
Vittorio Cristini[1,9] · Zhihui Wang[1,9]

Published online: 4 April 2019

Author's Personal Note My favorite memory with Mauro is from our very first interactions. I had received an email advertising a conference on cancer nanotechnology he was running in San Diego perhaps a decade and a half ago. I was a young assistant professor trying to push forward my ideas-at an embryonal stage then-on the possibility of developing clinically applicable mathematical models of cancer progression and tumor drug delivery. I decided to write to Mauro and introduce myself, describing these dreams I had of impacting cancer using mathematics. To my incredulity, he responded enthusiastically. Perhaps the only enthusiastic response my ideas received until then. We met in person and spent a few hours discussing research and "exchanging dreams." These dreams gave birth to many innovations, including the novel field of physical oncology, and a new impactful way to understand this disease. Mauro has helped me since: a constant point of reference in the uneasy seas of academia and translational research. Dear Mauro, congratulations on all your impressive accomplishments! For your help and support to my constant battle to push my ideas, my papers, my proposals forward, and for whatever success I have had or will ever have, big or small, I am forever grateful.

Vittorio Cristini

Abstract

Cancer continues to be among the leading healthcare problems worldwide, and efforts continue not just to find better drugs, but also better drug delivery methods. The need for delivering cytotoxic agents selectively to cancerous cells, for improved safety and efficacy, has triggered the application of nanotechnology in medicine. This effort has provided drug delivery systems that can potentially revolutionize cancer treatment. Nanocarriers, due to their capacity for targeted drug delivery, can shift the balance of cytotoxicity from healthy to cancerous cells. The field of cancer nanomedicine has made significant progress, but challenges remain that impede its clinical translation. Several biophysical barriers to the transport of nanocarriers to the tumor exist, and a much deeper understanding of nano-bio interactions is necessary to change the status quo. Mathematical modeling has been instrumental in improving our understanding of the physicochemical and physiological underpinnings of nanomaterial behavior in biological systems. Here, we present a comprehensive review of literature on mathematical modeling works that have been and are being employed towards a better understanding of nano-bio interactions for improved tumor delivery efficacy.

Keywords Agent-based modeling · Cancer treatment · Drug transport · Mechanistic modeling · Multiscale · Pharmacokinetics and pharmacodynamics

1 Introduction

Leveraging on the improvements in nanotechnology, nanomaterials can be meticulously engineered to obtain reproducible and custom biological behavior, which has catalyzed progress in the field of cancer nanomedicine through targeted delivery of biopharmaceutical agents to solid tumors. Nanoparticle[1] (NP)-based formulations (nanocarriers) are used to package and deliver cargos that are too toxic, insoluble, rapidly cleared, or unstable for delivery as free molecules (e.g., chemotherapeutic drugs, silencing RNAs, and contrast agents). Nanocarriers have been demonstrated to facilitate the delivery of such agents specifically to the tumor site by *passive accumulation* or *active targeting* (Sykes et al. 2014a), thereby overcoming potential drug-resistance mechanisms (Brocato et al. 2014). Whereas the former phenomenon depends on the leakiness of tumor vessels (defined as enhanced permeability and retention or EPR effect) (Jain and Stylianopoulos 2010), the latter involves surface functionalization of NPs with ligands specific to receptors

[1] Microscopic particles with aerodynamic diameter < 100 nm.

✉ Zhihui Wang
zwang@houstonmethodist.org

1 Mathematics in Medicine Program, The Houston Methodist Research Institute, HMRI R8-122, 6670 Bertner Ave, Houston, TX 77030, USA

2 Department of Mathematics, California State University, Northridge, CA 91330, USA

3 Department of Chemical Materials and Industrial Production Engineering, University of Naples Federico II, 80125 Naples, Italy

4 Department of Nanomedicine, Houston Methodist Research Institute, Houston, TX 77030, USA

5 Center for Micro-Engineered Materials, University of New Mexico, Albuquerque, NM 87131, USA

6 Chemical and Biological Engineering, University of New Mexico, Albuquerque, NM 87131, USA

7 Cancer Research and Treatment Center, Molecular Genetics and Microbiology, University of New Mexico, Albuquerque, NM 87131, USA

8 Self-Assembled Materials Department, Sandia National Laboratories, Albuquerque, NM 87185, USA

9 Department of Imaging Physics, University of Texas MD Anderson Cancer Center, Houston, TX 78701, USA

on the tumor cell surface. A prolonged circulation half-life is generally regarded as a common prerequisite for either process, and NPs have been demonstrated to alter the pharmacokinetics (PK) of the cargo and enhance its circulation half-life, thereby promoting delivery specifically to the tumor and reducing collateral damage to healthy cells (Chow et al. 2011). Additionally, nanocarriers can be engineered to allow triggered release of cargo, i.e., the payload remains encapsulated inside NPs unless a particular physiological or external stimulus is presented, e.g., pH, enzymes, temperature, or electromagnetic radiation. To date, a diverse variety of nanomaterials have been developed for tumor delivery applications. They include organic (e.g., polymeric, liposomal, solid lipid) and inorganic (e.g., silica, gold, iron, titanium) NPs, which can be spherical, rod-like, or disc-like in shape (among others), can be synthesized in a range of sizes, and possess a net surface charge (zeta potential). While remarkable progress has been made in the field of cancer nanomedicine over the past 50 years (Fig. 1), and some nanocarriers are already in clinical use (Table 1), there are significant challenges that remain unanswered, impeding the clinical translation of a majority of nanocarriers (Shi et al. 2016).

Based on a meta-analysis of literature in cancer nanomedicine over the past decade (Wilhelm et al. 2016), it has been noted that on average only 0.7% of the injected dose of NPs reaches the tumor. This is generally ascribable to the *biological barriers to transport of NPs that span across multiple scales and pose a serious threat to the clinical translation of upcoming nanoformulations*, making it imperative to investigate the underlying nano-bio interactions responsible for the suboptimal tumor deliverability of nanocarriers. While certain physicochemical properties of NPs may correlate with higher tumor deliverability in vitro or in vivo (e.g., size <100 nm, neutral charge, and rod-shape (Albanese et al. 2012; Wilhelm et al. 2016)), an improved understanding of biological barriers faced by nanocarriers can provide NP design guidelines for optimized tumor delivery, and foster their clinical translatability. To this end, mathematical modeling has been an important tool in supporting cancer nanomedicine, and also for many other cancer research fields (Wang and Deisboeck 2014; Wang et al. 2015; Wang and Maini 2017). For example, modeling has provided mechanistic understanding of phenomenological observations based on physical principles and helped establish important quantitative relationships. Modeling efforts in cancer nanomedicine have been applied to phenomena spanning several biologically relevant scales in time and space (Fig. 2). Here, we review various mathematical modeling techniques that have been helpful to the field of cancer nanomedicine to answer questions related to nano-bio interactions and to provide insights into the problem of low tumor deliverability of NPs.

2 The voyage of nanoparticles

Before we discuss any specific modeling work, it is critical to understand the journey of NPs from the site of injection to the site of action, which can be broadly divided into three phases: i) vascular, ii) transvascular, and iii) interstitial, to reach the cancerous cells (Fig. 3) (Jain and Stylianopoulos 2010; Chauhan et al. 2011; Nichols and Bae 2012). NPs are faced with physical and physiological challenges during these stages of transport that strongly influence their biological fate.

Immediately after NPs are injected into the blood stream, they are exposed to a high concentration of plasma proteins (60–80 $g \cdot l^{-1}$), such as albumins, apolipoproteins, and opsonins, which adsorb on the particle surface and form a surrounding biomolecular corona that redefines their chemical identity and provides them with the so-called *biological identity*, and ultimately plays an important role in governing

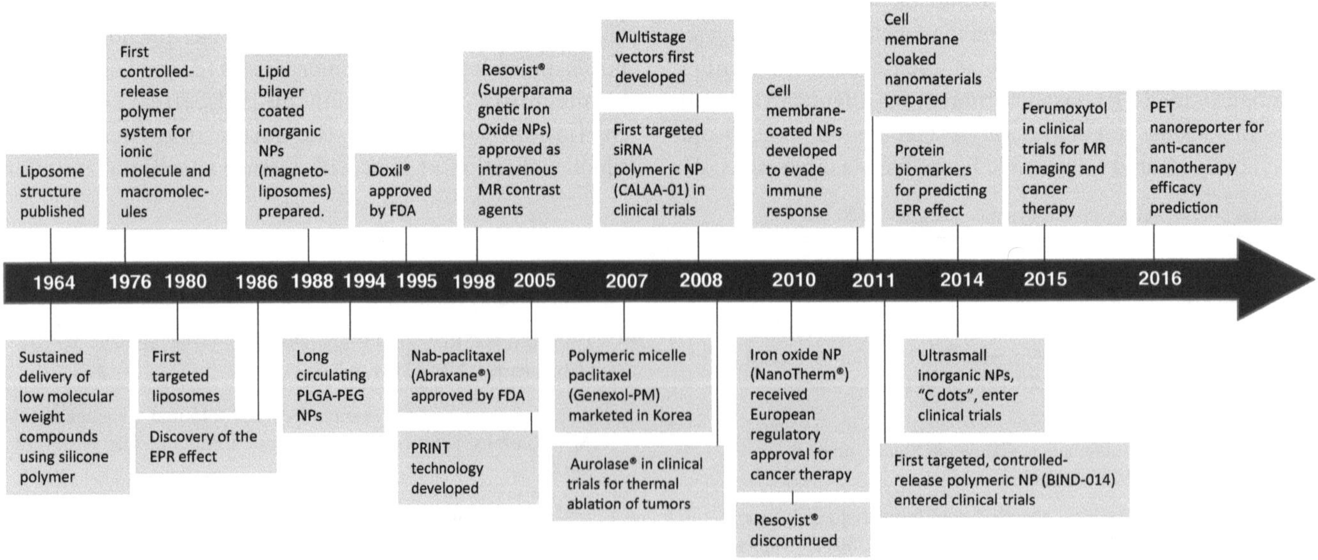

Fig. 1 Timeline. A historical timeline of the major advancements in cancer nanomedicine (Shi et al. 2016; Hassan et al. 2017)

Table 1 Anti-cancer nanomedicines in the clinic

Drug/Agent	Formulation	Product name/Company	Applications	NP size (nm)	Half-life (h)	References
Asparaginase	Polymeric conjugates	Oncaspar® (PEG[a])/Baxalta	Acute lymphoblastic leukemia	50–200	145–189.6	(Douer et al. 2007; Bawa et al. 2016)
Cytarbine	Liposome	Depocyt®/Pacira Pharmaceuticals	Lymphomatous leukemia	$10–20 \times 10^3$	82.4	(Pillai 2014; Bulbake et al. 2017)
Daunorubicin	Liposome	DaunoXome®/Ga-len	HIV-related Kaposi's sarcoma	45	4–5.6	(Fumagalli et al. 2000; Bellott et al. 2001; Immordino et al. 2006; Bulbake et al. 2017; Praça et al. 2017)
Doxorubicin	Liposome	Doxil®/ Caelyx®/Janssen	Kaposi's sarcoma Multiple Myeloma Ovarian cancer	87.3 ± 8.5	50–60	(Gabizon et al. 1993; Gabizon 2001; Gabizon et al. 2003; Gabizon et al. 2006; Soundararajan et al. 2009)
	Liposome	Myocet®/Teva UK	Metastatic breast cancer	150–250	2–3	(Immordino et al. 2006; Bulbake et al. 2017)
Irinotecan	Liposome	Onivyde®/Merri-mack Pharmaceuticals	Metastatic pancreatic cancer	110	19.4–22.3	(Drummond et al. 2006; Agency 2016)
Iron oxide	SPION[b]	NanoTherm®/M-ag-Force AG	Glioblastoma	12–20	N/A	(Bellizzi and Bucci 2018)
Mifamurtide	Liposome	Mepact®/Takeda Pharmaceutical	Osteocarcoma	$1–5 \times 10^3$	2.03–2.27	(Ando et al. 2011; Venkatakrishnan et al. 2014; Yu and Li 2014)
Paclitaxel	Polymeric micelles	Genexol®-PM/Samyang Biopharmaceutic-als	Non-small cell lung cancer Breast cancer Ovarian cancer	20–50	11–12.7	(Kim et al. 2004; Werner et al. 2013)
Paclitaxel	Albumin-bound	Abraxane®/Celg-ene	Advanced metastatic pancreatic cancer Advanced metastatic breast cancer Advanced non-small cell lung cancer	130	20.5–21.6	(Di Costanzo et al. 2009; Miele et al. 2009)
Styrine maleic anhydride neocarzinostatin	Polymer protein conjugate	Zinostatin Stimalamer	Unresectable hepatocellular carcinoma	N/A	7.1	(Toge et al. 1991)
Vincristine	Liposome	Marqibo®/Spectr-um Pharmaceuticals	Acute lymphoblastic leukemia	90–140	2.4–4.9	(Talon Therapeutics, Inc., 2012; Fenske and Cullis 2008; Silverman and Deitcher 2013; Yu and Li 2014)

[a] PEG (polyethylene glycol) is a hydrophilic polymer that is used to preclude plasma protein adsorption on NP surface through steric hindrance

[b] Superparamagnetic iron oxide nanoparticles

Fig. 2 Classification of mathematical models. Mathematical models in cancer nanomedicine can be classified based on the characteristic spatiotemporal scale of the system under consideration

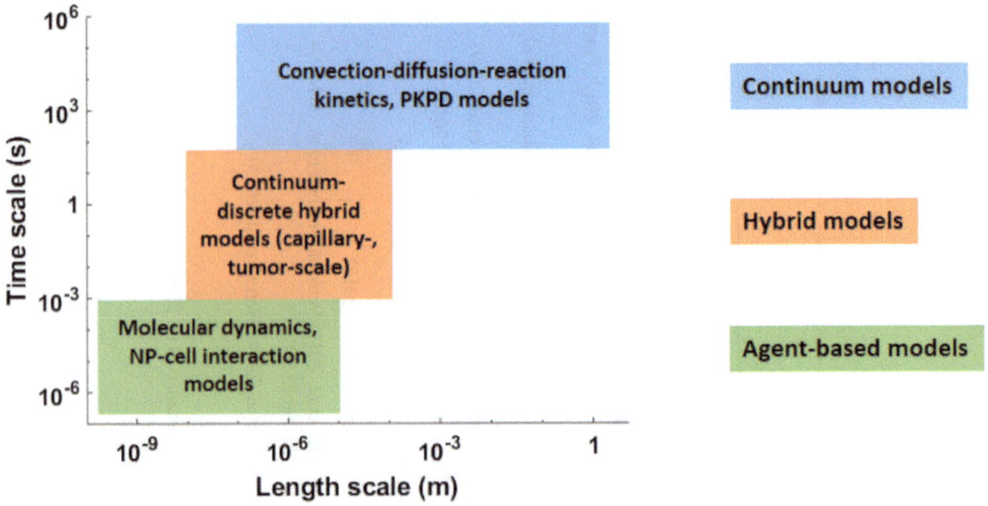

nano-bio interactions (Monopoli et al. 2012; Caracciolo et al. 2017). The bimolecular corona around the particles is a dynamic entity and, depending upon the relative affinity of the proteins for the NP surface, the coronas at steady state can have varied compositions. The nature of the corona can impact nano-bio interactions by affecting the hydrodynamic size, surface charge, and immunogenicity of the NPs, thereby affecting their cellular internalization, biodistribution, and circulation half-life (Aggarwal et al. 2009).

Simultaneously, NPs are transported across the body via the vascular network, and upon arrival at the finest blood vessels, i.e., capillaries, particles are faced with special anatomical, physiological, and hemodynamic conditions that strongly influence their fate. Healthy capillaries are broadly classified as: i) continuous, ii) fenestrated, and iii) sinusoidal, depending upon the upper limit of pore size in the vessel walls. Continuous capillaries have pore sizes <5 nm (e.g., brain, lungs, muscles, skin), fenestrated capillaries haves pores <15 nm (e.g., kidneys), and sinusoids in liver have pores <200 nm, while those in spleen are ~5 μm (Sarin 2010). Thus, the NP to pore size ratio becomes a determining factor (besides surface charge) in the extravasation of NPs into tissue interstitium (Stylianopoulos et al. 2013) or excretion in kidneys (Choi et al. 2007). In addition, the presence of resident macrophages in the lumen of capillaries causes NPs, which are already opsonized,[2] to be imminently phagocytosed and removed from circulation, thereby affecting their circulation time in the body. Kupffer cells of the liver (Tsoi et al. 2016) and splenic macrophages (Cataldi et al. 2017) have been recognized to contribute significantly to this mechanism of NP clearance. Given the significant decrease of blood flowrate in capillaries (~1 $mm \cdot s^{-1}$), compared to larger vessels (>10 $cm \cdot s^{-1}$), hemodynamic conditions exist in capillaries conducive for NP interaction with vessel wall pores or immune cells (Jain and Stylianopoulos 2010; Tsoi et al. 2016). However, it is important as a prerequisite that NPs gain near-wall access via a size-, shape-, and hematocrit[3]-dependent phenomenon referred as *margination*[4] (Lee et al. 2013; Müller et al. 2014). As a result, a complex interplay between these microscopic interactions inside blood capillaries defines the global biodistribution and clearance of NPs from the body, thereby strongly influencing the tumor delivery of NPs (Dogra et al. 2018).

Once NPs enter into the tumor interstitium following capillary extravasation, they are subject to a hostile microenvironment that weakens convective transport; thus diffusion becomes the primary means of transport for NPs. This significantly limits the penetration distance and delivery of cargo to cancerous cells distant from the tumor-feeding capillaries (Deisboeck et al. 2011; Pascal et al. 2013b; Wang et al. 2016; Cristini et al. 2017). *Readers are referred to the following review for a detailed discussion on intra-tumoral transport barriers responsible for chemotherapy resistance* (Brocato et al. 2014).

3 Mathematical modeling in cancer nanomedicine

We now discuss mathematical modeling works that focus on the above mentioned biophysical processes for optimizing NP design towards improved tumor delivery and efficacy. The following sections are organized based on the problem they seek to investigate (Table 2). Upon entering the blood stream, the *formation of an enveloping protein corona* changes the biochemical properties of NPs, a process which is best

[2] Adsorbed with plasma proteins of the complement system (opsonins) that cause recognition of xenobiotics by the immune system.

[3] Volume percentage of red blood cells in blood.

[4] Lateral drift of particles towards the vessel well.

Fig. 3 Nanoparticle transport in tumors. Biophysical barriers involved in the delivery of NPs to tumors via microcirculation

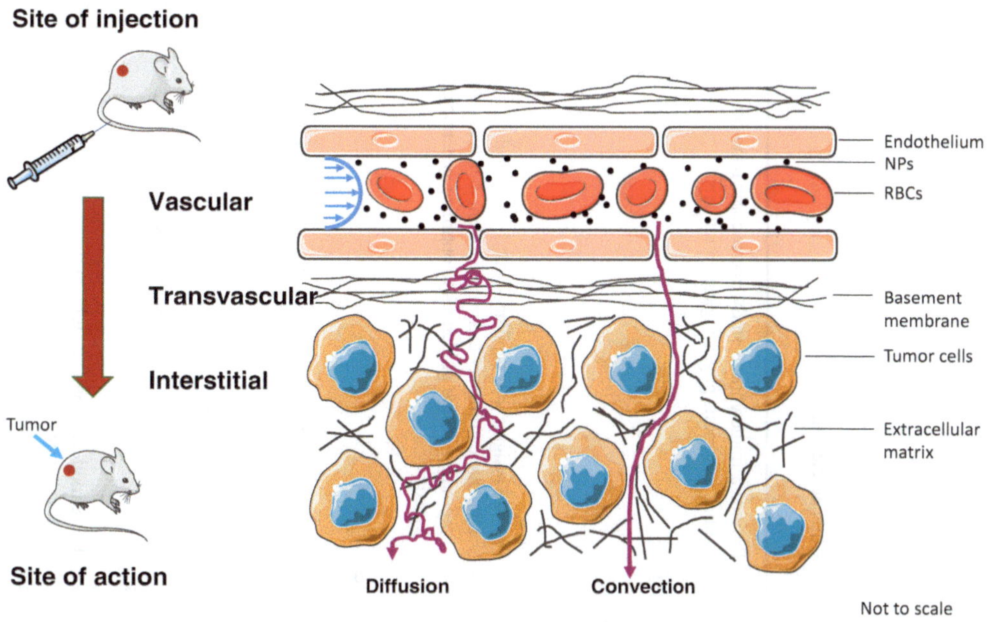

described mathematically by kinetic models or coarse-grained molecular dynamics simulations. Subsequent to corona formation, the biochemically altered NPs are transported via blood to tumor and various organs. The *processes during transport* are modeled by discrete, hybrid, or continuum models at the microscopic or mesoscopic length scales. Once NPs arrive at the capillary wall or extravasate into the extravascular space, *cellular uptake* of NPs is modeled using discrete modeling approaches. The emergent *whole-body distribution and clearance* of NPs due to microscopic nano-bio interactions is described by pharmacokinetic models, while *tumor deliverability* of particles is studied using hybrid modeling methods. Finally, *nanotherapy efficacy and toxicity* are evaluated with pharmacodynamic models.

3.1 Biomolecular corona formation

Kinetic modeling In a first attempt to model the NP-protein interactions during corona formation, Dell'Orco et al. developed a mathematical model to describe the kinetics of competitive interaction of human serum albumin (HSA), high density lipoprotein (HDL), and fibrinogen for binding with a 70 nm-sized copolymer NP (Dell'Orco et al. 2010). The model, based on the law of mass action, is formalized as a system of three linear first order differential equations describing the interaction of each protein with the NPs, represented as:

$$\frac{d[NP{\cdot}pro]}{dt} = n_{pro}k_{pro}^{on}[NP][pro]-k_{pro}^{off}[NP{\cdot}pro], \qquad (1)$$

where $[NP]$, $[pro]$, and $[NP \cdot pro]$ represent the concentration of NPs, protein, and NP-protein complex, respectively, and k_{pro}^{on} and k_{pro}^{off} are the association and dissociation rate constants of NP-protein binding. n_{pro}, which is the number of available binding sites on NP surface for a given NP-protein pairing, is approximated by the ratio of surface area of the NP-protein complex and cross-sectional area of protein molecules, assuming a spherical shape, given by $n_{pro} = \frac{4\pi\left(r_{NP}+r_{pro}\right)^2}{\pi r_{pro}^2}$ protein radius. This model has been used to simulate the temporal evolution of the corona composition that seems to depend on the association and dissociation rate constants of the proteins that govern the time required to reach steady state. Simulation results show that over time, the high affinity HDL displaces HSA from the corona, despite the much higher plasma concentration of the latter, primarily because of a relatively lower affinity of HSA for the NPs (Vroman effect). They also extended the model to study the effect of corona formation on the binding of NPs to target cell surface receptors, taking into account the kinetic, stoichiometric, and geometric variables of the system (Dell'Orco et al. 2012). Global sensitivity analysis revealed that successful NP-receptor binding is more strongly dependent on NP size than the number and size of peptides bound to NP surface.

Sahneh et al. made further improvements to the model by using population balance equations to derive the ordinary differential equations (ODEs) for *successive* binding of HSA, HDL, and fibrinogen to NPs, i.e. binding of a protein when the NP is already covered with different proteins, thereby representing a more realistic physiological scenario (Sahneh et al. 2013). Briefly, Sahneh et al. revised the NP-protein binding biochemical reactions such that Eq. (1) is modified to:

$$\frac{d[NP{\cdot}pro]}{dt} = n_{pro}k_{pro}^{on}[pro]Y-k_{pro}^{off}[NP{\cdot}pro] \qquad (2)$$

Table 2 A summary of key mathematical modeling approaches in cancer nanomedicine

Biological problem	Modeling-type	References	Major findings	Clinical relevance
Biomolecular corona formation	Kinetic modeling	(Dell'Orco et al. 2010; Dell'Orco et al. 2012; Sahneh et al. 2013)	During corona formation, high affinity proteins displace low affinity proteins (Vroman effect), and the corona evolves from a metastable to a stable state. NP size is more important than number and size of peptides bound to NP surface in governing successful NP-cell surface receptor binding.	Provide a framework to study microscopic nano-bio interactions in various physiological conditions.
	Coarse-grained molecular dynamics simulations	(Lopez and Lobaskin 2015; Tavanti et al. 2015a)	Protein adsorption energies for NP-protein interaction are primarily affected by NP size, while surface charge only has a small effect.	
Microvascular transport, margination, and binding	Continuum modeling	(Gentile et al. 2008)	An increase in hematocrit or vessel permeability reduces the effective diffusion coefficient of NPs, highlighting implications to intravascular transport of NPs.	Provide insights into capillary-scale biophysical interactions of NPs that can impact their macroscopic behavior, thereby providing design guidelines to optimize systemic circulation kinetics.
		(Tsoi et al. 2016)	NP sequestration in liver sinusoid is jointly affected by hemodynamic conditions and NP characteristics.	
	Hybrid modeling	(Lee et al. 2013; Müller et al. 2014; Fullstone et al. 2015)	Larger NP size correlates with greater margination, which is further promoted by discoidal NP shape and higher hematocrit.	
Cellular internalization	Discrete modeling	(Gao et al. 2005; Decuzzi and Ferrari 2008; Yuan and Zhang 2010)	A minimal particle size and ligand density are necessary for effective endocytosis.	Provide mechanistic understanding of the cellular uptake of NPs, which has implications in drug delivery or NP clearance by immune cells.
Whole-body biodistribution and clearance	PK modeling	(Dogra et al. 2018)	Small NP size correlates with longer systemic circulation and lower accumulation in mononuclear phagocytic system (MPS) organs, irrespective of route of injection. Positive charge supports excretion, and surface exposure of charged molecules increases the vulnerability to sequestration in MPS organs.	Provide a mechanistic description of whole-body phenomenological observations important for quantifying structure-activity relationships of NPs.
Tumor deliverability	Hybrid modeling	(Chauhan et al. 2012; Hendriks et al. 2012; Frieboes et al. 2013; Stapleton et al. 2013; Sykes et al. 2016)	Interplay between NP physicochemical properties (especially, size, surface charge) and vascular characteristics affects EPR-based accumulation and delivery of NPs to cancerous cells in the tumor interstitium.	These models provide insight on the intra-tumoral transport of NPs and provide critical design guidelines for improved tumor deliverability.
Nanotherapy efficacy and toxicity	PD modeling	(Pascal et al. 2013a; Leonard et al. 2016; Wang et al. 2016; Miller and Frieboes 2018)	Time integrated NP uptake by cancerous cells governs therapy efficacy. The outcome however is non-trivially affected by patient-specific tumor-perfusion heterogeneities.	Provide predictive tools that can be employed prospectively in the clinic to design personalized-nanomedicine regimens.
		(Laomettachit et al. 2017)	NP toxicity to healthy liver cells is dose-dependent, and while the effect of small exposures can be reversed due to cell proliferation, tissue damage due to higher dose exposures are generally irreversible.	Such studies are critical in assessing the toxicity potential of nanocarries in clinical doses, thereby providing guidelines for safe exposure limits.

where $Y = \left([NP]_0 - \sum_{j=1}^{m} \frac{[pro_j \cdot NP]}{n_{pro_j}}\right)$ is the concentration of available binding sites, as governed by the presence of already bound proteins on NP surface. This model reveals the temporal evolution of corona formation progressing from an early *metastable* composition to a late *stable* composition representing the hard corona, thus providing a mathematical underpinning to the phenomenon which can be used to make projections about NP behavior in varied physiological conditions. As a further extension of these modeling efforts, Zhdanov and Cho incorporated irreversible protein reconfiguration (denaturation) into the kinetic equations of adsorption and desorption of proteins on the NP surface (Zhdanov and Cho 2016). They also studied the role of protein diffusion during the adsorption process, given that the diffusion rate constant diminishes at the solution-solid interface compared to the bulk, and ultimately revealed that the rate limitations imposed by protein diffusion are negligible.

In the models discussed above, independent, single binding sites are assumed such that protein-protein interactions and cooperative effects are considered insignificant. To investigate this assumption, Angioletti-Uberti et al. developed a model based on the dynamic density functional theory (DDFT) which describes the density evolution of systems undergoing Brownian dynamics, to include the effects of inter-particle interactions on the adsorption kinetics of proteins (Angioletti-Uberti et al. 2014). Specifically, the free-energy function in their DDFT model is composed of the translational free energy of the proteins, free energy of protein-NP interaction, and free energy of protein-protein interactions. The model was used to understand the temporal evolution of the density profile of proteins around the NP-solution interface and protein adsorption on NP surface. In comparison to a reference model with ideal diffusion alone and no protein interactions, this model resulted in closer agreement with experimental findings. Thus, by incorporating fluxes due to different chemical potential gradients in the system, DDFT gives a more comprehensive picture than a simple diffusion-based model.

Coarse-grained molecular dynamics simulations While advancement of technology has made great strides to expand the ability to computationally simulate the atomic interactions of biochemical processes involving a handful of proteins for several nanoseconds, the feasibly-modeled size and time scales for such all-atom models are still well below what is relevant to the formation of a biomolecular corona around a NP (Tozzini 2005; Lopez and Lobaskin 2015). To reduce the degrees of freedom so that simulations of corona formation can be effectively carried out, coarse-graining techniques (CG) have been developed and employed in recent computational studies (Fig. 4). The CG approximation has been previously adopted to simulate phenomena associated with protein adsorption onto flat surfaces, such as the Vroman effect (Vilaseca et al. 2013). In a protein, atoms within each amino acid may be coarse-grained and thus represented by one or a few beads, allowing the interconnecting force fields between these beads to be incorporated. There are two major classes of CG protein models: elastic network models and Go-type models. In an elastic network model, the force fields are approximated as springs between beads, suitable for including thermal fluctuations and analyzing principal modes (Tozzini 2005). For corona formation, thermal fluctuations are negligible compared to the adsorption energies and ruggedness of protein folding energy landscapes, and Go-type models are usually adopted (Tozzini 2005; Tavanti et al. 2015a; Shao and Hall 2016). In a canonical Go model, a bead is placed at

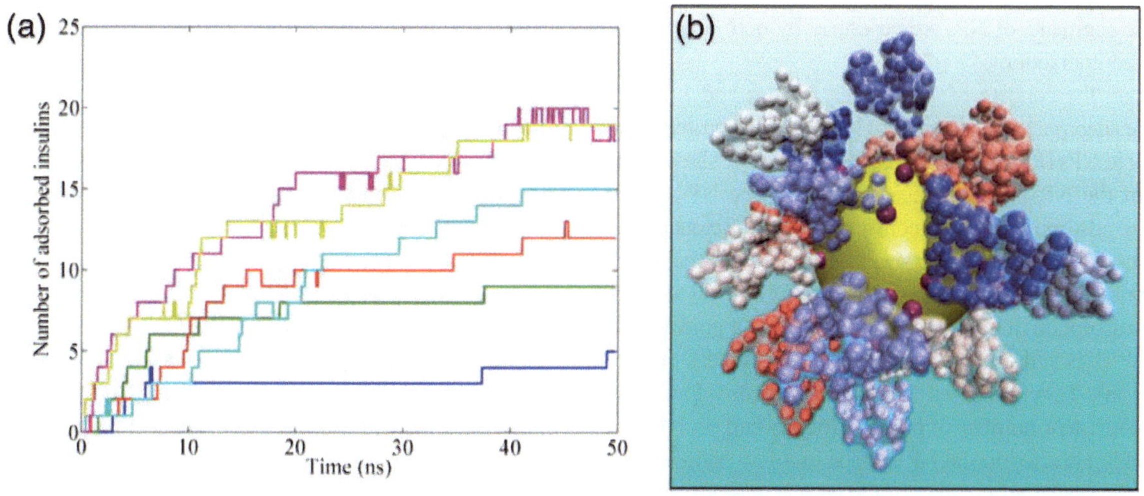

Fig. 4 Biomolecular corona formation studied with CG modeling. a Moving average (25 time steps; each time step represents 1 fs in real time) of number of insulin proteins adsorbed on citrate-coated gold NP over time. Colors represent different number of insulin molecules in the solution: blue, 10; green, 20; red, 34; cyan, 50; purple, 70; and brown, 100. **b** Snapshot, from a model simulation of a NP with 70 insulin molecules in solution taken at 45 ns shows the surrounding corona formation. Reproduced with permission from (Tavanti et al. 2015b)

the α-carbon position of each amino acid residue that collectively forms the crystal structure of a protein. Such a "one-bead" model can further be refined by placing one or more beads to represent the positions of the side chains if more details of the protein structure are desired and computationally feasible. The simulation of corona formation is carried out by discontinuous molecular dynamics, for which the protein structure can be retrieved from the Protein Data Bank (PDB) (Berman et al. 2003), and force fields governing the dynamics are evaluated between adjacent amino acid residues and between the surface of NPs through CG models. Typically, the force fields, expressed as interaction energies, consist of (Tavanti et al. 2015a, b)

$$U = \sum U_{\text{bonded}} + \sum U_{\text{non-bonded}}, \tag{3}$$

where the bonded force fields are

$$\sum U_{\text{bonded}} = \sum_{i,j} U_{\text{bonds}}(i,j) + \sum_{i,j,k} U_{\text{angles}}(i,j,k) + \sum_{i,j,k,l} U_{\text{dihedrals}}(i,j,k,l). \tag{4}$$

The first term on the right-hand side is the bonding energy between two consecutive amino acids; the second term describes the bending energy due to the angle between two consecutive bonds, hence three consecutive amino acids; and the last term describes the bending energy due to the dihedral angle of four consecutive amino acids. The non-bonded forces typically involve van der Waals interactions U_{VdW} and electrostatic interactions U_{el} between two arbitrary non-bonded amino acids:

$$\sum U_{\text{non-bonded}} = \sum_{i,j} U_{\text{VdW}}(i,j) + \sum_{i,j} U_{\text{el}}(i,j), \tag{5}$$

based on the pairwise Hamaker approach. For the process of corona formation, the interactions between amino acids and the surface segments of NPs are described by both bonded and non-bonded interactions.

Based on the concept of CG models, Lopez and Lobaskin studied the adsorption of six common blood plasma proteins onto hydrophobic NPs (Lopez and Lobaskin 2015). They showed that the size of the NP, and thus the curvature of the NP surface, affected the adsorption energies considerably for different protein orientations. Comparatively, NP surface charges had small effect on adsorption energies. Tavanti et al. used a CG model to investigate the adsorption of ubiquitin onto the surface of gold NPs for various NP sizes and coating conditions of NP surfaces (Tavanti et al. 2015a). They observed a two-phase corona formation, where a consistent number of proteins (depending on NP size) were quickly attracted to the NP surface in a few nanoseconds; however, the final configuration of the corona was not achieved until a slow reorientation, which could even be prolonged if NPs were coated with citrates, as ubiquitin competed with citrates for binding sites. Similar competition was also observed in the CG simulations between the adsorption of insulin and fibrinogen onto gold NPs (Tavanti et al. 2015b). In such models, force field parameters may be experimentally measured using monocomponent solutions of each protein. For example, Vilanova et al. parameterized their model by individually evaluating the binding affinities of three proteins, HSA, transferrin, and fibrinogen, and their model prediction of the prolonged adsorption due to competition agreed well with experiments (Vilanova et al. 2016). Shao and Hall used CG simulations to examine seven known isothermal adsorption formulas estimating the equilibrium adsorption coverage with varying protein concentration (Shao and Hall 2016). They found that for the two proteins (Trp-cage and WW domain) that they studied, four of the adsorption formulas (Langmuir, Freundlich, Temkin, and Kiselev) described the results reasonably well, while the remaining three (Elovich, Fowler-Guggenheirm, and Hill-de Boer) performed relatively worse.

Machine learning For the purpose of constructing a predictive model of corona formation to guide the engineering of NP designs, machine learning is a conceptually similar approach to coarse-graining in that it trades accuracy for efficiency. Instead of incorporating mechanical descriptions of the physiochemical processes, machine learning *statistically* infers observable features from controllable physiochemical properties. For example, using random forest classification, Findlay et al. constructed a decision tree learning model to predict the corona composition from biophysicochemical properties of proteins, NPs, and the surrounding solution (Findlay et al. 2018). Through model training, they were also able to estimate the weight of each biophysicochemical property, which could potentially be useful for guiding the design of NPs.

3.2 Microvascular transport, margination, and binding

Continuum models Decuzzi and Ferrari et al. studied the longitudinal transport of nanocarriers in non-permeable and permeable capillaries using the Taylor-Aris theory of shear dispersion (Decuzzi et al. 2006; Gentile et al. 2008). The transport of solute under fluid flow is governed by a convection-diffusion equation:

$$\frac{\partial C}{\partial t} = D_m \nabla^2 C - u \cdot \nabla C \tag{6}$$

where C is the local concentration of the solute, u is the fluid velocity field, and D_m is the diffusion coefficient[5] of the solute. This equation suggests that NPs in blood are transported

[5] Stokes-Einstein equation provides the diffusion coefficient for spherical particles in low Reynolds number. $D_m = \frac{k_B T}{6\pi\eta r}$, where k_B is Boltzmann's constant, T is the absolute temperature, η is dynamic viscosity, and r is the radius of NP.

under the influence of pressure gradient driven blood flow (advection) and their inherent Brownian motion (diffusion). Taylor (Taylor 1953) and Aris (Aris 1956) introduced the concept of an effective diffusion coefficient (D_{eff}) as a solution to the above equation averaged over the cross-section of a cylindrical tube of radius R_e and mean fluid velocity U, thus combining the contribution of both convection and diffusion in the longitudinal dispersion of NPs such that

$$D_{\text{eff}} = D_m\left[1 + P_e^2/48\right], \tag{7}$$

where P_e is the Peclet number ($P_e = R_e U/D_m$). Using the Taylor-Aris approach, Decuzzi and Ferrari et al. obtained an analytical formulation for the effective diffusion coefficient of NPs in a blood capillary (approximated as a cylindrical tube), incorporating the effects of vessel permeability and blood rheology. Eq. (7) is thus modified to:

$$D_{\text{eff}} = D_m\left[1 + \frac{P_{e_0}^2}{48}\frac{\left(\cosh\left(\tilde{z}\Gamma(\xi_c)\right) - \Omega\cosh(\Gamma(\xi_c)) - \tilde{z}\Gamma(\xi_c)\right)^2 \cdot G(\xi_c)}{1 - \Omega\cosh(\Gamma(\xi_c))} \cdot G(\xi_c)\right], \tag{8}$$

where P_{e_0} is the Peclet number at vessel inlet, $\tilde{z}$ is the dimensionless longitudinal coordinate, Γ is the permeability parameter, Ω is the pressure parameter, ξ_c is the rheological parameter (ratio between the plug and the vessel radii assuming a non-Newtonian Casson fluid velocity profile), and G is an expression that depends upon ξ_c. With this model, they found that an increase in hematocrit or vessel permeability causes a reduction in the effective diffusion coefficient of NPs, highlighting implications to intravascular transport of NPs.

In order to establish the mechanism of clearance of hard nanomaterials by the liver, Tsoi et al. implemented a minimal model of the liver sinusoid in order to study the role of flow dynamics and NP physicochemical properties on sequestration of NPs in the sinusoids (Tsoi et al. 2016). The sinusoid is modeled as a cylindrical channel of length L and radius r_0, with the inner wall of the channel lined by cells capable of sequestering NPs, which move in the channel under the influence of pressure gradient-driven advection along the longitudinal axis and Brownian motion along the radial axis. The model is expressed as a partial differential equation (PDE) that defines the temporal evolution of NP density along the length of the channel. The PDE is then solved to obtain an expression for the probability P of NP sequestration in the channel, given by:

$$P = \sum_{i=1}^{\infty}\left[1 - \exp\left(-\frac{DL}{Ur_0^2}\lambda_i\right)\right]b_i, \tag{9}$$

where D is NP diffusivity, U is the average flow velocity, and λ_i and b_i are numerical coefficients that depend upon flow profile and boundary conditions. In this expression, $\frac{DL}{Ur_0^2}$ resembles the inverse of the Peclet number ($1/P_e$) and is the ratio of the rate of diffusion to the rate of advection that controls the extent of NP interaction with the channel walls. Further, to account for the effect of NP-cell pair properties on cellular internalization of NPs and under the assumption that the cells do not function as perfect traps, a sticking coefficient (ratio of dissociation constant to association constant) is introduced into the model, making the model predictions more reliable. Model analysis demonstrated that hemodynamic conditions and particle properties that lead to $\frac{DL}{Ur_0^2} \gg 1$ favor greater NP sequestration in the channel, explaining the underlying mechanism for high NP sequestration in liver.

Discrete models In order to study the margination of NPs in blood vessels, Decuzzi et al. developed a mathematical model of a spherical NP of radius R circulating freely in blood at distance z from the endothelial wall (Decuzzi et al. 2005). The model considers buoyancy, hemodynamic, van der Waals, electrostatic, and steric force interactions acting on the particle and takes the form of the following non-linear differential equation:

$$a_1\tilde{h}(\eta)\frac{d\eta}{d\tau} - \frac{a_3}{(\eta-1)^4} + a_2 e^{-\tilde{k}(\eta-1)} + a_4 e^{-\frac{\eta-1}{\tilde{R}_g}} = 1, \tag{10}$$

where the parameter a_1 gives the ratio between hemodynamic force and buoyancy; a_2 gives the ratio between the repulsive electrostatic force and buoyancy; a_3 gives the ratio between the attractive van der Waals interaction and buoyancy, and a_4 gives the ratio between the repulsive steric force and buoyancy. η is the dimensionless distance of the particle from the endothelial wall, $\tilde{h}$ accounts for the hemodynamic forces acting on the particle, $\tilde{k}$ is the dimensionless Debye length, and $\tilde{R}_g$ is the dimensionless radius of gyration of the polymer chains on the NP surface. Numerical solution of the model determines the trajectory of the NP in the blood stream, in addition to its margination velocity and the time needed to contact the endothelium. Although the model did not account for the presence of erythrocytes, it revealed some interesting results. For example, it found that as the size of the particle is reduced, the time needed to reach the wall increases up to a maximum, beyond which the time to reach the wall decreases as the radius is further reduced. NP margination was further explored in a more comprehensive manner by Decuzzi and Lee et al. using a computational model of erythrocyte and NP transport in blood capillaries based on the immersed finite element method (discussed in next section) (Lee et al. 2013).

Furlani and Ng developed an analytical model specifically focused on magnetic NPs, to study their microvascular transport and capture under the influence of an external magnetic field (Furlani and Ng 2006). The model primarily accounts for magnetic and viscous forces acting on the particles and solves for the trajectory of the NPs in microvessels. By inputting magnetic force (obtained by the "effective" dipole moment method[6]) and fluidic force (obtained from Stoke's law[7]) into Newton's second law of motion, they arrived at an analytical expression for the trajectory of the NPs:

$$\overline{x}^3 - 3\overline{x} + 3f\left(\overline{x}_0, \overline{z}_0, \overline{z}\right) = 0, \tag{11}$$

where $\overline{x}$ and $\overline{z}$ are the dimensionless position coordinates of the NP along the x- and z-axis, respectively; f is a function of $\overline{z}$ and the normalized initial position coordinates of the NP $\overline{x}_0, \overline{z}_0$. Since magnetic force has no y-component, the model assumes NP motion in the x-z plane only. Their model demonstrates the suitability of paramagnetic NPs for delivering payload specifically to tumors under the influence of externally applied magnetic fields. Their model also determined the minimum radius of the particle required for successful capture in the capillary, by showing that the NP critical size is a function of the distance of the magnet from the capillary.

Hybrid models Hydrodynamic modeling hybridized with agent-based descriptions of NPs and blood cells has shed valuable insights into the dynamics of NP movement within capillary flow and shed new light on how NPs may be designed to increase extravasation. This is accomplished through a Navier-Stokes description of blood plasma flow within the capillary, which is solved numerically to obtain flow dynamics around agent representations of blood cells and NPs; these in turn move within the capillary due to forces applied as determined by the solution of the Navier-Stokes equation at their location (Fig. 5). Using these techniques, Lee et al. showed that NPs ≤ 100 nm diameter tended to remain in circulation with the blood bulk and have reduced incidences with and adhesion to the vasculature wall, while particles in the 500–1000 nm range were more likely to be displaced by red blood cells towards the vasculature walls (margination) (Lee et al. 2013). This increased rate of incidence with the vasculature endothelium likely increases extravasation rates and contributes to higher tumor accumulation in particles ≥100 nm than those ≤100 nm. Previously, Lee et al. had also studied the effect of particle shape on the margination propensity in linear laminar flow, although without incorporating RBCs into the model, and found that particles with discoidal shape and low aspect ratio have the highest propensity to marginate (Lee et al. 2009a, 2009b).

Interestingly, Müller et al. found similar effects of particle size on NP radial distribution within the capillary, and further demonstrated that particle shape, hematocrit, and capillary diameter also play key roles in NP distribution and dynamics within the capillary, as well as on NP extravasation (Müller et al. 2014). They found that increased hematocrit, smaller capillary diameter, and NP diameter ≥ 250 nm all promote increased margination; but, unlike the observations of Lee et al. discussed above, in the presence of RBCs, particles with aspect ratios approaching unity (spherical shapes) marginate better.

The above findings were also supported by results obtained from two other studies, which provided further evidence that red blood cells play a key role in plasma flow dynamics, with higher hematocrit increasing NP margination within the capillary (Tan et al. 2012; Fullstone et al. 2015). Tan et al. further demonstrated that small, rod-shaped particles have greater binding capacity than spherical ones due to reduced drag and greater contact area (Tan et al. 2013).

The adhesion of NPs with the endothelium during the transit of particles through microvasculature has been extensively investigated using a mathematical modeling approach by Decuzzi and Ferrari et al. (Decuzzi and Ferrari 2006; Van De Ven et al. 2012). They explored the binding of spherical and non-spherical particles under the influence of dislodging hemodynamic forces and adhesive nonspecific or specific interactions at the NP-endothelial cell interface. A force balance between hemodynamic forces, specific receptor-ligand interactions, and non-specific forces (van der Waals, electrostatic, steric) acting on the NP governs the probability of adhesion P_a of NPs to vasculature, expressed as:

$$P_a = m_r m_l K_a^0 A_c e^{-\frac{\lambda F_{dis}}{k_B T m_r A_c}}, \tag{12}$$

where m_r and m_l are the surface density of receptors on cells and ligands on NPs, respectively; K_a^0 is the association constant at zero load of ligand-receptor pair; A_c is the area of interaction between the particle and the substrate; λ is a characteristic length of the ligand–receptor bond typically on the order of 1 Å; $k_B T$ is the Boltzman thermal energy; and F_{dis} is the total dislodging force. Their results suggest that particles with an oblate shape tend to have a higher propensity and strength of adhesion under laminar flow compared to spherical particles.

3.3 Cellular internalization

Discrete models Depending upon the size of the particle and its surface chemistry, different pathways may be invoked for cellular internalization of NPs, with receptor-mediated endocytosis (clathrin-, caveolin-dependent and -independent) being the most common pathway for particles in the nano-sized range (Fig. 6) (Zhang et al. 2015). To this end, Gao et al. developed a mathematical model to understand the mechanism of cell membrane wrapping around a spherical or

[6] Magnetized NP is substituted by an equivalent point dipole with an effective dipole moment $m_{p,\ \mathrm{eff}}$ such that the magnetic force F_m on the dipole is, $F_m = \mu_f(m_{p,\ \mathrm{eff}} \cdot \nabla)H_a$, where μ_f is the permeability of the fluid, and H_a is the external magnetic field intensity.

[7] The drag F_d on a spherical object of radius r, moving through a viscous fluid of viscosity η at a relative velocity v, is, $F_d = -6\pi\eta r v$.

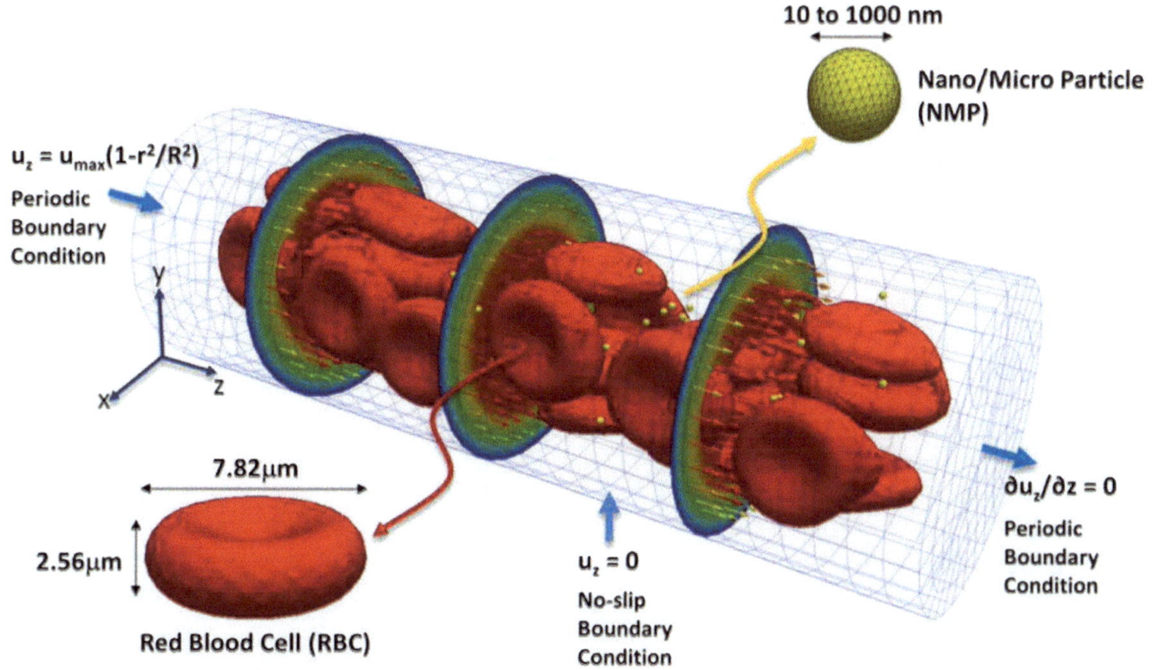

Fig. 5 Computational domain for hydrodynamic simulations of NP transport in microcirculation. NP transport was studied in a capillary of length 60 μm and diameter 20 μm in the presence of deformable red blood cells. Periodic boundary conditions are imposed at the inlet and outlet of the capillary, and a parabolic velocity profile with a maximum velocity of 100 $\mu m \cdot s^{-1}$ is imposed at the inlet. Reproduced with permission from (Lee et al. 2013)

cylindrical particle during receptor-mediated endocytosis (Gao et al. 2005). The model assumes that the particle surface is uniformly coated with immobile surface ligands that are complementary to mobile receptors on the cell membrane surface. Given that cells are much larger than NPs, the problem is simplified by assuming NP interaction with a flat membrane. Following initiation of contact and receptor-ligand binding, the diffusive receptors on cell membrane are drawn towards the site of contact to match receptor density with ligand density, thereby increasing the area of contact over time; this process continues until the area of contact equals the area of the particle. The power balance between the rate of free energy reduction gained from the wrapping process (receptor-ligand binding and membrane cytoskeleton bending) and the rate of energy dissipation consumed during receptor migration yields the particle wrapping time τ:

$$\tau = \left(\frac{\pi R}{2\alpha\sqrt{D}}\right)^2, \text{(for cylindrical particles)} \tag{13}$$

$$\tau = \left(\frac{R}{\alpha\sqrt{D}}\right)^2, \text{(for spherical particles)} \tag{14}$$

where R is the radius of the particle, α is the speed factor, and D is the diffusivity of membrane receptors. Their model highlights the importance of particle size in receptor-mediated endocytosis.

Yuan and Zhang developed a model based on the energy balance between receptor-ligand binding energy and work done in membrane wrapping around the particle (Yuan and Zhang 2010). Their model provided analytical expressions for minimal particle size for a given ligand density ξ_l (Eq. 15) and minimal ligand density for a given particle size of radius R (Eq. 16) below which receptor-mediated endocytosis cannot occur:

$$R_{min} = \sqrt{2B/[\xi_l(\varepsilon + ln\, \xi_0)]}, \tag{15}$$

$$\xi_{l,min} = 2B/\left[R^2(\varepsilon + ln\, \xi_0)\right], \tag{16}$$

where, ξ_0 is the receptor density on cell surface, B is the membrane bending modulus, and ε is the receptor-ligand binding energy. Their findings suggest that both NP size and ligand density are critical in governing the kinetics of endocytosis, and are important in guiding the rational design of NPs.

In another model, which was experimentally validated using iron oxide NPs and human macrophages, Lunov et al. ignored receptor diffusion on cell membrane surface, and instead employed pit formation followed by membrane wrapping around the spherical particle as the mechanism of receptor-mediated endocytosis (Lunov et al. 2011). By equating the mechanical work performed by cytoskeletal motor proteins for pit formation with the elastic energy of the membrane, the wrapping time τ is obtained as:

$$\tau = \frac{4\pi R^2 \sigma}{p}, \tag{17}$$

where R is the particle radius, σ is the surface tension of the

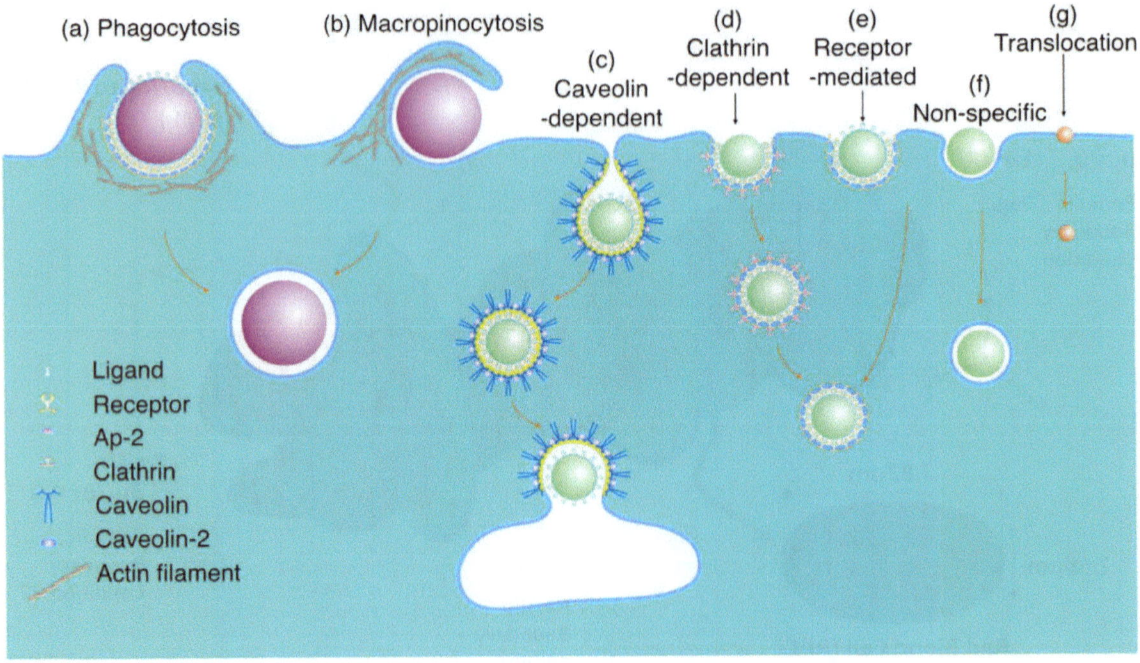

Fig. 6 Cellular internalization pathways. Phagocytosis (**a**) or micropinocytosis (**b**) may be involved in the internalization of micrometer-sized particles. Caveolin-dependent (**c**) or clathrin-dependent (**d**) endocytosis occurs by receptor-ligand binding leading to the formation of flask-shaped caveolae and clathrin-coated pits, respectively, on the cytosolic side of the cell membrane. Although receptor-mediated endocytosis may also be clathrin- and caveolin-independent (**e**). Non-specific interactions maybe involved in endocytosis of NPs without conjugated ligands (**f**). Small NPs and molecules (<1 nm) may enter the cell by diffusion (translocation) through the plasma membrane (**g**). Reproduced with permission from (Zhang et al. 2015)

membrane, and p is the power of motor proteins. Their analysis provides quantification of important parameters, including rate of NP uptake, number of NPs uptaken per cell at saturation, and mean uptake time.

Sorrell et al. investigated endocytosis of NPs by cancerous cells using a kinetic model based on a system of differential equations (Sorrell et al. 2014). The model accounts for the rate of change of unbound, bound, and internalized NPs. Key model parameters include association and dissociation rate constants of NP-receptor binding, the recycling rate constant of receptors, and the internalization rate constant of NPs. The results suggested that the rate of NP uptake depends on the number of receptors engaged by the particle.

Decuzzi and Ferrari modified the model by (Gao et al. 2005) assuming a non-constant radius of curvature of the cylindrical particle, i.e., the curvature depends on the particle-cell area of contact and thus varies with time (Decuzzi and Ferrari 2008). As a result, the contribution of the elastic bending energy of membrane wrapping around the particle varies as the area of contact changes. Their model reveals the occurrence of a critical size and aspect ratio for cylindrical particles: for circular cylindrical particles, radius below ~40 nm are energetically unfavorable for internalization, and for elliptical cylindrical particles there exists an optimal range of aspect ratio values, below or above which internalization is incomplete.

3.4 Whole-body biodistribution and clearance

Pharmacokinetic models For successful clinical translation of nanomedicine, it is important to establish the pharmacokinetics (PK) of NPs, and to this end, mathematical modeling approaches are routinely used; these are referred to as pharmacokinetic models. PK modeling typically operates at the macroscopic (organ) scale and involves a phenomenological description of the ADME (absorption, distribution, metabolism, and excretion) of NPs. PK modeling can be roughly classified into *classical* and *physiological* modeling approaches. The classical PK approach deconstructs the body into a system of compartments, and often contains a central compartment that may be connected via rate constants to one or more peripheral compartments (Fig. 7). The central compartment is a lumped compartment that contains the blood pool of the body and highly perfused organs, like heart, lungs, liver, and kidneys. Similarly, the peripheral compartment is generally formed by lumping the poorly perfused or slowly equilibrating tissues, like fat, bones, and muscle. First-order kinetics is typically assumed for mass transfer between compartments and for elimination from compartments. ODEs are used to describe various PK processes, which are fit to the concentration-time data of the NPs in plasma, urine, or other tissues, to obtain estimates for numerical coefficients and relevant PK parameters like half-life, clearance, volume of distribution, and mean residence time. Being simplistic and

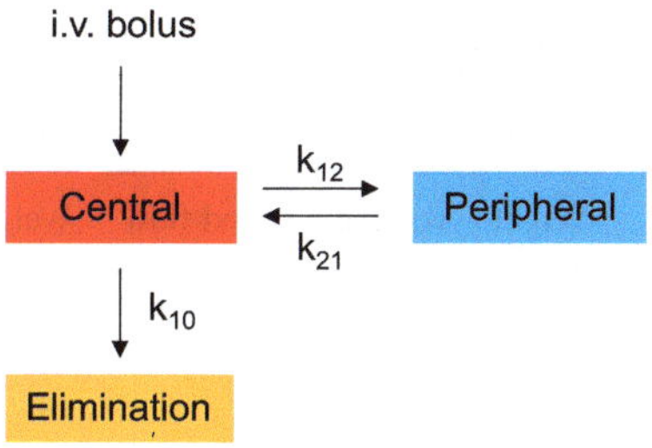

Central compartment:

$$\frac{dC_c}{dt} = k_{21} \cdot C_p - (k_{10}+k_{12}) \cdot C_c, \qquad C_c(0) = C_0$$

Peripheral compartment:

$$\frac{dC_p}{dt} = k_{12} \cdot C_c - k_{21} \cdot C_p, \qquad C_p(0) = 0$$

Fig. 7 Schematic of a classical pharmacokinetic model for i.v. bolus administration. A representative two-compartment PK model with a central compartment and a peripheral compartment is shown. Elimination is restricted to the central compartment. Master equations for the two compartments are also shown, where C_c and C_p are NP concentrations in the central and peripheral compartments, respectively; k_{12} and k_{21} are the rates of transfer of NPs between central and peripheral compartments, and k_{10} is the rate of elimination of NPs from the central compartment. C_0 is the concentration in the central compartment at time zero. All mass transfer processes are assumed to follow first-order kinetics, and solution of the coupled ordinary differential equations provides the temporal evolution of NP concentration in the given compartments

empirical (data-driven), such models have great clinical relevance, e.g. dosage regimen design, but their predictive capacity is very limited since they lack an underlying physiological and mechanistic reference (Gerlowski and Jain 1983; Gabrielsson and Weiner 2001; Jones and Rowland-Yeo 2013).

This is where physiological PK models fill the gap. Usually referred to as "physiologically based pharmacokinetic (PBPK) models," they have a similar underlying framework as classical PK models but have anatomically based compartments connected via physiological blood flow rates (Fig. 8). The models are thus parameterized with anatomical and physiological variables (e.g., tissue weights, tissue volumes, blood flow rates (Table 3)) and physicochemical information of the xenobiotics (e.g., blood-tissue partition coefficients), obtained from literature or data fitting (Gerlowski and Jain 1983; Li et al. 2010; Jones and Rowland-Yeo 2013). Each compartment representing a relevant organ or tissue is sub-compartmentalized into vascular, interstitial, and cellular spaces. Based on the physicochemical properties of the xenobiotic under investigation, PBPK models are classified into: i) perfusion or blood flow rate-limited and ii) diffusion or permeability-limited (Khalil and Läer 2011). Perfusion rate-limited models assume that xenobiotic transfer between the vascular space and interstitial space of an organ is not limited by capillary permeability but is only governed by the blood flow rate to the organ. Conversely, permeability-limited models assume capillary permeability to be the limiting factor. When applied to nanomedicine, PBPK models serve to provide a mechanistic description of the concentration time-course of NPs in any given tissue or plasma. Like classical PK, linear kinetics is usually assumed for mass transport across the system and coupled ODEs model the NP concentration in tissues. Because the underlying framework is physiologically meaningful, these models can help with dose extrapolation from animals to humans, or healthy volunteers to diseased patients, based on differences in physiological parameters (Sharma and McNeill 2009; Jones and Rowland-Yeo 2013). They are also valuable in predicting dose adjustments for special populations, like pregnant women (Gaohua et al. 2012) and pediatrics (Hornik et al. 2017). PBPK has also emerged as a critical tool in addressing regulatory questions about the effect of intrinsic (e.g., organ impairment, age, and genetics) and extrinsic (e.g., drug-drug interactions) factors on the PK and pharmacodynamics (PD) of drugs (Huang and Rowland 2012).

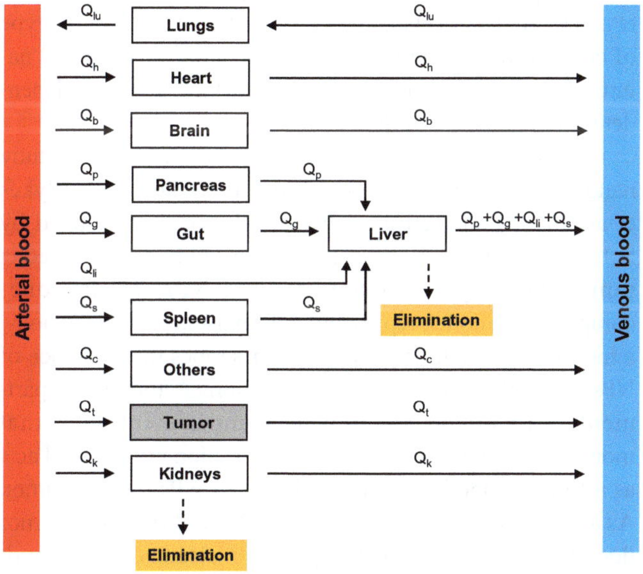

Fig. 8 Schematic of a physiologically based pharmacokinetic model. The various compartments are connected to the systemic circulation compartments (arterial and venous) via physiological blood flow rates (Q_i). Liver and kidneys are the compartments responsible for clearance. Liver has a dual blood supply as shown (hepatic artery and portal vein). "Others" represents the tissues not explicitly modeled and may include fat, muscle, and bones. Lymph flow is not depicted in this model

Like free drugs, PK modeling of NPs requires quantification of NP concentration (usually expressed as percent of injected dose per gram of tissue ($\%ID \cdot g^{-1}$)) over time in plasma, urine, and other tissues. For this purpose, the most commonly employed techniques include plasma/urine

Table 3 List of a priori physiological parameter values for lab animals and humans for organ- or whole-body-scale models

	Mouse (0.02 kg)			Rat (0.25 kg)			Rhesus monkey (5 kg)			Human (70 kg)		
	Weight (g)	Volume (ml)	Blood flow rate (ml/min)	Weight (g)	Volume (ml)	Blood flow rate (ml/min)	Weight (g)	Volume (ml)	Blood flow rate (ml/min)	Weight (g)	Volume (ml)	Blood flow rate (ml/min)
Brain	0.36	0.48	0.46	1.8	1.2	1.3	90	94	72	1400	1450	700
Heart	0.08	0.095	0.28	1	1.2	3.9	18.5	17	60	330	310	240
Lung	0.12	0.1	≈ C.O.	1.5	2.1	≈ C.O.	33	35.7	≈ C.O.	1000	1170	≈ C.O.
Liver	1.75	1.3	1.8	10	19.6	13.8	150	100	218	1800	1690	1450
Gut	1.5	1.5	1.5	6.3	11.3	7.5	230	230	125	2100	1650	1100
Spleen	0.1	0.1	0.09	0.75	1.3	0.63	8	5.95	21	180	192	77
Pancreas	0.2	0.09	0.27	1	1	1	5	12.5	10.2	80	104	41
Kidneys	0.32	0.34	1.3	2	3.7	9.2	25	30	138	310	280	1240
Muscle	–	10	0.91	–	245	7.5	–	2500	90	–	35,000	750
Fat	–	1.98	–	–	10	0.4	–	154	20	–	10,000	260
Skin	–	2.9	0.41	–	40	5.8	–	500	54	–	7800	300
Blood	–	1.7	–	–	13.5	–	–	367	–	–	5200	–
Plasma	–	1	–	–	7.8	–	–	224	–	–	3000	–
Hepatic artery	–	–	0.35	–	–	2	–	–	51	–	–	300
Portal vein	–	–	1.45	–	–	9.8	–	–	167	–	–	1150
Cardiac output (C.O.)	–	–	8	–	–	74	–	–	1086	–	–	5600

Source: (Davies and Morris 1993; Gabrielsson and Weiner 2001; Peters 2012; Shah and Betts 2012). Lymph flow rates are ~500 times lesser than blood flow rates (Shah and Betts 2012)

radioactivity quantification (Kommareddy and Amiji 2007; Xu et al. 2013), whole-body radioactivity imaging and quantification (e.g. single photo emission computed tomography (SPECT) (Woodward et al. 2007; Chrastina and Schnitzer 2010; Patil et al. 2011; Black et al. 2015; Li et al. 2016; Ming et al. 2017; Dogra et al. 2018), positron emission tomography (PET) (Schluep et al. 2009; Kumar et al. 2010; Phillips et al. 2014; Chen et al. 2016)), fluorescent imaging (Kumar et al. 2010; Vasquez et al. 2011), magnetic resonance imaging (MRI) (Neubauer et al. 2008), inductively coupled plasma mass spectroscopy (ICPMS) (Crayton et al. 2012), and accelerator mass spectrometry (AMS) (Malfatti et al. 2012). The most suitable compartmental model is then fit to the time course of NP concentration obtained using one or more of these experimental techniques. Most in vivo studies (Yu et al. 2012) report a biphasic decline in plasma (or mediastinum (Schluep et al. 2009), a substitute for plasma when imaging techniques are used) concentration of NPs, following intravenous (i.v.) injection. Thus, a two-compartment model is most commonly employed to describe systemic NP disposition (Fig. 7). However, it is important to note that the frequency of experimental data collected can impact the nature of concentration time course, and a biexponential decline will not always be the case. Hence, some studies report using a one compartment model (for monophasic decline) (Sykes et al. 2014b; Dogra et al. 2018). Thus, the selection of the type of compartmental model depends solely on the observed nature of the concentration time course, which in turn is dependent on the method of data collection.

Classical PK approaches have also been employed to study nanomedicine delivery to solid tumors. Sykes et al. (Sykes et al. 2014b) adapted a mathematical model, developed by Schmidt and Dane Wittrup (2009) for antibodies, to predict tumor delivery efficiency of gold NPs. In this one-compartment model, representing the plasma pool of the body, a monoexponential decay function describes the clearance of NPs from blood following i.v. injection. A tumor compartment is linked to the plasma compartment such that monoexponential decay of NP concentration in the blood acts as a forcing function to govern influx of NPs into the tumor. As a simplification, it is assumed that NP influx into the tumor does not influence the monoexponential clearance behavior of NPs in the blood. Once inside the tumor, permeability of tumor vessels governs the extravasation of NPs into the tumor interstitium. Following extravasation, NPs diffuse through the available volume of tumor interstitium, bind or unbind from cell surfaces, and are eventually endocytosed and degraded by tumor cells. With all reactions assumed to follow first order kinetics, the resulting biexponential function describes the NP-associated, dose-normalized, fluorescence concentration time-course (% ID · ml^{-1}, percent of injected dose per ml of tumor) in the tumor compartment. The model was then used to predict the influence of plasma clearance rate and tumor cell binding affinity of NPs on their tumor accumulation (represented by concentration-time area under the curve (AUC)). The model can thus be employed as an in silico platform to test NP design configurations for their impact on tumor accumulation of NPs.

The application of the more mechanistic PBPK models in nanomedicine is in only a premature stage to date, and most studies involve the application of PBPK models in their canonical form (Li et al. 2010; Li et al. 2012; Bachler et al. 2013; Li et al. 2013), i.e., models developed traditionally for free drugs. However, fundamental differences between physicochemical characteristics of NPs and free drugs demand modifications to the structure of conventional PBPK models to better fit the purpose of modeling NP disposition kinetics (Riviere et al. 2013).

Our group has recently demonstrated the application of PK modeling to disposition kinetics data of mesoporous silica NPs (MSNs) acquired through SPECT/CT imaging in vivo (Fig. 9) (Dogra et al. 2018). Guided by the data and built on our prior work (Das et al. 2013; Pascal et al. 2013b; Koay et al. 2014; Frieboes et al. 2015), we developed simple master equations in closed form to model the kinetics of MSN biodistribution and clearance with the goal of establishing the structure-activity relationships of MSNs. By systematically varying MSN physicochemical variables (including size, surface charge, and surface coatings) in the therapeutically relevant size range of ~25–150 nm, we examined the effect of size, zeta potential, and surface chemistry on in vivo disposition of hydrodynamically stable, monodisperse, non-targeted MSNs administered via intravenous (i.v.) or intraperitoneal (i.p.) injection. Our analysis showed that (1) smaller MSN size results in a higher systemic bioavailability, irrespective of the route of administration; (2) positive charge favors greater excretion; and (3) surface exposed charged molecules (amines) increase vulnerability to sequestration in liver and spleen. Importantly, a consistent mathematical function between one key PK parameter ($AUC_{0-24\ h}$; area under the curve—a parameter that evaluates the extent of xenobiotic exposure of organs) and MSN core diameter (*size*) was identified in the form of $AUC_{0-24\ h} = \lambda \cdot size^{-n}$ for systemic circulation and source-like organs (e.g., heart and lungs) in both i.v. and i.p. cases; conversely, for sink-like organs (e.g., spleen and liver), the function identified is $AUC_{0-24\ h} = \lambda \cdot size^{n}$, where λ and n are fitted parameters. This finding is highly significant because these newly discovered, consistent mathematical functions simplify and facilitate the understanding of the relationships between MSNs' physicochemical properties and their PK behaviors in vivo.

3.5 Tumor deliverability

Hybrid models We now discuss relevant modeling works that focus on the penetrability of NPs in the tumor tissue. These models prioritize various aspects of NP transport and help understand the relationships between physicochemical properties of NPs and their deliverability to the tumor tissue.

Chauhan et al. developed a mathematical model of tumor vasculature (Chauhan et al. 2012) based on *percolation theory* (Gazit et al. 1995). The model is cast in the form of a two-dimensional percolation network representing tumor vasculature with one inlet and one outlet to study the effect of vascular normalization on tumor delivery efficiency of nanomedicine. The network consists of a series of interconnected nodes representing blood vessel segments, with each segment endowed with vessel wall pores obtained from a distribution

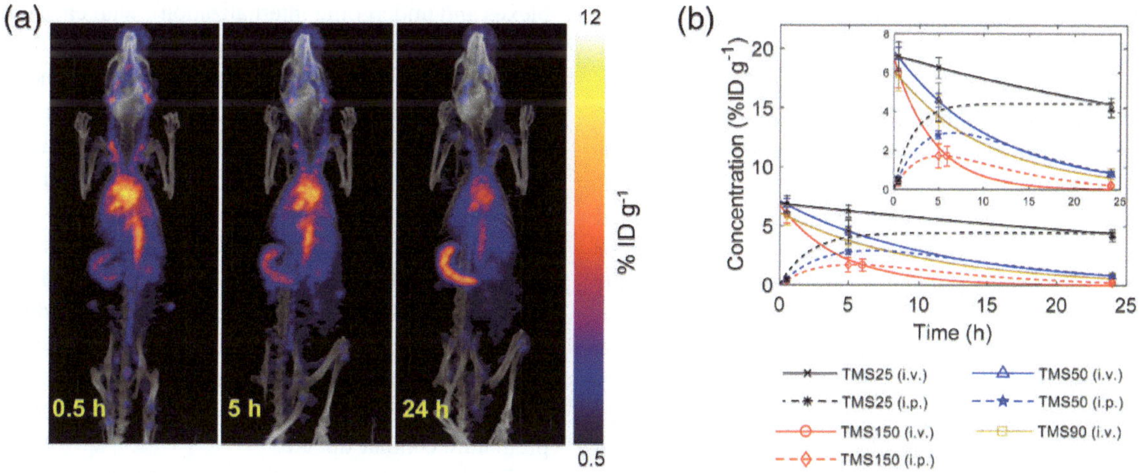

Fig. 9 **Imaging-based pharmacokinetics. a** Representative SPECT/CT images of a rat injected with radiolabeled, 25 nm-sized mesoporous silica NPs (MSNs). **b** Concentration kinetics obtained from fitting a phenomenological PK model to concentration versus time data for different types of MSNs injected i.v. or i.p. Reproduced from (Dogra et al. 2018)

of pore sizes. It is assumed that blood flow in the vessels follows Poiseuille's law,[8] transvascular fluid exchange occurs according to Starling's approximation,[9] and Darcy's law[10] governs interstitial fluid transport. Further, the transport of NPs across vessel wall pores is modeled using "pore theory" to determine barriers to diffusion and convection across cylindrical pores. By capturing critical transport phenomena at the tissue scale, this model provides insights into the significance of vascular normalization in cancer nanomedicine. Further, they also studied the effect of NP surface charge on transvascular flux of NPs, and demonstrated the superiority of cationic particles in crossing into the tumor interstitium (Stylianopoulos et al. 2013) using a mathematical model of tumor vasculature based on a previously developed algorithm of tumor-induced angiogenesis governed by vascular endothelial growth factors and fibronectin gradients (McDougall et al. 2006).

Hendricks et al. developed a multiscale kinetic model to study the tumor delivery efficiency of doxorubicin via drug-loaded liposomal formulations (Hendriks et al. 2012). The model consists of a one-compartment PK model of liposome disposition kinetics, integrated with a two-compartment PK model of liposome-released doxorubicin disposition kinetics, in plasma. The PK models are connected to a physiologically based tumor model that incorporates transvascular flow, interstitial transport, and cellular uptake of liposomes and free drug. Their model reveals strong dependence of delivery efficiency on liposomal PK and tumor vascular permeability to liposomes, thereby highlighting the significance of these patient-specific parameters in determining the success of nanotherapy.

Frieboes at al. developed a computational model to predict accumulation of NPs in tumor vasculature (Frieboes et al. 2013). It combines a previously developed two-dimensional model of tumor growth and angiogenesis (Anderson and Chaplain 1998; Cristini et al. 2003; McDougall et al. 2006; Macklin et al. 2009; Wu et al. 2013) with a mesoscopic model of NP adhesion to the tumor neovasculature. Tumor dynamics are primarily governed by cell division, cell death, cell migration, and cell-cell and cell-extracellular matrix adhesion. These processes are accounted for in the tumor growth compartment as a mathematical model based on conservation principles (mass, momentum) and transport phenomena (diffusion, convection). This tumor growth model is then coupled with a model of tumor-induced angiogenesis based on tumor angiogenic factor (TAF) gradients. This component accounts for tumor blood flow, non-Newtonian effects, vascular leakage, and vasculature remodeling due to shear stress and mechanical stress because of a growing tumor, thus simulating a pathophysiologically relevant scenario to test neovascular accumulation of systemically injected NPs based on a mesoscopic mathematical formulation (Eq. (12)) (Decuzzi and Ferrari 2006). Thus, by integrating the evolution of a growing tumor with NP-cell interactions, the model demonstrates the dependence of vascular accumulation of NPs based on the tumor growth stage, in addition to the importance of NP vascular affinity in controlling total accumulation of NPs and their spatial distribution inside a tumor (Fig. 10).

Hauert et al. (2013) modeled a representative section of tumor tissue to study the design characteristics of targeted NPs responsible for binding site barriers (Fujimori et al. 1990), and proposed guidelines to overcome such barriers. The modeled section represents a hypoperfused, near-necrotic region of the tumor where NPs extravasating from the microvessels diffuse into surrounding tumor tissue, and may bind to cell surfaces and subsequently be internalized. Reaction-diffusion kinetics thus forms the basis of such a model, and is represented by the following biochemical reaction: $NP_F + R \longleftrightarrow^{k_a,k_d} C \to^{k_i} NP_I + R$, where NP_F represents free NPs, R refers to cell-surface receptors, C are the NP-receptor complexes, NP_I represents internalized NPs, and k_a, k_d, and k_i are the association, dissociation, and internalization rate constants, respectively. This system has been modeled using both deterministic and stochastic approaches. The deterministic modeling approach is comprised of coupled reaction-diffusion PDEs governing the spatiotemporal evolution of species of interest (NP_F, R, C, NP_I) in a one-dimensional domain. The stochastic modeling approach discretizes the spatial domain into cubes with side S, assuming a well-mixed volume in each unit. NPs diffuse across these sub-volumes while interacting with surface receptors (potentially forming complexes and undergoing internalization), with each event being governed by a dynamic probability. Unlike the deterministic model, the stochastic model captures the randomness and fluctuations of a biochemical system; however, the former is a better choice for systems with large populations and relatively small fluctuations, since the SSC might be computationally expensive for such systems. Thus, in this work the authors used the deterministic model to simulate different experiments, and only validated key findings using the stochastic model. Their work suggests that delaying the binding of NPs to target cells could avoid binding site barriers on the tumor periphery and allow them to penetrate deeper into the tumor tissue, which can be achieved by engineering NPs to avoid premature cellular uptake.

Stapleton et al. developed a mathematical model to specifically investigate the transport of liposomes in solid tumors through the EPR effect (Stapleton et al. 2013). The

[8] $Q_{vasc} = -\frac{\pi d^4}{128\mu}\frac{\Delta p_v}{\Delta x}$ where Q_{vasc} is the blood volumetric flow rate in a vessel of length Δx and diameter d, for vascular pressure difference Δp_v and blood viscosity μ.

[9] $Q_{transv} = L_p S(p_v - p_i)$ where Q_{transv} is the volumetric fluid flow rate across a vessel wall of hydraulic conductivity L_p, vessel surface area S, and interstitial fluid pressure p_i.

[10] $Q_{inter} = -k_t A_c \frac{\Delta p_i}{\Delta x}$ where Q_{inter} is the volumetric fluid flow rate in interstitial space with hydraulic conductivity k_t, cross-sectional area A_c, tissue length Δx, and interstitial pressure difference Δp_i.

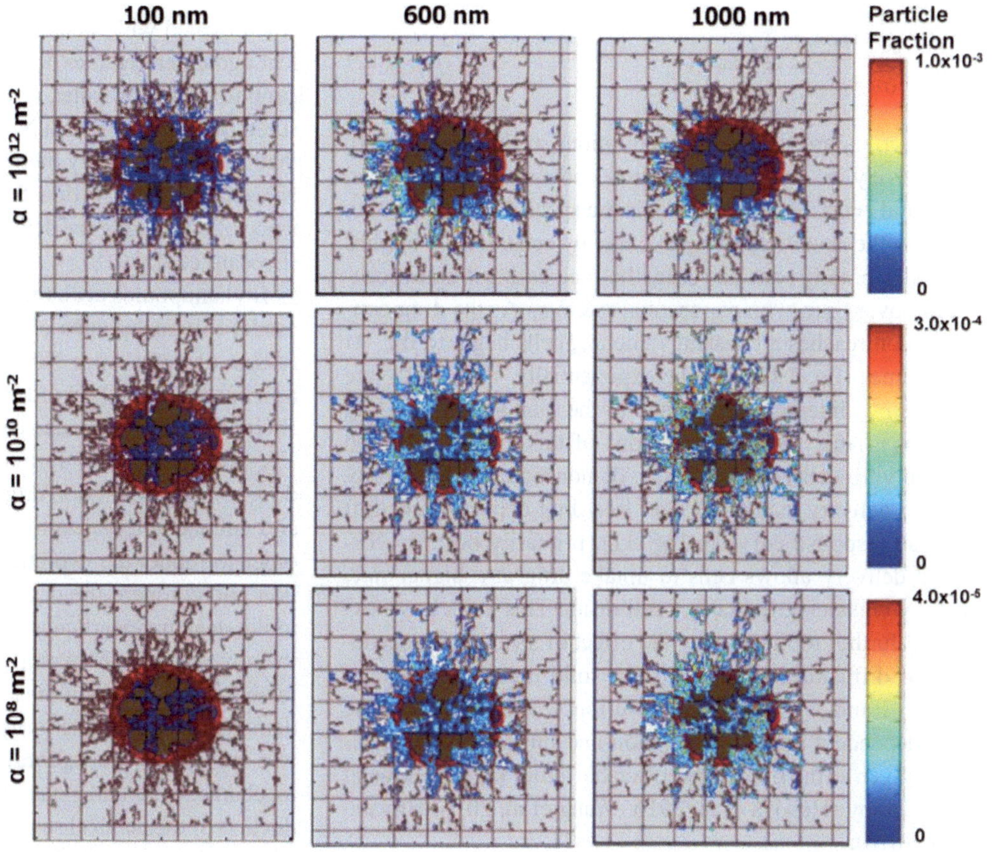

Fig. 10 Model predictions of NP tumor delivery efficiency. A multiscale tumor growth model is used to predict the delivery efficiency of nanocarriers of different sizes (100 nm, 600 nm, 1000 nm) and at different levels of intratumor vasculature receptor expression (α) 100 min after injection on day 18 of tumor growth. Reproduced with permission from (Frieboes et al. 2013)

model was formulated as a differential equation, and accounts for the transvascular and interstitial convective flux to determine the rate of liposome accumulation in the tumor interstitium. The extravasation of NPs from the vessels is primarily driven by the difference between microvascular pressure and interstitial fluid pressure, while interstitial transport is driven by an interstitial pressure gradient estimated by Darcy's law. The model was validated against experimental data and predicted inter-subject and intra-tumoral variations in the EPR effect based on biophysical properties of the tumor microenvironment.

Sykes at al. performed Monte Carlo simulations to study diffusion of NPs through tumor extracellular matrix (ECM) (Sykes et al. 2016), based on the approach of (Stylianopoulos et al. 2010). They modeled the tumor ECM in three-dimensions as an anisotropically oriented network of collagen fibers (the most abundant protein in ECM) to study the mobility of NPs through matrices of different collagen densities. Collagen fibers were approximated as immobile cylinders and NP-fiber collision was assumed to be elastic. NP movement was simulated as a discrete random walk following the Stokes-Einstein relation for diffusion of spherical particles in a fluid with low Reynolds number. They also modeled in 2D the microscopic collagen matrix pores to study the effect of NP size and matrix pore size on the frequency of NP-fiber collisions to provide a mechanistic explanation to the results of the 3D model. In this model, a pore was represented by a square bounded by collagen fibers, a NP underwent 2D Brownian motion inside the square, and particles were tracked for their collisions with the wall (fibers). Their models helped elucidate the mechanisms underlying particle size-dependent retention of NPs in clinically relevant tumor conditions.

3.6 Nanotherapy efficacy and toxicity

Pharmacodynamic models We now discuss the modeling approaches employed to investigate nanotherapy efficacy, which is the ultimate determinant of success of cancer nanomedicine. (Pascal et al. 2013a) evaluated the cytotoxicity of NP-loaded doxorubicin to hepatocellular carcinoma cells using a mathematical model based on cell and drug mass conservation. The model is based on the hypothesis that cell death rate is a function of the total amount of drug taken up by the cells over time, which is represented by the following equations that predict the asymptotic behaviors of cell and drug concentrations for short and long drug exposure times ($\lambda n_0 t$), respectively:

$$\begin{cases} \frac{n}{n_0} \approx 1 - \frac{1}{2}\lambda_A \sigma_0 \lambda n_0 t^2 \\ \frac{\sigma}{\sigma_0} = e^{-\lambda n_0 t} \end{cases}, \quad (\lambda n_0 t \ll 1) \tag{18}$$

$$\begin{cases} \frac{n}{n_0} \sim e^{-\lambda_A(\sigma_0-\sigma_\infty)t} \\ \sigma \approx \sigma_\infty \end{cases}, (\lambda n_0 t \gg 1) \quad (19)$$

where, n, n_0 are the cell concentrations at time t and $t=0$, respectively, σ, σ_0, σ_∞ are the drug concentrations at time t, $t=0$, and $t=\infty$, respectively, λ_A is the specific cell death rate, and λ is the specific drug uptake rate by cells. Eq. 18 implies that cells initially consume drug, thus decreasing drug concentration over time at an exponential rate λn_0. As the drug concentration reaches a constant value σ_∞, cells begin to die. At large times (Eq. 19), cells die exponentially, with death rate $\lambda_A(\sigma_0 - \sigma_\infty)$ linearly proportional to the total amount of drug $(\sigma_0 - \sigma_\infty)$. Analysis and predictions of this model, validated with in vitro cell viability assays, demonstrate the superiority of drug delivery using NPs over free-drug delivery (left shift in dose response curve in Fig. 11), primarily because NP-based delivery allows cells to uptake drug at a higher rate, thereby enhancing the rate of cell death. This model was extended further to incorporate the effect of spatial heterogeneities of diffusion and perfusion in solid tumors on therapy efficacy, and a closed-form solution was derived for the time-dependent drug concentration and tumor volume equations, given as the fraction f_{kill} of viable tumor killed following nanotherapy (Hosoya et al. 2016; Wang et al. 2016; Brocato et al. 2018):

$$f_{\text{kill}} = \frac{F\lambda_k}{2V_{T,0}}t^2 \quad (20)$$

where F is the flux of drug from the NPs, F is a function of NP size, λ_k is the cell death rate per unit cumulative drug concentration, and $V_{T,0}$ is the initial tumor volume. Note that there is a quadratic increase in treatment efficacy (represented by f_{kill}) with time. The model predicts faster and greater tumor death when treated with nanocarriers than with free drug, in excellent agreement with experimental observations under various conditions. In particular, using this model and by linking measured tumor growth with NP distribution, Brocato et al. found that treatment efficacy increases exponentially with increased NP accumulation within the tumor, highlighting the importance of optimizing the delivery efficiency of NPs to the tumor (Brocato et al. 2018).

Stylianopoulos et al. further developed their tumor delivery model (Stylianopoulos et al. 2013) to incorporate controlled release of drugs from NPs and their binding and internalization into tumor cells in order to study efficacy of NP-delivered chemotherapy on cancer cells (Stylianopoulos et al. 2015). They mathematically formulated the interstitial processes of drug release from NPs, drug diffusion, cell surface bonding, and cellular internalization in the form of coupled PDEs in order to model the evolution of internalized drug concentration and compare the cell-kill induced by a two-stage and a multi-stage nanoparticulate system with conventional chemotherapy. The model explored the effects of NP size, drug release kinetics, and binding affinity on therapy efficacy, and demonstrated the superiority of multistage nanocarriers over two-stage NPs. Papageorgis et al. used the same model to mechanistically understand the effect of Tranilast-induced reduction of mechanical stresses in tumors on improved efficacy of nanotherapeutics, thereby proposing the repurposing of tranilast as mechanical-stress alleviating agent (Papageorgis et al. 2017).

Leonard et al. (2016) employed the model developed by Frieboes et al. ((Frieboes et al. 2013) discussed previously) to investigate the efficacy of macrophage-encapsulated NP albumin-bound-paclitaxel in simulated patient breast cancer liver metastasis. Their model demonstrates the potential of this novel formulation for clinical translation based on the

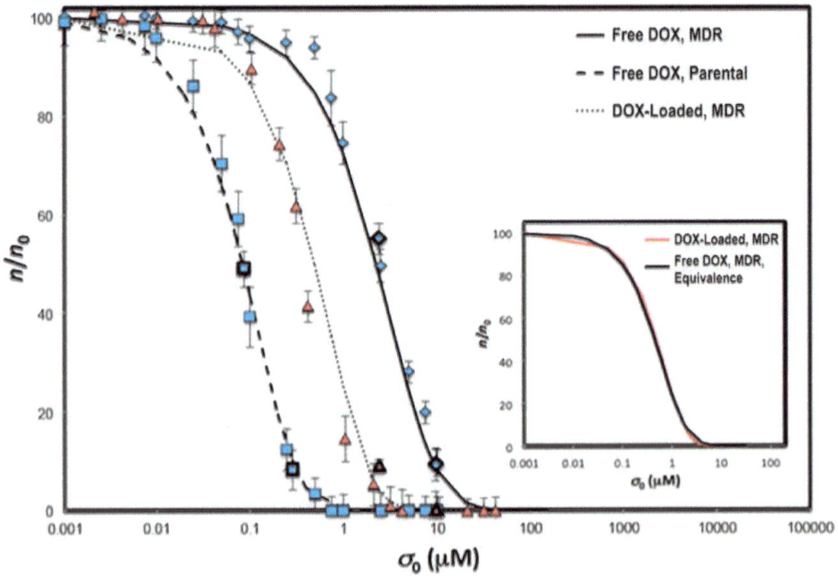

Fig. 11 Predictions of dose-response curve from a mathematical model. Numerical solutions of the model are tested against experimental observations of the fraction of viable cells 24 h after incubation with free-doxorubicin and doxorubicin-loaded NPs at variable drug concentrations. NP-mediated delivery causes a left-shift (lowered drug IC_{50}) in dose-response curves of multi-drug resistant hepatocellular carcinoma cells. Reproduced with permission from (Pascal et al. 2013a)

histological evidence of tumor biopsies. Miller and Frieboes employed the same model to study the effects of vasculature-induced heterogeneity and drug strength (IC_{50}) on therapy efficacy of drug-loaded gold NPs (Miller and Frieboes 2018). The results indicate that in the case of drugs with high $IC_{50,}$ the intra-tumoral vascular density should be above a threshold for optimal transport of nanocarriers for satisfactory efficacy, while efficacy depends less strongly on vessel density for NPs loaded with drugs with lower IC_{50}.

A relatively less studied aspect that can potentially stall nanomedicine from clinical translation is NP-induced toxicity. Recently, a multiscale hybrid model was developed by Laomettachit et al. to assess NP toxicity in human liver (Laomettachit et al. 2017). The model uses a standard whole-body PBPK framework to predict hepatic accumulation of administered NPs, and feeds the whole-body model-derived information into a tissue scale "cell-response model" of liver tissue. The cell-response model accounts for cell division and cellular uptake of NPs to predict the impact of NPs on hepatocyte viability. This multiscale platform demonstrates the dose-dependent toxicity of titanium dioxide NPs in liver, i.e., the tissue damage from a low dose of NPs is negligible and reversible due to compensation by cell proliferation, while high exposure can cause irreversible tissue damage unless a large fraction of cells undergoes cell division to renew the damaged tissue mass.

4 Conclusions

Cancer nanomedicine has been inspired by Paul Ehrlich's notion of the "magic bullet", which visualizes the design of therapeutic agents that selectively attack diseased cells while sparing the healthy ones. Due to the advances in nanotechnology, progress has been made towards specifically delivering cytotoxic agents to cancerous cells, despite the complexities inherent to a cancerous system. Although much of this technology still remains to be translated into the clinic, the investigation to understand and overcome the often-limited performance of nanomedicine-mediated cancer therapy continues. In a nutshell, the challenges associated with successful delivery of nanomedicine to the cancerous tissue range from sub-microscopic NP-protein interactions to microscopic NP-cell interactions, leading to an emergent behavior at the macroscopic scale in the form of whole-body biodistribuiton and clearance from the body. It is imperative to study these interactions in isolation, as well as in unification, to better understand the fundamental principles that lie underneath the problem. Moving forward, the intergration of mathematical modeling with experimental investigation of NP kinetics, efficacy, and toxicity in order to better under the structure-activity relationships of nanocarriers for their clinical translatability will become increasingly important. This review has explored many of the mathematical modeling approaches developed in the field of cancer nanomedicine that seek to expand our capability to fully explore the nano-bio interactions. In the future, we believe nanomedicine will bring notable improvements in patient survival and quality of life.

Acknowledgements This research has been supported in part by the National Science Foundation Grant DMS-1716737 (VC, ZW), the National Institutes of Health (NIH) Grants 1U01CA196403 (VC, ZW), 1U01CA213759 (VC, ZW), 1R01CA226537 (CJB, VC, ZW), 1R01CA222007 (VC, ZW), U54CA210181 (VC, ZW), and the University of Texas System STARS Award (VC). The funders had no role in study design, data collection and analysis, decision to publish, or preparation of the manuscript. CJB acknowledges additional support from the Sandia National Laboratories LDRD program. Sandia National Laboratories is a multimission laboratory managed and operated by National Technology & Engineering Solutions of Sandia, LLC, a wholly owned subsidiary of Honeywell International Inc., for the U.S. Department of Energy's National Nuclear Security Administration under contract DE-NA0003525. This paper describes objective technical results and analysis. Any subjective views or opinions that might be expressed in the paper do not necessarily represent the views of the U.S. Department of Energy or the United States Government.

Compliance with ethical standards

Competing interests The authors have declared that no competing interests exist.

References

Agency, European Medicines. 2016. "Assessment report: Onivyde." In, edited by Commitee for Medicinal Products for Human Use (CHMP)

P. Aggarwal, J.B. Hall, C.B. McLeland, M.A. Dobrovolskaia, S.E. McNeil, Nanoparticle interaction with plasma proteins as it relates to particle biodistribution, biocompatibility and therapeutic efficacy. Adv. Drug Deliv. Rev. **61**, 428–437 (2009)

A. Albanese, P.S. Tang, W.C.W. Chan, The effect of nanoparticle size, shape, and surface chemistry on biological systems. Annu. Rev. Biomed. Eng. **14**, 1–16 (2012)

A.R.A. Anderson, M.A.J. Chaplain, Continuous and discrete mathematical models of tumor-induced angiogenesis. Bull. Math. Biol. **60**, 857–899 (1998)

K. Ando, K. Mori, N. Corradini, F. Redini, D. Heymann, Mifamurtide for the treatment of nonmetastatic osteosarcoma. Expert. Opin. Pharmacother. **12**, 285–292 (2011)

S. Angioletti-Uberti, M. Ballauff, J. Dzubiella, Dynamic density functional theory of protein adsorption on polymer-coated nanoparticles. Soft Matter **10**, 7932–7945 (2014)

R. Aris, On the dispersion of a solute in a fluid flowing through a tube. Proc. R. Soc. Lond. A **235**, 67–77 (1956)

G. Bachler, N. von Goetz, K. Hungerbühler, A physiologically based pharmacokinetic model for ionic silver and silver nanoparticles. Int. J. Nanomedicine **8**, 3365 (2013)

R. Bawa, G.F. Audette, and I. Rubinstein, *Handbook of clinical nanomedicine: Nanoparticles, imaging, Therapy and Clinical Applications* (Pan Stanford Publishing Pte. Ltd.) (2016)

G. Bellizzi, and O.M. Bucci, *'Emerging Electromagnetic Technologies for Brain Diseases Diagnostics, Monitoring and Therapy.'* in L. Crocco, I. Karanasiou, M. James and R.C. Conceição (eds.) (Springer) . (2018)

R. Bellott, A. Auvrignon, T. Leblanc, Y. Perel, V. Gandemer, Y. Bertrand, F. Mechinaud, P. Bellenger, J. Vernois, G. Leverger, A. Baruchel, J. Robert, Pharmacokinetics of liposomal daunorubicin (DaunoXome) during a phase I-II study in children with relapsed acute lymphoblastic leukaemia. Cancer Chemother. Pharmacol. **47**, 15–21 (2001)

H. Berman, K. Henrick, H. Nakamura, Announcing the worldwide protein data bank. Nat. Struct. Mol. Biol. **10**, 980 (2003)

K.C.L. Black, W.J. Akers, G. Sudlow, B. Xu, R. Laforest, S. Achilefu, Dual-radiolabeled nanoparticle SPECT probes for bioimaging. Nanoscale **7**, 440–444 (2015)

T. Brocato, P. Dogra, E.J. Koay, A. Day, Y.-L. Chuang, Z. Wang, V. Cristini, Understanding drug resistance in breast Cancer with mathematical oncology. Curr. Breast Cancer Rep., 1–11 (2014)

T.A. Brocato, E.N. Coker, P.N. Durfee, Y.-S. Lin, J. Townson, E.F. Wyckoff, V. Cristini, C.J. Brinker, Z. Wang, Understanding the connection between nanoparticle uptake and Cancer treatment efficacy using mathematical modeling. Sci. Rep. **8**, 7538 (2018)

U. Bulbake, S. Doppalapudi, N. Kommineni, and W. Khan, 'Liposomal Formulations in Clinical Use: An Updated Review', Pharmaceutics, 9 (2017)

G. Caracciolo, O.C. Farokhzad, M. Mahmoudi. 'Biological identity of nanoparticles in vivo: Clinical implications of the protein corona', Trends Biotechnol. **35**, 257–264 (2017)

M. Cataldi, C. Vigliotti, T. Mosca, M.R. Cammarota, D. Capone, Emerging role of the spleen in the pharmacokinetics of monoclonal antibodies, nanoparticles and exosomes. Int. J. Mol. Sci. **18**, 1249 (2017)

V.P. Chauhan, T. Stylianopoulos, Y. Boucher, R.K. Jain, Delivery of molecular and nanoscale medicine to tumors: Transport barriers and strategies. Annu. Rev. Chem. Biomol. Eng. **2**, 281–298 (2011)

V.P. Chauhan, T. Stylianopoulos, J.D. Martin, Z. Popović, O. Chen, W.S. Kamoun, M.G. Bawendi, D. Fukumura, R.K. Jain, Normalization of tumour blood vessels improves the delivery of nanomedicines in a size-dependent manner. Nat. Nanotechnol. **7**, 383–388 (2012)

Chen, Feng, Shreya Goel, Reinier Hernandez, Stephen A Graves, Sixiang Shi, Robert J Nickles, and Weibo Cai. 2016. 'Dynamic positron emission tomography imaging of renal clearable gold nanoparticles', Small, 12: 2775–2782

H.S. Choi, W. Liu, P. Misra, E. Tanaka, J.P. Zimmer, B.I. Ipe, M.G. Bawendi, J.V. Frangioni, Renal clearance of quantum dots. Nat. Biotechnol. **25**, 1165 (2007)

Chow, Edward K, Xue-Qing Zhang, Mark Chen, Robert Lam, Erik Robinson, Houjin Huang, Daniel Schaffer, Eiji Osawa, Andrei Goga, and Dean Ho. 2011. 'Nanodiamond therapeutic delivery agents mediate enhanced chemoresistant tumor treatment', Sci. Transl. Med., 3: 73ra21

A. Chrastina, J.E. Schnitzer, Iodine-125 radiolabeling of silver nanoparticles for in vivo SPECT imaging. Int. J. Nanomedicine **5**, 653 (2010)

D. Costanzo, S.G. Francesco, V. Rotella, F. Di Costanzo, Targeted delivery of albumin bound paclitaxel in the treatment of advanced breast cancer. OncoTargets Ther **2**, 179–188 (2009)

S.H. Crayton, D.R. Elias, A. Al Zaki, Z. Cheng, A. Tsourkas, ICP-MS analysis of lanthanide-doped nanoparticles as a non-radiative, multiplex approach to quantify biodistribution and blood clearance. Biomaterials **33**, 1509–1519 (2012)

V. Cristini, J. Lowengrub, Q. Nie, Nonlinear simulation of tumor growth. J. Math. Biol. **46**, 191–224 (2003)

V. Cristini, E. Koay, Z. Wang, *An Introduction to Physical Oncology: How Mechanistic Mathematical Modeling Can Improve Cancer Therapy Outcomes* (CRC Press). (2017)

H. Das, Z. Wang, M.K.K. Niazi, R. Aggarwal, J. Lu, S. Kanji, M. Das, M. Joseph, M. Gurcan, V. Cristini, Impact of diffusion barriers to small cytotoxic molecules on the efficacy of immunotherapy in breast Cancer. PLoS One **8**, e61398 (2013)

B. Davies, T. Morris, Physiological parameters in laboratory animals and humans. Pharm. Res. **10**, 1093–1095 (1993)

A.L. Van De Ven, K. Pilhan, H. O'Hara, J.R. Fakhoury, G. Adriani, J. Schmulen, P. Moloney, F. Hussain, M. Ferrari, X. Liu, S.-H. Yun, P. Decuzzi. 'Rapid tumoritropic accumulation of systemically injected plateloid particles and their biodistribution', J. Control. Release, **158**, 148–155 (2012)

P. Decuzzi, M. Ferrari, The adhesive strength of non-spherical particles mediated by specific interactions. Biomaterials **27**, 5307–5314 (2006)

P. Decuzzi, M. Ferrari, The receptor-mediated endocytosis of nonspherical particles. Biophys. J. **94**, 3790–3797 (2008)

P. Decuzzi, S. Lee, B. Bhushan, M. Ferrari, A theoretical model for the margination of particles within blood vessels. Ann. Biomed. Eng. **33**, 179–190 (2005)

P. Decuzzi, F. Causa, M. Ferrari, P.A. Netti, The effective dispersion of nanovectors within the tumor microvasculature. Ann. Biomed. Eng. **34**, 633–641 (2006)

T.S. Deisboeck, Z. Wang, P. Macklin, V. Cristini, Multiscale cancer modeling. Annu. Rev. Biomed. Eng. **13**, 127–155 (2011)

D. Dell'Orco, M. Lundqvist, C. Oslakovic, T. Cedervall, S. Linse, Modeling the time evolution of the nanoparticle-protein corona in a body fluid. PLoS One **5**, e10949 (2010)

D. Dell'Orco, M. Lundqvist, T. Cedervall, S. Linse, Delivery success rate of engineered nanoparticles in the presence of the protein corona: A systems-level screening. Nanomedicine **8**, 1271–1281 (2012)

P. Dogra, N.L. Adolphi, Z. Wang, Y.-S. Lin, K.S. Butler, P.N. Durfee, J.G. Croissant, A. Noureddine, E.N. Coker, E.L. Bearer, V. Cristini, C. Jeffrey Brinker, Establishing the effects of mesoporous silica nanoparticle properties on in vivo disposition using imaging-based pharmacokinetics. Nat. Commun. **9**, 4551 (2018)

D. Douer, H. Yampolsky, L.J. Cohen, K. Watkins, A.M. Levine, A.P. Periclou, V.I. Avramis, Pharmacodynamics and safety of intravenous pegaspargase during remission induction in adults aged 55 years or younger with newly diagnosed acute lymphoblastic leukemia. Blood **109**, 2744–2750 (2007)

D.C. Drummond, C.O. Noble, Z. Guo, K. Hong, J.W. Park, D.B. Kirpotin, Development of a highly active nanoliposomal irinotecan using a novel intraliposomal stabilization strategy. Cancer Res. **66**, 3271–3277 (2006)

D.B. Fenske, P.R. Cullis, Liposomal nanomedicines. Expert Opin Drug Deliv **5**, 25–44 (2008)

M.R. Findlay, D.N. Freitas, M. Mobed-Miremadi, K.E. Wheeler, Machine learning provides predictive analysis into silver nanoparticle protein corona formation from physicochemical properties. Environ. Sci. Nano **5**, 64–71 (2018)

H.B. Frieboes, M. Wu, J. Lowengrub, P. Decuzzi, V. Cristini, A computational model for predicting nanoparticle accumulation in tumor vasculature. PLoS One **8**, e56876 (2013)

H.B. Frieboes, B.R. Smith, Z. Wang, M. Kotsuma, K. Ito, A. Day, B. Cahill, C. Flinders, S.M. Mumenthaler, P. Mallick, Predictive modeling of drug response in non-hodgkin's lymphoma. PLoS One **10**, e0129433 (2015)

K. Fujimori, D.G. Covell, J.E. Fletcher, J.N. Weinstein, A modeling analysis of monoclonal antibody percolation through tumors: A binding-site barrier. J. Nucl. Med. **31**, 1191–1198 (1990)

G. Fullstone, J. Wood, M. Holcombe, G. Battaglia. 'Modelling the Transport of Nanoparticles under Blood Flow Using an Agent-Based Approach', *Scientific reports*, 5 (2015)

L. Fumagalli, M. Zucchetti, I. Parisi, M.G. Vigano, B. Zecca, A. Careddu, M. D'Incalci, A. Lazzarin, The pharmacokinetics of liposomal encapsulated daunorubicin are not modified by HAART in patients with HIV-associated Kaposi's sarcoma. Cancer Chemother. Pharmacol. **45**, 495–501 (2000)

E.P. Furlani, K.C. Ng, Analytical model of magnetic nanoparticle transport and capture in the microvasculature. Phys. Rev. E **73**, 061919 (2006)

A.A. Gabizon, Pegylated liposomal doxorubicin: Metamorphosis of an old drug into a new form of chemotherapy. Cancer Investig. **19**, 424–436 (2001)

A.A. Gabizon, Y. Barenholz, M. Bialer, Prolongation of the circulation time of doxorubicin encapsulated in liposomes containing a polyethylene glycol-derivatized phospholipid: Pharmacokinetic studies in rodents and dogs. Pharm. Res. **10**, 703–708 (1993)

A. Gabizon, H. Shmeeda, Y. Barenholz, Pharmacokinetics of pegylated liposomal doxorubicin: Review of animal and human studies. Clin. Pharmacokinet. **42**, 419–436 (2003)

A.A. Gabizon, H. Shmeeda, S. Zalipsky, Pros and cons of the liposome platform in cancer drug targeting. J. Liposome. Res. **16**, 175–183 (2006)

J. Gabrielsson, D. Weiner, *Pharmacokinetic and pharmacodynamic data analysis: concepts and applications* (CRC Press) (2001)

G. Huajian, W. Shi, L.B. Freund, 'Mechanics of receptor-mediated endocytosis', Proc. Natl. Acad. Sci. **102**, 9469–9474 (2005)

L. Gaohua, K. Abduljalil, M. Jamei, T.N. Johnson, A. Rostami-Hodjegan, A pregnancy physiologically based pharmacokinetic (p-PBPK) model for disposition of drugs metabolized by CYP1A2, CYP2D6 and CYP3A4. Br. J. Clin. Pharmacol. **74**, 873–885 (2012)

Y. Gazit, D.A. Berk, M. Leunig, L.T. Baxter, R.K. Jain, Scale-invariant behavior and vascular network formation in normal and tumor tissue. Phys. Rev. Lett. **75**, 2428 (1995)

F. Gentile, M. Ferrari, P. Decuzzi, The transport of nanoparticles in blood vessels: The effect of vessel permeability and blood rheology. Ann. Biomed. Eng. **36**, 254–261 (2008)

L.E. Gerlowski, R.K. Jain, Physiologically based pharmacokinetic modeling: Principles and applications. J. Pharm. Sci. **72**, 1103–1127 (1983)

S. Hassan, G. Prakash, A.B. Ozturk, S. Saghazadeh, M.F. Sohail, J. Seo, M.R. Dokmeci, Y.S. Zhang, A. Khademhosseini, Evolution and clinical translation of drug delivery nanomaterials. Nano Today **15**, 91–106 (2017)

S. Hauert, S. Berman, R. Nagpal, S.N. Bhatia, A computational framework for identifying design guidelines to increase the penetration of targeted nanoparticles into tumors. Nano Today **8**, 566–576 (2013)

B.S. Hendriks, J.G. Reynolds, S.G. Klinz, E. Geretti, H. Lee, S.C. Leonard, D.F. Gaddy, C.W. Espelin, U.B. Nielsen, T.J. Wickham, Multiscale kinetic modeling of liposomal doxorubicin delivery quantifies the role of tumor and drug-specific parameters in local delivery to tumors. CPT Pharmacometrics Syst. Pharmacol. **1**, 1–11 (2012)

C.P. Hornik, H. Wu, A.N. Edginton, K. Watt, M. Cohen-Wolkowiez, D. Gonzalez, 'Development of a Pediatric Physiologically-Based Pharmacokinetic Model of Clindamycin Using Opportunistic Pharmacokinetic Data', Clin. Pharmacokinet. 1–11 (2017)

H. Hosoya, A.S. Dobroff, W.H.P. Driessen, V. Cristini, L.M. Brinker, F.I. Staquicini, M. Cardó-Vila, S. D'Angelo, F. Ferrara, B. Proneth, Integrated nanotechnology platform for tumor-targeted multimodal imaging and therapeutic cargo release. Proc. Natl. Acad. Sci. **113**, 1877–1882 (2016)

S.-M. Huang, M. Rowland. 'The role of physiologically based pharmacokinetic modeling in regulatory review', Clin. Pharmacol. Ther. **91**, 542–549 (2012)

M.L. Immordino, F. Dosio, L. Cattel, Stealth liposomes: Review of the basic science, rationale, and clinical applications, existing and potential. Int. J. Nanomedicine **1**, 297–315 (2006)

R.K. Jain, T. Stylianopoulos, Delivering nanomedicine to solid tumors. Nat. Rev. Clin. Oncol. **7**, 653 (2010)

H.M. Jones, K. Rowland-Yeo, Basic concepts in physiologically based pharmacokinetic modeling in drug discovery and development. CPT Pharmacometrics Syst. Pharmacol. **2**, e63 (2013)

F. Khalil, S. Läer, 'Physiologically Based Pharmacokinetic Modeling: Methodology, Applications, and Limitations with a Focus on its Role in Pediatric Drug development'. BioMed Res. Int. (2011)

T.Y. Kim, D.W. Kim, J.Y. Chung, S.G. Shin, S.C. Kim, D.S. Heo, N.K. Kim, Y.J. Bang, Phase I and pharmacokinetic study of Genexol-PM, a cremophor-free, polymeric micelle-formulated paclitaxel, in patients with advanced malignancies. Clin. Cancer Res. **10**, 3708–3716 (2004)

E.J. Koay, M.J. Truty, V. Cristini, R.M. Thomas, R. Chen, D. Chatterjee, Y.'a. Kang, P.R. Bhosale, E.P. Tamm, C.H. Crane, Transport properties of pancreatic cancer describe gemcitabine delivery and response. J. Clin. Invest. **124**, 1525–1536 (2014)

S. Kommareddy, M. Amiji, Biodistribution and pharmacokinetic analysis of long-circulating thiolated gelatin nanoparticles following systemic administration in breast cancer-bearing mice. J. Pharm. Sci. **96**, 397–407 (2007)

R. Kumar, I. Roy, T.Y. Ohulchanskky, L.A. Vathy, E.J. Bergey, M. Sajjad, P.N. Prasad, In vivo biodistribution and clearance studies using multimodal organically modified silica nanoparticles. ACS Nano **4**, 699–708 (2010)

T. Laomettachit, I.K. Puri, M. Liangruksa, 'A Two-Step Model of TiO 2 Nanoparticle Toxicity in Human Liver tissue'. Toxicol. Appl. Pharmacol. (2017)

S.-Y. Lee, M. Ferrari, P. Decuzzi, Design of bio-mimetic particles with enhanced vascular interaction. J. Biomech. **42**, 1885–1890 (2009a)

S.-Y. Lee, M. Ferrari, P. Decuzzi, Shaping nano−/micro-particles for enhanced vascular interaction in laminar flows. Nanotechnology **20**, 495101 (2009b)

T.-R. Lee, M. Choi, A.M. Kopacz, S.-H. Yun, W.K. Liu, P. Decuzzi, On the near-wall accumulation of injectable particles in the microcirculation: Smaller is not better. Sci. Rep. **3** (2013)

F. Leonard, L.T. Curtis, P. Yesantharao, T. Tanei, J.F. Alexander, M. Wu, J. Lowengrub, X. Liu, M. Ferrari, K. Yokoi, Enhanced performance of macrophage-encapsulated nanoparticle albumin-bound-paclitaxel in hypo-perfused cancer lesions. Nanoscale **8**, 12544–12552 (2016)

M. Li, K.T. Al-Jamal, K. Kostarelos, J. Reineke, Physiologically based pharmacokinetic modeling of nanoparticles. ACS Nano **4**, 6303–6317 (2010)

M. Li, Z. Panagi, K. Avgoustakis, J. Reineke, Physiologically based pharmacokinetic modeling of PLGA nanoparticles with varied mPEG content. Int. J. Nanomedicine **7**, 1345 (2012)

D. Li, G. Johanson, C. Emond, U. Carlander, M. Philbert, Olivier Jolliet, 'Physiologically Based Pharmacokinetic Modeling of Polyethylene Glycol-Coated Polyacrylamide Nanoparticles in rats'. Nanotoxicology 1–10 (2013)

X. Li, Z. Xiong, X. Xu, L. Yu, C. Peng, M. Shen, X. Shi, 99mTc-labeled multifunctional low-generation dendrimer-entrapped gold nanoparticles for targeted SPECT/CT dual-mode imaging of tumors. ACS Appl. Mater. Interfaces **8**, 19883–19891 (2016)

H. Lopez, V. Lobaskin, Coarse-Grained Model of Adsorption of Blood Plasma Proteins onto nanoparticles. J. Chem. Phys. **143**, 12B620_1 (2015)

O. Lunov, V. Zablotskii, T. Syrovets, C. Röcker, T. Kyrylo, G. Ulrich Nienhaus, T. Simmet, Modeling receptor-mediated endocytosis of polymer-functionalized iron oxide nanoparticles by human macrophages. Biomaterials **32**, 547–555 (2011)

P. Macklin, S. McDougall, A.R.A. Anderson, M.A.J. Chaplain, V. Cristini, J. Lowengrub, Multiscale modelling and nonlinear simulation of vascular tumour growth. J. Math. Biol. **58**, 765–798 (2009)

M.A. Malfatti, H.A. Palko, E.A. Kuhn, K.W. Turteltaub, Determining the pharmacokinetics and long-term biodistribution of SiO2 nanoparticles in vivo using accelerator mass spectrometry. Nano Lett. **12**, 5532–5538 (2012)

S.R. McDougall, A.R.A. Anderson, M.A.J. Chaplain, Mathematical modelling of dynamic adaptive tumour-induced angiogenesis: Clinical implications and therapeutic targeting strategies. J. Theor. Biol. **241**, 564–589 (2006)

E. Miele, G.P. Spinelli, E. Miele, F. Tomao, S. Tomao, Albumin-bound formulation of paclitaxel (Abraxane(®) ABI-007) in the treatment of breast cancer. Int. J. Nanomedicine **4**, 99–105 (2009)

H.A. Miller, H.B. Frieboes, 'Evaluation of Drug-Loaded Gold Nanoparticle Cytotoxicity as a Function of Tumor Vasculature-Induced Tissue Heterogeneity'. Ann. Biomed. Eng. 1–15 (2018)

H. Ming, L. Fang, J. Gao, C. Li, Y. Ji, Y. Shen, Y. Hu, N. Li, J. Chang, W. Li, Antitumor effect of nanoparticle 131I-labeled arginine-glycine-aspartate–bovine serum albumin–Polycaprolactone in lung Cancer. Am. J. Roentgenol. **208**, 1116–1126 (2017)

M.P. Monopoli, C. Åberg, A. Salvati, K.A. Dawson. 'Biomolecular coronas provide the biological identity of nanosized materials', Nat. Nanotechnol. **7**, 779 (2012)

K. Müller, D.A. Fedosov, G. Gompper,. 'Margination of micro-and nanoparticles in blood flow and its effect on drug delivery'. Sci. Rep. 4 (2014)

A.M. Neubauer, H. Sim, P.M. Winter, S.D. Caruthers, T.A. Williams, J. David Robertson, D. Sept, G.M. Lanza, S.A. Wickline, Nanoparticle pharmacokinetic profiling in vivo using magnetic resonance imaging. Magn. Reson. Med. **60**, 1353–1361 (2008)

J.W. Nichols, Y.H. Bae, Odyssey of a cancer nanoparticle: From injection site to site of action. Nano Today **7**, 606–618 (2012)

P. Papageorgis, C. Polydorou, F. Mpekris, C. Voutouri, E. Agathokleous, C.P. Kapnissi-Christodoulou, and T. Stylianopoulos, 'Tranilast-Induced Stress Alleviation in Solid Tumors Improves the Efficacy of Chemo-and Nanotherapeutics in a Size-Independent manner'. Sci. Rep. 7 (2017)

J. Pascal, C.E. Ashley, Z. Wang, T.A. Brocato, J.D. Butner, E.C. Carnes, E.J. Koay, C.J. Brinker, and V. Cristini, 'Mechanistic modeling identifies drug-uptake history as predictor of tumor drug resistance and Nano-carrier-mediated Response'. ACS Nano (2013a)

J. Pascal, E.L. Bearer, Z. Wang, E.J. Koay, S.A. Curley, V. Cristini, Mechanistic patient-specific predictive correlation of tumor drug response with microenvironment and perfusion measurements. Proc. Natl. Acad. Sci. **110**, 14266–14271 (2013b)

R.R. Patil, J. Yu, S.R. Banerjee, Y. Ren, D. Leong, X. Jiang, M. Pomper, B. Tsui, D.L. Kraitchman, H.-Q. Mao, Probing in vivo trafficking of polymer/DNA micellar nanoparticles using SPECT/CT imaging. Mol. Ther. **19**, 1626–1635 (2011)

S.A. Peters, *Physiologically-based pharmacokinetic (PBPK) modeling and simulations: principles, methods, and applications in the pharmaceutical industry* (John Wiley & Sons) (2012)

E. Phillips, O. Penate-Medina, P.B. Zanzonico, R.D. Carvajal, P. Mohan, Y. Ye, J. Humm, M. Gönen, H. Kalaigian, H. Schöder, Clinical translation of an ultrasmall inorganic optical-PET imaging nanoparticle probe. Sci. Transl. Med. **6**, 260ra149–260ra149 (2014)

G. Pillai, 'Nanomedicines for Cancer Therapy: An Update of FDA Approved and those under Various Stages of Development'. SOJ Pharm. Pharm. Sci. 1 (2014)

F.S.G. Praça, H. Susana Marinho, M. Bárbara F. Martins, Rogério Gaspar, M. Luísa Corvo, and W.S.G. Medina, 'Chapter 27 - Current aspects of breast cancer therapy and diagnosis based on a nanocarrier approach.' in Anton Ficai and Alexandru Mihai Grumezescu (eds.), *Nanostructures for Cancer Therapy* (Elsevier) (2017)

J.E. Riviere, C. Scoglio, F.D. Sahneh, N.A. Monteiro-Riviere, Computational approaches and metrics required for formulating biologically realistic nanomaterial pharmacokinetic models. Comput. Sci. Discovery **6**, 014005 (2013)

F.D. Sahneh, C. Scoglio, J. Riviere, Dynamics of nanoparticle-protein corona complex formation: Analytical results from population balance equations. PLoS One **8**, e64690 (2013)

H. Sarin, Physiologic upper limits of pore size of different blood capillary types and another perspective on the dual pore theory of microvascular permeability. J. Angiogenes. Res. **2**, 14 (2010)

T. Schluep, J. Hwang, I.J. Hildebrandt, J. Czernin, C.H.J. Choi, C.A. Alabi, B.C. Mack, M.E. Davis, Pharmacokinetics and tumor dynamics of the nanoparticle IT-101 from PET imaging and tumor histological measurements. Proc. Natl. Acad. Sci. **106**, 11394–11399 (2009)

M.M. Schmidt, K. Dane Wittrup, A modeling analysis of the effects of molecular size and binding affinity on tumor targeting. Mol. Cancer Ther. **8**, 2861–2871 (2009)

D.K. Shah, A.M. Betts, Towards a platform PBPK model to characterize the plasma and tissue disposition of monoclonal antibodies in preclinical species and human. J. Pharmacokinet. Pharmacodyn. **39**, 67–86 (2012)

Q. Shao, C.K. Hall, Protein adsorption on nanoparticles: Model development using computer simulation. J. Phys. Condens. Matter **28**, 414019 (2016)

V. Sharma, J.H. McNeill, To scale or not to scale: The principles of dose extrapolation. Br. J. Pharmacol. **157**, 907–921 (2009)

J. Shi, P.W. Kantoff, R. Wooster, O.C. Farokhzad, Cancer nanomedicine: Progress, challenges and opportunities. Nat. Rev. Cancer **17**, 20 (2016)

J.A. Silverman, S.R. Deitcher, Marqibo(®) (vincristine sulfate liposome injection) improves the pharmacokinetics and pharmacodynamics of vincristine. Cancer Chemother. Pharmacol. **71**, 555–564 (2013)

I. Sorrell, R.J. Shipley, V. Hearnden, H.E. Colley, M.H. Thornhill, C. Murdoch, S.D. Webb, Combined mathematical modelling and experimentation to predict polymersome uptake by oral cancer cells. Nanomedicine **10**, 339–348 (2014)

A. Soundararajan, A. Bao, W.T. Phillips, R. Perez, B.A. Goins, (186)re-liposomal doxorubicin (Doxil): In vitro stability, pharmacokinetics, imaging and biodistribution in a head and neck squamous cell carcinoma xenograft model. Nucl. Med. Biol. **36**, 515–524 (2009)

S. Stapleton, M. Milosevic, C. Allen, J. Zheng, M. Dunne, I. Yeung, D.A. Jaffray. 'A mathematical model of the enhanced permeability and retention effect for liposome transport in solid tumors', PLoS One, 8: e81157 (2013)

T. Stylianopoulos, B. Diop-Frimpong, L.L. Munn, R.K. Jain, Diffusion anisotropy in collagen gels and tumors: The effect of fiber network orientation. Biophys. J. **99**, 3119–3128 (2010)

T. Stylianopoulos, K. Soteriou, D. Fukumura, R.K. Jain. 'Cationic nanoparticles have superior transvascular flux into solid tumors: Insights from a mathematical model', Ann. Biomed. Eng. **41**, 68–77 (2013)

T. Stylianopoulos, E.-A. Economides, J.W. Baish, D. Fukumura, R.K. Jain, Towards optimal design of cancer nanomedicines: Multi-stage nanoparticles for the treatment of solid tumors. Ann. Biomed. Eng. **43**, 2291–2300 (2015)

E.A. Sykes, J. Chen, G. Zheng, W.C.W. Chan. 'Investigating the impact of nanoparticle size on active and passive tumor targeting efficiency', ACS Nano, **8**, 5696–5706 (2014a)

E.A. Sykes, J. Chen, G. Zheng, W.C.W. Chan, 'Investigating the impact of nanoparticle size on active and passive tumor targeting efficiency' (2014b)

E.A. Sykes, Q. Dai, C.D. Sarsons, J. Chen, J.V. Rocheleau, D.M. Hwang, G. Zheng, D.T. Cramb, K.D. Rinker, W.C.W. Chan. 'Tailoring nanoparticle designs to target cancer based on tumor pathophysiology', Proc. Natl. Acad. Sci. **113**, E1142-E1E51 (2016)

Talon Therapeutics, Inc. 'MARQIBO (VINCRISTINE SULFATE LIPOSOMES INJECTION) FOR THE TREATMENT OF ADVANCED RELAPSED AND/OR REFRACTORY PHILADELPHIA CHROMOSOME NEGATIVE (PH-) ADULT ACUTE LYMPHOBLASTIC LEUKEMIA' (2012)
J. Tan, A. Thomas, Y. Liu, Influence of red blood cells on nanoparticle targeted delivery in microcirculation. Soft Matter **8**, 1934–1946 (2012)
J. Tan, S. Shah, T. Antony, H. Daniel Ou-Yang, Y. Liu, The influence of size, shape and vessel geometry on nanoparticle distribution. Microfluid. Nanofluid. **14**, 77–87 (2013)
F. Tavanti, A. Pedone, M.C. Menziani, A closer look into the ubiquitin corona on gold nanoparticles by computational studies. New J. Chem. **39**, 2474–2482 (2015a)
F. Tavanti, A. Pedone, M.C. Menziani, Competitive binding of proteins to gold nanoparticles disclosed by molecular dynamics simulations. J. Phys. Chem. C **119**, 22172–22180 (2015b)
G.I. Taylor, Dispersion of soluble matter in solvent flowing slowly through a tube. Proc. R. Soc. Lond. A **219**, 186–203 (1953)
T. Toge, K. Jinushi, A. Sawamura, K. Matsuki, Study on the intra-tumor tissue concentration of zinostatin stimalamer in the liver following intra-arterial injection of zinostatin stimalamer (YM881) suspension. Gan To Kagaku Ryoho **18**, 1677–1680 (1991)
V. Tozzini, Coarse-grained models for proteins. Curr. Opin. Struct. Biol. **15**, 144–150 (2005)
K.M. Tsoi, MacParland S.A., X.-Z. Ma, V.N. Spetzler, J. Echeverri, B. Ouyang, S.M. Fadel, E.A. Sykes, N. Goldaracena, J.M. Kaths, 'Mechanism of Hard-Nanomaterial Clearance by the liver'. Nat. Mater. (2016)
K.O. Vasquez, C. Casavant, J.D. Peterson, Quantitative whole body biodistribution of fluorescent-labeled agents by non-invasive tomographic imaging. PLoS One **6**, e20594 (2011)
K. Venkatakrishnan, Y. Liu, D. Noe, J. Mertz, M. Bargfrede, T. Marbury, K. Farbakhsh, C. Oliva, A. Milton, Pharmacokinetics and pharmacodynamics of liposomal mifamurtide in adult volunteers with mild or moderate renal impairment. Br. J. Clin. Pharmacol. **77**, 986–997 (2014)
O. Vilanova, J.J. Mittag, P.M. Kelly, S. Milani, K.A. Dawson, J.O. Rädler, G. Franzese, Understanding the kinetics of protein–nanoparticle corona formation. ACS Nano **10**, 10842–10850 (2016)
P. Vilaseca, K.A. Dawson, G. Franzese, Understanding and modulating the competitive surface-adsorption of proteins through coarse-grained molecular dynamics simulations. Soft Matter **9**, 6978–6985 (2013)
Z. Wang, T.S. Deisboeck, Mathematical modeling in cancer drug discovery. Drug Discov. Today **19**, 145–150 (2014)
Z. Wang, P.K. Maini, Editorial special section on multiscale cancer modeling. IEEE Trans. Biomed. Eng. **64**, 501–503 (2017)
Z. Wang, J.D. Butner, R. Kerketta, V. Cristini, T.S. Deisboeck, Simulating cancer growth with multiscale agent-based modeling. Semin. Cancer Biol. **30**, 70–78 (2015)
Z. Wang, R. Kerketta, Y.-L. Chuang, P. Dogra, J.D. Butner, T.A. Brocato, A. Day, R. Xu, H. Shen, E. Simbawa, Theory and experimental validation of a spatio-temporal model of chemotherapy transport to enhance tumor cell kill. PLoS Comput. Biol. **12**, e1004969 (2016)
M.E. Werner, N.D. Cummings, M. Sethi, E.C. Wang, R. Sukumar, D.T. Moore, A.Z. Wang, Preclinical evaluation of Genexol-PM, a nanoparticle formulation of paclitaxel, as a novel Radiosensitizer for the treatment of non-small cell lung Cancer. Int. J. Radiat. Oncol. Biol. Phys. **86**, 463–468 (2013)
S. Wilhelm, A.J. Tavares, Q. Dai, S. Ohta, J. Audet, H.F. Dvorak, W.C.W. Chan, Analysis of nanoparticle delivery to tumours. Nat. Rev. Mater. **1**, 16014 (2016)
J.D. Woodward, S.J. Kennel, S. Mirzadeh, S. Dai, J.S. Wall, T. Richey, J. Avenell, A.J. Rondinone, In vivo SPECT/CT imaging and biodistribution using radioactive Cd125mTe/ZnS nanoparticles. Nanotechnology **18**, 175103 (2007)
M. Wu, H.B. Frieboes, S.R. McDougall, M.A.J. Chaplain, V. Cristini, J. Lowengrub, The effect of interstitial pressure on tumor growth: Coupling with the blood and lymphatic vascular systems. J. Theor. Biol. **320**, 131–151 (2013)
J. Xu, F. Gattacceca, M. Amiji, Biodistribution and pharmacokinetics of EGFR-targeted thiolated gelatin nanoparticles following systemic administration in pancreatic tumor-bearing mice. Mol. Pharm. **10**, 2031–2044 (2013)
L.X. Yu, B. Li, 'FDA Bioequivalence Standards.' in, *FDA Bioequivalence Standards* (Springer) (2014)
T. Yu, D. Hubbard, A. Ray, H. Ghandehari, In vivo biodistribution and pharmacokinetics of silica nanoparticles as a function of geometry, porosity and surface characteristics. J. Control. Release **163**, 46–54 (2012)
H. Yuan, S. Zhang, Effects of particle size and ligand density on the kinetics of receptor-mediated endocytosis of nanoparticles. Appl. Phys. Lett. **96**, 033704 (2010)
S. Zhang, H. Gao, G. Bao, Physical principles of nanoparticle cellular endocytosis. ACS Nano **9**, 8655–8671 (2015)
V.P. Zhdanov, N.-J. Cho, Kinetics of the formation of a protein corona around nanoparticles. Math. Biosci. **282**, 82–90 (2016)

Publisher's note Springer Nature remains neutral with regard to jurisdictional claims in published maps and institutional affiliations.

Biomedical Microdevices (2019) 21: 33
https://doi.org/10.1007/s10544-019-0368-y

Coupling tumor growth and bio distribution models

Raffaella Santagiuliana [1] · Miljan Milosevic [2,3] · Bogdan Milicevic [2] · Giuseppe Sciumè [4] · Vladimir Simic [2] · Arturas Ziemys [5] · Milos Kojic [2,5,6] · Bernhard A. Schrefler [1,5,7]

Published online: 25 March 2019

Author's Personal Note

I owe to Mauro the fact that I work since ten years in the field of computational transport oncophysics. I met Mauro personally for the first time in 2009 during a lecture he gave at the Faculty of Medicine of the University of Padua but I was then already well aware of what he was doing. He invited me to visit him in Houston, at that time at the University of Texas Health Center. In 2010 I took a sabbatical and spent some time in Austin at ICES of the University of Texas, with a fellowship of Professor Tinsley Oden. Professor Oden showed me a paper on tumor growth he was preparing, and I realized that the equations he was solving where very similar to those I was solving since over ten years in the field of porous media mechanics. His model however was mainly based on diffusion and the flow aspect was missing. After my stay in Austin I visited Mauro in Houston and told him I could prepare a more complete model for tumor growth and drug delivery which would include also flow aspects. He believed me and hired me and I started my regular and most fruitful research stays at HMRI where Mauro in the meantime had transferred his whole research group. I enjoyed the stays at HMRI enormously, met many friends and colleagues and learned a lot. I will always be grateful to Mauro for the trust he put in me and for the excellent collaboration.

Bernhard A. Schrefler

Abstract

We couple a tumor growth model embedded in a microenvironment, with a bio distribution model able to simulate a whole organ. The growth model yields the evolution of tumor cell population, of the differential pressure between cell populations, of porosity of ECM, of consumption of nutrients due to tumor growth, of angiogenesis, and related growth factors as function of the locally available nutrient. The bio distribution model on the other hand operates on a frozen geometry but yields a much refined distribution of nutrient and other molecules. The combination of both models will enable simulating the growth of a tumor in a whole organ, including a realistic distribution of therapeutic agents and allow hence to evaluate the efficacy of these agents.

Keywords Modeling · Multiphase · Biodistribution · Code coupling · Angiogenesis · Smearded finite element

1 Introduction

Cancer is an extraordinarily complex disease. It is now recognized that methods commonly used in physics can help reducing the complexity of cancer to a manageable set of underlying principles and phenomena (Michor et al. 2011; Moore et al. 2011). Among the fields to which oncophysics can contribute belong the evolution and evolutionary theory of cancer, information coding and transfer in cancer, deconvolution of cancer complexity, and transport oncophysics. Transport oncophysics views cancer as a disease of multiscale mass transport dysregulation involving biological barriers. In fact each one of us is the sum of a multiplicity of transport differentials (i.e., gradients of metabolites, chemotherapy, oxygen) between cellular compartments (e.g., DNA, organelles, cytosol, cellular membranes, extracellular matrix, and vasculature) and these multiscale mass transport differentials distinguish malignant from normal cells and tissues (Koay and Ferrari 2014). Understanding the disorganization of these differentials in cancer opens a realm of possibilities in all aspects of oncological research. Transport aspects further control delivery of therapeutic agents (e.g. chemotherapeutics or molecularly targeted therapeutics such as T cells, antibodies, particles) which must pass through different and heterogeneous tumor and healthy compartments (e.g. vascular, stroma) with distinct physical properties (Ferrari 2013, 2010). Delivery of drugs is an extremely complex procedure involving different spatial and temporal scales and taking place over several levels ranging from the organism to the intercellular environment. The underlying transport phenomena at individual tumor compartments may act as transport barriers possibly contributing to poor survival rates in cancer therapy (Ferrari 2013; Freyer et al. 1997).

In this paper we address a computational tool necessary for simulating simultaneously different aspects involved in transport oncophysics: tumor growth within the local tumor environment including angiogenesis, and bio distribution of nutrients, interstitial fluid flow and blood flow in the vasculature. This tool, together with imaging, analysis and quantification, will help to understand and predict cancer development and

Dedicated to Mauro Ferrari in occasion of his 60th birthday

✉ Raffaella Santagiuliana
santagiuliana.raffaella@gmail.com

1 Department of Civil, Environmental and Architectural Engineering, University of Padova, via Marzolo 9, 35131 Padova, Italy

2 Bioengineering Research and Development Center BioIRC Kragujevac, Prvoslava Stojanovica 6, Kragujevac 34000, Serbia

3 Belgrade Metropolitan University, Tadeuša Košćuška 63, Belgrade 11000, Serbia

4 Institut de Mécanique et d'Ingénierie (I2M, CNRS UMR 5295), University of Bordeaux, Bordeaux, France

5 The Department of Nanomedicine, Houston Methodist Research Institute, 6670 Bertner Ave., R7 117, Houston, TX 77030, USA

6 Serbian Academy of Sciences and Arts, Knez Mihailova 35, Belgrade 11000, Serbia

7 Institute for Advanced Study, Technische Universität München, Lichtenbergstrasse 2a, D-85748 Garching b. München, Germany

the efficacy of therapy, with the aim to design patient specific solutions for this complex disease.

Before presenting our tumor growth model we recall that the most advanced computational models for tumor growth prediction belong mainly to two types: multi-parameter and multi-phase models. Multi-parameter models are based on mixture theory (Cowin and Cardoso 2012) where the relevant balance equations are written directly at the level of interest and the thermodynamic consistency is satisfied at the same level. The evolution of phases and species within multi-parameter models is obtained either by use of phase field approach (Hawkins-Daarud et al. 2013; Lima et al. 2015, 2016; Oden et al. 2010, 2013, 2016; Rahman et al. 2017; Rocha et al. 2018; Vilanova et al. 2018) or of Volterra-Lotta (predator/prey like) equations (Carotenuto et al. 2018; Fraldi and Carotenuto 2018). Recent multiphase models (Kremheller et al. 2018; Sciumè et al. 2013, 2014a) are based on the Thermodynamically Constrained Averaging Theory (TCAT) (Gray and Miller 2014) where the model derivation proceeds systematically from known microscale relations to mathematically and physically consistent larger scale relations. This is accomplished by the use of averaging theorems. The thermodynamic analysis is consistent between scales, in the definitions of variables at different scales and in satisfying the entropy inequality. The closure relationships are obtained from a Simplified Entropy Inequality (SEI). Interfaces between constituents arise naturally from the solution of an initial-boundary value problem that must comprise the mass balance equations of all phases involved.

Our model is of the multiphase type with a deformable solid matrix (ECM) pervaded by three fluid phases, tumor cells (TCs), healthy cells (HCs) and interstitial fluid (IF), (Santagiuliana et al. 2016; Sciumè et al. 2013, 2014a, b). In this model the extracellular matrix is a porous solid which may undergo remodeling and the fluid phases fill its pores. Tumor cells partition into living cells and necrotic cells. Healthy cells are in homeostasis. The IF transports chemical species such as nutrients, oxygen, signaling molecules like Tumor Angiogenic Factor (TAF), cytokines, etc.. Transport within extravascular space takes place by convection and diffusion. The model is able to simulate growth, hypoxia, necrosis and lysis of the TCs, migration of cells through the ECM, invasion and infiltration of the TCs into the healthy tissue, different stiffness of cell population with respect to the ECM, deposition and remodeling of ECM, build-up of cortical tension between HCs and TCs, adhesion of the cells to their ECM as well as adhesion among cells, and possible detachment. Angiogenesis is modeled as follows (Santagiuliana et al. 2016): when the oxygen mass fraction is lower than a critical value the tumor living cells produce TAF in response to hypoxia. TAF diffuses in the surrounding tissue and creates a chemical gradient between the tumor and any existing vasculature. Endothelial cells (ECs), lining the blood vessels, respond to the TAF. In our model endothelial cells are a species transported in the IF, which diffuse following their own gradient and that of TAF. The density of endothelial cells represents the concentration of capillary sprouts formed by accumulation of endothelial cells which are recruited from the parent vessel. This smeared representation of the neovasculature is an ideal intersection with the bio distribution model described next. The choice of a smeared representation instead of a discrete one is justified by the fact that the neovasculature is extremely chaotic and that there is no relationship between vessel diameter and flow velocity (Dewhirst and Secomb 2017; Discher et al. 2005).

The model for the bio distribution introduced by the group of Prof. Kojic, developed at the Houston Methodist Research Institute and Bioengineering R&D Center Bioirc in Serbia, simulates the diffusion of molecules, oxygen, and the smeared capillary network in the tumor environment. The basic idea of the smeared concept consists in: 1) transformation of the 1D mass transport into the equivalent continuum form; 2) use of a standard continuum representation with the pressure and concentration fields within the particular domains of the composite finite elements; and 3) formulation of connectivity finite elements for membrane (capillary and cell walls) transport (Kojic et al. 2017a, b, 2018; Milosevic et al. 2018). The composite smeared finite element (CSFE) includes continuum domains occupying the corresponding volume fraction and also connectivity elements at each node. This concept simplifies the model generation of complex biological media, including tumors, and still provides satisfactory accuracy, particularly investigated in (Milosevic et al. 2018).

In sections 2 we recall briefly the governing equations of the tumor growth and in section 3 the basic relations of the CSFE formulation. Section 4 deals with the coupling of both models including operational aspects. The results are reported in section 5 where two simulations are shown: the first one is a 2D example of a tumor growing within a square domain, and the second one deals with melanoma growth in an axisymmetric setting. Finally the conclusions highlight the capability of the new code achieved by the connection of two original models.

2 The tumor growth model: the governing equations

The tumor growth model is a continuum model where the TCs and HCs are presented as" adhesive" fluids within a porous matrix (ECM). TCs may become necrotic upon exposure to low nutrient concentrations or excessive mechanical stresses. The HCs are in homeostasis. The IF is an aqueous solution of biological molecules, cell nutrients, oxygen and waste products and the Representative Elementary Volume (REV) is schematically shown in Fig. 1.

As already mentioned the model is built within the TCAT framework which is used to transform known microscale relations to mathematically and physically consistent macroscale relations by using averaging theorems (Gray and Miller 2014; Gray et al. 2013). These relations are adequate and sufficient to describe tumor development while filtering out high frequency spatial variability. The governing equations of the model are closed by introducing constitutive relations in the macroscale equations.

The ECM is a deformable porous solid with porosity ε. The volume fraction of the solid phase is $\varepsilon_s = 1 - \varepsilon$. The other phases, tumor cells (ε^t), healthy cells (ε^h) and interstitial fluid (ε^l), occupy the rest of the volume. The volume fractions for all phases add up to unity

$$\varepsilon^s + \varepsilon^h + \varepsilon^t + \varepsilon^l = 1 \tag{1}$$

The saturation degree of a fluid phase α is: $S^\alpha = \varepsilon^\alpha/\varepsilon$. Using porosity, ε, and volume fraction, ε^α, (1) yields

$$S^h + S^t + S^l = 1 \tag{2}$$

The macroscopic mass and momentum balance equations of phases and species have been derived in (Sciumè et al. 2013) and their transformation to take the differential pressures as primary variables has been obtained in (Sciumè et al. 2014a). The model has been enhanced in (Santagiuliana et al. 2015) to include ECM deposition by the tumor cells through new mass exchange terms, $\overset{l\to s}{\underset{ECM}{M}}$, and lysis, $\overset{t\to l}{\underset{lysis}{M}}$.

Below the general form of the governing equations for each phase is shown. ρ^α is the density and $\mathbf{v}^\alpha$ is the velocity of the phase α, see Appendix A of (Sciumè et al. 2014a), while the $\mathbf{v}^s$ is the velocity of the solid phase (ECM) that is the time derivative of the solid phase displacements u^s. The final form of the governing equations is then obtained from the general forms by introducing some simplifications and closure relationships such as a generalized Darcy's equation for flow of the fluid phases (Santagiuliana et al. 2016; Sciumè et al. 2013).

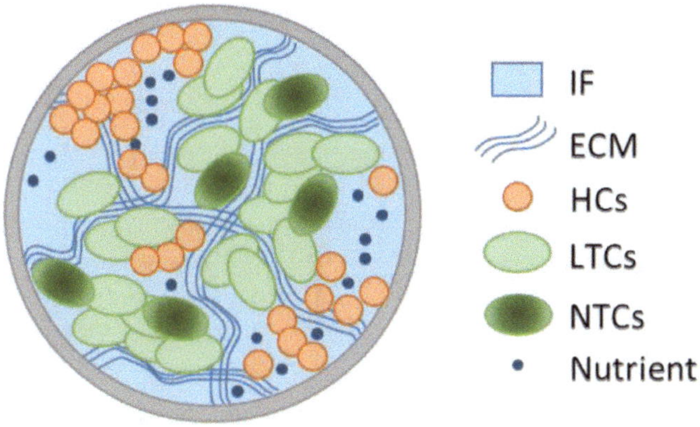

Fig. 1 Representative elementary volume (REV)

The mass balance equation of the ECM is

$$\frac{\partial \varepsilon}{\partial t} = \nabla\cdot\mathbf{v}^{\bar{s}} + \frac{(1-\varepsilon)}{\rho^s}\frac{\partial \rho^s}{\partial t} - \nabla\cdot\left(\varepsilon\mathbf{v}^{\bar{s}}\right) - \frac{\overset{l\to s}{\underset{ECM}{M}}}{\rho^s} \tag{3}$$

The mass balance equation of TCs reads

$$\frac{\partial(\varepsilon^t\rho^t)}{\partial t} + \nabla\cdot(\varepsilon^t\rho^t\mathbf{v}^t) = \overset{l\to t}{\underset{growth}{M}} - \overset{t\to l}{\underset{lysis}{M}} \tag{4}$$

where $\overset{l\to t}{\underset{growth}{M}}$ is an inter-phase exchange of mass between the phases l and t, and represents the mass of IF consumed by tumor cell growth.

The mass balance equation of HCs is

$$\frac{\partial(\varepsilon^h\rho^h)}{\partial t} + \nabla\cdot(\varepsilon^h\rho^h\mathbf{v}^h) = 0 \tag{5}$$

There are no mass exchange terms in Eq. (5) because the HCs are in homeostasis.

The mass balance equation of IF reads

$$\frac{\partial(\varepsilon^l\rho^l)}{\partial t} + \nabla\cdot(\varepsilon^l\rho^l\mathbf{v}^l) = \overset{t\to l}{\underset{lysis}{M}} - \overset{l\to t}{\underset{growth}{M}} - \overset{l\to s}{\underset{ECM}{M}} \tag{6}$$

where the mass exchange terms are the opposite of those seen in previous equations.

Similar are the mass balance equations for the species, presented below. Remind that the tumor cells are a phase composed of two species, the viable tumor cells and the necrotic ones. The governing equation for necrotic cells is

$$\frac{\partial(\varepsilon^t\rho^t\omega^{Nt})}{\partial t} + \nabla\cdot(\varepsilon^t\rho^t\omega^{Nt}\mathbf{v}^t) - \varepsilon^t r^{Nt} + \overset{t\to l}{\underset{lysis}{M}} = 0 \tag{7}$$

where ω^{Nt} is the mass fraction of TCs, $\varepsilon^t r^{Nt}$ is a reaction term, that is the death rate of tumor cells or rate of generation of necrotic cells. $M_{lysis}{}^{t\to l}$ takes into account of mass exchange between the necrotic tumor cells and the IF phase due to lysis.

The mass balance equation of the nutrient, by including a Fickian type equation for the diffusion of species, reads

$$\frac{\partial(\varepsilon^l\rho^l\omega^{nl})}{\partial t} + \nabla\cdot(\varepsilon^l\rho^l\omega^{nl}\mathbf{v}^l) - \nabla\cdot(\varepsilon^l\rho^l D^{nl}_{\text{eff}}\nabla\omega^{nl}) + \overset{nl\to t}{M} = 0 \tag{8}$$

where $D^{\overline{nl}}_{eff}$ is the effective diffusivity of the nutrient species in the extracellular space, ω^{nl} the mass fraction of nutrient species n and $M^{nl\to t}$ is the mass of nutrient consumed by tumor cells via metabolism and growth.

The other species considered are TAF and ECs. Since we have only four phases at disposal in Castem (the code referenced below), we model the endothelial cells as a transported species in the IF (Santagiuliana et al. 2016). The respective mass balance equations are

$$\frac{\partial(\varepsilon^l\rho^l\omega^{TAF})}{\partial t} + \nabla\cdot(\varepsilon^l\rho^l\omega^{TAF}\mathbf{v}^l) - \left(\varepsilon^l\rho^l D^{TAF}_{eff}\nabla\omega^{TAF}\right) - \overset{TAF\to t}{M} = 0 \tag{9}$$

Fig. 2 Diffusion from capillary to tissue through elementary capillary wall surface dA_{cap} which corresponds to the capillary volume dV_{cap} and total volume dV; dV_{tissue} is the volume occupied by tissue

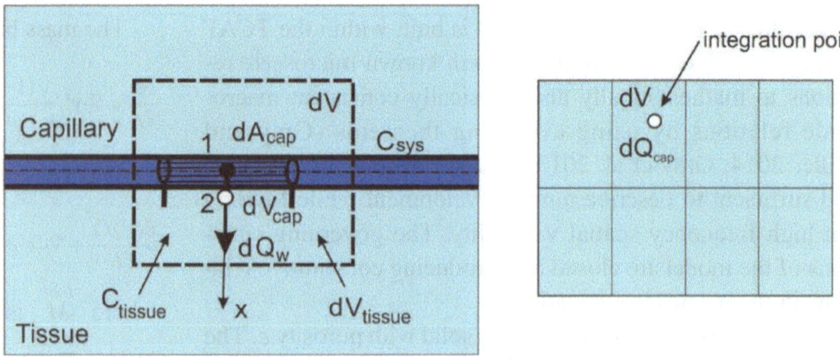

and

$$\frac{\partial(\varepsilon^l \rho^l \omega^{EC})}{\partial t} + \nabla\cdot(\varepsilon^l \rho^l \omega^{EC} \mathbf{v}^l) - \left(\varepsilon^l \rho^l D^{EC}_{eff} \left(\nabla\omega^{EC} - \nabla\omega^{TAF}\right)\right) - \overset{EC\to t}{M} = 0 \tag{10}$$

where D^{TAF}_{eff} and D^{EC}_{eff} are the effective diffusivity of the TAF and ECs in the extracellular space, ω^{TAF} and ω^{EC} the mass fractions of TAF and ECs.

The last governing equation is the linear momentum balance of the solid phase expressed in rate form as

$$\nabla\cdot\left(\frac{\partial \mathbf{t}^{\bar{\bar{s}}}_{eff}}{\partial t} - \overline{\alpha}\frac{\partial p^s}{\partial t}\mathbf{1}\right) = 0 \tag{11}$$

where the interaction between the solid and the three fluid phases is accounted for through the effective stress $\mathbf{t}^{\bar{\bar{s}}}_{eff}$ in the sense of porous media mechanics

$$\mathbf{t}^{\bar{\bar{s}}}_{eff} = \mathbf{t}^{\bar{\bar{s}}} + \overline{\alpha} p^s \mathbf{1} \tag{12}$$

$\mathbf{1}$ is the unit tensor, $\mathbf{t}^{\bar{\bar{s}}}$ is the total stress tensor in the solid phase, $\overline{\alpha}$ is the Biot's coefficient $\overline{\alpha} = 1 - K/K_s$, with K the compressibility of the empty ECM. In the modeled problem, K/K_s tends to zero hence we can assume a Biot's coefficient equal to 1. The solid pressure p^s is given as in (Gray and Schrefler 2007)

$$p^s = S^h p^h + S^t p^t + S^l p^l = p^l + \left(1 - S^l\right) p^{hl} + S^t p^{th} \tag{13}$$

where the Bishop parameter of each fluid phase (solid surface fraction in contact with the phase) has been taken equal to its own degree of saturation; p^t, p^t, p^l are respectively the tumor cells pressure, the host cells pressure and the interstitial fluid pressure; p^{hl} and p^{th} are the pressure differences between, respectively, host cells and interstitial fluid, and tumor cells and host cells. A large deformation regime is assumed for the elasto-visco-plastic ECM (solid phase).

The governing equations have been introduced in the finite elements code CAST3M or Castem (http://www-cast3m.cea.fr) of the French Atomic Energy Commission.

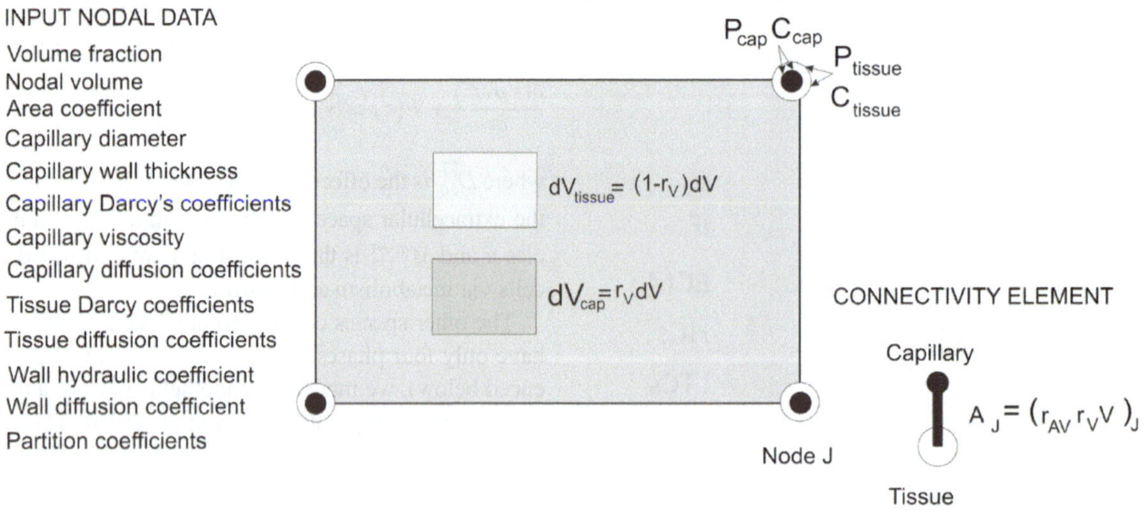

Fig. 3 Composite smeared finite element (CSFE) with capillary and tissue domain and connectivity elements at each node; list of nodal parameters is given in the figure

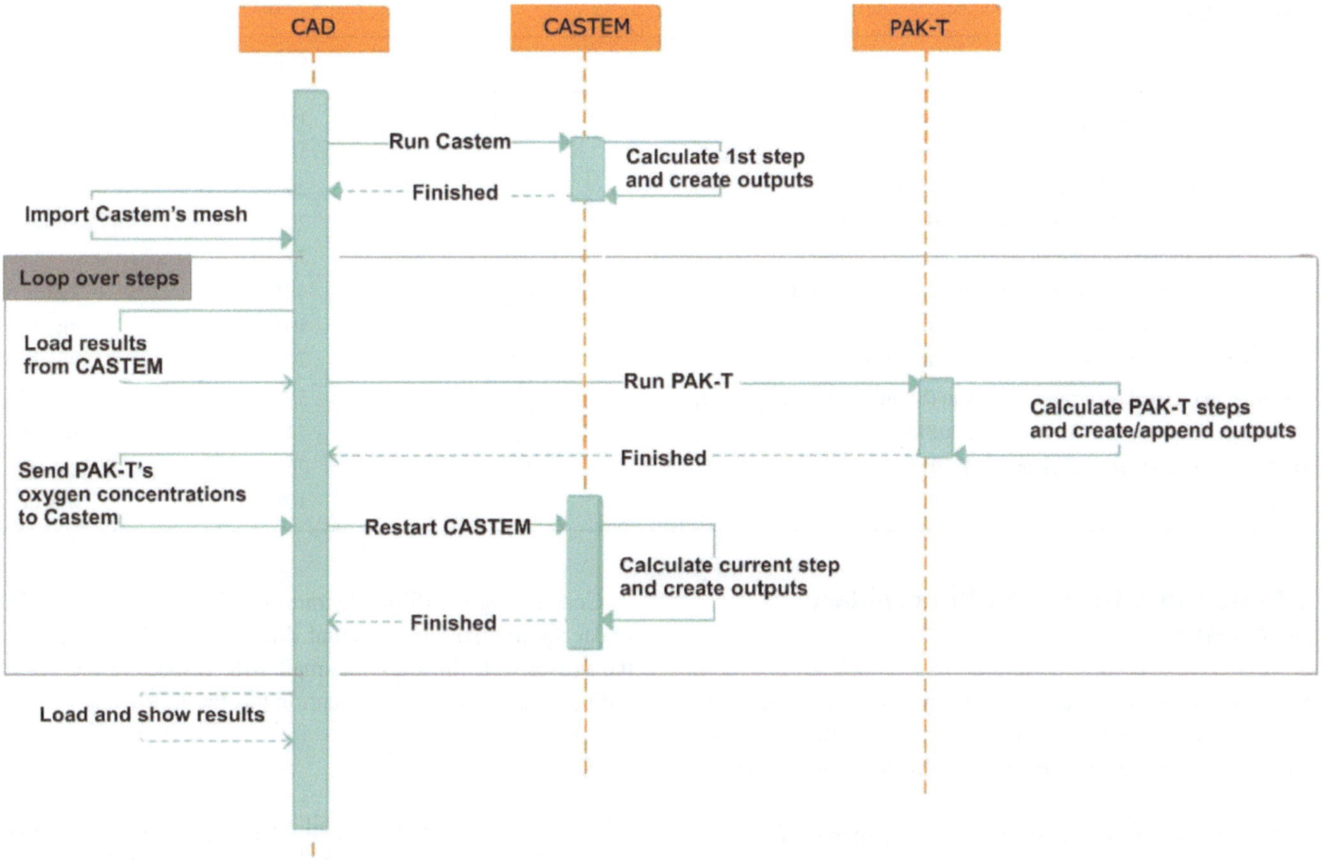

Fig. 4 Sequence diagram of automated exchange between Castem and PAKT, with using CAD user interface

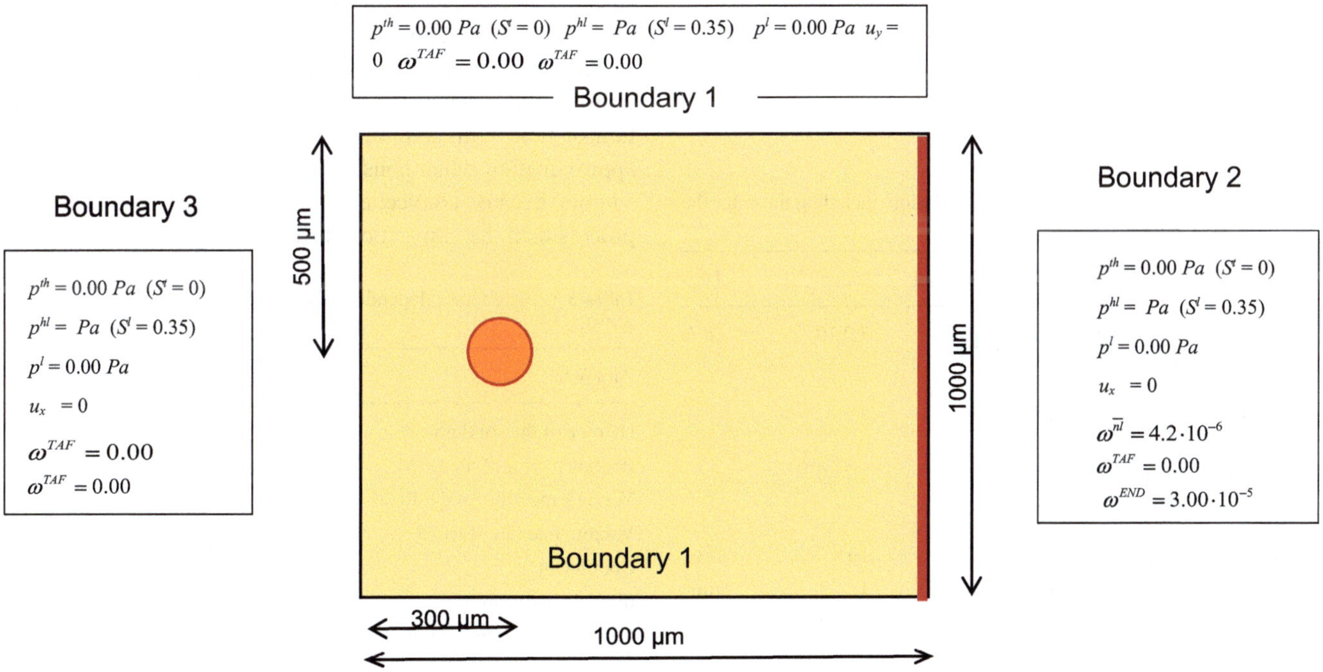

Fig. 5 Geometry of the problem with boundary conditions

Table 1 Parameters

Parameter	Symbol	Value	Unit
Density of the three fluid phases ($\alpha = h$, t and l)	ρ^{α}	1000	kg/m^3
Dynamic viscosity of IF (Sciumè et al. 2014b)	μ^{l}	$1 \cdot 10^{-2}$	Pa·sec
Dynamic viscosity of TC (Sciumè et al. 2014b)	μ^{t}	20	Pa·sec
Dynamic viscosity of HC (Sciumè et al. 2014b)	μ^{h}	20	Pa·sec
Critical mass fraction of oxygen	$\omega^{\overline{nl}}_{crit}$	$2.0 \cdot 10^{-6}$	–
Growth coefficient of tumor cells (Sciumè et al. 2013, 2014a)	γ^{t}_{growth}	$4 \cdot 10^{-2}$	kg/(m^3·s)
Necrosis coefficient (Sciumè et al. 2013, 2014a)	$\gamma^{t}_{necrosis}$	$1 \cdot 10^{-2}$	kg/(m^3·s)
Consumption related to growth in eqn (Sciumè et al. 2013, 2014a)	$\gamma^{\overline{nl}}_{growth}$	$2 \cdot 10^{-4}$	kg/(m^3·s)
Consumption related to metabolism in eqn (Sciumè et al. 2013, 2014a)	$\gamma^{\overline{nl}}_{0}$	$3 \cdot 10^{-4}$	kg/(m^3·s)
HC-IF interfacial tension (Sciumè et al. 2014b)	σ_{hl}	72	mN/m
TC-HC interfacial tension (Sciumè et al. 2014b)	σ_{th}	36	mN/m
TC-IF interfacial tension (Sciumè et al. 2014b)	σ_{tl}	108	mN/m

3 Coupling diffusion within capillary and tissue

We summarize now the methodology, which will be further used as the basis for the development of the composite smeared finite element to model diffusion within capillary network and tissue.

We write the balance equation for diffusion in a form used in our smeared formulation. Hence, according to Eq. (8) we have

$$-\frac{\partial \omega^{nl}}{\partial t} - \frac{\partial \omega^{nl}}{\partial x_i} v_i + \frac{\partial}{\partial x_i}\left(D_{ij}\frac{\partial \omega^{nl}}{\partial x_j}\right) + \frac{1}{\varepsilon^{l}\rho^{l}}\overset{nl\rightarrow t}{M}$$
$$= 0, \quad \text{sum on } i, j; \quad i = 1, 2, 3 \qquad (14)$$

where D_{ij} are diffusion tensor coefficients ($D^{\overline{nl}}_{eff}$ of a nutrient in extracellular space in (8)).

Table 2 Parameters related to oxygen diffusion, including those for the smeared model

Parameter	Symbol	Value	Unit
Diffusion coefficient of oxygen in interstitial fluid (Sciumè et al. 2013, 2014a)	$D^{\overline{nl}}_{0}$	$3.2 \cdot 10^{-9}$	m^2/s
Coefficient δ (Sciumè et al. 2013, 2014a)	δ	2	–
Normal mass fraction of oxygen in tissue (Sciumè et al. 2013, 2014a)	$\omega^{\overline{nl}}_{env}$	$4.2 \cdot 10^{-6}$	–
Mean capillary diameter	D_{cap}	10	μm
Thickness of the endothelial layer	δ_{EC}	1	μm
Diffusion coefficient of oxygen through endothelial layer	D_{wall}	$8.73 \cdot 10^{-10}$	m^2/s
Mass fraction of oxygen in capillaries	ω_{sys}	$0.2451 \cdot 10^{-3}$	–

Considering the diffusive transport through the wall, a 1D linear approximation for radial diffusion will be used, since the vessel wall thickness is small with respect to the vessel radius. The mass balance equation has the following form,

$$-\frac{\partial \omega^{nl}}{\partial t} + D_w \frac{\partial^2 \omega^{nl}}{\partial x^2} = 0 \qquad (15)$$

where D_{wall} is the capillary wall diffusive coefficient; convection through the wall is neglected which in reality is small. Mass transport through capillary walls does not have a mass exchange term because we assume that consumption is present in the tissue domain but not in capillary walls. For simplicity of writing, consistent with the references related to the smeared concept, we will further use $c \equiv \omega^{nl}$ as the nutrient (oxygen) mass fraction called also concentration.

We further assume that concentration within capillaries is uniform and given as the systemic concentration $C_{sys}(t)$ function of time. This assumption can be taken as a reasonable approximation when considering a small tissue domain, e.g. a tumor, because convection within capillaries is a much faster process than the convection-diffusion within tissue.

Table 3 Parameters depending for ECM taken from (Sciumè et al. 2014b)

Parameter	Symbol	Value	Unit
Density of the solid phase	ρ^{s}	$1 \cdot 10^{3}$	kg/m^3
Poisson's ratio of the ECM	ν	0.4	–
Young's modulus of the ECM	E	$2.0 \cdot 10^{2}$	Pa
Volume fraction of ECM	ε^{s}	0.2	–
Coefficient a	a	590	Pa
Intrinsic permeability	k	$1.8 \cdot 10^{-15}$	m^2
Yield effective stress limit	$t^{s}_{eff,y}$	$0.5 \cdot 10^{1}$	Pa
Viscosity	η	5	Pa·sec
Hardening modulus	H	$1.0 \cdot 10^{2}$	Pa

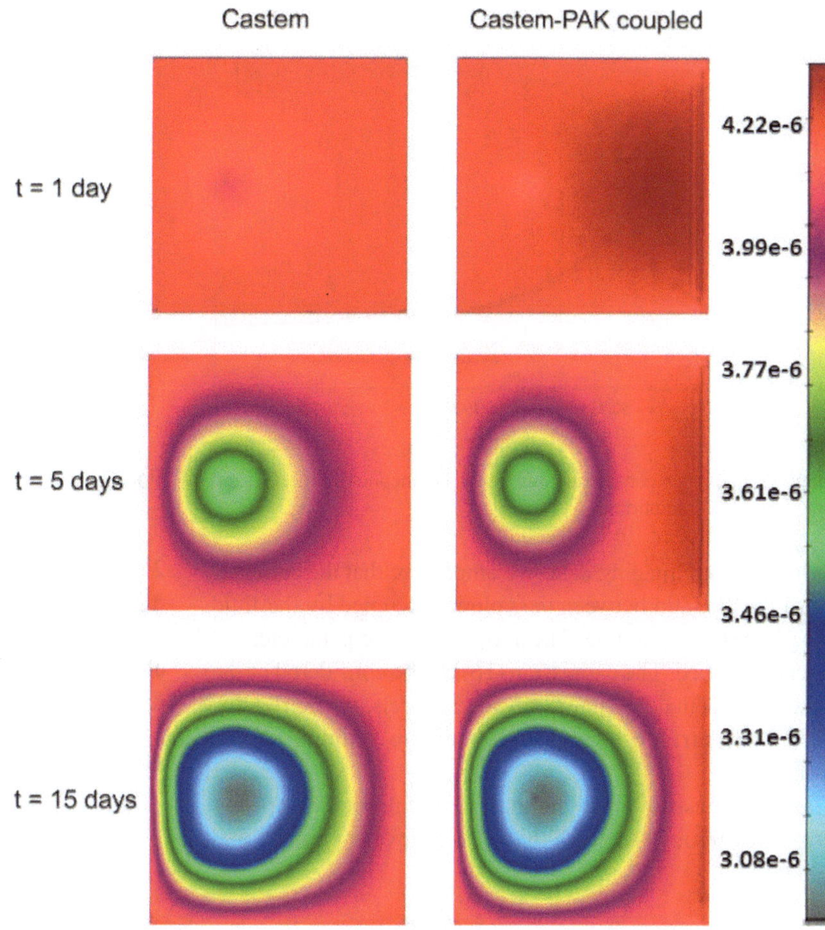

Fig. 6 Oxygen concentration (mass fraction) field obtained by using Castem and Castem-PAK, for t = 1, 5, 15 days

Consider diffusion through a capillary wall as schematically shown in Fig. 2. First, the elementary area of the surface of the internal wall dA_{cap} can be related to the elementary volume dV_{cap} and further to the elementary total volume dV, as follows:

$$dA_{cap} = r_{AV} dV_{cap} = r_{AV} r_V dV \tag{16}$$

where r_{AV} is the capillary area-to volume ratio (called further surface ratio), and r_V is the capillary volumetric (mass) ratio within tissue, or capillary density; the volume of tissue is $(1-r_v)dV$. Note that in case of a straight capillary, the surface ratio is $r_{AV} = 4/D_{cap}$ where D_{cap} is the capillary internal diameter; in case of complex geometries r_{AV} can be different, and can be

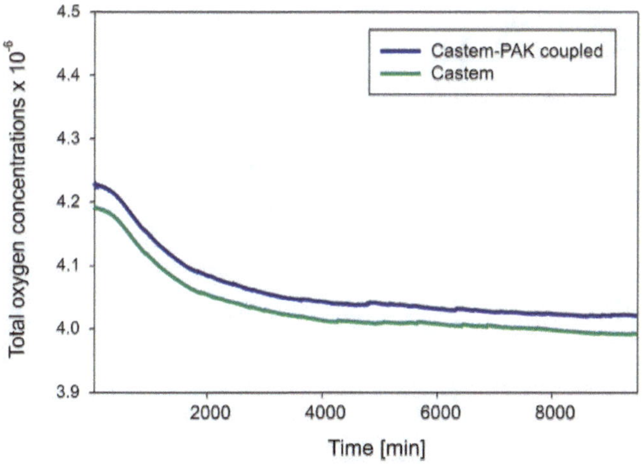

Fig. 7 Mean oxygen concentrations in whole domain as function of time, Castem and Castem-PAK solutions

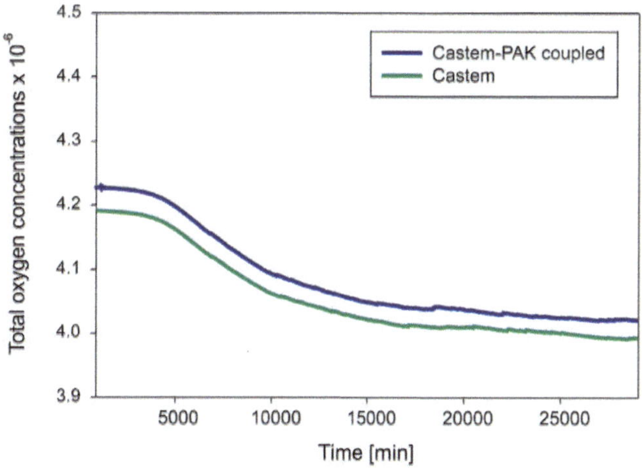

Fig. 8 Mean oxygen concentrations in tissue as function of time, Castem and Castem-PAK solutions

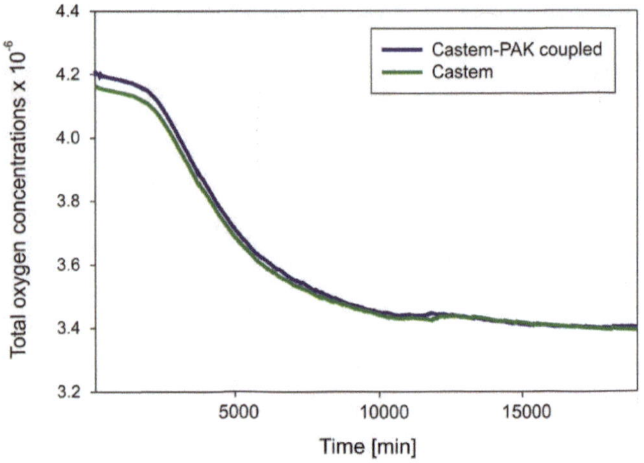

Fig. 9 Mean oxygen concentration in tumor as function of time, Castem and Castem-PAK solutions

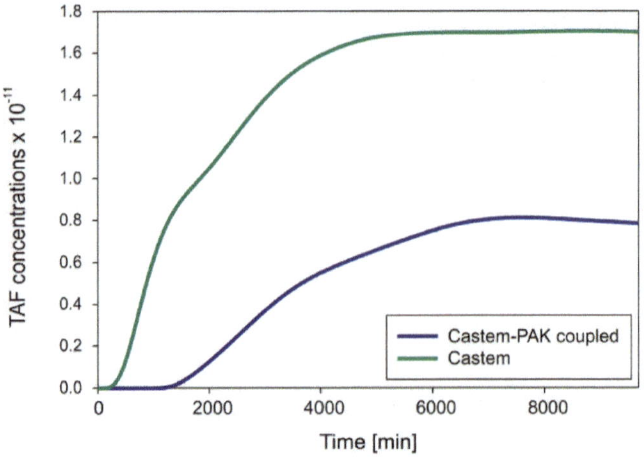

Fig. 11 Mean TAF concentration as function of time, Castem and Castem-PAK solutions

evaluated from imaging data. We emphasize that the capillary density is the ratio between the volume occupied by the fluid (blood) and total volume. The above ratios are parameters of the capillary bed. The expression (16) can be considered as the most fundamental one in our smeared models, where the discrete wall surface is smeared over the volume of the continuum.

Next, we assume that the mass concentration is linearly distributed through the wall thickness (between points 1 and 2 in Fig. 2), which is acceptable for thin capillary walls; this is in accordance with (15). Then, the flux through the wall at point 2, corresponding to the elementary surface dA_{cap}, including partitioning P_1 and P_2 at the internal and external capillary surface, can be expressed as

$$dQ_w = \left[D_{wall}\left(P_1 C_{sys} - P_2 C_{tissue}\right) - \frac{hP}{6\Delta t}\left(P_1 C - C^t\right)_{sys} - \frac{h}{3\Delta t}\left(P_2 C - C^t\right)_{tissue}\right] r_{AV} r_V dV \quad (17)$$

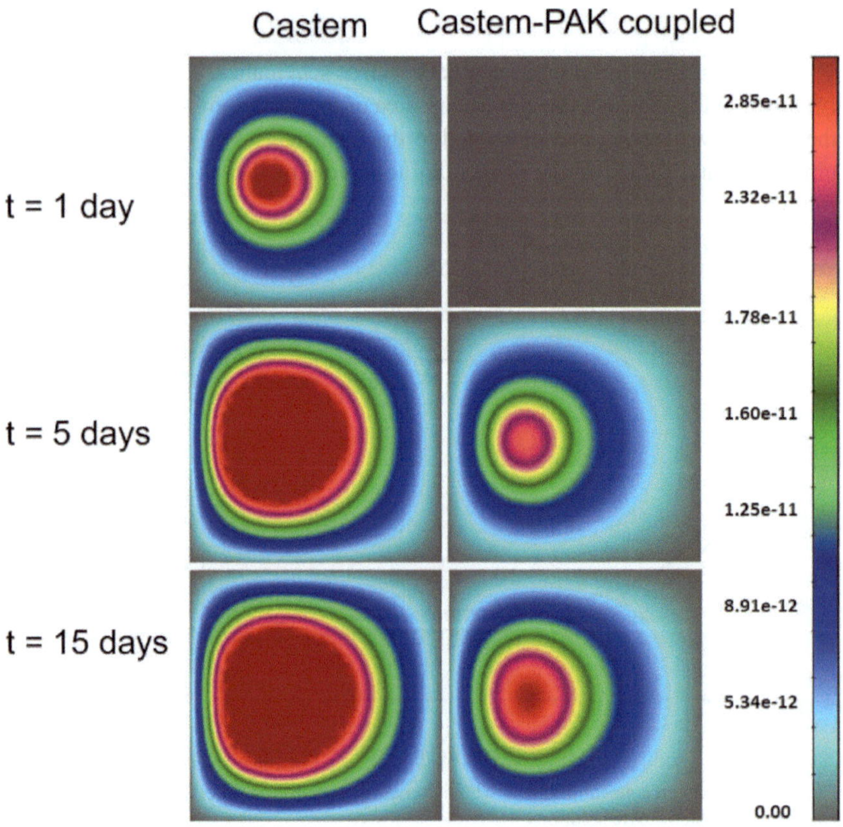

Fig. 10 TAF concentrations using CASTEM and CASTEM-PAK, for t = 1, 5 and 15 days

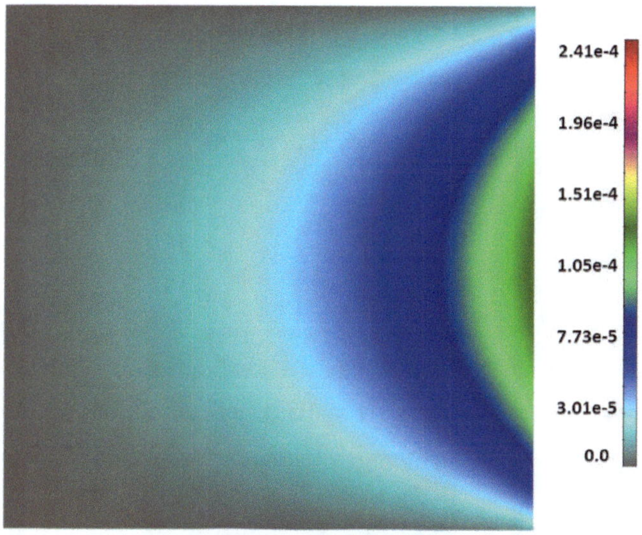

Fig. 12 Endothelial cells mass fraction, after 5 days

where C_{sys}, C^t_{sys}, C_{tissue}, C^t_{tissue} are the concentration in the lumen, and concentration within tissue at the end and start of time step, respectively, and h is the wall thickness. Note that D_{wall} represents the overall transport coefficient of the wall (with pores, fenestrations, etc.); it can be related to the diffusion coefficient of the wall porous material coefficient $D_{material}$ as $D_{wall} = h\ D_{material}$. Partitioning coefficient P is used as a measure of repelling or attraction of molecules at the boundary between two media which produces a discontinuity at the common surface ($P=C_1/C_2$ where C_1 and C_2 are concentrations at two sides of the surface). As can be seen from Fig. 2, point 1 is in lumen and point 2 in tissue domain, hence P1 is partitioning at lumen/capillary wall interface, and P2 is partitioning at capillary wall/tissue interface.

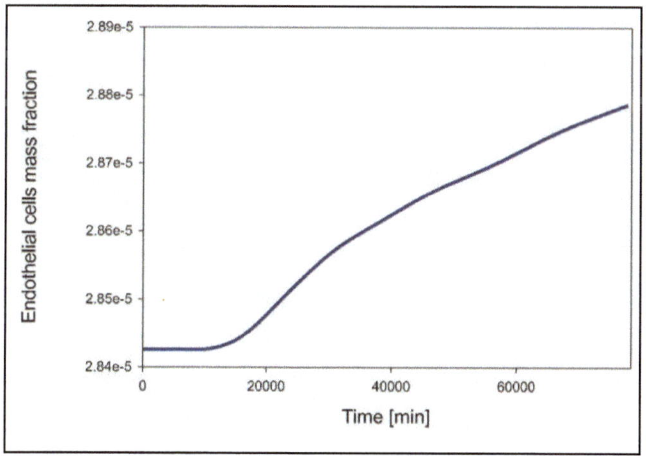

Fig. 13 Mean mass fraction of endothelial cells as a function of time

We have now the tissue continuum within which capillaries are distributed and are producing the source of mass according to (17). Therefore, the nodal fluxes of a continuum finite element are

$$Q_{wI} = \int_V N_I dQ_w = \int_V N_I(\ldots)(1-r_V)dV \tag{18}$$

where terms within the parenthesis (…) follow from (17), and N_I are the continuum interpolation functions of the element with volume V. When evaluating the integral (18), concentration C_{tissue} is the current concentration within the tissue at an integration point. Note that the factor (1-r_V) is used since the volume of tissue is reduced due to presence of capillaries.

Instead of using source terms at FE integration points, connectivity elements can be assigned at each continuum node. Then, the balance equation for the connectivity element at a continuum node I can be written as

$$\left(\frac{1}{\Delta t}M_{22} + K_{22}\right)P_2\Delta C_I = -\left(K_{21} + \frac{1}{\Delta t}M_{21}\right)P_1 C_{sys} \tag{19}$$

$$-\left(\frac{1}{\Delta t}M_{22} + K_{22}\right)P_2 C_I + \frac{1}{\Delta t}M_{21}C^t_{sys} + \frac{1}{\Delta t}M_{22}C^t_I$$

where

$$M_{22} = \frac{1}{3}P_2 A_{capI} h_I, \quad M_{21} = \frac{1}{6}P_2 A_{capI} h_I \qquad K_{22} = -K_{21} = A_{capI} P_2 D_{(wall)I} \tag{20}$$

and C_I and C^t_I are concentrations at node I at end and start of time step, respectively. Also, P_1 and P_2 are partitioning coefficients as in (17); $D_{(wall)I}$ is the wall diffusion coefficient, h_I is the wall thickness at node I; and A_{capI} is the wall surface area belonging to the node I, which is

$$A_{capI} = (r_{AV} r_V)_I V_I \tag{21}$$

with $(r_V)_I$, $(r_{AV})_I$ and V_I the volumetric ratio, the area coefficient and the volume of the continuum which belongs to the node, respectively. The volume V_I can numerically be evaluated as

$$V_I = \sum_{elements} \int_V N_I dV \tag{22}$$

where summation includes all elements containing the node I. We found that convergence was improved by applying the concept of these connectivity elements at nodes instead of continuously distributed source terms.

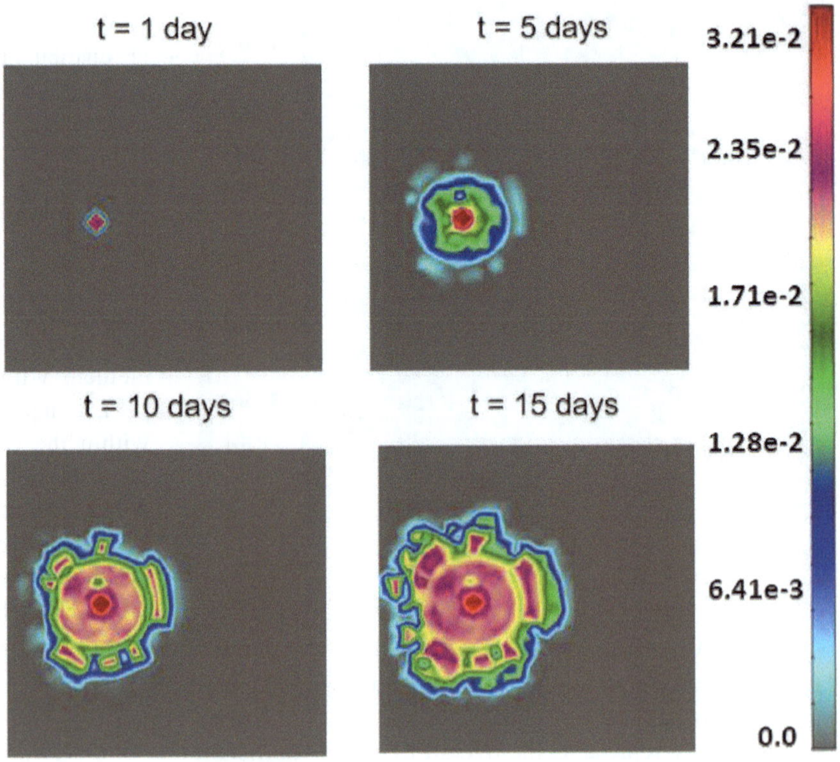

Fig. 14 Tumor growth for t = 1, 5, 10 and 15 days. Results are almost the same for Castem and Castem-PAK model

It can be concluded from the above that diffusive transport between capillaries and tissue can be performed by discretizing the continuum (tissue) only. The parameters of the model, assigned to each continuum node I include geometrical data (the volumetric ratio of capillaries $(r_V)_I$, the surface ratio $(r_{AV})_I$, the wall thickness (h_I)) and material data of capillaries consisting of wall diffusion coefficient $(D_{wall})_I$ and partition coefficients P_{1I} and P_{2I} at the capillary surfaces. Effects of partitioning are neglected in this study.

4 Coupling the models

4.1 Numerical solution and computational procedure for the tumor growth model

The weak form of the governing equations is obtained by means of the standard Galerkin procedure and is then discretized in space by means of the finite element method. Integration in the time domain is carried out by the Finite Difference Method adopting a quasi-Crank–Nicolson scheme (θWilson method with θ = 0.52). Within each time step the equations are linearized by the Newton-Raphson method.

Five computational units are used in the staggered scheme: the first is for the nutrient mass fraction $\overline{\omega}^{nl}$, the second for the TAF mass fraction $\overline{\omega}^{TAF}$, the third is for the endothelial cells mass fraction $\overline{\omega}^{EC}$, the fourth to compute p^{th}, p^{hl} and p^{l}, the difference of pressure between TCs and HCs, HCs and IF respectively and the pressure of IF, and the fifth is used to obtain the displacement vector $\overline{\mathbf{u}}^{s}$. Within each iteration the mass fraction of NTC, $\omega^{N\bar{t}}$, is updated using (7).

The final system of equations can be expressed in matrix form as follows,

$$\mathbf{C}_{ij}(\mathbf{x})\frac{\partial \mathbf{x}}{\partial t} + \mathbf{K}_{ij}(\mathbf{x})\mathbf{x} = \mathbf{f}_i(\mathbf{x}) \tag{23}$$

with

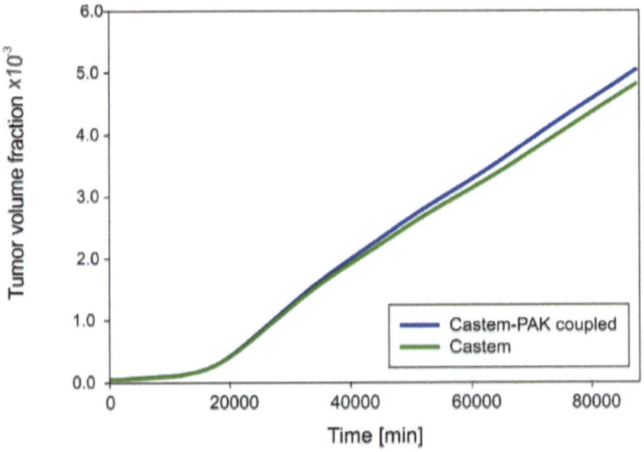

Fig. 15 Mean volume fraction of tumor as function of time, for Castem and Castem-PAK

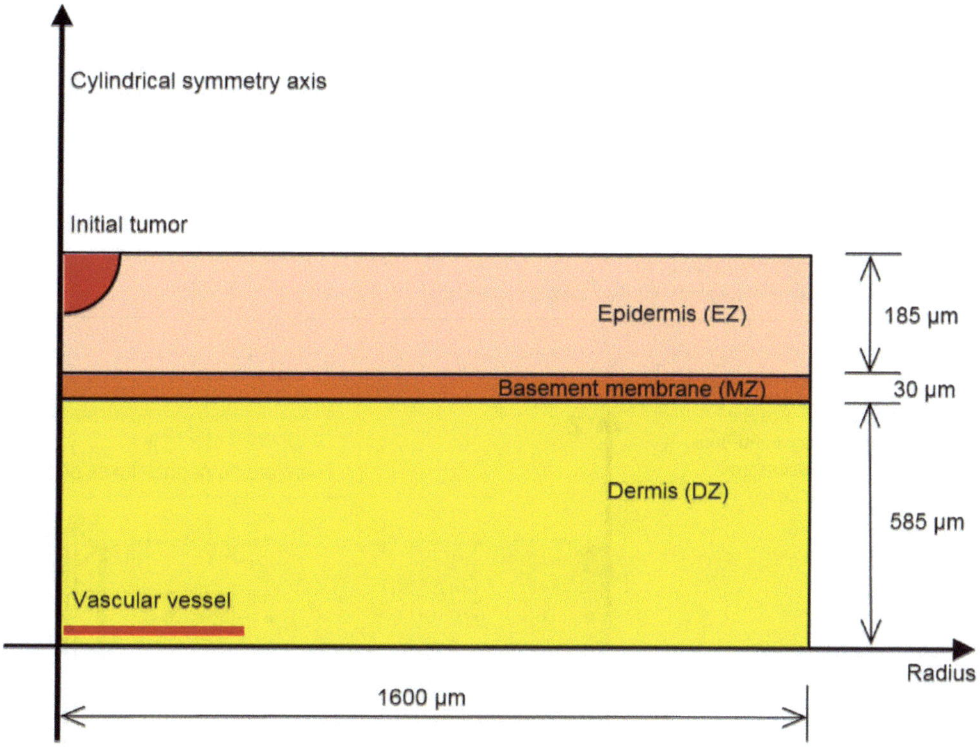

Fig. 16 Skin structure and geometry of the modeled case, redrawn with permission from (Santagiuliana et al. 2016)

Table 4 Parameters depending on ECM type taken from (Sciumè et al. 2014b), redrawn with permission from (Santagiuliana et al. 2016)

Parameter	Symbol	Value	Unit
Density of the solid phase	ρ^s	1000	kg/m^3
Poisson' ratio of the ECM	ν	0.4	–
	E_{EZ}	$2{\cdot}10^2$	
Young's modulus of the ECM	E_{MZ}	$3{\cdot}10^2$	Pa
	E_{DZ}	$1{\cdot}10^2$	
	ε^s_{EZ}	0.2	–
Volume fraction of ECM (initial)	ε^s_{MZ}	0.3	–
	ε^s_{DZ}	0.1	–
	a_{EZ}	590	
Coefficient a in Eq. (23)	a_{MZ}	516	Pa
	a_{DZ}	664	
	k_{EZ}	$1.80{\cdot}10^{-15}$	
Intrinsic permeability	k_{MZ}	$1.21{\cdot}10^{-15}$	m^2
	k_{DZ}	$2.56{\cdot}10^{-15}$	
Yield effective stress limit	$t^s_{\mathrm{eff},y}$	$0.5{\cdot}10^1$	Pa
Viscosity	η	5	Pa·sec
Hardening modulus	H	$1.0{\cdot}10^2$	Pa

Table 5 Parameters for TAF end Endothelial cells, redrawn with permission from (Santagiuliana et al. 2016)

Parameter	Symbol	Value	Unit
Limit mass fraction of oxygen for hypoxia	$\omega_{hyp}^{\overline{nl}}$	$4.0 \cdot 10^{-6}$	–
Diffusion coefficient of TAF in interstitial fluid (Anderson and Chaplain 1998)	D_0^{TAF}	$3.5 \cdot 10^{-4}$	m^2/s
Diffusion coefficient of endothelial cells in interstitial fluid (Eikenberry et al. 2009)	D_0^{end}	$1.29 \cdot 10^{-12}$	m^2/s
coefficient for uptake of TAF by endothelial cells	ν_{TAF}	$1.00 \cdot 10^{-15}$	–
degradation rate coefficient for TAF demise	β_{TAF}	$8.00 \cdot 10^{-13}$	–
coefficient for TAF and EC production	c	$1.00 \cdot 10^{-3}$	–
coefficient for new oxygen brought by the new capillary network	ϕ	$1.00 \cdot 10^{-8}$	–

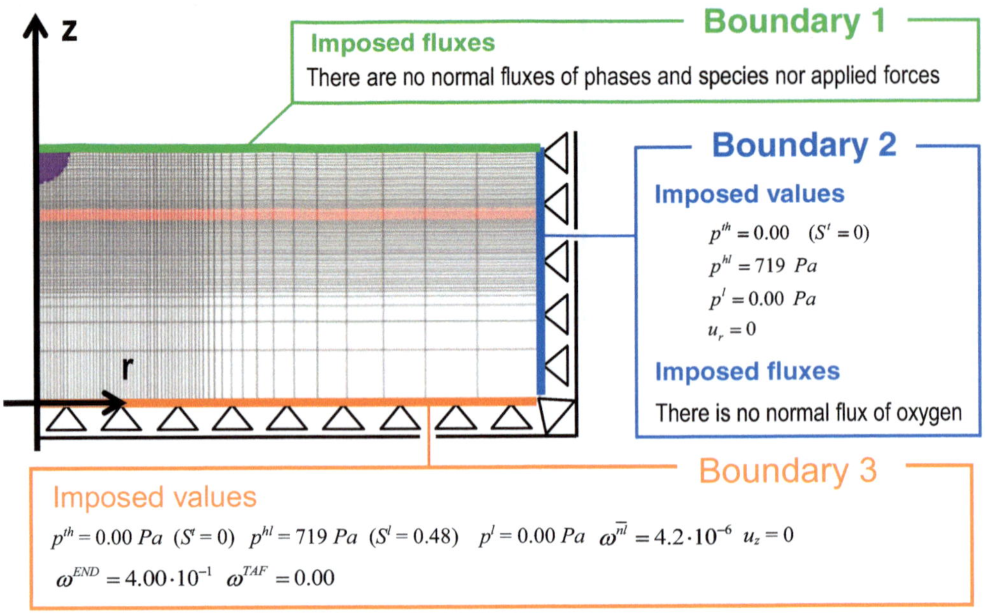

Fig. 17 Discretization of the domain and boundary conditions, redrawn with permission from (Santagiuliana et al. 2016)

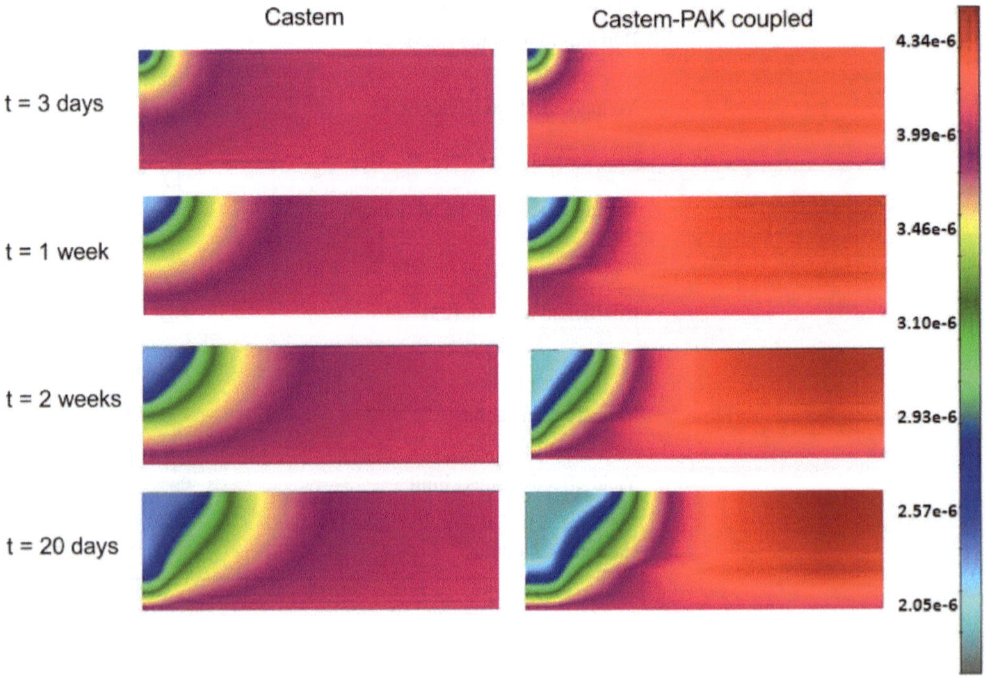

Fig. 18 Oxygen field concentrations obtained using Castem and Castem-PAK, for t = 3 days, 1 week, 2 weeks and 20 days

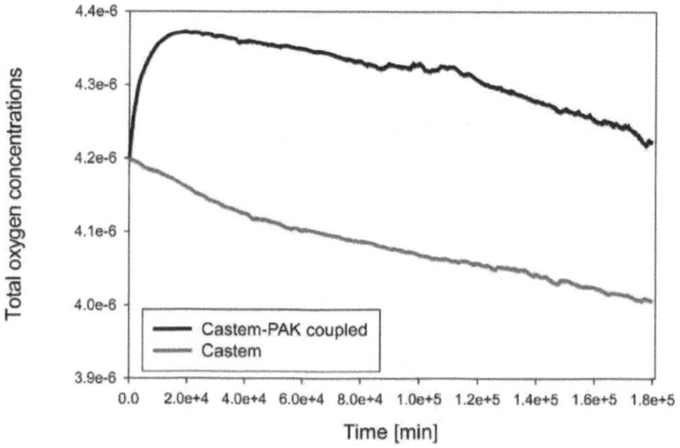

Fig. 19 Mean oxygen concentrations in tumor and tissue as function of time, for Castem and Castem-PAK coupled model

$$\mathbf{C}_{ij} = \begin{pmatrix} \mathbf{C}_{nn} & 0 & 0 & 0 & 0 & 0 & 0 \\ 0 & \mathbf{C}_{TT} & 0 & 0 & 0 & 0 & 0 \\ 0 & 0 & \mathbf{C}_{EE} & 0 & 0 & 0 & 0 \\ 0 & 0 & 0 & \mathbf{C}_{tt} & \mathbf{C}_{th} & \mathbf{C}_{tl} & 0 \\ 0 & 0 & 0 & \mathbf{C}_{ht} & \mathbf{C}_{hh} & \mathbf{C}_{tl} & 0 \\ 0 & 0 & 0 & \mathbf{C}_{lt} & \mathbf{C}_{lh} & \mathbf{C}_{ll} & 0 \\ 0 & 0 & 0 & 0 & 0 & 0 & \mathbf{C}_{uu} \end{pmatrix}$$

$$\mathbf{K}_{ij} = \begin{pmatrix} \mathbf{K}_{nn} & 0 & 0 & 0 & 0 & 0 & 0 \\ 0 & \mathbf{K}_{TT} & 0 & 0 & 0 & 0 & 0 \\ 0 & 0 & \mathbf{K}_{EE} & 0 & 0 & 0 & 0 \\ 0 & 0 & 0 & \mathbf{K}_{tt} & \mathbf{K}_{th} & \mathbf{K}_{tl} & 0 \\ 0 & 0 & 0 & \mathbf{K}_{ht} & \mathbf{K}_{hh} & \mathbf{K}_{hl} & 0 \\ 0 & 0 & 0 & \mathbf{K}_{lt} & \mathbf{K}_{lh} & \mathbf{K}_{ll} & 0 \\ 0 & 0 & 0 & 0 & 0 & 0 & 0 \end{pmatrix}, \quad \mathbf{f}_i = \begin{pmatrix} \mathbf{f}_n \\ \mathbf{f}_T \\ \mathbf{f}_E \\ \mathbf{f}_t \\ \mathbf{f}_h \\ \mathbf{f}_l \\ \mathbf{f}_u \end{pmatrix} \tag{24}$$

where $\mathbf{x}^T = \left\{ \overline{\boldsymbol{\omega}}^{nl}, \overline{\boldsymbol{\omega}}^{TAF}, \overline{\boldsymbol{\omega}}^{EC}, \overline{\mathbf{p}}^{th}, \overline{\mathbf{p}}^{hl}, \overline{\mathbf{p}}^{l}, \overline{\mathbf{u}}^{s} \right\}$. The nonlinear coefficient matrices $\mathbf{C}_{ij}(\mathbf{x})$, $\mathbf{K}_{ij}(\mathbf{x})$ and $\mathbf{f}_i(\mathbf{x})$ are given in (Santagiuliana et al. 2016).

The endothelial cells diffusion dependence on TAF concentration has been taken into account in the code by adding

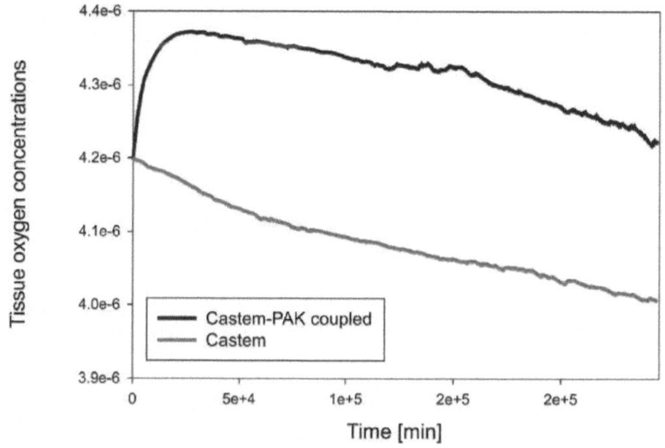

Fig. 20 Mean oxygen concentrations in tissue as function of time, for Castem and Castem-PAK coupled model

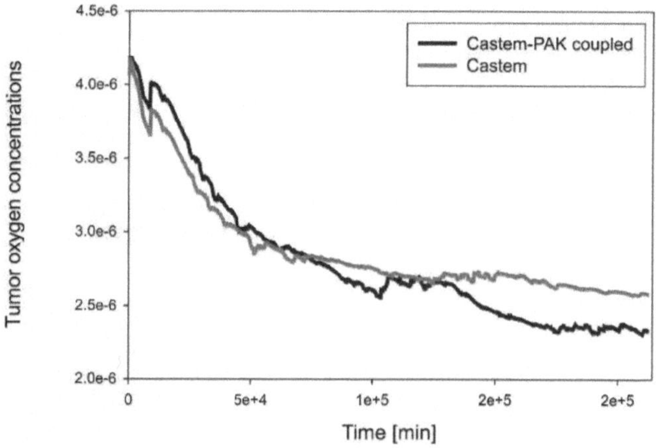

Fig. 21 Mean oxygen concentrations in tumor as function of time, for Castem and Castem-PAK

a coupling matrix for the endothelial cells and the TAF as a condition on the diffusion of the endothelial cells.

4.2 Formulation of the smeared finite element to model mass transport in the capillary-tissue system

Here we summarize the basic relations in formulating the composite smeared finite element (CSFE), according to (Kojic et al. 2017a). The basic requirement of the smeared concept for modeling transport within the capillary-tissue system, is that the transport characteristics of the system should appropriately be preserved in the smeared model.

A schematic representation of this element is shown in Fig. 3. The domains and the nodal variables used here are those relevant for this tumor growth model – capillary and tissue, while extension to include different cell groups and intracellular transport with cytosol and organelles (Kojic et al. 2018) is not considered.

The capillary network is represented by the fraction r_V of the FE volume (mass) and the nodal variables are pressure and concentrations. The fundamental step now is to transform 1D fluid flow and diffusion to the adequate continuum form. This is achieved by introducing the corresponding Darcy and diffusion tensors. The Darcy tensor can be derived in the form (Kojic et al. 2017a)

$$k_{Dij} = \frac{1}{A_{tot}} \sum_K k_{pK} \ell_{Ki} \ell_{Kj} = \frac{\pi}{128 \mu A_{tot}} \sum_K d_K^4 \ell_{Ki} \ell_{Kj} \tag{25}$$

where d_K are capillary diameters, ℓ_{Ki} are directional cosines, μ is the fluid viscosity, and A_{tot} is the total cross-sectional area of capillaries at a considered spatial point vicinity, $A_{tot} = \sum_K A_K = \frac{\pi}{4} \sum_K d_K^2$. Here, the Hagen-Poisseuile law is assumed for

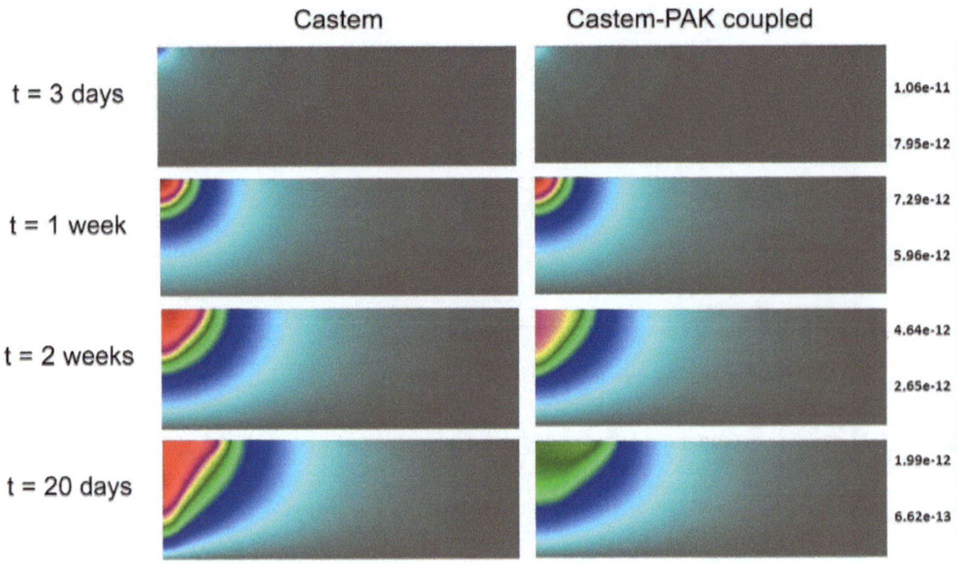

Fig. 22 TAF concentrations obtained using Castem and Castem-PAK models, for t = 3 days, 1 week, 2 weeks and 20 days

flow within capillaries. Likewise, the diffusion tensor can be expressed as

$$D_{ij} = \frac{1}{A_{tot}} \sum_{K} D_{capK} A_K \ell_{Ki} \ell_{Kj} \tag{26}$$

where D_{capK} are diffusion coefficients within capillaries and A_K are the cross-sectional areas. Therefore, we model the capillary network by a continuum; for convenience and efficiency we first evaluate the continuum volume V_J belonging to each FE node, (22) and use the area-to-volume ratio r_{AV} to evaluate the capillary wall surface A_J needed for nodal connectivity elements. The balance equations for connectivity elements are given by (19). The CSFE is built into our FE code PAK (Kojic et al. 2010).

4.3 Computational procedure for coupling the models

We will use three different modules to couple the models: Castem, CAD, and PAKT. Castem is the FE software for numerical simulation in structural mechanics and other types of scientific problems. CAD is indoor user interface for pre- and post-processing results for the package PAKT, and PAKT (a module of the PAK software package) is the indoor FE solver for concentration field which incorporates the smeared finite element.

The complete process of coupling the models is shown in the sequence diagram of Fig. 4, and consist of the following steps:

1. Run first time step in Castem and create outputs.
2. Import Castem's mesh into CAD software.
3. Load data from Castem into CAD and run PAKT simulation. Data loaded from Castem are: ε, ε^t,$\overline{\varpi}^{nl}$, $\overline{\varpi}^{TAF}$,$\overline{\varpi}^{EC}$, $\omega^{N\bar{t}}$ and p^{th}.
4. Pick up PAKT oxygen nodal mass fractions (concentrations) and send them to Castem.
5. If there are no more time steps to calculate go to step 6, otherwise go back to step 3.
6. Load results into CAD for post-processing.

5 Results

In the next paragraphs two examples of simulations are shown. The first one is a 2D example with an initial circular tumor growing within a squared domain with a blood vessel on a side. The second one is an example of the cutaneous melanoma growth in axial-symmetric conditions. Both simulations are performed with the coupled codes above described CAST3M and PAKT, and are compared with results of the

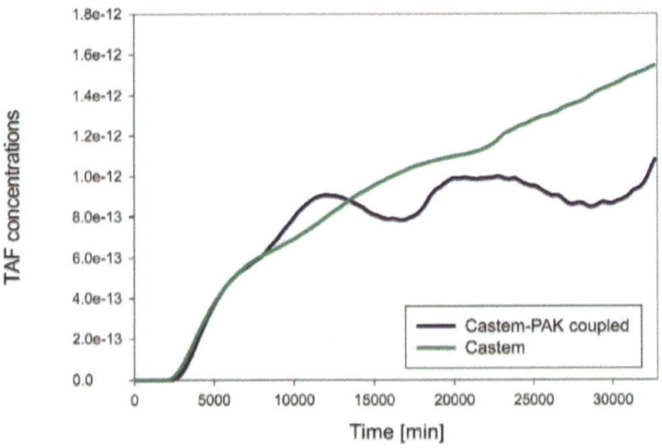

Fig. 23 Mean TAF concentrations as function of time, Castem and Castem-PAK models

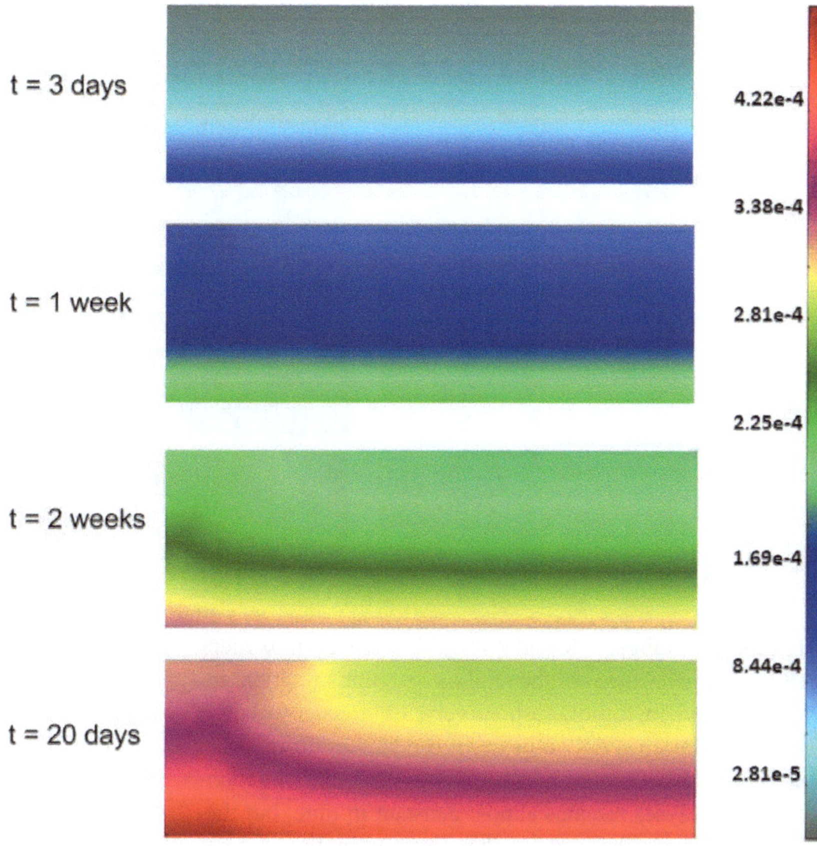

Fig. 24 Endothelial cells mass fraction for t = 3 days, 1 week, 2 weeks and 20 days, Castem model

simulations achieved with the code CAST3M only, i.e. just with the tumor growth model.

5.1 Simulation of a 2D example

A circular tumor with diameter 50 μm is situated in a square domain 1000 × 1000 μm^2 with a blood vessel on the right side, see Fig. 5. HCs, ECM and IF are present in the whole domain. The oxygen diffuses in the IF. As initial conditions the oxygen mass fraction is fixed at 4.2 10^{-6} in the whole domain, the ECM volume fraction is 0.2, the TCs volume

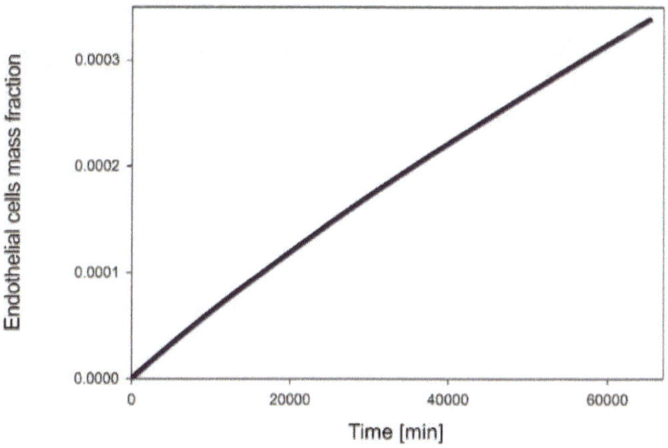

Fig. 25 Mean mass fraction of endothelial cells as function of time, Castem model

fraction is 0.02, the HCs volume fraction is 0.45. The endothelial cells are present on the blood vessel on the right side of the domain, with initial mass fraction 3 10^{-5}. The boundary conditions are imposed on all the sides of the square for oxygen and TAF; for the endothelial cells zero concentration is prescribed in the upper, bottom and left sides. The parameters for the growth of tumor cells, for the ECM, for the HCs and for the diffusion of oxygen are listed in Tables 1, 2, 3 (Sciumè et al. 2014b). The whole domain is discretized with triangular elements.

The fields for the oxygen mass fraction, obtained by the Castem and Castem-PAK solvers are shown in Fig. 6. A higher consumption of oxygen appears in the first analysis (Castem). In the coupled analysis (Castem-PAK) the oxygen mass is larger because the supply from the capillary network is taken into account. The capillary domain in the smeared model is only present at FE nodes where the endothelial cell mass fraction (determined by the Castem) is different from zero. The nodal mass fraction of capillaries $\overline{\omega}_{cap}$ (capillary density) is determined as

$$\overline{\omega}_{cap} = D_{cap}\overline{\omega}^{\overline{EC}}/(4\delta_{EC}) \tag{27}$$

where D_{cap} and δ_{EC} are capillary diameter and thickness of the endothelial layer (see Table 2).

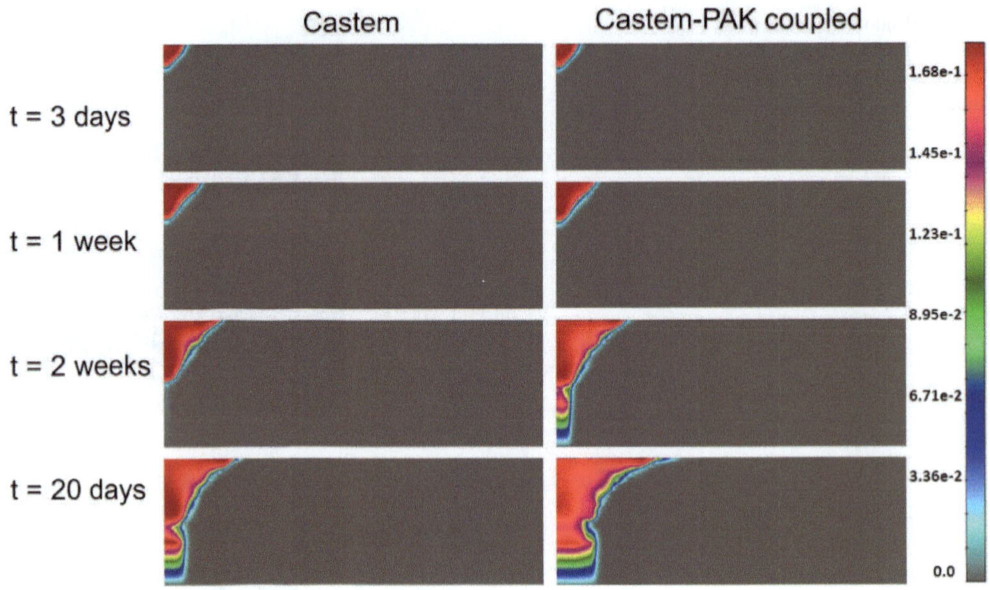

Fig. 26 Tumor growth according to Castem and Castem-PAK models, for t = 3 days, 1 week, 2 weeks and 20 days

The graphs in Figs. 7, 8, and 9 show differences in oxygen concentration over time for the total oxygen mass fraction, in the tissue, and in tumor, respectively, obtained by using Castem and Castem-PAK software. It can be seen from Figs. 7 and 8 that the difference between the oxygen mass fraction of the two analyses remains almost constant in time, but the coupled solution curve is above that obtained by the Castem only. In the zone of the tumor (Fig. 9) this difference tends to zero as time progresses because the oxygen consumption increases with the tumor growth. By comparing Figs. 7 and 8 it appears that most of the oxygen is in the tissue.

When the oxygen concentration is lower than a fixed threshold, the living tumor cells produce the tumor angiogenic factor (TAF) that diffuses from the tumor into the IF. Solutions for the TAF concentration obtained by Castem and Castem-PAK, for three different time stations, are shown in Figs. 10 and 11. It can be seen that lower concentration in the Castem solution produces larger TAF concentrations.

As response to the TAF the endothelial cells start moving from the vessel based on the right side of the domain towards the tumor as shown in Fig. 12 after 5 days for the coupled analysis Castem-PAK. Note that PAK doesn't change endothelial cells mass fraction, hence in fact only Castem results are shown.

In Fig. 13 the graph for the endothelial cells mass fraction over time is depicted. After the initial response to the TAF where endothelial cells concentration remains constant, there is an almost linear increase as time progresses.

Figure 14 shows tumor growth after 1 day, 5, 10 and 15 days. Note how the tumor rim becomes irregular as the tumor grows. For this example, Castem and Castem-PAK produce very similar results of the tumor size and shape (this also can be seen in Fig. 15). Hence only Castem results are shown.

Figure 15 shows the tumor growth according to the two analyses. With the coupled codes the final tumor growth is larger because of the higher oxygen concentration due to supply. The difference between the two models is not significant because angiogenesis is low (small endothelial mass fraction, Fig. 13).

5.2 Simulation of melanoma growth

Cutaneous melanoma growth with angiogenesis has been simulated in (Santagiuliana et al. 2016). Here the simulation is repeated by using the coupled models and results are compared. A detailed description of the physical problem can be found in (Santagiuliana et al. 2016) together with the geometry, boundary and initial conditions. We recall that the melanoma first expands radially, then a penetration through the basement membrane occurs and angiogenesis begins as described above.

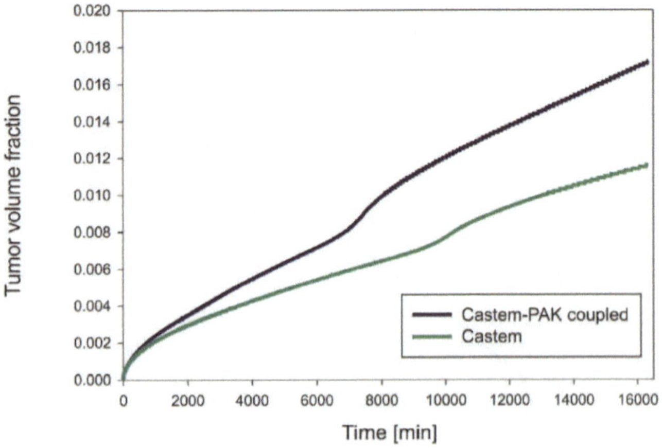

Fig. 27 Mean volume fraction of tumor as function of time, for Castem and Castem-PAK

The geometry of the skin is drawn in Fig. 16.

The parameters used are indicated in Table 4.

The parameters in for the four phases and the oxygen are those of Tables 1 and 2. For the ECM, the parameters are in Table 4 (Sciumè et al. 2014b). The parameters for TAF and Endothelial cells in Table 5 stem from literature, (Anderson and Chaplain 1998; Breward et al. 2003; Eikenberry et al. 2009) and from sensitivity analysis.

The problem of Fig. 16 is simulated in axial symmetry and the domain is discretized by 3720 plane four node elements. The boundary conditions are described in Fig. 17.

In the whole domain the initial interstitial fluid pressure (IFP), p^l, is set equal to 0 Pa, while the HC-IF pressure difference, p^{hl}, is set equal to 719 Pa, which correspond to a saturation degree of IF, in EP, MZ and DZ respectively equal to 0.44, 0.40, and 0.48. At time t = 0 h, all four phases coexist in the purple region of Fig. 17 (it has a radius of 40 μm), while in the remaining regions of the domain no TCs are present. Within the purple region, the initial saturation degree of the tumor cells (TC) is set to 0.125, corresponding to $p^{th} \approx 59$ Pa. Oxygen is here the sole nutrient species, and in the whole domain its mass fraction is initially set equal to $4.2.10^{-6}$, which corresponds to the average of the dissolved oxygen in the plasma of a healthy individual. Initial TAF concentration is zero. When the oxygen is less than a critical value, fixed for this simulation as $3.0.10^{-6}$, i.e. when there is hypoxia, the living tumor cells produce TAFs that diffuse. The boundary condition for TAF is zero concentration at the base of dermis. The endothelial cells are initially present at the base of the dermis, in the blood vessel with mass fraction equal to $4.0.10^{-1}$. At the upper boundary (B1), there are no fluxes of phases and species nor applied forces. At the right boundary (B2) the primary variables p^l, p^{hl}, p^{th} and u_r are fixed in time (Dirichlet boundary conditions) and zero flux is imposed for oxygen. At the lower boundary (B3), the primary variables p^l, p^{hl}, p^{th} and u_z are fixed in time (see Fig. 17). The boundary conditions at the z axis are assumed respecting cylindrical symmetry.

The results of this simulation are now shown. Figure 18 depicts the oxygen concentration fields at four time instants, according to the tumor growth model alone and the coupled model. It can be seen that oxygen concentrations are larger when the capillary network is included within the smeared model.

In Figs. 19 and 20 mean oxygen concentrations are shown, respectively for the total oxygen concentration in tumor and tissue, and the oxygen concentration in tissue alone. After an initial short period the evolution is quite similar and it can be seen that most of oxygen is in the tissue.

Tumor oxygen concentrations are shown in Fig. 21. As time goes on there is a bit less oxygen in the tumor when the coupled model Castem-PAK is used as compared to the Castem solution. Tumor consumes oxygen and this seems to be the predominant behavior.

TAF concentrations from Castem and Castem-PAK are shown in Fig. 22 for four different time instants. It appears, see also Fig. 23, that lower oxygen concentration in the Castem model produces larger TAF.

The endothelial cells mass fraction determined by the Castem model is shown in Fig. 24. Over time concentration of endothelial cells grows. It can be seen that after 2 weeks endothelial cells migrate towards tumor because of the gradient of the TAF. Mean mass fraction of ECs is depicted in Fig. 25. The increase of ECs is almost linear. Tumor growth, with the tumor domains progressing with time, is shown in Fig. 26 for both models and for four different time instants.

The mean volume fraction of tumor, which represents the tumor growth, is drawn against time in Fig. 27. The tumor volume is larger when using the coupled models (Castem-PAK) instead of just the tumor growth model alone (Castem). There is more oxygen available with the presence of capillaries, see Figs. 19 and 20. Comparing the tumor growth in the two examples, Figs. 15 and 27, it also appears that the effect of oxygen supply from capillaries on tumor growth is larger in the melanoma tumor; this is due to difference in the evolution rate of endothelial cells (around two order of magnitude larger in melanoma tumor, see Figs. 13 and 25) and therefore of the rate of angiogenesis.

6 Conclusions

A very general multiphase porous media model for tumor growth has been coupled with a bio distribution model. This coupled code allows to simulate a more realistic dynamics of molecules involved in proliferating tumors. The tumor growth model yields as a function of the available nutrient, evolution of tumor cell population, both viable and necrotic, of the differential pressure between cell populations, of porosity of ECM, of consumption of nutrients due to tumor growth, of angiogenesis and related growth factors. On the other hand the bio distribution model integrates the tumor growth model with the diffusion of molecules, oxygen and the smeared capillary network in the tumor environment. The two models and their connection have been presented in detail. The code coupling is the new aspect of this paper. Two examples have been shown: a 2D example of tumor growth in a square domain and an axisymmetric example of melanoma growth. In both cases comparison has been carried out between the new results from the coupled codes and the result from the tumor growth code alone. The model connection offers a better insight into regulation of the oxygen diffusion and of possible other molecules including therapeutic agents e.g. chemotherapeutics or molecularly targeted therapeutics such as T cells, antibodies for immunotherapy, and of nanoparticles and multistage platforms. The combined models appear as an appropriate tool for simulating the growth of a tumor in a whole organ allowing at the same time to evaluate the efficacy of the therapeutic agents.

Acknowledgements B.A.S. gratefully acknowledges the support of the Technische Universität München - Institute for Advanced Study, funded by the German Excellence Initiative and the TUV SÜD Foundation. The authors acknowledge CITO Award, Houston Methodist Research Institute, Houston, NCI U54 CA210181. The authors affiliated to Serbian institutions also acknowledge support from Ministry of Education and Science of Serbia, grants OI 174028 and III 41007, and City of Kragujevac.

References

A.R. Anderson, M.A. Chaplain, Bull. Math. Biol. **60**, 857 (1998)

C.J.W. Breward, H.M. Byrne, C.E. Lewis, A multiphase model describing vascular tumour growth. Bull. Math. Biol. **65**, 609–640 (2003). https://doi.org/10.1016/S0092-8240(03)00027-2

A.R. Carotenuto, A. Cutolo, A. Petrillo, R. Fusco, C. Arra, M. Sansone, D. Larobina, L. Cardoso, M. Fraldi, J. Mech. Behav. Biomed. Mater. **86**, 55 (2018)

S.C. Cowin, L. Cardoso, Mech. Mater. **44**, 47 (2012)

M.W. Dewhirst, T.W. Secomb, Nat. Rev. Cancer **17**, 738 (2017)

D.E. Discher, P. Janmey, Y.-L. Wang, Science **310**, 1139 (2005)

S. Eikenberry, C. Thalhauser, Y. Kuang, PLoS Comput. Biol. **5**, e1000362 (2009)

M. Ferrari, Trends Biotechnol. **28**, 181 (2010)

M. Ferrari, Int. J. Non. Linear. Mech. **56**, 3 (2013)

M. Fraldi, A.R. Carotenuto, J. Mech. Phys. Solids **112**, 345 (2018)

G. Freyer, B. Ligneau, B. Tranchand, C. Ardiet, F. Serre-Debeauvais, V. Trillet-Lenoir, Cancer Treat. Rev. **23**, 153 (1997)

W. Gray, C. Miller, *Introduction to the Thermodynamically Constrained Averaging Theory for Porous Medium Systems* (Springer International Publinshing, Cham, 2014)

W. Gray, B. Schrefler, Int. J. Numer. Anal. Methods Geomech. **31**, 541 (2007)

W.G. Gray, C.T. Miller, B.A. Schrefler, Adv. Water Resour. **51**, 123 (2013)

A. Hawkins-Daarud, S. Prudhomme, K.G. van der Zee, J.T. Oden, J. Math. Biol. **67**, 1457 (2013)

E.J. Koay, M. Ferrari, Phys. Biol. **11**, 60201 (2014)

M. Kojic, M. Milosevic, V. Simic, E.J. Koay, N. Kojic, A. Ziemys, M. Ferrari, Multiscale smeared finite element model for mass transport in biological tissue: from blood vessels to cells and cellular organelles. Comput. Biol. Med. **99**, 7–23 (2018). https://doi.org/10.1016/j.compbiomed.2018.05.022

M. Kojic, M. Milosevic, V. Simic, E.J. Koay, J.B. Fleming, S. Nizzero, N. Kojic, A. Ziemys, M. Ferrari, Comput. Methods Appl. Mech. Eng. **324**, 413 (2017a)

M. Kojic, M. Milosevic, V. Simic, E.J. Koay, N. Kojic, A. Ziemys, M. Ferrari, J. Serbian Soc. Comput. Mech. **11**, 108 (2017b)

M. Kojic, R. Slavkovic, M. Zivkovic, N. Grujovic, N. Filipovic, M. Milosevic, PAK-Finite element program for linear and nonlinear analysis; software patent. Univ Kragujevac and R&D Center for Bioengineering, Kragujevac, (2010)

J. Kremheller, A.T. Vuong, L. Yoshihara, W.A. Wall, B.A. Schrefler, Comput. Methods Appl. Mech. Eng. **340**, 657 (2018)

E.A.B.F. Lima, R.C. Almeida, J.T. Oden. Analysis and numerical solution of stochastic phase-field models of tumor growth. Numerical methods for partial differential equations. **31**, 552–574 (2015). https://doi.org/10.1002/num.21934

E.A.B.F. Lima, J.T. Oden, D.A. Hormuth 2nd, T.E. Yankeelov, R.C. Almeida, Math. Models Methods Appl. Sci. **26**, 2341 (2016)

F. Michor, J. Liphardt, M. Ferrari, J. Widom, hat does physics have to do with cancer? Nat. Rev. Cancer. **11**, 657–670 (2011). https://doi.org/10.1038/nrc3092

M. Milosevic, V. Simic, B. Milicevic, E. Koay, M. Ferrari, A. Ziemys, M. Kojic, Comput. Methods Appl. Mech. Eng. **338**, 97 (2018)

N.M. Moore, N. Kuhn, S.E. Hanlon, J.S.H. Lee, L.A. Nagahara, Phys. Biol. **8**, 10302 (2011)

J.T. Oden, A. Hawkins, S. Prudhomme, Math. Model. Methods Appl. Sci. **20**, 477 (2010)

J.T. Oden, E.A.B.F. Lima, R.C. Almeida, Y. Feng, M.N. Rylander, D. Fuentes, D. Faghihi, M.M. Rahman, M. DeWitt, M. Gadde, J.C. Zhou, Arch. Comput. Methods Eng. **23**, 735 (2016)

J.T. Oden, E.E. Prudencio, A. Hawkins-Daarud, Math. Model. Methods Appl. Sci. **23**, 1309 (2013)

M.M. Rahman, Y. Feng, T. Yankeelov, J.T. Oden, Comput. Methods Appl. Mech. Eng. **320**, 261 (2017)

H.L. Rocha, R.C. Almeida, E.A.B.F. Lima, A.C.M. Resende, J.T. Oden, T.E. Yankeelov, Math. Model. Methods Appl. Sci. **28**, 61 (2018)

R. Santagiuliana, M. Ferrari, B.A. Schrefler, Comput. Methods Appl. Mech. Eng. **304**, 197 (2016)

R. Santagiuliana, C. Stigliano, P. Mascheroni, M. Ferrari, P. Decuzzi, B.A. Schrefler, Adv. Model. Simul. Eng. Sci. **2**, 19 (2015)

G. Sciumè, W.G. Gray, F. Hussain, M. Ferrari, P. Decuzzi, B.A. Schrefler, Comput. Mech. **53**, 465 (2014a)

G. Sciumè, R. Santagiuliana, M. Ferrari, P. Decuzzi, B.A. Schrefler, Phys. Biol. **11**, 65004 (2014b)

G. Sciumè, S. Shelton, W.G. Gray, C.T. Miller, F. Hussain, M. Ferrari, P. Decuzzi, B.A. Schrefler, New J. Phys. **15**, 15005 (2013)

G. Vilanova, M. Burés, I. Colominas, H. Gomez, J. R. Soc. Interface **15**, 20180415 (2018)

Author's Personal Note

I first met Dr. Ferrari in June 2006, when he attended the First South-East European Conference in Kragujevac, Serbia. Due to flight cancellations he arrived at 2 am (instead of 2 pm) and still managed to give a phenomenal general lecture early next morning. We were impressed how he was handling everything so graciously under the circumstances, but later we got to know that this was classic "Mauro working style". After spending time in Serbia, we formed a special bond, which resulted in me joining the Methodist Research Institute and since then myself and my Serbia team were privileged to have his unwavering support.

As a result of our close collaboration, we not only published numerous papers, but also helped advance computational mechanics in Kragujevac and in Serbia in general. Specifically, Dr. Ferrari directly impacted the growth of the newly-formed Bioengineering R&D Center in Kragujevac. For his tremendous contributions, the City of Kragujevac bestowed its highest honor – the Plaque of St. George, to Dr. Ferrari in 2008.

Furthermore, the Serbian Academy of Sciences and Arts was honored to elect Dr. Ferrari as its prestigious Foreign Member in 2018.

Lastly, as an illustration of Dr. Ferrari's incredible personality, I wanted to share the following anecdote. On the day he received the Plaque of St. George, as we were entering the ceremonial hall, he asked me what were the names of the key City of Kragujevac dignitaries that were going to be present. I told him: Stevanovic (mayor), Milenic (vice-mayor) and Milovanovic (president of the city parliament). One hour later, towards the end of the ceremony, when Dr. Ferrari was concluding his acceptance speech, he thanked all three individuals by perfectly pronouncing their not-so-easy-to-pronounce names!

Milos Kojic

Biomedical Microdevices (2019) 21: 44
https://doi.org/10.1007/s10544-019-0390-0

A quantitative approach for determining the role of geometrical constraints when shaping mesenchymal condensations

Valentina Onesto[1,2] · William B. Barrell[3] · Mary Okesola[3] · Francesco Amato[1] · Francesco Gentile[4] · Karen J. Liu[3] · Ciro Chiappini[3]

Published online: 8 April 2019

Abstract

In embryogenesis, mesenchymal condensation is a critical event during the formation of many organ systems, including cartilage and bone. During organ formation, mesenchymal cells aggregate and undergo compaction while activating developmental programmes. The final three-dimensional form of the organ, as well as cell fates, can be influenced by the size and shape of the forming condensation. This process is hypothesized to result from multiscale cell interactions within mesenchymal microenvironments; however, these are complex to investigate *in vivo*. Three-dimensional *in vitro* models that recapitulate key phenotypes can contribute to our understanding of the microenvironment interactions regulating this fundamental developmental process. Here we devise such models by using image analysis to guide the design of polydimethylsiloxane 3D microstructures as cell culture substrates. These microstructures establish geometrically constrained micromass cultures of mouse embryonic skeletal progenitor cells which influence the development of condensations. We first identify key phenotypes differentiating face and limb bud micromass cultures by linear discriminant analysis of the shape descriptors for condensation morphology, which are used to guide the rational design of a micropatterned polydimethylsiloxane substrate. High-content imaging analysis highlights that the geometry of the microenvironment affects the establishment and growth of condensations. Further, cells commit to establish condensations within the first 5 h; condensations reach their full size within 17 h; following which they increase cell density while maintaining size for at least 7 days. These findings elucidate the value of our model in dissecting key aspects of mesenchymal condensation development.

Keywords 3D *in vitro* models · Mesenchymal condensation · High-content imaging · Microtopography · Stem cells · Embryogenesis

1 Introduction

In the embryo, the formation of bone and cartilage is induced by a mesenchymal condensation process in which previously dispersed mesenchymal cells gather together into a tightly packed cell mass (Hall and Miyake 2000). The size and shape of the condensed cell mass define the final three-dimensional form of the tissue. Multiscale cellular interactions play a critical role in establishing the permissive microvenironment conditions to induce condensations, which initiate chondrogenic differentiation during embryonic development (Stott et al. 1999). Although advances in molecular biology have allowed a systematic investigation of the role of morphogens in regulating the differentiation and growth of chondrogenic cells and tissues (Ichinose et al. 2005; Niswander 2003; Daniels et al. 1996), the role of biophysical and extracellular matrix microenvironmental cues are less well understood.

It has been appreciated for nearly a century that skeletal morphogenesis involves interactions between cell intrinsic factors (e.g. genetic cues) and extrinsic factors including environmental constraints (Murray and Huxley 1925). Most parameters that regulate the reaction-diffusion of chemical factors and the mechanochemical interaction between cells and extracellular matrix on chondrogenic patterning (Oster et al. 1983) are based on *in vivo* observations. While

✉ Ciro Chiappini
ciro.chiappini@kcl.ac.uk

1 Department of Experimental and Clinical Medicine, University of Magna Graecia, 88100 Catanzaro, Italy

2 Center for Advanced Biomaterials for HealthCare@CRIB, Istituto Italiano di Tecnologia, Largo Basanti e Matteucci 53, 80125 Naples, Italy

3 Centre for Craniofacial and Regenerative Biology, Dental Institute, King's College London, London SE1 9RT, UK

4 Department of Electrical Engineering and Information Technology, University Federico II, Naples, Italy

theoretical models have highlighted their contributions, they have yet found little experimental validation (Glimm et al. 2012). Chondrogenic morphogenesis *in vivo* is a markedly three-dimensional process and the presence of surrounding tissues provides physical constraints to the cell population, which plays a role in the establishment of the condensation (Klumpers et al. 2014; Cukierman et al. 2001).

In vitro systems can provide a degree of experimental validation in a controlled environment; but their two-dimensional nature and the lack of scope to modulate biophysical interactions restricts the quality of information they can provide into the mechanisms of chondrogenic morphogenesis (Rottmar et al. 2010). A more three-dimensional system, such as the micromass model, can better mimic *in vivo* conditions and can also be used to test cell intrinsic and extrinsic parameters required for morphogenesis (Archer et al. 1985). This approach has since been used to determine the relative contributions of signaling factors and mechanical compression during differentiation of the tooth, where it has been shown that artificial compaction is sufficient to induce cell fate switches (Mammoto et al. 2011). 3D substrates for cell culture have been extensively used to achieve directional alignment or migration of mesenchymal cells (Rashidi et al. 2014), to study the role of mechanical stiffness (Fuhrer et al. 2013) and the influence of surface properties including charge (Webb et al. 1998) and wettability (Webb et al. 1998; Arima and Iwata 2007) on adhesion (Bacakova et al. 2011), phenotype and functionality of mesenchymal stem cells (Curran et al. 2005). Therefore, we set out to adapt our micromass cultures for use with our synthetic three-dimensional environments.

To understand what role geometric boundary conditions can play on condensation patterning, we used image analysis of mouse embryonic skeletal progenitor cells (ESPCs) in standard micromass culture to extract shape descriptors of condensation morphology, and used them to direct the design of 3D polydimethylsiloxane (PDMS) microenvironments. We identified the covalent binding of (3-aminopropyl)triethoxy silane (APTES) followed by fibronectin physisorption as the ideal strategy for long-term cell adhesion to PDMS, enabling the establishment of patterned 3D mesenchymal condensations for up to 7 days. High content live imaging analysis enabled determining the characteristic features of the early stages of condensation and to relate them to the degree of geometric constraining. In our system, ESPC condensations established within 5 h and grew in size up to 17 h, following which they increased in cell density up to 7 days. Uniaxial geometric constraint through microgrooves highlighted a groove-size dependent effect on condensation growth.

2 Results

2.1 Identifying key descriptors of condensation morphology

Condensation morphology is dependent on the local environment and represents an important large-scale manifestation of the global organizational and physiological state of cells across body sites. Comparing morphological information can provide guidelines when designing environmental cues that direct the establishment and development of condensations. To this end, we developed a Matlab® script for analysis of brightfield images (Fig. 1a). The script locates and isolates individual condensations (Fig. 1b-d) in micromass cultures of ESPCs isolated from the facial prominences and the limb buds. The analysis of the condensations extracted 11 individual shape descriptors used to characterize the micromass culture (Table 1). Linear discriminant analysis identified that *area, perimeter, major axis* and *roundness* were the most prominent morphological differences between face and limb condensations (Fig. 1e). In particular, limb bud condensations displayed more rounded and shorter shapes (i.e. lower aspect ratios) while facial condensations presented higher branching.

2.2 Establishing a three-dimensional model for ESPC culture

Having identified that the *axis length* and *roundness* are significant contributors to site-specific variance in mesenchymal condensation, we employed microgrooves as an ad-hoc model to uniaxially constrain condensations, and modulate these key features *in vitro,* in order to dissect their role in skeletal shape generation. Thus, we generated PDMS microgrooves with 100 μm depth and variable width, selected at 25, 50, 100, 200 and 300 μm. To overcome the limited ability of PDMS to sustain cell adhesion and proliferation, we explored biofunctionalization strategies, including direct physisorption of poly-L-lysine (PLL), poly-D-lysine (PDL), gelatin, collagen I, fibronectin, and physisorption of fibronectin to a covalently bound (APTES) monolayer (Fig. 2a-c). It is known that the formation of a cationic coating layer on PDMS, despite the presence of positive charges at the cell interface, can improve the cytocompatibility of the surface in comparison to umodified PDMS (Nourmohammadi et al. 2015; Kuddannaya et al. 2013). The APTES + fibronectin surface modification approach outperformed the competing strategies in improving the attachment and proliferation of ESPCs. Fibronectin efficiently adsorbed on the APTES layer, enabling maintenance of ESPCs in culture for up to 7 days. In comparison, the alternative approaches exhibited significant detachment of

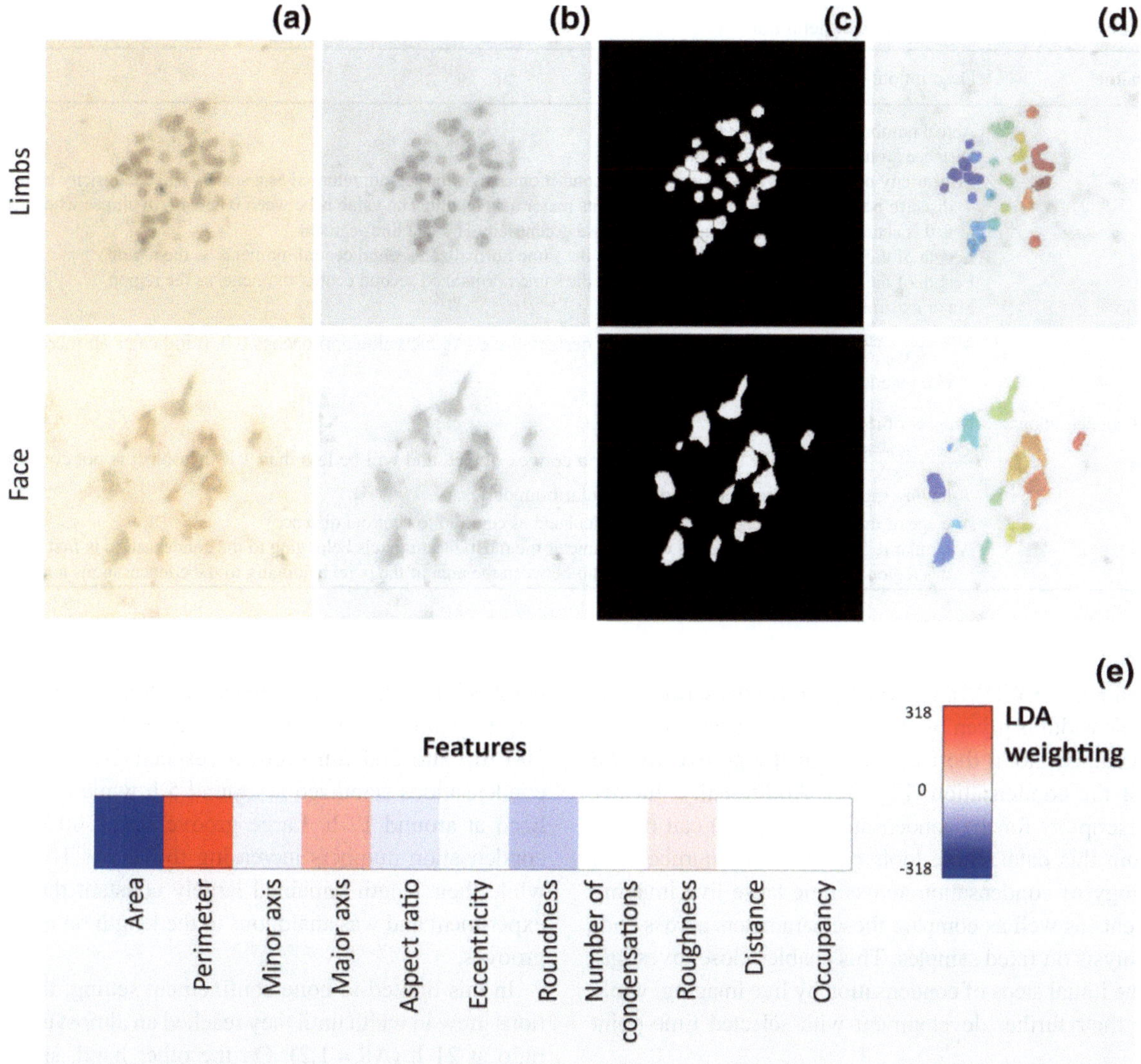

Fig. 1 Identification of morphological features discriminating between limb buds and face condensations. (**a**) Brightfield stereoscopic images of condensations are (**b**) converted to grayscale, (**c**) binarized and (**d**) watershed segmented to identify individual objects. (**e**) Weighting coefficients for morphological descriptors of individual condensation arising from linear discriminant analysis between face and limb condensations. Higher absolute values indicate most significant descriptors

embryonic mouse cells starting from day 3 in culture (Fig. 2c). The ESPCs cultured within this system assembled into three-dimensional cell bundles within each groove, establishing condensations therein (Fig. 2d).

2.3 Developing a tool for high content imaging analysis of mesenchymal condensations

Quantifying key morphological features of the mesenchymal condensations within the 3D *in vitro* model is crucial to understanding the role of geometric constraints. High content live imaging analysis provides unprecedented power in dissecting the transition between cell seeding and the establishment of condensations. To this aim we developed a custom analytical tool capable of identifying individual condensations and track their development across a time-series, in order to determine the evolution of key shape descriptors.

The script isolates individual grooves throughout a time series (Fig. 3a). Within each groove it identifies condensations at the final timepoint and tracks them backwards to their initial establishment. The condensations are detected by the local increase in cell density observed through the increased fluorescence intensity of Hoechst live nuclear staining. The script averages Hoechst intensity along the minor axis of the groove, to generate an intensity line plot along the major axis (Fig. 3b). Condensations are defined by the peaks identified in this plot. Their length, defined as their size along the major axis of the groove, is determined by the full width at half maximum of the

Table 1 List of shape descriptors and their definition

Property name	Description
Area	Actual number of pixels in the region
Perimeter	Distance around the boundary of the region
Eccentricity	Eccentricity of the ellipse that has the same second-moments as the region, returned as a scalar. The eccentricity is the ratio of the distance between the foci of the ellipse and its major axis length. The value is between 0 and 1. An ellipse whose eccentricity is 0 is actually a circle, while an ellipse whose eccentricity is 1 is a line segment
Minor axis	Length of the minor axis of the ellipse that has the same normalized second central moments as the region
Major axis	Length of the major axis of the ellipse that has the same normalized second central moments as the region.
Aspect Ratio	Major axis/minor axis ratio.
Roundness	$4 \times \frac{[\text{Area}]}{\pi \times [\text{Major axis}]^2}$. A value of 1.0 indicates a perfect circle. As the value approaches 0.0, it indicates an increasingly elongated shape
Number of condensations	Number of regions in the image.
Roughness	$\frac{[\text{Convex Perimeter}]}{[\text{Perimeter}]}$. It will take the value of 1 for a convex object, and will be less than 1 if the object is not convex. Objects having high roughness values show irregular boundaries.
Distance	Average of the 4 nearest-neighbor distance (calculated as centroid to centroid distance)
Occupancy	A circular region having as extremes of the diameter the most distant pixels belonging to the condensation is first selected. In this region the occupancy is given by the ratio between the area of the pixel belonging to the condensations and the total area.

associated peak (FWHM). For each peak/condensation identified, their width is given by the FWHM of Hoechst fluorescence averaged along the major axis of the groove for the length of the condensation (Fig. 3c). Additionally, further shape descriptors for the condensations (Table 1) can be derived from this data. These tools can track the number and morphology of condensation across time-lapse live imaging experiments as well as compare these parameters across end-point analysis on fixed samples. This enables close investigation of the initial steps of condensation by live imaging, while tracking their further development with selected time-point analysis.

2.4 Influence of geometric constrains on mesenchymal condensations

Combining the microtopography of the substrate with high content imaging analysis allowed investigating the role of geometric constraint on the development of mesenchymal condensations. We live tracked the establishment of condensations in limb bud ESPCs over 36 h (Fig. 4a-d). From this analysis we identified three different constraint regimes provided by small (25 and 50 μm), medium (100 and 200 μm) and large grooves (300 μm). Cells confined in small grooves showed a limited increase in the number and size of condensations. Indeed, condensation number increased to 1.2 and 1.3 per 100 μm while their length, which is the unconstrained dimension, increased only by 49% and 56% over the 36 h for 25 μm and 50 μm grooves respectively, indicating that the strong geometric constraint imposed by this topography highly limited the establishment of the condensations. In contrast, for intermediate size grooves (100 and 200 μm), condensations significantly grew in numbers and size. The number of condensations reached 2.6 and 2.4 per 100 μm length, while their length increased by 192% and 190% for 100 μm and 200 μm grooves respectively. The number of condensations stabilized at around 5 h while their size stabilized at around 17 h. Large groove size (300 μm) showed condensation numbers increasing to 3.6 per 100 μm length while their length remained largely constant throughout the experiment and was analogous to the length on medium sized grooves.

In this limited-to-none confinement setting, the condensations grew in width until they reached an almost unitary aspect ratio at 21 h (AR = 1.2). On the other hand, small grooves induced elongated condensations with aspect ratios greater than 2, while medium grooves showed intermediate aspect ratio values (AR ≈ 1.5). The constrained dimensions of the condensations, their width, also reached a steady-state value within 15 h for the small and medium grooves while the large grooves steadied at 21 h. The width was characteristic of the constraint regime. Small grooves had a 20–35 μm width, medium grooves a 70–80 μm width and large grooves a 125 μm width. This data overall indicates that microtopography does impose constraints on mesenchymal condensation which fall within three separate regimes. Small grooves tightly confined the establishment of condensations, limiting their length as well as their width, while medium grooves only limited the width of condensations without affecting length and the large grooves did not impose perceivable geometric constraints.

This view was confirmed in the analysis of long-term condensation maturation performed on fixed and stained samples at 2, 5 and 7 days (Fig. 5a-c). The length, width and overall aspect ratio for both face and limb bud condensations did not increase past day 2, agreeing with the live imaging observation that following 17 h the condensations have been established.

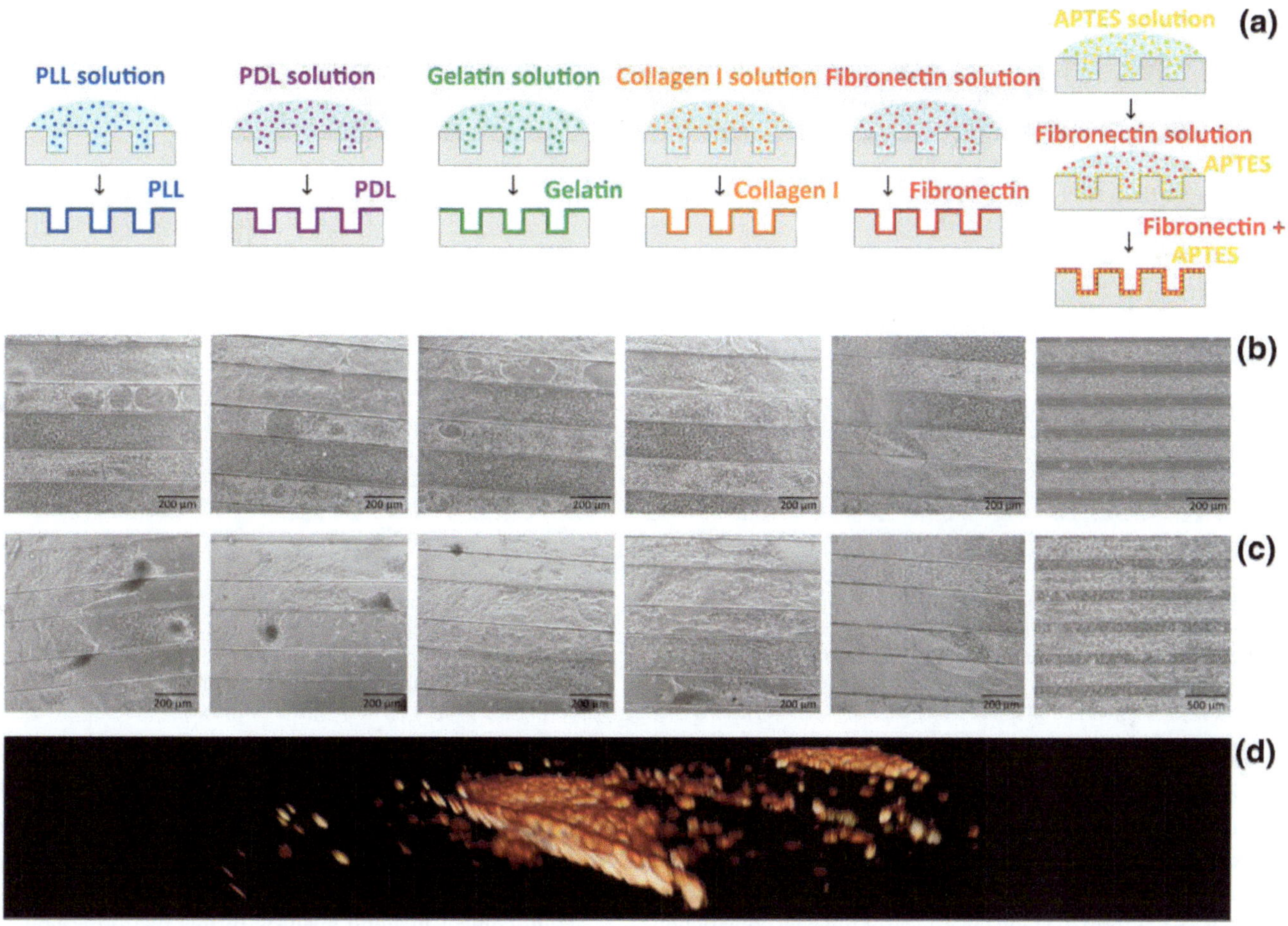

Fig. 2 Establishment of microgroove culture system for ESPCs. **a** The PDMS substrates are functionalised with different coating strategies. **b-c** Phase-contrast images of ESPCs cultured on substrates with different coating strategies at (**b**) 1 DIV and (**c**) 4 DIV. Cell detachment and aggregation is visible at day 4 except for the APTES+fibronectin strategy. **d** Volumetric reconstruction of a confocal stack shows the 3D shape of the condensations in the microgrooves

A markedly different behavior instead was observed for peak prominence, a measure of the density of cells within the condensation (Fig. 5d). Limb bud condensations displayed a rapid increase of peak prominence between day 2 and day 5, which was further sustained until day 7. This increase in cell compaction was supported by confocal imaging displaying an increased cell density as time progressed (Fig. 5e) and suggests a continued proliferation within the condensation, indicating a more mature state. Face condensations displayed a similar, but slower, increase in prominence between 2 and 7 days (Fig. 5d), which also suggest progression towards maturation and well agrees with independent reports of a slower maturation kinetics for face vs limb condensations. To further support the maturation of condensations we identified that all cells within limb condensations, for all groove types were Sox9 positive (Fig. 6a-b), a marker for chondrogenic differentiation, starting from at least day 2.

3 Conclusion

The establishment of mesenchymal condensations is a crucial first step in the differentiation of a wide range of embryonic organs including bone, cartilage, somites and kidneys. In this process, cells aggregate to form tightly packed groups which predict the final size and shape of the organ. Traditionally, high density micromass cultures of pre-chondrogenic mesenchyme have been used to mimic the *in vivo* condensation process. Relevant to tissue engineered bone and cartilage, micromass experiments have suggested that the ultimate shape and size of skeletal elements can be modulated by cell-cell or cell-matrix interactions within 3D spatial constraints. Therefore, we established an approach for the rapid assessment of the role of geometrical constraints during the establishment and maturation of micromass condensations. In the long term, our approach will be broadly applicable to testing and modifying multiple parameters in organogenesis.

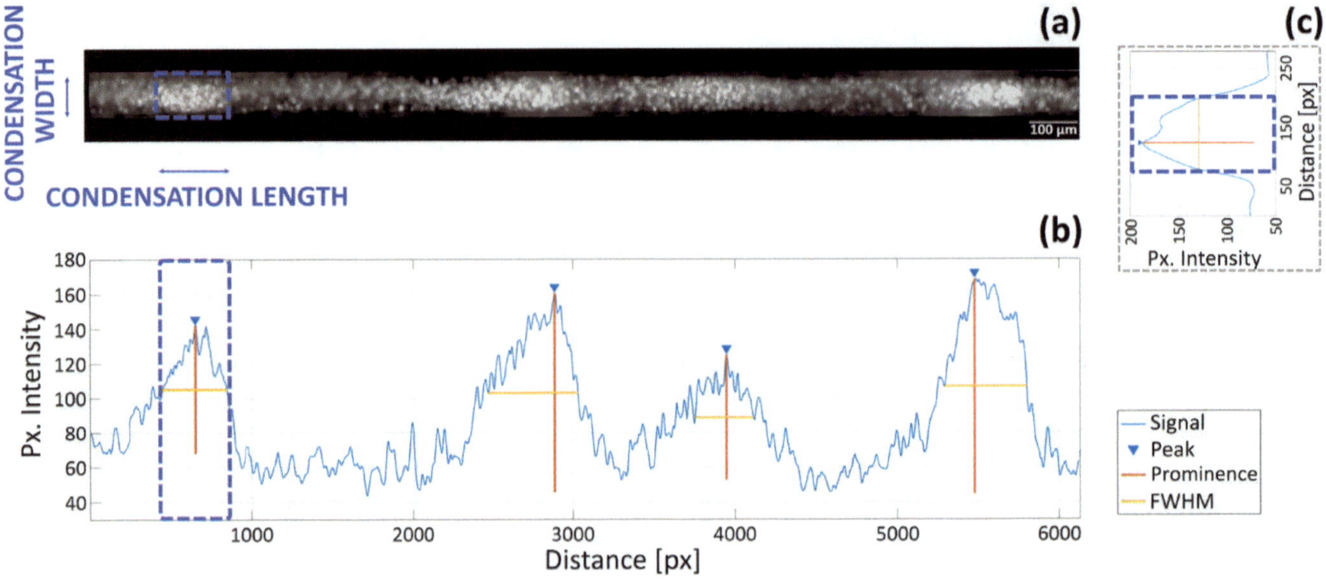

Fig. 3 Strategy for the analysis of condensation morphology by high content live imaging. **a** Individual grooves are extracted from the image. **b** Plot of the fluorescence intensity along the minor axis of the groove for the highlighted condensation. Fluorescence intensity is averaged over the major axis of the groove within the blue box. **c** Plot of the fluorescence intensity along the major axis of the groove. Fluorescence intensity is averaged over the minor axis of the groove. The light and dark bands in the fluorescence image correspond to peaks and valley in the graph, respectively. Individual condensations are identified by peaks in the plot. Their length is defined by the FWHM of the peak

First, it is crucial to have a standardized quantitative measure of the relevant cellular descriptors. On the basis of shape changes extracted from the segmentation of conventional brightfield micromass images of face and limb bud cells, we designed a 3D PDMS platform to study the effects of uniaxial constriction on condensation formation. We chose 'linearisation' via geometrical constraint of these cellular condensations in order to allow simplification and expansion of existing mathematical models of mesenchymal condensation. To improve the biocompatibility of PDMS substrates and to maintain long-term cultures, we optimised the combination of adhesive proteins and antifouling treatments.

High content imaging analysis of this platform, alongside confirmatory confocal and molecular analysis revealed two distinct regimes of uniaxial constriction on the establishment of mesenchymal condensations. For constriction between

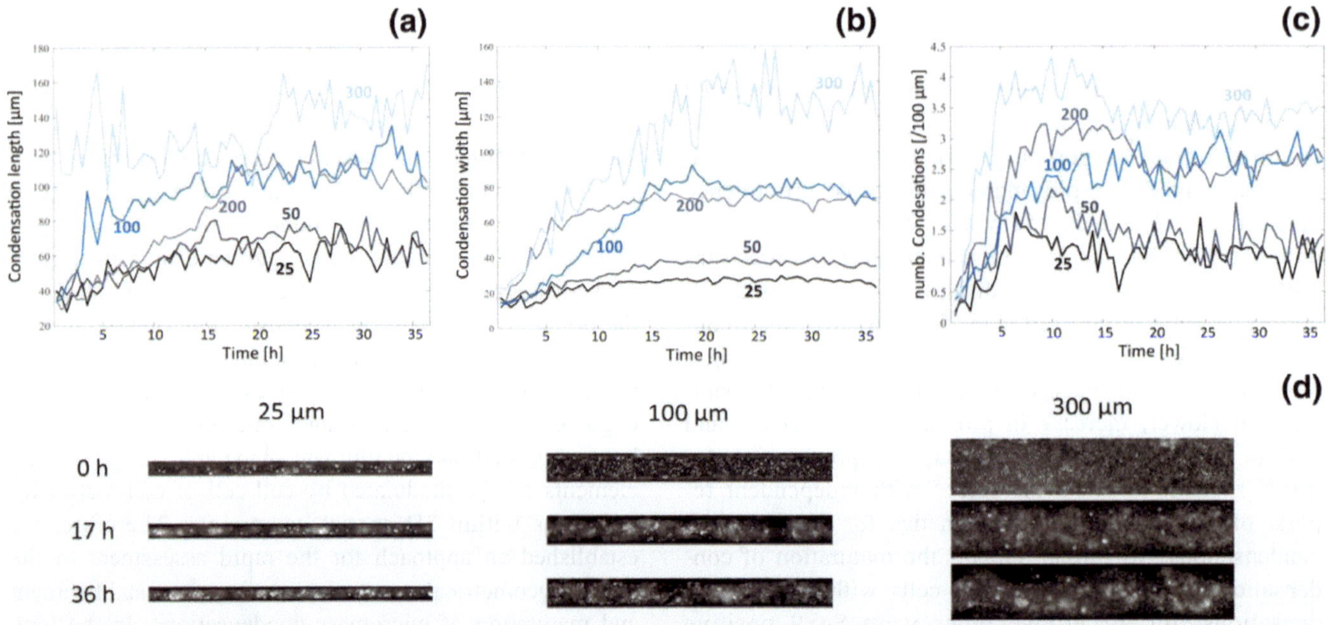

Fig. 4 High content live imaging analysis of mesenchymal condensations. Evolution of **(a)** length, **(b)** width and **(c)** number of condensations measured over the 36 h of the live imaging for different groove sizes (25–300 µm). **(d)** Snapshot of individual grooves at representative time points

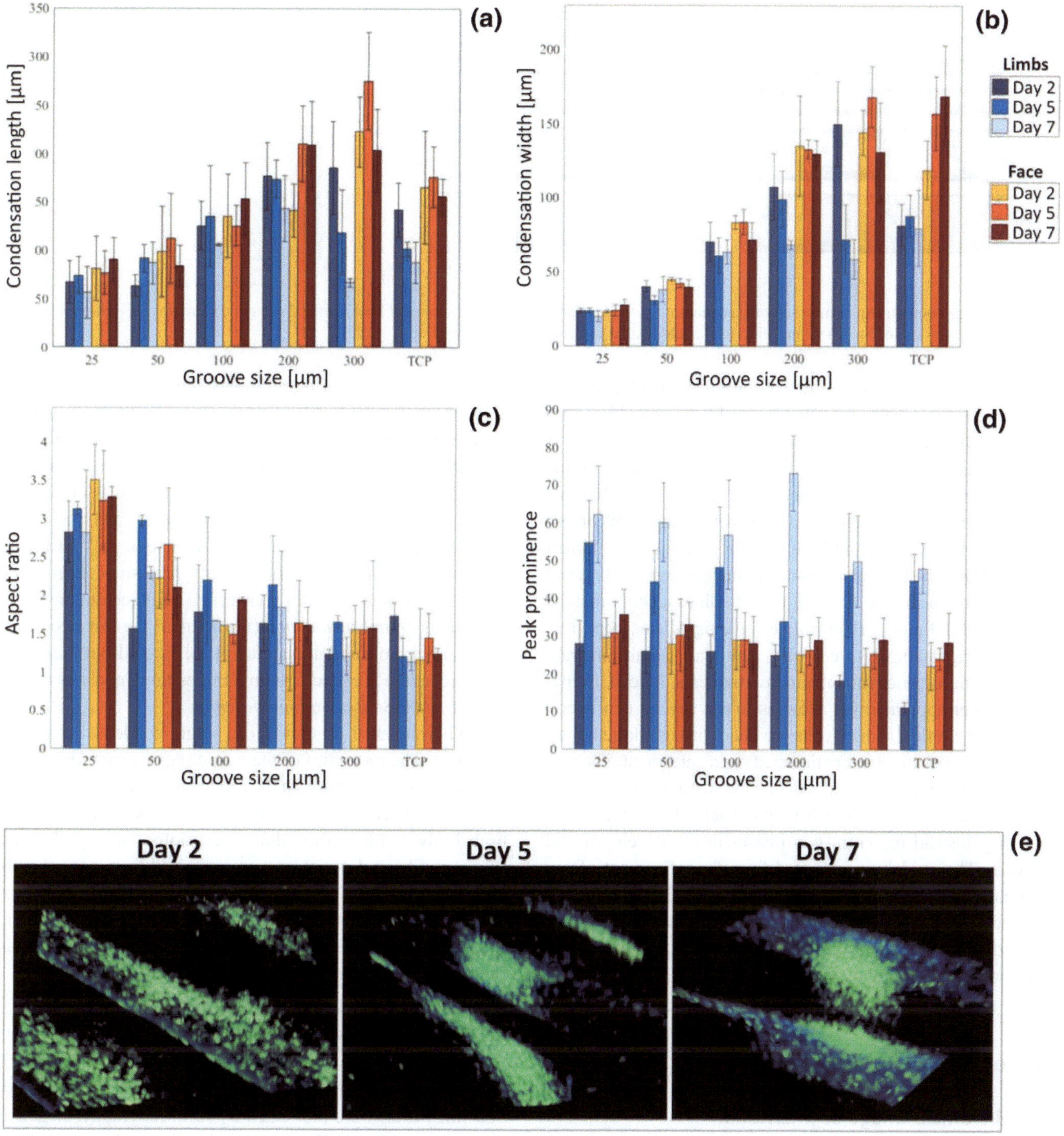

Fig. 5 Long-term analysis of mesenchymal condensations. **a** Mesenchymal condensation width, (**b**) length, (**c**) aspect ratio and (**d**) peak prominence of limb buds (blue) and face (red) ESPCs at 2, 5, 7 days. **e** Confocal images of condensations at 2, 5 and 7 days

25 µm and 50 µm, the constraint affected the elongation of the condensation along the free dimension, while for constrictions between 100 µm and 200 µm elongation along the free dimension was unconstrained, leading to increased aspect ratios. At 300 µm there were no constrictions as the condensations evolved isotropically with an aspect ratio of 1, analogous to what observed in unconstrained systems. Regardless of the constriction regimes all condensations reached their final dimensions by 17 h post-seeding, then until 7d they increased in cell density without significantly altering their size. Throughout this process the cells present within limb condensations were committed to chondrogenic differentiation. This phenotypic evolution leads us to hypothesise the presence of two distinct stages in the development of mesenchymal

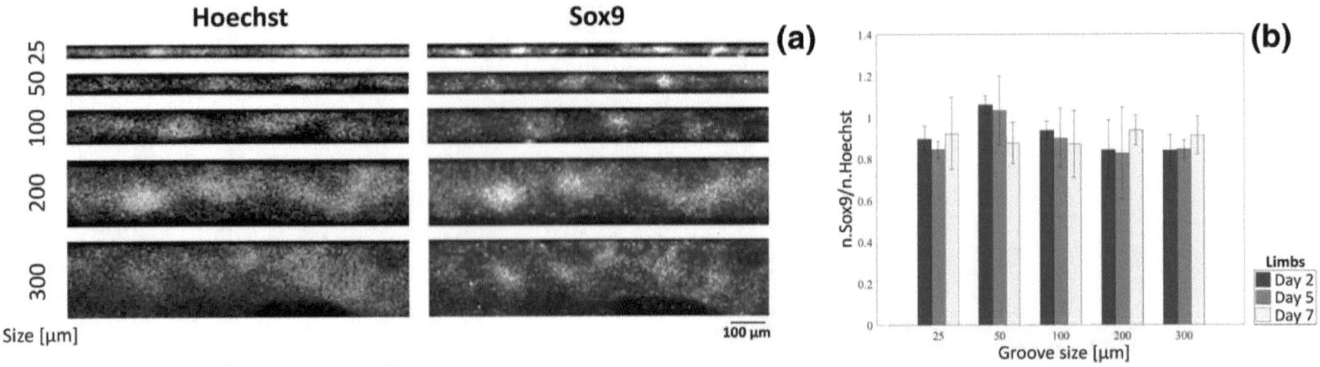

Fig. 6 Lineage commitment of mesenchymal condensations. **a** Hoechst and Sox9 co-staining of ESPCs after 2 days in culture, for different groove sizes (25–300 μm). **b** Ratio of Sox9 + ve nuclei to total nuclei within mesenchymal condensations

condensations: an initiation phase during the 0-17 h period where cells are locally recruited to establish a condensation, followed by a maturation phase lasting at least until day 7 during which cells within the condensation continue to proliferate, increasing its density rather than its size. This hypothesis requires further confirmation in a study combining high content analysis with molecular and cell biology investigations, aimed at identifying the biological processes underlying the observed evolution of phenotype. Overall, these findings identify that the established 3D model is suitable to investigate emergent phenotypes originating from complex multiscale cellular interactions within mesenchymal microenvironments. Indeed, refinement of our approach will allow us to test several conflicting models in the field, where cells are either brought together by attractive forces to form a compacted aggregate, or, alternatively, short range repulsive forces define the size and density of the aggregates. These two models may not actually be in conflict, but may instead represent two phases in the developmental process. The guidelines extracted from the refinement of our approach can provide design principles for scaffold tailored to the specific stages of mesenchymal condensation that we have here identified, allowing developing strategies to dynamically support skeletal morphogenesis.

4 Methods

4.1 Micromass image segmentation

Brightfield images of face and limb buds cells (Fig. 1a) at 20 days in normal micromass culture were analysed with Matlab® (2017b) to extract condensation shape descriptors. Images were first converted to grayscale (Fig. 1b), preprocessed to enhance contrast by contrast-limited adaptive histogram equalization (CLAHE) and low-pass filtered to remove constant power additive noise before being binarized (Fig. 1c) with Otsu's method. Morphological opening was performed to remove any small white noises in the image, and morphological closing to remove any small holes in the object. All connected components that had fewer than 24 pixels were removed and structures that were lighter than their surroundings and connected to the image border were suppressed. The images were segmented by a watershed transformation (Fig. 1d) and a distance transform was used as segmentation function to split out the regions. Watershed transform is known for its tendency to "oversegment" an image because each local minimum, no matter how small, becomes a "catchment basin". For this reason we (i) filtered out tiny local minima using *imextendedmin* (ii) modified the distance transform so that no minima occured at the filtered-out locations. This procedure is called "minima imposition" and was implemented via the function *imimposemin*.

With watershed segmentation, single condensations were identified. The properties of the image regions were directly or indirectly calculated through the *regionprops* function (Table 1).

Linear Discriminant Analysis (LDA) between face and limb buds condensation shape descriptors was performed (Fig. 1e). Shape descriptors were extracted from experiments with three technical repeats with three biological replicate for each one. The obtained values were standardized for all of the predictors (shape descriptors) and the regression model was fitted with the *fitcdiscr* function of Matlab®.

4.2 3D microwells fabrication

Polydimethylsiloxane (PDMS) grooves (25, 50,100, 200, 300 μm width) were fabricated by soft lithography. Negative photo resist SU-82100 was deposited onto clean and dried silicon wafers with a thickness of 100 μm and patterned by exposure to UV light through a transparency photomask. The stamp mold was silanized with vapour phase deposition of trichloro (1H, 1H, 2H, 2H-perfluorooctyl) silane (Sigma-Aldrich) to allow for an easy peel-off of the PDMS. PDMS prepolymer base and curing agent from a Sylgard 184 kit were mixed in a 10:1 ratio. The bubbles were removed by degassing in a vacuum desiccator. The PDMS was spin-coated on the patterned silicon wafer (400 rpm for 40s with acceleration of 1000 rpm/s) and cured at 95 °C for 10 min on a hotplate. After cooling, PDMS was peeled off the mold and the stamps were cut in 6 mm diameter discs and stored in 100% ethanol.

4.3 Surface functionalization

To chemically functionalize the surface of the material, different coating strategies were tested. First, the PDMS microstructures were rendered hydrophilic by exposing them to oxygen plasma treatment (3 min 70 W at 0.4mBar, with an O_2 flow rate of 20 sccm; Diener Zepto-W6). Thereafter, the substrates were coated with Poly-D-lysine (PDL, Sigma-Aldrich, 1 mg/10 ml in sterile water), Poly-L-lysine (PLL, Sigma-Aldrich, 1 mg/10 ml in sterile water), Gelatin (0.1% in PBS), Collagen type 1 (Roche, #111791179001, 5 mg/ml in PBS), or Fibronectin (5μg/cm^2 in PBS, Sigma-Aldrich). The solutions were incubated for 2 h, the excess of liquid was aspirated, and the PDMS discs were left to become dry overnight. Coatings were washed twice with PBS prior to cell plating.

A coating strategy with (3-aminopropyl)triethoxy silane (APTES, Sigma-Aldrich) to immobilize adhesive proteins (Fibronectin) was developed (Fig. 2a). The PDMS microstructures were rendered hydrophilic by exposing them to oxygen plasma treatment (3 min 70 W at 0.4mBar, with an O_2 flow rate of 20 sccm; Diener Zepto-W6). The substrates were coated with APTES solution (5% in 100% ethanol (*v*/v)) for 2 h at RT, washed three times with 100% ethanol and washed three times with PBS. Each mould was then coated with 5μg/cm^2 fibronectin in PBS for 2 h at RT and washed two times with PBS.

4.4 Micromass culture

Animal husbandry was carried out in accordance with King's College London ethical guidelines and UK Home Office Licence P8D5E2773. Facial prominences (including the frontonasal process, maxilla and mandible) and limb buds were collected from E12.5 embryos (CD1 strain). Pooled facial and limb tissues were triturated. Cells were then enzymatically dissociated in a water bath using Dispase (Roche, 1unit/ml in 10% FBS) for 1 h 15 min shaking (70 rpm) at 37 °C. Gentle vortexing was carried out every 15mins. Dispase was neutralised with an equal volume of micromass media (50:50 DMEM:F12 (Sigma), 10% FBS (Gibco), 1X ABAM (Sigma), 2 mM L-glutamine (Sigma), 5 mM Beta-glycerophosphate (Sigma) and 0.1mg/ml Ascorbic acid (Sigma)) and passed through a 40 μm cell strainer (Flowmi, Belart). Limb and facial prominence plating densities were 200,000 and 1,000,000 cells per well respectively (in a 20ul droplet). Droplets were incubated for 1-2 h in 37 °C with 5% CO_2. Humidity was maintained by placing PBS in the inter-well space of the plate. Wells were inspected for cell attachment then media was added and cells were incubated for the stated time.

4.5 Staining

Cells were fixed using 4% PFA for 10 mins at room temperature. A 5 min permeabilisation step was performed (0.5% TritonX-100 (Sigma) in 1X PBS) at room temperature. Then, blocking solution was applied (3% BSA (Sigma) in 1X PBS, 0.01% Tween20 (Sigma)) for 1-2 h at room temperature. Primary anti-SOX9 (Millipore, 1:500) in blocking solution was incubated overnight at 4 °C. Samples were washed three times (1% BSA, 0.01%Tween20) and then incubated in 1:500 far red anti-Rabbit (Alexafluor), 20μg/ml Hoechst 33342 (Sigma) and 1:500 Phalloidin-488 (Life Technologies) in blocking solution for 1-2 h. Samples were washed three times and then plates were imaged using a Leica DMi8 or Nikon Spinning Disc Confocal.

4.6 Live imaging

Micromass media was supplemented with 0.02μg/ml Hoescht 33342 (Sigma) as a DNA stain during live imaging. Cells were maintained at 37 °C with 5%CO_2 in an environment chamber (Solent scientific). Brightfield and Hoescht fluorescence images were collected every 30 mins for a total of 36 h using a Leica DMi8.

4.7 Stained images and high content live imaging analysis

To quantify condensation characteristics, images of Hoechst nuclear staining at 36 h, 2, 5 and 7 days of culture were analysed using Matlab® (2017b). Condensations were identified with fluorescence intensity analysis. First, single grooves were selected (Fig. 3a). Mesenchymal condensation length is the size of the condensation along the major axis of the groove and is given by full width at half maximum (FWHM) for peaks of Hoechst fluorescence averaged along the minor axis of the groove (Fig. 3c). Peak prominences were extracted as well (with the function *findpeaks*) to obtain the peak intensity values of the single condensations. For each peak identified, mesenchymal condensation width is given by FWHM for peaks of Hoechst fluorescence averaged along the major axis of the groove (Fig. 3b). Aspect ratio was calculated as ratio between condensation length and width (except for 300 μm grooves and tissue culture plate (TCP) where, due to the fact that condensation sizes are not confined, it is calculated as ratio between the longest and the shortest size).

In order to observe the evolution of mesenchymal condensation parameters over 36 h within each groove, condensations were identified at the final timepoint and tracked backwards to their initial establishment.

All parameters were measured per condensation and then averaged over a range of replicate condensation.

Author Contributions VO, WBB, KJL and CC conceived of this project and designed experiments. VO, WBB, MO performed experiments, analysed and presented data. VO and CC wrote the manuscript with support from WBB and KJL. F.G. discussed the results and commented on the manuscript. F.A. commented on the manuscript. All authors reviewed and approved the manuscript.

Funding WBB funded by the EPSRC, KJL is funded by BBSRC, MRC and BHF. CC recognizes support from the European Research Agency (ERC StG EnBiON 759,577). MO is funded by The Rosetrees Trust.

References

C.W. Archer, P. Rooney, C.P. Cottrill, Cartilage morphogenesis *in vitro*. J Embryol Exp Morphol. **90**, 33–48 (1985)

Y. Arima, H. Iwata, Effect of wettability and surface functional groups on protein adsorption and cell adhesion using well-defined mixed self-assembled monolayers. Biomaterials **28**, 3074–3082 (2007)

L. Bacakova, E. Filova, M. Parizek, Modulation of cell adhesion, proliferation and differentiation on materials designed for body implants. Biotechnol. Adv. **29**, 739–767 (2011)

E. Cukierman, R. Pankov, D.R. Stevens, K.M. Yamada, Science **294**, 1708–1712 (2001)

J.M. Curran, R. Chen, J.A. Hunt, Controlling the phenotype and function of mesenchymal stem cells *in vitro* by adhesion to silane-modified clean glass surfaces. Biomaterials **26**, 7057–7067 (2005)

K. Daniels, R. Reiter, M. Solursh, Micromass cultures of limb and other mesenchyme. Methods Cell Biol. **51**, 237–247 (1996)

R. Fuhrer, S. Hofmann, N. Hild, et al., Pressureless mechanical induction of stem cell differentiation is dose and frequency dependent. PLoS One **8**, e81362 (2013)

T. Glimm, D. Headon, M.A. Kiskowski, Computational and mathematical models of chondrogenesis in vertebrate limbs. Birth Defects Res C Embryo Today. **96**, 176–192 (2012)

B.K. Hall, T. Miyake, All for one and one for all: Condensations and the initiation of skeletal development. Bioessays **22**, 138–147 (2000)

S. Ichinose, M. Tagami, T. Muneta, I. Sekiya, Morphological examination during *in vitro* cartilage formation by human mesenchymal stem cells. Cell Tissue Res. **322**(2), 217–226 (2005)

D.D. Klumpers, A.S. Mao, T.H. Smit, D.J. Mooney, Linear patterning of mesenchymal condensations is modulated by geometric constraints. J. R. Soc. Interface **11**, 20140215 (2014)

S. Kuddannaya, Y.J. Chuah, M.H.A. Lee, N.V. Menon, Y. Kang, Y. Zhang, Surface chemical modification of poly(dimethylsiloxane) for the enhanced adhesion and proliferation of mesenchymal stem cells. ACS Appl. Mater. Interfaces **5**(19), 9777–9784 (2013)

T. Mammoto, A. Mammoto, Y.S. Torisawa, T. Tat, A. Gibbs, R. Derda, R. Mannix, M. de Bruijn, C.W. Yung, D. Huh, D.E. Ingber, Mechanochemical control of mesenchymal condensation and embryonic tooth organ formation. Dev. Cell **21**(4), 758–769 (2011)

P.D. Murray, J.S. Huxley, Self-differentiation in the grafted limb-bud of the Chick. J. Anat. **59**(Pt 4), 379–384 (1925)

L. Niswander, Pattern formation: Old models out on a limb. Nat Rev Genet. **4**(2), 133–143 (2003)

J. Nourmohammadi, T. Hajibabaei, G. Amoabediny, S.H. Jafari, N. Salehi-Nik, Aminosilane layer formation inside the PDMS tubes improves wettability and Cytocompatibility of human endothelial cells. Trends Biomater Artif Organs. **29** (2015)

G.F. Oster, J.D. Murray, A.K. Harris, Mechanical aspects of mesenchymal morphogenesis. J Embryol Exp Morphol. **78**, 83–125 (1983)

H. Rashidi, J. Yang, K.M. Shakesheff, Surface engineering of synthetic polymer materials for tissue engineering and regenerative medicine applications. Biomater. Sci. **2**, 1318 (2014)

M. Rottmar, M. Håkanson, M. Smith, et al., Stem cell plasticity, osteogenic differentiation and the third dimension. J Mater Sci Mater Med **21**, 999–1004 (2010)

N.S. Stott, T.X. Jiang, C.M. Chuong, Successive formative stages of precartilaginous mesenchymal condensations *in vitro*: Modulation of cell adhesion by Wnt-7A and BMP-2. J. Cell. Physiol. **180**(3), 314–324 (1999)

K. Webb, V. Hlady, P. Tresco, Relative importance of surface wettability and charged functional groups on NIH 3T3 fibroblasts attachment, spreading, and cytoskeletal organisation. J. Biomed. Mater. Res. **41**, 422–430 (1998)

MIX
Papier aus verantwortungsvollen Quellen
Paper from responsible sources
FSC® C105338

If you have any concerns about our products,
you can contact us on
ProductSafety@springernature.com

In case Publisher is established outside the EU,
the EU authorized representative is:
Springer Nature Customer Service Center GmbH
Europaplatz 3, 69115 Heidelberg, Germany

Printed by Libri Plureos GmbH
in Hamburg, Germany